Engineering mathematics
through applications

KULDEEP SINGH

Senior Lecturer in Mathematics,
Department of Mathematics
University of Hertfordshire

palgrave
macmillan

Dedication

— To Bibi Chanan Kaur

First published 2003 by
PALGRAVE MACMILLAN
Houndmills, Basingstoke, Hampshire RG21 6XS and
175 Fifth Avenue, New York, N.Y. 10010
Companies and representatives throughout the world

PALGRAVE MACMILLAN is the global academic imprint of the Palgrave
Macmillan division of St. Martin's Press, LLC and of Palgrave Macmillan Ltd.
Macmillan® is a registered trademark in the United States, United Kingdom
and other countries. Palgrave is a registered trademark in the European
Union and other countries.

ISBN 0–333–92224–7

This book is printed on paper suitable for recycling and made from fully
managed and sustained forest sources.

A catalogue record for this book is available from the British Library.

10   9   8   7   6   5   4   3   2   1
12   11   10   09   08   07   06   05   04   03

Printed and bound in Great Britain by
Ashford Colour Press Ltd, Gosport

Contents

Note to the Student

This book is ideally suited for anyone doing mathematics as part of their engineering or science undergraduate studies. Very little mathematical knowledge is assumed.

The heart of the book is the wealth of engineering examples drawn from a wide range of disciplines such as aerospace, building services, civil, control, electrical, manufacturing, mechanical, etc. In this respect the book is unique. The inclusion of these engineering disciplines will show from the outset the many applications of the mathematics you are studying. Understanding why you need to learn the mathematics and using it in real engineering examples is a real motivating factor.

These examples are not formed on the basis of prior engineering knowledge. However you will be asked to reflect on your final answer, i.e. how does your result relate to the problem? Reflection is a good way of learning mathematics and therefore frequent use of question and answer is made throughout the book. A question mark icon represents the questions. The question mark icon will help you to stop and think rather than pressing on and reading the answer on the next line.

The key to learning mathematics is to do exercises. If you want to become familiar with mathematical language and understand the theory there is no substitute for practice. There are exercises associated with each section and a miscellaneous exercise at the end of each chapter. The exercises are an integral part of the book and not just a bolt on. Moreover every section begins with a list of objectives and ends with a summary. The chapter also begins with a list of objectives. Doing the exercises and checking your answers are the only way you can meet these objectives.

Complete solutions to *all* exercises are provided on the book's website at www.palgrave.com/science/engineering/singh. This makes the book suitable for students working by themselves or on a distance learning course, as well as for lecture-based courses. These complete solutions provide a thorough step-by-step guide through each problem. Also on the website is a link through to dozens of test questions which offer immediate feedback and scores.

The proof of mathematical results is kept to a minimum. Instead the emphases of the book are on you learning by investigating results, observing patterns, visualizing graphs, answering questions, using technology, etc.

You need to understand the first two chapters thoroughly, because basic arithmetic and algebra are vital ingredients for the remaining chapters. For example you cannot do any calculus unless you have a firm footing in basic algebra. The content of these chapters should eventually become second nature to you. Recent news articles have stated that there are fewer and fewer students applying for engineering courses at university, which is a serious problem for industry. One of the reasons is that the mathematical nature of engineering is perceived as difficult. This book addresses this problem by using an everyday language step-by-step guide through each example. For example if an expansion of brackets is carried out at a particular step in the manipulation then this is stated as [Expanding] adjacent to the step.

Every formula has been given a reference number. For example reference number 2.1 means that it is the first formula in Chapter 2. If a particular formula is used then it is either stated in the main text or placed in the footnote on the page so that you do not need to search for it.

Major figures in the history of engineering mathematics are discussed. Of course these are not central to the mathematics but they do provide motivation for that particular topic.

There are occasionally some examples that are more difficult to follow. These are highlighted.

We assume you have access to a basic scientific calculator. Use is also made of a computer algebra system called MAPLE* and a graphical calculator. However you can use any computer algebra system. The book does not try to teach MAPLE but uses it to enlighten examples. The use of this software is not integral to understanding the mathematics.

Finally good luck with your work through this book, and enjoy it!

Kuldeep Singh
University of Hertfordshire
February 2003

---

* MAPLE is a registered trademark of Waterloo Maple Inc., http://www.maplesoft.com

Preface

For some time now I have felt a need for a textbook for teaching mathematics to undergraduates beginning engineering and science degrees which really grasped the nettle and taught the required mathematics in an engineering context right from the very start. We know, from our own experience and from what we read in the news and in professional papers, that students are starting out on degree and pre-degree courses with an increasingly low level of mathematics. What is worse is that students feel that mathematics is hard and abstract, unrelated to what they are doing elsewhere on their courses. Although many textbooks add occasional examples or indeed occasional chapters showing mathematics in its engineering contexts, none does this in every chapter right through the book. This is what I have done in this book and I know that students will find this use of engineering applications interesting and above all motivating, which should help to overcome some of these problems. I have included a wide range of applications in as realistic an array of examples as possible.

Several other factors have driven me to write this book. I wanted to see a textbook with:

- an accessible entry level
- full worked solutions to every exercise (these are on the companion website)
- use of mathematical tools
- a student-friendly style.

Accessible entry level

We all recognize that there is an increasing need to start a textbook at an entry level which will make the book suitable for students from a wide range of backgrounds and previous mathematical experience. The first two chapters provide a thorough revision of basic arithmetic and algebra, and should provide a useful launch pad from which to move on to the rest of the material. These chapters can of course be skipped by the more able or more advanced student.

Full worked solutions to every exercise

I feel very strongly that if students are to become confident and comfortable with the mathematics, then they need to have full worked solutions for *all* the exercises in the book. To my knowledge no other similar textbook does this. These are accessible on the companion website to the book, which is at www.palgrave.com/science/engineering/singh. Here you will also find a link through to dozens of test questions which offer students immediate feedback and scores.

Use of mathematical tools

Students are being asked to become familiar with calculators and mathematical software from early on in their courses. Textbooks should reflect this fact. I have assumed student access to a basic scientific calculator. Use is also made of MAPLE and a graphical calculator. The book in no way tries to teach MAPLE but uses it to illustrate examples. The use of this software is not integral to understanding the mathematics presented in the book, and any other similar software could be used in place of MAPLE. Similarly, instructions for using calculators have been kept generic.

Student-friendly style

Finally, none of these features will help the student if they do not feel at ease with the language in the book. My aim has been to speak to the student direct, and I have made use of a question/answer technique, marked by an icon, which encourages students to stop and think, and not simply plough on to the next line and the next without reflecting on what they are doing.

It has taken me six years to write this book and thanks must go to my long suffering family for my frequent anti-social behaviour over these years. Special thanks go to Helen Bugler at Palgrave Macmillan for her support and enthusiasm throughout the years and for overseeing the whole project. Thanks too to Jon Hopkins for encouraging me to take the project to Palgrave Macmillan. My appreciation also goes to Bill Perry for his careful attention to detail. Finally I thank the many anonymous reviewers of the initial manuscript, particularly a couple who gave a detailed review of the later drafts.

I do hope that this book will suit the needs of you and your students. If you have any comments on the book, please contact me via the companion website at www.palgrave.com/science/engineering/singh.

Kuldeep Singh
University of Hertfordshire
February 2003

Acknowledgements

The author and publishers would like to thank the following for the use of copyright material:

Waterloo Maple Inc.
Cambridge University Press for the table on page 821, taken from J. C. P. Miller & F. C. Powell, *The Cambridge Elementary Mathematical Tables*, 2nd edn, 1979. Reproduced with permission.

INTRODUCTION

Arithmetic for Engineers

SECTION A **Whole numbers**

By the end of this section you will be able to:
▶ understand the symbols $<$, $\leq$, $>$ and $\geq$
▶ evaluate the arithmetic of positive and negative numbers
▶ use a calculator

A1 **Directed numbers**

Figure 1 shows the number line. The negative numbers are to the left of zero and have a negative symbol, $-$, in front of them.

The positive numbers are to

Fig. 1

the right of zero but do **not** have an associated $+$ symbol. For example $+3 = 3$.

Numbers to the right are greater than numbers to the left, and greater than is denoted by the symbol $>$.

5 is greater than 2 is written $5 > 2$.

Similarly -1 is greater than -3 because -1 is to the right of -3, see Fig. 1.

? **How can we write this in symbol format?**

$$-1 > -3$$

The symbol $\geq$ means greater than or equal to.

Conversely numbers to the left are less than numbers to the right and less than is denoted by the symbol $<$.

? **We know 1 is less than 4, so how can we write this in terms of symbols?**

$$1 < 4$$

? **In the same manner, -4 lies to the left of -1, so how can we write this?**

$$-4 < -1$$

The symbol $\leq$ represents less than or equal to.

Example 1

Fill in the most appropriate symbol $>$, $\geq$, $<$, $\leq$ for $\square$ in the following:

a $2 \square 3$ **b** $-1 \square -3$ **c** $-1 \square -1$ **d** $11 \square -11$

e $-11 \square 11$ **f** $-5 \square -100$ **g** $-7 \square 0$

Example 1 *continued*

Solution

a Since 2 is less than 3 we have $2 < 3$.
b On the number line -1 is to the right of -3 so -1 is greater than -3:

$$-1 > -3$$

c -1 is equal to -1 so either $\leq$ or $\geq$ will do. Thus

$$-1 \leq -1 \text{ or } -1 \geq -1$$

Both of these are correct. However we normally write this as

$$-1 = -1$$

d 11 is to the right of -11 so 11 is greater than -11. Hence $11 > -11$.
e This is the reverse of **d**, so $-11 < 11$.
f -5 is to the right of -100 on the number line (Fig. 2).
-5 is greater than -100, therefore $-5 > -100$.

Fig. 2 (number line showing -100, -5, 0)

g -7 is to the left of 0 so it is less than 0. Thus $-7 < 0$.

We can add ($+$), subtract ($-$), multiply ($\times$) and divide ($\div$) numbers. To add and subtract negative numbers (numbers left of zero) we need to apply the following rules:

Rule **1**. Adding a negative number is the same as subtracting a positive number.

Rule **2**. Subtracting a negative number is the same as adding a positive number.

Example 2

Evaluate the following: **a** $5 + (-3)$ **b** $-3 + (-4)$ **c** $8 - (-4)$
    d $-7 - (-5)$ **e** $6 + (-6)$

Solution

a Using Rule **1** gives

$$5 + (-3) = 5 - 3 = 2$$

b Using Rule **1** again gives

$$-3 + (-4) = -3 - 4$$

From -3 we move 4 steps to the left (Fig. 3).

$$-3 - 4 = -7$$

Fig. 3 (number line showing -7, -6, -5, -4, -3, 0)

Example 2 *continued*

c Applying Rule **2** gives

$$8 - (-4) = 8 + 4 = 12$$

d Applying Rule **2** again gives

$$-7 - (-5) = -7 + 5 = -2$$

(From -7 you are moving 5 steps to the right.)

e Applying Rule **1** gives

$$6 + (-6) = 6 - 6 = 0$$

Next we consider the multiplication and division of negative numbers. It's worth remembering

'Minus times minus makes a plus.

The reason for this we need not discuss.'

For multiplication and division of two non-zero numbers, the rule is that if we have the same sign for both numbers then the result is a positive number. If the signs are different we get a negative number.

For example

$$9 \times (-3) = -27$$
$$(-9) \times (-3) = 27$$
$$9 \div (-3) = -3$$
$$(-9) \div (-3) = 3$$

For multiplication:

$$\underbrace{(+) \times (+)}_{\text{same}} = (+)$$

$$\underbrace{(+) \times (-)}_{\text{different}} = (-)$$

$$\underbrace{(-) \times (+)}_{\text{different}} = (-)$$

$$\underbrace{(-) \times (-)}_{\text{same}} = (+)$$

Similarly for division:

$$\underbrace{(+) \div (+)}_{\text{same}} = (+)$$

$$\underbrace{(+) \div (-)}_{\text{different}} = (-)$$

$$\underbrace{(-) \div (+)}_{\text{different}} = (-)$$

$$\underbrace{(-) \div (-)}_{\text{same}} = (+)$$

Example 3

Evaluate the following:

a $(-6) \times 2$ **b** $(-3) \times (-4)$ **c** $8 \div (-2)$

d $(-15) \div (-3)$ **e** $(-2) \times (-3) \times (-4)$

Solution

a Since -6 is a negative number multiplied by a positive number 2, the result is a negative number:

$$(-6) \times 2 = -12$$

b Minus times minus makes a plus. So $(-3) \times (-4) = 12$.

c Positive number, 8, divided by a negative number, -2, gives a negative number:

$$8 \div (-2) = -4$$

d Both numbers, -15 and -3, have the same sign so the result is a positive number:

$$(-15) \div (-3) = 5$$

e Minus times minus gives a plus, so $(-2) \times (-3) = 6$.
Then $6 \times (-4) = -24$. Hence

$$\underbrace{(-2) \times (-3)}_{= 6} \times (-4) = 6 \times (-4) = -24$$

It's worth remembering that if the number of negative numbers in the multiplication is odd then the final result is a negative number. Conversely if the number of negative numbers is even then the final result is a positive number:

$$\underbrace{(-) \times (-) \times \ldots \times (-)}_{\substack{\text{odd number of} \\ \text{negative numbers}}} = (-)$$

$$\underbrace{(-) \times \ldots \times (-)}_{\substack{\text{even number of} \\ \text{negative numbers}}} = (+)$$

We can represent multiplication in several ways, such as $\times$, $\cdot$ or no space between the two numbers. For example

$$3 \times 2 = 3 \cdot 2 = (3)(2)$$

Division can also be represented in several ways. The number $4 \div 2$ can also be written as $\dfrac{4}{2}$ or 4/2. Thus

$$4 \div 2 = \frac{4}{2} = 4/2$$

The evaluations with negative numbers can be implemented on a calculator by using the $\boxed{(-)}$ button or the $\boxed{+/-}$ button.

To compute $(-1729) \times (343)$ on a calculator, PRESS

| (−) | 1729 | × | 343 | = | which should show -593047.

Whole numbers are called **integers**. There are numbers between whole numbers which will be discussed in subsequent sections.

Fig. 4

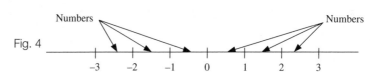

Figure 4 shows a portion of the real number line indicating some of the whole numbers. Any number on this line is a **real number**.

SUMMARY

The symbols $<$ and $\le$ mean less than and less than or equal to respectively. Similarly $>$ and $\ge$ mean greater than and greater than or equal to respectively.

For adding and subtracting, remember the following rules:

Adding a negative number is the same as subtracting a positive number.

Subtracting a negative number is the same as adding a positive number.

For division and multiplication:

If both numbers have the same sign then the result is positive, otherwise it is negative.

Exercise Intro(a)

Solutions at end of book. Complete solutions available at www.palgrave.com/science/engineering/singh

1 Fill in the most appropriate symbol $<$, $\le$, $>$ or $\ge$ in place of $\square$ for the following:

a $17 \,\square\, -17$ **b** $-17 \,\square\, 17$
c $-2 \,\square\, -1$ **d** $-5 \,\square\, -5$
e $-2 \,\square\, 0$

2 Complete the correct symbol, $+$, $-$, $\div$, $\times$, in place of $\square$ for the following:

a $3 \,\square\, 2 = 5$ **b** $3 \,\square\, -2 = 5$
c $9 \,\square\, 3 = 3$ **d** $-7 \,\square\, (-11) = 77$
e $-6 \,\square\, (-5) = -1$

3 Evaluate the following:

a $7 - 12$ **b** $-3 + 1$
c $-3 - (-1)$ **d** $-11 + (-11)$
e $-11 - (-11)$

4 Calculate

a $(-6) \div (-2)$ **b** $(-6) \times 2$
c $(-6) \times (-2)$ **d** $(-6) \div 2$
e $(-6) \div (-2)$ **f** $6 \div (-2)$
g $(-1) \times (-2) \times (-8)$

5 By using your calculator, or otherwise, compute the following:

a $(-343) \times 343$
b $(-343) \times (-343)$
c $(-729) \div 81$
d $\dfrac{(-729)}{81}$
e $(-666) - (-1945)$
f $(-2) \times (-5) \times (-7) \times (-10)$

SECTION B **Indices**

By the end of this section you will be able to:
▶ understand the terms **index**, **power** and **indices**
▶ evaluate indices
▶ evaluate square roots, cube roots etc.

B1 **Powers and roots**

[?] Instead of writing $\underbrace{2 \times 2 \times 2 \times 2 \times 2}_{\text{5 copies}}$, can you remember a shorter way of writing this number?

It can be written as 2 with superscript 5, 2^5. The superscript is called the index or the power, so in 2^5 the 5 is the **power** or **index**. The plural of index is **indices**. We can write 7×7 as 7^2 pronounced '7 squared', that is

$$7 \times 7 = 7^2$$

[?] How can we write $7 \times 7 \times 7$ in this index format?

$$\underbrace{7 \times 7 \times 7}_{\text{3 copies}} = 7^3$$

7^3 is pronounced '7 cubed'. We can write repetitive multiplication of the same number with an index (or power).

[?] What is 9^7 (9 to the index 7) equal to?

$$\underbrace{9 \times 9 \times 9 \times 9 \times 9 \times 9 \times 9}_{\text{7 copies}}$$

To evaluate this we use a calculator. PRESS

| 9 | | x^y | | 7 | | = |, which should show 4782969. Therefore

$$9^7 = 4782969$$

On some calculators you might have to press the ⃞ ∧ key instead of the ⃞ x^y key.
See the handbook of your calculator.

Powers of numbers can be easily evaluated on a calculator. Let's do an example.

Example 4

By using your calculator, or otherwise, compute the following:

a 3^4 **b** 88^2 **c** 20^3 **d** 20^1 **e** 7^0 **f** $(-2)^3$

Solution

Use the x^y or ∧ button on your calculator.

a $3^4 = \underbrace{3 \times 3 \times 3 \times 3}_{\text{4 copies}} = 81$

Example 4 *continued*

b $88^2 = 88 \times 88 = 7744$
c $20^3 = 20 \times 20 \times 20 = 8000$

?

d How can we write 20^1?

$$20^1 = 20$$

Any number to the index 1 is the number itself because it is **not** multiplied by itself again.

e Use your calculator to find 7^0:

$$7^0 = 1$$

There is a general result which says that: any number, apart from zero, to the index 0 gives 1.

f Odd number of negatives multiplied gives a negative number:

$$(-2)^3 = \underbrace{(-2) \times (-2)}_{=4} \times (-2)$$
$$= 4 \times (-2)$$
$$= -8$$

? Since $9^2 = 81$, can we somehow extract 9 from the number 81?

We move in the opposite direction, from 81 to 9. The 9 is the square root of 81. The square root is represented by the symbol $\sqrt{}$. Thus we write it as

$$\sqrt{81} = 9$$

? Can you find another square root of 81 or is 9 the only one?

? What is $(-9) \times (-9) = (-9)^2$ equal to?

$$(-9) \times (-9) = 81$$

because minus times minus gives a plus. So -9 is also a square root of 81.
$\sqrt{81} = 9$ or -9, sometimes written as

$$\sqrt{81} = \pm 9$$

where $\pm$ is the plus or minus symbol. The square root of a positive number gives you two numbers, one positive and the other negative.

? What is $\sqrt{64}$ equal to?

$$\sqrt{64} = \pm 8 \ (+8 \text{ or } -8)$$

We will generally use the following notation:

$$\sqrt{64} = 8 \text{ (only positive root)}$$
$$\pm\sqrt{64} = \pm 8$$

Most calculators will only give the positive square root. For the other square root we just place a minus sign in front.

In this Introductory chapter when we refer to roots we mean **only** the **real** roots.

We also have the cube root of a number. For example $12^3 = 1728$, so a cube root of 1728 is 12. The symbol for the cube root is $\sqrt[3]{\ }$. Hence

$$\sqrt[3]{1728} = 12$$

There is **only one** real cube root.

? **What is the cube root of 27?**

A number multiplied by itself three times gives 27:

$$(\text{number}) \times (\text{number}) \times (\text{number}) = 27$$

Use a calculator. There should be a $\sqrt[3]{\ }$ button on your calculator.

? **What is the answer?**

It's 3, thus

$$\sqrt[3]{27} = 3$$

Remember there is only one cube root. Similarly we can define other roots, for example

$$3^4 = 81$$

? **What is the 4th root of 81, $\sqrt[4]{81}$, equal to?**

$$\sqrt[4]{81} = 3 \text{ because } 3^4 = 81$$

In this case -3 is also a root. So

$$\pm \sqrt[4]{81} = \pm 3$$

In general, the **even** root of a number results in two **numbers**, one positive and the other negative. For **odd** roots we only obtain **one** root.

Example 5

Calculate

a $\pm\sqrt{25}$ **b** $\pm\sqrt{1444}$ **c** $\sqrt{19600}$ **d** $\sqrt[3]{8}$ **e** $\sqrt[3]{-8}$ **f** $\pm\sqrt[4]{625}$

Solution

a Since $5 \times 5 = 25$ and $(-5) \times (-5) = 25$ we have

$$\pm\sqrt{25} = \pm 5 \ (5 \text{ or } -5)$$

b Use the $\sqrt{\ }$ button on your calculator:

$$\pm\sqrt{1444} = \pm 38$$

c Similarly $\sqrt{19600} = 140$ (only the positive root)

? **d** **What is the cube root of 8, $\sqrt[3]{8}$, equal to?**

We know $2 \times 2 \times 2 = 8$, so $\sqrt[3]{8} = 2$ (only one real root).

e Also $(-2) \times (-2) \times (-2) = -8$. Thus

$$\sqrt[3]{-8} = -2$$

Try this on your calculator by using the $\sqrt[3]{\ }$ button.

Example 5 *continued*

f To find $\sqrt[4]{625}$ we use a calculator. PRESS

| 4 | | SHIFT | | x^y | | 625 | | = | , which should show 5. Hence

$$\sqrt[4]{625} = 5$$

On your calculator you may not have the SHIFT key but an INV or 2nd F key. Press this in place of the SHIFT key for the above evaluation. Also some calculators may have the $\sqrt[y]{x}$ key. In this case PRESS 4 | $\sqrt[y]{x}$ | 625 | =

Since we want to find the 4th (even) roots of 625, there is another root, −5. So

$$\pm\sqrt[4]{625} = \pm 5$$

The square root is most useful in various fields of engineering such as electronics, structures, vibrations, control etc. In the next example we examine the properties of the square root, $\sqrt{\ }$.

Example 6

Compute the following:

a $\sqrt{25 - 16}$ **b** $\sqrt{64 + 36}$ **c** $\sqrt{36 \times 81}$ **d** $\sqrt{\dfrac{256}{16}}$

Solution

a We first carry out the subtraction and then take the square root of the result:

$$\sqrt{25 - 16} = \sqrt{9} = 3$$

It is **critical** that you know

$$\sqrt{25 - 16} \neq \sqrt{25} - \sqrt{16} \quad \text{[Not equal]}$$

b Similarly $\sqrt{64 + 36} = \sqrt{100} = 10$. [Addition first and then root]

c With multiplication under the square root, it doesn't matter about the order:

$$\sqrt{36 \times 81} = \underset{=6}{\sqrt{36}} \times \underset{=9}{\sqrt{81}}$$
$$= 6 \times 9 = 54$$

Also by using a calculator we could have

$$\sqrt{36 \times 81} = \sqrt{2916} = 54$$

However for the first approach we didn't need to use a calculator.

d Division is similar to multiplication, so it doesn't matter about the order:

$$\sqrt{\frac{256}{16}} = \frac{\sqrt{256}}{\sqrt{16}} = \frac{16}{4} = 4$$

Note that for addition (and subtraction) the arithmetic under the square root is executed first. For multiplication (and division) you can apply the square root first and then multiply or vice versa.

The same properties apply to the cube root, 4th roots, etc.

SUMMARY

A number squared is a number with superscript 2 and means it is multiplied by itself:

$$(\text{number})^2 = (\text{number}) \times (\text{number})$$

The superscript is called the index or power. We also have indices of 3, 4, 5, etc.

Square root, $\sqrt{}$, is the reverse of squaring:

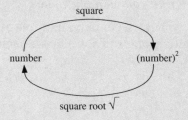

The cube root $\left(\sqrt[3]{}\right)$ and the 4th root $\left(\sqrt[4]{}\right)$ are defined in a similar way. The even real root of a given positive number results in two numbers, one positive and the other negative ($\pm$ number).

If we have addition (or subtraction) under the square root sign then we have to implement the addition (or subtraction) first. For multiplication (and division), order does not matter.

Exercise **Intro(b)**

Solutions at end of book. Complete solutions available at www.palgrave.com/science/engineering/singh

1 Evaluate the following:
- **a** 8^2
- **b** 8^3
- **c** 8^8
- **d** 45^2
- **e** 264^2
- **f** 20001^0
- **g** $(-3)^5$

2 Compute

- **a** $\pm\sqrt{49}$
- **b** $\pm\sqrt{1681}$
- **c** $\sqrt{183184}$
- **d** $\sqrt[3]{2197}$
- **e** $\sqrt[3]{729}$
- **f** $\pm\sqrt[4]{16}$
- **g** $\sqrt[4]{2401}$
- **h** $\pm\sqrt[8]{256}$
- **i** $\sqrt[5]{243}$
- **j** $\sqrt[5]{(-243)}$

3 Calculate the following:

- **a** $\sqrt{100 - 36}$
- **b** $\sqrt{129 + 40}$
- **c** $\sqrt{7225 - 5929}$
- **d** $\sqrt[4]{7225 - 5929}$
- **e** $\sqrt[3]{500 + 500}$
- **f** $\sqrt[3]{(-500 - 500)}$

4 Evaluate

- **a** $\sqrt{16 \times 25}$
- **b** $\sqrt{144 \times 81}$
- **c** $\sqrt{\dfrac{144}{36}}$
- **d** $\sqrt{\dfrac{225}{25}}$
- **e** $\sqrt[5]{\dfrac{7776}{243}}$
- **f** $\sqrt[5]{-\left(\dfrac{7776}{243}\right)}$

SECTION C **Numbers**

By the end of this section you will be able to:
▶ evaluate the lowest common multiple, LCM, of numbers
▶ understand what is meant by a prime number
▶ find the prime factors of a whole number

C1 **Lowest common multiple – LCM**

[?] **What do you think the term *lowest common multiple, LCM*, of two numbers means?**

It's the lowest positive number which is a common multiple of both numbers. For example to find the lowest common multiple of 4 and 6, we first list some of the multiples of these numbers.

[?] **What are the multiples of 4?**

Multiples of 4 are 4, 8, 12, 16, 20, 24, 28, ...

[?] **What are the multiples of 6?**

Multiples of 6 are 6, 12, 18, 24, 30, ...

Amongst these two lists we look for numbers which are common to both these lists.

[?] **Can you spot a number in both these lists?**

There are two numbers, 12 and 24, which lie in both lists. Both these numbers, 12 and 24, are common multiples of 4 and 6.

[?] **Which one of these is the lowest common multiple of 4 and 6?**

It's 12 because it is the smaller number. We say the lowest common multiple of 4 and 6 is 12. Let's try another example.

Example 7

Find the lowest common multiple of 7 and 8.

Solution

[?] **What do we need to do first?**

List the multiples of 7 and 8.

The multiples of 7 are 7, 14, 21, 28, 35, 42, 49, 56, 63, 70, ...

The multiples of 8 are 8, 16, 24, 32, 40, 48, 56, 64, ...

[?] **What is the smallest number common to both lists?**

56. Hence the lowest common multiple of 7 and 8 is 56.

C2 **Prime factors**

This method of finding the lowest common multiple (LCM) can be laborious, even for simple numbers. There is another technique used to find the LCM which involves finding the prime factors.

? **What do we mean by prime factors?**

In **Example 7** we say 7 and 8 are factors of 56 – whole numbers which divide into 56.

? **Can you think of any factors of 12?**

>3 and 4

Of course there are other factors of 12:

>1, 2, 3, 4, 6 and 12

Factors are whole numbers that divide exactly into a given number.

? **What are the factors of 20?**

>1, 2, 4, 5, 10 and 20. That is, all the whole numbers that divide into 20 exactly.

? **What are the factors of 7?**

>7 and 1

? 7 and 1 are the only factors of 7. **Can you remember the specific name given to a number whose only factors are the number itself and 1?**

It's called a **prime number**. The only factors of a prime number are 1 and itself. For example, prime numbers less than 20 are 2, 3, 5, 7, 11, 13, 17 and 19. Note that 1 is not a prime number.

? **What's so special about prime numbers?**

Every whole number can be written as a prime or a multiple of its prime factors.

? **What do we mean by prime factors?**

These are factors of a number which are prime. For example, 10 can be written as

>$10 = 2 \times 5$

where 2 and 5 are prime. Thus the prime factors of 10 are 2 and 5.

? **How can we write 20 in terms of prime factors?**

>$20 = 2 \times 10$

and 10 can be written as 2×5, so we have

>$20 = 2 \times 2 \times 5$

Also we can write 2×2 as 2^2:

>$20 = 2^2 \times 5$

We say $2^2 \times 5$ is the prime decomposition of 20.

Example 8

Find the prime decomposition of the numbers:

a 55   **b** 1000   **c** 31

Solution

a We know that $5 \times 11 = 55$ and since 5 and 11 are primes, so the prime decomposition of 55 is 5×11.

b We know $100 \times 10 = 1000$ and we can write 100 as 10×10. We have

$$1000 = \underbrace{100 \times 10}_{=10 \times 10} = 10 \times 10 \times 10$$

What are the prime factors of 10?

$$10 = 5 \times 2$$

Hence:

$$1000 = 10 \times 10 \times 10$$
$$= (5 \times 2) \times (5 \times 2) \times (5 \times 2)$$
$$= \underbrace{(5 \times 5 \times 5)}_{3 \text{ copies}} \times \underbrace{(2 \times 2 \times 2)}_{3 \text{ copies}}$$
$$1000 = 5^3 \times 2^3$$

c 31 is a prime number, so $31 = 31 \times 1$.

We can use this idea of writing a number in its prime factors to find the lowest common multiple. If you write down the prime factors of each number, then the LCM is the multiplication of the highest power of each prime factor.

Example 9

Obtain the lowest common multiple of 9, 15 and 40.

Solution

Writing the prime factors of each number gives:

$$9 = 3 \times 3 = 3^2$$
$$15 = 5 \times 3$$
$$40 = 8 \times 5$$
$$= (4 \times 2) \times 5$$
$$= \underbrace{(2 \times 2 \times 2)}_{3 \text{ copies}} \times 5$$
$$40 = 2^3 \times 5$$

So all the prime factors are 2, 3 and 5.

Example 9 *continued*

What is the highest power of each prime factor?

We have 2^3, 3^2 and 5. So the LCM $= 2^3 \times 3^2 \times 5$

$$= 8 \times 9 \times 5$$

$$= 360$$

The lowest common multiple of 9, 15 and 40 is 360.

Note that we can find the LCM for more than two numbers. The initial method for finding the LCM was to write down the multiples of 9, 15 and 40 and then spot the lowest number common among the list of all three multiples. This would have been tedious. Try it!

The lowest common multiple is used in the addition and subtraction of fractions covered later in this Introductory chapter.

SUMMARY

The lowest common multiple (LCM) of a collection of numbers is the smallest positive number which is common to all the multiples of the numbers.

A factor is a whole number which divides exactly into another number.

A prime number is a whole number with only 1 and itself as factors.

Every whole number can be written as a prime or a multiple of prime factors. The LCM of numbers can be evaluated by multiplying the highest power of each prime factor.

Exercise **Intro(c)**

Solutions at end of book. Complete solutions available at www.palgrave.com/science/engineering/singh

1 By listing the multiples, find the lowest common multiple of

 a 5, 6

 b 6, 14

 c 6, 27

 d 3, 7, 42

2 Write the following numbers in terms of its prime factors:

 a 90

 b 144

 c 94

 d 495

3 Find the lowest common multiple of

 a 24, 54

 b 8, 27, 64

 c 121, 125, 144

SECTION D Fractions

By the end of this section you will be able to:
▶ understand the different types of fractions
▶ write a fraction in its simplest form

D1 Fractions

? **What does the word fraction mean?**

$$\frac{\text{whole number}}{\text{whole number}}$$

A fraction is the division of one number by another. Examples are $\frac{1}{2}, \frac{1}{4}, \frac{17}{18}$, etc.

These numbers, $\frac{1}{2}, \frac{1}{4}, \frac{17}{18}$, are examples of **proper fractions**, that is the top number is smaller than the bottom number. A fraction is normally denoted by

$$\frac{\text{top number}}{\text{bottom number}}$$

The top number is called the **numerator** and the bottom number is called the **denominator**:

$$\frac{\text{numerator}}{\text{denominator}}$$

A proper fraction, $\frac{\text{numerator}}{\text{denominator}}$, is where the numerator is always less than the denominator.

? **What is an *improper fraction*?**

This is when the numerator is greater than the denominator, such as $\frac{8}{7}, \frac{4}{3}, \frac{377}{120}$, etc.

An improper fraction is also called a **top-heavy** fraction.

? **What does the term *mixed fraction* mean?**

This is a whole number with a proper fraction such as $2\frac{1}{2}, 3\frac{7}{11}, 7\frac{1}{8}$, etc.

Considering a fraction as a division, for example $\frac{22}{7}$, is the same as $22 \div 7$. In this way we can express top-heavy fractions as mixed fractions by dividing the numerator by the denominator. For example

$$\frac{22}{7} = 22 \div 7 \text{ which gives 3 remainder 1}$$

So we can write $\frac{22}{7}$ as 3 whole ones and $\frac{1}{7}$ left over, thus

$$\frac{22}{7} = 3\frac{1}{7}$$

Also $\frac{21}{7}$ is equal to 3.

? Going the other way, a mixed fraction can be expressed as a top-heavy fraction. **How do we write $2\frac{7}{9}$ as a top-heavy fraction?**

$$2 = \frac{18}{9} = \frac{2 \times 9}{9}$$

Thus we have

$$2\frac{7}{9} = \frac{(2 \times 9) + 7}{9}$$
$$= \frac{18 + 7}{9} = \frac{25}{9}$$

We can also reduce a fraction to its simplest terms if the numerator and denominator have a common factor between them. Otherwise the fraction is already in its simplest form.

For example $\frac{8}{21}$ is in its simplest form because there is **no** whole number which goes into both 8 and 21. However $\frac{8}{18}$ is not in its simplest form because 2 divides into 8 and 2 divides into 18. Hence

$$\frac{8}{18} = \frac{4 \times 2}{9 \times 2} = \frac{4}{9} \qquad \text{[Cancelling 2's]}$$

If the same number appears in the numerator and denominator then we can cancel this number. In this example we had $\frac{2}{2} = 1$.

Example 10

Express $\frac{36}{48}$ in its simplest form.

Solution

? **Which whole number goes into 36 and 48?**

12, that is $3 \times 12 = 36$ and $4 \times 12 = 48$. Thus

$$\frac{36}{48} = \frac{3 \times 12}{4 \times 12} = \frac{3}{4} \qquad \text{[Cancelling 12's]}$$

? **Conversely we can move from $\frac{3}{4}$ to $\frac{36}{48}$, how?**

By multiplying the numerator and denominator by 12.

$\frac{3}{4}$ and $\frac{36}{48}$ are called **equivalent fractions**. You can also write a fraction in its simplest terms by using a calculator. To write $\frac{512}{640}$ in its simplest form on a calculator, PRESS

[512] [a b/c] [640] [=], which should show [4] [⌐] [5]. Thus

$$\frac{512}{640} = \frac{4}{5}$$

SUMMARY

There are different types of fractions – proper fractions, top-heavy fractions, mixed fractions and equivalent fractions.

We can write a top-heavy fraction as a mixed fraction and vice versa.

We can express a fraction in its simplest form by cancelling a number common to the numerator and the denominator.

Exercise **Intro(d)**

Solutions at end of book. Complete solutions available at www.palgrave.com/science/engineering/singh

1 Express the following numbers as mixed fractions:

$$\frac{355}{113}, \quad \frac{213}{71}, \quad \frac{878}{323}, \quad \frac{577}{408}, \quad \frac{64}{7}$$

2 Write the following fractions in their simplest form:

a $\dfrac{7}{21}$ **b** $\dfrac{8}{20}$ **c** $\dfrac{72}{100}$

d $\dfrac{56}{75}$ **e** $\dfrac{272}{272}$

3 Which of the following pairs are equivalent fractions?

a $\dfrac{16}{64}, \dfrac{1}{4}$ **b** $\dfrac{26}{65}, \dfrac{1}{5}$ **c** $\dfrac{49}{89}, \dfrac{4}{8}$

4 Write the following mixed fractions as top-heavy fractions:

a $3\dfrac{1}{3}$ **b** $1\dfrac{70}{69}$ **c** $9\dfrac{87}{100}$

SECTION E **Arithmetic of fractions**

By the end of this section you will be able to:
▶ add and subtract fractions
▶ multiply fractions
▶ divide fractions

E1 **Addition and subtraction of fractions**

We can only add and subtract the numerators of fractions if they have the same denominator. For example

$$\frac{2}{7} + \frac{3}{7} = \frac{2+3}{7} = \frac{5}{7}$$

$$\frac{2}{7} - \frac{3}{7} = \frac{2-3}{7} = \frac{-1}{7} = -\frac{1}{7}$$

Note that −1 over 7 is the same as putting the negative sign in front, $-\dfrac{1}{7}$.

? **How can we add fractions with different denominators?**

[?] **We have to rewrite the fractions with a common denominator first, how?**

We find the lowest common multiple, LCM, of the denominators of the fractions concerned. The procedure for adding or subtracting fractions is

1 Find the LCM of the denominators.
2 Express each fraction with the LCM as the denominator.
3 Add or subtract the numerators.
Let's try an example.

Example 11

Find: **a** $\dfrac{1}{4} + \dfrac{3}{10}$ **b** $\dfrac{22}{7} - \dfrac{25}{8}$

Solution

[?] **a How do we find the LCM of 4 and 10?**

We first find the multiples of 4 and 10:
Multiples of 4 are 4, 8, 12, 16, 20, 24, 28, ...
Multiples of 10 are 10, 20, 30, 40, ...

[?] **Among these two lists, is there a number in both lists?**

Yes, 20. This is the lowest common multiple of 4 and 10. We need to place each fraction with 20 as the denominator. Since 4×5 makes 20, we multiply the numerator and denominator of $\dfrac{1}{4}$ by 5:

$$\frac{1}{4} = \frac{1 \times 5}{4 \times 5} = \frac{5}{20}$$

Similarly 2×10 makes 20 so we multiply the numerator and denominator of $\dfrac{3}{10}$ by 2:

$$\frac{3}{10} = \frac{3 \times 2}{10 \times 2} = \frac{6}{10}$$

We can now add the numerators:

$$\frac{1}{4} + \frac{3}{10} = \frac{5}{20} + \frac{6}{20}$$
$$= \frac{5 + 6}{20}$$
$$= \frac{11}{20}$$

b The multiples of 7 and 8 are as follows:

7, 14, 21, 28, 35, 42, 49, <u>56</u>, 63, 70 ···

8, 16, 24, 32, 40, 48, <u>56</u>, 64 ···

The LCM is 56, so we need to write each fraction under the common denominator of 56. Remember it is 7×8 which makes 56 so we multiply the numerator and denominator of $\dfrac{22}{7}$ by 8:

$$\frac{22}{7} = \frac{22 \times 8}{7 \times 8} = \frac{176}{56}$$

Example 11 *continued*

Similarly multiply the numerator and denominator of $\dfrac{25}{8}$ by 7:

$$\frac{25}{8} = \frac{25 \times 7}{8 \times 7} = \frac{175}{56}$$

Thus subtracting the numerators:

$$\frac{22}{7} - \frac{25}{8} = \frac{176}{56} - \frac{175}{56}$$

$$= \frac{176 - 175}{56}$$

$$= \frac{1}{56}$$

Note that both fractions in **b** are top-heavy fractions. The evaluation in **Example 11 b** of $\dfrac{22}{7} - \dfrac{25}{8}$ can also be carried out by writing each fraction as a mixed number:

$$\frac{22}{7} = 22 \div 7 = 3 \text{ remainder } 1$$

So $\dfrac{22}{7} = 3\dfrac{1}{7}$. Also $\dfrac{25}{8} = 3$ remainder 1, thus $\dfrac{25}{8} = 3\dfrac{1}{8}$. Hence

$$\frac{22}{7} - \frac{25}{8} = 3\frac{1}{7} - 3\frac{1}{8}$$

$$= (3-3) + \left(\frac{1}{7} - \frac{1}{8}\right)$$

$$= 0 + \underbrace{\left(\frac{1}{7} - \frac{1}{8}\right)}_{= \,^1/_{56}}$$

$$= 0 + \frac{1}{56}$$

$$\frac{22}{7} - \frac{25}{8} = \frac{1}{56}$$

(Both $\dfrac{22}{7}$ and $\dfrac{25}{8}$ are approximations for π.)

The evaluations with fractions can also be executed on a calculator. For example to evaluate $\dfrac{22}{7} - \dfrac{25}{8}$ on a calculator, PRESS

| 22 | | a b/c | | 7 | | (−) | | 25 | | a b/c | | 8 | | = |, which should show | 1⌐ 56 |

E2 **Multiplication and division of fractions**

Multiplication and division of fractions is effortless compared to addition and subtraction. For multiplication, multiply the numerators and multiply the denominators.

Example 12

Evaluate **a** $\dfrac{2}{9} \times \dfrac{4}{5}$ **b** $1\dfrac{1}{5} \times 1\dfrac{5}{12}$

Solution

a This is straightforward:

$$\frac{2}{9} \times \frac{4}{5} = \frac{2 \times 4}{9 \times 5} = \frac{8}{45}$$

b $1\dfrac{1}{5}$ and $1\dfrac{5}{12}$ are mixed fractions so we can write them as top-heavy fractions:

$$1\frac{1}{5} = \frac{(1 \times 5) + 1}{5} = \frac{6}{5}$$
$$1\frac{5}{12} = \frac{(1 \times 12) + 5}{12} = \frac{17}{12}$$

? **How do we multiply these fractions?**

As part **a**, multiply the numerators and multiply the denominators:

$$1\frac{1}{5} \times 1\frac{5}{12} = \frac{6}{5} \times \frac{17}{12}$$
$$= \frac{6 \times 17}{5 \times 12}$$
$$= \frac{102}{60}$$

? **The fraction $\dfrac{102}{60}$ is not in its simplest form, how can we place this in its simplest form?**

The factor 6 is common between 102 and 60 so $\dfrac{102}{60} = \dfrac{17 \times \cancel{6}}{10 \times \cancel{6}} = \dfrac{17}{10} = 1\dfrac{7}{10}$.

Hence $1\dfrac{1}{5} \times 1\dfrac{5}{12} = 1\dfrac{7}{10}$.

The evaluation **b** can be made somewhat simpler as follows:

$$1\frac{1}{5} \times 1\frac{5}{12} = \frac{6}{5} \times \frac{17}{12} = \frac{\cancel{6}}{5} \times \frac{17}{2 \times \cancel{6}} \underset{\substack{\text{cancelling} \\ \text{the 6's}}}{=} \frac{1 \times 17}{5 \times 2} = \frac{17}{10} = 1\frac{7}{10}$$

We cancel the 6's before multiplication.

For division of fractions we turn the second fraction upside down and multiply. Turning a fraction upside down is generally referred to as **inverting** the fraction.

Example 13

Determine **a** $\dfrac{5}{8} \div \dfrac{2}{9}$ **b** $2\dfrac{18}{25} \div 2\dfrac{71}{100}$

Solution

a $\dfrac{5}{8} \div \dfrac{2}{9} = \dfrac{5}{8} \times \underbrace{\dfrac{9}{2}}_{\substack{\text{turning the second} \\ \text{fraction upside down}}} = \dfrac{5 \times 9}{8 \times 2} = \dfrac{45}{16}$

We can write $\dfrac{45}{16}$ as a mixed fraction, how?

$45 \div 16 = 2$ remainder 13, so $\dfrac{45}{16} = 2\dfrac{13}{16}$. Hence $\dfrac{5}{8} \div \dfrac{2}{9} = 2\dfrac{13}{16}$.

b We can write both, $2\dfrac{18}{25}$ and $2\dfrac{71}{100}$, as top-heavy fractions:

$$2\dfrac{18}{25} = \dfrac{(2 \times 25) + 18}{25} = \dfrac{68}{25}$$

$$2\dfrac{71}{100} = \dfrac{(2 \times 100) + 71}{100} = \dfrac{271}{100}$$

Substituting these gives

$$2\dfrac{18}{25} \div 2\dfrac{71}{100} = \dfrac{68}{25} \div \dfrac{271}{100}$$

$$= \dfrac{68}{25} \times \dfrac{100}{271}$$

How can we evaluate $\dfrac{68}{25} \times \dfrac{100}{271}$?

Since 25 is common to 25 and 100 we can cancel:

$$\dfrac{68}{25} \times \dfrac{100}{271} = \dfrac{68}{25} \times \dfrac{4 \times 25}{271} \underset{\substack{\text{cancelling} \\ \text{25's}}}{=} \dfrac{68}{1} \times \dfrac{4}{271} = \dfrac{68 \times 4}{1 \times 271} = \dfrac{272}{271} = 1\dfrac{1}{271}$$

Hence $2\dfrac{18}{25} \div 2\dfrac{71}{100} = 1\dfrac{1}{271}$.

We can also divide and multiply fractions on a calculator. For example to do $2\frac{18}{25} \div 2\frac{71}{100}$ on a calculator, PRESS

| 2 | a b/c | 18 | a b/c | 25 | ÷ | 2 | a b/c | 71 | a b/c | 100 | = | which should show

| 1 ⌐ 1 ⌐ 271 |.

Get familiar with your calculator to compute fractions.

SUMMARY

To add and subtract fractions we first have to find the lowest common multiple (LCM) of the denominators and put each fraction with this new denominator. It is then just a matter of adding or subtracting the numerators.

For multiplication of fractions we simply multiply the numerators and denominators separately. Cancelling a common factor first simplifies the multiplication.

For division, invert the second fraction and then multiply the fractions.

Exercise **Intro(e)**

Solutions at end of book. Complete solutions available at www.palgrave.com/science/engineering/singh

1 Determine the following:

a $\frac{1}{2} + \frac{1}{3}$ **b** $\frac{1}{4} + \frac{1}{9}$ **c** $\frac{1}{6} + \frac{1}{60}$

d $\frac{1}{5} + \frac{1}{7}$ **e** $\frac{2}{3} + \frac{6}{8}$

2 Evaluate

a $\frac{1}{2} - \frac{1}{3}$ **b** $\frac{22}{7} - \frac{16}{9}$

c $\frac{41}{29} - \frac{204}{145}$ **d** $\frac{3}{31} - \frac{2}{41}$

3 Calculate

a $\frac{1}{2} + \frac{1}{3} + \frac{1}{6}$ **b** $\frac{1}{2} + \frac{1}{3} - \frac{1}{4} + \frac{1}{5}$

c $\frac{1}{7} - \frac{1}{3} + \frac{2}{8}$

4 Compute the following and write your final answers in their simplest form:

a $\frac{1}{2} \times \frac{1}{3}$ **b** $\frac{17}{12} \times \frac{84}{60}$

c $\frac{235}{19} \times \frac{38}{5}$

5 Write the following as a single fraction in its simplest form:

a $\frac{1}{2} \div \frac{1}{3}$ **b** $\frac{99}{70} \div \frac{22}{7}$

c $\frac{235}{19} \div \frac{5}{38}$

For question 6 use your calculator.

6 Evaluate the following:

a $1 + \frac{24}{60} + \frac{51}{3600}$

b $\frac{1}{56} + \frac{1}{679} + \frac{1}{776}$

c $\frac{1351}{780} - \frac{265}{153}$ **d** $\frac{377}{120} - \frac{333}{106}$

e $\frac{577}{408} \times \frac{17}{12}$ **f** $\left(\frac{17}{12}\right)^2$

g $\frac{41}{29} \div \frac{99}{70}$ **h** $\frac{377}{120} \div \frac{22}{7}$

SECTION F **Decimals**

By the end of this section you will be able to
▶ understand the decimal number system
▶ round off numbers correct to a number of decimal places or significant figures
▶ understand the term irrational numbers

F1 **Decimal number system**

You must have seen numbers in decimal format such as 253.679.

? **What does the point (.) in 253.679 represent?**

Its called the decimal point and is the cut-off position between whole numbers and fractions of whole numbers.

? **What does the 253 signify in 253.679?**

253 is made up from 2 hundreds, 5 tens and 3 units:

Hundreds (100)	Tens (10)	Units (1)
2	5	3

Thus

$$253 = (2 \times 100) + (5 \times 10) + (3 \times 1)$$

? **What is the value of .679 in 253.679?**

.679 is 6 tenths, 7 hundredths and 9 thousandths:

tenths $\left(\dfrac{1}{10}\right)$	hundredths $\left(\dfrac{1}{100}\right)$	thousandths $\left(\dfrac{1}{1000}\right)$
6	7	9

Thus

$$.679 = \left(6 \times \frac{1}{10}\right) + \left(7 \times \frac{1}{100}\right) + \left(9 \times \frac{1}{1000}\right)$$

Table 1 shows the place value of certain numbers with respect to the decimal point.

Number	Hundreds (100)	Tens (10)	Units (1)	Decimal point	Tenths $\left(\dfrac{1}{10}\right)$	Hundredths $\left(\dfrac{1}{100}\right)$	Thousandths $\left(\dfrac{1}{1000}\right)$
53.76		5	3	.	7	6	
0.213			0	.	2	1	3
25.001		2	5	.	0	0	1
829.705	8	2	9	.	7	0	5

TABLE 1

? **What is the place value of 1 in 25.001?**

Since it is in the 3rd position after the decimal point, its value is 1 thousandth, or $\dfrac{1}{1000}$.

Let's investigate decimal places.

F2 Decimal place

In countless numbers of cases the calculator does not give you an accurate answer but is only correct to the number of digits on your display. We can approximate numbers by stating them accurate to a number of decimal places (d.p.). For example

$$25.3691 = 25.369 \text{ (correct to 3 d.p.)}$$

$$= 25.37 \text{ (correct to 2 d.p.)}$$

$$= 25.4 \text{ (correct to 1 d.p.)}$$

For 3 decimal places you examine the fourth number after the decimal point:

25.369 1
     fourth
     number

If this number is 5 or larger we increase the third number after the decimal point by 1, otherwise we leave the third number intact:

25.36 9 1
    third
    number

Since we are interested in writing the number correct to 3 d.p. we erase the fourth number after the decimal point. In this case

$$25.3691 = 25.369 \text{ (correct to 3 d.p.)}$$

Similarly, for 2 decimal places, since 9 is greater than 5 we increase the second number after the decimal point to 7 and erase the numbers to the right:

$$25.3691 = 25.37 \text{ (correct to 2 d.p.)}$$

For 1 decimal place we examine the second position after the decimal point:

25.3 6 91
   second
   number

and since this is bigger than 5 we increase the first number after the decimal point to 4 and erase the remaining numbers to the right:

$$25.3691 = 25.4 \text{ (correct to 1 d.p.)}$$

Example 14

Write the following numbers correct to 2 d.p.:

a 0.534   **b** 7.697
c 0.0461   **d** 0.005

Solution

a Since the 4 in 0.534 is less than 5 we leave the second number after the decimal place intact:

$$0.534 = 0.53 \text{ (2 d.p.)}$$

b Since the last 7 on the right in 7.697 is bigger than 5 we increase the 9 to 10. Thus

$$7.697 = 7.70 \text{ (2 d.p.)}$$

Notice that 69 becomes 70 after the decimal point.

c Similarly 0.0461 = 0.05 (2 d.p.).

d Since we have 5 in the third position after the decimal point in 0.005 we increase the 0 before 5 by 1. Thus

$$0.005 = 0.01 \text{ (2 d.p.)}$$

(2 d.p.) means the answer is correct to 2 decimal places.

F3 Significant figures

We can also approximate numbers by using significant figures. The following illustrates the significance of each of the digits in 386:

3 8 6

The 3 in 386 has a value of 300, the 8 a value of 80 and the 6 a value of 6. Note that the 3 is the most significant figure although it is the smallest digit in 386. Rounding 386 to 1 significant figure (1 s.f.) gives 400 because the 8 next to the 3 is bigger than 5, so we increase 3 to 4 and place zeros after that. Writing a number to 1 significant figure means that at most one digit is not zero, hence 386 rounds to 400. **What is 386 to 2 significant figures?**

Since the 6 is larger than 5 we increase the 8 to 9 and place a zero in the remaining right position:

$$386 = 390 \text{ (2 s.f.)}$$

(2 s.f.) means that the answer is correct to 2 significant figures.

Example 15

Write the following correct to the number of significant figures stated in the brackets:

a 7.526 (2 s.f.)   **b** 186 895 (3 s.f.)   **c** 0.000 617 (1 s.f.)

d 9.99 (1 s.f.)   **e** 11.399 (2 s.f.)

Solution

a In 7.526, the 2 in the 3rd position is less than 5 so
$$7.526 = 7.5 \ (2 \text{ s.f.})$$

b Similarly 186 895 = 187 000 (3 s.f.)

c In this case we examine the second digit after the zeros:
$$0.000\ 6 \quad \underline{1} \quad 7$$
$$\text{this digit}$$
Since 1 is less than 5 we leave the 6 intact, thus
$$0.000\ 617 = 0.000\ 6 \ (1 \text{ s.f.})$$

d Similarly 9.99 = 10.00 (1 s.f.)

e Since 3 is less than 5 in 11.399, therefore
$$11.399 = 11.000 \ (2 \text{ s.f.})$$

F4 Irrational numbers

There are some fundamental numbers in engineering such as π (pi), $\sqrt{2}$ and e. You might remember π from the area of a circle (πr^2).

What is the exact value of π?

If we evaluate π on a 10-digit calculator it shows

3.141592654

Of course this is **only** correct to 10 s.f., π goes on for an infinite number of decimal places. Today π has been approximated to more than a billion decimal places. Yes, I said approximated because it is **not** the exact value of π. We cannot write down the exact fraction. The only way to write π exactly is to leave it as π.

There is another number called e which like π cannot be expressed exactly. The number e correct to 12 d.p. is

$$e = 2.718281828459$$

The number e crops up in many fields of engineering such as mechanics, aerodynamics, electronics, control etc. We will examine this number further in **Chapter 5**. Another notable number is $\sqrt{2}$. On a 10-digit calculator

$$\sqrt{2} = 1.414213562 \text{ (correct to 9 d.p.)}$$

Again we cannot evaluate the correct value of $\sqrt{2}$ in decimal or fraction format. The only way to write $\sqrt{2}$ exactly is to leave it as $\sqrt{2}$.

A number which we cannot write as a whole number or a fraction is called an **irrational number.**

These numbers, π, e and $\sqrt{2}$, are all examples of irrational numbers.

There are an infinite number of irrational numbers on the real line, such as $\sqrt{3}$, $\sqrt{5}$, 3π, etc.

SUMMARY

We can round off numbers to so many decimal places (d.p.) or significant figures (s.f.).

Irrational numbers often used in engineering are π, e and $\sqrt{2}$.

Exercise **Intro(f)**

Solutions at end of book. Complete solutions available at www.palgrave.com/science/engineering/singh

1 Write the numbers correct to the number of decimal places stated:

a 1.618 034	**i** 3 d.p.	**ii** 2 d.p.
b 4.6692	**i** 4 d.p.	**ii** 2 d.p.
c 2.502 907 875	**i** 3 d.p.	**ii** 1 d.p.
d 0.373 96	**i** 3 d.p.	**ii** 2 d.p.

2 Write the following numbers correct to the number of significant figures stated:

a 1.618 034	**i** 3 s.f.	**ii** 2 s.f.
b 2.973 214	**i** 5 s.f.	**ii** 2 s.f.
c 0.110 001	**i** 4 s.f.	**ii** 1 s.f.
d 9.869	**i** 2 s.f.	**ii** 1 s.f.

3 Write the following numbers to 2 s.f.:

a 1729	**b** 99 954
c 107 928 278 317	

4 Write π, e and $\sqrt{2}$ to

a 2 d.p.   **b** 2 s.f.

SECTION G **Powers of 10**

By the end of this section you will be able to:
▶ write numbers in standard form
▶ examine the SI units and their symbols
▶ write physical quantities in the most appropriate symbol format
▶ estimate basic arithmetical results

G1 **Standard form**

? **What is the value of 10^2, 10^3 and 10^4?**

These are all powers of 10 and can be expanded as

$$10^2 = 10 \times 10 = 100$$

$$10^3 = 10 \times 10 \times 10 = 1000$$

$$10^4 = 10 \times 10 \times 10 \times 10 = 10\ 000$$

? We have $10^2 = 100$, $10^3 = 1000$ and $10^4 = 10\,000$. **What do you notice?**

The index number (superscript number) is the same as the number of zeros after 1.

We will use this concept in this section:

$$10^{\text{index number}} = 1\ \underbrace{0...0}_{\substack{\text{number of zeros} \\ = \text{ index number}}}$$

? **However, can you remember how you multiplied numbers by 10?**

We moved all the figures one place to the left, for example

$$7.530 \times 10 = 75.30$$

? **How did you multiply by 100?**

We moved all the figures two places to the left, for example

$$7.530 \times 100 = 753.0\acute{}$$

? **What about multiplying by 1000?**

We moved all the figures three places to the left, for example

$$7.530 \times 1000 = 7530$$

Writing a number in standard form involves going the other way, that is we are given a number and we write this as a number between 1 and 10 multiplied by a power of 10:

$$\text{number} = (\text{number between 1 and 10}) \times 10^{\text{index}}$$

For example 9801 can be written in standard form as

$$9801 = 9.801 \times 1000 = 9.801 \times 10^3$$

? **What's the point? Why can't we leave it as 9801?**

Well we can but this particular notation (standard form) allows us to express very large numbers in a more compact form. For example the distance of the earth from the sun is 93 million miles, or 93 000 000 miles, and writing this number in standard form gives
? a more concise representation. **How can we write 93 000 000 in standard form?**
93 times a million, and since a million has 6 zeros:

$$93\,000\,000 = 93.0 \times 10^6$$

The number is not in standard form. Remember the first number should be between 1 and 10 but we have 93.0.

We have to move the decimal point in 93.0 one place to the left which means multiplying by another 10.

The index of 10 is 7 because it is $10^6 \times 10$. So we have

$$93\,000\,000 = 9.3 \times 10^7$$

Let's do some more examples.

Example 16

Write the following numbers in standard form:

a 333 000
b 133 000 000 000 000 000 000
c 1054.35

Solution

We need to write each number between 1 and 10 multiplied by a power of 10.

a **The number between 1 and 10 is 3.33, so how many places does the decimal point have to move?**

$$333\ 000 = \underbrace{333\ 000.0}_{5\ \text{places}}$$

5 places to the left. Thus $333\ 000 = 3.33 \times 10^5$.

b **The number of zeros in this large number is 18 but then we have 133 which we need to write as 1.33×100. Hence there are 2 more places to move, which makes 20 altogether. We have**

$$\underbrace{133\ 000\ 000\ 000\ 000\ 000\ 000}_{20\ \text{places}} = 1.33 \times 10^{20}$$

Notice how such a large number can be written in compact form.

c **What is the number between 1 and 10 for 1054.35?**

It's 1.05435, so how many places do we shift the decimal point?

$$1\underbrace{054.35}$$

3 places to the left. Hence

$$1054.35 = 1.05435 \times 10^3$$

We need to use the following rule for the next part:

$$\frac{1}{10^{\text{index}}} = 10^{-\text{index}}$$

Small numbers can also be represented in standard form. By small numbers we mean numbers between 0 and 1. For small numbers we multiply by $\frac{1}{10}, \frac{1}{100}, \frac{1}{1000}$, etc. which is the same as dividing by 10, 100, 1000, etc. respectively.

? **What is the result of multiplying 753.67 by $\frac{1}{10}$ (or dividing by 10)?**

All the figures move to the right by 1 place:

$$753.67 \times \frac{1}{10} = 75.367$$

Similarly, multiplying by $\frac{1}{100}$ moves all the figures to the right by 2 places:

$$753.67 \times \frac{1}{100} = 7.5367$$

? **What is the consequence of multiplying by $\frac{1}{1000}$?**

All the figures move to the right by 3 places:

$$753.67 \times \frac{1}{1000} = 0.75367$$

Using $10^1 = 10$, $10^2 = 100$ and $10^3 = 1000$, we can write $\frac{1}{10}$, $\frac{1}{100}$ and $\frac{1}{1000}$ in terms of negative indices because $\frac{1}{10^{index}} = 10^{-index}$:

$$\frac{1}{10} = \frac{1}{10^1} = 10^{-1}$$

$$\frac{1}{100} = \frac{1}{10^2} = 10^{-2}$$

$$\frac{1}{1000} = \frac{1}{10^3} = 10^{-3}$$

? **How can we write 0.007 78 in standard form?**

Remember numbers in standard form can be written as

$$(\text{number between 1 and 10}) \times 10^{index}$$

? **For 0.007 78 the number between 1 and 10 is 7.78, so how many places do we have to move the decimal point?**

$$0.\underbrace{007}_{3 \text{ places}} 78, \text{ we move 3 places to the right}$$

Hence $0.007\,78 = 7.78 \times 10^{-3} (= 7.78 \div 10^3)$.

Notice the negative index, that's because we are shifting the decimal point to the right.

Example 17

Write each of the following numbers in standard form:

a 0.000 072 92   **b** 0.000 000 000 066 73   **c** $2 \div 10^5$

Solution

a **The number between 1 and 10 is 7.292, so how many places do we shift the decimal point?**
0.000 072 92, 5 places to the right. Hence

$$0.000\ 072\ 92 = 7.292 \times 10^{-5}$$

b **What number do we write down between 1 and 10 for 0.000 000 000 066 73?**

6.673

We shift the decimal point by 11 places to the right:

0.000 000 000 066 73

Hence $0.000\ 000\ 000\ 066\ 73 = 6.673 \times 10^{-11}$.

c We can write 10^5 as $\dfrac{10^5}{1}$ because dividing by 1 gives the same number:

$$2 \div 10^5 = 2 \div \frac{10^5}{1} = 2 \times \frac{1}{10^5} = 2 \times 10^{-5} \text{ (turning the second fraction upside down)}$$

We can use a calculator to enter numbers in standard form. Use the $\boxed{\text{EXP}}$ (or $\boxed{\text{E}}$ or $\boxed{\text{EE}}$) button on your calculator. For example to enter 1.6×10^{-19} on a calculator, PRESS $\boxed{1.6}$ $\boxed{\text{EXP}}$ $\boxed{(-)}$ $\boxed{19}$ $\boxed{=}$, which should show 1.6^{-19}.

Standard form is also known as exponential notation, or scientific notation, that's the reason for the $\boxed{\text{EXP}}$ button. Most calculators can place numbers in standard form. See the handbook of your calculator.

G2 SI units

In engineering, some indices of 10 are more popular than others, particularly the multiples of 3: 10^3, 10^6, 10^{-3}, etc. Moreover these powers of 10 have a particular name, for example 10^3 is kilo. When we say Milton Keynes is 85 kilometres from London we mean

$$85 \times 10^3 \text{ metres or 85 000 metres}$$

The kilo is a prefix. Table 2 shows the prefix for some powers of 10 and their associated symbols.

Power of 10	Prefix	Symbol for prefix
10^{18}	exa	E
10^{15}	peta	P
10^{12}	tera	T
10^{9}	giga	G
10^{6}	mega	M
10^{3}	kilo	k
10^{-3}	milli	m
10^{-6}	micro	μ
10^{-9}	nano	n
10^{-12}	pico	p
10^{-15}	femto	f
10^{-18}	atto	a

Generally we will use the SI (*Système International*) units of measurement. Table 3 shows some primary physical quantities with SI units and their associated symbol for the SI unit.

TABLE 3

Quantity	SI unit	Symbol for SI unit
Mass	kilogram	kg
Length	metre	m
Time	second	s
Resistance	ohm	Ω
Capacitance	farad	F
Current	ampere	A
Inductance	henry	H
Potential	volt	V

Quantity	SI unit	Symbol for SI unit
Power	watt	W
Frequency	hertz	Hz
Pressure	pascal	Pa
Energy	joule	J
Temperature	kelvin	K
Force	newton	N

TABLE 3 CONTINUED

The SI units also allow the use of some secondary units such as hour, day, etc. We can use Table 2 in conjunction with Table 3 to express a physical quantity. For example the distance 85 thousand metres is better expressed as 85 kilometres, or in symbol format 85 km.

Example 18 *electrical principles*

Write the following electrical quantities in the most appropriate symbol format:

a 0.005 amp   **b** 0.000 003 5 farad   **c** 15 000 000 ohms

Solution

a $0.005 \text{ amp} = 5 \times 10^{-3} \text{ amp} = 5 \underbrace{\text{milli}}_{\text{by Table 2}} \text{ amps} = 5 \text{ mA}$

b $0.000\ 003\ 5 \text{ farad} = 3.5 \times 10^{-6} \text{ farad} = 3.5 \underbrace{\text{micro}}_{\text{by Table 2}} \text{ farads} = 3.5\ \mu\text{F}$

c $15\ 000\ 000 \text{ ohms} = 15 \times 10^{6} \text{ ohms} = 15 \underbrace{\text{mega}}_{\text{by Table 2}} \text{ ohms} = 15\ \text{M}\Omega$

Note that the number does not have to be in standard form.

Example 19

Convert the following into the appropriate units of Table 3:

a 85 km   **b** 15 mH   **c** 15 GHz   **d** 3 ns   **e** 100 pF   **f** 2.54 mm

Solution

a 85 km is 85 kilometres, so we need to convert this into metres. **What does kilo represent?**

Kilo = 10^3, hence 85 km = 85×10^3 m

Example 19 *continued*

b 15 mH is 15 millihenrys. We need to convert 15 mH into henrys because that is the unit of inductance in Table 3. Remember that milli = 10^{-3}, so we have

$$15 \text{ mH} = 15 \times 10^{-3} \text{ H}$$

c **What is 15 GHz?**

15 gigahertz. **What does giga represent?**

Giga is 10^9, so

$$15 \text{ GHz} = 15 \times 10^9 \text{ Hz}$$

d 3 ns is 3 nanoseconds. By Table 2, nano = 10^{-9}, hence

$$3 \text{ ns} = 3 \times 10^{-9} \text{ s}$$

e 100 pF is 100 picofarads. **What does pico represent?**

By Table 2, pico is 10^{-12}:

$$100 \text{ pF} = 100 \times 10^{-12} \text{ F}$$

f 2.54 mm is 2.54 millimetres. Hence

$$2.54 \text{ mm} = 2.54 \times 10^{-3} \text{ m}$$

G3 Estimation and accuracy

You often find that people go to extreme lengths to find an accurate answer to a numerical engineering problem. However in many cases this is not required. For example in some electronics industries you can only obtain certain resistance values such as 10, 12, 15.... Moreover each resistor has a tolerance limit and a $10\,\Omega$ resistor may vary between say $9\,\Omega$ and $11\,\Omega$. Thus the resistor value lies between two limits. So evaluating a resistance value correct to 10 decimal places in an electrical problem is laborious and superfluous. Generally there is widespread use of tolerance limits in the field of engineering.

So why do we go to these extraordinary measures to evaluate answers accurate to the number of digits on our calculator?

Most students copy down the numbers from the display of their calculator. It's very straightforward to plug numbers into your calculator and then reproduce the results on paper. Some students have adopted this habit from their primary school years and find difficulty in abandoning this procedure.

It's valuable practice to estimate your answer first and then evaluate it on a calculator. Let's investigate by an example.

Example 20

Estimate the following:

a 705×0.83 **b** $70.4 \div 0.73$ **c** $\dfrac{510 \times 0.62}{0.33}$

Solution

a We can round 705 to 700 and 0.83 to 0.8. Thus

$$705 \times 0.83 \approx 700 \times 0.8 = 70 \times 8 = 560$$

So 705×0.83 is approximately 560. The notation $\approx$ is used for approximation.

b We can round 70.4 to 70 and 0.73 to 0.7. We have

$$70.4 \div 0.73 \approx 70 \div 0.7 = 700 \div 7 = 100$$

c Similarly rounding 510 to 500, 0.62 to 0.6 and 0.33 to 0.3 we have

$$\frac{510 \times 0.62}{0.33} \approx \frac{500 \times 0.6}{0.3}$$

$$= \frac{50 \times 6}{0.3}$$

$$= \frac{300}{0.3}$$

$$= \frac{3000}{3}$$

$$= 1000$$

So $\dfrac{510 \times 0.62}{0.33}$ is approximately 1000. Compare these results with the evaluation on your calculator.

Estimation gives you a rough guide to your final result and can be used as a check on your calculator output.

The calculator or computer are very useful tools but we need to be cautious. There is a grave danger that you may bypass the thinking phase and carry out evaluations on autopilot. I have heard of students implementing division by 1 on a calculator.

SUMMARY

A number in standard form is written as

(number between 1 and 10) $\times 10^{index}$

Large numbers have a positive index and small numbers have a negative index.

In engineering, indices of multiples of 3 are popular and are given a particular name.

The SI units in Table 3 are the units for physical quantities.

In many engineering problems you do not require exact evaluations but a close approximation will do.

Exercise **Intro(g)**

Solutions at end of book. Complete solutions available at www.palgrave.com/science/engineering/singh

1 Write the following numbers in standard form:

  a 186 000   **b** 1 392 000
  c 136 000   **d** 0.000 000 034 39
  e 0.000 000 095 1   **f** 0.009 29
  g 0.000 025 8   **h** 14.96×10^{6}
  i 273.15   **j** 7.6×10^{2}

2 The following numbers are in standard form. Write them in conventional form:

  a 6.4×10^{6} **b** 3.3×10^{-9}
  c 7.292×10^{-5} **d** 3×10^{8}

3 Arrange the following numbers in order of size, smallest first:

  a 12 750, 12.75×10^{2}, 12.75×10^{-3} and 12.75
  b 3.14×10^{3}, $3.14 \div 10^{3}$ and 3.14×10^{-2}

4 Evaluate the following on a calculator and give your final answer correct to 2 decimal places:

  a $\dfrac{1.25 \times 10^{3} \times 0.15 \times 348}{15 \times 10^{5}}$

  b $0.5 \times 9.11 \times 10^{-31} \times (1.86 \times 10^{5})^{2} \times 10^{20}$

  c $2\pi\sqrt{\dfrac{6 \times 10^{-2}}{9.81}}$

5 [*electrical principles*]   Write the following electrical quantities in the most appropriate symbols:

  a 100×10^{-12} farads
  b 30 000 ohms
  c 0.000 3 amps

6 [*mechanics*]   Write the following mechanical quantities in the most appropriate symbol format:

  a 8536 N   **b** 75 000 000 W
  c 0.2×10^{12} Pa

7 Rewrite the following quantities in terms of the SI units in Table 2 (in standard form):

  a 3000 mm   **b** 573 kN
  c 25 MJ   **d** 12 ps
  e 25 mW

8 Estimate the following:

  a $\dfrac{22}{7}$ **b** $\dfrac{333}{106}$

  c 99×99 **d** $\dfrac{714 \times 0.63}{14.45}$

> ## SECTION H **Conversion**
>
> By the end of this section you will be able to:
> ► convert SI units
> ► convert imperial to SI units

H1 **Conversion of units**

Generally we use SI units throughout the book. Conversion of units is best understood by examples.

Example 21 *mechanics*

An object has a velocity of 5 m/s. Find the velocity in km/h, where h = hour.

Solution

What does 5 m/s mean?

5 m/s means the object has covered a distance of 5 m in one second.

How many metres will it cover in one hour?

Since there are 60×60 seconds (s) in one hour, the object covers

$$5 \times 60 \times 60 = 18\ 000$$

18 000 metres in one hour.

How can we write this in km/h (km per hour)?

Divide the result by 1000 because 1000 m = 1 km:

$$\frac{18000}{1000} = 18$$

Thus 5 m/s = 18 km/h.

Example 22 *fluid mechanics*

The density of mercury is 13.55 g/cm³. Write this in units of kg/m³ (g = gram, cm = centimetre and 100 cm = 1 m).

Solution

13.55 g/cm³ means 1 cm³ of mercury has a mass of 13.55 g. We need to find the mass for 1 m³.

Example 22 *continued*

What is 1 m³ in terms of (centimetre)³?

$$1 \text{ m}^3 = 1 \text{ m} \times 1 \text{ m} \times 1 \text{ m} = 100 \text{ cm} \times 100 \text{ cm} \times 100 \text{ cm}$$

How many grams is 1 m³ of mercury?

$$13.55 \times 100 \times 100 \times 100 = 13\,550\,000$$

So 1 m³ of mercury has a mass of 13 550 000 grams. **Since we need kg/m³, we have to write 13 550 000 grams in kg. How?**

Divide by 1000 because 1000 g = 1 kg:

$$\frac{13\,550\,\cancel{000}}{1\cancel{000}} = 13\,550$$

So 1 m³ has a mass of 13 550 kg, hence 13.55 g/cm³ = 13 550 kg/m³.

Example 23

Given that 1 mile = 1.6 km, convert 70 miles per hour to m/s.

Solution

We have

$$70 \text{ miles} = 70 \times 1.6 \text{ km} = 112 \text{ km}$$

What is 112 km in metres (m)?

$$112 \text{ km} = 112 \times \underset{\text{kilo} = 1000}{\underline{1000}} \text{ m} = 112\,000 \text{ m}$$

70 miles per hour = 112 000 metres per hour, but we need to write this in metres per second.

How many seconds are there in 1 hour?

$$1 \text{ hour} = 60 \text{ minutes} = 60 \times 60 \text{ s} = 3600 \text{ s (s = seconds)}$$

How do we convert 112 000 metres per hour into m/s?

Divide by 3600:

$$70 \text{ miles per hour} = \frac{112\,000}{3600} \text{ m/s} = 31.11 \text{ m/s (2 d.p.)}$$

The answer is correct to 2 decimal places.

The unit of miles per hour (mph) is an example of imperial units. There are other imperial units such as feet per second, inches, ft³, etc.

SUMMARY

For SI units we look at the prefix of the unit and then multiply or divide accordingly. For converting imperial units into SI units we multiply (or divide) by the relevant factor.

Exercise Intro(h)

Solutions at end of book. Complete solutions available at www.palgrave.com/science/engineering/singh

Use the following data (correct to 3 d.p.) for this exercise:

1 foot = 0.305 m, 1 km = 0.621 mile and 1 mile = 1.609 km

1 [mechanics] An object has a velocity of 31 m/s. Convert this into km/h.

2 [mechanics] Acceleration due to gravity is 32.174 feet/sec^2. Write this in m/s^2.

3 [mechanics] The velocity of sound is 342 m/s. Express this in miles per hour.

4 [fluid mechanics] The density of air is 1.206 kg/m^3. Convert this into g/cm^3.

5 [materials] The second moment of area of a body is 0.63 cm^4. Convert this into m^4.

6 Convert 5 feet^2 to m^2.

7 Convert the following into m/s:
   a 30 miles per hour   **b** 80 km/h
   c 186 000 miles per second

8 [materials] Modulus of elasticity is 0.2×10^{11} N/m^2. Write this in N/cm^2.

SECTION I Arithmetical operations

By the end of this section you will be able to:
▶ use BROIDMAS to understand the priority of arithmetical operations
▶ evaluate arithmetical problems using BROIDMAS

I1 Sequence of operations

The sequence of operations such as $+, \times, \sqrt{\ }$, etc. is critical in evaluating arithmetical problems. **For example what is the value of $5 + 2 - 3 \times 6$?**

Which operation is carried out first? The addition, multiplication or subtraction?

We use the mnemonic BROIDMAS to remember the priority of operations:

Brackets   } First

ROots
Indices   } Second

$$\left.\begin{array}{l}\underline{\text{D}}\text{ivision}\\ \underline{\text{M}}\text{ultiplication}\end{array}\right\}\text{Third}$$

$$\left.\begin{array}{l}\underline{\text{A}}\text{ddition}\\ \underline{\text{S}}\text{ubtraction}\end{array}\right\}\text{Fourth}$$

Combining the underlined letters makes the word **BROIDMAS**.

 To evaluate $5 + 2 - 3 \times 6$, which operation do we implement first?

Multiplication, because it has priority over addition and subtraction:

$$5 + 2 - 3 \times 6 = 5 + 2 - 18$$

At this point, $5 + 2 - 18$, we work from left to right because addition and subtraction have the same priority. Hence

$$\underset{=7}{\underline{5+2}} - 18 = 7 - 18 = -11$$

Let's proceed with another example.

Example 24

Compute

(†) $\left(5 + \sqrt{25 - 16}\,\right) \div 2$

Solution

 What do we evaluate first in $\left(5 + \sqrt{25 - 16}\,\right) \div 2$?

Brackets have first priority, so we initially need to calculate

$$5 + \sqrt{25 - 16}$$

Within this, the square root has priority over addition (B<u>RO</u>IDM<u>A</u>S). We have to first carry out the subtraction and then take the square root:

$$\sqrt{25 - 16} = \sqrt{9} = 3$$

Hence

$$5 + \sqrt{25 - 16} = 5 + 3 = 8$$

Replacing $5 + \sqrt{25 - 16}$ with 8 in (†) gives

$$8 \div 2 = 4$$

Therefore the final result is 4. Note that we could write the above as

(*) $\left(5 + \sqrt{25 - 16}\ \div\ 2\right) = \dfrac{5 + \sqrt{25 - 16}}{2} = 4$

Even if there are no brackets around the numerator (the middle expression in (*)) we still divide the whole numerator by 2. We will encounter these types of arithmetical problems in subsequent chapters.

Example 25

Compute the following:
$$6^3 - 5 \times 5^3 + 10 \times 4^3 - 10 \times 3^3 + 5 \times 2^3 - 1$$

Solution

Which operation do we employ first?

BRO**I**D**M**AS discloses that the indices have preference over multiplication, so we first evaluate the indices by using a calculator. Hence
$$6^3 = 216, \quad 5^3 = 125, \quad 4^3 = 64, \quad 3^3 = 27 \text{ and } 2^3 = 8$$
Substituting these into the original problem gives
$$216 - \underbrace{5 \times 125}_{=625} + \underbrace{10 \times 64}_{=640} - \underbrace{10 \times 27}_{=270} + \underbrace{5 \times 8}_{=40} - 1$$
$$= 216 - 625 + 640 - 270 + 40 - 1$$
$$= 0$$
Hence
$$6^3 - 5 \times 5^3 + 10 \times 4^3 - 10 \times 3^3 + 5 \times 2^3 - 1 = 0$$
Note that 5×5^3 can written as 5^4 because it is $5 \times \underbrace{5 \times 5 \times 5}_{3 \text{ copies}}$.

Example 26

Compute $\dfrac{4}{3} \pi \times 5^3$ correct to 2 decimal places.

Solution

If we want our final answer correct to 2 d.p. then we have to work to 3 d.p. and round off at the end.

How do we calculate $\dfrac{4}{3} \pi \times 5^3$?

Employing BRO**I**D**M**AS tells us that indices have preference over multiplication, so we first calculate 5^3:
$$5^3 = 125$$
We have
$$\frac{4}{3} \pi \times 5^3 = \frac{4}{3} \pi \times 125$$
$$= \frac{4}{3} \pi \times 125$$
Multiplication and division have the same priority so we can multiply $4 \times \pi \times 125$, and then divide the result by 3; or we can divide first, $4 \div 3$, and then multiply the result by π and 125:
$$\frac{4}{3} \pi \times 125 = (4 \times \pi \times 125) \div 3 = 523.60 \ (2 \text{ d.p.})$$

SUMMARY

BROIDMAS is used as a priority list to evaluate arithmetical problems.

Exercise **Intro(i)**

Solutions at end of book. Complete solutions available at www.palgrave.com/science/engineering/singh

1 Evaluate the following:

  a $(1 + 3 + 5) \times (1 \times 3 \times 5)$
  b $(1 + 4 + 4) \times (1 \times 4 \times 4)$
  c $(49 \times 1) + (49 \times 3) + (49 \times 5)$
  d $1 + 7^2 + 5^3$
  e $(2 \times 4^3) + (4 \times 3^3) + (3 \times 1)$
  f $(2 \times 3 \times 5 \times 7 \times 11) + 1$

2 Calculate

  a $2^{11} - 1$ **b** $\dfrac{1 - 5}{2}$

  c $\dfrac{1}{2} - \dfrac{1}{3 \times 2^3} + \dfrac{1}{5 \times 2^5}$

3 Compute

  a $\dfrac{1 + \sqrt{5}}{2} + \dfrac{1 - \sqrt{5}}{2}$

  b $\dfrac{5 + \sqrt{25 - 24}}{2}$

  c $\dfrac{5 \pm \sqrt{(-5)^2 - (4 \times 1 \times 6)}}{2}$

  d $\dfrac{-4 \pm \sqrt{(4)^2 - 4 \times (-5)}}{2}$

  e $\dfrac{-65 \pm \sqrt{(-65)^2 + (4 \times 14 \times 25)}}{28}$

4 Evaluate the following correct to 2 decimal places:

  a $\pi \times 7^2 \times 3$

  b $\dfrac{\pi}{3} \times 4^2 \times 5$ **c** $\dfrac{4\pi \times 5^3}{3}$

  d $2\pi \times 7(5 + 7)$

  e $2\pi \sqrt{\dfrac{3^2 + 4^2}{2}}$

5 Evaluate

  a $2^4(2^5 - 1)$ **b** $2^{2^3} + 1$

SECTION J **Percentages**

By the end of this section you will be able to:
▶ convert percentages to a fraction or a decimal
▶ convert a fraction or a decimal to a percentage
▶ apply percentages to engineering problems

J1 **Percentages, fractions and decimals**

? What does the word *percentage* mean?

It means out of 100 and is denoted by the symbol %. For example 10% means 10 out of 100 or $\dfrac{10}{100}$ ($= 10 \div 100$):

$$10\% = \frac{10}{100} = \frac{1}{10} = 0.1$$

These are **all** equivalent. Hence we can write a percentage as a fraction or a decimal. Let's investigate percentages by doing some examples.

Example 27

Convert the following percentages to their simplest fractions:

a 15%   **b** $8\frac{1}{3}\%$

Solution

a What does 15% mean?

15 out of 100:

$$15\% = \frac{15}{100} = \frac{5 \times 3}{5 \times 20} = \frac{3}{20} \qquad \text{[Cancelling 5's]}$$

Hence $15\% = \frac{3}{20}$.

b Similarly $8\frac{1}{3}\% = \frac{8^{1}/_{3}}{100}$. **How else can we write $8\frac{1}{3}$?**

As a top-heavy fraction: $8\frac{1}{3} = \frac{(8 \times 3) + 1}{3} = \frac{25}{3}$

$$\frac{8^{1}/_{3}}{100} = \frac{25/3}{100} = \frac{25}{3} \div 100$$

$$= \frac{25}{3} \div \frac{100}{1} \qquad \left[\text{Because } 100 = \frac{100}{1} \right]$$

$$= \frac{25}{3} \times \frac{1}{100} \qquad \left[\begin{array}{l} \text{Inverting the} \\ \text{second fraction} \end{array} \right]$$

$$= \frac{25}{3} \times \frac{1}{4 \times 25}$$

$$= \frac{1}{3 \times 4} = \frac{1}{12} \qquad \text{[Cancelling 25's]}$$

Hence $8\frac{1}{3}\% = \frac{1}{12}$.

To convert a percentage into a decimal number is straightforward, for example 16.8% means 16.8 out of 100 which can also be represented by

$$\frac{16.8}{100} = 16.8 \div 100 = 0.168$$

(We move the decimal point 2 places to the left.) Hence

$$16.8\% = 0.168$$

Conversely we can write decimals and fractions as a percentage. To convert a decimal or a fraction into a percentage, multiply by 100.

Example 28

Transform the following into percentages:

a $\dfrac{9}{20}$ **b** $\dfrac{7}{15}$ **c** 0.6735

Solution

Multiply each of these by 100:

a $\dfrac{9}{2\cancel{0}} \times 10\cancel{0} = \dfrac{9 \times 10}{2} = \dfrac{90}{2} = 45$. Thus $\dfrac{9}{20} = 45\%$.

b Similarly

$$\dfrac{7}{15} \times 100 = \dfrac{7}{\cancel{5} \times 3} \times (\cancel{5} \times 20) = \dfrac{7 \times 20}{3} = \dfrac{140}{3} = 140 \div 3 = 46.67$$

Hence $\dfrac{7}{15} = 46.67\%$ (2 d.p.)

c $0.6735 \times 100 = 67.35$. We have $0.6735 = 67.35\%$.

J2 Evaluation of percentages

It is often easier to use a calculator to evaluate percentages of quantities. There should be a percentage symbol, %, on your calculator. To evaluate $17\frac{1}{2}\%$ of 250 on a calculator, PRESS

$\boxed{250}$ $\boxed{\text{x}}$ $\boxed{17.5}$ $\boxed{\%}$ or $\boxed{250}$ $\boxed{\text{x}}$ $\boxed{17.5}$ $\boxed{\text{SHIFT}}$ $\boxed{=}$, which should show 43.75.

If you want to add this figure, 43.75, to the initial 250, you press the $\boxed{+}$ button after the $\boxed{=}$ button in the above. Similar for subtraction. For other percentage calculations see the handbook of your calculator.

Let's attempt some percentages with engineering examples.

Example 29 *electrical principles*

A resistor has a value of 1300 Ω. If the tolerance limits of the resistor are 3% find the smallest and largest value of the resistor.

Solution

We need to evaluate 3% of 1300.

What does 3% mean?

3 out of 100, $\dfrac{3}{100}$. Hence

$$3\% \text{ of } 1300 = \dfrac{3}{10\cancel{0}} \times 13\cancel{00}$$
$$= \dfrac{3 \times 13}{1} = 39$$

Example 29 *continued*

The resistor value lies between $1300 - 39$ to $1300 + 39$ (or 1300 ± 39). The smallest and largest values are given by $1261\ \Omega$ and $1339\ \Omega$ respectively. We sometimes say the range of the resistor is $1261\ \Omega$ to $1339\ \Omega$.

Example 30 *heat transfer*

A rod has a length of 0.583 m. After heating, it is expanded by 0.23%. Find, correct to 3 d.p., the new length.

Solution

First we want to find 0.23% of 0.583. How?

$$0.23\% = \frac{0.23}{100}, \text{ hence}$$

$$0.23\% \text{ of } 0.583 = \frac{0.23}{100} \times 0.583 = 0.0023 \times 0.583$$

$$= 1.34 \times 10^{-3}$$

The new length $= 0.583 + (1.34 \times 10^{-3}) = 0.584$ m (correct to 3 d.p.).

The percentage error, % error, for an experiment is defined as

$$\boxed{I.1} \qquad \% \text{ error} = \frac{\text{experimental value} - \text{exact value}}{\text{exact value}} \times 100$$

$\boxed{I.1}$ is a reference number which we refer to later on.

Example 31 *heat transfer*

The experimental temperature value at which water freezes is 275.16 K; the exact temperature is 273.15 K.

Find the percentage error in the experimental value.

Solution

Substitute 275.16 for the experimental value and 273.15 for the exact value into $\boxed{I.1}$:

$$\% \text{ error} = \frac{\text{experimental value} - \text{exact value}}{\text{exact value}} \times 100$$

We have

$$\% \text{ error} = \left(\frac{275.16 - 273.15}{273.15} \right) \times 100$$

$$= \left(\frac{2.01}{273.15} \right) \times 100$$

$$= 0.74\% \ (2 \text{ d.p.})$$

SUMMARY

Percentage, %, means out of a 100.

To transform a percentage to a fraction, write the percentage figure over 100 and simplify.

To express a fraction or decimal in terms of a percentage, multiply by 100.

Use your calculator for percentage calculations.

I.1 $\% \text{ error} = \dfrac{\text{experimental value} - \text{exact value}}{\text{exact value}} \times 100$

Exercise Intro(j)

Solutions at end of book. Complete solutions available at www.palgrave.com/science/engineering/singh

1 Convert the following to percentages:

a $\dfrac{1}{10}$ b $\dfrac{1}{12}$

c $\dfrac{7}{8}$ d $\dfrac{3}{13}$

e 0.167   f 2.583

2 Transform the following percentages to fractions in their simplest form:

a 4%   b 9%

c 17.5%   d 2.5%

3 [electrical principles]   The following are tolerance limits of various resistors. Calculate the range of each resistor:

a 900 Ω ± 10%

b 1200 Ω ± 5%

c 27 kΩ ± 1%

d 19 kΩ ± 3%

e 5 MΩ ± 0.15%

4 A speedometer has an error of 4.5%. If the speed indicated is 120 km/h, find the range of the speed.

5 [heat transfer]   The length of a bar is 3.567 m. If the bar expands by 0.15%, what is the new length correct to 4 s.f.?

6 The exact value of sound velocity is 342 m/s. If the experimental value produces 345 m/s, find the percentage error.

7 [thermodynamics]   The volume of a gas in a piston is 0.005 00 m^3. If pressure is applied then the volume of the gas is compressed by 0.57%. Find the new volume.

8 [heat transfer]   A rod of length 0.35 m is expanded to a length of 0.38 m. Find the expansion as a percentage of the original length.

9 [electrical principles]   An ammeter is specified to have a tolerance limit of ± 4.3%. Find the highest value of a current reading of 25 mA.

SECTION K **Ratios**

By the end of this section you will be able to:
▶ understand the term **ratio**
▶ simplify and evaluate ratios

K1 **Ratio**

? **What does the word *ratio* mean?**

It's a relationship that tells us the amount to which one set of objects is compared to another. For example the ratio of boys to girls in a classroom. Consider another example. Concrete is made by mixing 3 bags of sand for every bag of cement and then adding water to the mixture. We say that the sand to cement ratio is 3 to 1 and it is denoted by 3:1. This notation 3:1 is equivalent to $3 \div 1$ or a fraction $\dfrac{3}{1}$. The ratio 3:1 is also the same as 9:3, that is

$$3:1 = 9:3$$

In the case of making concrete it means we need 9 bags of sand for every 3 bags of cement. Multiplying and dividing (not by zero) by the same number leaves the ratio unaltered. Thus

$$3:1 = 6:2 = 9:3 = 27:9 = 1:\frac{1}{3} \text{ etc.}$$

We say 3:1 is the simplified ratio because it is placed into the simplest whole numbers.

Example 32

Simplify the following ratios:

a 7:14   **b** 100:5   **c** $\dfrac{1}{5}:20$ **d** $\dfrac{1}{5}:\dfrac{1}{4}$ **e** 0.36:0.06

Solution

? **a What do 7 and 14 have in common?**

Both have a common factor of 7. Dividing both, 7 and 14, by 7 gives 1 and 2 respectively, hence

$$7:14 = 1:2$$

? **b What is a common factor of 5 and 100?**

5, so dividing both, 100 and 5, by 5 gives 20 and 1 respectively. Hence

$$100:5 = 20:1$$

For **c**, **d** and **e** we need to use whole numbers rather than fractions or decimals.

Example 32 *continued*

c Since we have $\dfrac{1}{5}$, it is better to multiply by 5. Why?

Because $5 \times \dfrac{1}{5} = 1$. We also need to multiply 20 by 5, thus

$$\dfrac{1}{5} : 20 = 1:100$$

d How can we simplify the ratio $\dfrac{1}{5} : \dfrac{1}{4}$?

Find the lowest common multiple of 4 and 5 and then multiply by this number. The lowest common multiple of 4 and 5 is 20, thus

$$20 \times \dfrac{1}{5} = 4$$
$$20 \times \dfrac{1}{4} = 5$$

We have

$$\dfrac{1}{5} : \dfrac{1}{4} = 4:5$$

e First we convert the decimals, 0.36 and 0.06, into whole numbers. How?

Multiply by 100:

$$0.36 \times 100 = 36$$
$$0.06 \times 100 = 6$$

We have

$$0.36:0.06 = 36:6$$

Can we simplify 36:6?

Yes, both 36 and 6 have a common factor of 6:

$$36 \div 6 = 6$$
$$6 \div 6 = 1$$

So 36:6 = 6:1, hence

$$0.36:0.06 = 6:1$$

We can write 0.36:0.06 directly as 6:1 because $6 \times 0.06 = 0.36$.

We can have a ratio of more than two quantities. For example say we want to divide £500 in the ratio 2:3:5. We need to find the amount of money contained in each part.

How many parts are there altogether?

$$\text{Total number of parts} = 2 + 3 + 5 = 10$$
$$\text{Money for each part} = £500 \div 10 = £50$$

The first part is 2 out of 10 so it contains $2 \times £50 = £100$. Similarly the second part has $3 \times £50 = £150$, and the third part has $5 \times £50 = £250$.

Example 33

Engineering students at a university are divided into civil, mechanical, electrical and aerospace in the ratio 2:3:5:1 respectively. If there are 1089 engineering students find the number of students in each of the disciplines.

Solution

Students are divided into $2 + 3 + 5 + 1 = 11$ parts.

How many students are there in one part?

$$1089 \div 11 = 99$$

Number of students in civil $= 99 \times 2 = 198$

Number of students in mechanical $= 99 \times 3 = 297$

Number of students in electrical $= 99 \times 5 = 495$

Number of students in aerospace $= 99 \times 1 = 99$

As a check, you can total the number of students:

$$198 + 297 + 495 + 99 = 1089$$

SUMMARY

A ratio between two objects makes a comparison of their relative sizes. Ratio is denoted by : and we can compare more than two objects.

Exercise **Intro(k)**

Solutions at end of book. Complete solutions available at www.palgrave.com/science/engineering/singh

1 Simplify the following ratios:
   a 100:10   **b** 69:23
   c 5:25:45   **d** 6:42:54:72
   e $\dfrac{2}{3}:\dfrac{5}{6}$

2 By using a calculator, or otherwise, simplify the following:
   a $1\dfrac{1}{5}:2\dfrac{1}{4}$ **b** $3\dfrac{1}{2}:2\dfrac{5}{12}$

3 Simplify the following ratios:
   a 0.4:0.5   **b** 0.52:0.72
   c $\sqrt{2}:\sqrt{8}$

4 A piece of metal of length 0.64 m is cut into three pieces in the ratio 2:3:5. Find the length of each piece.

5 A metal brass is a mixture of copper and zinc where the copper to zinc ratio is 8:3. If the total mass of the brass is 66 kg, find the mass of copper and zinc.

6 A metal alloy is made from copper, zinc and nickel in the ratio 3:4:6. If the alloy has a mass of 45.5 kg, find the mass of each metal.

7 Engineering students at a university are divided into manufacturing, building services, vehicle and control in the ratio 2:3:5:4 respectively. If there are 1260 engineering students, find the number of students in each discipline.

Miscellaneous exercise **Intro**

1 Convert 1 mm^3 to m^3.

2 Evaluate

a 5^3 **b** $\sqrt{100}$ **c** $\sqrt[3]{-8}$

d $\sqrt{\dfrac{196}{49}}$ **e** $\sqrt{12^2 + 5^2}$

3 Calculate the following and write your answer in its simplest form:

a $\dfrac{2}{3} + \dfrac{3}{5}$ **b** $\dfrac{2}{3} \times \dfrac{4}{5}$ **c** $\dfrac{2}{3} \div \dfrac{3}{5}$

4 Write the following to the number of decimal places (d.p.) or significant figures (s.f.) stated:

a 3.141 5926   **i** 3 d.p.   **ii** 3 s.f.

b $\sqrt{10}$ **i** 2 d.p.   **ii** 3 d.p

c 1.6449   **i** 2 d.p.   **ii** 3 d.p

d 16^5 **i** 1 s.f.   **ii** 3 s.f.

e $2^{2^4} + 1$ **i** 3 s.f.   **ii** 2 s.f.

5 Compute the following correct to 2 decimal places:

a $\dfrac{\pi}{4} + 1$

b $\dfrac{-(-7) \pm \sqrt{(-7)^2 - (4 \times 1 \times 12)}}{2}$

c $\dfrac{2}{3}\sqrt{(30 \times 5) + 1}$

d $\left(\dfrac{5}{2.718}\right)^5 \sqrt{10\pi}$

6 Evaluate the following:

a $\dfrac{1}{2\pi \times 50 \times 3 \times 10^{-6}}$

b $\dfrac{(5 \times 10^6) \pm \sqrt{(5 \times 10^6)^2 - (16 \times 10^{12})}}{2}$

7 [electrical principles]   Write the following electrical quantities in symbol notation of Table 2 and Table 3:

a 378 000 V   **b** 0.000 01 A   **c** 1300 Ω

8 [electrical principles]   Compute the range of values of the following resistors:

a 100 Ω ± 5%   **b** 5 kΩ ± 2.5%
c 13 MΩ ± 0.1%

9 [electrical principles]   The input to output current ratio of an amplifier is given by $1\dfrac{1}{2} : 10\dfrac{3}{4}$. Simplify this ratio.

10 Evaluate 15% of $0.5 \times 10^{-3} \times 50 \times 10^3$.

11 [thermodynamics]   Thermal efficiency of a plant is given by

$$\text{thermal efficiency} = \dfrac{3.5 \times 10^6}{24 \times 10^6}$$

Write this as a percentage.

12 [electrical principles]   The power produced by an electrical generator is 20 kW. If it loses 6% of its power, find the power loss.

13 Calculate the following, correct to 2 d.p., by using your calculator:

a $\dfrac{20 \times 10^{11}}{1.5 \times 10^6}$ **b** $\dfrac{115 \times 10^3}{(15 + 1.8)^2}$

c $500 + \dfrac{100^2 - 157^2}{3 \times 10^3} - 160$

d $\dfrac{(2 \times 10^6 \times 0.015) - (2 \times 10^5 \times 0.075)}{1.44 - 1}$

e $\dfrac{(20 \times 10^9)(1 \times 10^3)\left[(1.5 \times 10^3)^2 - (0.5 \times 10^3)^2\right]^{\frac{3}{2}}}{25 \times 10^{21}}$

CHAPTER 1
Engineering Formulae

In this chapter we look at the applications of basic algebra to engineering problems.

The word *algebra* comes from the Arabic '*al-jabr*' which occurs in Al-Khwarizmi's book '*Hisab al-jabr w'al-muqabala*' written in the early ninth century. *Al-jabr* means restoration (or transpose to remove the negative quantities of an equation, e.g. $3x + 1 = 8 - 4x$ goes to $7x + 1 = 8$). Al-Khwarizmi (780–850 AD) was born in Khwarizm, now called Khiva, a town located in Uzbekistan (a former Soviet republic which became independent in 1991).

SECTION A **Substitution and transposition**

By the end of this section you will be able to:
▶ evaluate formulae using BROIDMAS
▶ solve equations
▶ transpose formulae

A1 **Evaluating formulae**

A **formula** is a general rule or law of mathematics. The plural of formula is **formulae**.

In evaluating formulae, the mnemonic BROIDMAS gives the order of operation (see Introductory chapter):

<u>B</u>rackets	} First
<u>RO</u>ots	⎫
<u>I</u>ndices	⎬ Second
<u>D</u>ivision	⎫
<u>M</u>ultiplication	⎬ Third
<u>A</u>ddition	⎫
<u>S</u>ubtraction	⎬ Last

It's imperative that you understand BROIDMAS because it tells us the rules of algebra and is used for evaluating and simplifying algebraic expressions. Moreover it can be useful for typing in an expression into a computer algebra package or a calculator.

In algebra, letters or symbols are used to represent numbers. These letters or symbols may be **constants**, that is fixed, or **variables**, which means they can take up various values.

No space between letters represents multiplication, for example

$$ab = a \times b = a \cdot b$$

So if $a = 3$ and $b = 7$ then $ab = 3 \times 7 = 21$. To evaluate a formula we substitute the given numbers in place of letters and then apply BROIDMAS to evaluate the arithmetical expression as in the Introductory chapter.

For example, evaluate $a(b + c) + \dfrac{c(a + b)^2}{b}$ where $a = 2$, $b = 3$ and $c = 5$:

$$2(3 + 5) + \frac{5(2 + 3)^2}{3} = (2 \times 8) + \left(\frac{5 \times 5^2}{3}\right)$$

$$= 16 + \left(\frac{5 \times 25}{3}\right)$$

$$= 16 + \frac{125}{3}$$

$$= 57\frac{2}{3}$$

Example 1

Pythagoras theorem gives the length of the longest side, c, in terms of the other two sides of a right-angled triangle, a and b, as

$$c = \sqrt{a^2 + b^2}$$

Evaluate c for $a = 5$ and $b = 12$.

Solution

Substituting $a = 5$ and $b = 12$ into $c = \sqrt{a^2 + b^2}$ gives

$$c = \sqrt{5^2 + 12^2}$$

$$= \sqrt{25 + 144}$$

$$= \sqrt{169} = 13$$

Hence $c = 13$.

A2 Transposition of formulae

In the formula $v = u + at$, we say v is the **subject** of the formula. If we want to make t the subject of the formula then we need to change the form to

$$t = \boxed{}$$

This process of changing the subject is called **transposition** of formulae.

When transposing we can

► add, subtract, multiply or divide by the same quantities on both sides of the formula (though we cannot divide by zero)

A3 **Transposition applied to equations**

An **equation** is a mathematical statement that says two expressions are equal. For example

$$x - 3 = 7$$

is an equation where x is a unknown variable. To solve this equation means we need to find the value (or values) of x so that

Left-Hand Side = Right-Hand Side

 Hence we need to make x the subject of $x - 3 = 7$. **That is we need to remove the 3 on the Left-Hand Side. How?**

Add 3. We need to add 3 to both sides because we have to maintain the **balance** of the equation:

$$x - 3 + 3 = 7 + 3$$

$$x - 0 = 10$$

$$x = 10$$

In this case $x = 10$ is a **solution**, or a **root**, of the above equation. The process of finding the value of x is the same as transposition of formulae. Let's try some examples in the field of electrical principles and mechanics.

Example 2 *electrical principles*

If the voltage, V, across a resistor $R = 100 \, \Omega$ is 10 volts, find the current I through the resistor, given that $V = IR$.

Solution

Substituting $V = 10$ and $R = 100$ into $V = IR$ gives

$$10 = 100I$$

 What are we trying to find?

 The value of I. **How do we find I?**

Divide both sides by 100:

$$\frac{100I}{100} = \frac{10}{100}$$

Cancelling the 100's on the Left-Hand Side:

$$I = \frac{10}{100} = 0.1 \text{ amp (A)}$$

The unit for current is amp and will generally be denoted by A.

As a check you can substitute $I = 0.1$ into $10 = 100I$, thus $10 = 100 \times 0.1$.

Example 3 *mechanics*

A vehicle's speed, v, is given by

$$v = 14 + 5t$$

where t is time. Find the time taken in seconds to reach a speed of 23 m/s.

Solution

Substituting $v = 23$ gives

$$14 + 5t = 23$$

We need to find t. How?

Subtract 14 from both sides:

$$5t = 23 - 14 = 9$$

How do we remove the 5 from the Left-Hand Side?

Divide both sides by 5:

$$\frac{5t}{5} = \frac{9}{5}$$

Cancelling 5's on the Left-Hand Side gives:

$$t = \frac{9}{5} = 1.8 \text{ s}$$

We use SI units throughout the book – see the Introductory chapter. For example, velocity is given in m/s, acceleration in m/s^2, time in s, etc.

The above equations, $14 + 5t = 23$ and $100I = 10$, are examples of **linear equations**.

A4 Transposition applied to engineering formula

Algebraic expressions can be simplified by adding, subtracting or cancelling like terms, for example

$$x + x = 2x, \quad x - x = 0, \quad x + 5x = 6x \quad \text{and} \quad \frac{x}{x} = 1 \quad [\text{provided } x \neq 0]$$

We can **only** add and subtract like terms. We cannot simplify the following:

$$x + y = x + y, \, x - y = x - y, \, x + 5y = x + 5y \quad \text{and} \quad \frac{x}{y} = \frac{x}{y}$$

The procedure in applying transposition to formulae is very similar to that used in solving equations. Let's try some engineering examples.

Example 4 *electronics*

The power, P, dissipated in a circuit is given by

$$P = IV$$

where I is current and V is voltage. Transpose the formula for V.

Solution

We want to get $V = - - -$. **How can we achieve this from $P = IV$?**

We need to remove I from the Right-Hand Side (RHS). How?

Divide both sides by I. Thus

$$\frac{P}{I} = \frac{IV}{I} \qquad \text{[the I's on the RHS cancel out]}$$

$$\frac{P}{I} = V \text{ or } V = \frac{P}{I}$$

In Example 4, how do we know that we need to divide both sides by I?

The subject that we want to obtain is V.

What does the formula $P = IV$ do to V?

It is multiplied by I.

We want to remove the I and find V on its own. **How can we remove I?**

We can divide by I.

One way of obtaining the subject in many cases is to see what the formula does to the subject and then do the opposite on the other side. In **Example 4** the subject (V) is multiplied by I so we need to divide the other side by I, thus obtaining $V = \frac{P}{I}$.

Example 5 *mechanics*

The velocity, v, of an object with an initial velocity u and constant acceleration a after time t is given by

$$v = u + at$$

Transpose to make t the subject of the formula.

Solution

We need to get from $v = u + at$ to $t = - - -$

Example 5 *continued*

We need to remove the *u* first. How?

Subtract *u* from both sides:

$$v - u = (u + at) - u$$

$$= \underbrace{u - u}_{=0} + at$$

$$v - u = at$$

How can we obtain *t* from *v* − *u* = *at*?

Divide both sides by *a*:

$$\frac{v - u}{a} = \frac{at}{a}$$

The *a*'s on the Right-Hand Side cancel out to give

$$t = \frac{v - u}{a}$$

We will assume that the variable we are dividing by is **not** zero because we cannot divide by zero. So in **Example 5**, *a* is not zero.

SUMMARY

In evaluating the formula we use the mnemonic BROIDMAS: Brackets, ROots, Indices, Division, Multiplication, Addition, Subtraction.

We can transpose a formula to make a certain variable the subject of the formula. Transposing involves arithmetical operations carried out on both sides of the formula. We use transposition to solve equations.

Exercise 1(a)

Solutions at end of book. Complete solutions available at www.palgrave.com/science/engineering/singh

1 Given that

$$c = \sqrt{a^2 + b^2}$$

evaluate *c* for $a = 24$ and $b = 7$.

2 [*electrical principles*] If the voltage, V, across a resistor $R = 1000 \ \Omega$ is 15 V, then find the current I given that
$$V = IR.$$

3 [*thermodynamics*] A gas is expanded from an initial pressure P_1 and volume V_1 of $5 \times 10^6 \ \text{N/m}^2$ and $2 \times 10^{-4} \ \text{m}^3$ respectively, to a final pressure $P_2 = 2 \times 10^7 \ \text{N/m}^2$. Find the new volume V_2 given that $P_1 V_1 = P_2 V_2$.

4 [*thermodynamics*] A gas has pressure $P = 5.6 \times 10^5 \ \text{N/m}^2$, volume $V = 0.015 \ \text{m}^3$ and is at a temperature $T = 312$ K. If there are $n = 34.6$ mole of gas, determine the mass, m, given that
$$PV = nmRT$$
where $R = 8.31 \ \text{J/(K mole)}$ (R is called the universal or molar gas constant).

Exercise **1(a) continued**

Solutions at end of book. Complete solutions available at www.palgrave.com/science/engineering/singh

5 🚗 [*mechanics*]   The distance, s, travelled in time t is related by

$$s = ut + \frac{1}{2}at^2$$

where u is the initial velocity and a is constant acceleration. Determine a, given that $s = 30$ m, $u = 2$ m/s and $t = 5$ s.

6 🔲 [*electrical principles*]   The resistance, R, of a wire at $t\,°C$ is given by

$$R = R_0(1 + \alpha t)$$

where R_0 is the resistance at $0\,°C$ and α is the temperature coefficient of resistance. Determine α, given that $R_0 = 33$ Ω, $R = 35$ Ω and $t = 89\,°C$. (The units of α are / °C.)

7 🔲 [*electrical principles*]   A battery with e.m.f. $E = 12$ V and an internal resistance $r = 1$ Ω is connected across a resistor $R = 20$ Ω. Find the voltage V across R, given that

$$E = \frac{V(R + r)}{R}$$

8 🚗 [*mechanics*]   The velocity, v, of an object is given by

$$v = u + at$$

where u is the initial velocity, a is constant acceleration and t is time. Transpose to make
  i u the subject
  ii a the subject.

9 🔲 [*electrical principles*]   Ohm's law states that

$$I = \frac{V}{R}$$

where I is current, V is voltage and R is resistance. Transpose to make R the subject of the formula.

10 🔲 [*thermodynamics*]   The characteristic equation of a perfect gas is given by

$$PV = mRT$$

(P = pressure, V = volume, R = gas constant, m = mass and T = temperature). Make T the subject.

11 🚗 [*mechanics*]   The slip, S (%), of a vehicle is given by

$$S = \left(1 - \frac{r\omega}{v}\right) \times 100$$

where r = radius of tyre, ω = angular velocity and v = velocity. Make ω the subject of the formula.

SECTION B **Transposing engineering formulae**

By the end of this section you will be able to:
▶ transpose complicated formulae
▶ evaluate formulae by substitution

Transposition and substitution are two very important concepts of algebra and it is critical that you fully understand the processes involved because the remaining chapters in this book rely upon these techniques.

This section is more difficult than Section A in the sense that the formulae involve roots and inverse.

B1 Formulae involving roots

As discussed in the Introductory chapter, the square root and the *n*th root are denoted by $\sqrt{}$ and $\sqrt[n]{}$ respectively. We can write these as

1.1 $\sqrt{a} = a^{1/2}$ [square root]

1.2 $\sqrt[n]{a} = a^{1/n}$ [*n*th root]

For example

$$\sqrt{49} = 7 \text{ (positive square root)}$$

$$\sqrt[3]{8} = 2 \text{ (because } 2 \times 2 \times 2 = 8 \quad \text{or} \quad 2^3 = 8; \sqrt[3]{} \text{ denotes the cube root)}$$

? **What is $8^{1/3}$ and $256^{1/4}$ equal to?**

$$8^{1/3} = \sqrt[3]{8} = 2$$

$$256^{1/4} = 4 \text{ (because } 4 \times 4 \times 4 \times 4 = 256)$$

Now let's take a look at roots where letters represent variables.

We also have:

$$\sqrt{a^2} = \left(\sqrt{a}\right)^2 = a$$
$$\sqrt[n]{a^n} = \left(\sqrt[n]{a}\right)^n = a$$

(These can be shown by using the rules of indices which are explored in the next section.)

✈ Example 6 *aerodynamics*

The lift force, *L*, on an aircraft is given by

$$L = \frac{1}{2}\rho v^2 A C$$

where ρ is density, v is speed, A is area and C is lift coefficient. Make v the subject of the formula.

Solution

? **How can we get $v = - - - $?**

We can first find v^2 and then take the square root of both sides.

? **How do we get $v^2 = - - - $?**

? **First we need to remove the $\frac{1}{2}$. How?**

Multiply both sides by 2: $2L = \rho v^2 A C$

? **Next we need to remove $\rho A C$ from the Right-Hand Side. How?**

Divide through by $\rho A C$:

$$\frac{2L}{\rho A C} = v^2$$

> ### Example 6 *continued*
>
>
>
> **How can we find v?**
>
> Take the square root of both sides (because $\sqrt{v^2} = v$):
>
> $$\sqrt{v^2} = \sqrt{\frac{2L}{\rho AC}}$$
>
> $$v = \sqrt{\frac{2L}{\rho AC}} \text{ which we may write as } \underbrace{\left(\frac{2L}{\rho AC}\right)^{1/2}}_{\text{by } \boxed{1.1}}$$

In the last line where we say 'by $\boxed{1.1}$' means the result follows by this reference quoted earlier and repeated at the bottom of this page.

We adopt this approach of quoting a reference number throughout the book and the formula itself will either be in the main text or on the bottom of the page below a horizontal line so that you do not need to flick over pages to find the reference.

> ### Example 7 *materials*
>
> The second moment of area, I, of a rectangle of height h and breadth b is given by
>
> $$I = \frac{1}{12}bh^3$$
>
> Make h the subject of the formula.
>
> ### Solution
>
> **First we need to remove $\dfrac{1}{12}$ from the Right-Hand Side. How?**
>
> Multiply both sides by 12: $12I = bh^3$
>
> **By what means can we find h?**
>
> We can initially obtain h^3 and then find h. So divide both sides by b:
>
> $$\frac{12I}{b} = h^3$$
>
> and now take the cube root, $\left(\sqrt[3]{}\right)$, of both sides:
>
> $$\sqrt[3]{\frac{12I}{b}} = h \left(\text{because } \sqrt[3]{h^3} = h\right)$$
>
> or
>
> $$h \underset{\text{by } \boxed{1.2}}{=} \left(\frac{12I}{b}\right)^{1/3}$$

$\boxed{1.1}$ $\sqrt{a} = a^{1/2}$ $\boxed{1.2}$ $\sqrt[3]{a} = a^{1/3}$

As discussed in the Introductory chapter, in many engineering examples it is sufficient to give your final answer to the smallest number of significant figures consistent with the data. The intermediate working has to be one more decimal point (d.p.) or significant figure (s.f.) than is needed. Thus in order to give your final answer to 1 d.p. (or 1 s.f.) you need to work to 2 d.p. (or 2 s.f.).

In the next example we use substitution and transposition of formulae to evaluate the capacitance, C. It is more difficult than the above examples.

Example 8 *electronics*

The impedance, Z, of a circuit containing a resistor R, capacitor C and inductor L is given by

$$Z = \sqrt{R^2 + (X_L - X_C)^2}$$

where $X_L = 2\pi f L$ and $X_C = \dfrac{1}{2\pi f C}$ (f represents frequency).

Determine C if $R = 100\,\Omega$, $Z = 104\,\Omega$, $L = 0.1$ henry and $f = 50$ Hz.

Solution

Substituting $f = 50$ and $L = 0.1$ into $X_L = 2\pi f L$ gives

$$X_L = 2\pi \times 50 \times 0.1 = 10\pi$$

Substituting $X_L = 10\pi$, $Z = 104$ and $R = 100$ into the given formula, $Z = \sqrt{R^2 + (X_L - X_C)^2}$, results in

$$104 = \sqrt{100^2 + (10\pi - X_C)^2}$$

What are we trying to find?

We need to determine C but first we find X_C and then obtain C.

Squaring both sides gives

$$104^2 = 100^2 + (10\pi - X_C)^2$$

Transposing

$$104^2 - 100^2 = 816 = (10\pi - X_C)^2$$

We have

$$(10\pi - X_C)^2 = 816$$

Taking square root of both sides:

$$10\pi - X_C = \sqrt{816} = 28.566$$

Hence $X_C = 10\pi - 28.566 = 2.850$

Since $X_C = \dfrac{1}{2\pi f C}$ we have

$$\frac{1}{2\pi f C} = 2.850$$

Example 8 continued

Transposing

$$C = \frac{1}{2\pi f \times 2.85}$$

$$\underset{\substack{\text{substituting}\\ f = 50}}{=} \frac{1}{2\pi \times 50 \times 2.85} = 0.0011$$

Hence $C = 0.0011$ farad or 1.1×10^{-3} farad $= 1.1$ millifarad (mF). Remember that the prefix milli, m, represents 10^{-3}.

B2 Formulae involving the inverse

The multiplicative inverse of x ($\neq 0$) is denoted by x^{-1} and defined as

1.3 $x^{-1} = \dfrac{1}{x}(= 1 \div x)$

Example 9

Show that

$$\left(\frac{a}{b}\right)^{-1} = \frac{b}{a} \quad (a \neq 0, b \neq 0)$$

Solution

We have

$$\left(\frac{a}{b}\right)^{-1} = \frac{1}{\left(\dfrac{a}{b}\right)} \quad \left[\text{by (1.3) with } x = \frac{a}{b} \right]$$

$$= 1 \div \frac{a}{b}$$

$$= 1 \times \frac{b}{a} = \frac{b}{a}$$

Remember $1 \div \dfrac{a}{b} = 1 \times \dfrac{b}{a}$ because when we divide fractions we turn the second fraction upside down and multiply.

We give this important result a reference number:

1.4 $$\left(\frac{a}{b}\right)^{-1} = \frac{b}{a}$$

We also have $\left(\frac{a}{b}\right)^2 = \frac{a^2}{b^2}$, but more of this in the next section.

Example 10 *thermodynamics*

A gas in a cylinder in state 1 with pressure P_1, temperature T_1 and volume V_1 expands to state 2 with pressure P_2, temperature T_2 and volume V_2. A formula relating these variables is given by $\dfrac{P_1 V_1}{T_1} = \dfrac{P_2 V_2}{T_2}$. Make T_1 the subject of the formula.

Solution

From $\dfrac{P_1 V_1}{T_1} = \dfrac{P_2 V_2}{T_2}$ we need $T_1 = - - -$. Taking the inverse, $(\ )^{-1}$, of both sides gives

$$\left(\frac{P_1 V_1}{T_1}\right)^{-1} = \left(\frac{P_2 V_2}{T_2}\right)^{-1}$$

$$\frac{T_1}{P_1 V_1} = \frac{T_2}{P_2 V_2} \quad [\text{by } \boxed{1.4}\,]$$

? **How can we find $T_1 = - - -$?**

Multiply both sides by $P_1 V_1$:

$$T_1 = \frac{P_1 V_1 T_2}{P_2 V_2}$$

SUMMARY

The square root, $\sqrt{\ }$, and the nth root, $\sqrt[n]{\ }$, are defined as

1.1 $\sqrt{a} = a^{1/2}$

1.2 $\sqrt[n]{a} = a^{1/n}$

The inverse of x ($\neq 0$) is defined as

1.3 $x^{-1} = \dfrac{1}{x}$

1.4 $\left(\dfrac{a}{b}\right)^{-1} = \dfrac{b}{a} \quad (a \neq 0, b \neq 0)$

It is well worth spending some time learning these, 1.1 to 1.4 , because they are used throughout the book.

Solutions at end of book. Complete solutions available at www.palgrave.com/science/engineering/singh

Exercise **1(b)**

1 [electrical principles] The power P dissipated in a resistor of resistance R is given by

$$P = \frac{V^2}{R} \quad [V \text{ is voltage}]$$

Make V the subject of the formula.

2 [acoustics] The speed, c, of sound in air is given by

$$c = \sqrt{\frac{\gamma P}{\rho}}$$

where γ is the specific heat ratio, P is the pressure and ρ is the density. Make P the subject of the formula.

3 [mechanics] The airflow over a vehicle causes drag D, which is given by

$$D = \frac{1}{2}\rho C v^2 A$$

where ρ is density, C is drag coefficient, v is velocity and A is the frontal area of the vehicle. Make A the subject of the formula.

4 [electrical principles] Evaluate the total resistance, R, in a circuit containing two resistors in parallel, $R_1 = 100\,\Omega$ and $R_2 = 270\,\Omega$, where

$$\frac{1}{R} = \frac{1}{R_1} + \frac{1}{R_2}$$

(Ω is the SI unit ohm used to measure electrical resistance).

5 [mechanics] The time, T, taken for a pendulum to make a complete swing is given by

$$T = 2\pi\sqrt{\frac{l}{g}}$$

where l = length of pendulum and $g = 9.81$ m/s². Determine l, if $T = 0.5$ s.

6 [electronics] The impedance, Z, of a circuit containing a resistor R, capacitor C and inductor L is given by

$$Z = \sqrt{R^2 + (X_L - X_C)^2}$$

where $X_L = 2\pi f L$ and $X_C = \dfrac{1}{2\pi f C}$

(f represents frequency).

Determine C if $R = 50\,\Omega$, $Z = 100\,\Omega$, $L = 1$ henry and $f = 50$ Hz.

7 [aerodynamics] The power, P, required to drive an air screw of diameter D is given by

$$P = 2\pi k \rho n^3 D^5$$

(k = torque coefficient, ρ = density, n = number of revolutions per second). Make D the subject of the formula.

8 [electronics] A system with feedback β and gain A has an input voltage v_{in} given by

$$v_{in} = \left(\frac{1}{A} - \beta\right)v_{out}$$

(v_{out} = output voltage). Show that

$$\frac{v_{out}}{v_{in}} = \frac{A}{1 - A\beta}$$

9 [materials] A cylinder of radius r is subject to a torque T at each end, which causes it to twist. The shear stress τ is given by

$$\tau = \frac{T}{\frac{1}{2}\pi r^3}$$

Make r the subject of the formula.

10 The following formulae occur in various engineering fields. Make the letter in the square brackets the subject of the formula.

a $V = \dfrac{ER}{R + r}$ $\qquad\qquad$ $[r]$

b $v^2 = u^2 + 2as$ $\qquad\qquad$ $[u]$

c $v = \left(\dfrac{K + 4a/3}{\rho}\right)^{1/2}$ $\qquad$ $[K]$

Exercise **1(b) continued**	Solutions at end of book. Complete solutions available at www.palgrave.com/science/engineering/singh

d $\eta = \dfrac{\pi P r^4 t}{8 v L}$ $[r]$ $\qquad$ **g** $W = \dfrac{P_1 V_1 - P_2 V_2}{n - 1}$ $[V_1]$

e $T = 2\pi\sqrt{\dfrac{l}{g}}$ $[l]$ $\qquad$ **h** $PV^n = C$ $[V]$

f $RT = \left(P + \dfrac{a}{V^2}\right)(V - b)$ $[P]$ $\qquad$ **i** $f = C\dfrac{W}{D}\sqrt{\dfrac{h}{u}}$ $[u]$

SECTION C **Indices**

By the end of this section you will be able to:
▶ use the laws of indices to simplify expressions
▶ use the laws of indices in applications of thermodynamics

? **Do you remember what 3^5 represents?**

It is $\underbrace{3 \times 3 \times 3 \times 3 \times 3}_{\text{5 copies}}$ which is equal to 243.

The 5 in 3^5 is called the **index** or the **power**. The plural of index is **indices**. In this section we will predominantly apply the rules of indices to letters rather than numbers.

The topic of indices is very important for engineers but many students do find this a difficult topic – invariably because they don't know the rules well enough.

C1 **Some rules of indices**

We have already stated some rules of indices in the last section, 1.1 to 1.4 . Other important rules of indices are

1.5 $a^m a^n = a^{m+n}$

1.6 $a^m \div a^n = \dfrac{a^m}{a^n} = a^{m-n}$ $(a \neq 0)$

1.7 $(a^m)^n = a^{m \times n}$

1.8 $a^0 = 1$ $(a \neq 0)$

1.9 $a^1 = a$

1.10 $a^{-n} = \dfrac{1}{a^n}$ $(a \neq 0)$

1.11 $(ab)^n = a^n b^n$

1.12 $\left(\dfrac{a}{b}\right)^n = \dfrac{a^n}{b^n}$ $(b \neq 0)$

Example 11

Simplify the following:

a $x^3 x^2$ **b** $\dfrac{x^3}{x^2}$ **c** $\dfrac{x}{\sqrt{x}}$ **d** $\left(\sqrt[3]{x}\right)^2 \sqrt[3]{x}$

Solution

Using the above rules we have

a $x^3 x^2 \underset{\text{by } \boxed{1.5}}{=} x^{3+2} = x^5$

b $\dfrac{x^3}{x^2} \underset{\text{by } \boxed{1.6}}{=} x^{3-2} = x^1 \underset{\text{by } \boxed{1.9}}{=} x$

c $\dfrac{x}{\sqrt{x}} = \dfrac{x^1}{x^{1/2}} = \underset{\text{by } \boxed{1.6}}{x^{1-1/2}}$

$= x^{1/2} = \sqrt{x} \qquad [\text{by } \boxed{1.1}]$

d $\left(\sqrt[3]{x}\right)^2 \sqrt[3]{x} \underset{\text{by } \boxed{1.2}}{=} \left(x^{1/3}\right)^2\left(x^{1/3}\right)$

$\underset{\text{by } \boxed{1.7}}{=} x^{2/3} x^{1/3} = \underset{\text{by } \boxed{1.5}}{x^{(2/3)+(1/3)}} = x^1 = x$

As **Example 11** shows, the rules of indices, $\boxed{1.1}$ to $\boxed{1.12}$, can be used to simplify algebraic expressions. We can also apply these to show results that we have already used, such as $\sqrt{a^2} = a$:

$$\sqrt{a^2} = (a^2)^{1/2} = a^{2 \times 1/2} = a^1 = a$$

Similarly we have

$$\sqrt[n]{a^n} = (a^n)^{1/n} = a^{n \times 1/n} = a^1 = a$$

Note that if $x^n = a$ then taking the nth root of both sides gives

$$(x^n)^{1/n} = a^{1/n}$$

Thus we have

$\boxed{\dagger} \qquad x = a^{1/n}$

We call this ($\dagger$) because we will refer to it later on.

Let's try an engineering example.

$\boxed{1.1}$ $\sqrt{x} = x^{1/2}$ $\boxed{1.2}$ $\sqrt[n]{a} = a^{1/n}$ $\boxed{1.5}$ $a^m a^n = a^{m+n}$

$\boxed{1.6}$ $\dfrac{a^m}{a^n} = a^{m-n}$ $\boxed{1.7}$ $(a^m)^n = a^{m \times n}$ $\boxed{1.9}$ $a^1 = a$

📖 Example 12 *thermodynamics*

A gas in a cylinder is compressed according to the law

$$P_1 V_1^{1.5} = P_2 V_2^{1.5}$$

where P is pressure and V is volume. If the gas has an initial volume of $V_1 = 0.16$ m³ and pressure of $P_1 = 140 \times 10^3$ N/m² and is then compressed to a pressure of $P_2 = 750 \times 10^3$ N/m², find the new volume, V_2.

Solution

Substituting $P_1 = 140 \times 10^3$, $P_2 = 750 \times 10^3$ and $V_1 = 0.16$ into

$$P_1 V_1^{1.5} = P_2 V_2^{1.5}$$

gives

$$(140 \times 10^3) \times (0.16)^{1.5} = (750 \times 10^3) \times V_2^{1.5}$$

$$V_2^{1.5} = \frac{(140 \times 10^3) \times (0.16)^{1.5}}{750 \times 10^3} \quad \text{[Dividing by } 750 \times 10^3\text{]}$$

$$= \frac{140 \times (0.16)^{1.5}}{750} \quad \text{[Cancelling } 10^3 \text{ 's]}$$

$$V_2^{1.5} = 0.0119$$

Applying the index $1/1.5$ to both sides and using ⊹ yields:

$$V_2 = (0.0119)^{1/1.5} = 0.052 \text{ m}^3 \text{ (2 s.f.)}$$

SUMMARY

We can use the rules of indices, 1.1 to 1.12 , to simplify algebraic expressions and in engineering applications such as those in thermodynamics.

Exercise 1(c)

Solutions at end of book. Complete solutions available at www.palgrave.com/science/engineering/singh

1 Simplify the following:

a $x^5 x^2$ **b** $x^{1/5} x^{1/2}$ **c** $\dfrac{x^3}{x^3}$ **d** $\dfrac{x^7}{x^9}$

e $\left(\sqrt[5]{x}\right)^2 \sqrt[3]{x}$

2 Simplify

a $(1 + y)^2 (1 + y)$ **b** $\dfrac{(1 + x^2)^5}{(1 + x^2)^3}$

c $\left(\sqrt[3]{x^2 + x + 1}\right)^5 \sqrt[3]{x^2 + x}$

d $\left(\sqrt[3]{x^2 + x + 1}\right)^5 \sqrt[3]{x^2 + x + 1}$

📖 Questions 3 to 5, inclusive, are on [*thermodynamics*]

3 A gas in an engine obeys the law

$$P_1 V_1^{1.45} = P_2 V_2^{1.45}$$

where P represents pressure and V represents volume.

Exercise **1(c) continued**

Solutions at end of book. Complete solutions available at www.palgrave.com/science/engineering/singh

If $P_1 = 2 \times 10^6$ N/m^2, $V_1 = 0.15$ m^3 and $P_2 = 2 \times 10^5$ N/m^2, find V_2.

4 The work done, W, on the face of a piston by a gas is given by

$$W = \frac{CV_2^{-0.35} - CV_1^{-0.35}}{-0.35}$$

where $C = P_1V_1^{1.35} = P_2V_2^{1.35}$ (P and V are pressure and volume respectively and C is a constant). Show that

$$-0.35W = P_2V_2 - P_1V_1$$

5 The state of a gas changes from P_1, V_1 and T_1 to P_2, V_2 and T_2 (pressure, volume and temperature respectively). The characteristic equation is given by

$$\frac{P_1V_1}{T_1} = \frac{P_2V_2}{T_2}$$

By the polytropic law we have

$$P_1V_1^n = P_2V_2^n$$

By using these formulae, show that

$$\frac{T_1}{T_2} = \left(\frac{P_1}{P_2}\right)^{1-\frac{1}{n}}$$

6 [aerodynamics] In aerodynamics the following equation holds:

$$\frac{\rho_2}{\rho_1} = \left(\frac{T_2}{T_1}\right)^{g/LC}\left(\frac{T_1}{T_2}\right)$$

where ρ_1, ρ_2, T_1 and T_2 represent the densities and temperatures at altitude 1 and 2 respectively. L is the rate of decrease of temperature with altitude. C is a constant and g is acceleration due to gravity. Show that

$$\frac{\rho_2}{\rho_1} = \left(\frac{T_2}{T_1}\right)^{\frac{g-LC}{LC}}$$

SECTION D **Dimensional analysis**

By the end of this section you will be able to:
► apply the rules of indices to check equations which involve physical quantities

D1 **Dimensional analysis**

There are **three** fundamental dimensions: Mass, Length and Time (M, L and T respectively). All mechanical quantities can be expressed in terms of powers of M, L and T. (Non-mechanical quantities such as electrical current can also be expressed in terms of M, L and T, but it is easier to introduce a fourth fundamental dimension – charge Q.)

We use the following notation:

[force] represents the dimension of force

Example 13

Obtain the fundamental dimensions of velocity (units m/s), acceleration (units m/s²) and force (=mass × acceleration).

Solution

We know the units of velocity are m/s so the dimensions are

$$\frac{\text{Length}}{\text{Time}} = \frac{L}{T} = L\left(\frac{1}{T}\right) \underset{\text{by } 1.3}{\equiv} LT^{-1} \text{ etc}$$

Similarly acceleration has units m/s² so the dimensions are

$$\frac{\text{Length}}{(\text{Time})^2} = \frac{L}{T^2} = L\left(\frac{1}{T^2}\right) \underset{\text{by } 1.10}{\equiv} LT^{-2}$$

What about force?

$$\text{force} = \text{mass} \times \text{acceleration}$$

$$[\text{force}] = M \times (LT^{-2})$$

$$= MLT^{-2}$$

Similarly we can evaluate the dimensions of the other quantities as shown in Table 1. Try verifying some of these in your own time.

TABLE 1

Quantity	Units	Dimensions
Area	m²	L^2
Volume	m³	L^3
Velocity	m/s	LT^{-1}
Acceleration	m/s²	LT^{-2}
Force	newton (N)	MLT^{-2}
Work (or energy)	joule (J)	$ML^2 T^{-2}$
Power	watt (W)	$ML^2 T^{-3}$
Pressure	N/m²	$ML^{-1} T^{-2}$
Density	kg/m³	ML^{-3}
Frequency	hertz (Hz)	T^{-1}

1.3 $\dfrac{1}{x} = x^{-1}$ 1.10 $\dfrac{1}{a^n} = a^{-n}$

Dimensional analysis is a method used in checking an equation by establishing the same dimension formula on each side of the equation, that is

[Left-Hand Side] = [Right-Hand Side]

Numbers with no units attached to them are dimensionless.

Example 14 *fluid mechanics*

Bernoulli's equation is given by

$$P + \frac{1}{2}\rho v^2 + \rho gz = \text{constant}$$

where P = pressure, ρ = density, v = velocity, z = height and g = acceleration due to gravity.

Find the dimensions of the constant.

Solution

Using Table 1 we have $\left(\text{remember } \dfrac{1}{2} \text{ is dimensionless}\right)$

$$\underbrace{ML^{-1}\,T^{-2}}_{P} + \underbrace{ML^{-3}}_{\rho}\underbrace{(LT^{-1})^2}_{v^2} + \underbrace{ML^{-3}}_{\rho}\underbrace{LT^{-2}}_{g}\underbrace{L}_{z}$$

$$= ML^{-1}T^{-2} + ML^{-3}\underbrace{L^2\,T^{-2}}_{\text{by }1.7} + M\underbrace{L^{-3+2}}_{\text{by }1.5}\,T^{-2}$$

$$= ML^{-1}\,T^{-2} + \underbrace{ML^{-1}T^{-2}}_{=\,L^{-3+2}} + ML^{-1}T^{-2}$$

Hence the constant has the dimensions $ML^{-1}T^{-2}$.

A physical requirement is that **dimensional homogeneity** holds, that is both sides of an equation have the same dimensions.

Example 15 *mechanics*

The period T of a pendulum of length l is given by

$$T = 2\pi\sqrt{\frac{l}{g}}$$

where g is acceleration due to gravity. Show that the formula has dimensional homogeneity.

1.5 $a^m\,a^n = a^{m+n}$ 1.7 $(a^m)^n = a^{m\times n}$

Example 15 *continued*

Solution

Remember 2π is dimensionless. By Table 1, g has the dimensions LT^{-2}. So we have

$$[T] = \sqrt{\frac{L}{LT^{-2}}} = \left(\frac{\cancel{L}}{\cancel{L}T^{-2}}\right)^{\frac{1}{2}} \underset{\text{by } 1.12}{=} \frac{1^{\frac{1}{2}}}{(T^{-2})^{\frac{1}{2}}} \underset{\text{by } 1.7}{=} \frac{1}{T^{-2\times\frac{1}{2}}} = \frac{1}{T^{-1}} = T$$

The last step is justified by

$$\frac{1}{T^{-1}} \underset{\text{by } 1.3}{=} (T^{-1})^{-1} \underset{\text{by } 1.7}{=} T^{(-1)\times(-1)} = T^{1} \underset{\text{by } 1.9}{=} T$$

Clearly period T has dimensions T.

SUMMARY

There are three fundamental dimensions – mass M, length L and time T. We can apply the rules of indices to check dimensional homogeneity.

Exercise **1(d)**

Solutions at end of book. Complete solutions available at www.palgrave.com/science/engineering/singh

 All questions in this exercise belong to [*dimensional analysis*].

1 Show that the dimensions of

a pressure $\left(= \dfrac{\text{force}}{\text{area}}\right)$ are $ML^{-1}T^{-2}$

b density $\left(= \dfrac{\text{mass}}{\text{volume}}\right)$ are ML^{-3}

c momentum ($=$ mass $\times$ velocity) are MLT^{-1}

d power ($=$ force $\times$ velocity) are $ML^{2}T^{-3}$

e impulse ($=$ force $\times$ time) are MLT^{-1}

f kinetic energy $\left(= \dfrac{1}{2} \times \text{mass} \times (\text{velocity})^2\right)$ are $ML^{2}T^{-2}$

g potential energy ($=$ mass $\times$ acceleration $\times$ height) are $ML^{2}T^{-2}$

2 The pressure, P, at a depth d of a fluid of density ρ is given by

$$P = \rho g d \; (g = \text{acceleration})$$

Show that the formula has dimensional homogeneity.

3 Which of the following are dimensionally correct (have dimensional homogeneity)?

a $F = mgl$ **b** $s = ut + \dfrac{1}{2} gt^2$

c $v^2 = u^2 + 2gs$ **d** $W = F \times v$

e $P = F \times l$

(m = mass, g = acceleration, l = length, t = time, s = distance, u and v = velocities, F = force, W = work and P = power).

1.3 $\dfrac{1}{x} = x^{-1}$ **1.7** $(a^m)^n = a^{m \times n}$ **1.9** $a^1 = a$ **1.12** $\left(\dfrac{a}{b}\right)^n = \dfrac{a^n}{b^n}$

Solutions at end of book. Complete solutions available at www.palgrave.com/science/engineering/singh

Exercise **1(d) continued**

4 The dynamic coefficient of viscosity μ (viscosity of a fluid) is found from

$$F = \frac{\mu A v}{d}$$

where v = velocity, d = distance, F = force and A = area. Find the dimensions of μ.

5 Show that the following are dimensionless parameters by checking that the dimensions of each are equal to 1:

a Reynolds Number $= \dfrac{\rho v l}{\mu}$

b Mach Number $= \dfrac{v}{c}$

c Euler Number $= \dfrac{p}{\rho v^2}$

d Froude Number $= \dfrac{v}{\sqrt{gl}}$

e Weber Number $= \dfrac{v^2 l \rho}{\sigma}$

(ρ is density, v is velocity, g is acceleration due to gravity, l is length, μ is viscosity, p is pressure, c is speed of sound and σ is surface tension whose units are N/m.)

SECTION E **Expansion of brackets**

By the end of this section you will be able to:
▶ expand brackets
▶ use expansion of brackets in engineering applications
▶ expand brackets of the type $(a + b)(c + d)$ using FOIL

E1 **Revision of brackets**

? **What does $5(x + 3)$ mean?**

All the terms inside the bracket are multiplied by 5:

$$5(x + 3) = (5 \times x) + (5 \times 3)$$
$$= 5x + 15$$

Let's do a few examples.

Example 16

Multiply out the brackets of the following:

a $5(2x + 1)$ **b** $3(3x - 2)$ **c** $-(x - 1)$ **d** $-2(-x - 4)$

Solution

a $5(2x + 1) = (5 \times 2x) + (5 \times 1) = 10x + 5$
b $3(3x - 2) = (3 \times 3x) - (3 \times 2) = 9x - 6$
c Remember that 'minus times minus equals plus':

Example 16 *continued*

$$-(x - 1) = -1(x - 1) = (-1 \times x) - \underbrace{[1 \times (-1)]}_{= -1} = -x + \underset{\substack{\text{because} \\ -(-1) = 1}}{\underline{1}} = 1 - x$$

The result of taking a negative sign inside a bracket is to change **all** the signs inside the bracket.

d $-2(-x - 4) = [-2 \times (-x)] - \underbrace{[2 \times (-4)]}_{= -8} = 2x + 8$

Example 17

Simplify the following:

a $3(x + 2) + 5(2x + 3)$ **b** $(x + 5) - 2(x - 1)$ **c** $-(2x + 3) + (2x + 3)$

Solution

We add all the like terms:

a $3(x + 2) + 5(2x + 3) = (3x + 6) + (10x + 15)$
$$= \underbrace{3x + 10x}_{\text{collecting all the } x \text{ terms}} + (6 + 15)$$
$$= 13x + 21$$

b Multiplying out the brackets gives
$$(x + 5) - 2(x - 1) = (x + 5) - (2 \times x) - (2 \times (-1))$$
$$= (x + 5) - 2x + 2$$
$$= x - 2x + (5 + 2)$$
$$= 7 - x$$

c $-(2x + 3) + (2x + 3) = 0$

Example 18 *structures*

The deflection, y, at a distance x from one end of a beam of length l is given by

$$y = \frac{wx^2}{6EI}(3l - x)$$

where w is the load per unit length and EI is the flexural rigidity of the beam. Remove the brackets of this expression.

Solution

We have
$$y = \frac{wx^2}{6EI}(3l - x) = \frac{wx^2}{6EI}3l - \frac{wx^2}{6EI}x$$
$$= \frac{3wx^2 l}{6EI} - \frac{wx^2 x}{6EI}$$
$$= \frac{wx^2 l}{2EI} - \frac{wx^3}{6EI}$$

E2 Using FOIL

How do we remove the brackets from an expression like $(x + 3)(x + 2)$?

Each term of the first bracket (x and 3) multiplies the second bracket ($x + 2$):

$$(x + 3)(x + 2) = x(x + 2) + 3(x + 2)$$
$$= (x \times x) + (x \times 2) + (3 \times x) + (3 \times 2)$$
$$= x^2 + \underbrace{2x + 3x}_{=5x} + 6$$
$$= x^2 + 5x + 6$$

Another way is to use FOIL, which is a mnemonic for $\underline{F}$irst, $\underline{O}$utside, $\underline{I}$nside and $\underline{L}$ast:

$$(x + 3)(x + 2) = \underbrace{(x \times x)}_{F} + \underbrace{(x \times 2)}_{O} + \underbrace{(3 \times x)}_{I} + \underbrace{(3 \times 2)}_{L}$$
$$= x^2 + 5x + 6$$

Multiply

> The $\underline{F}$irst terms in each bracket
>
> The $\underline{O}$utside terms
>
> The $\underline{I}$nside terms
>
> The $\underline{L}$ast terms

The process of multiplying brackets is also known as **expanding brackets**.

Example 19

Expand the following:

a $(x + 4)(x + 5)$ **b** $(x + 5)(x - 1)$ **c** $(2x + 3)(3x + 5)$ **d** $(3x - 1)(4x - 2)$

Solution

Using FOIL in each case gives

a $(x + 4)(x + 5) = \underbrace{(x \times x)}_{F} + \underbrace{(x \times 5)}_{O} + \underbrace{(4 \times x)}_{I} + \underbrace{(4 \times 5)}_{L}$

$$= x^2 + \underbrace{5x + 4x}_{=9x} + 20$$

$$= x^2 + 9x + 20$$

b $(x + 5)(x - 1) = \underbrace{(x \times x)}_{F} + \underbrace{(x \times (-1))}_{O} + \underbrace{(5 \times x)}_{I} + \underbrace{(5 \times (-1))}_{L}$

$$= x^2 - x + 5x - 5$$

$$= x^2 + \underbrace{4x}_{=5x-x} - 5$$

Example 19 *continued*

c $(2x + 3)(3x + 5) = \underbrace{(2x \times 3x)}_{\text{F}} + \underbrace{(2x \times 5)}_{\text{O}} + \underbrace{(3 \times 3x)}_{\text{I}} + \underbrace{(3 \times 5)}_{\text{L}}$

$$= 6x^2 + \underbrace{10x + 9x}_{=19x} + 15$$

$$= 6x^2 + 19x + 15$$

d $(3x - 1)(4x - 2) = \underbrace{(3x \times 4x)}_{\text{F}} + \underbrace{(3x \times (-2))}_{\text{O}} + \underbrace{((-1) \times 4x)}_{\text{I}} + \underbrace{((-1) \times (-2))}_{\text{L}}$

$$= 12x^2 \underbrace{- 6x - 4x}_{=-10x} + 2$$

$$= 12x^2 - 10x + 2$$

E3 Important expansions

Important expansions are $(a + b)^2$ and $(a - b)^2$. Let's use FOIL to expand these.

We have

$$(a + b)^2 = (a + b)(a + b) = \underbrace{(a \times a)}_{\text{F}} + \underbrace{(a \times b)}_{\text{O}} + \underbrace{(b \times a)}_{\text{I}} + \underbrace{(b \times b)}_{\text{L}}$$

$$= a^2 + \underbrace{ab + ba}_{=2ab} + b^2$$

$$= a^2 + 2ab + b^2$$

Similarly we have

$$(a - b)^2 = (a - b)(a - b) = \underbrace{(a \times a)}_{\text{F}} - \underbrace{(a \times b)}_{\text{O}} - \underbrace{(b \times a)}_{\text{I}} + \underbrace{(b \times b)}_{\text{L}}$$

$$= a^2 - ab - ba + b^2$$

$$= a^2 - 2ab + b^2$$

Note these results:

$$(a + b)^2 \neq a^2 + b^2 \quad \text{[Not equal]}$$

$$(a - b)^2 \neq a^2 - b^2 \quad \text{[Not equal]}$$

The symbol '$\neq$' means 'does not equal'.

It is useful to remember these results:

| 1.13 | $(a + b)^2 = a^2 + 2ab + b^2$ |
| 1.14 | $(a - b)^2 = a^2 - 2ab + b^2$ |

For example, by using 1.13 with $a = 2x$ and $b = 3$ we have

$$(2x + 3)^2 = (2x)^2 + (2 \times 2x \times 3) + 3^2$$

$$= 2^2 x^2 + (4x \times 3) + 9$$

$$= 4x^2 + 12x + 9$$

Similarly, using $\boxed{1.14}$ with $a = 5x$ and $b = 2$ we have

$$(5x - 2)^2 = (5x)^2 - (2 \times 5x \times 2) + 2^2$$
$$= 5^2x^2 - (10x \times 2) + 4$$
$$= 25x^2 - 20x + 4$$

Another important result which will be discussed in **Exercise 1(e)** is

$\boxed{1.15}$ $\qquad (a - b)(a + b) = a^2 - b^2$

Expansions of the type $\boxed{1.13}$, $\boxed{1.14}$ and $\boxed{1.15}$ are prevalent in many fields of engineering and are worth learning until they become second nature to you.

Example 20 *electrical principles*

The impedance, Z, of a circuit is given by

$$Z^2 = R^2 + \left(\omega L - \frac{1}{\omega C} \right)^2$$

where R is resistance, L is inductance, C is capacitance and ω is angular frequency. Expand the brackets and simplify.

Solution

Substituting $a = \omega L$ and $b = \dfrac{1}{\omega C}$ into $(a - b)^2 = a^2 - 2ab + b^2$ produces

$$\left(\omega L - \frac{1}{\omega C} \right)^2 = (\omega L)^2 - \left(2 \times \cancel{\omega} L \times \frac{1}{\cancel{\omega} C} \right) + \left(\frac{1}{\omega C} \right)^2$$
$$= \omega^2 L^2 - \left(\frac{2 \times L \times 1}{C} \right) + \frac{1^2}{(\omega C)^2}$$
$$= \omega^2 L^2 - \frac{2L}{C} + \frac{1}{\omega^2 C^2} \qquad \text{[Simplifying]}$$

Substituting this into the original formula gives

$$Z^2 = R^2 + \omega^2 L^2 - \frac{2L}{C} + \frac{1}{\omega^2 C^2}$$

SUMMARY

Expand brackets of the form $(a + b)(c + d)$ by using FOIL (First, Outside, Inside, Last). Important expansions are

$\boxed{1.13}$ $\qquad (a + b)^2 = a^2 + 2ab + b^2$

$\boxed{1.14}$ $\qquad (a - b)^2 = a^2 - 2ab + b^2$

$\boxed{1.15}$ $\qquad (a + b)(a - b) = a^2 - b^2$

Solutions at end of book. Complete solutions available at www.palgrave.com/science/engineering/singh

Exercise 1(e)

1 Multiply out the brackets and simplify:
   a $2(3x + 1)$ **b** $-(2x + 1)$
   c $-3(5y + 1)$ **d** $x(3x + 5)$
   e $3(y - 1) - (2y + 1)$
   f $x(x - 3) + x(3x + 2)$

2 [*structures*]   Remove the brackets from the following and simplify:

   a $y = \dfrac{w}{2EI}(Lx^3 - x^4)$

   b $y = \dfrac{wx^3}{8EI}(2L - 3x)$

   c $y = \dfrac{wx^2}{48EI}(3L^2 - 2x^2)$

   d $y = -\dfrac{w}{12EI}\left(Lx^3 - \dfrac{x^4}{2} - \dfrac{L^3x}{2}\right)$

   (L is length of beam, x is distance along the beam, EI is the flexural rigidity, y is deflection of the beam and w is the load per unit length).

3 Expand the following brackets and simplify:
   a $(x + 1)(x + 2)$
   b $(2x + 3)(3x + 5)$
   c $(2x - 1)^2$
   d $(a + b)^2 - (a - b)^2$
   e $(xy + 1)^2 - x(y + 1)$

4 By expanding brackets show that
   a $(x - 5)(x + 5) = x^2 - 25$
   b $(2x - 3)(2x + 3) = 4x^2 - 9$
   c $(9x - 7)(9x + 7) = 81x^2 - 49$

? **What do you notice about the above results?**

In general

   1.15 $(a + b)(a - b) = a^2 - b^2$

This is known as the difference between two squares.

5 [*electrical principles*]   Expand and simplify the following:

   a $(R + \omega L)(R - \omega L)$

   b $\dfrac{1}{R^2} + \left(\omega C - \dfrac{1}{\omega L}\right)^2$

   (R is resistance, ω is angular frequency, L is inductance and C is capacitance).

6 Find

   $(x - a)(x - b)(x - c) \ldots (x - z)$

   where … means ($x -$ number represented by next letter of the Roman alphabet).

SECTION F **Factorization**

By the end of this section you will be able to:
▶ factorize simple expressions
▶ factorize quadratic expressions

F1 **Factorizing expressions**

? We investigated factors in the Introductory chapter. **What are the factors of 10?**

          5 and 2 because $5 \times 2 = 10$

Of course there are other factors of 10:1 and 10.

Similarly $2 \times 5 \times 7 = 70$ and we say that 2, 5 and 7 are factors of 70.

In this section we look at factors of algebraic expressions.

Example 21

Factorize $5x + 5y + 5z$.

Solution

What do you notice about $5x + 5y + 5z$?

The number 5 is common to all the terms in $5x + 5y + 5z$. We write

$$5x + 5y + 5z = 5(x + y + z)$$

and say that 5 and $x + y + z$ are **factors** of $5x + 5y + 5z$.

Factorization is the reverse process of expansion discussed in the previous section.

How do we factorize an expression like

$$5x - 4x^2?$$

We know x is common in both terms because $x^2 = xx$, thus

$$5x - 4x^2 = 5x - 4xx = x(5 - 4x)$$

How do we factorize an engineering expression such as

$$y = \frac{wx^2}{EI} - \frac{wx^3}{EI}?$$

We know from the rules of indices that x^3 can be written as x^2x, so we have:

$$y = \frac{wx^2}{EI} - \frac{wx^2 \, x}{EI}$$

What is common between the two terms on the Right-Hand Side?

Clearly it is $\dfrac{wx^2}{EI}$. So we can take out this common factor and write y as

$$y = \frac{wx^2}{EI}1 - \frac{wx^2}{EI}x = \frac{wx^2}{EI}(1 - x)$$

Let's do another example.

Example 22 *structures*

The deflection y of a beam of length L at distance x is given by

$$y = \frac{wx^2 \, L^2}{16EI} + \frac{wx^4}{16EI}$$

where w is the load per unit length and EI is the flexural rigidity. Factorize this expression.

Example 22 *continued*

Solution

From the rules of indices we have $x^4 = x^2x^2$, so we can write y as

$$y = \frac{wx^2\,L^2}{16EI} + \frac{wx^2\,x^2}{16EI}$$

$\dfrac{wx^2}{16EI}$ is common to both terms on the Right-Hand Side, so we can take this factor out:

$$y = \frac{wx^2}{16EI}(L^2 + x^2)$$

The next example is a lot more difficult because it involves an algebraic fraction with different denominators.

Example 23 *structures*

The deflection, y, of a beam of length L at distance x is given by

$$y = \frac{wx^2\,L^2}{8EI} - \frac{wx^4}{24EI}$$

where w is the load per unit length and EI is the flexural rigidity. Factorize this expression.

Solution

From the rules of indices we have $x^4 = x^2x^2$, so we can write y as

$$y = \frac{wx^2\,L^2}{8EI} - \frac{wx^2x^2}{24EI}$$

$\dfrac{wx^2}{EI}$ is common to both terms on the Right-Hand Side, so we can take this factor out:

$$y = \frac{wx^2}{EI}\left(\frac{L^2}{8} - \frac{x^2}{24}\right)$$

Can we factorize this further?

Yes. The bracket term $\dfrac{L^2}{8} - \dfrac{x^2}{24}$ is an example of an algebraic fraction. It is dealt with in the same way as an arithmetic fraction.

How do you evaluate $\dfrac{1}{8} - \dfrac{1}{24}$?

We need a common denominator, 24. Hence

$$\frac{1}{8} - \frac{1}{24} = \underbrace{\frac{3}{24} - \frac{1}{24}}_{=1/8}$$

Example 23 *continued*

Similarly we have

$$\frac{L^2}{8} - \frac{x^2}{24} = \frac{3L^2}{24} - \frac{x^2}{24} = \frac{3L^2 - x^2}{24}$$

(Of course we cannot simplify $3L^2 - x^2$ any further because they are not like terms.)

Substituting $\dfrac{L^2}{8} - \dfrac{x^2}{24} = \dfrac{3L^2 - x^2}{24}$ into ⬛ * ⬛ gives

$$y = \frac{wx^2}{EI}\left(\frac{3L^2 - x^2}{24}\right)$$

$$= \frac{wx^2}{24EI}(3L^2 - x^2)$$

In **Example 23** the examination of the fraction, $\dfrac{1}{8} - \dfrac{1}{24}$, might seem like a diversion, but to deal with the algebraic fraction, $\dfrac{L^2}{8} - \dfrac{x^2}{24}$, we need to consider the arithmetic fraction.

F2 **Factorizing quadratics** $(ax^2 + bx + c)$

An expression of the form $ax^2 + bx + c$ (where a is not zero) is called a **quadratic.**
Expand $(x + 2)(x + 5)$.

We can use FOIL:

$$(x + 2)(x + 5) = \underbrace{(x \times x)}_{F} + \underbrace{(x \times 5)}_{O} + \underbrace{(2 \times x)}_{I} + \underbrace{(2 \times 5)}_{L}$$

$$= x^2 + 5x + 2x + 10$$

$$= x^2 + 7x + 10$$

Remember in this section we go in the opposite direction.

How can we obtain $(x + 2)(x + 5)$ given the quadratic $x^2 + 7x + 10$? (Or how do we factorize $x^2 + 7x + 10$?)

Let's assume we don't know the factors of $x^2 + 7x + 10$. We know
$x^2 + 7x \underset{\uparrow}{+} 10 = (x \pm *)(x \pm \bullet)$ because $x \times x$ gives x^2.

1 If the sign in front of 10 is

$+$ then $\pm$ and $\pm$ in the brackets are the same sign
$-$ then $\pm$ and $\pm$ in the brackets are different signs

In this example, $\pm$ and $\pm$ are the same sign but we have to establish which sign.

2 If the signs are the same then

$$x^2 \underset{\uparrow}{\pm} 7x + 10 = (x \pm *)(x \pm \bullet)$$

   this first sign tells you what the sign is, hence

$$x^2 + 7x + 10 = (x + *)(x + \bullet)$$

3 Now we look at the factors of 10 (because 10 is the only term in the quadratic which does not contain an x). **What are the factors of 10?**

$$10 \text{ and } 1 \text{ or } 5 \text{ and } 2$$

We have a $7x$ on the Left-Hand Side, therefore

$$7x = x\bullet + \ast x \qquad \text{[Expanding]}$$
$$= (\bullet + \ast)x \qquad \text{[Factorizing]}$$

Since we want 7 in the middle, $\ast$ must be 5 and $\bullet$ must be 2 (or vice versa). So we have

$$x^2 + 7x + 10 = (x + 5)(x + 2)$$

Let's do another example.

Example 24

Factorize $x^2 - 2x - 3$.

Solution

Using the above procedure we have

$$x^2 - 2x \underset{\uparrow}{-} 3 = (x + \ )(x - \ )$$
$$\text{Because of this, the signs are different. Next we look at the factors of 3.}$$

What are the factors of 3?

$$1 \text{ and } 3$$

Hence we have

$$(x + 1)(x - 3) \text{ or } (x + 3)(x - 1)$$

Since we want -2 in the middle it is -3 and $+1$. Thus

$$x^2 - 2x - 3 = (x + 1)(x - 3)$$

How do we factorize $x^2 + 5x - 3$?

Since the only factors of 3 are 1 and 3 we can only have

$$(x + 1)(x - 3) \text{ or } (x - 1)(x + 3)$$

Multiplying out either of these does not give

$$x^2 + 5x - 3$$

Where have we made a mistake?

There is **no** mistake. Simply, **not** all quadratics, $ax^2 + bx + c$, can be factorized into whole numbers. The actual factorization is

$$x^2 + 5x - 3 = \left(x + \frac{5 - \sqrt{37}}{2}\right)\left(x + \frac{5 + \sqrt{37}}{2}\right)$$

Of course this looks horrendous and you are not expected to attempt this factorization in this chapter. The quadratic $x^2 + 5x - 3$ cannot be factorized into simple whole numbers.

Example 25

Factorize $2x^2 + 7x - 15$.

Solution

We know:

$$2x^2 + 7x - 15 = (2x \pm *)(x \pm \bullet) \qquad \text{[Because we want } 2x^2\text{]}$$
$$\uparrow$$

This sign tells us that the signs in the middle ($\pm$ and $\pm$) are different.

So we have

$$(2x - *)(x + \bullet) \text{ or } (2x + *)(x - \bullet)$$

Let's consider the case $(2x - *)(x + \bullet)$. The factors of 15 are 15 and 1 or 5 and 3. We need a 7 in the middle (the x term). In this example we need to be careful because the middle term is obtained by

$$
\overbrace{}^{\text{only the middle term}}
$$
$$(2x - *)(x + \bullet) = \ldots \underbrace{2x \times \bullet}_{\text{outside}} + \underbrace{x \times (-*)}_{\text{inside}} \ldots$$

Clearly the factors 15 and 1 are useless because we will never get 7. They need to be 5 and 3 because $(2 \times 5) - 3 = 7$. So $\bullet$ is 5 and $*$ is 3, that is the x term is made from $(2 \times 5) - 3 = 7$. We have

$$2x^2 + 7x - 15 = (2x - 3)(x + 5)$$

You can always check your result; expanding $(2x - 3)(x + 5)$ gives $2x^2 + 7x - 15$. Also note that if you change the signs such that we have $(2x + 3)(x - 5)$ then you get $-7x$ in the middle and not $+7x$ as required. You can only judge the placement of signs by practising a number of factorizations. It's good practice to expand your final factorization to check your result.

F3 Important factorization

How do we factorize $x^2 - 25$?

It is a quadratic because the highest power term is x^2 and it doesn't matter if there is no x. Remember a quadratic is $ax^2 + bx + c$ where a is **not** zero but b or c may be zero.

How do we factorize this, $x^2 - 25$?

We can use

1.15 $\qquad a^2 - b^2 = (a + b)(a - b)$

It's easier than the above, so we have

$$x^2 - 25 = x^2 - 5^2$$
$$\underset{\text{by } 1.15}{=} (x + 5)(x - 5)$$

Similarly we have

$$x^2 - 9 = (x + 3)(x - 3)$$
$$x^2 - 16 = (x + 4)(x - 4)$$
$$x^2 - 5 = x^2 - \left(\sqrt{5}\right)^2 \qquad \left[\text{Rewriting } 5 = \left(\sqrt{5}\right)^2\right]$$
$$= \left(x + \sqrt{5}\right)\left(x - \sqrt{5}\right)$$

Let's investigate a more challenging example.

Example 26 *mechanics*

The equation of an object falling in air is given by

$$ma = mg - kv^2$$

where m ($\neq 0$) is the mass of the object, a is acceleration, v is velocity, g is acceleration due to gravity and k is a constant. Show that

$$a = \left(\sqrt{g} - cv\right)\left(\sqrt{g} + cv\right) \text{ where } c = \sqrt{\frac{k}{m}}$$

Solution

Dividing the initial equation, $ma = mg - kv^2$, by m gives

$$a = \frac{mg}{m} - \frac{k}{m}v^2 = g - \frac{k}{m}v^2$$

From earlier work we know that $g = \left(\sqrt{g}\right)^2$ and $\dfrac{k}{m} = \left(\sqrt{\dfrac{k}{m}}\right)^2$ so we have

$$a = \left(\sqrt{g}\right)^2 - \left(\sqrt{\frac{k}{m}}\right)^2 v^2$$

$$= \left(\sqrt{g}\right)^2 - c^2 v^2 \qquad \text{where } c = \left(\sqrt{\frac{k}{m}}\right)$$

$$= \left(\sqrt{g}\right)^2 - \underbrace{(cv)^2}_{\text{by } \boxed{1.11}}$$

How can we place $\left(\sqrt{g}\right)^2 - (cv)^2$ **into two brackets?**

Use $\boxed{1.15}$, hence

$$a = \left(\sqrt{g}\right)^2 - (cv)^2 = \left(\sqrt{g} - cv\right)\left(\sqrt{g} + cv\right)$$

$\boxed{1.11}$ $a^n b^n = (ab)^n$ $\boxed{1.15}$ $a^2 - b^2 = (a - b)(a + b)$

SUMMARY

To factorize $ax^2 + bx + c$ we need to look at factors of a and c and signs inside the expression. A common factorization is

$$\boxed{1.15} \qquad a^2 - b^2 = (a - b)(a + b)$$

You need to know this result in both directions, that is from left to right and right to left.

Exercise 1(f)

Solutions at end of book. Complete solutions available at www.palgrave.com/science/engineering/singh

1 Factorize the following:
  a $4x + 4y + 4z$ **b** $8x + 8xy$
  c $2x - 4y$ **d** $3x - 2x^2$
  e $x^2 - xy$

2 [*mechanics*] The following formulae occur in mechanics. Factorize each of them.

  a $s = ut + \dfrac{1}{2}at^2$

  b $F = \dfrac{mv_2}{t} - \dfrac{mv_1}{t}$

  c $F = \rho A v_2 v_1 - \rho A v_1^2$

3 The surface area, S, of a cone of radius r and height h is given by
$$S = \pi r^2 + \pi r (r^2 + h^2)^{1/2}$$
Factorize this formula.

4 Factorize the following:
  a $x^2 + 7x + 10$ **b** $x^2 + 5x + 4$
  c $x^2 - 5x + 4$ **d** $x^2 - 4x - 12$
  e $2x^2 + x - 1$ **f** $x^2 - 3x - 4$
  g $21x^2 + 29x - 10$

5 [*electrical principles*] Factorize the following:

  a $Z^2 - R^2$ **b** $\omega^2 L^2 - \dfrac{1}{\omega^2 C^2}$

(Z is impedance, R is resistance, L is inductance, C is capacitance and ω is angular frequency).

6 [*aerodynamics*] The Froude efficiency, F, of a propulsive system is given by
$$F = \frac{2(VV_s - V^2)}{V_s^2 - V^2}$$
[V_s and V are velocities].

Show that
$$F = \frac{2V}{V_s + V}$$

7 [*structures*] Factorize the following:

  a $y = \dfrac{3wLx^2}{6EI} - \dfrac{wx^3}{6EI}$

  b $y = \dfrac{wLx^3}{4EI} - \dfrac{3wx^4}{8EI}$

  c $y = \dfrac{wx^4}{24EI} - \dfrac{wLx^3}{12EI} + \dfrac{wL^2 x^2}{24EI}$

where y is the deflection at a distance x along a beam of length L, w is load per unit length and EI is flexural rigidity.

8 [*mechanics*] The acceleration, a, of an object in vibration is given by
$$a = g - k^2 \omega^2$$
where g is acceleration due to gravity, ω is angular frequency and k is a constant.

Show that $a = \left(\sqrt{g} - k\omega \right)\left(\sqrt{g} + k\omega \right)$.

9 [*structures*] The deflection, y, of a beam of length l at a distance x from one end is given by
$$y = \frac{wx^3}{12EI} - \frac{lx^2 w}{8EI} + \frac{l^2 wx}{24EI}$$
where EI is flexural rigidity and w is load per unit length on the beam. Show that
$$y = \frac{wx}{24EI}(2x - l)(x - l)$$

Exercise **1(f) continued**

Solutions at end of book. Complete solutions available at www.palgrave.com/science/engineering/singh

10 [electrical principles] Show that

a If $N = \dfrac{Z_0 + \frac{1}{2}Z_1}{Z_0 - \frac{1}{2}Z_1}$ then

$$Z_1 = 2Z_0\left(\dfrac{N - 1}{N + 1}\right)$$

b If $Z_1(N - 1)^2 + 2Z_0(N^2 - 1)$

$$= Z_1(N + 1)^2$$

then $Z_1 = Z_0\left(\dfrac{N^2 - 1}{2N}\right)$

(Z_1, Z_0 are impedances and N is a number).

SECTION G **Quadratic equations**

By the end of this section you will be able to:
▶ solve some quadratic equations of the form $ax^2 = b$
▶ solve some quadratic equations by factorization
▶ solve all quadratic equations by formula

In Section A3 we considered linear equations. In this section we consider the different methods involved in solving quadratic equations.

A **quadratic equation** is an equation with the unknown variable to the second power. It has the form

$$ax^2 + bx + c = 0 \qquad [a \neq 0]$$

where x is the unknown variable. In **Example 27** below both equations are quadratics.

G1 **Solving quadratics using factorization**

We use the process of factorization described in the previous section to solve quadratic equations.

We know from the Introductory chapter that if the result of multiplying two numbers is zero then one of the numbers must be zero. This can be stated as:

If A and B are numbers and $A \times B = 0$ then $A = 0$ or $B = 0$.

We use this to solve various equations, for example to solve $x^2 - 2x = 0$.

Since x is common in both terms we can factorize, thus

$$x^2 - 2x = xx - 2x = 0$$

$$x(x - 2) = 0 \qquad [\text{Factorizing}]$$

$$x = 0 \text{ or } x - 2 = 0$$

$$x = 0 \text{ or } x = 2$$

Example 27

Solve the following equations:

a $x^2 - x - 6 = 0$ **b** $27x^2 - 6x - 5 = 0$

Solution

a **What are we trying to find?**

The value(s) of x satisfying $x^2 - x - 6 = 0$. **Can we factorize $x^2 - x - 6$?**

$$x^2 - x - 6 = (x + 2)(x - 3)$$

So we have

$$(x + 2)(x - 3) = 0$$

What can we say about $(x + 2)(x - 3) = 0$?

$$(x + 2) = 0 \quad \text{or} \quad (x - 3) = 0$$

Hence we have

$$x + 2 = 0 \quad \text{or} \quad x - 3 = 0$$

$$x = -2 \quad \text{or} \quad x = 3$$

b Factorizing $27x^2 - 6x - 5$ is more difficult but it can factorized into whole numbers:

$$27x^2 - 6x - 5 = (3x + 1)(9x - 5)$$

$$(3x + 1)(9x - 5) = 0$$

which gives

$$3x + 1 = 0 \quad \text{or} \quad 9x - 5 = 0$$

$$3x = -1 \quad \text{or} \quad 9x = 5$$

$$x = \frac{-1}{3} = -\frac{1}{3} \quad \text{or} \quad x = \frac{5}{9}$$

Remember that **not all** quadratics **can** be factorized into simple whole numbers.

Example 28 *mechanics*

A body of mass $m = 35$ kg has kinetic energy, $KE = 3500$ joule (J). Find the speed v given that

$$KE = \frac{1}{2}mv^2$$

Solution

Substituting $m = 35$ and $KE = 3500$ into $\frac{1}{2}mv^2 = KE$ gives

$$\frac{1}{2}35v^2 = 3500$$

Example 28 *continued*

How do we find v?

We first find v^2 and then take the square root.

So we need to remove the $\dfrac{1}{2}$ and 35 from the Left-Hand Side. How?

Multiply both sides by 2:

$$35v^2 = 3500 \times 2 = 7000$$

Divide both sides by 35:

$$v^2 = \frac{7000}{35} = 200$$

Taking the square root of both sides and using a calculator gives

$$v = \sqrt{200} = 14.14 \text{ (2 d.p.)} = 14 \text{ m/s (2 s.f.)}$$

G2 Solving quadratics using formula

The formula for solving a quadratic equation

$$ax^2 + bx + c = 0$$

where x is a unknown variable is given by

1.16
$$x = \frac{-b \pm \sqrt{b^2 - 4ac}}{2a}$$

Subsequently we will show this result in Chapter 2. (See question **8** in **Exercise 2(d)**.)

If the factorization is difficult or impossible then we use 1.16 . Generally students prefer to use this formula rather than factorization even when they shouldn't, for example to solve $x^2 - 2x = 0$.

Example 29 *structures*

The bending moment, M, of a beam is given by

$$M = 0.3x^2 + 0.35x - 2.6$$

where x is the distance (in m) along a beam from one end. Find the value of x for which $M = 0$.

Solution

We have

$$0.3x^2 + 0.35x - 2.6 = 0$$

Example 29 *continued*

It is not easy to factorize this, so we use formula 1.16 to determine x. For the formula, a is the number next to x^2, b is the number next to x and c is the number without any x attached to it. Hence

$$a = 0.3, \quad b = 0.35 \quad \text{and} \quad c = -2.6$$

Substituting this, $a = 0.3$, $b = 0.35$ and $c = -2.6$, into

1.16 $$x = \frac{-b \pm \sqrt{b^2 - 4ac}}{2a}$$

gives

$$x = \frac{-0.35 \pm \sqrt{0.35^2 - (4 \times 0.3 \times (-2.6))}}{2 \times 0.3}$$

$$= \frac{-0.35 \pm \sqrt{3.243}}{0.6}$$

$$= \frac{-0.35 \pm 1.801}{0.6}$$

$$x = \frac{-0.35 + 1.801}{0.6} \quad \text{or} \quad \frac{-0.35 - 1.801}{0.6}$$

$$x = 2.42 \text{ (2 d.p.)} \quad \text{or} \quad x = -3.59 \text{ (2 d.p.)}$$

Since we cannot have a distance of -3.59 m on the beam, the bending moment $M = 0$ is at $x = 2.42$ m.

SUMMARY

For a quadratic equation, $ax^2 + bx + c = 0$, first seek factorization. If this fails then try the formula

1.16 $$x = \frac{-b \pm \sqrt{b^2 - 4ac}}{2a}$$

Exercise 1(g)

Solutions at end of book. Complete solutions available at www.palgrave.com/science/engineering/singh

1 Solve the following equations:

a $2x - 1 = 0$

b $x^2 + 5x + 6 = 0$

c $x^2 - 10x + 21 = 0$

d $6x^2 - 13x - 5 = 0$

e $5x^2 + 14x - 3 = 0$

2 [mechanics] A vehicle with velocity v and constant acceleration a is related by

$$v^2 = u^2 + 2as$$

where s is the distance and u is the initial velocity. If $a = 3.2$ m/s², $s = 187$ m and $v = 35$ m/s, find u.

Solutions at end of book. Complete solutions available at www.palgrave.com/science/engineering/singh

3 [structures]   The maximum deflection of a beam occurs at x satisfying

$$2x^2 - 3xL + L^2 = 0$$

where L is the length of the beam and x is the distance along the beam from one end. Find x at maximum deflection.

4 [mechanics]   The displacement, s, of a particle is given by

$$s = 1.9t + 4.3t^2 \quad (t \geq 0)$$

where t is time. Find the time taken for a displacement of 50 m.

5 A rectangular conservatory has length l and its width is 5 m shorter then its length. Given that the area of the floor is 84 m^2, find the dimensions of the floor.

6 [structures]   The bending moment, M, of a beam is given by

$$M = 3000 - 500x - 20x^2$$

where x is the distance along the beam from one end. At what distance is the bending moment $M = 0$.

7 [structures]   A simply supported beam has the bending moment, M, given by

$$M = \frac{15}{8}x - \frac{29}{4}\left(x - \frac{1}{2}\right)^2$$

where x is the distance along the beam from one support. Find the value(s) of x for $M = 0$.

8 [mechanics]   The height h (above the ground level) of a ball thrown vertically upwards is given by

$$h = -4.9t^2 + 55t + 12$$

where t is time. Find the time taken to reach the ground.

9 [mechanics]   A ball is thrown vertically upwards from a height h_0. The height h above ground level is given by

$$h = h_0 + ut - \frac{1}{2}gt^2$$

where t is time and u is initial velocity. Find an expression of t for the ball to reach the ground.

10 [aerodynamics]   The following equation occurs in aerodynamics:

$$\frac{-T}{2wL^{3/2}} = \frac{4kL^{5/2} - 3(DL^{1/2} + kL^{5/2})}{2L^3}$$

$$(L \neq 0, w \neq 0)$$

where T is thrust, w is weight, L is lift coefficient, D is drag coefficient and k ($\neq 0$) is a constant. Show that

$$L = \frac{-T \pm \sqrt{T^2 + 12kDw^2}}{2kw}$$

SECTION H Simultaneous equations

By the end of this section you will be able to:
► solve a pair of linear simultaneous equations

H1 Solving simultaneous linear equations

Simultaneous means occurring together. **Simultaneous equations** are a set of equations such that the unknown variables $x, y, z\ldots$ have the same values satisfying each equation. In this section we solve two simultaneous linear equations.

Example 30

Solve the simultaneous equations:

$$150x + 140y = 10.4$$

$$150x + 100y = 10$$

Solution

What are we trying to find?

The values of x and y satisfying the above equations:

| † | $150x + 140y = 10.4$ |

| †† | $150x + 100y = 10$ |

If we have one equation with one unknown then it is easy, we can apply a simple transposition of formulae. **Can we possibly get one equation with one unknown from** † **and** †† **?**

If we subtract these equations, † − †† :

| † | $150x + 140y = 10.4$ |
| †† | $-(150x + 100y = 10)$ |

we get $\quad\quad 0 + 40y = 0.4$

Can we solve $40y = 0.4$?

This is just a **linear equation** with one unknown, y :

$$40y = 0.4$$

$$y = \frac{0.4}{40} = 0.01$$

Have we completed this problem?

No, we need to find x. **How do we find x?**

Substitute $y = 0.01$ into †† (or †):

$$150x + (100 \times 0.01) = 10$$
$$150x + 1 = 10$$
$$150x = 9$$
$$x = \frac{9}{150} = 0.06$$

Hence

$$x = 0.06 \text{ and } y = 0.01$$

Example 30 *continued*

We can check our solution by substituting $x = 0.06$ and $y = 0.01$ into the original equations:

$$150x + 140y = 10.4$$
$$150x + 100y = 10$$

We get

$$(150 \times 0.06) + (140 \times 0.01) = 10.4$$
$$(150 \times 0.06) + (100 \times 0.01) = 10$$

The procedure outlined in the above example is a process of **elimination**. We eliminate one of the unknown variables and then solve for the remaining unknown variable.

Example 31 *mechanics*

The distance, s, travelled by an object is given by

$$s = ut + \frac{1}{2}at^2$$

where a is constant acceleration, u is initial velocity and t is time.

An experiment produces the following results: After times of 3 s and 5 s the distances travelled by the object are 66 m and 160 m respectively.

Determine the values of u and a.

Solution

Substituting $t = 5$ and $s = 160$ into $ut + \frac{1}{2}at^2 = s$ gives

$$5u + \underbrace{\frac{1}{2}5^2}_{=12.5}a = 160$$

Substituting $t = 3$ and $s = 66$ into $ut + \frac{1}{2}at^2 = s$ gives

$$3u + \underbrace{\frac{1}{2}3^2}_{=4.5}a = 66$$

Rewriting these as

† $5u + 12.5a = 160$

†† $3u + 4.5a = 66$

How can we get one equation with one unknown from † **and** †† **?**

In the previous example we had the same number of x's so, when we subtracted, the x's vanished. **Can we remove the u's from** † **and** †† **?**

Yes, we need to make the numbers in front of u (the coefficients of u) to be the same. In † we have $5u$ and in †† we have $3u$.

Example 31 *continued*

If we multiply $5u$ by 3 we get $15u$ and if we multiply $3u$ by 5 we also get $15u$.
Thus multiplying [†] by 3 gives

$$(3 \times 5u) + (3 \times 12.5a) = 3 \times 160$$

[*] $$15u + 37.5a = 480$$

and multiplying [††] by 5 yields

[**] $$15u + 22.5a = 330$$

Why?

Because when we subtract, [*] – [**], the u's are eliminated:

[*] $\quad 15u + 37.5a = 480$
[**] $\quad \underline{-(15u + 22.5a = 330)}$
$$0 + 15a = 150$$

$$a = \frac{150}{15}$$

$$a = 10$$

Substituting $a = 10$ into $3u + 4.5a = 66$ gives the linear equation

$$3u + (4.5 \times 10) = 66$$

$$3u + 45 = 66$$

$$3u = 66 - 45 = 21$$

$$u = \frac{21}{3} = 7$$

We have $a = 10$ m/s^2 and $u = 7$ m/s. You can check your result by plugging these
numbers into the original equations.

Why do you think we remove the u's in the above example?

It is straightforward to find a common multiple of 3 and 5, that is 15, rather than find a
common multiple of 4.5 and 12.5. We say that the 5 of $5u$ is the coefficient of u.

SUMMARY

For two simultaneous linear equations, eliminate one of the unknown variables and the
result is a linear equation with one unknown. Solve for this unknown, substitute this
value into one of the original equations and solve for the remaining unknown.

Exercise 1(h)

1 Solve the simultaneous equations:

$$8x + 5y = 13$$
$$x + 5y = 6$$

2 [mechanics]   A lifting machine obeys the law

$$E = aW + b$$

where E is effort force, W is load and a, b are constants. An experiment produces the following results: Effort forces of 45.5 N and 53 N lift loads of 70 N and 120 N respectively. Find the values of the constants a and b.

3 [mechanics]   The displacement, s, of a body is given by

$$s = ut + \frac{1}{2}at^2$$

where t is time, u is initial velocity and a is acceleration. If at $t = 2$ s then $s = 33$ m and at $t = 3$ s then $s = 64.5$ m, find the initial velocity (u) and acceleration (a).

4 [electrical principles]   By applying Kirchhoff's law in a circuit we obtain

$$25(I_1 - I_2) + 56I_1 = 2.225$$
$$17I_2 - 3(I_1 - I_2) = 1.31$$

where I_1 and I_2 represent currents. Find I_1 and I_2.

5 [materials]   The length, ℓ, of an alloy varies with temperature t according to the law

$$\ell = \ell_0(1 + \alpha t)$$

where ℓ_0 is the original length of the alloy and α is the coefficient of linear expansion. An experiment produces the following results:

At $t = 55\,^{\circ}\text{C}$ $\ell = 20.11$ m

At $t = 120\,^{\circ}\text{C}$ $\ell = 20.24$ m

Determine ℓ_0 and α. (The units of α are /$^{\circ}$C.)

(*Hint*: Eliminate ℓ_0 by division.)

6 [electrical principles]   Resistors R_1 and R_2 are parallel in a circuit and satisfy

$$\frac{1}{R_1} + \frac{1}{R_2} = 1.2 \times 10^{-3}$$

$$\frac{5}{R_1} + \frac{8}{R_2} = 6.6 \times 10^{-3}$$

Determine R_1 and R_2.

(*Hint*: Work in terms of $1/R_1$ and $1/R_2$.)

7 [dimensional analysis]   The force, F, of a jet is a function of density ρ, area A and velocity v. By assuming

$$F = K\rho^a A^b v^c$$

and dimensional homogeneity, find a, b and c and express F in terms of ρ, A and v. (K, a, b and c are real numbers. *Hint*: Use the fact that the equations must be dimensionally homogeneous to write three simultaneous equations by using Table 1.)

8 [dimensional analysis]   The power, P, required to drive an air screw depends on the diameter D, the number n of revolutions per second and density ρ. Assume

$$P = K\rho^a n^b D^c$$

where K, a, b and c are real numbers. Using dimensional analysis, or otherwise, determine a, b and c and write down the equation relating P, ρ, n and D. (Take the dimensions of n to be T^{-1}.)

Solutions at end of book. Complete solutions available at www.palgrave.com/science/engineering/singh

Miscellaneous exercise 1

1 [aerodynamics]   The pressure coefficient C is defined by

$$C = \frac{\frac{1}{2}\rho(v^2 - u^2)}{\frac{1}{2}\rho v^2}$$

where u, v are velocities and ρ is density. Simplify this formula.

2 [fluid mechanics]   The pressures P_1 and P_2 at depths d_1 and d_2 respectively are given by

$$P_1 = \rho g(d - d_1)$$
$$P_2 = \rho g(d - d_2)$$

where d is depth of the fluid, ρ is the density of fluid and g is acceleration due to gravity. Show that

$$P_2 - P_1 = -\rho g(d_2 - d_1)$$

3 [fluid mechanics]   The head loss, h, of a fluid in a pipe is given by

$$h = \frac{v_2}{g}(v_2 - v_1) - \frac{v_2^2 - v_1^2}{2g}$$

(g is acceleration due to gravity and v_1, v_2 are velocities of fluid). Show that

$$h = \frac{(v_1 - v_2)^2}{2g}$$

4 Evaluate $x^2 + x + 41$ for $x = 0, 1, 2, 3, 4$ and 5. What do your results have in common?

5 [electronics]   The resonant frequency, f_0, of a tuned circuit is given by

$$f_0 = \frac{1}{2\pi\sqrt{LC}}$$

a Evaluate f_0 for $L = 5 \times 10^{-3}$ henry and $C = 1 \times 10^{-6}$ farad.
b If $f_0 = 1000$ Hz and $L = 1 \times 10^{-3}$ henry then determine C.

6 [electrical principles]   The total resistance, R, of two resistors, R_1 and R_2, in parallel is given by

$$\frac{1}{R} = \frac{1}{R_1} + \frac{1}{R_2}$$

Show that

$$R = \frac{R_1 R_2}{R_1 + R_2}$$

7 [electrical principles]   Find the total resistance, R, of a circuit consisting of three resistors, $R_1 = 10$ kΩ, $R_2 = 15$ kΩ and $R_3 = 1.2$ kΩ, connected in parallel. (The total resistance, R, is given by

$$\frac{1}{R} = \frac{1}{R_1} + \frac{1}{R_2} + \frac{1}{R_3}.)$$

8 [aerodynamics]   An aircraft's drag, D, at speed v in a medium of density ρ is given by

$$D = \frac{1}{2}\rho v^2 AC_D + \frac{1}{2}\rho v^2 AkC_L^2$$

where C_D, C_L are drag coefficients, A is area and k is a constant. Transpose to make v the subject of the formula.

9 [mechanics]   The excess energy, E, of an engine between the points of maximum speed, v_1, and minimum speed, v_2, is given by

$$E = \frac{1}{2}Iv_1^2 - \frac{1}{2}Iv_2^2$$

where I is the moment of inertia. Make I the subject of the formula.

10 [fluid mechanics]   The flow of liquid from location 1 to location 2 can be described by Bernoulli's equation:

$$\frac{p_1}{\rho g} + \frac{v_1^2}{2g} + h_1 = \frac{p_2}{\rho g} + \frac{v_2^2}{2g} + h_2$$

Miscellaneous exercise **1 continued**

Solutions at end of book. Complete solutions available at www.palgrave.com/science/engineering/singh

where v is flow velocity, p is pressure, h is height, g is acceleration due to gravity and ρ is density. Make v_1 the subject of the formula.

11 Solve the following equations:

  a $5x - 1 = 0$ **b** $3x + 2 = 8$

  c $(x - 1)(x + 2) = 0$

  d $(3x - 1)(2x + 3) = 0$

12 [mechanics]   The vertical displacement, y, of a projectile in motion is given by

$$y = ut - \frac{1}{2}gt^2$$

where u is the initial velocity of the projectile. Find t for $y = 10$ m and $u = 14$ m/s. (Take $g = 9.8$ m/s².)

13 [thermodynamics]   The exit velocity, u, of a fluid from a nozzle is given by

$$u = \left\{ \frac{2\gamma P_1 V_1}{\gamma - 1} \left[1 - \frac{P_2 V_2}{P_1 V_1} \right] \right\}^{\frac{1}{2}}$$

where P_1, V_1 represent the entrance pressure and specific volume respectively and P_2, V_2 represent the exit pressure and specific volume respectively. γ is the ratio of specific heat capacities. Given that

$$P_1 V_1^\gamma = P_2 V_2^\gamma$$

show that

$$u^2 = \frac{2\gamma P_1 V_1}{\gamma - 1} \left[1 - \left(\frac{P_2}{P_1} \right)^{1 - 1/\gamma} \right]$$

Find u (correct to 1 d.p.) given that $\gamma = 1.39$, $P_1 = 5.2 \times 10^6$ N/m², $V_1 = 3.1 \times 10^{-3}$ m³/kg and $V_2 = 5 \times 10^{-3}$ m³/kg.

14 [electrical principles]   In an electrical circuit a resistor R satisfies

$$R = 1 + \frac{3(9 + R)}{12 + R}$$

Determine R.

15 Solve the following equations:

  a $x^2 - 7x + 10 = 0$

  b $x^2 - 1 = 0$

  c $2x^2 - 3x + 1 = 0$

  d $15x^2 - x - 2 = 0$

  e $-100x^2 + 400x - 300 = 0$

16 [dimensional analysis]   The following coefficients occur in aerodynamics:

Lift coefficient $C_L = \dfrac{L}{PA}$

Drag coefficient $C_D = \dfrac{D}{PA}$

Moment coefficient $C_M = \dfrac{M}{PA\ell}$

where L is the lift (in N), D is the drag (in N), P is the pressure, A is the area, M is the moment (in Nm) and l is the length. Show that C_L, C_D and C_M are dimensionless.

17 [structures]   The deflection, y, of a beam of length L is given by

$$y = \frac{wx^4}{36EI} - \frac{wLx^3}{8EI} + \frac{wL^4}{36EI}$$

where w is the load per unit length, EI is the flexural rigidity and x is the distance along the beam from one end. Factorize this expression.

18 [structures]   The critical load, P, of a steel column can be obtained from

$$L\sqrt{\frac{P}{EI}} = n\pi$$

where L is the length, EI is flexural rigidity and n is a positive whole number.

Miscellaneous exercise **1 continued**

Solutions at end of book. Complete solutions available at www.palgrave.com/science/engineering/singh

i Transpose to make P the subject of the formula.

ii Determine P (correct to 2 d.p.) for $n = 1$, $E = 0.2 \times 10^{12}$ N/m^2, $I = 6.95 \times 10^{-6}$ m^4 and $L = 1.07$ m.

19 Solve the following equations:

a $x^2 + 3x + 1 = 0$

b $x^2 + 4x + 2 = 0$

c $5x^2 + 2x - 1 = 0$

d $1 - 3x - 2x^2 = 0$

20 [vibrations]   A constant, C, in a vibrational problem is defined as

$$C = \frac{F_0}{k - m\alpha^2} \qquad \left(\alpha \neq \sqrt{k/m}\right)$$

where F_0 is the magnitude of the forcing function, k is the spring stiffness, m is the mass and α is the angular frequency.

If $\omega = \sqrt{\dfrac{k}{m}}$ and $r = \dfrac{\alpha}{\omega}$ then show that

$$C = \frac{F_0/k}{1 - r^2}$$

21 [electrical principles]   Applying Kirchhoff's law to a circuit gives

$$12(I_1 + I_2) + 67I_2 = 5.794$$
$$3I_1 - 5(I_1 - I_2) = 0.306$$

where I_1 and I_2 represent currents. Determine I_1 and I_2.

22 [fluid mechanics]   The acoustic velocity, v, is given by

$$v = (\gamma k \rho^{\gamma - 1})^{1/2}$$

Using $k\rho^\gamma = P$ and $\dfrac{P}{\rho} = RT$, show that

$$v = \sqrt{\gamma RT}$$

(k, R are constants, T is temperature, P is pressure, ρ is density and γ is the specific heat capacity ratio).

23 [fluid mechanics]   The ratio of depths $\dfrac{d_2}{d_1} \left(\dfrac{\text{upstream depth}}{\text{downstream depth}} \right)$ of water flowing through a channel can be derived from

$$d_1^2 - d_2^2 = \frac{2Q}{g}\left(\frac{Q}{d_2} - \frac{Q}{d_1} \right)$$
$$(d_1 \neq 0, d_2 \neq 0, d_1 \neq d_2)$$

where Q is the flow rate and g is acceleration due to gravity. Given that the Froude number, F, is defined as

$$F = \frac{Q}{\sqrt{gd_1^3}}$$

show that

$$\frac{d_2}{d_1} = \frac{1}{2}\left[-1 \pm \sqrt{1 + 8F^2} \right]$$

24 Solve the following simultaneous equations:

a $2x + 3y = 5$
   $x + 2y = 3$

b $3x + 8y = -18$
   $5x + 5y = 25$

c $3x + 2y = 7$
   $x + 5y = 6$

25 [vibrations]   The natural frequency, ω, of a flywheel is given by

$$\omega^2 = \frac{JG}{IL}$$

where I and J are moments of inertia, G is shear modulus of elasticity and L is length.

If mass m is placed at a distance r from the centre then the natural frequency, α, of the flywheel becomes

$$\alpha^2 = \frac{JG}{(I + 2mr^2)L}$$

From these two formulae, show that

$$I = \frac{2mr^2\alpha^2}{\omega^2 - \alpha^2}$$

Miscellaneous exercise **1 continued** Solutions at end of book. Complete solutions available at www.palgrave.com/science/engineering/singh

26 ⓨ [*vibrations*]   When looking at vibrational problems, we often need to solve the quadratic equation

⬛ * $$mx^2 + \zeta x + k = 0$$

where m is the mass, ζ is the damping coefficient and k is the spring constant. Show that

$$x = \frac{\zeta}{2m}\left(-1 \pm \sqrt{1 - \frac{4mk}{\zeta^2}}\right)$$

For critical damping, we need $b^2 - 4ac = 0$ of the quadratic $ax^2 + bx + c = 0$. For what value of ζ (> 0) in (*) does critical damping occur.

27 🩺 [*thermodynamics*]   The specific heat at constant volume c_v and the specific heat at constant pressure c_p are related by

$$c_p - c_v = R$$

where R is the gas constant. If $k = \frac{c_p}{c_v}$, show that

a $c_v = \dfrac{R}{k-1}$ **b** $c_p = \dfrac{Rk}{k-1}$

28 ⓨ [*vibrations*]   The four natural frequencies, ω_1, ω_2, ω_3 and ω_4, of a system are given by the roots of the equation

$$\omega^4 - 402\omega^2 + 800 = 0$$

Determine ω_1, ω_2, ω_3 and ω_4.

29 ⓨ [*vibrations*]   The natural frequencies, ω_1, ω_2, ω_3 and ω_4, of two pendulums connected by a spring can be obtained from the equation

$$\omega^4 - 2\left(\frac{g}{\ell} + \frac{kx^2}{m\ell^2}\right)\omega^2 + \frac{g^2}{\ell^2} + \frac{2kx^2 g}{m\ell^3} = 0$$

where ℓ, x are lengths, m is mass, k is spring stiffness and g is acceleration due to gravity. Show that

$$\omega_1 = \sqrt{\frac{g}{\ell}}, \quad \omega_2 = -\omega_1,$$

$$\omega_3 = \sqrt{\frac{g}{\ell} + \frac{2kx^2}{m\ell^2}} \text{ and } \omega_4 = -\omega_3$$

30 ✈ [*aerodynamics*]   Minimum drag occurs when the lift coefficient, L, satisfies

$$kL^2 = Z$$

where Z is the zero lift coefficient and k is a constant. The velocity, v, of an aircraft satisfies

$$w = \frac{1}{2}\rho v^2 LA$$

where w is weight, ρ is density and A is area. Show that

$$v^4 = \left(\frac{2w}{\rho A}\right)^2 \cdot \frac{k}{Z}$$

31 🌉 [*structures*]   The deflection, y, of a beam of length, L, is given by

$$EIy = \frac{w(L-x)L}{6} - \frac{w}{6L}(L-x)^3$$

$$[EI \neq 0, w \neq 0]$$

where w is the load per unit length, EI is the flexural rigidity and x is the distance along the beam. Determine the value(s) of x for which the deflection is zero.

CHAPTER 2
Visualizing Engineering Formulae

SECTION A **Graphs**

By the end of this section you will be able to:
▶ draw graphs of the form $y = mx + c$
▶ define the gradient and intercept of a straight line graph
▶ observe what effect the change in m and c has on the graph of $y = mx + c$

A1 **Linear graphs**

? **What does the term *'linear equation'* mean?**

It is an equation which gives a straight line when plotted on a graph.

? **What does the word *graph* mean?**

It is a set of points connected together which shows a relationship between two variables.

Here is an example to remind you of the terminology of graphs.

Example 1

Plot the graph of $y = 2x + 1$ for $x = -2, -1, 0, 1$ and 2.

Solution

We first find the y values by substituting the given x values into $y = 2x + 1$ as shown in Table 1.

	x	$y = 2x + 1$
TABLE 1	-2	$[2 \times (-2)] + 1 = -3$
	-1	$[2 \times (-1)] + 1 = -1$
	0	$[2 \times 0] + 1 = 1$
	1	$[2 \times 1] + 1 = 3$
	2	$[2 \times 2] + 1 = 5$

Next we plot the points on a graph. For example the first point $(-2, -3)$ means -2 horizontal (i.e. 2 to the left) and -3 vertical (i.e. 3 down). Similarly the last point $(2,5)$ means 2 horizontal (i.e. 2 to the right) and 5 vertical (i.e. 5 up).

Example 1 *continued*

The *x* values are plotted horizontally and the *y* values vertically. To obtain a graph we connect all the points in the table as the graph in Fig. 1 shows.

Fig. 1 Portion of the graph of $y = 2x + 1$

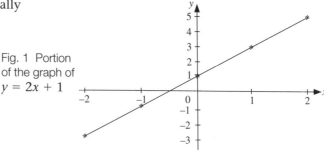

Generally the horizontal axis is called the *x* axis and the vertical axis is called the *y* axis. The point (0, 0) is called the *origin*. This is known as the rectangular Cartesian co-ordinate system, *x*−*y* axes. We say *x* is the **independent variable** and *y* is the **dependent variable** (depends on *x*). Generally the independent variable is plotted along the horizontal axis and the dependent variable along the vertical axis.

The name *Cartesian* is derived from the French philosopher, **René Descartes** born in 1596. Descartes attended a Jesuit college and because of his poor health he was allowed to remain in bed until 11 o'clock in the morning, a habit he continued until his death in 1650.

? **What does the term *gradient* mean?**

It gives a figure to the steepness of a hill but in terms of graphs it's a value which describes the slope of a line on a graph.

? **How do we measure this slope?**

It is the ratio between the vertical distance between two points to the horizontal distance between them (Fig. 2).

Fig. 2

$$\text{Gradient} = \frac{\text{Vertical}}{\text{Horizontal}}$$

? **What is the gradient of the straight line $y = 2x + 1$ shown in Fig. 1?**

The first step is to select any two points. Let's choose the two end points $(-2, -3)$ and $(2, 5)$. Then we draw horizontal and vertical lines as shown in Fig. 3.

Fig. 3 Portion of the graph of $y = 2x + 1$

$$\text{Gradient} = \frac{\text{Vertical}}{\text{Horizontal}} = \frac{5 + 3}{2 + 2} = 2$$

So the gradient of the line $y = 2x + 1$ is 2.

? **Is there any relationship between $y = 2x + 1$ and the gradient, 2?**

Consider other examples:

i   ii   iii

i The gradient of $y = 3x + 5$ is 3.

ii The gradient of $y = -\dfrac{1}{2}x + 7$ is $-\dfrac{1}{2}$.

iii The gradient of $y = -\pi x - 10$ is $-\pi$.

In fact every relationship of the form $y = mx + c$ is a straight line and the gradient of the general straight line

$$y = mx + c$$

is m, the number next to the x variable. This number is called the **x coefficient.**

? **Consider another aspect of the graph. At what value of y does the graph $y = 2x + 1$ cross the y axis?**)

At 1. This is known as the y-intercept. Similarly

i The y-intercept of $y = 3x + 5$ is 5, see graph **i.**

ii The y-intercept of $y = -\dfrac{1}{2}x + 7$ is 7, see graph **ii.**

iii The y-intercept of $y = -\pi x - 10$ is -10, see graph **iii.**

? **What rule can you gather from these examples?**

The y-intercept is the value of the constant, c, of the graph $y = mx + c$.

So summarizing the above, we have the graph of the straight line

2.1 $y = mx + c$

where m is the gradient and c is the y-intercept.

Next we look at how a change in m and c changes the graph. **Example 2** illustrates the effect of changing the gradient on a straight line graph. It also shows that we can deal with other variables besides x and y. In this example the horizontal and vertical axes are x and f respectively.

Example 2 *mechanics*

The force, f, required to produce an extension, x, of a spring is given by

$$f = kx$$

where k is the spring stiffness. On the same axes, sketch the graphs of f against x for $0 \leq x \leq 0.2$ and

Example 2 continued

i $k = 10$ N/m   **ii** $k = 20$ N/m   **iii** $k = 40$ N/m

What force, f, is required for values of k in parts i, ii and iii to produce an extension of 0.1 m?

Solution

Putting $k = 10$, 20 and 40 into $f = kx$ gives the equations

$$f = 10x, f = 20x \text{ and } f = 40x$$

In each case the y-intercept is 0 and the gradients of $f = 10x$, $f = 20x$ and $f = 40x$ are 10, 20 and 40 respectively. We have the graphs shown in Fig. 4.

Fig. 4
Graphs of force against extension

Notice that as the gradient increases, the line gets steeper. For an extension of 0.1 m we read the values from the graph where $x = 0.1$, which gives

$$k = 10 \text{ N/m}, f = 1 \text{ N}$$

$$k = 20 \text{ N/m}, f = 2 \text{ N}$$

$$k = 40 \text{ N/m}, f = 4 \text{ N}$$

This example also shows that the scales on each axis do not need to be the same. The scale on the x axis is 0.05 whilst that on the f axis is 1.

Example 3 fluid mechanics

The streamlines of a fluid flow are given by the equation

$$y - x = C$$

where C is a constant. On the same axes, sketch the graphs for $C = 0$, $1, -1$, $2, -2$, 3 and -3.

Do you notice any relationship between the graph and the value for C?

Solution

Fig. 5 Portion of the graphs of $y = x + C$

By rearrangement we have $y = x + C$, which is a straight line with gradient equal to 1 since the x coefficient is 1. (Remember that $x = 1x$.) For the given C values we have the straight line graphs shown in Fig. 5. The y-intercept is the value of C.

From **Examples 2** and **3**, observe how a change in m and c alters the straight line graph $y = mx + c$. As m gets larger, the line gets steeper and c is the point where the line cuts the y axis.

? **What effect does a negative gradient (negative m value) have on the straight line $y = mx + c$?**

A negative gradient means that the line slopes downwards as you move to the right, that is in the direction of increasing x.

Example 4 illustrates the impact of a negative gradient. In this example, the horizontal and vertical axes are t and a respectively.

Example 4 *mechanics*

The acceleration, a, of a particle is given by

$$a = 1 - \frac{t}{3} \qquad 0 \le t \le 3$$

Sketch the graph of a against t.

Solution

The vertical a-intercept is 1 and the gradient is $-\frac{1}{3}$ because this is the number next to the t variable. Since the gradient is negative the line slopes downwards to the right, that is as t increases so a decreases.

Substituting $a = 0$ gives $0 = 1 - \frac{t}{3}$. We have $\frac{t}{3} = 1$ which gives $t = 3$.

Hence when $t = 3$, $a = 0$, that is the graph touches the t axis at 3. Joining the points on the axes gives the straight line graph shown in Fig. 6.

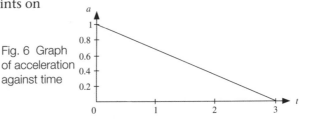

Fig. 6 Graph of acceleration against time

If $m = 0$ in $y = mx + c$ then the graph of $y = c$ is a horizontal line with value c.
If $m = \pm\infty$ (plus or minus infinity) then we have the vertical line $x = c$.

SUMMARY

A straight line graph is defined by

2.1 $\qquad y = mx + c$

where m is the gradient and c is the y-intercept. Negative m means the line slopes downwards as you move to the right.

Solutions at end of book. Complete solutions available at www.palgrave.com/science/engineering/singh

Exercise 2(a)

1 Find the gradient, m, and y-intercept, c, of the following graphs, and sketch them:

   a $y = x + 1$ **b** $y = x - 1$

   c $y = 3x + 5$ **d** $y = \dfrac{1}{2}x - 7$

   e $y - \pi = 0$ **f** $y + \pi - x = 0$

2 Sketch the following graphs:

   a $y = 3$
   b $x = 3$
   c $2x + 2y = 6$
   d $x + 2y = -3$
   e $x - y - 3 = 0$
   f $2x + 7y - 5 = 0$

3 [mechanics]   Hooke's law states that the force F (N) exerted on a spring for an extension x (m) is given by

$$F = kx$$

where k is the spring stiffness (constant) (N/m). Plot the graphs of F against x for $k = 10$, 50 and 100 N/m. What effect does the spring stiffness k have? (*Hint*: Consider an extension of 0.5 m.)

4 [fluid mechanics]   The streamlines of a fluid flow are given by

$$y = 3x + c$$

where c is a constant. On the same axes, sketch the graphs for $c = -10, -5, -1, 0, 1, 5$ and 10.

5 [electrical principles]   The voltage, V, across a capacitor is given by

$$V = \begin{cases} 2t & 0 \le t < 1 \\ 4 - 2t & t \ge 1 \end{cases}$$

Sketch the graph of V against t for $0 \le t \le 2$.

(*Hint*: The graph of V is given in two parts. You sketch $V = 2t$ for $0 \le t < 1$ and then sketch $V = 4 - 2t$ for $t \ge 1$.)

SECTION B **Applications of graphs**

By the end of this section you will be able to:
▶ find the acceleration and displacement from the $v - t$ graph

B1 **Velocity–time graphs**

There are many engineering applications of graphs but in this section we only examine the area of mechanics. We state how you can obtain the acceleration and displacement from a velocity–time graph. You may have covered the material on a mechanics module.

Fig. 7 ————————————
            O   P

Consider a particle which starts from O and moves in a straight line as shown in Fig. 7.

The distance O to P travelled by the particle is called the **displacement** of the particle. If we plot the velocity–time graph of the particle then the area under the graph gives the total displacement of the particle and the gradient of the graph gives the acceleration of the particle.

For example, let the graph in Fig. 8 be the velocity–time, v–t, graph of a particle moving along the horizontal line shown in Fig. 7. From this graph we can obtain the acceleration and total displacement. These quantities are given by

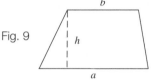

Fig. 8
Graph of velocity against time

| 2.2 | acceleration = gradient of the v–t graph |
| 2.3 | total displacement = area under the graph |

For the next example we need to know the area of a trapezium. Figure 9 shows a trapezium.

Fig. 9

| * | Area of the trapezium $= \dfrac{1}{2}(a + b)h$ |

Example 5 *mechanics*

Consider the v–t graph of Fig. 8.
a Determine the acceleration for
  i $0 \le t \le 2$
  ii $2 < t \le 10$
  iii $10 < t \le 13$
b Determine the total displacement.
c Obtain an equation for v in terms of t.

Solution

By examining the graph of Fig. 8 we have
a i For $0 \le t \le 2$, the gradient $= \dfrac{3}{2}$, hence acceleration $= \dfrac{3}{2}$ m/s².

  ii For $2 < t \le 10$, there is no slope, therefore the gradient = 0, hence acceleration = 0 m/s².

  iii For $10 < t \le 13$, the line slopes downwards to the right so we have a negative gradient. The gradient $= -\dfrac{3}{3} = -1$, thus acceleration = −1 m/s².

b Total displacement = Area of trapezium

$$= \frac{1}{2}(13 + 8) \times 3 \quad [\text{by} \boxed{*}]$$

$$= 31.5 \text{ m}$$

c Remember that m of 2.1 is the gradient = acceleration in this example and c is the v-intercept. Since the graph consists of three straight lines, we can write each line in the form $v = mt + c$ where m is the gradient and c is the v-intercept. Let's find the equation of each line separately.

The first line is when t is between 0 and 2 seconds, $0 \le t \le 2$.

What is the value of the gradient, m?

$$m = \frac{3}{2} \quad [\text{ from } \boxed{\textbf{a i}} \ ]$$

Example 5 *continued*

Where does the line cut the *v* axis?

At zero, hence $c = 0$. Thus substituting $m = \dfrac{3}{2}$ and $c = 0$ into $v = mt + c$ gives

$$v = \frac{3}{2}t \quad \text{for} \quad 0 \le t \le 2$$

The second line is when *t* is between 2 and 10 seconds, $2 < t \le 10$.

What is the gradient, *m*, equal to in this case?

$$m = 0 \; [\,\text{from} \;\; \boxed{\textbf{a ii}}\;]$$

The *v*-intercept is at $c = 3$ because if the horizontal line is extended to touch the *v* axis then it will cross the axis at 3. Substituting $m = 0$, $c = 3$ into $v = mt + c$ gives

$$v = 3 \quad \text{for} \quad 2 < t \le 10$$

Consider the last line for *t* between 10 and 13 seconds, $10 < t \le 13$.

What is the gradient, *m*, equal to?

$$m = -1 \; [\,\text{from} \;\; \boxed{\textbf{a iii}}\;]$$

What is the v-intercept, *c*, equal to?

We cannot read off the *v*-intercept from the graph so at the moment we only have

$$v = -t + c \qquad \boxed{\dagger}$$

How can we find *c*?

We know that when $t = 13$, $v = 0$ because this is where the line cuts the *t* axis. Substituting $t = 13$, $v = 0$ into $\boxed{\dagger}$ gives

$$0 = -13 + c$$

Adding 13 to make *c* the subject, establishes

$$c = 13$$

Hence the equation of the third line is

$$v = -t + 13 \text{ or } v = 13 - t \qquad 10 < t \le 13$$

Assembling these three equations together gives

$$v = \begin{cases} \dfrac{3}{2}t & 0 \le t \le 2 \\ 3 & 2 < t \le 10 \\ 13 - t & 10 < t \le 13 \end{cases}$$

This is the equation for *v* in terms of *t*.

SUMMARY

From a v–t graph, the acceleration and total displacement are given by

| 2.2 | acceleration = gradient of the v–t graph |
| 2.3 | total displacement = area under the graph |

Exercise **2b**

Solutions at end of book. Complete solutions available at www.palgrave.com/science/engineering/singh

🚗 All questions in this exercise are in the field of [*mechanics*].

1 The velocity, v, of a vehicle is given by

$$v = \begin{cases} \dfrac{3}{2}t & 0 \le t \le 2 \\ 3 & 2 < t \le 100 \\ 103 - t & 100 < t \le 103 \end{cases}$$

Sketch the v–t graph.

a Determine the acceleration for $2 < t \le 100$.

b Determine the total displacement of the vehicle.

2 Figure 10 shows a velocity–time graph for a vehicle.

Fig. 10

Write the equation of the graph.

3

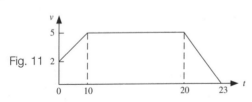

Fig. 11

Figure 11 shows a $v - t$ graph for a train travelling along a track. Determine an equation for v in terms of t.

4 The velocity, v, of an object is given by

$$v = \begin{cases} 2t & 0 \le t \le 2 \\ 4 & 2 < t \le 5 \\ 9 - t & 5 < t \le 9 \end{cases}$$

a Sketch the graph of v against t.

b The acceleration, a, can be obtained by evaluating the gradient of the velocity, v. Sketch the graph of a against t.

5 An object has an initial velocity of 20 m/s at $t = 0$. For the first 10 seconds it has **no** acceleration and then it has a constant acceleration of $a = -5$ m/s^2.

i Sketch the velocity–time graph for $0 \le t \le 15$.

ii At what time t is the velocity equal to zero?

6 Figure 12 shows a velocity–time graph for an object.

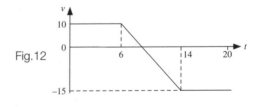

Fig.12

a State the durations for which the acceleration $a = 0$.

b What is the acceleration of the object for $6 < t < 14$?

Solutions at end of book. Complete solutions available at www.palgrave.com/science/engineering/singh

Exercise **2b continued**

7

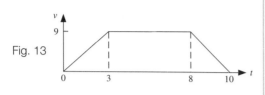

Fig. 13

Figure 13 shows a *v–t* graph of a particle moving in a straight line. Sketch the corresponding *a–t* graph.

SECTION C **Quadratic graphs**

By the end of this section you will be able to:
▸ sketch graphs of the form $y = kx^2 + c$
▸ see what effect changes in k and c have on the graph of $y = kx^2 + c$

C1 **Quadratic graph $y = x^2$**

We looked at quadratic equations in Chapter 1. In this section we examine quadratic curves. First we plot the simplest quadratic curve, $y = x^2$.

Example 6

Plot the graph of $y = x^2$ for x between -3 and $+3$.

Solution

Considering whole numbers between -3 and $+3$ gives the values shown in Table 2.

	x	$y = x^2$
TABLE 2	-3	$(-3)^2 = 9$
	-2	$(-2)^2 = 4$
	-1	$(-1)^2 = 1$
	0	$0^2 = 0$
	1	$1^2 = 1$
	2	$2^2 = 4$
	3	$3^2 = 9$

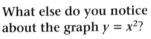

Connecting the points $(-3, 9), (-2, 4), \ldots$ and $(3, 9)$ gives the graph of $y = x^2$ (Fig. 14).

Note that the graph of $y = x^2$ is not a straight line.

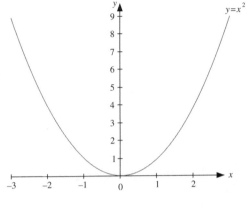

Fig. 14 Graph of $y = x^2$ for x between -3 and 3

What else do you notice about the graph $y = x^2$?

  i It is symmetrical about the y axis; for negative and positive values of x it gives the same y value. For example $x = 2$ and $x = -2$ gives $y = 4$.

 ii It goes through the origin, $(0, 0)$.

Property **i** gives the definition of an **even** function, but we will discuss this in Chapter 3 (**Miscellaneous exercise 3**).

C2 **Graph of $y = x^2 + c$**

Next we look at the graph of $y = x^2 + c$ where c is a constant. This graph is similar to $y = x^2$ but shifted up or down according to the value of c.

Example 7 *fluid mechanics*

The streamlines of fluid flow are given by:

$$y = x^2 + c$$

where c is a constant. Sketch the streamlines for $c = 0, -1, 1, -2, 2, -3$ and 3.

Solution

The graphs of $y = x^2 + c$ for $c = 0, -1, 1, -2, 2, -3$ and 3 are $y = x^2, y = x^2 - 1, y = x^2 + 1, y = x^2 - 2, y = x^2 + 2, y = x^2 - 3$ and $y = x^2 + 3$ respectively and are shown in Fig. 15.

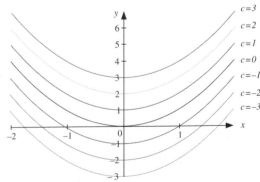

Fig. 15 Portion of the graphs of $y = x^2 + c$

Notice how the graph of $y = x^2 + c$ varies as c changes. The c is where the curve cuts the y axis.

C3 **Graph of $y = kx^2$**

Now we look at the graph of $y = kx^2$ ($k > 0$) and see what effect k has on the graph of $y = x^2$. Comparing $y = x^2$ and $y = 2x^2$ (Fig. 16), the y value in $y = 2x^2$ is twice the y value in $y = x^2$.

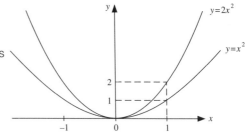

Fig. 16 Graphs of $y = x^2$ and $y = 2x^2$

Example 8 *electrical principles*

The power dissipated, P, in a resistor, R, is given by

$$P = i^2R$$

where i is the current flowing through the resistor. Sketch on the same axes the graphs of P against i for $0 \le i \le 1$ and

a $R = 1\,\Omega$ **b** $R = 5\,\Omega$ **c** $R = 10\,\Omega$ **d** $R = 20\,\Omega$

What happens to the power dissipated for an increase in resistance R?

Solution

The formulae of the graphs for the given values of R are obtained by substituting $R = 1$, $R = 5$, $R = 10$ and $R = 20$ into $P = i^2R$:

$$P = i^2, \quad P = 5i^2, \quad P = 10i^2 \text{ and } P = 20i^2$$

These graphs are the quadratic curves, with each curve getting steeper as the coefficient of i^2 gets larger (Fig. 17).

As R increases, the power dissipated in the resistor also increases. The power dissipated in a $20\,\Omega$ resistor is 20 times more than the power dissipated in a $1\,\Omega$ resistor.

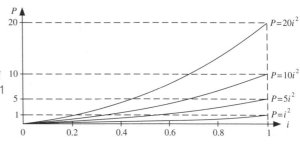

Fig. 17 Graphs of $P = i^2 R$ for i between 0 and 1

Comparing $P = 20i^2$ and $P = i^2$, the P value in $P = 20i^2$ is 20 times the P value for $P = i^2$.

Notice how the graphs of $y = kx^2$ (for positive k) are stretched by a factor of k.

The graph of $y = -x^2$ is the graph of $y = x^2$ inverted (Fig. 18).

Notice that with $y = -x^2$, the curve has a peak, ∩, whilst $y = x^2$ gives a trough, ∪. We say the graph $y = -x^2$ has a maximum value at the peak, ∩, and the graph $y = x^2$ has a

minimum value at the trough, ∪.
If the number next to x^2 is positive
then we have a trough, ∪,
and if the x^2 number is
negative then we have a
peak, ∩. The number next
to x^2 is called the **coefficient**
of x^2.

Fig. 18 Portion
of the graphs of
$y = x^2$ and
$y = -x^2$

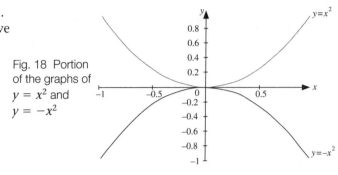

C4 **Graph of $y = kx^2 + c$**

We can sketch the graph of $y = kx^2 + c$ by locating the points where it cuts the axes and
then drawing a smooth curve through these points.

? **Where does $y = kx^2 + c$ cut the y axis?**

> At c

? **Where does $y = kx^2 + c$ cut the x axis?**

> This is where $y = 0$, that is $kx^2 + c = 0$

We try to find real values of x which satisfy $kx^2 + c = 0$.

For example $y = x^2 - 16$ cuts the y axis at -16, and the x axis when

> $$x^2 - 16 = 0$$

Solving for x gives

> $$x^2 = 16$$
> $$x = \pm\sqrt{16}$$
> $$x = 4, -4$$

Hence $y = x^2 - 16$ cuts the x axis at -4 and $+4$.
The graph of $y = x^2 - 16$ is essentially the graph
of $y = x^2$ but shifted down by 16 (Fig. 19).

Fig. 19

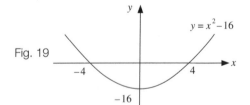

? **What does the graph of $y = x^2 + 16$
look like?**

It is basically the graph of $y = x^2$
but shifted up by 16 (Fig. 20).

Fig. 20

Note that this graph, $y = x^2 + 16$,
does **not** cut the x axis.

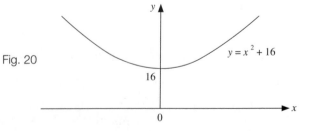

SUMMARY

The graph of $y = x^2$ is a curve with a trough ∪.

The graph of $y = -x^2$ is a curve with a peak ∩.

The graph of $y = x^2 + c$ is similar to the graph of $y = x^2$ but shifted up or down according to the value of c. If $c > 0$ then it is shifted up and if $c < 0$ then it is shifted down.

The graph of $y = kx^2$ is stretched by a factor of k ($k > 0$).

The graph of $y = kx^2 + c$ cuts the y axis at c.

Not all graphs of the form $y = kx^2 + c$ cut the x axis, as seen in Fig. 20.

However if there are real values of x which satisfy $kx^2 + c = 0$ then the graph $y = kx^2 + c$ cuts the x axis at these real values.

Exercise 2(c)

Solutions at end of book. Complete solutions available at www.palgrave.com/science/engineering/singh

1 [*fluid mechanics*]   The following table shows the velocities of fluid between two parallel plates 100 mm apart:

Distance x (mm)	0	10	20	30	40	50
Velocity v (m/s)	0	15	18	21	25	29

Distance x (mm)	60	70	80	90	100
Velocity v (m/s)	26	20	16	12	0

Plot the graph of v against x.

2 [*fluid mechanics*]   The velocity, v, of a fluid through a pipe is given by

$$v = x^2 - 9 \qquad -3 \le x \le 3$$

Sketch the graph of v against x.

3 [*electrical principles*]   The power dissipated, P, in a resistor, R, is given by

$$P = \frac{V^2}{R}$$

where V is the voltage across the resistor. Plot on the same axes the graphs of P against V for $0 \le V \le 5$ and

a $R = 1\,\Omega$ **b** $R = 5\,\Omega$ **c** $R = 10\,\Omega$
d $R = 20\,\Omega$

4 [*fluid mechanics*]   The velocity, v, of a fluid through a pipe is given by

$$v = \frac{1}{40}(400 - x^2) \qquad -20 \le x \le 20$$

where x is the distance from the central axis. Sketch the graph of v against x.

5 [*mechanics*]   The velocity, v, of a particle is given by

$$v = 48 - 3t^2 \qquad \text{where } 0 \le t \le 4$$

Sketch the graph of v against t. What can you conclude about the acceleration of the particle?

6 [*electrical principles*]   The voltage, V, of a circuit is given by

$$V = \begin{cases} 0.1t^2 & 0 \le t \le 5 \\ 2.5 & t > 5 \end{cases}$$

Sketch the graph of V against t.

7 [*mechanics*]   The position, s, of a particle along a line is given by

$$s = \begin{cases} t^2 + 1 & 0 \le t < 2 \\ 2t + 1 & 2 \le t < 10 \end{cases}$$

Sketch the graph of s against t for $0 \le t < 10$.

SECTION D **Quadratics revisited**

By the end of this section you will be able to:
▶ complete the square on the expression $ax^2 + bx + c$
▶ sketch the graph of $y = Ax^2 + Bx + C$, locating the minimum, maximum and the points where the graph cuts the axes

---

D1 **Completing the square**

Completing the square is a challenging topic and you may have to spend time to consolidate the algebra involved in the process. It might seem like a huge leap from the earlier sections but completing the square can be used to solve quadratic equations and to sketch quadratic curves, and is an important technique in more advanced engineering mathematics.

Example 9

Expand $(x + 1)^2 + 5$.

Solution

Opening up the brackets and simplifying gives

$$(x + 1)^2 + 5 = \underbrace{(x^2 + 2x + 1)}_{\text{by } \boxed{1.13}} + 5$$

$$= x^2 + 2x + 6$$

By looking at **Example 9**, note that $x^2 + 2x + 6$ can be written as $(x + 1)^2 + 5$. In this section we are given the expanded version $x^2 + 2x + 6$ and we try to rewrite this as $(x + 1)^2 + 5$. In other words we travel from the solution to the question of **Example 9**:

$$x^2 + 2x + 6 = (x + 1)^2 + 5$$

This is an example of an **identity.** The term *identity* means that two mathematical expressions are equal for **all** values of the variables. Other examples of identities are

$$x^2 + 4x + 1 = (x + 2)^2 - 3$$

$$x^2 - 5x + 2 = \left(x - \frac{5}{2}\right)^2 - \frac{17}{4}$$

$$x^2 + 7x + 13 = \left(x + \frac{7}{2}\right)^2 + \frac{3}{4}$$

$\boxed{1.13}$ $(a + b)^2 = a^2 + 2ab + b^2$

? **Do you notice any relationship between the Left-Hand Side and the Right-Hand Side of these examples?**

Look at the x coefficient (the number next to the x) on the left and the number in the bracket on the right.

The number within the brackets is half the x coefficient on the left (Table 3).

	x coefficient	Number within the bracket
TABLE 3	4	2
	-5	$-5/2$
	7	$7/2$

Let's consider $x^2 + 4x + 1 = (x + 2)^2 - 3$.

? **The 2 in the brackets is half of 4 but where does -3 on the Right-Hand Side materialize from?**

By applying $(a + b)^2 = a^2 + 2ab + b^2$ with $a = x$ and $b = 2$ we have

$$(x + 2)^2 = x^2 + 2x(2) + 2^2 = x^2 + 4x + 4$$

So we have

$$x^2 + 4x = (x + 2)^2 - 4$$

We need to subtract 4 because we have produced an extra 4 within the $(x + 2)^2$ term. Hence

$$x^2 + 4x + 1 = (x + 2)^2 - 4 + 1$$
$$= (x + 2)^2 - 3$$

In a similar manner we have

$$x^2 - 5x + 2 = \left(x - \frac{5}{2}\right)^2 - \underbrace{\left(\frac{5}{2}\right)^2 + 2}$$

$$= \left(x - \frac{5}{2}\right)^2 - \underbrace{\frac{17}{4}}_{= -\left(\frac{5}{2}\right)^2 + 2}$$

and

$$x^2 + 7x + 13 = \left(x + \frac{7}{2}\right)^2 - \left(\frac{7}{2}\right)^2 + 13$$

$$= \left(x + \frac{7}{2}\right)^2 + \underbrace{\frac{3}{4}}_{= -\left(\frac{7}{2}\right)^2 + 13}$$

This method is called **completing the square** and results in an algebraic identity. To understand completing the square you have to ensure that the following result is second nature to you:

1.13 $(a + b)^2 = a^2 + 2ab + b^2$

Let's try some more examples.

Example 10

Complete the square on

a $x^2 - 8x + 3$ **b** $2x^2 - 5x + 3$ **c** $10 - 6x - 3x^2$

Solution

a $x^2 - 8x + 3 = \underbrace{(x - 4)}_{\text{half of } -8}{}^2 - 16 + 3$

$\quad\quad = (x - 4)^2 - 13$

With **b** and **c** we need to take out the x^2 coefficient.

b We first take a factor of 2 out:

$$2x^2 - 5x + 3 = 2\left[x^2 - \frac{5}{2}x + \frac{3}{2}\right]$$

$$= 2\left[\underbrace{\left(x - \frac{5}{4}\right)^2}_{\text{half of } 5/2} - \left(\frac{5}{4}\right)^2 + \frac{3}{2}\right]$$

$$= 2\left[\left(x - \frac{5}{4}\right)^2 - \underbrace{\frac{1}{16}}_{= -(5/4)^2 + 3/2}\right]$$

$$= 2\left(x - \frac{5}{4}\right)^2 - \underbrace{\frac{1}{8}}_{= 2/16} \quad\quad\text{[Expanding]}$$

c Factorizing the -3 in the last two terms gives

$$10 - 6x - 3x^2 = 10 - 3[2x + x^2]$$

$$= 10 - 3\left[(x + 1)^2 - 1\right] \quad\text{[Completing the square]}$$

$$= 10 - 3(x + 1)^2 + 3 \quad\quad\text{[Expanding]}$$

$$= 13 - 3(x + 1)^2$$

D2 Applications of completing the square

We can use the procedure of completing the square to solve quadratic equations and to sketch quadratic curves.

Example 11

Solve the quadratic equation

$$10 - 6x - 3x^2 = 0$$

Solution

We have already completed the square in **Example 10(c)** so we have the identity

$$10 - 6x - 3x^2 = 13 - 3(x + 1)^2$$

Putting the Right-Hand Side to zero and solving for x we have

$$13 - 3(x + 1)^2 = 0$$
$$3(x + 1)^2 = 13 \qquad \text{[Transposing]}$$
$$(x + 1)^2 = \frac{13}{3} \qquad \text{[Dividing by 3]}$$
$$(x + 1) = \pm\sqrt{\frac{13}{3}} \qquad \text{[Taking square root]}$$
$$x = -1 \pm \sqrt{\frac{13}{3}}$$

The roots of $10 - 6x - 3x^2 = 0$ are $x = -1 + \sqrt{\dfrac{13}{3}}$ or $x = -1 - \sqrt{\dfrac{13}{3}}$

Generally if the quadratic expression can be factorized easily then it is not worth investing your effort into solving the quadratic equation by completing the square because the solution of the factorized equation is straightforward. However most quadratic equations, $ax^2 + bx + c = 0$, cannot be easily factorized, so we are compelled to use the completing-the-square procedure or the quadratic equation formula

1.16 $\qquad x = \dfrac{-b \pm \sqrt{b^2 - 4ac}}{2a}$

as discussed in Chapter 1. You are asked to show result 1.16 in **Exercise 2(d).**

D3 Sketching quadratics

One advantage of the completing-the-square procedure over the quadratic formula is in sketching quadratic curves. Completing the square tells us where the minimum or the maximum point of the quadratic curve occurs. For example, say we want to sketch the quadratic $y = (x - a)^2 + b$. It is basically a x^2 graph but shifted horizontally by a units and vertically by b units (Fig. 21).

Similarly the graph of $y = b - (x - a)^2$ is shown in Fig. 22.

Fig. 21

In both Fig. 21 and Fig. 22 we have assumed a and b to be positive. This transformation of graphs is more thoroughly analysed in Chapter 3.

Thus we have

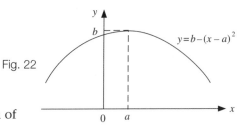

| 2.4 | The quadratic $y = (x - a)^2 + b$ has a minimum value of b at $x = a$ |

Fig. 22

| 2.5 | The quadratic $y = b - (x - a)^2$ has a maximum value of b at $x = a$ |

We use 2.4 or 2.5 to help us sketch the graph of $y = Ax^2 + Bx + C$.

Note that if the x^2 coefficient is positive then we have a minimum, but if the x^2 coefficient is negative then we have a maximum.

We execute the following to sketch the general quadratic curve $y = Ax^2 + Bx + C$:

i Mark C on the y axis; this is the point where the graph cuts the y axis.

ii If we can find real values of x which satisfy $Ax^2 + Bx + C = 0$ then the graph $y = Ax^2 + Bx + C$ cuts the x axis at these real values. The graph may not cut the x axis.

iii Complete the square to ascertain the points of minimum or maximum by using 2.4 or 2.5 . Mark these on the x–y plane.

Combining **i**, **ii** and **iii** gives us a good sketch of $y = Ax^2 + Bx + C$.

Example 12 *aerodynamics*

The height, h, of an aircraft during a manoeuvre is given by

$$h = t^2 - 10t + 500$$

where t is time. Find the minimum height reached by the aircraft and sketch the graph of h against t.

Solution

By completing the square we have

$$h = t^2 - 10t + 500 = (t - 5)^2 - 25 + 500$$
$$= (t - 5)^2 + 475$$

Hence by 2.4 the minimum height reached by the aircraft is 475 m.

The graph, $h = t^2 - 10t + 500$, cuts the h axis at 500 but where does it cut the t axis?

Since the minimum value is 475, the graph cannot cut the t axis. We can show this algebraically. We need to solve

$$(t - 5)^2 + 475 = 0$$
$$(t - 5)^2 = -475$$

Example 12 *continued*

There are no real values of t which satisfy
█ █ because when we take the
square root of a negative number
(-475) we do not get a real number.   Fig. 23
Therefore we can conclude that the
graph does not cut the t axis.

By 2.4 the minimum value of h occurs at
$t = 5$ and is 475 (Fig. 23).

SUMMARY

To complete the square on a quadratic expression we consider half the x coefficient and then subtract the extra square term produced by completing the square. If the x^2 coefficient is not 1 then we need to factorize this out first. This method gives an algebraic identity.

To sketch $y = Ax^2 + Bx + C$ we first mark C on the y axis then we solve the quadratic equation for x:

$$Ax^2 + Bx + C = 0$$

These x values are where the graph cuts the x axis. Not all quadratics cut the horizontal axis as in Fig. 23. Next we complete the square on $Ax^2 + Bx + C$ and then use 2.4 or 2.5 to locate the maximum or minimum.

Exercise 2(d)

Solutions at end of book. Complete solutions available at www.palgrave.com/science/engineering/singh

1 Complete the square on the following:

a $x^2 - 4x + 3$ **b** $9 + 8x - x^2$

c $x^2 + 7x + 1$ **d** $2x^2 + 7x + 1$

2 Solve the following equations:

a $x^2 - 4x + 3 = 0$ **b** $9 + 8x - x^2 = 0$

c $x^2 + 7x + 1 = 0$ **d** $2x^2 + 7x + 1 = 0$

3 [electrical principles]   The voltage, V, of a circuit is defined by:

$$V = t^2 - 5t + 6 \quad (t \geq 0)$$

Sketch the graph of V against t, indicating the minimum value of V.

4 [electrical principles]   The voltage, V, of a circuit is given by

$$V = t^2 - t \quad (t \geq 0)$$

Sketch the graph of V against t.

5 [mechanics]   The height h of a projectile fired from the ground with respect to time t is given by

$$h = 12t - t^2$$

Sketch this curve. Find the maximum height reached by the projectile.

6 [mechanics]   The height h (m) of a projectile fired vertically upwards is given by

$$h = 200t - 4.9t^2$$

2.4   The quadratic $y = (x - a)^2 + b$ has a minimum value of b at $x = a$

Solutions at end of book. Complete solutions available at www.palgrave.com/science/engineering/singh

Exercise **2(d) continued**

where t is time in seconds. Evaluate the maximum height reached by the projectile and sketch the curve.

7 [materials] Sketch the graph of M against x, where

$$M = \frac{W}{24EI}(Lx - x^2)$$

indicating the maximum or minimum value of M.

(M is the bending moment of the beam, L is the length of the beam, EI (> 0) is the flexural rigidity, x is the distance along the beam and W is the uniform load per unit length.)

8 By completing the square, show that the solution of the general quadratic equation $ax^2 + bx + c = 0$ is given by

1.16
$$x = \frac{-b \pm \sqrt{b^2 - 4ac}}{2a}$$

SECTION E **Further graphs**

By the end of this section you will be able to:

▶ understand the term 'asymptote'

▶ sketch the non-linear graph $\dfrac{1}{x}$

▶ sketch graphs involving variations of $\dfrac{1}{x}$, x^3, x^4...

▶ plot more complicated graphs using a symbolic manipulator or a graphical calculator

E1 **Non-linear graphs**

Non-linear graphs are graphs which do not have the shape of a straight line, thus the quadratic graphs $y = ax^2 + bx + c$ just considered are examples of non-linear graphs. First we look at the graph of $y = \dfrac{1}{x}$.

Example 13 *electrical principles*

The relationship between frequency, f, and periodic time, T, is defined as

$$f = \frac{1}{T} \qquad\qquad (T > 0)$$

Sketch the graph of f against T.

Solution

We have $f = \dfrac{1}{T}$. Note that as T increases, f decreases. For example if $T = 2$, $f = \dfrac{1}{2}$ and if $T = 10$, $f = \dfrac{1}{10}$. Consequently f gets smaller and smaller as T gets larger and larger.

Example 13 *continued*

How is this reflected in the graph of $f = \dfrac{1}{T}$?

It means that f gets closer and closer to the T axis

for large values of T without ever touching it

(Fig. 24). It never touches the T axis   Fig. 24 Graph

because $f = \dfrac{1}{T}$ cannot equal zero. We   of $f = \dfrac{1}{T}$

say that the T axis is an **asymptote** to

the graph $f = \dfrac{1}{T}$.

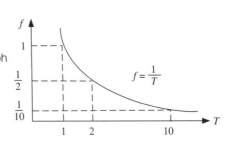

Similarly for small positive values of T, f is large and the graph gets steeper and steeper.

For example if $T = \dfrac{1}{2}$, $f = \dfrac{1}{1/2} = 2$ and if $T = \dfrac{1}{10}$, $f = 10$. Moreover the graph gets

closer to the f axis. Hence the f axis is

also an asymptote to the graph

$f = \dfrac{1}{T}$. The resulting graph, $f = \dfrac{1}{T}$,   Fig. 25 Graph

is shown in Fig. 25   of f against T
for $T > 0$

Example 14

Sketch the graph of $y = \dfrac{1}{x}$.

Solution

For positive values of x, $x > 0$, we   Fig. 26 Graph
have the graph of Fig. 25 with the   of y against x
T, f axes labelled as x, y respectively.   for $x < 0$
For negative values of x, $x < 0$, we
obtain the graph shown in Fig. 26:

Note that the graph of $y = \dfrac{1}{x}$ is not defined at $x = 0$.

Why not?

Because $\dfrac{1}{0}$ does not equal a real number.   Fig. 27
Combining the graphs of Fig. 25 and   Graph of y
Fig. 26 gives that shown in Fig. 27.   against x

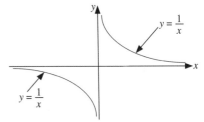

Other non-linear graphs which occur in engineering, such as x^3, $\sqrt{x}$ and x^4, are sketched in Figs 28, 29 and 30 respectively.

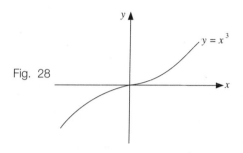

Fig. 28

Fig. 29

Next we sketch graphs involving variations of these terms.

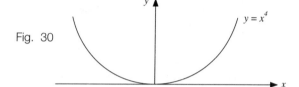

Fig. 30

Example 15 *fluid mechanics*

The velocity, v, of a fluid through a pipe is given by

$$v = 1 - x^4 \quad (-1 \le x \le 1)$$

where x is the distance from the central axis. Sketch this graph. (In Fluid mechanics, this graph is called the *velocity profile*.)

Solution

By Fig. 30 we know the graph of x^4.

? What then does the graph of $-x^4$ look like?

Fig. 31 Graph of $v = -x^4$

The graph of $-x^4$ is the inverted graph of x^4 (Fig. 31).

? How does the graph $v = 1 - x^4$ differ from $v = -x^4$?

The graph of $1 - x^4$ is the same shape as $-x^4$ but shifted up by 1 unit. Since v cuts the x axis at $v = 0$ we have

$$1 - x^4 = 0$$
$$x^4 = 1 \text{ which gives the}$$

real roots $x = 1, -1$

Fig. 32 Graph of $v = 1 - x^4$

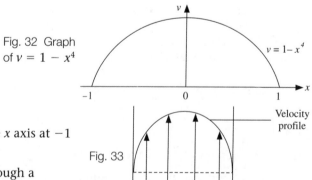

Hence the graph $v = 1 - x^4$ cuts the x axis at -1 and 1, as shown in Fig. 32.

The velocity profile, $v = 1 - x^4$, through a pipe can be drawn as shown in Fig. 33.

Fig. 33

Similarly the graph of $x^3 + C$ will have the same shape as the graph of x^3, but shifted up or down depending on the value of C. Next we plot graphs by evaluation.

E2 **Plotting graphs**

Example 16 *mechanics*

A particle moves along a horizontal line according to $x = t^3 - 6t^2 + 7t - 1$ where x is the displacement and t is time. Plot the graph of x against t for $0 \le t \le 5$.

Solution

We evaluate x by substituting whole numbers for t between 0 and 5 into $x = t^3 - 6t^2 + 7t - 1$ (Table 4).

t	$x = t^3 - 6t^2 + 7t - 1$
0	$0^3 - (6 \times 0^2) + (7 \times 0) - 1 = -1$
1	$1^3 - (6 \times 1^2) + (7 \times 1) - 1 = 1$
2	$2^3 - (6 \times 2^2) + (7 \times 2) - 1 = -3$
3	$3^3 - (6 \times 3^2) + (7 \times 3) - 1 = -7$
4	$4^3 - (6 \times 4^2) + (7 \times 4) - 1 = -5$
5	$5^3 - (6 \times 5^2) + (7 \times 5) - 1 = 9$

TABLE 4

Using this table we plot the points and connect them to give us the graph of $x = t^3 - 6t^2 + 7t - 1$ (Fig. 34).

Fig. 34 Graph of $x = t^3 - 6t^2 + 7t - 1$ for t between 0 and 5

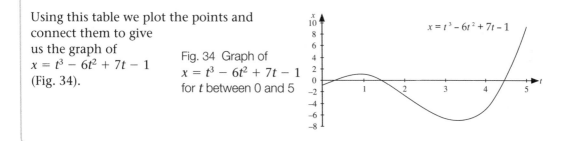

Graphs are a particular fruitful topic for computer algebra packages and graphical calculators. By changing the coefficients of variables we can visualize how the graph changes.

The remaining examples show the plotting of graphs on a symbolic manipulator (MAPLE) and a graphical calculator. If you have not used MAPLE then you might find MAPLE syntax difficult to follow, so try these examples on a package you have available (Derive, Matlab, Mathcad etc.).

Example 17 *vibrations*

In the theory of vibrations, the magnification factor M is defined as

$$M = \frac{1}{\left[\left(1 - r^2\right)^2 + \left(2\zeta r\right)^2\right]^{\frac{1}{2}}}$$

where r is the frequency ratio and ζ (zeta) is the damping ratio.

Plot, on the same axes, the graphs of M versus r for the following values of ζ: $\zeta = 0.1$, 0.2 and 0.3 and for r between 0 and 2.

Solution

The MAPLE output for this example is shown below. The % symbol in MAPLE refers to the previous output and %% refers to the penultimate output.

```
> M:= 1/sqrt((1 − r^2)^2 + (2*zeta*r)^2);
```

$$M: = \frac{1}{\sqrt{1 - 2r^2 + r^4 + 4\zeta^2 r^2}}$$

```
> M[1]:= subs (zeta = 0.1,%);
```

$$M_1: = \frac{1}{\sqrt{1 - 1.96\, r^2 + r^4}}$$

```
> M[2]:= subs (zeta = 0.2,%%);
```

$$M_2: = \frac{1}{\sqrt{1 - 1.84\, r^2 + r^4}}$$

```
> M[3]:= subs (zeta = 0.3, %%%);
```

$$M_3: = \frac{1}{\sqrt{1 - 1.64\, r^2 + r^4}}$$

```
> plot ({M[1], M[2], M[3]}, r = 0..2, color = [black, gold, gray]);
```

Fig. 35

Example 18 *reliability engineering*

The failure density function, f, is given by

$$f = \frac{0.5}{(1 + kt)^5} \qquad t \geq 0$$

where t is in years and $k \geq 0$. Plot, on the same axes, the graphs of f against t for $t \geq 0$ with

i $k = 0.01$ **ii** $k = 0.1$ **iii** $k = 1$

What happens as k gets large?

Solution

The MAPLE commands are similar to **Example 17**. The portability of the graphical calculator is a great advantage over a computer algebra system. Try this example on your graphical calculator. Your screen should show something similar to Fig. 36.

Fig. 36 Graphs of $f = \dfrac{0.5}{(1 + kt)^5}$ for different values of k

If the screen fails to show graphs of Fig. 36 then adjust the range. Also the calculator screen may not show the labels attached to the graph.

Notice that the graph decays more rapidly for large k.

SUMMARY

Use the assistance of a computer algebra system or a graphical calculator to plot complicated graphs.

Graphs can also be plotted by evaluating in a table.

Graphs of $y = x^3$, $y = \sqrt{x}$ and $y = x^4$ are shown in Figs 28–30.

Exercise **2(e)**

Solutions at end of book. Complete solutions available at www.palgrave.com/science/engineering/singh

1 [*mechanics*] The displacement, s, of an object is given by
$s = t^3 - 2t^2 + 5$.
Plot the graph of s against t for t between 0 and 3.

2 [*mechanics*] The velocity, v, of a particle is given by

$$v = t^3 - t \qquad (t \geq 0)$$

Plot the graph of v against t.

3 [*electrical principles*] The current, I, through a variable load resistor, R, is given by $I = \dfrac{10}{5 + R}$. Plot the graph of I versus R for $0 \leq R \leq 10$.

4 [*electrical principles*] The voltage, V, across a variable load resistor, R, is given by $V = \dfrac{10R}{5 + R}$. Plot the graph of V versus R for $0 \leq R \leq 10$.

Exercise **2(e) continued**

Solutions at end of book. Complete solutions available at www.palgrave.com/science/engineering/singh

5 ≋ [*fluid mechanics*]   The velocity profile, v, of a fluid is given by

$$v = 3x^2 - x^3$$
$$(x \text{ is between 0 and 3})$$

where x represents distance. Plot the graph of v against x, considering $x = 0, 0.5, 1, 1.5, 2.0, 2.5$ and 3.0.

6 ≋ [*fluid mechanics*]   The following equations are the velocity distributions v for a fluid:

a $v = 2x^2 - x^3$

b $v = 4x^2 - 2x^3$

where x is the distance and is between 0 and 2. Sketch the graphs of v against x on the same axes.

7 ≋ [*fluid mechanics*]   The flow rate, Q, of a liquid through a discharge valve is given by

$$Q = kP^{\frac{1}{2}}$$

where P is the pressure across the valve and k is a constant. Sketch the graphs, on the same axes, of Q versus P for the following values of k:

  i $k = 0.5$

 ii $k = 0.75$

iii $k = 1$

 iv $k = 2$

  v $k = 10$

8 ≋ [*fluid mechanics*]   The velocity, v, of a fluid flow is given by $v = \dfrac{k}{x}$. Sketch, on the same axes, the graphs of v against x for x between 0 and 10 and

a $k = 1$

b $k = 10$

c $k = 100$

What happens as k increases?

9 🌡 [*heat transfer*]   The thermal resistance, R, of a material is defined as

$$R = \frac{t}{kA}$$

where t is the thickness, A is the cross-sectional area and k is the thermal conductivity of the material. For $t = 0.01$ m, $A = 5$ m², sketch the graph of R against k for $k > 0$. What happens to the thermal resistance, R, as k increases? What value does R tend to as k goes to infinity?

10 🌡 [*heat transfer*]   For $t = 0.05$ m, $k = 5 \times 10^{-3}$ (W/m K), sketch the graph of the thermal resistance R against area A for $A > 0$ and $R = \dfrac{t}{kA}$. What effect does large area, A, have on the thermal resistance, R?

11 ≋ [*fluid mechanics*]   The velocity, v, of a fluid through a pipe is given by

$$v = 64 - x^4 \; \left(-64^{\frac{1}{4}} \le x \le 64^{\frac{1}{4}} \right)$$

where x is the distance from the central axis. Sketch the graph of v versus x.

12 ≋ [*fluid mechanics*]   The velocity, v, of a fluid through a pipe is given by

$$v = \sqrt{25 - x^2} \;, -5 \le x \le 5$$

where x is the distance from the central axis. Sketch the profile of v.

💻 **For the remaining questions use a symbolic manipulator (MAPLE) or a graphical calculator.**

13 ⊞ [*electrical principles*]   For the following, P represents power and R represents variable loads in an electrical circuit. Plot the graphs (on different axes) of P versus R for the corresponding R values.

Exercise **2(e) continued**

a $P = \dfrac{400R}{(10 + R)^2}$, $0 \leq R \leq 20$

b $P = \dfrac{500\,R}{[(10 \times 10^3) + R]^2}$,
$0 \leq R \leq 20 \times 10^3$

c $P = \dfrac{2500R}{(1500 + R)^2}$, $0 \leq R \leq 3000$

By using the various facilities on your graphical calculator or symbolic manipulator, determine the R values for maximum power P in each of the above cases. What do you notice about your results?

14 [electrical principles]   The voltage, V, across a variable load resistor, R_L, is given by

$$V = \frac{ER_L}{R + R_L}$$

where E is the source e.m.f. and R is the source resistance. Plot the graphs (on different axes) of V versus R_L for the corresponding values:

a $E = 60$ volts, $R = 10\ \Omega$ for $0 \leq R_L \leq 20$

b $E = 15$ volts, $R = 3 \times 10^3\ \Omega$
for $0 \leq R_L \leq 6 \times 10^3$

c $E = 10$ volts, $R = 15 \times 10^3\ \Omega$
for $0 \leq R_L \leq 30 \times 10^3$

From each of your graphs, determine the value of V at $R_L = R$. Do you notice any relationship between V and E at $R_L = R$? Show algebraically that if

$R_L = R$ then $V = \dfrac{E}{2}$.

SECTION F **Binomial expansion**

By the end of this section you will be able to:
▶ expand $(1 + x)^n$ where n is a small positive whole number
▶ use Pascal's triangle to find coefficients
▶ expand $(a + b)^n$ where n is a small positive whole number

F1 **Expansion of** $(1 + x)^n$

The prefix bi in binomial refers to two or twice, and nomial is from the Greek word *nomen* meaning name or term. Thus the word binomial means two terms.

? **What does binomial expansion refer to?**

It's expanding an expression with two terms; expanding expressions such as $(1 + x)^2$, $(5 + 2x)^7$, $(3 - 7x)^{10}$ etc. Let's examine $(1+x)^n$ where n is a positive whole number.

? **How do we expand $(1 + x)^2$?**

By using $(a + b)^2 = a^2 + 2ab + b^2$ with $a = 1$ and $b = x$ as in Chapter 1.

$$(1 + x)^2 = (1 + x)(1 + x)$$
$$= 1 + 2x + x^2$$

By expanding further we can also obtain the following:

$$(1 + x)^3 = 1 + 3x + 3x^2 + x^3$$

$$(1 + x)^4 = 1 + 4x + 6x^2 + 4x^3 + x^4$$

$$(1 + x)^5 = 1 + 5x + 10x^2 + 10x^3 + 5x^4 + x^5$$

Remember $(1 + x)^5 = (1 + x)(1 + x)(1 + x)(1 + x)(1 + x)$

? **Can you see a pattern in the index powers of x in the final expansion above?**

The x index increases by 1 each time you move across the terms. First term is $x^0 = 1$, 2nd term involves $x^1 = x$, 3rd term involves x^2 etc, up to x^5:

$$(1 + x)^5 = 1 + 5x + 10x^2 + 10x^3 + 5x^4 + x^5$$

The numbers in front of x, the **coefficients** of x, can be determined by Pascal's triangle, which is shown in Fig. 37.

Fig. 37

```
                              n = 0                1
                       n = 1              1                1
                 n = 2            1              2               1
                                        1 + 2 = 3
           n = 3          1              3               3              1
                                       3 + 3 = 6
      n = 4        1            4              6              4             1
   n = 5      1          5             10            10            5           1
 n = 6    1        6            15            20           15           6          1
n = 7  1      7          21           35          35          21          7         1
```

Each row begins and ends with 1. Every other number in a row is obtained by adding the two numbers in the row above which are immediately to the left and right of it. For example, the 5 underlined in Fig. 37 is determined by adding 1 and 4. Similarly the 20 is established by summing 10 and 10. Furthermore each of the rows gives the coefficients for the binomial expansion of the corresponding values of n. For example, $n = 5$ gives 1, 5, 10, 10, 5 and 1. Hence

$$(1 + x)^5 = 1 + (5 \times x) + (10 \times x^2) + (10 \times x^3) + (5 \times x^4) + (1 \times x^5)$$

$$= 1 + 5x + 10x^2 + 10x^3 + 5x^4 + x^5$$

This method is painless compared to multiplying out:

$$(1 + x)(1 + x)(1 + x)(1 + x)(1 + x)$$

? **By using Pascal's triangle can you find the expansion of $(1 + x)^7$?**

? **The x index increases by 1 but how do we find the coefficients of x?**

Looking at the row $n = 7$ in Pascal's triangle (Fig. 37) gives 1, 7, 21, 35, 35, 21, 7 and 1:

> **Blaise Pascal** was a French mathematician born in 1623. Pascal developed the first digital calculator in 1645. He also worked on hydrostatics, projective geometry and probability theory. Pascal suffered from poor health and died at the young age of 39 in 1662. However he has been called the great 'might have been' mathematician.

$$(1 + x)^7 = 1 + 7x + 21x^2 + 35x^3 + 35x^4 + 21x^5 + 7x^6 + x^7$$

Note the symmetrical nature of Pascal's triangle.

F2 **Expanding** $(a + b)^n$

We investigate the expansion of the general binomial expression $(a + b)^n$. By using Pascal's triangle for $n = 5$ we have

$$(a + b)^5 = a^5 + 5a^4 b + 10a^3 b^2 + 10a^2 b^3 + 5ab^4 + b^5$$

The coefficients in front of the a, b are obtained from Pascal's triangle.

The a, b powers are determined by reducing a by an index of 1, giving $a^5, a^4, a^3, \ldots, a^0$; and increasing b by an index of 1, giving $b^0, b, b^2, \ldots, b^5$.

Generally

$\boxed{2.6}$ $\qquad (a + b)^n = C_n a^n + C_{n-1} a^{n-1} b + C_{n-2} a^{n-2} b^2 + \ldots + C_0 b^n$

where $C_n, C_{n-1}, \ldots$ are coefficients found from Pascal's triangle. Note that the two end coefficients $C_n = C_0 = 1$. Also note that the coefficients next to the ends, C_{n-1} and C_1, are both equal to n. Let's consider some examples.

Example 19

Expand $\quad$ **a** $(2 + x)^3$ $\qquad$ **b** $(5 - x)^4$ $\qquad$ **c** $(3 + 2x)^5$

Solution

a From Fig. 37 with $n = 3$ we have the coefficients 1, 3, 3 and 1. Using $\boxed{2.6}$ with $a = 2, b = x$ and $n = 3$ gives

$$(2 + x)^3 = (1 \times 2^3) + (3 \times 2^2 \times x) + (3 \times 2 \times x^2) + (1 \times x^3)$$

$$= 8 + 12x + 6x^2 + x^3$$

b The coefficients in the expansion of $(5 - x)^4$ are 1, 4, 6, 4 and 1. Putting $a = 5, b = -x$ and $n = 4$ into $\boxed{2.6}$ gives

$$(5 - x)^4 = (5 + (-x))^4$$

$$= (1 \times 5^4) + [4 \times 5^3 \times (-x)] + [6 \times 5^2 \times (-x)^2]$$
$$+ [4 \times 5 \times (-x)^3] + [1 \times (-x)^4]$$

$$= 625 + [500 \times (-x)] + [150 \times x^2] + [20 \times (-x^3)] + [1 \times x^4]$$

$$= 625 - 500x + 150x^2 - 20x^3 + x^4$$

Example 19 *continued*

c The coefficients in the expansion of $(3 + 2x)^5$ are 1, 5, 10, 10, 5 and 1. (From row $n = 5$ in Fig. 37.) Putting $a = 3$, $b = 2x$ and $n = 5$ into 2.6 gives

$$(3 + 2x)^5 = [1 \times 3^5] + [5 \times 3^4 \times 2x] + \left[10 \times 3^3 \times \underbrace{(2x)^2}_{= 4x^2} \right]$$

$$+ \left[10 \times 3^2 \times \underbrace{(2x)^3}_{= 8x^3} \right] + \left[5 \times 3 \times \underbrace{(2x)^4}_{= 16x^4} \right] + [1 \times (2x)^5]$$

$$= 243 + (5 \times 3^4 \times 2)x + (10 \times 3^3 \times 4)x^2$$

$$+ (10 \times 3^2 \times 8)x^3 + (5 \times 3 \times 16)x^4 + 2^5 x^5$$

$$= 243 + 810x + 1080x^2 + 720x^3 + 240x^4 + 32x^5$$

$$= 32x^5 + 240x^4 + 720x^3 + 1080x^2 + 810x + 243$$

We use Pascal's triangle to find the binomial expansion if n is a small positive whole number.

? **Why can't we use Pascal's triangle to expand terms such as $(a + b)^{20}$?**

It's a laborious effort to evaluate the row $n = 20$ of Pascal's triangle. The binomial expansion for other values of n will be covered in Chapter 7.

SUMMARY

Binomial expansion is the expansion of two terms to a given index.

For a small positive whole number, n, we have

2.6 $(a + b)^n = C_n a^n + C_{n-1} a^{n-1} b + C_{n-2} a^{n-2} b^2 + \cdots + C_0 b^n$

where $C_n, C_{n-1}, \ldots$ represent the coefficients determined by Pascal's triangle. Also $C_n = C_0 = 1$; $C_{n-1} = C_1 = n$.

Exercise 2(f)

Solutions at end of book. Complete solutions available at www.palgrave.com/science/engineering/singh

1 Expand   **a** $(x + 1)^6$ **b** $(a + b)^6$

2 Expand   **a** $(5 + x)^4$ **b** $(2 + 3x)^5$

  c $(4 - 3x)^6$

3 [communications]   The amplitudes of a n-element array are determined by

the coefficients of the binomial expansion of

$$(1 + x)^{n-1}$$

Find the amplitudes for a 10-element array.

4 Determine the expansion of $\left(\dfrac{w}{4} - \dfrac{x}{3} \right)^7$.

Solutions at end of book. Complete solutions available at www.palgrave.com/science/engineering/singh

Miscellaneous exercise 2

1 [mechanics] The displacement, s, of a vehicle is given by $s = t^2$. Sketch the s–t graph.

2 [fluid mechanics] The velocity, v, of a fluid is given by

$$v = 36 - x^2 \qquad -6 \leq x \leq 6$$

where x is the distance. Sketch the graph of v against x.

3 [thermodynamics] The pressure, P, is related to the specific volume, V, by

$$P = \frac{C}{V}$$

where C is a constant. Sketch the graphs on the same axes for $V > 0$ and
a $C = 90$ **b** $C = 120$
c $C = 150$ **d** $C = 180$

4 [fluid mechanics] The velocity distribution, v, of a fluid varies between $v = \sqrt{x}$ and $v = \sqrt{2(x-3)}$. Sketch the velocity profile.

5 [fluid mechanics] The velocity profile, v, of a fluid through a pipe is given by

$$v = 5\sqrt{1 - \frac{x^2}{9}}$$

where $-3 \leq x \leq 3$

(x is the distance from the central axis). Sketch the velocity profile.

6 [mechanics]

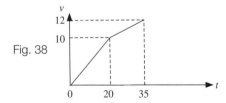

Fig. 38

Figure 38 shows a velocity–time (v–t) graph for a vehicle moving along a straight track. Find the equation giving v in terms of t.

7 [mechanics] Figure 39 shows a v–t graph of a car travelling along a straight road. Sketch the a–t graph.

Fig. 39

8 [mechanics] Figure 40 shows a displacement–time (s–t) graph of a car moving along a straight road. Find the equation of s in terms of t. Sketch the v–t and a–t graphs for $0 \leq t \leq 20$.

Fig. 40

9 By comparing coefficients of $\sqrt{-2}$ in

$$z + \sqrt{-2} = \left(x + y\sqrt{-2}\right)^3$$

show that $1 = y(3x^2 - 2y^2)$.

10 Sketch the graphs of $y = \dfrac{x^2}{a^2}$, on the same axes, for $a = 1$, 5 and 10. (Take your x values to be between -10 and 10.) What effect does a have on y?

11 [reliability engineering] The failure distribution function, Q, is described as in Fig. 41.

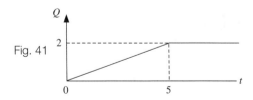

Fig. 41

Write an expression for Q.

12 [reliability engineering]   The hazard function, h, of a component is described as in Fig. 42.

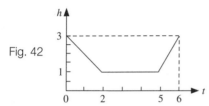

Fig. 42

Write an expression for h.

13 [control engineering]   The amplitude ratio, r, of the output and the input of a system is given by

$$r = \frac{1}{\sqrt{1 + \dfrac{\omega^2}{9}}}$$

where ω is the angular frequency. Plot the graph of r against ω for $0 \le \omega \le 10$.

For the remaining questions use a computer algebra system (MAPLE) or a graphical calculator.

14 [vibrations]   The magnification factor, M, in the theory of vibrations is defined as

$$M = \begin{cases} \dfrac{1}{1 - r^2} & r < 1 \\ \dfrac{1}{r^2 - 1} & r > 1 \end{cases}$$

where r is the frequency ratio. Plot the graph of M versus r for $r \ge 0$.

15 [thermodynamics]   The pressure, P, and volume, V, are related by

$$P = \frac{100}{V^n}$$

where n is a constant. Plot the graphs of P against V on the same axes for $V = 0.1$ to 0.9 and

a $n = 1$ **b** $n = 1.3$ **c** $n = 1.6$.

16 [thermodynamics]   The pressure, P, and volume, V, of a gas are related by

$$PV^n = 1000, \qquad 0.01 \le V \le 0.1$$

Taking V to be the horizontal axis, plot the graphs, on the same axes, of P against V for $n = 1.2, 1.3, 1.4$ and 1.5.

17 [fluid mechanics]   For hydrogen gas, the viscosity, η, is determined by

$$\eta = (0.7 \times 10^{-6}) \left[\frac{T^{1.5}}{T + 70} \right]$$

where T is the temperature in Kelvin. Plot the graph of η against T for $273 \le T \le 373$.

18 [fluid mechanics]   The viscosity, η, of a fluid at temperature, θ, is given by

$$\eta = \frac{1.8 \times 10^{-3}}{1 + (34 \times 10^{-3})\theta + (22 \times 10^{-6})\theta^2}$$

Plot the graph of η against θ for $\theta = 0$ to 1000.

19 [fluid mechanics]   The viscosity, η, of water at temperatures between 0 °C and 100 °C is given by

$$\eta = \frac{1.8 \times 10^{-3}}{1 + (22 \times 10^{-3})\theta + (22 \times 10^{-6})\theta^2}$$

Plot the graph of η against θ for this range.

CHAPTER 3
Functions in Engineering

SECTION A **Concepts of functions**

By the end of this section you will be able to:
▶ understand the terms function, domain, range, one–one and many–one
▶ evaluate functions
▶ find simplified expressions for functions

A1 **Definition of a function**

? **In general, what does the word 'function' mean in everyday life?**

It's an aim of a person or an object, for example switching on a television is a function. We are interested in mathematical examples.

? **Can you think of any mathematical examples of functions?**

Adding 5, subtracting 2, dividing by 10, etc.

Adding 5 can be represented by

$$2 \xrightarrow{+5} 7$$

$$3 \xrightarrow{+5} 8$$

$$4 \xrightarrow{+5} 9$$

? **Where does adding 5 take the variable x?**

$$x \xrightarrow{+5} ?$$

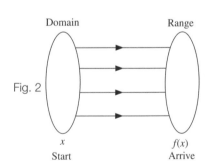

Fig. 1

$x + 5$. We say that the function is to add 5 for any given number. This can be drawn as shown in Fig. 1.

The function, adding 5 to any given number, is normally denoted by f as follows:

$$f(x) = x + 5$$

? **What does $f(x) = x + 5$ mean?**

It means that the function, f, takes x to $x + 5$. The aim is to add 5 to any given value. The function $f(x)$ is pronounced 'f of x'. The set A of Fig. 1 where we start from is called the **domain** of the function. The set B of Fig. 1 where we arrive is called its **range** (Fig. 2). *Set* means a collection of objects.

Fig. 2

The domain and range do not just consist of a few selected points shown in Fig. 1 but could be all the real numbers.

? The variable x is called the **independent** variable because $f(x)$ depends on x. **What does the term 'function' mean in mathematics?**

A **function** is a rule that gives every x in the domain **only** one value in the range, $f(x)$.

A2 Evaluating functions

Example 1

Let $f(x) = x^2 + 1$, and evaluate the following:

$$f(-3), f(3), f(-2), f(2), f(-1) \text{ and } f(1)$$

Solution

The function squares each value and then adds 1, thus

$$f(-3) = (-3)^2 + 1 = 10$$
$$f(3) = 3^2 + 1 = 10$$
$$f(-2) = (-2)^2 + 1 = 5$$
$$f(2) = 2^2 + 1 = 5$$
$$f(-1) = (-1)^2 + 1 = 2$$
$$f(1) = 1^2 + 1 = 2$$

Example 1 can be represented as shown in Fig. 3.

$f(x) = x^2 + 1$ is an example of a many-to-one function. It indicates that many (i.e. at least 2) from the start arrive at only one destination. This is sometimes denoted by many → one. Let's investigate an example of a one-to-one function, one → one.

Fig. 3

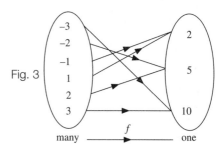

Example 2

Let $g(x) = 3x$ and determine $g(0)$, $g(1)$, $g(-1)$, $g(2)$ and $g(-2)$.

Solution

The function, g, trebles (multiplies by 3) each given value:

$$g(0) = 3 \times 0 = 0$$
$$g(1) = 3 \times 1 = 3$$
$$g(-1) = 3 \times (-1) = -3$$
$$g(2) = 3 \times 2 = 6$$
$$g(-2) = 3 \times (-2) = -6$$

Note that $g(x)$ depends on the given values of x.

This can be drawn as shown in Fig. 4.

One element from the start arrives at only one element (or one input has only one output). Remember that functions can only have one destination (or one output).

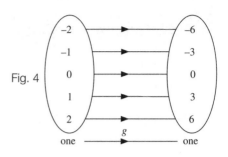

Fig. 4

So functions can be many → one or one → one but **not** one → many or many → many. The essential property is that you must arrive at only one station (or one output) for any given input. An input always has only **one** output if it is a function.

A3 **Engineering applications of functions**

? **Why do engineering and mathematics books use *f*(x) to denote functions?**

Because '*f*' is the first letter of the word 'function' and *x* normally represents an independent variable. In the next example we use the Greek letter 'ϕ', pronounced 'fi', and *t* as the independent variable representing time.

Example 3 *mechanics*

The displacement, $\phi(t)$, of a particle at time *t* is given by

$$\phi(t) = 2t^3 + t^2 - 10t + 10$$

a Evaluate $\phi(2)$, $\phi(3)$ and $\phi(5)$.
b Find simplified expressions for
  i $\phi(t^2)$ **ii** $\phi(t + 1)$

Solution

a We have

$$\phi(2) = (2 \times 2^3) + 2^2 - (10 \times 2) + 10 = 10$$
$$\phi(3) = (2 \times 3^3) + 3^2 - (10 \times 3) + 10 = 43$$
$$\phi(5) = (2 \times 5^3) + 5^2 - (10 \times 5) + 10 = 235$$

b i To find $\phi(t^2)$ we replace *t* with t^2 in $\phi(t) = 2t^3 + t^2 - 10t + 10$:

$$\phi(t^2) = 2 \times (t^2)^3 + (t^2)^2 - 10t^2 + 10$$
$$= \underline{2t^6} + \underline{t^4} - 10t^2 + 10$$
by 1.7 by 1.7

ii To find $\phi(t + 1)$ we replace *t* with $t + 1$ in ⋆ :

† $$\phi(t + 1) = 2(t + 1)^3 + (t + 1)^2 - 10(t + 1) + 10$$

1.7 $(a^m)^n = a^{m \times n}$

Example 3 *continued*

How do we find $(t + 1)^3$ of ┼ ?

Use the binomial expansion:

$$(t + 1)^3 = t^3 + 3t^2 + 3t + 1$$

How do we find $(t + 1)^2$?

Use:

1.13 $(a + b)^2 = a^2 + 2ab + b^2$
$(t + 1)^2 = t^2 + 2t + 1$

Substituting these into ┼ gives

$$\phi(t + 1) = 2(t^3 + 3t^2 + 3t + 1) + (t^2 + 2t + 1) - 10t - 10 + 10$$
$$= 2t^3 + 6t^2 + 6t + 2 + t^2 + 2t + 1 - 10t$$
$$= 2t^3 + \underbrace{6t^2 + t^2}_{= 7t^2} + \underbrace{6t + 2t - 10t}_{= -2t} + 2 + 1$$
$$= 2t^3 + 7t^2 - 2t + 3$$

In **Example 3**, the displacement, $\phi(t)$, depends on the time t, hence t is the independent variable.

SUMMARY

A *function* is a rule which gives only **one** output for every given input. The set where we start from is called the *domain* and the set where we finish is called the *range*. There are one → one and many → one functions. There is **no** such thing as one → many or many → many functions.

Exercise **3(a)**

Solutions at end of book. Complete solutions available at www.palgrave.com/science/engineering/singh

1 Let $f(x) = x^2$ and find $f(1)$, $f(2)$, $f(3)$, $f(-1)$, $f(-2)$ and $f(-3)$. What type, one → one or many → one, of function is $f(x)$?

2 Let $f(t) = \dfrac{9}{5}t + 32$. This function converts temperature from °C to °F. Evaluate $f(0)$, $f(100)$ and $f(24)$.

3 Let $g(x) = x^3$ and find $g(1)$, $g(-1)$, $g(2)$, $g(-2)$, $g(3)$ and $g(-3)$. What do you notice about your results?

4 Let $g(t) = -4.9t^2$, evaluate $g(1)$, $g(\pi)$, $g\left(\dfrac{1}{\sqrt{4.9}}\right)$.

5 Is $f(h) = \pm\sqrt{20h}$ a function? If it is a function then is it a one → one or a many → one function?

6 [mechanics]   The velocity, $v(t)$, of an object is given by

$$v(t) = 9.8t$$

a Evaluate $v(0)$, $v(1)$, $v(2)$ and $v(5)$.
b Find simplified expressions for $v(t^2)$ and $v(t + 1)$.

7 [mechanics]   The displacement, $s(t)$, of a particle is given by

$$s(t) = t^3 - t^2 + 5$$

Exercise **3(a) continued**

Solutions at end of book. Complete solutions available at www.palgrave.com/science/engineering/singh

a Evaluate $s(0)$, $s(1.5)$ and $s(2.5)$.
b Simplify $s(t + 1)$.

8 🔲 [*electrical principles*]   The power dissipated, $P(V)$, by a resistor is given by

$$P(V) = \frac{V^2}{R}$$

Find $P(IR)$.

9 🔲 [*electrical principles*]   The voltage, $V(R)$, of a battery is given by

$$V(R) = \frac{ER}{R + r}$$

where r is the internal resistance and E is the electromotive force. Find $V(r)$ and $V(2r)$.

10 🚗 [*mechanics*]   The height, $h(t)$, of an object is given by

$$h(t) = 200t - 4.9t^2$$

a Determine t for $h(t) = 0$.
b Simplify $h(t + 2)$.

SECTION B **Inverse functions**

By the end of this section you will be able to:
▶ understand what is meant by 'inverse functions'
▶ find inverse functions

B1 **Inverse of a function**

? **What does the word 'inverse' mean in everyday life?**

Reverse or opposite.

? **What is the inverse of:**

a **turning a light on?**
b **united?**
c **multiplying by 3?**
d **adding 2 and then multiplying by 3?**

The inverses are

a turning a light off.
b divided.
c dividing by 3.
d dividing by 3 and then subtracting 2.

Inverse functions simply unlock what you have done. Note the inverse of **d**. You undo the last operation first.

In the next example the notation {2, 3, 4} means that the set contains only the numbers 2, 3 and 4.

Example 4

Consider $f(x) = 2x$ with domain {2,3,4}. Determine the inverse function.

Solution

? f is the function 'multiply by 2'. **Where does f take 2, 3 and 4?**

? f takes $2 \rightarrow 4$, $3 \rightarrow 6$ and $4 \rightarrow 8$ (Fig. 5). **What should the inverse function do?**

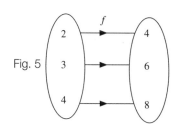

Fig. 5

Since the inverse function undoes f, it should take $4 \rightarrow 2$ (Fig. 6), $6 \rightarrow 3$ and $8 \rightarrow 4$.

The opposite of multiplying by 2 is dividing by 2, thus

$$\text{inverse}\quad f(x) = \frac{x}{2}$$

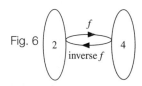

Fig. 6

Instead of writing inverse $f(x)$, we can write this as $f^{-1}(x)$ so $f^{-1}(x) = \frac{x}{2}$. f^{-1} does not mean to the power of -1; it is the conventional notation for inverse functions.

There are various ways of finding the inverse function – one method is to transpose formulae. To find the inverse function, $f^{-1}(x)$, we introduce a new variable, y, and let $f(x) = y$. Then we transpose to make x the subject, $x = \dots$. Let's do an example.

Example 5

If $f(x) = 2x + 3$, find $f^{-1}(x)$.

Solution

Let $f(x) = y$, then we have

$$y = 2x + 3$$

? **We need to make x the subject, $x = \dots$. How?**

? **We need to remove 3 and 2 from the Right-Hand Side. How?**

Subtract 3:

$$y - 3 = 2x$$

Divide by 2:

$$\frac{y - 3}{2} = x$$

We replace the y with x to obtain $f^{-1}(x)$ (a function of x, not y):

$$f^{-1}(x) = \frac{x-3}{2}$$

Example 6

Consider the domain of f to be real numbers. If $f(x) = \dfrac{x^3}{12} + 5$, find $f^{-1}(x)$ and the real number $f^{-1}(23)$.

Solution

Let $y = f(x) = \dfrac{x^3}{12} + 5$. Therefore

$$\frac{x^3}{12} + 5 = y$$

[?] We need to transpose this formula to obtain $x = \ldots$. **How?**

First we remove the $+5$ from the Left-Hand Side, by subtracting 5 from both sides:

$$\frac{x^3}{12} = y - 5$$

[?] **Next how do we remove the divide by 12?**

Multiply both sides by 12:

$$x^3 = 12(y - 5)$$

[?] **By what means do we remove the power 3 on the x term?**

Take the cube root, $\sqrt[3]{\ }$, of both sides:

$$x = \sqrt[3]{12(y - 5)}$$

We replace the y with x to obtain the inverse function, $f^{-1}(x)$:

$$f^{-1}(x) = \sqrt[3]{12(x - 5)}$$

The domain of f^{-1} is also the real numbers.

[?] **How do we evaluate $f^{-1}(23)$?**

We replace x with 23 in $f^{-1}(x) = \sqrt[3]{12(x-5)}$:

$$
\begin{aligned}
f^{-1}(23) &= \sqrt[3]{12(23-5)} \\
&= \sqrt[3]{12 \times 18} \\
&= \sqrt[3]{216} \\
f^{-1}(23) &= 6
\end{aligned}
$$

This can be represented by Fig. 7.

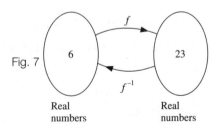

Fig. 7

We need to be careful with statements such as

$$f(x) = \frac{1}{x - 2}$$

[?] This f is not a function because it is not defined at $x = 2$. **Why not?**

Because the denominator is zero at $x = 2$ and of course we cannot divide by zero. So we write

$$f(x) = \frac{1}{x - 2} \quad \text{(provided } x \neq 2\text{)}$$

Now f is a function with domain of all the real numbers apart from 2.

Similarly, for inverse functions we cannot divide by zero. The inverse of f is

$$f^{-1}(x) = 2 + \frac{1}{x} \quad \text{(provided } x \neq 0\text{)}$$

More functions of this nature occur in **Exercise 3(b)**.

Not **all** functions have an inverse function. It is only **one → one** functions which have an inverse function.

SUMMARY

Inverse functions, denoted by $f^{-1}(x)$, undo $f(x)$ (reverse process). To find $f^{-1}(x)$ of $f(x)$, let $y = f(x)$ and then obtain $x = \ldots$. Finally, replace the y with x to give $f^{-1}(x)$.

Exercise 3(b)

Solutions at end of book. Complete solutions available at www.palgrave.com/science/engineering/singh

Let the domain of all the functions be real numbers with the restriction shown in brackets.

1 Obtain the inverse function, $f^{-1}(x)$, in each of the following cases:

  a $f(x) = 7x - 2$

  b $f(x) = \dfrac{1}{x}$ (provided $x \neq 0$)

  c $f(x) = 1 - x$

  d $f(x) = \dfrac{3}{5 - x}$ (provided $x \neq 5$)

2 Find the inverse function of the following:

  a $g(t) = \dfrac{t}{t + 2}$ (provided $t \neq -2$)

  b $h(x) = \dfrac{x - 1}{2 - x}$ (provided $x \neq 2$)

  c $f(x) = \dfrac{2x - 1}{x + 1}$ (provided $x \neq -1$)

3 i If f is defined as $f(x) = \dfrac{x + 1}{3x + 1}$ then show that

$$f^{-1}(x) = \frac{1 - x}{3x - 1} \quad \text{(provided } 3x - 1 \neq 0\text{)}$$

  ii If $g(t) = \dfrac{1 - t}{3t - 1}$ $\left(\text{provided } t \neq \dfrac{1}{3}\right)$ then find $g^{-1}(t)$.

4 i Let $f(x) = \dfrac{x^3 + 3}{2 - x^3}$ (provided $2 - x^3 \neq 0$). Show that the inverse function, $f^{-1}(x)$, is given by

$$f^{-1}(x) = \left(\frac{2x - 3}{1 + x}\right)^{\frac{1}{3}} \quad \text{(provided } x \neq -1\text{)}$$

  ii Let $g(t) = \left(\dfrac{2t - 3}{1 + t}\right)^{\frac{1}{3}}$ (provided $t \neq -1$), then find $g^{-1}(t)$.

5 What do you notice about your results to $g^{-1}(t)$ in questions **3 ii** and **4 ii**?

SECTION C **Graphs of functions**

By the end of this section you will be able to:
▶ plot graphs of functions
▶ sketch the inverse function, f^{-1}, from the function f
▶ sketch graphs by transformation

C1 **Graphing functions**

Graphs of functions are drawn on the rectangular Cartesian co-ordinate system, x–y axes. Normally the domain is plotted on the horizontal axis and the range on the vertical axis, as shown in Fig. 8.

Fig. 8

Remember that x is the independent variable and $f(x)$ is the dependent variable (depends on x).

Much of this section is similar to the work on graphs in Chapter 2. However in this section we examine graphs from a function viewpoint.

Example 7

Plot the graph of the function $f(x) = x^2 - x + 5$ with the domain between -3 and 3.

Solution

We can substitute whole numbers for x between -3 and $+3$ into the function $x^2 - x + 5$ (see Table 1).

TABLE 1

x	$f(x) = x^2 - x + 5$
-3	$(-3)^2 - (-3) + 5 = 17$
-2	$(-2)^2 - (-2) + 5 = 11$
-1	$(-1)^2 - (-1) + 5 = 7$
0	$0^2 - 0 + 5 = 5$
1	$1^2 - 1 + 5 = 5$
2	$2^2 - 2 + 5 = 7$
3	$3^2 - 3 + 5 = 11$

Example 7 *continued*

Thus the points on the graph are $(-3, 17)$, $(-2, 11)$, $(-1, 7)$, $(0, 5)$, $(1, 5)$, $(2, 7)$ and $(3, 11)$. Connecting these points gives the graph of the function $y = f(x)$ – see Fig. 9.

We plot like the normal x–y graph with $y = f(x) = x^2 - x + 5$.

Fig. 9 Graph of $f(x) = x^2 - x + 5$ for x between -3 and $+3$.

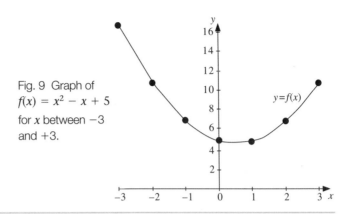

C2 Plotting inverse functions

Example 8

Plot the line $y = x$ and $f(x) = 2x$. Find the inverse function, $f^{-1}(x)$, and plot it on the same axes.

Solution

Since the function f multiplies by 2, the inverse function, f^{-1}, divides by 2. So we have

$$f^{-1}(x) = \frac{x}{2}$$

What sort of graphs are the functions $f(x) = 2x$, $y = x$ **and** $f^{-1}(x) = \frac{x}{2}$ **?**

They are all straight lines going through the origin with gradient 2, 1 and $\frac{1}{2}$ respectively. (See Fig. 10.)

Fig. 10 Portion of the graphs of $f(x) = 2x$, $y = x$ and $f^{-1}(x) = \frac{x}{2}$

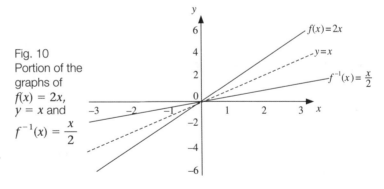

Example 9

Using a graphical calculator, plot the graph of the function $f(x) = x^3$ between $x = -1$ and $x = 1$. On the same diagram plot $y = x$. The inverse of $f(x)$ is

$$f^{-1}(x) = \sqrt[3]{x}$$

Plot $f^{-1}(x)$ on the same axes.

Example 9 *continued*

Solution

On a graphical calculator plot all three functions. It should show the plots given in Fig. 11.

Fig. 11
Graphs of
$y = x$,
$f(x) = x^3$ and
$f^{-1}(x) = \sqrt[3]{x}$
for x between
-1 and 1

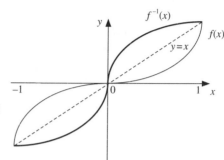

By observing the graphs of Figs 10 and 11, can you see a relationship between f, f^{-1} and $y = x$?

3.1   The inverse function, f^{-1}, is the reflection of f in the line $y = x$

So if f is a function with inverse f^{-1} then f^{-1} is obtained by reflecting the graph of f in the line $y = x$ (Fig. 12).

Fig. 12
Graphs of $f(x)$,
$y = x$ and
$f^{-1}(x)$

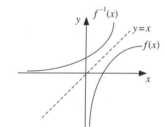

C3 Transformation of graphs

We have sketched graphs of some important engineering functions in Chapter 2 such as $y = x^2$, $y = x^3$, $y = \dfrac{1}{x}$ etc. These can also be written as $f(x) = x^2$, $f(x) = x^3$, $f(x) = \dfrac{1}{x}$ etc. respectively. From these basic functions we can sketch more complicated functions as was discussed in Chapter 2.

In the following cases take c to be a positive constant.

From the graph of $f(x)$ we can sketch the graph of $f(x) + c$ and $f(x) - c$.

Figure 13 shows how the graph of $f(x) + c$ is shifted upwards by c from the graph of $f(x)$. Similarly the graph of $f(x) - c$ is shifted downwards by c.

Fig. 13
Graphs of $f(x)$,
$f(x) + c$ and
$f(x) - c$

Example 10 *fluid mechanics*

The streamline, $f(x)$, of a fluid flow is sketched in Fig. 14.

Sketch, on the same axes, the graphs of $f(x) + 2$, $f(x) - 2$ and $f(x) - 4$.

Fig. 14

Solution

The graph of $f(x) + 2$ is shifted upwards by 2 units from the graph of $f(x)$. Similarly $f(x) - 2$ is shifted down by 2 units and $f(x) - 4$ is shifted down by 4 units (Fig. 15).

Fig. 15
Graphs of $f(x)$, $f(x) + 2$, $f(x) - 2$ and $f(x) - 4$

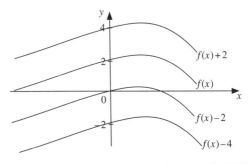

The graph of $f(x + c)$ is the same as the graph of $f(x)$ but translated to the left by c units (Fig. 16).

Fig. 16
Graphs of $f(x + c)$ and $f(x)$

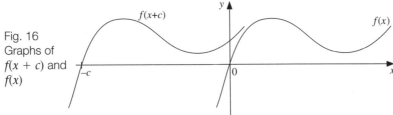

Which way does the graph $f(x - c)$ shift in relation to $f(x)$?

It is translated to the right by c units, as shown in Fig. 17.

Fig. 17
Graphs of $f(x - c)$ and $f(x)$

Example 11 *mechanics*

The displacement, $x(t)$, of a particle is given by

$$x(t) = (t - 3)^2$$

i Sketch the graph of displacement versus time.
ii At what time(s) is $x(t) = 0$?

Example 11 *continued*

Solution

i How do we sketch the graph of $x(t) = (t-3)^2$?
It is the same shape as the quadratic graph t^2 but shifted to the right by 3 units (Fig. 18).

Fig. 18
Graph of
$x(t) = (t-3)^2$

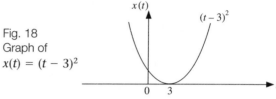

ii $x(t) = 0$ at $t = 3$.

For $c > 1$, the graph of $cf(x)$ is stretched vertically by a factor of c (Fig. 19).

For $0 < c < 1$, the graph of $cf(x)$ is squashed vertically by a factor of $1/c$ (Fig. 20).

Fig. 19

Fig. 20

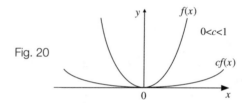

Example 12

Sketch, on the same axes, the graphs of $f(x) = x^3$ and $\dfrac{1}{8}f(x)$.

Solution

From Chapter 2, we have the graph of $f(x) = x^3$. **How do we sketch $\dfrac{1}{8}x^3$?**

It is the graph of x^3 but squashed vertically by a factor of 8 (Fig. 21).

Fig. 21
Graphs of x^3
and $\dfrac{1}{8}x^3$

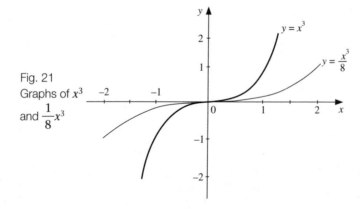

The function $\dfrac{1}{8}x^3$ has a value of one-eighth of $f(x) = x^3$.

The next example is more challenging than the previous examples because of the algebra involved in tackling the problem.

Example 13 *mechanics*

The displacement, $x(t)$, of a particle is given by

$$x(t) = 2t^2 - 4t - 1$$

Sketch the graph of $x(t)$ indicating where it cuts the t axis.

Solution

How do we sketch the graph of $x(t) = 2t^2 - 4t - 1$?

We rewrite $2t^2 - 4t - 1$ as

$$2t^2 - 4t - 1 = 2[t^2 - 2t] - 1$$

We first complete the square on $t^2 - 2t$. We take half the t coefficient, half of -2, and then adjust the arithmetic:

$$2[t^2 - 2t] - 1 = 2[(t - 1)^2 - 1] - 1$$
$$= 2(t - 1)^2 - 2 - 1$$
$$= 2(t - 1)^2 - 3$$

How do we sketch the graph of $x(t) = 2(t - 1)^2 - 3$?

We know it has the same shape as the basic graph of t^2.

How do we adjust the basic graph to make it into $2(t - 1)^2 - 3$?

The $2(t - 1)^2$ shifts the graph of t^2 to the right by 1 unit and stretches it vertically by a factor of 2, and the -3 shifts it down by 3 units. Moreover the graph crosses the t axis at $x(t) = 0$. We have

$$2(t - 1)^2 - 3 = 0$$
$$2(t - 1)^2 = 3 \qquad \text{[Adding 3]}$$
$$(t - 1)^2 = \frac{3}{2} \qquad \text{[Dividing by 2]}$$
$$t - 1 = \pm\sqrt{\frac{3}{2}} \qquad \text{[Taking the square root]}$$
$$t = 1 \pm \sqrt{\frac{3}{2}} \qquad \text{[Adding 1]}$$
$$t = 1 + \sqrt{\frac{3}{2}} \quad \text{or} \quad t = 1 - \sqrt{\frac{3}{2}}$$

Example 13 *continued*

The graph crosses the t axis at $t = 1 + \sqrt{\dfrac{3}{2}}$, $t = 1 - \sqrt{\dfrac{3}{2}}$ (Fig. 22).

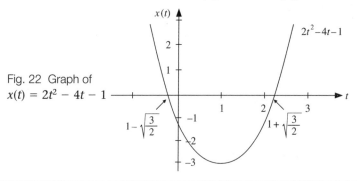

Fig. 22 Graph of
$x(t) = 2t^2 - 4t - 1$

SUMMARY

Graphs are plotted with domain on the horizontal axis and range on the vertical axis.
Inverse functions, f^{-1}, of f can be determined by reflecting f in the line $y = x$.
From the graph of $f(x)$ we can sketch the following (c is a positive constant):

▶ $f(x) + c$ shifts $f(x)$ up by c units
▶ $f(x) - c$ shifts $f(x)$ down by c units
▶ $f(x + c)$ translates $f(x)$ to the left by c units
▶ $f(x - c)$ translates $f(x)$ to the right by c units
▶ for $c > 1$, the graph of $cf(x)$ is stretched vertically by a factor of c
▶ for $0 < c < 1$, the graph of $cf(x)$ is squashed vertically by a factor of $1/c$
▶ for $c > 1$, the graph of $f(cx)$ is squashed horizontally by a factor of c
▶ for $0 < c < 1$, the graph of $f(cx)$ is stretched horizontally by a factor of $1/c$

Examples of the last two points will be given in the next chapter.

Exercise 3(c)

Solutions at end of book. Complete solutions available at www.palgrave.com/science/engineering/singh

1 Plot, on different diagrams, each of the following functions:
  a $f(x) = 2x + 3$ with domain between -5 and 5
  b $f(I) = 2.54I$ with domain between -5 and 5
  c $f(t) = -9.8t$ with domain between 0 and 5
  d $g(f) = 2\pi f$ with domain between 0 and 50
  e $f(C) = \dfrac{9}{5}C + 32$ with domain between 0 and 100.

2 Sketch the following functions with domain between -1 and 1:
  i $f(x) = x^2$ **ii** $g(x) = -x^2$

Do you notice any relationship between **i** and **ii**?

Exercise **3(c) continued**

Solutions at end of book. Complete solutions available at www.palgrave.com/science/engineering/singh

3 Sketch the following functions:

  i $f(x) = x^2 + 2x + 1$

  ii $f(x) = (x + 1)^2$

4 Sketch the following functions on the same axes:

  a $f(r) = \pi r^2$ and $g(r) = r^2$ with domain $0 \le r \le 10$

  b $f(v) = \dfrac{v^2}{20}$ and $g(v) = v^2$ with domain $0 \le v \le 10$

  c $f(T) = \dfrac{10}{T}$ and $g(T) = \dfrac{1}{T}$ with domain $0 < T \le 10$.

  Can you see any connection between the graphs of f and g?

5 The following diagrams show the graphs of $f(x)$. Sketch these and their inverse function, $f^{-1}(x)$, on the same axes.

  a

  b

  c

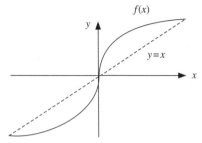

6 🔲 [*electrical principles*] The energy stored, $w(i)$, of an inductor,

L, is given by $w(i) = \dfrac{1}{2}Li^2$. Sketch the graph of $w(i)$ versus i for $L = 0.1$ H and $i \ge 0$.

7 🔲 [*electronics*] The current, $i(v)$, through a semiconductor with voltage, v, is given by

$$i(v) = 2v^2 - 3$$

Sketch the graph of $i(v)$ against v, labelling the points where the graph cuts the axes.

8 🔲 [*electrical principles*] The current, $i(R)$, through a resistor, R, is given by

$$i(R) = \dfrac{1}{R + 5}$$

Sketch the graph of $i(R)$ against R.

9 🔲 [*mechanics*] The displacements, $x(t)$, of a particle are given by

  a $x(t) = (t - 1)^3$

  b $x(t) = 5(t - 1)^3$

  c $x(t) = \dfrac{1}{2}(t - 1)^3$

  Sketch the displacement–time (t) function in each case on the same axes.

SECTION D **Combinations of functions**

By the end of this section you will be able to:
▶ understand composite functions
▶ simplify functions using combinations such as addition, multiplication etc.
▶ apply combination rules to engineering examples such as reliability and control engineering

D1 **Composition of functions**

Let f and g be functions, then $f \circ g$ (f composite g) is called the **composite** function and is defined as

$$3.2 \qquad f \circ g = f(g(x))$$

First apply the function g to x and then apply function f to $g(x)$.

Example 14

If $f(x) = 2x$ (double) and $g(x) = x^2$ (squared), then find

i $f \circ g$ **ii** $g \circ f$ **iii** $f \circ f$

Solution

i $f \circ g = f(g(x))$

$\qquad = f(\underbrace{x^2}_{\text{squared}})$

$f \circ g \underset{\text{double}}{=} 2x^2$

Hence you first apply g which squares x, then you apply f which doubles your result of squaring.

ii $g \circ f = g(f(x))$

$\qquad \underset{\text{double}}{=} g(2x)$

$\qquad \underset{\substack{\text{squares} \\ \text{the result}}}{=} (2x)^2$

$g \circ f = 4x^2$

iii $f \circ f = f(f(x))$

$\qquad \underset{\text{double}}{=} f(2x)$

$\qquad \underset{\substack{\text{double} \\ \text{the result}}}{=} 4x$

In part **i** we square first and then double. In part **ii** we double first and then square.

? **What can you conclude from your answers i and ii of Example 14?**

The result **i** $f \circ g$ is different from **ii** $g \circ f$:

$$f \circ g \neq g \circ f$$

In general, we have the property

3.3 $f \circ g \neq g \circ f$ [Not equal]

D2 Other combinations of functions

We have

3.4 $(f \pm g)(x) = f(x) \pm g(x)$

3.5 $\dfrac{f}{g}(x) = \dfrac{f(x)}{g(x)}$ provided $g(x) \neq 0$

3.6 $f \cdot g(x) = f(x) \times g(x)$

Example 15

Let f and g be functions defined by
$$f(x) = x - 1$$
$$g(x) = 3x + 5$$

Determine

i $f + g$ **ii** $f - g$ **iii** $g - 5$ **iv** $f \cdot g$ **v** $\dfrac{f}{g}$

Solution

i Using 3.4 :
$$f + g = (f + g)(x)$$
$$= f(x) + g(x)$$
$$= (x - 1) + (3x + 5)$$
$$= (x + 3x) + (5 - 1)$$
$$f + g = 4x + 4$$

ii Using 3.4 :
$$f - g = (f - g)(x)$$
$$= f(x) - g(x)$$
$$= (x - 1) - (3x + 5)$$
$$= (x - 3x) + (-1 - 5)$$
$$f - g = -2x - 6$$

Example 15 *continued*

iii Using 3.4 with a constant function, 5, taken away from g:

$$g - 5 = g(x) - 5$$
$$= (3x + 5) - 5$$
$$= 3x$$

iv $f \cdot g$ means

$$f \cdot g(x) = f(x) \times g(x) \qquad [\text{by } 3.6\;]$$
$$= (x - 1)(3x + 5)$$
$$\underset{\text{FOIL}}{=} (x \times 3x) + (x \times 5) + (-1 \times 3x) + (-1 \times 5)$$
$$= 3x^2 + \underset{= 2x}{\underbrace{5x - 3x}} - 5$$
$$f \cdot g(x) = 3x^2 + 2x - 5$$

v Using 3.5 :

$$\frac{f}{g} = \frac{f(x)}{g(x)}$$
$$= \frac{x - 1}{3x + 5} \qquad (\text{provided } 3x + 5 \neq 0)$$

D3 Engineering applications

We apply the above combinations of functions to reliability and control engineering. The reliability example is straightforward but the control example is long and difficult. You need to use your algebraic skills to deal with the control problem.

Example 16 *reliability engineering*

The failure density function, $f(t)$, for a component is given by

$$f(t) = \frac{1}{8} \text{ where } 0 < t < 8 \text{ years}$$

Find $F(t)$, $R(t)$ and $h(t)$ where these are defined as

$$F(t) = tf(t) \qquad \text{(Failure distribution function)}$$
$$R(t) = 1 - F(t) \qquad \text{(Reliability function)}$$
$$h(t) = \frac{f(t)}{R(t)} \qquad \text{(Hazard rate function)}$$

and $0 < t < 8$ years.

Example 16 *continued*

Solution

We have

$$F(t) = t\left(\frac{1}{8}\right) = \frac{t}{8}$$

$$R(t) = 1 - F(t) = 1 - \frac{t}{8}$$

$$h(t) = \frac{f(t)}{R(t)} = \frac{1/8}{1 - \dfrac{t}{8}}$$

$$= \frac{1}{8 - t} \quad \begin{bmatrix} \text{Multiplying numerator} \\ \text{and denominator by 8} \end{bmatrix}$$

Example 17 *control engineering*

The transfer function, $G(s)$, of a system is given by

$$G(s) = \frac{1/(2s + 1)}{7/(s^2 + 5s + 2)}$$

Simplify $\dfrac{G(s)}{1 + G(s)N(s)}$ where $N(s) = 10s$.

Solution

We need to first simplify $G(s)$. How?

$$G(s) = \frac{1}{(2s + 1)} \div \frac{7}{(s^2 + 5s + 2)}$$

$$= \frac{1}{(2s + 1)} \times \frac{(s^2 + 5s + 2)}{7} \quad \begin{bmatrix} \text{Turning the second} \\ \text{fraction upside down} \\ \text{and multiplying} \end{bmatrix}$$

$$= \frac{1 \times (s^2 + 5s + 2)}{7(2s + 1)}$$

$$G(s) = \frac{s^2 + 5s + 2}{14s + 7}$$

Putting $G(s) = \dfrac{s^2 + 5s + 2}{14s + 7}$ and $N(s) = 10s$ into

$$\frac{G(s)}{1 + N(s)G(s)} = \frac{(s^2 + 5s + 2)/(14s + 7)}{1 + 10s \cdot \left(\dfrac{s^2 + 5s + 2}{14s + 7}\right)}$$

Example 17 *continued*

How do you simplify [*]?

Multiply numerator and denominator by $14s + 7$:

$$\frac{G(s)}{1 + G(s)N(s)} = \frac{s^2 + 5s + 2}{(14s + 7) + 10s(s^2 + 5s + 2)}$$

$$= \frac{s^2 + 5s + 2}{14s + 7 + \underbrace{10s \cdot s^2 + (10s \times 5s) + (2 \times 10s)}_{\text{expanding } 10s(s^2 + 5s + 2)}}$$

$$= \frac{s^2 + 5s + 2}{14s + 7 + 10s^3 + 50s^2 + 20s}$$

$$\frac{G(s)}{1 + G(s)N(s)} = \frac{s^2 + 5s + 2}{10s^3 + 50s^2 + \underbrace{34s + 7}_{= 20s + 14s}}$$

Notice how we only need to use the algebra of Chapter 1 to simplify $\dfrac{G(s)}{1 + G(s)N(s)}$ of **Example 17**.

SUMMARY

The composition of functions $f(x)$ and $g(x)$ denoted by $f \circ g$ is defined as

3.2 $f \circ g = f\big(g(x)\big)$

First apply g, then f. Also in general

3.3 $f \circ g \neq g \circ f$

Other combinations are

3.4 $(f \pm g)(x) = f(x) \pm g(x)$

3.5 $\dfrac{f}{g}(x) = \dfrac{f(x)}{g(x)}$ [provided $g(x) \neq 0$]

3.6 $f \cdot g(x) = f(x) \times g(x)$

Exercise 3(d)

Solutions at end of book. Complete solutions available at www.palgrave.com/science/engineering/singh

Let the domain of all the functions be real numbers with any restriction in brackets.

1 Let f and g be functions defined by
$$f(x) = x + 1, \quad g(x) = 2x + 3$$
Find

i $f \circ g$ **ii** $f \cdot g$

iii $g - 1$ **iv** $f + g$

v $f - g$ **vi** $\dfrac{f}{g}$ ($g(x) \neq 0$)

2 Let f be the function $f(x) = 2x + 3$.
Find

i $f \circ f$ **ii** $f \circ f \circ f$ **iii** $f\big(f(f(-3))\big)$

Exercise **3(d) continued**

Solutions at end of book. Complete solutions available at www.palgrave.com/science/engineering/singh

3 Let $f(x) = ax^2 + bx + c$ where a, b and c are non-zero constants.

Find and simplify

  i $f(0)$

  ii $f(x + 1)$

  iii $f(x + 1) - f(x)$

4 Let f and g be functions defined by
$$f(x) = x^2, \quad g(x) = x + 1$$

Find

  i $g \circ g$

  ii $f \circ g$

  iii $g \circ f$

5 If $f(x) = x^2 - 1$ and $g(x) = x$ (identity function), find

  i $f \circ g$

  ii $g \circ f$

  iii What do you notice about the results of **i** and **ii**?

Determine **iv** $g \circ g$ **v** $f \circ f$

6 The function f is defined by
$$f(x) = \frac{6}{3 - x} \quad (x \neq 3)$$

Find

  i f^{-1}

  ii $f \circ f^{-1}$

  iii $f^{-1} \circ f$

7 If $f(x) = \dfrac{x^3 + 3}{2 - x^3} \quad (2 - x^3 \neq 0)$ then

$f^{-1}(x) = \left(\dfrac{2x-3}{1 + x}\right)^{\frac{1}{3}} \quad (1 + x \neq 0)$.

Find

  i $f \circ f^{-1}$

  ii $f^{-1} \circ f$

8 What do your results to questions **6 ii, iii** and **7 i, ii** have in common?

9 Let $f(x) = x - 1$ and $g(x) = \sqrt{x} \ (x \geq 0)$ be functions. Express the following in terms of f and g:

  i $\sqrt{x} - 1$ **ii** $\sqrt{x-1}$ **iii** $\sqrt{x-1} + 7$

  iv $\sqrt{x^2 - 2x + 1}$

10 ⬛ [*reliability engineering*]   The failure density function, $f(t)$, for a system is given by

$$f(t) = \frac{1}{5} \text{ where } 0 < t < 5 \text{ years}$$

Find $F(t)$, $R(t)$ and $h(t)$ where these functions are defined as in **Example 16**.

11 🔧 [*control engineering*]   A system transfer function, $T(s)$, is defined by

$$T(s) = \frac{G(s)}{1 + N(s)G(s)}$$

Simplify $T(s)$ for the following (assume the denominator $\neq 0$):

  a $G(s) = \dfrac{k}{s(s + 1)}$ where k is a constant and $N(s) = 0.01$

  b $G(s) = \dfrac{1}{s + k_1}$, $N(s) = k_2$ where k_1 and k_2 are constants

  c $G(s) = \dfrac{s + 1}{s^2 + 3s + 2}$ and $N(s) = 0.3$.

12 🔧 [*control engineering*]   A transfer function, $G(s)$, of a system is given by

$$G(s) = \frac{10}{(s - 2)} \times \frac{s}{(s^2 + 2s - 5)}$$

Simplify $\dfrac{G(s)}{1 + G(s)H(s)}$ where

$$H(s) = s + 3.$$

SECTION E **Limits of functions**

By the end of this section you will be able to:
▶ evaluate limits of functions
▶ apply limits of functions to engineering problems

Limits of functions is a complex and difficult topic. This section might seem like a colossal jump from the previous sections but functions and limits are powerful concepts in engineering mathematics, particularly calculus.

E1 Finding limits of functions

The phrase 'x tends to zero' is denoted in mathematical terms by
$$'x \rightarrow 0'$$
which means x can be made as close as possible to zero.

Example 18

By using your calculator, investigate to 5 decimal places
$$f(x) = 2^x \text{ as } x \rightarrow 0$$

Solution

How can we find an approximation to $f(x) = 2^x$ as x goes to zero?

We can substitute certain values of x.

What values of x do we substitute into f?

We can use x values close to zero such as $x = 1, 0.1, 1 \times 10^{-3}, 1 \times 10^{-6}$ and 1×10^{-9}.
Let's establish a table of values for $f(x) = 2^x$ (see Table 2).

TABLE 2	x	$f(x) = 2^x$
	1	$2^1 = 2$
	0.1	$2^{0.1} = 1.071773$
	1×10^{-3}	$2^{1 \times 10^{-3}} = 1.000693$
	1×10^{-6}	$2^{1 \times 10^{-6}} = 1.000000$
	1×10^{-9}	$2^{1 \times 10^{-9}} = 1.000000$

From the last two rows, notice that as x gets closer and closer to zero, $f(x)$ gets closer and closer to 1.00000. Trying even smaller values of x, you will find $f(x)$ is 1.00000 rounded to 5 d.p.

We say the function $f(x)$ has limit 1.00000 as $x \rightarrow 0$ and write this as
$$\lim_{x \rightarrow 0} f(x) = 1.00000 \quad \text{(5 d.p.)}$$

Similarly we can write 'x tends to infinity' as '$x \to \infty$', which means x can be made as large as possible. Let's examine a function for large values of x, $(x \to \infty)$.

Example 19

By using your calculator, investigate to 5 decimal places

$$g(x) = \left(1 + \frac{1}{x}\right)^x \quad \text{as } x \to \infty$$

Solution

How do we find an approximation for $g(x)$ as x goes to infinity?

We can substitute certain values of x.

What values of x do we substitute into g?

We can use $x = 1, 10, 1000, 1 \times 10^6$ and 1×10^{10}. To input 1×10^6 on a calculator we PRESS

| 1 | EXP | 6 | or | 1 | EE | 6 |

Let's establish a table of values for $g(x)$ – see Table 3.

TABLE 3	x	$g(x) = \left(1 + \dfrac{1}{x}\right)^x$
	1	$\left(1 + \dfrac{1}{1}\right)^1 = 2$
	10	$\left(1 + \dfrac{1}{10}\right)^{10} = 2.593742$
	1000	$\left(1 + \dfrac{1}{1000}\right)^{1000} = 2.716924$
	1×10^6	$\left(1 + \dfrac{1}{1 \times 10^6}\right)^{1 \times 10^6} = 2.718280$
	1×10^{10}	$\left(1 + \dfrac{1}{1 \times 10^{10}}\right)^{1 \times 10^{10}} = 2.718282$

From the last two rows, notice that as x gets larger and larger, $g(x)$ gets closer and closer to 2.71828. Trying even larger values of x, you will find $g(x)$ is 2.71828 rounded to 5 d.p.

We say the function $g(x)$ has limit 2.71828 as $x \to \infty$ and write this as

$$\lim_{x \to \infty} g(x) = 2.71828 \qquad \text{(5 d.p.)}$$

Actually the function $g(x)$ closes in on an irrational number called e ($= 2.718281828\ldots$). Thus

$$\lim_{x \to \infty} \left(1 + \frac{1}{x}\right)^x = e$$

Remember that an irrational number means we cannot write it as a fraction of two whole numbers. Moreover it has non-repeating decimals:

$$e = 2.71828182846\ldots$$

So any technological gadget cannot evaluate the exact value of numbers like e. Computer algebra packages and calculators use a rounded value of e. We revisit this important number, e, in Chapter 5.

We can plot the graph of $\left(1 + \dfrac{1}{x}\right)^x$ for large x on a computer algebra system. In MAPLE we use the commands shown in Fig. 23 and notice how the graph gets closer and closer to $2.71828\ldots$.

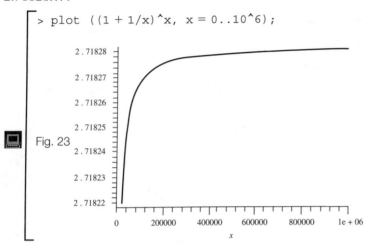

Fig. 23

In general, the notation

$$\lim_{x \to a} f(x) = L$$

means as x gets close to a, $f(x)$ gets close to the limit L.

Let's examine what happens to the function

$$f(x) = \frac{x^2 - 4}{x - 2}$$

as x gets close to 2.

Example 20

By using your calculator, investigate to 1 decimal place

$$f(x) = \frac{x^2 - 4}{x - 2} \text{ as } x \to 2$$

Solution

How do you evaluate $f(x)$ as x gets close to 2?

Substitute x values near to 2 into $f(x)$.

Example 20 *continued*

Which values should we substitute for *x*?

We could try 1.9, 1.99, 1.999, 2.001, 2.01, 2.1, etc. Substituting these *x* values into $f(x) = \dfrac{x^2 - 4}{x - 2}$ and rounding to 2 d.p. we have the results shown in Table 4.

TABLE 4		
	x	$f(x) = \dfrac{x^2 - 4}{x - 2}$
	1.9	$\dfrac{1.9^2 - 4}{1.9 - 2} = 3.90$
	1.99	$\dfrac{1.99^2 - 4}{1.99 - 2} = 3.99$
	1.999	$\dfrac{1.999^2 - 4}{1.999 - 2} = 4.00$
	2.001	$\dfrac{2.001^2 - 4}{2.001 - 2} = 4.00$
	2.01	$\dfrac{2.01^2 - 4}{2.01 - 2} = 4.01$
	2.1	$\dfrac{2.1^2 - 4}{2.1 - 2} = 4.1$

We can see from the table that the function $f(x)$ gets closer and closer to 4 as $x \to 2$. Trying even closer values to 2, you will find $f(x)$ goes to 4. We say the function $f(x)$ has limit 4 as $x \to 2$. This is generally written as

$$\lim_{x \to 2} f(x) = 4.0 \qquad \text{(1 d.p.)}$$

We can plot the function $f(x) = \dfrac{x^2 - 4}{x - 2}$ for x close to 2 in MAPLE. The output from MAPLE is shown in Fig. 24.

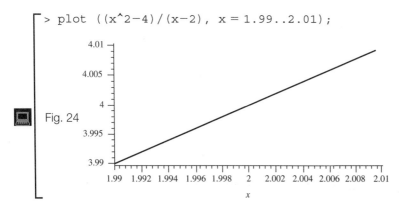

```
> plot ((x^2-4)/(x-2), x = 1.99..2.01);
```

Fig. 24

Actually the function does not exist at $x = 2$ because we have $\dfrac{0}{0}$ which is not defined.

? **Why does the MAPLE output show a value of 4 at $x = 2$ for $f(x) = \dfrac{x^2 - 4}{x - 2}$?**

MAPLE is a powerful computer algebra package but it can be misleading, and the value 4 shown for $f(x)$ is for x values close to 2 and not at $x = 2$.

E2 Employing algebra to find limits

You can also find limits of functions algebraically. We show the above result

$$\lim_{x \to 2} f(x) = 4$$

by using algebra in the next example.

Example 21

Evaluate $\lim\limits_{x \to 2} \dfrac{x^2 - 4}{x - 2}$ by using algebra.

Solution

? **How can we find the value of $\dfrac{x^2 - 4}{x - 2}$ close to 2?**

Remember that we cannot substitute $x = 2$ because we end up with $\dfrac{0}{0}$ which is not defined. We can rearrange some algebra on the function $\dfrac{x^2 - 4}{x - 2}$.

? **What can we use?**

Apply

1.15 $a^2 - b^2 = (a - b)(a + b)$

to factorize the numerator:

$$x^2 - 4 = x^2 - 2^2 = (x - 2)(x + 2)$$

Thus

$$\lim_{x \to 2} \frac{x^2 - 4}{x - 2} = \lim_{x \to 2} \frac{(x - 2)(x + 2)}{(x - 2)}$$

$$\underset{\substack{\text{cancelling}\\ x - 2}}{\dagger} = \lim_{x \to 2}(x + 2)$$

We can only cancel $x - 2$ because in the limit x never equals 2.

Example 21 *continued*

As x gets close to 2, what does $(x + 2)$ get close to?

As x approaches 2 $(x \to 2)$ so $x + 2$ approaches 4 $(x + 2 \to 4)$. Thus

$$\lim_{x \to 2}(x + 2) = 4$$

Substituting this into ⊤ gives

$$\lim_{x \to 2}\frac{x^2 - 4}{x - 2} = \lim_{x \to 2}(x + 2) = 4$$

Example 22 *control engineering*

The steady-state error, e_{ss}, of a system is given by

$$e_{ss} = \lim_{s \to 0}\frac{1}{s[1 + G(s)]}$$

Evaluate e_{ss} for $G(s) = \dfrac{5(s + 1)}{s(s + 2)}$.

Solution

Substituting $G(s) = \dfrac{5(s + 1)}{s(s + 2)}$ into e_{ss} gives

$$e_{ss} = \lim_{s \to 0}\frac{1}{s\left[1 + \dfrac{5(s + 1)}{s(s + 2)}\right]}$$

$$= \lim_{s \to 0}\frac{1}{\left[s + s\dfrac{5(s + 1)}{s(s + 2)}\right]} = \lim_{s \to 0}\frac{1}{\left[s + \dfrac{5(s + 1)}{(s + 2)}\right]}$$

As s gets close to 0, s goes to 0 and $\dfrac{5(s + 1)}{(s + 2)}$ goes to $\dfrac{5(1)}{(2)} = \dfrac{5}{2}$. Therefore

$$e_{ss} = \frac{1}{0 + {}^5/_2} = \frac{2}{5}$$

SUMMARY

The notation $x \to 0$ means x can be made as close as possible to zero. The notation

$$\lim_{x \to a} f(x) = L$$

means $f(x)$ tends to L as x goes to a.

Exercise **3(e)**

Solutions at end of book. Complete solutions available at www.palgrave.com/science/engineering/singh

1 By using your calculator, estimate the following limits:

a $\lim\limits_{x \to 0} 5^x$

b $\lim\limits_{x \to \infty} 2.71828^{-x}$

2 🚗 [mechanics]   The terminal velocity, v, of an object falling in air is given by

$$v = \lim_{t \to \infty} 10\left(1 - 2.71828^{-0.5t}\right)$$

By using your calculator, estimate v.

3 Employ algebra to evaluate the following limits:

a $\lim\limits_{x \to 3} \dfrac{x^2 - 9}{x - 3}$

b $\lim\limits_{x \to -\frac{1}{2}} \dfrac{2x^2 - x - 1}{2x + 1}$

c $\lim\limits_{t \to 1} \dfrac{t - 1}{\sqrt{t} - 1}$

4 🌾 [control engineering]   The steady-state error, e_{ss}, of a closed loop system is given by

$$e_{ss} = \lim_{s \to 0} \frac{1}{s\left(1 + G(s)\right)}$$

For $G(s) = \dfrac{3(s + 1)}{s(s + 2)}$, evaluate e_{ss}.

5 🌾 [control engineering]   The steady-state error, e_{ss}, of a closed loop system is given by

$$e_{ss} = \lim_{s \to 0} \frac{s}{1 + G(s)}$$

Evaluate e_{ss} for $G(s) = \dfrac{5(s + 2)}{s(s + 5)}$.

6 🌾 [control engineering]   The steady-state error, e_{ss}, of a system is given by

$$e_{ss} = \lim_{s \to 0} \frac{1}{s\left(1 + \dfrac{k}{s}\right)}$$

where k is a positive constant. Evaluate e_{ss}.

7 🌾 [control engineering]   The steady-state error, e_{ss}, of a system is given by

$$e_{ss} = \lim_{s \to 0} \frac{M}{1 + G(s)}$$

where M is a constant. Given that

$$G(s) = \left(\frac{k_1 + k_2}{s}\right) \times \left(\frac{1}{1 + s\tau}\right)$$

where k_1, k_2 and τ are non-zero constants, evaluate e_{ss}.

Employ algebra to find the limits of questions **8** and **9**.

8 Let $f(x) = x^2$. Find $\lim\limits_{h \to 0} \dfrac{f(x + h) - f(x)}{h}$.

9 Let $g(x) = x^3$, determine

$$\lim_{h \to 0} \frac{g(x + h) - g(x)}{h}$$

SECTION F **Modulus function**

By the end of this section you will be able to:
▶ understand and sketch the modulus function

F1 **Definition and sketching**

The modulus function, denoted $|x|$, is defined as

3.7 $$|x| = \begin{cases} x & \text{if } x \geq 0 \\ -x & \text{if } x < 0 \end{cases} \qquad \begin{array}{l} \text{(if } x \text{ is positive or zero)} \\ \text{(if } x \text{ is negative)} \end{array}$$

The modulus function gives the size of a number, and it is always positive or zero. We say $|3|$, pronounced mod 3, is equal to 3.

? **What is the value of $|-3|$?**

We use the second line of 3.7 given above with $x = -3$:

$$|-3| = -(-3) = 3$$

Thus $|-3| = 3$.

? **What are the values of $|5|$ and $|-5|$?**

In the same way we have $|5| = 5$ and $|-5| = 5$. Let's plot the graph of $|x|$.

Example 23

Plot the graph of $|x|$ for x between -3 and 3.

Solution

Substituting whole numbers between -3 and 3 into 3.7 gives the results shown in Table 5.

TABLE 5	x	$|x|$
	-3	$|-3| = -(-3) = 3$
	-2	$|-2| = -(-2) = 2$
	-1	$|-1| = -(-1) = 1$
	0	$|0| = 0$
	1	$|1| = 1$
	2	$|2| = 2$
	3	$|3| = 3$

Example 23 *continued*

Connecting the points $(-3, 3)$, $(-2, 2)$,...,$(3, 3)$ gives the graph of Fig. 25.

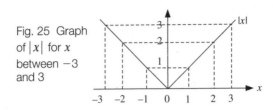

Fig. 25 Graph
of $|x|$ for x
between -3
and 3

Example 24

Sketch, on different axes, the graphs of

$$g(x) = |x - 1| \quad \text{and} \quad h(x) = |x + 3| - 2$$

Solution

We know the graph of $g(x) = |x - 1|$ is the same shape as $|x|$ but it has been modified in some way.

What adjustments do we need to make for the graph of $|x-1|$?

The graph of $|x|$ is shifted to the right by 1 unit to give $|x - 1|$ (see Fig. 26).

Fig. 26 Graph
of $|x - 1|$

The $|x + 3|$ in $h(x)$ shifts the graph $|x|$ to the left by 3 units and the -2 drops it down by 2 units. Thus the graph of $h(x) = |x + 3| - 2$ is as shown in Fig. 27.

Fig. 27
Graph of
$|x + 3| - 2$

SUMMARY

The modulus function, $|x|$, is defined by

3.7 $\quad |x| = \begin{cases} x & \text{if} \quad x \geq 0 \\ -x & \text{if} \quad x < 0 \end{cases}$

Solutions at end of book. Complete solutions available at www.palgrave.com/science/engineering/singh

Exercise **3(f)**

1 On the same axes plot the graphs of $|1 - x|$ and $|x - 1|$ for x values between -3 and $+3$. What do you notice about your results?

2 ≋ [*fluid mechanics*]   The streamlines of a fluid flow are given by

$$f(x) = |x| + c$$

Sketch the streamlines for $c = -2, -1, 0, 1$ and 2.

3 Sketch the graph of
$$|x - 3| + 1$$

Miscellaneous exercise **3**

Solutions at end of book. Complete solutions available at www.palgrave.com/science/engineering/singh

Let the domain of all the functions be real numbers with any restriction in brackets.

1 If $f(x) = \dfrac{2x - 1}{1 - x}$ (assume the denominator of $f(x) \neq 0$), find

i $\left(f\left(\dfrac{1}{2}\right) \right)^2$ **ii** $\dfrac{6f(2) - 3f(5)}{f(5)}$ **iii** $f \circ f$

2 Let f be the function defined by
$$f(t) = 4.9t^2$$
Determine the following with $h \neq 0$:

i $f(t + h)$

ii $\dfrac{f(t + h) - f(t)}{h}$

iii $\dfrac{f(1 + h) - f(1)}{h}$

iv $\dfrac{f(t + h) - f(t)}{h}$ for $t = 1$ and $h = 10^{-20}$

v $\lim\limits_{h \to 0} \dfrac{f(t + h) - f(t)}{h}$

3 Let $\phi(t) = t^3$. Find expressions for the following with $h \neq 0$:

i $\dfrac{\phi(t + h) - \phi(t)}{h}$

ii $\lim\limits_{h \to 0} \dfrac{\phi(t + h) - \phi(t)}{h}$

(Hint: $(a + b)^3 = a^3 + 3a^2 b + 3ab^2 + b^3$)

4 Sketch: **i** $|x - 3|$ **ii** $|x + 1|$

5 Let $f(x) = 2x^4 - x^2 + 1001$. Show that $f(-a) = f(a)$. (A function with this property is called an **even** function.)

6 Let $g(t) = t^5 - t^3 - t$. Show that $g(-a) = -g(a)$. (A function with this property is called an **odd** function.)

7 Let $f(x) = x^2 + 5$ and $g(x) = 6x$. Find the roots of the equation $f(x) = g(x)$.

8 Let $f(x) = x^3 - 2x^2 - x - 0.625$.
Evaluate **i** $f(2)$ **ii** $f(3)$

Without plotting the graph of f, what can you say about the graph of f between $x = 2$ and $x = 3$. Use a computer algebra package (symbolic manipulator) to plot $f(x)$ and find a root of the equation $f(x) = 0$.

9 Sketch the graph $f(x) = \dfrac{1}{x}$ $(x \neq 0)$.
Find its inverse f^{-1} and sketch it on the same axes.

10 Let ϕ be the function defined by
$$\phi(x) = \dfrac{ax - b}{cx - a}   (cx - a \neq 0)$$
Show that $\phi^{-1}(x) = \phi(x)$. If
$f(x) = \dfrac{2x - 3}{5x - 2}$ then find $f^{-1}(x)$.
(Assume the denominator $5x - 2 \neq 0$.)

Solutions at end of book. Complete solutions available at www.palgrave.com/science/engineering/singh

You may use a computer algebra system, or a graphical calculator, for questions 11 and 12.

11 Plot, on the same axes, the graphs of $f(x) = x^3$ and $g(x) = 4 + 4x - x^2$. Find the roots of $f(x) - g(x) = 0$.

12 Let $f(x) = 2.718281828^x$. For the following, h is not zero:

  i Plot $f(x)$ for x between -2 and 2.

  ii Estimate $\lim\limits_{h \to 0} \dfrac{f(h) - 1}{h}$.

  iii Find $\lim\limits_{h \to 0} \dfrac{f(x + h) - f(x)}{h}$.

13 [*mechanics*]   The displacement, $x(t)$, of a particle is given by

$$x(t) = t^2 - t - 2$$

Sketch the graph of $x(t)$ against t, marking the points where the graph crosses the axes.

14 [*control engineering*]   The transfer function, $G(s)$, of a system is given by

$$G(s) = \left(\frac{3}{s + 2}\right) \times \left(\frac{s + 1}{s^2 + 5s + 6}\right)$$

Show that

$$G(s) = \frac{3s + 3}{s^3 + 7s^2 + 16s + 12}$$

15 [*control engineering*]   The poles of a system occur where the denominator of the transfer function, $G(s)$, is zero. That is, if

$$G(s) = \frac{Y(s)}{X(s)}$$

then the system poles are the values of s which satisfy $X(s) = 0$.

Find the system poles for

a $G(s) = \dfrac{s^2 + 1}{s^2 + 4s + 3}$

b $G(s) = \dfrac{s + 1}{s^2 + 5s - 7}$

c $G(s) = \dfrac{s}{3}$

16 [*reliability engineering*]   The failure density function, $f(t)$, for a system is given by

$$f(t) = \frac{1}{10} \quad 0 < t < 10 \text{ years}$$

Find $F(t)$, $R(t)$ and $h(t)$ where

$F(t) = tf(t)$ (Failure density function)

$R(t) = 1 - F(t)$ (Reliability function)

$h(t) = \dfrac{f(t)}{R(t)}$ (Hazard rate function)

Plot, by using a symbolic manipulator or otherwise, on different axes, $F(t)$, $R(t)$ and $h(t)$ for $0 < t < 10$.

17 [*reliability engineering*]   In reliability engineering we have the following relationship:

$$R(t) = 1 - Q(t)$$

where $R(t)$ is called the reliability function, $Q(t)$ is called the failure distribution function and t is time in years. For a set of components, the failure distribution function, $Q(t)$, is given by

$$Q(t) = 0.3t - 0.015t^2$$

Sketch $Q(t)$ for $t \geq 0$. Also find $R(t)$.

Use a computer algebra system for question 18.

18 [*control engineering*]   The overall transfer function, $T(s)$, of a system is given by

$$T(s) = \frac{G(s)}{1 + G(s)H(s)}$$

Simplify $T(s)$ for

$$G(s) = \frac{2}{s + 1} \times \frac{s}{(s + 2)(s + 5)}$$

and $H(s) = 0.001$

CHAPTER 4

Trigonometry and Waveforms

The word *trigonometry* is a combination of the Greek words *trigono* meaning triangle and *metro* meaning to measure. It was the early Greeks who laid the foundation stones of trigonometry around 140 BC. However the sine of an angle was developed by Aryabhata, a Hindu.

SECTION A **Trigonometric functions**

By the end of this section you will be able to:
▶ find sin, cos and tan of angles
▶ solve right-angled triangles
▶ calculate inverse trigonometric functions
▶ find cosec, sec and cot of angles

A1 **Review of trigonometric ratios**

Consider the right-angled triangle ABC (Fig. 1) with an angle θ at A.

A right-angled triangle is a triangle that has one angle equal to 90° and this is represented by a square, □, as seen in Fig. 1.

We define

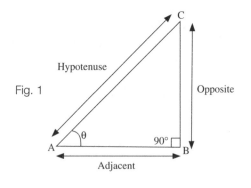

Fig. 1

adjacent = side AB, **adjacent** to the angle θ

opposite = side BC, **opposite** to the angle θ

hypotenuse = side AC, the **longest** side of the right-angled triangle

We characterize sin (short for sine), cos (short for cosine) and tan (short for tangent) as follows:

4.1
$$\sin(\theta) = \frac{\text{opposite length}}{\text{hypotenuse length}}$$

4.2
$$\cos(\theta) = \frac{\text{adjacent length}}{\text{hypotenuse length}}$$

4.3
$$\tan(\theta) = \frac{\text{opposite length}}{\text{adjacent length}}$$

4.1 to 4.3 are called the trigonometric ratios and as each of sin, cos and tan are functions of the angle θ, they are generally referred to as the **trigonometric functions**. Let *opp* = opposite, *adj* = adjacent and *hyp* = hypotenuse, then by using Fig. 1 we can rewrite these trigonometric functions in terms of *opp*, *adj* and *hyp* (Fig. 2):

4.1 $$\sin(\theta) = \frac{opp}{hyp}$$

4.2 $$\cos(\theta) = \frac{adj}{hyp}$$

Fig. 2

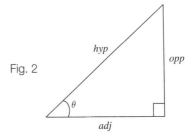

4.3 $$\tan(\theta) = \frac{opp}{adj}$$

We can use these ratios to find an unknown angle or length of a side of any right-angled triangle. Note that we **must** have a right-angled triangle. Before we proceed to an example, we need to recall Pythagoras's theorem. **By considering Fig. 2 what does Pythagoras tell us about the relationship between the lengths *adj*, *opp* and *hyp*?**

P $$hyp^2 = adj^2 + opp^2$$

We give this formula reference (P) because it is central to this chapter and you should remember it.

In order to find *hyp* we take the square root of both sides. By transposing formula (P) we can find *adj* and *opp*. Often by using this theorem we can find the length of a side we need.

Example 1

Find the length AC of Fig. 3.

Without using a calculator, evaluate $\sin(45°)$, $\cos(45°)$ and $\tan(45°)$.

Fig. 3

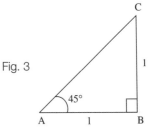

Solution

How can we find AC?

Since opposite = 1 and adjacent = 1 we can apply Pythagoras:

$$AC^2 = 1^2 + 1^2 = 2$$

Take the square roots of both sides: $AC = \sqrt{2}$

Thus hypotenuse = $\sqrt{2}$. With *adj* = 1, *opp* = 1 and *hyp* = $\sqrt{2}$ we have

$$\sin(45°) = \frac{opp}{hyp} = \frac{1}{\sqrt{2}}$$

$$\cos(45°) = \frac{adj}{hyp} = \frac{1}{\sqrt{2}}$$

$$\tan(45°) = \frac{opp}{adj} = \frac{1}{1} = 1$$

Example 2

Consider the right-angled triangle shown in Fig. 4.

Without using a calculator, find the values of

i sin(60°), cos(60°) and tan(60°)

ii sin(30°), cos(30°) and tan(30°).

Fig. 4

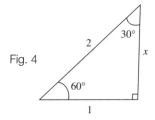

Solution

i Considering the 60° case we have the situation shown in Fig. 5.

What is sin(60°) equal to?

Fig. 5

For 60°, opposite = x, hypotenuse = 2 and adjacent = 1.

By ⬛4.1⬛ **, sin(60°) =** $\dfrac{x}{2}$ **but what is the value of** x**?**

Rearranging Pythagoras:

$$x^2 = 2^2 - 1^2 = 3 \text{ which gives } x = \sqrt{3}$$

Applying ⬛4.1⬛ , ⬛4.2⬛ and ⬛4.3⬛ with $opp = \sqrt{3}$, $adj = 1$ and $hyp = 2$ respectively, gives

$$\sin(60°) = \frac{\sqrt{3}}{2}, \ \cos(60°) = \frac{1}{2} \text{ and } \tan(60°) = \frac{\sqrt{3}}{1} = \sqrt{3}$$

ii For the 30° case we have the situation shown in Fig. 6.

Now with reference to 30°, adjacent = $\sqrt{3}$, opposite = 1 and hypotenuse = 2:

Fig. 6

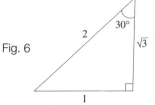

$$\sin(30°) = \underbrace{\frac{1}{2}}_{\text{by } 4.1}, \ \cos(30°) = \underbrace{\frac{\sqrt{3}}{2}}_{\text{by } 4.2} \text{ and } \tan(30°) = \underbrace{\frac{1}{\sqrt{3}}}_{\text{by } 4.3}$$

The values obtained in **Examples 1** and **2** are the exact values of the corresponding trigonometric ratios. Thus combining these with trigonometric ratios for angles 0°, 90° we obtain Table 1.

⬛4.1⬛ $\sin(\theta) = opp/hyp$ ⬛4.2⬛ $\cos(\theta) = adj/hyp$ ⬛4.3⬛ $\tan(\theta) = opp/adj$

TABLE 1		$\theta = 0°$	$\theta = 30°$	$\theta = 45°$	$\theta = 60°$	$\theta = 90°$
	$\sin(\theta)$	0	$\dfrac{1}{2}$	$\dfrac{1}{\sqrt{2}}$	$\dfrac{\sqrt{3}}{2}$	1
	$\cos(\theta)$	1	$\dfrac{\sqrt{3}}{2}$	$\dfrac{1}{\sqrt{2}}$	$\dfrac{1}{2}$	0
	$\tan(\theta)$	0	$\dfrac{1}{\sqrt{3}}$	1	$\sqrt{3}$	undefined

This table shows the exact trigonometric ratios for some special angles.

The angles are read from the top row. Note that $\sin(0°) = 0$, $\cos(0°) = 1$ and $\tan(0°) = 0$. To evaluate any of the trigonometric functions, sin, cos or tan, we can use a calculator. For example, to evaluate $\sin(30°)$, place the calculator into degrees mode first and then PRESS $\boxed{\sin}$ $\boxed{30}$. The calculator should show 0.5. On some older calculators you enter the argument [30] first and then the function [sin]. Hence $\sin(30°) = 0.5 = \dfrac{1}{2}$. Similarly by using a calculator we have

$$\cos(22°) = 0.927, \ \tan(15°) = 0.268,$$
$$\sin(53.1°) = 0.800, \text{ etc.}$$

Note that these are **not** exact values but rounded to 3 decimal places.

Fig. 7

Opposite

Adjacent

θ

? **Why is tan(90°) not defined?**

$$\tan(\theta) = \frac{\text{opposite}}{\text{adjacent}}$$

? **If $\theta = 90°$ in Fig. 7 then what is the size of the adjacent length?**

Adjacent must be zero and so

$$\tan(90°) = \frac{\text{opposite}}{0}$$

Of course we cannot divide by zero. This is why $\tan(90°)$ is **not** defined.

A2 Solving right-angled triangles

Transposing the standard formulae, 4.1 to 4.3 , allows us to evaluate many other unknowns of the right-angled triangle. By transposing these we have the following:

4.4	$opp = hyp \times \sin(\theta)$

| 4.5 | $adj = hyp \times \cos(\theta)$ |

| 4.6 | $opp = adj \times \tan(\theta)$ |

| 4.7 | $hyp = \dfrac{opp}{\sin(\theta)}$ |

4.8

$$hyp = \frac{adj}{\cos(\theta)}$$

4.9

$$adj = \frac{opp}{\tan(\theta)}$$

Example 3

Figure 8 shows a symmetrical roof truss.

Span BF is of length 8 m. Rafters AB and AC are inclined at angles 40° and 65° respectively. Find the lengths AF and DE.

Fig. 8

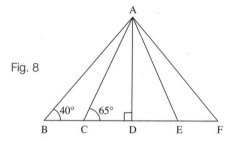

Solution

Since the roof truss is symmetrical, angle $F = 40°$ and DF = 4 m (Fig. 9).

How can we find length AF?

AF is the hypotenuse and we know the adjacent = 4.

So what formula can we use?

Use $hyp = \frac{adj}{\cos(\theta)}$:

Fig. 9

$$AF = \frac{4}{\cos(40°)} = \frac{4}{0.7660} = 5.222 \text{ m (3 d.p.)}$$

For length DE, consider triangle ADE. We know by symmetry that angle $E = 65°$ and length AD can be found from Fig. 9. In Fig. 9, AD is the opposite, so using $opp = adj \times \tan(\theta)$ we have

$$AD = 4 \times \tan(40°)$$

$$= 4 \times 0.8391 = 3.356$$

(see Fig. 10)

Fig. 10

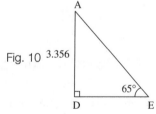

How can we find length DE?

DE is adjacent to 65° and we have $opp = 3.356$, so using $adj = \frac{opp}{\tan(\theta)}$ gives

$$DE = \frac{3.356}{\tan(65°)} = \underbrace{\frac{3.356}{2.1445}}_{\text{by calculator}} = 1.565 \text{ m (3 d.p.)}$$

A3 Inverse trigonometric functions

Let the sine of angle A be x, that is

$$\sin(A) = x$$

then the angle A is given by

$$A = \sin^{-1}(x)$$

where $\sin^{-1}$ denotes **inverse sine.** Remember from Chapter 3 that the inverse function is represented by the index -1. This is why the inverse of the sine function is denoted by $\sin^{-1}$. **What is $\sin^{-1}\left(\dfrac{1}{\sqrt{2}}\right)$ equal to?**

From Table 1 we know that $\sin(45°) = \left(\dfrac{1}{\sqrt{2}}\right)$, thus

$$\sin^{-1}\left(\frac{1}{\sqrt{2}}\right) = 45°$$

We can also use a calculator to determine inverse sin. For example, to find angle A from $\sin(A) = 0.791$ we PRESS $\boxed{\textbf{SHIFT}}\ \boxed{\textbf{SIN}}\ \boxed{\textbf{0.791}}\ \boxed{\textbf{=}}$, which should show 52.27906149. (Make sure you put your calculator into degrees mode.)

On many basic scientific calculators, the $\sin^{-1}$ is a secondary function so we need to press the $\boxed{\textbf{SHIFT}}$ or $\boxed{\textbf{INV}}$ or $\boxed{\textbf{2nd FUN}}$ button prior to entering the sin button. (This is just like pressing the $\boxed{\textbf{SHIFT}}$ key on a computer keyboard to access capital letters.) Likewise, the inverse cos is denoted by $\cos^{-1}$ and inverse tan by $\tan^{-1}$ and defined similarly.

On some calculators and computer algebra systems, the inverse sin, $\sin^{-1}$, is denoted by arcsin. Similarly

$$\cos^{-1} = \arccos \text{ and } \tan^{-1} = \arctan$$

Example 4

Evaluate the following:

$$\sin^{-1}(0.23),\ \cos^{-1}(0.75) \text{ and } \tan^{-1}(1.73)$$

Solution

By using a calculator and rounding to 2 decimal places we have

$$\sin^{-1}(0.23) = 13.30°,\ \cos^{-1}(0.75) = 41.41° \text{ and } \tan^{-1}(1.73) = 59.97°$$

Example 5

Figure 11 shows the symmetrical part of a bridge. Point X is the midpoint of ABCD. Find the size of angle θ if CD = 10 m and AC = 18 m.

Solution

Consider triangle AXY where Y is the midpoint of AB.

? What is the length of AY and XY?

Since X is the midpoint, AY = 5 and XY = 9.

? How can we find θ?

Angle AXY is half of θ (Fig. 12).

? How can we find $\theta/2$?

We know with reference to angle $\theta/2$ that opposite = 5 and adjacent = 9, so we can use 4.3 :

$$\tan\left(\frac{\theta}{2}\right) = \frac{5}{9}$$ ★

? How do we extract $\theta/2$ from ★ ?

Take inverse tan of both sides:

$$\frac{\theta}{2} = \tan^{-1}\left(\frac{5}{9}\right) = 29.055° \text{ [by calculator]}$$

So $\theta = 2 \times 29.055° = 58.11°$

Fig. 11

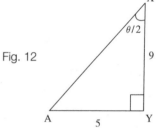

Fig. 12

A4 **Additional trigonometric functions**

The other trigonometric functions are called cosecant (cosec), secant (sec) and cotangent (cot), and are determined by

4.10 $$\text{cosec}(\theta) = \frac{1}{\sin(\theta)}$$ (provided $\sin(\theta) \neq 0$)

4.11 $$\sec(\theta) = \frac{1}{\cos(\theta)}$$ (provided $\cos(\theta) \neq 0$)

4.12 $$\cot(\theta) = \frac{1}{\tan(\theta)}$$ (provided $\tan(\theta) \neq 0$)

4.3 $\tan(x) = opp/adj$

We can evaluate these new trigonometric functions, cosec, sec and cot, by using a calculator. Some calculators evaluate these directly, on others you have to write them in terms of the basic trigonometric functions and then evaluate them.

Example 6

Evaluate the following to 3 decimal places:

$$\text{cosec}(22°), \text{sec}(19.61°) \text{ and } \cot(89°)$$

Solution

By using a calculator we have

$$\text{cosec}(22°) = \frac{1}{\sin(22°)} = \frac{1}{0.3746} = 2.669 \text{ (3 d.p.)}$$

$$\sec(19.61°) = \frac{1}{\cos(19.61°)} = \frac{1}{0.9420} = 1.062 \text{ (3 d.p.)}$$

$$\cot(89°) = \frac{1}{\tan(89°)} = \frac{1}{57.2900} = 0.018 \text{ (3 d.p.)}$$

SUMMARY

The basic trigonometric functions are defined as

4.1	$\sin(\theta) = opp/hyp$
4.2	$\cos(\theta) = adj/hyp$
4.3	$\tan(\theta) = opp/adj$

Inverse sin is denoted by $\sin^{-1}$.

Exercise 4(a)

Solutions at end of book. Complete solutions available at www.palgrave.com/science/engineering/singh

1 Evaluate the following to 2 decimal places:

$\sin(35°), \cos(47°)$ and $\tan(21°)$

2 By using Table 1, find the exact values of the following:

i $\cos(45°)\sin(45°)$

ii $\tan(60°) + \tan(30°)$

iii $\sin(60°) + \tan(60°)$

iv $\dfrac{\sin(45°)}{\tan(45°)\cos(45°)}$

v $\cos(45°) + \sin(60°)$

vi $\sin(60°)\cos(30°) + \cos(60°)\sin(30°)$

Exercise **4(a) continued**

Solutions at end of book. Complete solutions available at www.palgrave.com/science/engineering/singh

3 Figure 13 shows a symmetrical roof truss, with rise $r = 4$ m and rafter AC inclined at 25°.

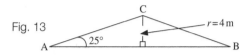

Fig. 13

Evaluate the length of the rafter AC and span AB.

4 By using $\sin(\theta) = \dfrac{opp}{hyp}$ and $\cos(\theta) = \dfrac{adj}{hyp}$

show that $\tan(\theta) = \dfrac{\sin(\theta)}{\cos(\theta)}$.

5 Figure 14 shows a garage roof with $r = 3$ m and $s = 6$ m. Find the size of the angle θ.

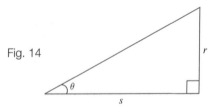

Fig. 14

6 Figure 15 shows a symmetrical part of a bridge lattice. Determine the length h, given that AD = 5 m, angle DAB = 55° and X is the midpoint of ABCD.

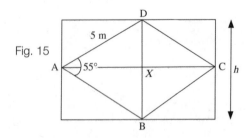

Fig. 15

7 Figure 16 shows a cross-section of a rain gutter with PQ = QR = RS = 180 mm

and angle $\theta = 60°$. Find the area of the cross-section PQRS.

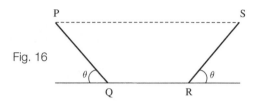

Fig. 16

8 Figure 17 shows a circular car tunnel with AB = 3 m and angle BAC = 60°. Assume angle BAZ = angle BCZ = 90°. Without using a calculator, find the diameter BZ of the tunnel.

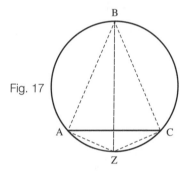

Fig. 17

9 Without using a calculator, find the values of

i $\cos(\theta)$ and $\tan(\theta)$, if $\sin(\theta) = \dfrac{1}{\sqrt{3}}$

ii $\sin(\theta)$ and $\tan(\theta)$, if $\cos(\theta) = \dfrac{2}{3}$

iii $\sin(\theta)$ and $\cos(\theta)$, if $\tan(\theta) = \dfrac{3}{4}$

10 Evaluate the following to 3 decimal places:

$\sec(36.4°)$, $\cot(29.17°)$

and $\operatorname{cosec}(44.44°)$

SECTION B **Angles and graphs**

By the end of this section you will be able to:
▸ find the trigonometric functions for angles of any size
▸ plot the graphs of these functions

B1 **Angles of any size**

Let P be the point (x, y) in the Cartesian co-ordinate system at a distance r from the origin O, at an angle θ. Consider the triangle in Fig. 18.

By the formula adjacent = hypotenuse $\times$ cos(θ) we have

$$x = r \times \cos(\theta)$$

By the formula opposite = hypotenuse $\times$ sin(θ) we have

$$y = r \times \sin(\theta)$$

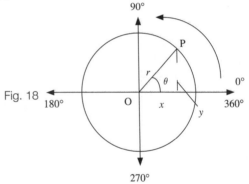

Fig. 18

If P rotates **anticlockwise** about the origin, then we define the trigonometric functions (sin, cos and tan) for any angle θ as follows:

4.13 $\qquad \sin(\theta) = \dfrac{y}{r}$

4.14 $\qquad \cos(\theta) = \dfrac{x}{r}$

4.15 $\qquad \tan(\theta) = \dfrac{y}{x}$

Here θ can be of any size (magnitude). One complete revolution is 360°, two complete revolutions are $2 \times 360° = 720°$. You can think of P in Fig. 18 as a point on the edge of a wheel, and θ as the angle made by P from the 0° position. Thus if the wheel makes 10 complete revolutions then it has completed 3600°.

By convention, the angle θ is normally measured from the positive x axis in an anticlockwise direction. The angle θ can be divided into four equal quadrants:

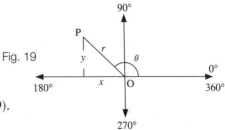

Fig. 19

0°–90°, 90°–180°, 180°–270° and 270°–360° (Fig. 19).

B2 **Graphs of sin, cos and tan**

By using a graphical calculator, or a computer algebra package, you can easily plot the graphs of sin, cos and tan between 0° and 360°. However on a more basic calculator you have to evaluate some points and connect them to make a graph, as the next example shows.

Example 7

Plot the graph of $y = \sin(x)$ for $x = 0°$ to 360°.

Solution

By taking intervals of 30° and evaluating the sine of each angle on a calculator, rounding to 3 d.p., we establish Table 2.

TABLE 2	x	0°	30°	60°	90°	120°	150°	180°	210°	240°	270°	300°	330°	360°
	$y = \sin(x)$	0	0.5	0.866	1	0.866	0.5	0	−0.5	−0.866	−1	−0.866	−0.5	0

Connecting the points gives the graph of $y = \sin(x)$ (Fig. 20).

Fig. 20 Graph of $y = \sin(x)$ for x between 0° and 360°

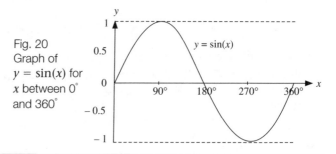

Similarly we can plot the graphs of $y = \cos(x)$ and $y = \tan(x)$ (Fig. 21) – see **Exercise 4(b).**

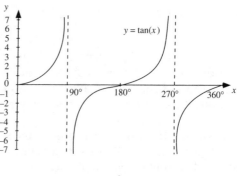

Fig. 21

Note that $\sin(x)$ and $\cos(x)$ lie between −1 and +1.

Also it is worth observing the symmetrical nature of the sin and cos graphs:

i the positive part of the sin(x) graph is symmetrical about $x = 90°$ and the negative part is symmetrical about $x = 270°$

ii the cos(x) graph is symmetrical about $x = 180°$.

Remember from the previous section that tan(90°) is not defined. Similarly tan(270°) is not defined. In general, if n is any whole number then $\tan[(2n + 1) \times 90°]$, that is the tan of an odd multiple of 90° is not defined.

If we extend the x-axis domain, it can be seen by plotting the graphs that the sin and cos graphs are repeated every 360° interval and the tan graph is repeated every 180° interval (Fig. 22).

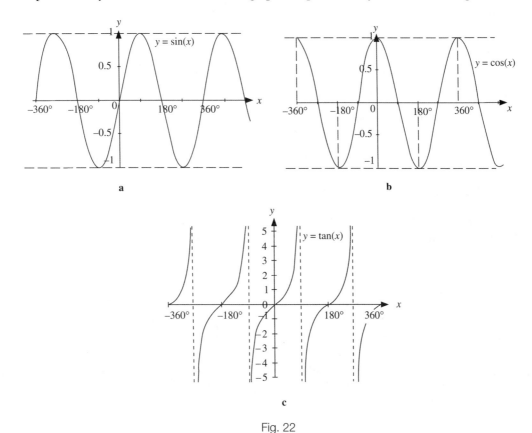

Fig. 22

B3 **Transformation of graphs**

We can apply the transformation of graphs covered in Chapter 3 to the transformation of trigonometric graphs.

Example 8

Sketch the graph of $y = \cos(x - 90°)$ for $0° \le x \le 360°$.

What do you notice about your sketch?

Example 8 *continued*

Solution

Let $f(x) = \cos(x)$ then
$f(x - 90°) = \cos(x - 90°)$, so we
shift the cos graph in Fig. 22**b** to
the right by 90° (Fig. 23). (Remember
from Chapter 3 that $f(x - c)$ is
the graph of $f(x)$ shifted to the right
by c units.)

Fig. 23

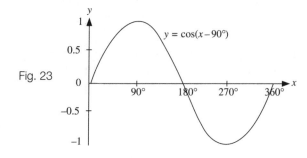

It is identical to the sine graph
(see Fig. 20). Actually for all x:

$$\cos(x - 90°) = \sin(x)$$

This is an example of a trigonometric identity. There are many other trigonometric
identities which we will examine later in this chapter.

Example 9

For $0° \le x \le 360°$, sketch the graph of $y = \sin(2x)$ and compare with the
graph of $y = \sin(x)$.

Solution

Let $f(x) = \sin(x)$ then $f(2x) = \sin(2x)$.
The graph of $\sin(2x)$ is basically
the graph of $\sin(x)$ but it is
squeezed by a factor of 2 as shown
in Fig. 24. (In the summary of Chapter
3 we pointed out that $f(cx)$ is the graph
of $f(x)$ but squashed horizontally by
a factor of c if $c > 1$.)

Fig. 24

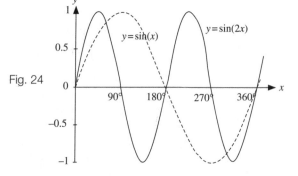

The graph $\sin(2x)$ has two cycles
between 0° and 360° compared to the
one cycle for $\sin(x)$. Similarly, a graph of $\sin(3x)$ will have three cycles
between 0° and 360°.

SUMMARY

We can define the trigonometric functions for angles of any size. The graphs of $\sin(x)$
and $\cos(x)$ lie between -1 and $+1$; note the symmetrical nature of these graphs. Also
$\tan(x)$ is not defined for odd whole number multiples of 90°.

Solutions at end of book. Complete solutions available at www.palgrave.com/science/engineering/singh

Exercise **4(b)**

1 Plot the graph of $y = \cos(x)$ for $x = 0°$ to $360°$.
2 Plot the graph of $y = \tan(x)$ for $x = 0°$ to $360°$.

3 Sketch, on different axes, the graphs of
   a $\cos(2x)$ **b** $\sin(x + 90°)$
   c $\tan(x - 180°)$ **d** $\sin(x + 360°)$

SECTION C **Trigonometric equations**

By the end of this section you will be able to:
▶ find particular solutions of trigonometric equations
▶ find general solutions of trigonometric equations

Solving trigonometric equations is a highly sophisticated and difficult topic. Finding the general solution of these equations requires special proficiency in algebra.

C1 **Particular solutions of trigonometric equations**

Example 10

Solve:

$$\sin(\theta) = 0.5$$

where θ is between $0°$ and $360°$.

Solution

What are we trying to find?

We need to find the angle θ which satisfies $\sin(\theta) = 0.5$. **How?**

Take inverse sin, $\sin^{-1}$, of both sides:

$$\theta = \sin^{-1}(0.5) = 30°$$

Ideally you should know $\sin(30°) = 0.5$ but you can use a calculator. We PRESS

SHIFT | SIN | 0.5 and the calculator should show 30. (On some calculators we press

INV or 2nd FUN instead of SHIFT)

By looking at the graph of $\sin(\theta)$, there is another θ value between $0°$ and $180°$ which satisfies $\sin(\theta) = 0.5$ (Fig. 25).

Example 10 *continued*

The other angle is obtained by subtracting
30° from 180° because the sine graph is
symmetrical about $x = 90°$ (Fig. 25). So

$$\theta = 30° \text{ or } \theta = 180° - 30° = 150°$$ Fig. 25

Checking our result, we have
$\sin(30°) = \sin(150°) = 0.5$.

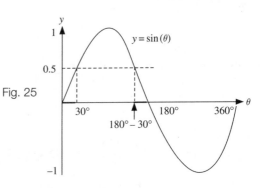

**Do we need to plot the graph of $\sin(\theta)$ to solve
$\sin(\theta) = 0.5$?**

No.

Next we examine another technique for
solving trigonometric equations like
$\sin(\theta) = 0.5$.

In the previous section we divided the
interval 0° to 360° into four equal quadrants:
$0°-90°$, $90°-180°$, $180°-270°$ and $270°-360°$.
By the graphs of sin, cos and tan
we have Fig. 26:

All [$\sin(\theta)$, $\cos(\theta)$ and $\tan(\theta)$] are positive,
if $0° < \theta < 90°$.

Only $\sin(\theta)$ is positive, if $90° < \theta < 180°$.

Only $\tan(\theta)$ is positive, if $180° < \theta < 270°$.

Only $\cos(\theta)$ is positive, if $270° < \theta < 360°$.

The mnemonic CAST can be
used to remember the signs of the
trigonometric functions as shown in
Fig. 27. By convention, θ is negative
if we measure clockwise from the
positive x axis.

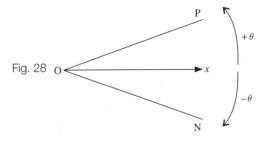

The equation $\sin(\theta) = 0.5$ from **Example 10**
can also be solved by looking at CAST (Fig. 29).

We know θ lies in quadrants ▨* and ▨**
because those are the only quadrants where
$\sin(\theta)$ is positive, $\sin(\theta) = +0.5$. Hence
$\theta = 30°$ or $150°$. One of the angles is found by
using your calculator and the other angle(s)
can be established using CAST.

Example 11

Solve:

$$2\cos^2(\theta) = 2 - 3\cos(\theta)$$

where $-180° \leq \theta \leq 180°$.

Solution

What are we trying to find?

The value of θ which satisfies the above equation. Rearranging gives

$$2\cos^2(\theta) + 3\cos(\theta) - 2 = 0 \qquad \boxed{\ \star\ }$$

The term $\cos^2(\theta) = [\cos(\theta)]^2$. **So how do we solve this equation?**

Let $x = \cos(\theta)$, so we have

$$2x^2 + 3x - 2 = 0$$

We need to find x. **How?**

Factorize:

$$(2x - 1)(x + 2) = 0$$

$$x = \frac{1}{2} \text{ or } x = -2 \qquad \text{[Solving]}$$

$$\cos(\theta) = \frac{1}{2} \text{ or } \cos(\theta) = -2 \quad \text{[Replacing } x = \cos(\theta)\text{]}$$

The result $\cos(\theta) = -2$ is impossible because $\cos(\theta)$ lies between -1 and $+1$, as can be seen by the graph of $\cos(\theta)$ in Fig. 30.

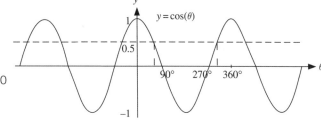

Fig. 30

So $\cos(\theta) = \frac{1}{2} = 0.5$.

How do we find θ?

Take inverse cos, $\cos^{-1}$, of both sides: $\theta = \cos^{-1}(1/2) = 60°$.

The other angle lies in the 4th quadrant because that is the only other quadrant where $\cos(\theta)$ is positive, $\cos(\theta) = +0.5$ (Fig. 31).

Since θ needs to be between $-180°$ and $180°$, according to the question, we measure the other angle clockwise from $0°$. The other angle is $-60°$.

Fig. 31

Therefore $\theta = 60°$ or $-60°$. Check your result by substituting $\theta = 60°$ and $\theta = -60°$ into $\boxed{\ \star\ }$.

C2 General solutions of trigonometric equations

Example 12

Find the general solution of

$$\sin(\theta) = 0.5$$

Solution

Here, there is no restriction on θ. By looking at the graph of $\sin(\theta)$, and using our results from **Example 10**, we find (Fig. 32) that

Fig. 32

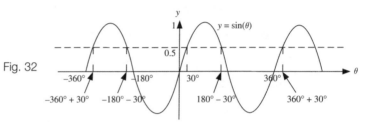

$$\theta = 30°, 180° - 30°, 360° + 30°, 540° - 30°, \ldots$$
$$-180° - 30°, -360° + 30°, -540° - 30°, \ldots$$

This can also be written as

$$\boxed{\quad * \quad} \qquad \theta = 30°, 180° - 30°, (2 \times 180)° + 30°, (3 \times 180)° - 30°, \ldots$$
$$(-1 \times 180)° - 30°, (-2 \times 180)° + 30°, (-3 \times 180)° - 30° \ldots$$

By looking at $\boxed{*}$, can you see a pattern?

The general solution of $\sin(\theta) = 0.5$ is

$$\theta = (n \times 180)° + [(-1)^n \times 30]°$$

where n is any whole number.

The $(-1)^n$ simply changes the sign alternately. **What is $[(-1)^n \times 30]°$ if $n = 1$?**

$$[(-1)^1 \times 30]° = -30°$$

Similarly if $n = 2$ then $[(-1)^2 \times 30]° = +30°$. For $n = 3$, we have $-30°$, and for $n = 4$ we have $+30°$ etc.

In general, if $\sin(\theta) = R$ (where R is a real number between -1 and 1 inclusive) then

$$\boxed{4.16} \qquad \theta = (180 \times n)° + (-1)^n \alpha$$

where α is the angle shown in Fig. 33.

$\boxed{4.16}$ is called the general solution of $\sin(\theta) = R$.

To find angle α, use your calculator.

Fig. 33

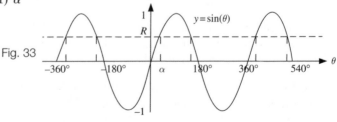

Similarly the general solution of $\cos(\theta) = R$ is

| 4.17 | $\theta = (360 \times n)° \pm \alpha$ |

where α is the angle shown in Fig. 34.

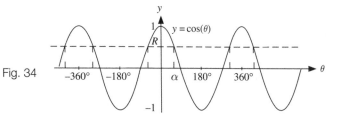

Fig. 34

The general solution of $\tan(\theta) = R$ (in this case R is any real number) is

| 4.18 | $\theta = (180 \times n)° + \alpha$ |

where α is the angle shown in Fig. 35.

Fig. 35

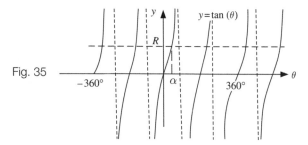

Example 13

Find the general solution of

$$\cos(\theta) = 0.927$$

Solution

By using 4.17 :

$$\theta = (360 \times n)° \pm \alpha$$

where $\alpha = \cos^{-1}(0.927) = 22.03°$. (This angle is obtained by using a calculator.) For any whole number n

$$\theta = (360 \times n)° \pm 22.03°$$

Example 14

Solve

$$\tan^2(\theta) = 2$$

where $0° \leq \theta \leq 720°$ and $\tan(\theta)$ is positive.

Solution

We have

$$\tan^2(\theta) = [\tan(\theta)]^2 = 2$$

> **Example 14** *continued*
>
> **?** **How do we find tan(θ)?**
>
> Take the square root of both sides:
>
> $$\tan(\theta) = \sqrt{2} \quad \text{[only positive root]}$$
>
> By 4.18 :
>
> $$\theta = (180 \times n)° + \alpha$$
>
> where $\alpha = \tan^{-1}\left(\sqrt{2}\right) = 54.74°$. Substituting this into θ for $n = 0, 1, 2$ and 3:
>
> $$\theta = (180 \times 0)° + 54.74°, (180 \times 1)° + 54.74°,$$
> $$(180 \times 2)° + 54.74°, (180 \times 3)° + 54.74°$$
> $$= 54.74°, 180° + 54.74°, 360° + 54.74°, 540° + 54.74°$$
> $$= 54.74°, 234.74°, 414.74°, 594.74°$$

The result is that you add 180° each time you want to find a new positive angle in **Example 14.**

SUMMARY

We can solve trigonometric equations by examining the appropriate graph or using CAST.
For general solutions of $\sin(\theta) = R$, $\cos(\theta) = R$ and $\tan(\theta) = R$ we apply 4.16 , 4.17
and 4.18 respectively.

Exercise 4(c)

Solutions at end of book. Complete solutions available at www.palgrave.com/science/engineering/singh

1 Solve the following for θ, where $0° \leq \theta \leq 360°$:

a $\sin(\theta) = 0.707$ **b** $\cos(\theta) = 0.5$

c $\tan(\theta) = \sqrt{3}$ **d** $\tan^2(\theta) - 3 = 0$

e $\cos^2(\theta) - \cos(\theta) + \dfrac{1}{4} = 0$

f $2\sin^2(\theta) - 3\sin(\theta) = 0$

g $7\sin(\theta) - 3 = 0$

2 Find the general solution (if possible) of
a $\tan(\theta) = 0.8$ **b** $\cos(\theta) = -0.125$

c $\sin(\theta) = 0.22$ **d** $\sin(3\theta) = 0.72$

e $\cos(2\theta + 60°) = 2$

f $\cos(2\theta + 60°) = \dfrac{1}{2}$

3 Solve (if possible) the following for θ in the given domain:

a $\tan^2(\theta) - 2\tan(\theta) = -1$

$$0° \leq \theta \leq 720°$$

b $\cos(\theta)\sin(\theta) + \cos(\theta) = 0$

$$0° \leq \theta \leq 360°$$

Solutions at end of book. Complete solutions available at www.palgrave.com/science/engineering/singh

Exercise **4(c) continued**

c $2\cos^2(\theta) - 4\cos(\theta) - 5 = 0$
$$0° \le \theta \le 180°$$

d $\tan^2(\theta) - \tan(\theta) - 1 = 0$
$$0° \le \theta \le 360°$$

e $\sin^2(\theta) - 6\sin(\theta) = -4$
$$0° \le \theta \le 720°$$

f $\cos^2(\theta) - 10\cos(\theta) + 23 = 0$
$$-360° \le \theta \le 360°$$

4 Find the exact values of θ, where $0° \le \theta \le 360°$, which satisfies the following equations.

a $2\cos^3(\theta) - \cos(\theta) = 0$

b $4\sin^2(\theta) + \sqrt{2}\sin(\theta) = \sqrt{6}\sin(\theta)$

c $16\cos^2(\theta)\sin^2(\theta) - 12\cos^2(\theta) - 12\sin^2(\theta) + 9 = 0$

SECTION D **Trigonometric rules**

By the end of this section you will be able to:
▶ apply the sine rule to solve a non-right-angled triangle
▶ apply the cosine rule to solve a non-right-angled triangle

? What property do the triangles of Section A have in common?

Each of these triangles contains a 90° angle. The formulae of that section **only** apply to a right-angled triangle. You cannot use the formulae for any triangle. In this section we examine some new formulae, called the sine and cosine rules, that can be applied to **any** triangle. These rules are used to find unknown sides and angles.

In this section we use the property:

The sum of the angles in a triangle is equal to 180°

D1 **Sine rule**

Consider any triangle ABC with **side** a opposite **angle** A, **side** b opposite **angle** B and **side** c opposite **angle** C (Fig. 36). Then we have

4.19
$$\frac{a}{\sin(A)} = \frac{b}{\sin(B)} = \frac{c}{\sin(C)}$$

4.19 is called the **sine rule.** The sine rule can be used provided we are given either:

i two angles and one side or
ii two sides and one angle but the angle is **not** between the two given sides.

Fig. 36

Note that we can apply the sine rule for any triangle; it doesn't need to be a right-angled triangle.

Example 15

Figure 37 shows part of a crane. Find the length of the inciined jib BC.

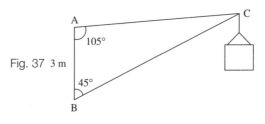

Fig. 37 3 m

Solution

Since we are given two angles and one side, we can use the sine rule 4.19 :

We need to find length a (Fig. 38). From 4.19 :

Fig. 38 $c = 3$ m

$$* \qquad \frac{a}{\sin(A)} = \frac{c}{\sin(C)}$$

We don't need to use the central part of 4.19 because length b is not required. **We know length $c = 3$, but what is the value of angle C?**

Since the sum of the angles of a triangle is 180° we have

$$\text{angle } C = 180° - (105° + 45°) = 30°$$

Substituting $A = 105°$, $C = 30°$ and $c = 3$ into $*$ gives

$$\frac{a}{\sin(105°)} = \frac{3}{\sin(30°)} = 6$$
$$a = 6 \times \sin(105°)$$
$$= 5.80 \text{ m}$$

Length BC = 5.8 m (1 d.p.).

D2 Ambiguous case of the sine rule

Example 16

Find angle B in the case shown in Fig. 39.

Fig. 39

Solution

Can we apply the sine rule?

Since we have two sides and an angle, we can use the sine rule. Labelling the sides and angles according to the sine rule we have $a = 5$, $b = 11$ and angle $A = 18°$. Using 4.19 :

$$\frac{a}{\sin(A)} = \frac{b}{\sin(B)}$$

Example 16 *continued*

with $a = 5$, $b = 11$ and A = 84° we have

$$\frac{5}{\sin(18°)} = \frac{11}{\sin(B)}$$

$$16.18 = \frac{11}{\sin(B)}$$

$$\sin(B) = \frac{11}{16.18} = 0.68$$

How do we find angle *B*?

Take the inverse sin, $\sin^{-1}$, of both sides:

$$B = \sin^{-1}(0.68) = 42.84° \text{[By calculator]}$$

By looking at Fig. 39 we can see angle *B* is much larger than 42.84°.

What mistake(s) have we made in our calculation?

There are no mistakes but this is **not** the **only** solution for angle *B*. We know from the previous section that the trigonometric equation

$$\sin(B) = 0.68$$

has many solutions. Figure 40 shows the sine graph.

Fig. 40

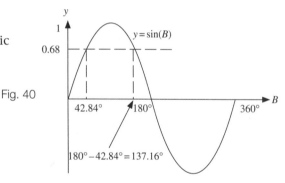

Examining Fig. 40, angle *B* can have two values:

$$B = 42.84° \text{ or } B = 180° - 42.84° = 137.16°$$

The triangle of Fig. 39 has angle $B = 137.16°$ (2 d.p.).

Remember that the sum of the angles in a triangle is 180°. Thus we only examine the sine graph between 0° and 180°.

In **Example 16** if you are **not** shown the triangle but just given the data, angle *A* is 18°, *a* is 5 and *b* is 11, then angle *B* could be either 42.84° or 137.16°. Thus we obtain two triangles (Fig. 41).

Note that angle *C* and length AB will depend on the size of angle *B*. We obtain two triangles because the general trigonometric equation

$$\sin(\theta) = \text{positive real number}$$

has two solutions between 0° and 180°.

This situation where two triangles are obtained is called the **'ambiguous case'** of the sine rule.

Fig. 41

D3 **Cosine rule**

With reference to Fig. 42, we also have

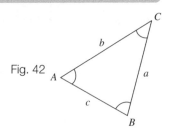

Fig. 42

| 4.20 | $a^2 = b^2 + c^2 - 2bc \cos(A)$ |

| 4.21 | $b^2 = a^2 + c^2 - 2ac \cos(B)$ |

| 4.22 | $c^2 = a^2 + b^2 - 2ab \cos(C)$ |

4.20 – 4.22 are called the **cosine rule.** The cosine rule can be used provided we are given either

i any three sides, or
ii two sides and an angle between the two given sides.

We can find any of the angles by transposing formulae 4.20 – 4.22 :

| 4.23 | $\cos(A) = \dfrac{b^2 + c^2 - a^2}{2bc}$ |

| 4.24 | $\cos(B) = \dfrac{a^2 + c^2 - b^2}{2ac}$ |

| 4.25 | $\cos(C) = \dfrac{a^2 + b^2 - c^2}{2ab}$ |

Again, the cosine rule can be used for any triangle.

Example 17

Figure 43 shows a crank mechanism. Crank arm OA, of length 150 mm, rotates clockwise about O. Connecting rod AB is of length 350 mm. If angle OAB = 75°, find length OB.

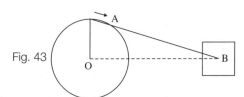

Fig. 43

Solution

We have the situation shown in Fig. 44.

Since we are given two sides and the angle between them, which rule do we employ?

Fig. 44

The cosine rule 4.20

If we let O be C then $c = 350$, $b = 150$ and angle $A = 75°$. We label side OB as a because OB is opposite angle A. We need to find length a. Substituting these values into $a^2 = b^2 + c^2 - 2bc \cos(A)$ gives

$$a^2 = 150^2 + 350^2 - [2 \times 150 \times 350 \times \cos(75°)]$$
$$= 117\,824$$

Taking the square root of both sides:

$$a = \sqrt{117\,824} = 343 \text{ mm} \quad (3 \text{ s.f.})$$

SUMMARY

If we define sides a, b and c to be the lengths opposite angles A, B and C respectively, then we have

| 4.19 | $\dfrac{a}{\sin(A)} = \dfrac{b}{\sin(B)} = \dfrac{c}{\sin(C)}$ | (sine rule) |

4.20	$a^2 = b^2 + c^2 - 2bc\,\cos(A)$	
4.21	$b^2 = a^2 + c^2 - 2ac\,\cos(B)$	(cosine rule)
4.22	$c^2 = a^2 + b^2 - 2ab\,\cos(C)$	

We can transpose 4.20 – 4.22 to find any of the angles.

Exercise 4(d)

Solutions at end of book. Complete solutions available at www.palgrave.com/science/engineering/singh

1 Figure 45 shows part of a car window with angle A = 110°, angle B = 40° and length AB = 105 mm. Find the lengths BC and AC.

Fig. 45

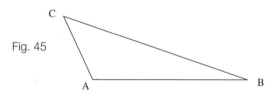

2 Figure 46 shows part of a hinge for a window. As the window is opened, part B moves down to the fixed point O and A moves to the right. When angle A = 22°, find length OB.

Fig. 46

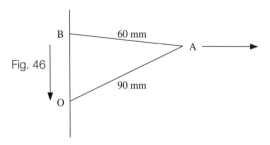

60 mm

90 mm

3 A crane consists of an inclined jib BC = 6.0 m, a vertical part AB = 2.0 m and a tie AC = 5.0 m. Find angle A.

4 Figure 47 shows a robot at C moving an object from A to B. Lengths AC = 1.40 m, BC = 1.90 m and angle A = 50.0°. Find the distance the robot moves the object.

Fig. 47

5 A roof (of a triangular shape) has a span (or width) of 10.0 m and the two sides make an angle of 40° and 45° at each end. Find the lengths of these sides.

6 Figure 48 shows a slider crank mechanism. Arm OA = 60 mm, connecting rod AB = 180 mm and length OB = 200 mm. Find the size of angle A.

Fig. 48

Exercise **4(d) continued**

Solutions at end of book. Complete solutions available at www.palgrave.com/science/engineering/singh

7 Let ABC be a triangle with AC = 1 m, AB = 2 m and angle C = 45°. Find the length BC.

8 A crank mechanism is shown in Fig. 49. It consists of an arm OA = 30 mm rotating anticlockwise about O and a connecting rod AB = 60 mm. B moves along the horizontal line OB. What is the distance OB when OA has rotated by $\frac{1}{8}$ of a revolution? (A complete revolution is 360°.)

Fig. 49

9 A surveyor wants to know the distance from location A to location B. She knows AC = 291 m, BC = 405 m and angle C = 79°, where C is another location. From these measurements, find the distance AB.

10 Figure 50 shows a symmetric framework about OC. Using the measurements shown, find the lengths CD and DE.

Fig. 50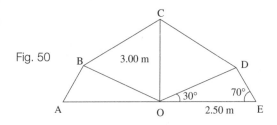

11 Find all the unknown sides and angles of a triangle ABC for the following:
   a AC = 10, BC = 6 and angle A = 30°
   b AB = 7, BC = 9 and angle C = 35°.

SECTION E **Radians**

By the end of this section you will be able to:
▶ convert angles from degrees to radians
▶ convert angles from radians to degrees

E1 **Radian measure**

Up until now we have been measuring angles in degrees, but the SI unit of angular measure is called the **radian** and is derived from the words radius angle, denoted by rad.

Let's recall some of the definitions connected with a circle.

The **radius**, r, is the distance from the centre of the circle to the edge.

Fig. 51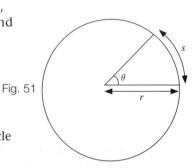

The **circumference** is the length of the perimeter of the circle and is equal to $2\pi r$.

The **arc** is part of the circle (or curve).

The angle θ of Fig. 51 in radians is defined as

$$\theta = \frac{\text{arc length}}{\text{radius}} = \frac{s}{r} \text{ radians}$$

If the line of the radius does one complete revolution, $\theta = 360°$, then

$$\theta = \frac{\text{circumference of circle}}{\text{radius}}$$

$$= \frac{2\pi r}{r} = 2\pi \text{ radians} \qquad [\text{Cancelling } r\text{'s}]$$

Thus we have

$$360° = 2\pi \text{ radians}$$

? **What is 180° in radians?**

For 180°, the radius line only completes half a revolution, so $180° = \pi$ radians.

? **What is 1° in radians?**

$$1° = \frac{\pi}{180} \text{ radians}$$

? **What is $\theta°$ in radians?**

$$\theta° = \theta \times \frac{\pi}{180} \text{ radians}$$

To convert from $\theta°$ to radians we use

$$\boxed{4.26} \qquad \theta° = \frac{\theta \times \pi}{180} \text{ radians}$$

Example 18

Express the following in radians (leaving π in your answer, wherever appropriate):

a 45° **b** 80° **c** 135° **d** 153.2°

Solution

Using $\theta° = \dfrac{\theta \times \pi}{180}$ gives

a $\quad 45° = \dfrac{45 \times \pi}{180} = \dfrac{\pi}{4} \text{ rad} \left(\text{because } \dfrac{45}{180} = \dfrac{1}{4}\right)$

b $\quad 80° = \dfrac{80 \times \pi}{180} = \dfrac{8\pi}{18} = \dfrac{4\pi}{9} \text{ rad}$

c $\quad 135° = \dfrac{135 \times \pi}{180} = \dfrac{3\pi}{4} \text{ rad} \left(\text{because } \dfrac{135}{180} = \dfrac{3}{4}\right)$

d $\quad 153.2° = \dfrac{153.2 \times \pi}{180} = 2.67 \text{ rad} \quad (2 \text{ d.p.})$

? **How do we convert θ radians to degrees?**

By transposing formula 4.26 :

4.27 $$\theta \text{ radians} = \left(\frac{\theta \times 180}{\pi}\right)^\circ$$

We use 4.27 to convert from radians to degrees.

Example 19

Express the following radians in degrees:

a $\dfrac{\pi}{2}$ **b** 2.5   **c** 1   **d** $\dfrac{3\pi}{2}$

Solution

Using $\theta = \left(\dfrac{\theta \times 180}{\pi}\right)^\circ$, by replacing θ with the appropriate value above, gives

a $\dfrac{\pi}{2} = \left(\dfrac{\pi \times 180}{2\pi}\right)^\circ = 90°$ [Cancelling π's]

b $2.5 = \left(\dfrac{2.5 \times 180}{\pi}\right)^\circ = 143.24°$ (2 d.p.)

c $1 = \left(\dfrac{1 \times 180}{\pi}\right)^\circ = 57.30°$ (2 d.p.)

? **What does 1 radian mean?**

This is the angle (57.3°) where the radius = arc length.

d $\dfrac{3\pi}{2} = \left(\dfrac{3\pi \times 180}{2\pi}\right)^\circ = 270°$ [Cancelling π's]

You may use your calculator to convert from degrees to radians or vice versa.

SUMMARY

To convert from degrees to radians we use

4.26 $$x° = \frac{x \times \pi}{180} \text{ radians}$$

To convert from radians to degrees we use

4.27 $$x \text{ radians} = \left(\frac{x \times 180}{\pi}\right)^\circ$$

Exercise **4(e)**

(You may use your calculator to convert.)

1 Express the following in radians:
  i 123°   **ii** 13°
  iii 131.67°   **iv** 333.3°

2 Convert the following into radians (leave π in your answers):
  i 90°   **ii** 30°   **iii** 330°   **iv** 22.5°
  v 10°   **vi** 27°   **vii** 3°   **viii** 144°

3 Express the following radians in degrees:
  i $\dfrac{\pi}{6}$ **ii** $\dfrac{3\pi}{10}$ **iii** 1.45   **iv** $\dfrac{\pi}{126}$

  v 200   **vi** $\dfrac{\pi}{9}$ **vii** $\dfrac{7\pi}{81}$ **viii** $\dfrac{13\pi}{120}$

4 Evaluate the following:

  i $\sin\left(\dfrac{\pi}{10}\right)$ **ii** $\cos\left(\dfrac{7\pi}{2}\right)$ **iii** $\dfrac{\cos(7\pi)}{2}$

5 Find the exact values of

  i $\dfrac{\pi}{2}\sin\left(\dfrac{\pi}{4}\right)$ **ii** $\dfrac{1}{\sqrt{3}}\tan\left(\dfrac{\pi}{3}\right)$

  iii $\sin\left(\dfrac{\pi}{4}\right)\cos\left(\dfrac{\pi}{4}\right)$

6 The following are car engine speeds in revolutions per minute. Convert these into radians per second.
  i 2500   **ii** 500   **iii** 6300

SECTION F **Wave theory**

By the end of this section you will be able to:
▶ find the amplitude, period, frequency and phase of a wave
▶ sketch graphs of waves
▶ determine the time displacement between two waves

F1 **Waves**

In many engineering applications, it is appropriate to write the angle θ in terms of time t seconds rather than degrees or radians. The graph of $y = \sin(t)$ is sometimes called the sine wave (Fig. 52).

Fig. 52

Notice that the graph of $y = \sin(t)$ looks like a wave and it does one complete cycle from 0 to 2π seconds.

Similarly the graph of $y = \cos(t)$ is sometimes called the cosine wave (Fig. 53).

Fig. 53

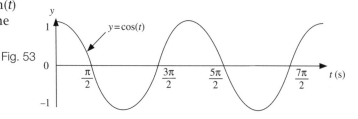

F2 Amplitude, period, frequency and phase of a wave

Consider a big wheel of radius R at a fairground rotating anticlockwise about O, as shown, with angular velocity, often called angular frequency, ω rad/s (Fig. 54).This means the wheel has swept out ω radians in 1 second.

After t seconds, OB has moved to OA and the angle swept out in this time is

Fig. 54

$$\omega t \ \text{radians}$$

Let x and y be the horizontal and vertical components of OA respectively (Fig. 55).

With reference to angle ωt, we have
x = adjacent, y = opposite and R = hypotenuse.

By 4.5 we have

Fig. 55

$$x = R\cos(\omega t)$$

and by 4.4 we have

$$y = R\sin(\omega t)$$

where x represents the horizontal distance of A from the centre O, and y represents the vertical distance of A from the centre line OC, the height of A from the centre line OC. Thus the horizontal distance, x, traces a cosine wave and the vertical distance, y, traces a sine wave. Consider the sine wave, $R \sin(\omega t)$.

The **period** of the big wheel is the time taken to complete one revolution and is denoted by T. This is also the time taken to complete one cycle of the sine wave or cosine wave. Sometimes the word *cycle* or *oscillation* is used for revolution.

If the big wheel does f revolutions in each second then f is called the **frequency** of the big wheel or frequency of the wave, $R \sin(\omega t)$:

Frequency f = number of revolutions occurring per second

Since the angular velocity of the big wheel is ω rad/s and there are 2π radians in each revolution, therefore

$$\text{frequency } f = \frac{\omega}{2\pi}$$

The SI unit of frequency is hertz, denoted by Hz.

4.4 opposite = hypotenuse $\times$ sin(θ)   4.5 adjacent = hypotenuse $\times$ cos(θ)

For the wave $R\sin(\omega t)$ we define the following:

| 4.28 | **amplitude** of $R\sin(\omega t)$ is R (it is the maximum displacement from rest position of the wave) |

| 4.29 | **period**, T, of $R\sin(\omega t)$ is $T = \dfrac{2\pi}{\omega}$ (time taken to complete 1 cycle) |

| 4.30 | **frequency**, f, of $R\sin(\omega t)$ is $f = \dfrac{\omega}{2\pi}$ or $f = \dfrac{1}{T}$ |

Multiplying both sides of $f = \dfrac{\omega}{2\pi}$ by 2π gives

| 4.31 | **angular velocity** $\omega = 2\pi f$ |

These can be shown for a general wave as in Fig. 56.

Fig. 56

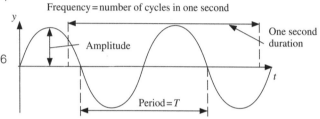

Frequency = number of cycles in one second

One second duration

Amplitude

Period = T

The above definitions also hold for $R\cos(\omega t)$.

Example 20

Find the exact values for the amplitude, frequency and period of

a $y = \dfrac{1}{5}\sin(3t)$ **b** $y = A\cos(100\pi t)$

Solution

a What is the amplitude of $y = \dfrac{1}{5}\sin(3t)$?

By 4.28 :   amplitude $= 1/5$

By 4.29 with $\omega = 3$:   period $T = \dfrac{2\pi}{\omega} = \dfrac{2\pi}{3}$ seconds

By 4.30 :   frequency $f = \dfrac{\omega}{2\pi} = \dfrac{3}{2\pi}$ Hz

b The amplitude of $y = A\cos(100\pi t)$ is A. Comparing $A\cos(100\pi t)$ with $R\cos(\omega t)$ gives $\omega = 100\pi$, and applying 4.29 and 4.30 we have

$$\text{period } T = \dfrac{2\pi}{100\pi} = \dfrac{1}{50} = 0.02\,\text{s or 20 milliseconds}$$

$$\text{frequency } f = \dfrac{100\pi}{2\pi} = 50\,\text{Hz} \qquad [\text{Cancelling } \pi\text{'s}]$$

Example 21

State the equation of the waves shown in Fig. 57.

Solution

a By looking at Fig. 57**a**, we know this is a sine wave. **What is the value of the amplitude?**

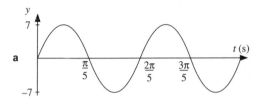

Fig. 57

Amplitude = 7

thus $y = 7\sin(\omega t)$. We need to find ω; the sine wave covers one cycle in $\dfrac{2\pi}{5}$ seconds.

Period, T, is the time taken to complete one cycle, hence

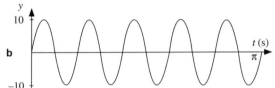

$$\text{period } T = \frac{2\pi}{\omega} = \frac{2\pi}{5}, \text{ giving } \omega = 5$$

Substituting $\omega = 5$ into $y = 7\sin(\omega t)$ gives the equation of the wave as $y = 7\sin(5t)$.

b Examining Fig. 57**b**, we have another sine wave. The amplitude of this graph is 10. The equation of the wave is of the form $y = 10\sin(\omega t)$, and we need to find the value of ω. By observing the graph, there are five cycles in π seconds so

$$T = \frac{\pi}{5} \qquad \text{[time taken to complete one cycle]}$$

By $T = 2\pi/\omega$ we have

$$\frac{2\pi}{\omega} = \frac{\pi}{5}, \text{ and transposing gives } \omega = 5 \times 2 = 10$$

Substituting $\omega = 10$ into $y = 10\sin(\omega t)$ gives the equation $y = 10\sin(10t)$.

F3 Phase and time displacement of waves

Example 22

Sketch on the same axes the graphs of:

$$y = 2\sin(t), \; y = 2\sin\left(t - \frac{\pi}{4}\right)$$

Example 22 *continued*

Solution

Let $f(t) = 2\sin(t)$, then $f\left(t - \dfrac{\pi}{4}\right) = 2\sin\left(t - \dfrac{\pi}{4}\right)$, replacing t with $t - \dfrac{\pi}{4}$. **How do we**

sketch the graph of $f\left(t - \dfrac{\pi}{4}\right) = 2\sin\left(t - \dfrac{\pi}{4}\right)$?

From Chapter 3 we know that $2\sin\left(t - \dfrac{\pi}{4}\right)$ is the graph of $2\sin(t)$ but shifted to the right by $\dfrac{\pi}{4}$ (Fig. 58).

Since $2\sin\left(t - \dfrac{\pi}{4}\right)$ reaches its peaks (and troughs) $\dfrac{\pi}{4}$ **after** $2\sin(t)$, we say $2\sin\left(t - \dfrac{\pi}{4}\right)$ **lags**

$2\sin(t)$ by a time of $\dfrac{\pi}{4}$.

Fig. 58

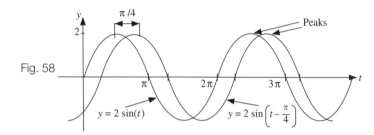

If you had difficulty in following **Example 22** it might help to revise the transformation of graphs section of Chapter 3. Remember that the graph of $f(t - c)$ is the graph of $f(t)$ but shifted to the right by c units.

What is the relationship between the graphs $f(t)$ and $f(t + c)$?

The graph of $f(t + c)$ is the graph of $f(t)$ but shifted to the left by c units.

We use this concept in the next example.

Example 23

Sketch on the same axes:

i $y = 5\sin(2t)$ **ii** $y = 5\sin\left(2t + \dfrac{\pi}{3}\right)$

Solution

We can rewrite **ii** $y = 5\sin\left(2t + \dfrac{\pi}{3}\right)$ as

$$y = 5\sin\left(2t + \dfrac{\pi}{3}\right)$$
$$= 5\sin\left(2t + \dfrac{2\pi}{6}\right) \quad \left[\text{Remember } \dfrac{1}{3} = \dfrac{2}{6}\right]$$
$$= 5\sin\left[2\left(t + \dfrac{\pi}{6}\right)\right] \quad \text{[Taking out a factor of 2]}$$

Example 23 *continued*

Let $f(t) = 5\sin(2t)$ then $f\left(t + \dfrac{\pi}{6}\right) = 5\sin\left[2\left(t + \dfrac{\pi}{6}\right)\right]$, replacing t with

$t + \dfrac{\pi}{6}$.

By the above statement, the graph of $f\left(t + \dfrac{\pi}{6}\right) = 5\sin\left[2\left(t + \dfrac{\pi}{6}\right)\right]$ is the graph of

$5\sin(2t)$ but shifted to the left

by $\dfrac{\pi}{6}$ (Fig. 59).

In this example

$5\sin\left(2t + \dfrac{\pi}{3}\right)$ reaches its

Fig. 59

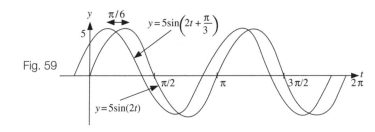

peak $\dfrac{\pi}{6}$ **before** $5\sin(2t)$, so

we say that

$$5\sin\left(2t + \dfrac{\pi}{3}\right) \text{ leads } 5\sin(2t) \text{ by a time of } \dfrac{\pi}{6}$$

By using a graphical calculator, or otherwise, you can find the following results:

a $\sin\left(5t + \dfrac{\pi}{6}\right)$ **leads** $\sin(5t)$ by $\dfrac{\pi/6}{5} = \dfrac{\pi}{5 \times 6} = \dfrac{\pi}{30}$

b $100\sin(7t + 2)$ **leads** $100\sin(7t)$ by $\dfrac{2}{7}$

This lead or lag time is called the **time displacement**.

? By examining the above examples, can you predict a general rule for the time
displacement of $R\sin(\omega t)$ and $R\sin(\omega t + \alpha)$?

4.32 $R\sin(\omega t + \alpha)$ **leads** $R\sin(\omega t)$ by $\dfrac{\alpha}{\omega}$

? What is the time displacement between $R\sin(\omega t)$ and $R\sin(\omega t - \alpha)$?

4.33 $R\sin(\omega t - \alpha)$ **lags** $R\sin(\omega t)$ by $\dfrac{\alpha}{\omega}$

4.34   The **angle** α in $R\sin(\omega t \pm \alpha)$ is generally known as the **phase angle** or just
the **phase**

These wave definitions also apply to the cosine wave and are commonly used in electrical
principles, as the following example shows.

Example 24 *electrical principles*

Find the amplitude, frequency and period of $i = 6.9\sin(100\pi t - 0.31)$ amps (t in s).
Also determine the time displacement between $6.9\sin(100\pi t - 0.31)$ and $6.9\sin(100\pi t)$.

Solution

Amplitude = 6.9 A. Comparing

$$6.9\sin(100\pi t - 0.31) \text{ with } R\sin(\omega t \pm \alpha)$$

gives $\omega = 100\pi$. Substituting this, $\omega = 100\pi$, into

4.30   frequency $f = \dfrac{\omega}{2\pi}$

gives

$$\text{frequency, } f = \frac{\omega}{2\pi} = \frac{100\pi}{2\pi} = 50 \text{ Hz}$$

$$\text{period} = \frac{1}{f} = \frac{1}{50} = 0.02 \text{ s}$$

By applying

4.33 $R\sin(\omega t - \alpha)$ lags $R\sin(\omega t)$ by $\dfrac{\alpha}{\omega}$

we find that $6.9\sin(100\pi t - 0.31)$ lags $6.9\sin(100\pi t)$ by:

$$\frac{0.31}{100\pi} = 9.87 \times 10^{-4} \text{ or } 0.987 \text{ ms (milliseconds)}$$

SUMMARY

If a wave has the equation $y = R\sin(\omega t \pm \alpha)$ then

R = amplitude

ω = angular velocity (rad/s)

4.30   frequency $f = \dfrac{\omega}{2\pi} = \dfrac{1}{T}$ (Hz)

α = phase angle of lag or lead compared with $R\sin(\omega t)$ in radians or degrees

$\dfrac{\alpha}{\omega}$ = time displacement

Exercise **4(f)**

Solutions at end of book. Complete solutions available at www.palgrave.com/science/engineering/singh

1 State the amplitude, period and frequency of the following waves:

a $10\sin(2t)$

b $\dfrac{1}{10}\cos\left(\dfrac{t}{10}\right)$

c $H\sin(\pi t)$

d $\dfrac{\cos(\pi^2 t)}{\pi}$

e $10\,000\sin\left(\dfrac{t}{\pi^2}\right)$

f $\dfrac{\cos(t/\pi e)}{\pi e}$

g $220\cos(1000\pi t)$

h $\sin\left(\dfrac{t}{\pi/6}\right)$

2 Determine the amplitude, period and phase (state lead or lag) of the following:

a $4\sin\left(t + \dfrac{\pi}{3}\right)$

b $\sin(314.159t - 0.25)$

c $20\sin\left[\left(100t + \dfrac{1}{5}\right)\pi\right]$

d $315\cos(7t + 30°)$

e $5 + 2\sin\left(2\pi t - \dfrac{\pi}{2}\right)$

3 Find the time displacement (stating lead or lag) of the following waves:

a $\sin(t)$

b $7\sin\left(t + \dfrac{\pi}{7}\right)$

c $3\cos\left(2t + \dfrac{\pi}{2}\right)$

d $\cos[(100t - 1)\pi]$

e $31\cos\left(t - \dfrac{\pi}{2}\right)$

f $\sin\left(\dfrac{t}{10} - \pi\right)$

g $\sin\left(\dfrac{t - \pi}{10}\right)$

h $\cos\left[\dfrac{1}{e\pi}\left(\dfrac{t + 1}{e\pi}\right)\right]$

4 What is the time displacement between the following waves?

a $A\sin\left(5t + \dfrac{\pi}{5}\right)$ and $A\sin\left(5t - \dfrac{\pi}{7}\right)$

b $\cos(7t - 0.26)$ and $5\cos(7t + \pi)$

c $\cos\left(2t + \dfrac{\pi}{2}\right)$ and $\cos\left(2t + \dfrac{\pi}{3}\right)$

5 State the equations of the waveforms shown in Fig. 60.

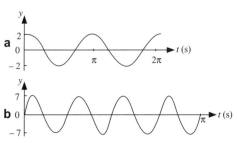

Fig. 60

6 ▨ [electrical principles]   Sketch the voltage waveform, $v = 5\cos(\omega t)$, as t varies over two complete cycles, 0 to $4\pi/\omega$, showing important points.

7 Sketch one cycle for the following graphs starting with $t = 0$. Show important points.

a $7\sin(3t)$

b $2\cos\left(\dfrac{t}{2}\right)$

c $3\sin\left(t + \dfrac{\pi}{3}\right)$

8 ▨ [electrical principles]   An alternating current $i = 5\sin(120\pi t + 0.52)$, ($i$ in mA, t in s). Find

a The amplitude, period, frequency and phase (in degrees) of i.

b The current i at: **i** $t = 0$ **ii** $t = \dfrac{1}{60}$ **iii** $t = \dfrac{1}{30}$ **iv** $t = \dfrac{1}{15}$

What do you notice about your results? Why is this?

c Find the time, t, when current, i, is first a maximum.

d Find the time displacement between $i = 5\sin(120\pi t + 0.52)$ and $i = 5\sin(120\pi t)$.

e Sketch one complete cycle of the waveform $i = 5\sin(120\pi t + 0.52)$.

SECTION G **Trigonometric identities**

By the end of this section you will be able to:
▶ establish some trigonometric identities
▶ use these identities to evaluate the sine, cosine and tangent of particular angles

In this section we state a lot of trigonometric identities because they are used in subsequent chapters. Some of these identities are shown in this section but you are asked to verify many of them in **Exercise 4(g)**.

G1 **Trigonometric identities**

In **Exercise 4(a)** we showed the following:

4.35 $$\tan(A) = \frac{\sin(A)}{\cos(A)}$$

This is an example of a trigonometric identity.

What do we mean by an identity?

It's two expressions which are equal for all values of the variables. Another example is

4.36 $$\cot(A) = \frac{\cos(A)}{\sin(A)}$$

Example 25

Show the identity $\cot(A) = \dfrac{\cos(A)}{\sin(A)}$.

Solution

From 4.12 we have $\cot(A) = \dfrac{1}{\tan(A)}$, thus

$$\cot(A) = \frac{1}{\tan(A)} = \underbrace{\frac{1}{\sin(A)/\cos(A)}}_{\text{by } 4.35}$$

$$= 1 \div \frac{\sin(A)}{\cos(A)}$$

$$= 1 \times \frac{\cos(A)}{\sin(A)} \quad \text{[Inverting the second fraction]}$$

$$= \frac{\cos(A)}{\sin(A)}$$

The following are more examples of trigonometric identities:

| 4.37 | $\sin(A + B) = \sin(A)\cos(B) + \cos(A)\sin(B)$ |

| 4.38 | $\sin(A - B) = \sin(A)\cos(B) - \cos(A)\sin(B)$ |

| 4.39 | $\cos(A + B) = \cos(A)\cos(B) - \sin(A)\sin(B)$ |

| 4.40 | $\cos(A - B) = \cos(A)\cos(B) + \sin(A)\sin(B)$ |

| 4.41 | $\tan(A + B) = \dfrac{\tan(A) + \tan(B)}{1 - \tan(A)\tan(B)}$ |

| 4.42 | $\tan(A - B) = \dfrac{\tan(A) - \tan(B)}{1 + \tan(A)\tan(B)}$ |

We can use these identities to establish other identities or to evaluate sin, cos and tan of an angle, as the following example shows.

Example 26

By using the above identities and Table 1 on page 171, find the exact value of $\sin(15°)$.

Solution

By examining Table 1 we only have the sin of angles $0°$, $30°$, $45°$, etc. So we need to write $15°$ in terms of these angles.

?   **How can we rewrite 15°?**

$$15° = 45° - 30°$$

Substituting this we have

$$\sin(15°) = \sin(45° - 30°) \underset{\text{using } 4.38}{=} [\sin(45°)\cos(30°)] - [\cos(45°)\sin(30°)]$$

$$\underset{\text{by Table 1}}{=} \left(\frac{1}{\sqrt{2}} \cdot \frac{\sqrt{3}}{2} \right) - \left(\frac{1}{\sqrt{2}} \cdot \frac{1}{2} \right)$$

$$\sin(15°) = \frac{1}{2\sqrt{2}} \left(\sqrt{3} - 1 \right) \qquad \left[\text{Factorizing } \frac{1}{2\sqrt{2}} \right]$$

Example 27

Show that $\cos(90° - A) = \sin(A)$.

Solution

?   We can expand $\cos(90° - A)$. **How?**

Use

4.38 $\sin(A - B) = \sin(A)\cos(B) - \cos(A)\sin(B)$

Example 27 *continued*

4.40 $\cos(A - B) = \cos(A)\cos(B) + \sin(A)\sin(B)$

$\cos(90° - A) = \left[\cos(90°)\cos(A)\right] + \left[\sin(90°)\sin(A)\right]$
$= \left[0 \times \cos(A)\right] + \left[1 \times \sin(A)\right]$
$= \sin(A)$

Similarly we can establish the following identities. Try verifying them.

4.43 $\sin(A) = \cos(90° - A)$

4.44 $\cos(A) = \sin(90° - A)$

4.45 $\sin(A) = \sin(180° - A)$

4.46 $\cos(A) = -\cos(180° - A)$

4.47 $\tan(A) = -\tan(180° - A)$

4.48 $\sin(A) = -\sin(360° - A)$

4.49 $\cos(A) = \cos(360° - A)$

4.50 $\sin(-A) = -\sin(A)$

4.51 $\cos(-A) = \cos(A)$

4.52 $\tan(-A) = -\tan(A)$

Example 28

Verify the trigonometric identity:

$\sin(2A) = 2\sin(A)\cos(A)$

Solution

We can rewrite $2A$ as $A + A$.

$\sin(2A) = \sin(A + A)$
$= \sin(A)\cos(A) + \cos(A)\sin(A)$
using 4.37
$= 2\sin(A)\cos(A)$

Example 28 is an example of the **double-angle formula**. Thus the following are examples of the double-angle formulae:

4.53 $\sin(2A) = 2\sin(A)\cos(A)$

4.54 $\cos(2A) = \cos^2(A) - \sin^2(A) = 2\cos^2(A) - 1 = 1 - 2\sin^2(A)$

4.55 $\tan(2A) = \dfrac{2\tan(A)}{1 - \tan^2 (A)}$

You are asked to verify the last two, 4.54 and 4.55 , in **Exercise 4(g)**.

4.37 $\sin(A + B) = \sin(A)\cos(B) + \cos(A)\sin(B)$

Example 29

Show that

$$2\sin(A)\cos(B) = \sin(A + B) + \sin(A - B)$$

Solution

Expanding the Right-Hand Side by applying 4.37 and 4.38 gives

$$\sin(A + B) + \sin(A - B) = \underbrace{\sin(A)\cos(B) + \cos(A)\sin(B)}_{\text{using } 4.37} + \underbrace{\sin(A)\cos(B) - \cos(A)\sin(B)}_{\text{using } 4.38}$$

$$= \sin(A)\cos(B) + \sin(A)\cos(B) = 2\sin(A)\cos(B)$$

Similarly we can establish the following:

4.56 $\qquad 2\sin(A)\cos(B) = \sin(A + B) + \sin(A - B)$

4.57 $\qquad 2\cos(A)\sin(B) = \sin(A + B) - \sin(A - B)$

4.58 $\qquad 2\cos(A)\cos(B) = \cos(A + B) + \cos(A - B)$

4.59 $\qquad 2\sin(A)\sin(B) = \cos(A - B) - \cos(A + B)$

4.60 $\qquad \sin(A) + \sin(B) = 2\sin\left(\dfrac{A + B}{2}\right)\cos\left(\dfrac{A - B}{2}\right)$

4.61 $\qquad \sin(A) - \sin(B) = 2\cos\left(\dfrac{A + B}{2}\right)\sin\left(\dfrac{A - B}{2}\right)$

4.62 $\qquad \cos(A) + \cos(B) = 2\cos\left(\dfrac{A + B}{2}\right)\cos\left(\dfrac{A - B}{2}\right)$

4.63 $\qquad \cos(A) - \cos(B) = 2\sin\left(\dfrac{A + B}{2}\right)\sin\left(\dfrac{B - A}{2}\right)$

Remember

$$\sin(A)\sin(A) = [\sin(A)]^2 = \sin^2(A)$$

Other trigonometric functions can be represented in a similar manner.

Example 30

Show that

$\qquad$ * $\qquad \cos^2(A) + \sin^2(A) = 1$

Solution

This is a **very** important identity which is used throughout the book. **How can we verify this identity?**

4.37 $\sin(A + B) = \sin(A)\cos(B) + \cos(A)\sin(B)$ $\qquad$ 4.38 $\sin(A - B) = \sin(A)\cos(B) - \cos(A)\sin(B)$

Example 30 *continued*

Remember that cos(0) = 1, so we rewrite the 1 in [*] as cos(0). **But how does that serve us?**

Well, cos(0) = cos(A − A), so by using

4.40 $\cos(A - B) = \cos(A)\cos(B) + \sin(A)\sin(B)$

we have

$$1 = \cos(A - A)$$
$$= \cos(A)\cos(A) + \sin(A)\sin(A)$$
$$= [\cos(A)]^2 + [\sin(A)]^2$$
$$1 = \cos^2(A) + \sin^2(A)$$

This identity can also be proved directly from Pythagoras's theorem. Consider a right-angled triangle and try proving it.

Example 31

Show that

$$1 + \tan^2(A) = \sec^2(A)$$

Solution

Dividing both sides of

$$\cos^2(A) + \sin^2(A) = 1$$

by $\cos^2(A)$ gives

$$\frac{\cos^2(A)}{\cos^2(A)} + \frac{\sin^2(A)}{\cos^2(A)} = \frac{1}{\cos^2(A)}$$

$$1 + \frac{\sin^2(A)}{\cos^2(A)} = \frac{1}{\cos^2(A)}$$

We can rewrite tan as $\dfrac{\sin}{\cos}$ and so $\tan^2(A) = \dfrac{\sin^2(A)}{\cos^2(A)}$. Also $\dfrac{1}{\cos(A)} = \sec(A)$

which implies $\dfrac{1}{\cos^2(A)} = \sec^2(A)$. Using these, we have

$$1 + \frac{\sin^2(A)}{\cos^2(A)} = \frac{1}{\cos^2(A)}$$
$$1 + \tan^2(A) = \sec^2(A)$$

We can establish the following trigonometric identities:

| 4.64 | $\cos^2(A) + \sin^2(A) = 1$ |

| 4.65 | $1 + \tan^2(A) = \sec^2(A)$ |

| 4.66 | $1 + \cot^2(A) = \operatorname{cosec}^2(A)$ |

| 4.67 | $\cos^2(A) = \dfrac{1}{2}\left[1 + \cos(2A)\right]$ |

| 4.68 | $\sin^2(A) = \dfrac{1}{2}\left[1 - \cos(2A)\right]$ |

The next three identities are examples of **multiple-angle formulae**:

| 4.69 | $\sin(3A) = 3\sin(A) - 4\sin^3(A)$ |

| 4.70 | $\cos(3A) = 4\cos^3(A) - 3\cos(A)$ |

| 4.71 | $\tan(3A) = \dfrac{3\tan(A) - \tan^3(A)}{1 - 3\tan^2(A)}$ |

Below, 4.72 to 4.74 are examples of **half-angle formulae**.

If $t = \tan(A/2)$ then

| 4.72 | $\sin(A) = \dfrac{2t}{1 + t^2}$ |

| 4.73 | $\cos(A) = \dfrac{1 - t^2}{1 + t^2}$ |

| 4.74 | $\tan(A) = \dfrac{2t}{1 - t^2}$ |

Many of these results will be shown in **Exercise 4(g)**.

You don't have to remember all these trigonometric identities, but some of these are more important than others and worth remembering. For example

$$\cos^2(A) + \sin^2(A) = 1$$

$$\tan(A) = \frac{\sin(A)}{\cos(A)}$$

$$\sin(2A) = 2\sin(A)\cos(A)$$

$$\cos(2A) = \cos^2(A) - \sin^2(A)$$

Sometimes these are referred to as **fundamental trigonometric identities**.

SUMMARY

We have established some trigonometric identities. Remember that an identity is an expression which is equal for all values of the variables.

Solutions at end of book. Complete solutions available at www.palgrave.com/science/engineering/singh

Exercise **4(g)**

1 By using the identities 4.37 – 4.42 and the values in Table 1, find the exact values of

  i $\sin(75°)$ **ii** $\sin(120°)$ **iii** $\cos(105°)$

  iv $\cos(150°)$ **v** $\tan(15°)$

2 Show that
  i $\sin(90° - A) = \cos(A)$
  ii $\sin(-A) = -\sin(A)$
  iii $\cos(-A) = \cos(A)$
  iv $\tan(-A) = -\tan(A)$

3 Show that
  i $\cos(2\theta) = \cos^2(\theta) - \sin^2(\theta)$

  ii $\tan(2\theta) = \dfrac{2\tan(\theta)}{1 - \tan^2(\theta)}$

4 Show that $1 + \cot^2(\theta) = \operatorname{cosec}^2(\theta)$.

5 Show that

  i $\dfrac{1}{2}\left[1 + \cos(2\theta)\right] = \cos^2(\theta)$

  ii $\dfrac{1}{2}\left[1 - \cos(2\theta)\right] = \sin^2(\theta)$

  [Can also by written as
$\cos(2\theta) = 2\cos^2(\theta) - 1 = 1 - 2\sin^2(\theta)$.]

6 Show that
  i $\sin(3\theta) = 3\sin(\theta) - 4\sin^3(\theta)$

  ii $\cos(3\theta) = 4\cos^3(\theta) - 3\cos(\theta)$

  iii $\tan(3\theta) = \dfrac{3\tan(\theta) - \tan^3(\theta)}{1 - 3\tan^2(\theta)}$

7 Show that
  i $\sin(A + B) - \sin(A - B)$
    $= 2\cos(A)\sin(B)$
  ii $\cos(A + B) + \cos(A - B)$
    $= 2\cos(A)\cos(B)$
  iii $\cos(A - B) - \cos(A + B)$
    $= 2\sin(A)\sin(B)$

8 Show that

  i $2\sin\left(\dfrac{A + B}{2}\right)\cos\left(\dfrac{A - B}{2}\right)$
    $= \sin(A) + \sin(B)$

  ii $2\cos\left(\dfrac{A + B}{2}\right)\sin\left(\dfrac{A - B}{2}\right)$
    $= \sin(A) - \sin(B)$

  iii $2\cos\left(\dfrac{A + B}{2}\right)\cos\left(\dfrac{A - B}{2}\right)$
    $= \cos(A) + \cos(B)$

  iv $2\sin\left(\dfrac{A + B}{2}\right)\sin\left(\dfrac{B - A}{2}\right)$
    $= \cos(A) - \cos(B)$

SECTION H **Applications of identities**

By the end of this section you will be able to:
▶ use identities of **Section G** to show various engineering results

This section might seem like a colossal jump from previous sections, but most of it involves applying algebra to trigonometric formulae.

H1 **Engineering applications of identities**

We can simplify or rewrite engineering trigonometric formulae by using the identities of the previous section, as the following two examples show.

Example 32 *electrical principles*

The power, p, in a circuit is given by

$$p = iv$$

where i and v are the a.c. current and voltage respectively. Show that, if $i = I\sin(\omega t - \phi)$ and $v = V\sin(\omega t)$ then

$$p = \frac{IV}{2}\left[\cos(\phi) - \cos(2\omega t - \phi)\right]$$

(I = current amplitude, V = voltage amplitude, ω = angular frequency and ϕ = phase). Don't let the Greek letters put you off.

Solution

Substituting $i = I\sin(\omega t - \phi)$ and $v = V\sin(\omega t)$ into $p = iv$ gives

†

$$p = IV\sin(\omega t - \phi)\sin(\omega t)$$

By putting $A = \omega t - \phi$ and $B = \omega t$ into

4.59

$$2\sin(A)\sin(B) = \cos(A - B) - \cos(A + B)$$

$$2\sin(\omega t - \phi)\sin(\omega t) = \cos(\omega t - \phi - \omega t) - \cos(\omega t - \phi + \omega t)$$

$$= \underbrace{\cos(-\phi)}_{=\cos(\phi)\ \text{by}\ 4.51} - \cos(2\omega t - \phi)$$

$$= \cos(\phi) - \cos(2\omega t - \phi)$$

Dividing both sides by 2 gives

$$\sin(\omega t - \phi)\sin(\omega t) = \frac{1}{2}\left[\cos(\phi) - \cos(2\omega t - \phi)\right]$$

Substituting this into the Right-Hand Side of † gives the required result:

$$p = \frac{IV}{2}\left[\cos(\phi) - \cos(2\omega t - \phi)\right]$$

Example 33 *mechanics*

A projectile is fired with a velocity u from an inclined angle of β. The projectile makes an angle of α with the horizontal.

i If $\left(u\sin(\alpha) - \dfrac{1}{2}gt\right)\cos(\beta) = u\sin(\beta)\cos(\alpha)$, show that

$$t = \frac{2u\sin(\alpha - \beta)\sec(\beta)}{g}$$

where g is acceleration due to gravity and t is time of flight.

4.51 $\cos(-A) = \cos(A)$

🚗 Example 33 *continued*

ii If $gR = u^2\sec^2(\beta)\left[1 - \sin(\beta)\right]$, show that $R = \dfrac{u^2}{g\left[1 + \sin(\beta)\right]}$ where R is the range of the projectile.

Solution

i Expanding the Left-Hand Side gives

$$u\sin(\alpha)\cos(\beta) - \frac{1}{2}gt\cos(\beta) = u\sin(\beta)\cos(\alpha)$$

$$u\sin(\alpha)\cos(\beta) - u\sin(\beta)\cos(\alpha) = \frac{1}{2}gt\cos(\beta) \quad \text{[Transposing]}$$

$$u\left[\sin(\alpha)\cos(\beta) - \cos(\alpha)\sin(\beta)\right] = \frac{1}{2}gt\cos(\beta) \quad \text{[Factorizing]}$$

$$u\underbrace{\sin(\alpha - \beta)}_{\text{by } 4.38} = \frac{1}{2}gt\cos(\beta)$$

$$\frac{2u\sin(\alpha - \beta)}{g\cos(\beta)} = t \qquad \text{[Transposing]}$$

Thus

$$t = \frac{2u\sin(\alpha - \beta)}{g\cos(\beta)} = \frac{2u\sin(\alpha - \beta)}{g}\cdot\underbrace{\left(\frac{1}{\cos(\beta)}\right)}_{=\sec(\beta)}$$

$$t = \frac{2u\sin(\alpha - \beta)}{g}\sec(\beta) = \frac{2u\sin(\alpha - \beta)\sec(\beta)}{g}$$

ii Examining the Right-Hand Side of $gR = u^2\sec^2(\beta)\left[1 - \sin(\beta)\right]$:

$$u^2\sec^2(\beta)\left[1 - \sin(\beta)\right] = \frac{u^2\left[1 - \sin(\beta)\right]}{\cos^2(\beta)} \quad \left[\text{Because } \sec^2(\beta) = \frac{1}{\cos^2(\beta)}\right]$$

$$= \frac{u^2\left[1 - \sin(\beta)\right]}{1 - \sin^2(\beta)} \qquad \text{[by } 4.64 \text{ on denominator]}$$

$$= \frac{u^2\left[1 - \sin(\beta)\right]}{\underbrace{\left[1 - \sin(\beta)\right]\left[1 + \sin(\beta)\right]}_{\text{by } 1.15}} \qquad \text{[Factorizing denominator]}$$

$$u^2\sec^2(\beta)\left[1 - \sin(\beta)\right] = \frac{u^2}{1 + \sin(\beta)} \qquad \left[\text{Cancelling } 1 - \sin(\beta)\right]$$

1.15 $a^2 - b^2 = (a - b)(a + b)$ 4.38 $\sin(A)\cos(B) - \cos(A)\sin(B) = \sin(A - B)$

4.64 $\sin^2(A) + \cos^2(A) = 1$

🚗 Example 33 *continued*

Substituting this into the original equation gives

$$gR = \frac{u^2}{\left[1 + \sin(\beta)\right]}$$

$$R = \frac{u^2}{g\left[1 + \sin(\beta)\right]} \qquad \text{[Dividing through by } g]$$

SUMMARY

We can use the identities of **Section G** to simplify or rewrite engineering trigonometric formulae.

Exercise **4(h)**

Solutions at end of book. Complete solutions available at www.palgrave.com/science/engineering/singh

1 🚗 [*mechanics*]   A vehicle is tested on a circular track with a bank at an angle θ. By doing the dynamic analysis we obtain the equations

$$N\cos(\theta) - mg = 0,$$

$$N\sin(\theta) - \frac{mv^2}{r} = 0$$

where N = normal reaction of road surface, m = mass of vehicle, r = track radius of curvature and v = neutral steer speed (the speed at which the car goes round the track without steering). Show that

$$v = \sqrt{rg\,\tan(\theta)}$$

2 📱 [*communications*]   An amplitude modulated (a.m.) signal, v, is given by

$$v = V_c\cos(\omega_c t)$$

$$+ V_m\cos(\omega_m t)\cos(\omega_c t)$$

where V_c = carrier amplitude, V_m = modulating signal amplitude, ω_c = angular frequency of carrier and ω_m = angular frequency of signal. Show that

$$v = V_c\left\{\cos(\omega_c t) + \frac{d}{2}\left(\cos\left[(\omega_m + \omega_c)t\right]\right.\right.$$

$$\left.\left. + \cos\left[(\omega_m - \omega_c)t\right]\right)\right\}$$

where $d = \dfrac{V_m}{V_c}$ (sometimes called the depth of modulation if given as a percentage).

3 📱 [*communications*]   An amplitude modulated signal, v, is given by

$$v = \left[V_c + V_m\sin(\omega_m t)\right]\sin(\omega_c t)$$

where V_c, V_m, ω_m, ω_c and d are defined as in the above question. Show that

$$v = \frac{V_c}{2}\left(2\sin(\omega_c t) + d\cos\left[(\omega_c - \omega_m)t\right]\right.$$

$$\left. - d\cos\left[(\omega_c + \omega_m)t\right]\right)$$

4 🔌 [*electrical principles*]   The power, p, in a circuit is given by $p = iv$ where i = current and v = voltage. Show the following:

a If $i = I\sin(\omega t)$ and $v = V\sin(\omega t)$ then

$$p = \frac{IV}{2}\left[1 - \cos(2\omega t)\right]$$

Solutions at end of book. Complete solutions available at www.palgrave.com/science/engineering/singh

Exercise **4(h) continued**

b If $i = I\sin(\omega t)$ and $v = V\cos(\omega t)$ then
$$p = \frac{IV}{2}\sin(2\omega t)$$

c If $i = I\sin(\omega t)$ and $v = V\sin\left(\omega t - \frac{\pi}{2}\right)$
then $p = -\frac{IV}{2}\sin(2\omega t)$

d If $i = I\cos(\omega t)$ and $v = V\cos\left(\omega t + \frac{\pi}{2}\right)$
then $p = \frac{IV}{2}\cos\left(2\omega t + \frac{\pi}{2}\right)$

e If $i = I\sin\left(\omega t - \frac{\pi}{2}\right)$ and $v = V\sin(\omega t)$
then $p = -\frac{IV}{2}\cos\left(2\omega t - \frac{\pi}{2}\right)$

f If $i = 12\sin\left(100\pi t - \frac{\pi}{3}\right)$ and
$v = 240\sin(100\pi t)$ then
$$p = 720 - 1440\cos\left(200\pi t - \frac{\pi}{3}\right)$$

SECTION I **Conversion**

By the end of this section you will be able to:
▶ rewrite the function $a\cos(\theta) + b\sin(\theta)$ as $R\cos(\theta - \beta)$
▶ sketch the graph of $a\cos(\theta) + b\sin(\theta)$
▶ solve engineering applications with $a\cos(\theta) + b\sin(\theta)$

Another challenging section. This section relies on **one** particular formula, 4.75 , to be described below. The most difficult part is verifying this formula which is carried out in **Example 35**. As an engineering student, you should at least be able to apply this formula.

I1 **The function $a\cos(\theta) + b\sin(\theta)$**

We can write $a\cos(\theta) + b\sin(\theta)$ just in terms of the cosine function, as the following examples show.

Example 34

Show that
$$2\cos(\theta) + 3\sin(\theta) = \sqrt{13}\cos(\theta - 56.31°)$$

Solution

Expanding the Right-Hand Side by using:

Example 34 *continued*

4.40 $\cos(A - B) = \cos(A)\cos(B) + \sin(A)\sin(B)$

we have

$$\sqrt{13}\cos(\theta - 56.31°) = \sqrt{13}\left[\cos(\theta)\cos(56.31°) + \sin(\theta)\sin(56.31°)\right]$$

$$= \underbrace{\sqrt{13}\cos(56.31°)}_{=2}\cos(\theta) + \underbrace{\sqrt{13}\sin(56.31°)}_{=3}\sin(\theta)$$

$$= 2\cos(\theta) + 3\sin(\theta)$$

In this section we are **not** given $\sqrt{13}\cos(\theta - 56.31°)$ but we need to find it from $2\cos(\theta) + 3\sin(\theta)$. Let's show the general case and then try some examples.

Example 35

Show that

★ $a\cos(\theta) + b\sin(\theta) = R\cos(\theta - \beta)$ where $R = \sqrt{a^2 + b^2}$ and $\beta = \tan^{-1}\left(\dfrac{b}{a}\right)$

Solution

We can expand the Right-Hand Side of ★ by using

4.40 $\cos(A - B) = \cos(A)\cos(B) + \sin(A)\sin(B)$

$Rcos(\theta - \beta) = R\left[\cos(\theta)\cos(\beta) + \sin(\theta)\sin(\beta)\right]$

$= R\cos(\beta)\cos(\theta) + R\sin(\beta)\sin(\theta)$

So we have

$a\cos(\theta) + b\sin(\theta) = \left[R\cos(\beta)\right]\cos(\theta) + \left[R\sin(\beta)\right]\sin(\theta)$

Equating coefficients of $\cos(\theta)$ gives

$a = R\cos(\beta)$

and equating coefficients of $\sin(\theta)$ gives

$b = R\sin(\beta)$

? **How can we show $R = \sqrt{a^2 + b^2}$?**

We can find $a^2 + b^2$:

$$a^2 + b^2 = \left[R\cos(\beta)\right]^2 + \left[R\sin(\beta)\right]^2$$

$$= R^2\cos^2(\beta) + R^2\sin^2(\beta)$$

$$\underset{\substack{\text{factorizing}\\ R^2}}{=} R^2\underbrace{\left[\cos^2(\beta) + \sin^2(\beta)\right]}_{=1 \text{ by } 4.64} = R^2$$

4.64 $\cos^2(A) + \sin^2(A) = 1$

Example 35 *continued*

Taking square root gives $R = \sqrt{a^2 + b^2}$. Next we show $\beta = \tan^{-1}\left(\dfrac{b}{a}\right)$.

$$\frac{b}{a} = \frac{R\sin(\beta)}{R\cos(\beta)} = \frac{\sin(\beta)}{\cos(\beta)} \underset{\text{by } \boxed{4.35}}{=} \tan(\beta)$$

Taking the inverse tan gives

$$\beta = \tan^{-1}\left(\frac{b}{a}\right)$$

Fig. 61

where β is the angle shown in Fig. 61.

The result of **Example 35** is useful in many engineering applications:

 $\quad a\cos(\theta) + b\sin(\theta) = R\cos(\theta - \beta)$ where $R = \sqrt{a^2 + b^2}$ and $\beta = \tan^{-1}\left(\dfrac{b}{a}\right)$

It is used in engineering fields because it converts $a\cos(\theta) + b\sin(\theta)$ into **amplitude–phase form**, $R\cos(\theta - \beta)$, where R is the amplitude and β is the phase.

We need to be careful in choosing β. The angle β can be found in two different ways:

1 Examining the signs of a and b and then plotting it in the correct quadrant (Fig. 62).

Fig. 62

2 This second method does not need a plot but relies on memory.

If $\cos(\beta)$ has the **same sign** as a or $\sin(\beta)$ has the **same sign** as b, then β is the angle found via a calculator. If the signs are different, **add** 180° to the angle found.

Example 36

Write the following in the form $R\cos(\theta \pm \beta)$:

i $2\cos(\theta) + 3\sin(\theta)$ **ii** $-5\cos(\theta) + 8\sin(\theta)$ **iii** $-\cos(\theta) - \sin(\theta)$

iv $-\sqrt{3}\sin(\theta) + \cos(\theta)$

$\boxed{4.35}$ $\sin(A)/\cos(A) = \tan(A)$

Example 36 *continued*

Solution

We use $\boxed{4.75}$ with a as the coefficient of $\cos(\theta)$ and b as the coefficient of $\sin(\theta)$:

i Substituting $a = 2$ and $b = 3$ into $\boxed{4.75}$ gives:

$$R = \sqrt{2^2 + 3^2} = \sqrt{13}$$

$$\beta = \tan^{-1}\left(\frac{3}{2}\right) = 56.31° \quad \text{[By calculator]}$$

Thus $2\cos(\theta) + 3\sin(\theta) = \sqrt{13}\cos(\theta - 56.31°)$. This is the result in **Example 34**.

ii Putting $a = -5$ and $b = 8$ into $\boxed{4.75}$:

$$R = \sqrt{(-5)^2 + 8^2} = \sqrt{89}$$

$$\text{angle} = \tan^{-1}\left(\frac{8}{-5}\right) = -57.99° \quad \text{[By calculator]}$$

$\cos(-57.99°)$ is positive but $a = -5$, so we add $180°$ to $-57.99°$:

$$\beta = -57.99° + 180° = 122.01°$$

Fig. 63

We can also check this angle by plotting (Fig. 63).

Using $R = \sqrt{89}$ and $\beta = 122.01°$ we have

$$-5\cos(\theta) + 8\sin(\theta) = \sqrt{89}\cos(\theta - 122.01°)$$

iii We have $a = -1$ and $b = -1$. Applying $\boxed{4.75}$:

$$R = \sqrt{(-1)^2 + (-1)^2} = \sqrt{2}$$

$$\text{angle} = \tan^{-1}\left(\frac{-1}{-1}\right) = 45°$$

$\cos(45°)$ is positive but $a = -1$. So we add $180°$:

$$\beta = 45° + 180° = 225°$$

Using $\boxed{4.75}$:

Fig. 64

$$-\cos(\theta) - \sin(\theta) = \sqrt{2}\cos(\theta - 225°)$$

Note that $225° = -135°$ (Fig. 64).

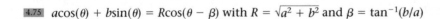

$\boxed{4.75}$ $a\cos(\theta) + b\sin(\theta) = R\cos(\theta - \beta)$ with $R = \sqrt{a^2 + b^2}$ and $\beta = \tan^{-1}(b/a)$

Example 36 *continued*

Thus we can also write $-\cos(\theta) - \sin(\theta)$ as

$$-\cos(\theta) - \sin(\theta) = \sqrt{2}\cos(\theta+135°)$$

iv In this case $b = -\sqrt{3}$ and $a = 1$. (Remember that a and b are the cosine and sine coefficients respectively.)

$$R = \sqrt{1^2 + (-\sqrt{3})^2} = 2$$

$$\beta = \tan^{-1}\left(-\sqrt{3}\right) = -60°$$

Fig. 65

The plot of $-\sqrt{3}\sin(\theta) + \cos(\theta)$ is shown in Fig. 65.

By 4.75 we have

$$-\sqrt{3}\sin(\theta) + \cos(\theta) = 2\cos\left[\theta - (-60°)\right] = 2\cos(\theta + 60°)$$

12 **Engineering applications**

Example 37 *electrical principles*

An a.c. current, i, is given by

$$i = 3\cos(\omega t) + 4\sin(\omega t)$$

Find the amplitude, and sketch one complete cycle for $\omega = 100\pi$ rad/s.

Solution

Using 4.75 with $a = 3$ and $b = 4$, we have

$$R = \sqrt{3^2 + 4^2} = 5$$

$$\beta = \tan^{-1}\left(\frac{4}{3}\right) = 0.9273$$

(Better to use radians since ω is in radians.) Hence

$$3\cos(\omega t) + 4\sin(\omega t) = 5\cos(\omega t - 0.9273)$$

What is the amplitude of the waveform?

Amplitude $= 5$ A

4.75 $a\cos(\theta) + b\sin(\theta) = R\cos(\theta - \beta)$ with $R = \sqrt{a^2 + b^2}$ and $\beta = \tan^{-1}(b/a)$

Example 37 *continued*

For sketching, we need to find period T and the time displacement from $5\cos(\omega t)$. **How do we find the period?**

Use 4.29 with $\omega = 100\pi$:

$$\text{Period } T = \frac{2\pi}{\omega} = \frac{2\pi}{100\pi} = 20 \times 10^{-3} = 20 \text{ ms}$$

The time displacement with $\omega = 100\pi$ and the phase $\alpha = 0.9273$ is found by using

4.33   time displacement $= \alpha/\omega$

$$\text{time displacement} = \frac{0.9273}{100\pi} = 2.95 \text{ ms}$$

So $5\cos(100\pi t - 0.9273)$ lags $5\cos(100\pi t)$ by 2.95 ms (Fig. 66).

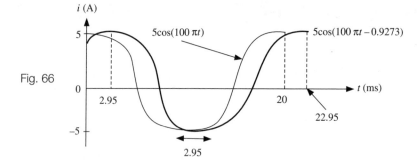

Fig. 66

Example 38 *vibrations*

Consider a spring of length L which is fixed at one end and has a mass m attached at the other end. If the spring is extended by a length x from its natural position then the motion of the mass can be described by

$$x = A\cos\left(\sqrt{\frac{\lambda}{Lm}}t\right) + B\sin\left(\sqrt{\frac{\lambda}{Lm}}t\right)$$

where t is the time after the mass, m, has been released, and λ, A, B are constants.

Find the amplitude and frequency of the vibrations, if $A = 3$ and $B = \sqrt{3}$.

Example 38 *continued*

Solution

Substituting $A = 3$ and $B = \sqrt{3}$ gives

$$x = 3\cos\left(\sqrt{\frac{\lambda}{Lm}}t\right) + \sqrt{3}\sin\left(\sqrt{\frac{\lambda}{Lm}}t\right)$$

Using 4.75 :

$$R = \sqrt{3^2 + (\sqrt{3})^2} = \sqrt{12} = \sqrt{4 \times 3} = 2\sqrt{3}$$

and

$$\beta = \tan^{-1}\left(\frac{\sqrt{3}}{3}\right) = \tan^{-1}\left(\frac{1}{\sqrt{3}}\right) = \frac{\pi}{6} \text{ (or 30°)}$$

Thus

$$x = 2\sqrt{3}\cos\left(\sqrt{\frac{\lambda}{Lm}}t - \frac{\pi}{6}\right)$$

What is the amplitude and frequency?

$$\text{Amplitude} = 2\sqrt{3}$$

We can find the frequency by using

4.30 $$f = \frac{\omega}{2\pi}$$

with $\omega = \sqrt{\dfrac{\lambda}{Lm}}$:

$$\text{frequency } f = \frac{\omega}{2\pi} = \frac{\sqrt{\lambda/Lm}}{2\pi} = \sqrt{\frac{\lambda}{4\pi^2 Lm}} \qquad \left[\text{because } (2\pi)^2 = 4\pi^2\right]$$

SUMMARY

We can convert $a\cos(\theta) + b\sin(\theta)$ as

4.75 $$a\cos(\theta) + b\sin(\theta) = R\cos(\theta - \beta) \text{ where } R = \sqrt{a^2 + b^2} \text{ and } \beta = \tan^{-1}\left(\frac{b}{a}\right)$$

Remember, to find β we have to check the signs of a and b.

We can use 4.75 to convert $a\cos(\theta) + b\sin(\theta)$ into amplitude–phase form and to sketch graphs.

Exercise **4(i)**

Solutions at end of book. Complete solutions available at www.palgrave.com/science/engineering/singh

1 Write the following in the form $R\cos(\omega t \pm \beta)$:

  i $10\cos(\omega t) + 10\sin(\omega t)$

  ii $\sqrt{2}\cos(\omega t) + 2\sin(\omega t)$

  iii $-2\sin(\omega t) + 5\cos(\omega t)$

  iv $\cos(\omega t) - \sin(\omega t)$

  v $-\cos(\omega t) - \sqrt{3}\sin(\omega t)$

  vi $-5\cos(\omega t) + 5\sin(\omega t)$

2 Sketch one complete cycle of the following waveforms:

  i $y = \sin(\theta) + \cos(\theta)$

  ii $y = \sqrt{3}\cos(\theta) - \sin(\theta)$

  iii $y = \sin(\theta) - 2\cos(\theta)$

3 [electrical principles] An a.c. current, i, is represented by

$$i = 3\cos(3t) - \sqrt{3}\sin(3t)$$

Find the amplitude and period.

4 [vibrations] The motion of a bob of a pendulum can be described by

$$\theta = A\cos(\omega t) + B\sin(\omega t)$$

where $\omega = \sqrt{\dfrac{g}{l}}$, g = acceleration due to gravity, l = length, θ the angle made by the bob and A, B are constants.

For $l = 1$ m, $g = 10$ m/s², $A = 1$, and $B = 0.5$, write θ in the form

$$R\cos(\omega t \pm \beta)$$

State amplitude, period of oscillation and phase.

5 [vibrations] In a spring–mass system, the motion of the mass is described by

$$x = A\cos(\omega t) + B\sin(\omega t)$$

where x is the distance of the mass from its natural position, ω is the natural frequency of vibration and A, B are constants. For $A = \sqrt{3}$, $B = 1$ and $\omega = 10$:

  i Write x in the form $R\cos(\omega t \pm \beta)$ and state the amplitude of x.

  ii Sketch one complete cycle of x.

6 [thermodynamics] Figure 67 shows a piston moving in a cylinder with rod AB. Point B is rotating clockwise with an angular velocity of ω.

Fig. 67

The motion of B can be described by

$$r = C\cos(\omega t) + D\sin(\omega t)$$

where r is the (horizontal + vertical) distance of B from O, and C, D are constants. For $C = D = 4$ cm and $\omega = 2$ rad/s, write r in the form $R\cos(\omega t \pm \beta)$ and state the amplitude and period of r. Sketch one complete cycle of r.

Solutions at end of book. Complete solutions available at www.palgrave.com/science/engineering/singh

Miscellaneous exercise 4

1 Figure 68 shows a sheet of metal which is symmetrical about the broken line. Length AB = x, length BC = y and angle BAE = α.

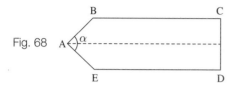

Fig. 68

Show that the area of the sheet ABCDE is

$$\frac{x}{2}\left[x\sin(\alpha) + 4y\sin\left(\frac{\alpha}{2}\right)\right]$$

2 [electrical principles] Let

$$i_1 = \sin\left(\omega t + \frac{\pi}{3}\right) \text{ and } i_2 = \cos\left(\omega t - \frac{\pi}{6}\right).$$

Show that

$$i_1 + i_2 = 2\cos\left(\omega t - \frac{\pi}{6}\right)$$

3 [signal processing] Let $f_1(t)$ and $f_2(t)$ be two signals defined by

$$f_1(t) = A_1\sin(\omega t) \text{ and}$$

$$f_2(t) = A_2\sin[\omega(t + \tau) + \phi]$$

where A_1 = amplitude of $f_1(t)$, A_2 = amplitude of $f_2(t)$, τ = time shift, ϕ = phase shift, ω = frequency and t = time. Show that

$$f_1(t)f_2(t) = \frac{A_1 A_2}{2}\left[\cos(\omega\tau + \phi) - \cos(2\omega t + \omega\tau + \phi)\right]$$

4 [electrical principles] The power in a circuit is given by $P = \dfrac{v^2}{R}$. If

$$v = V\cos\left(\omega t + \frac{\pi}{4}\right) \text{ show that}$$

$$P = \frac{V^2}{2R}\left[1 - \sin(2\omega t)\right]$$

5 [electrical principles] Let

$$v = 4\sin\left(\omega t + \frac{\pi}{4}\right) \text{ and } i = \sin(\omega t) \text{ in a}$$

circuit. Show that power, $P = iv$, is given by

$$P = \sqrt{2}\left[2\sin^2(\omega t) + \sin(2\omega t)\right]$$

6 [mechanics] A projectile fired at an angle α has its horizontal (x) and vertical (y) distances given by

$$x = ut\cos(\alpha) \text{ and}$$

$$y = ut\sin(\alpha) - \frac{gt^2}{2} \text{ respectively}$$

where u is the initial velocity of the projectile, g is the acceleration due to gravity and t is time. Show that

$$y = \frac{1}{2u^2}\left[2u^2 x\tan(\alpha) - gx^2\sec^2(\alpha)\right]$$

7 Find the lengths of the two unknown sides in Fig. 69.

Fig. 69

8 Three sides of a triangle are 14, 16 and 18. Find all the angles of the triangle.

9 By using the identity $\sin^2(\theta) + \cos^2(\theta) = 1$, deduce

$$1 + \cot^2(\theta) = \operatorname{cosec}^2(\theta)$$

10 Let x be an angle between 0° and 90°, with

$$\tan(x) = \sqrt{2} - 1$$

Without using a calculator, find the value of x.

$$\left(\text{Hint: Use } \tan(2x) = \frac{2\tan(x)}{1 - \tan^2(x)}\right)$$

11 Solve, for x between 0 and 2π, the equations

a $\cos(2x) = 0$

b $\cos\left(x + \dfrac{\pi}{6}\right) = \sin(x)$

c $\tan(2x) \cdot \tan(x) = 1$

d $4\sin(x)\cos(x) - 4\cos^2(x) = -2$

12 [electronics] An a.c. voltage, v, across a circuit is given by

$$v = 20\sin(50\pi t + 0.6)$$

i Find the amplitude, period and frequency of v.

ii Calculate the time displacement of v with respect to $20\sin(50\pi t)$.

iii Sketch one complete cycle of v.

13 [electronics] An a.c. current through a circuit is given by

$$i = 200\sin\left(\omega t + \dfrac{\pi}{3}\right)$$

Sketch one complete cycle of i against t.

14 [vibrations] Let x_1 and x_2 represent the horizontal displacement of two simple pendulums from their central position. Let

$$x_1 = 0.5\sin\left(\dfrac{\pi t}{5}\right) \text{ and } x_2 = 0.1\sin\left(\dfrac{\pi t}{4}\right)$$

where t is in seconds. If they both start together at $t = 0$, find the time taken for them to meet at $x_1 = x_2 = 0$ again.

15 [vibrations] As in the above question, let

$$x_1 = 0.3\sin\left(\dfrac{\pi t}{3}\right) \text{ and } x_2 = 0.3\cos\left(\dfrac{\pi t}{3}\right)$$

At $t = 0$ they are not together. Find the time taken for them to meet for the first time.

16 [structures] The displacement curve of a uniform pin-ended strut is given by

$$y = A\cos(kx) + B\sin(kx)$$
where k is a constant

By using the conditions $x = 0$, $y = e$ and at $x = L$, $y = e$, find A and B. Also show that

$$y = e\left(\sin(kx)\tan\left(\dfrac{kL}{2}\right) + \cos(kx)\right)$$

Show that the central displacement y_c at $x = \dfrac{L}{2}$ is $y_c = e\sec\left(\dfrac{kL}{2}\right)$.

$$\left(Hint: \dfrac{1 - \cos(x)}{\sin(x)} = \tan\left(\dfrac{x}{2}\right)\right)$$

17 Solve, if possible, the following equations for x where $0° \le x \le 360°$:

i $\cos(x) - \sec(x) = 1$

ii $\cos(x) + \sec(x) = 1$

iii $\cos^2(x) = 2\cos(x) + \tan^2(x)$

18 Find, by using Table 1, the exact values of

a $\dfrac{\tan(40°) + \tan(20°)}{1 - \tan(40°)\tan(20°)}$

b $\cos(15°) - \sin(15°)$

c $\cos(75°) + \sin(75°)$

19 Find the general solution of

$$\tan(2x) + \tan(x) = 0$$

Logarithmic, Exponential and Hyperbolic Functions

SECTION A **Indices revisited**

By the end of this section you will be able to:
▶ use the laws of indices
▶ apply these rules to thermodynamics

A1 **Review of indices**

We first visited indices in Chapter 1. The rules of indices, stated below, are **critical** in most engineering disciplines. We will use these rules throughout this chapter.

5.1	$a^m a^n = a^{m+n}$
5.2	$\dfrac{a^m}{a^n} = a^{m-n} \quad (a \neq 0)$
5.3	$(a^m)^n = a^{m \times n}$
5.4	$\dfrac{1}{a^n} = a^{-n} \quad (a \neq 0)$
5.5	$a^{\frac{1}{n}} = \sqrt[n]{a}$
5.6	$a^{\frac{m}{n}} = \sqrt[n]{a^m}$
5.7	$a^0 = 1 \quad (a \neq 0)$
5.8	$a^1 = a$

Let's look at some applications of indices in the field of thermodynamics.

Example 1 *thermodynamics*

A gas in a cylinder expands according to the law

$$P_1(V_1)^{1.35} = P_2(V_2)^{1.35}$$

where P_1, P_2 represent pressure and V_1, V_2 represent volume. Find an expression for V_2.

Solution

We have

$$P_2(V_2)^{1.35} = P_1(V_1)^{1.35}$$

$$(V_2)^{1.35} = \frac{P_1(V_1)^{1.35}}{P_2} \quad \text{[Dividing by } P_2\text{]}$$

† $\quad (V_2)^{1.35} = \frac{P_1}{P_2} \cdot (V_1)^{1.35}$

Example 1 *continued*

 How do we remove the power of V_2, 1.35?

By taking both sides to the power of $\dfrac{1}{1.35}$ because

$$\left(V_2^{1.35}\right)^{\frac{1}{1.35}} \underset{\text{by } 5.3}{=} V_2^{1.35 \times \frac{1}{1.35}} = V_2^1 \underset{\text{by } 5.8}{=} V_2$$

Applying the power of $\dfrac{1}{1.35}$ to both sides of ▮ † gives

$$V_2 = \left[\frac{P_1}{P_2} \cdot V_1^{1.35}\right]^{\frac{1}{1.35}}$$

$$= \left[\frac{P_1}{P_2}\right]^{\frac{1}{1.35}} V_1$$

Example 2 *thermodynamics*

A gas in a piston has an initial volume of 0.01 m^3 at a pressure of 250 kN/m^2. The gas is then compressed to a pressure of 800 kN/m^2 according to the law $P(V)^{1.2} = C$, where C is a constant. Find the final volume.

Solution

We have $P_1(V_1)^{1.2} = P_2(V_2)^{1.2}$ and we need to find V_2.
Substituting $V_1 = 0.01$, $P_1 = 250 \times 10^3$ and $P_2 = 800 \times 10^3$ gives

$$250 \times 10^3 \times (0.01)^{1.2} = 800 \times 10^3 \times (V_2)^{1.2}$$

 How do we find V_2?

First we obtain $V_2^{1.2}$:

$$(V_2)^{1.2} = \frac{250 \times 10^3}{800 \times 10^3} \times (0.01)^{1.2} \quad \text{[Dividing by } 800 \times 10^3\text{]}$$

$$= \frac{25}{80} \times (0.01)^{1.2} \quad \text{[Cancelling]}$$

As in **Example 1**, taking both sides to the power of $\dfrac{1}{1.2}$ gives

$$V_2 = \left[\frac{25}{80}\right]^{\frac{1}{1.2}} \times 0.01 = 3.79 \times 10^{-3}$$

Hence the final volume is 0.0038 m^3 (2 s.f.).

SUMMARY

We can apply the rules of indices, 5.1 to 5.8, to engineering applications such as those shown in **Examples 1** and **2**.

Exercise 5(a)

Solutions at end of book. Complete solutions available at www.palgrave.com/science/engineering/singh

1 [thermodynamics] A gas with an initial volume $V_1 = 0.3$ m^3 is compressed from $P_1 = 100$ kPa to $P_2 = 350$ kPa.
From the formula

$$P_1(V_1)^{5/3} = P_2(V_2)^{5/3}$$

evaluate the final volume V_2.
(Pascal, Pa = N/m^2, is the named SI unit of pressure.)

2 [thermodynamics] A gas in a cylinder with an initial volume $V_1 = 0.25$ m^3 and pressure $P_1 = 100$ kPa is compressed to a final pressure $P_2 = 450$ kPa.
Given that

$$P_1(V_1)^{1.33} = P_2(V_2)^{1.33}$$

determine the final volume V_2.

3 [thermodynamics] A volume, 0.1 m^3, of gas at a pressure of 200 kN/m^2 is compressed to a pressure of 632 kN/m^2. If the gas obeys the law

$$P(V)^{1.5} = C \quad (C = \text{constant})$$

then find the new volume.

4 [thermodynamics] For an ideal gas, the PVT equations are given by

$$P_1(V_1)^k = P_2(V_2)^k \text{ and } \frac{P_1 V_1}{T_1} = \frac{P_2 V_2}{T_2}$$

where P, V, T are pressure, volume, temperature respectively and k is a constant. Show that

i $\dfrac{T_2}{T_1} = \left(\dfrac{V_1}{V_2}\right)^{k-1}$

ii $\dfrac{T_1}{T_2} = \left(\dfrac{V_2}{V_1}\right)^{k-1}$

5 Show that $x^{\frac{1}{\Psi}}\left(y^{\frac{\Psi-1}{\Psi}} - x^{\frac{\Psi-1}{\Psi}}\right)$ can be

simplified to $x\left[\left(\dfrac{y}{x}\right)^{\frac{\Psi-1}{\Psi}} - 1\right]$.

6 [fluid mechanics] The velocity, v, of a fluid in a channel of slope s and radius r is defined as $v = \dfrac{r^{1/6}}{\eta}\sqrt{rs}$ where η is a constant. Show that

$$v = \frac{r^{2/3}s^{1/2}}{\eta}$$

SECTION B **The exponential function**

By the end of this section you will be able to:
▶ use a calculator to evaluate exponential expressions
▶ plot the graphs of e^x and e^{-x}
▶ apply exponential functions to engineering problems

B1 **Exponential function**

We sometimes hear the statement – the population is growing exponentially.

? What does this mean?

It means that the population is rapidly increasing. For example, if the population starts from 100 and doubles every year, we have the situation shown in Table 1.

TABLE 1

Year (x)	Population (y)
0	100
1	200
2	400
3	800
4	1600

We can predict the future population, y, by using the equation, $y = 100 \times (2^x)$. The initial population is 100 and it doubles every year, (2^x).

? However if the population trebles every year then what equation can we use?

$$y = 100 \times (3^x)$$

TABLE 2

Year (x)	Population (y)
0	100
1	300
2	900
3	2700
4	8100

This doubling and trebling of population is known as **exponential growth.** A particular exponential growth is defined as $y = e^x$ where

$$e = 2.71828182846 \ldots$$

This number e is an irrational number (like π) and so we cannot write down the exact value of e.

? **Can you recall where you have seen this number before?**

In the Introductory chapter and in Chapter 3 under **Section E** (Limits of functions) where it was defined as

$$\lim_{x \to \infty} \left(1 + \frac{1}{x}\right)^x = e$$

? **Why is e^x important?**

The function e^x occurs in many engineering disciplines such as aerodynamics, mechanics, electrical principles, etc. It also crops up in natural sciences, for example population growth.

The population growth for the exponential function $y = 100 \times (e^x)$ is given in Table 3.

TABLE 3	Year (x)	Population (y) (nearest whole number)
	0	100
	1	272
	2	739
	3	2009
	4	5460

The exponential function, e^x, more than doubles but does not quite treble over a period. Compare Tables 1, 2 and 3.

Plots of the general exponential functions $y = 2^x$, 3^x and e^x are shown in Fig. 1.

Fig. 1

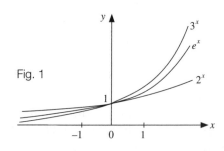

> The exponential function, e^x, is sometimes denoted by $\exp(x)$. It was the Swiss mathematician **Euler** who introduced the notation e^x in 1727. This function, e^x, is extensively used throughout engineering.

Example 3

Complete Table 4.

TABLE 4	x	−2	−1	0	1	2	3	4
	$y = e^x$							

Hence plot the function $y = e^x$.

Example 3 *continued*

Solution

To find the y values we use our calculator. Most calculators have the exponential button, e^x; it might be a secondary function under the ln button. For example to find e^2, PRESS:

$\boxed{\text{SHIFT}}$ $\boxed{\text{ln}}$ $\boxed{2}$ $\boxed{=}$ and the calculator should show 7.389...

On some calculators we need to replace the $\boxed{\text{SHIFT}}$ button with the $\boxed{\text{INV}}$ or $\boxed{\text{2nd FUN}}$ button (see the handbook of your calculator).

We obtain Table 5.

TABLE 5	x	-2	-1	0	1	2	3	4
	$y = e^x$	0.135	0.368	1.000	2.718	7.389	20.086	54.598

Plotting these points gives the graph shown in Fig. 2.

Fig. 2

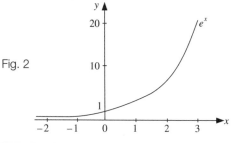

Example 4 *aerodynamics*

The pressure, P, of a gas at an altitude, Z, and temperature, T, is given by

$$P = Ae^{-\left(\frac{g}{RT}\right)Z}$$

Find P given that $A = 20 \times 10^3$ N/m^2, $R = 287$ J/(kg K), $Z = 5 \times 10^3$ m, $T = 200$ K and $g = 9.81$ m/s^2.

Solution

Substituting these values into the above formula with $e = \exp$ and using a calculator gives

$$P = (20 \times 10^3)\exp\left(-\frac{9.81}{(287 \times 200)} \times (5 \times 10^3)\right)$$

$$= (20 \times 10^3)\exp(-0.8545)$$

Example 4 *continued*

The screen of the calculator should show 8509.918. Remember that k = kilo = 10^3 and the unit of pressure, P, is N/m². Thus

$$P = 8510 \text{ N/m}^2 = 8.51 \text{ kN/m}^2 \text{ (3 s.f.)}$$

B2 Properties of the exponential function

We have the following results for the exponential function:

5.9 $e^0 = 1$ [This is 5.7 with $a = e$]

5.10 $\dfrac{1}{e^x} = e^{-x}$ [This is 5.4 with $a = e$]

The graph of $y = e^{-x}$ is a decaying graph (Fig. 3).

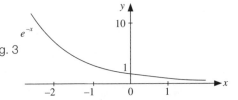

Fig. 3

? **Can you think of a decaying population?**

Possibly black rhinos or giant pandas.

As x becomes positively large then e^{-x} goes to zero. That is, as $x \to \infty$, $e^{-x} \to 0$. We can write this as

5.11 $\lim\limits_{x \to \infty}(e^{-x}) = 0$

Example 5 *electronics*

The voltage, v, across a capacitor in a *RC* circuit is given by

$$v = 100e^{-t/\tau}$$

where τ is the time constant. Find the percentages of voltage, v, at $t = \tau, 2\tau, 3\tau, 4\tau$ and 5τ of the initial voltage at $t = 0$. Sketch the graph of v versus t for $0 \le t \le 5\tau$.

Solution

Some engineering students find difficulty in following this example because of the Greek letter τ pronounced 'taw'. Generally in electronics and electrical principles τ is used to represent the time constant.

Substituting $t = 0$ into the given equation, $v = 100e^{-t/\tau}$, yields

$$v = 100e^0 = 100$$

The following results are correct to 1 d.p.

Example 5 *continued*

$$\text{At } t = \tau: \quad v = 100e^{-\frac{\tau}{\tau}} = 100e^{-1} = 36.8$$

$$\text{At } t = 2\tau: \quad v = 100e^{-\frac{2\tau}{\tau}} = 100e^{-2} = 13.5$$

$$\text{At } t = 3\tau: \quad v = 100e^{-\frac{3\tau}{\tau}} = 100e^{-3} = 5.0$$

$$\text{At } t = 4\tau: \quad v = 100e^{-\frac{4\tau}{\tau}} = 100e^{-4} = 1.8$$

$$\text{At } t = 5\tau: \quad v = 100e^{-\frac{5\tau}{\tau}} = 100e^{-5} = 0.7$$

The percentages of voltage after $t = \tau, 2\tau, 3\tau, 4\tau$ and 5τ are 36.8%, 13.5%, 5.0%, 1.8% and 0.7% respectively (Fig. 4).

Fig. 4

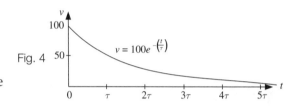

Note that after 5 time constants (5τ), the voltage v is approximately 0 volts.

How can we simplify expressions like $e^{-2t}e^{-4t}$?

We can apply the rule of indices, $a^m a^n = a^{m+n}$ with $a = e$:

$$e^{-2t}e^{-4t} = e^{-2t-4t} = e^{-6t}$$

Simplify $(e^{-200t})^2$.
Applying $(a^m)^n = a^{m \times n}$ with $a = e$ gives

$$(e^{-200t})^2 = e^{-200t \times 2} = e^{-400t}$$

The following example is more **demanding** because it involves a number of algebraic topics such as the rules of indices, expansion of brackets, substitution etc.

Example 6 *electrical principles*

The energy, w, of an inductor, L, is defined as

$$w = \frac{1}{2}Li^2$$

For $L = 2 \times 10^{-3}$ H and $i = 5(e^{-100t} - e^{-200t})$ A, show that

$$w = 0.025(e^{-200t} - 2e^{-300t} + e^{-400t})$$

Example 6 *continued*

Solution

Substituting $L = 2 \times 10^{-3}$ and $i = 5(e^{-100t} - e^{-200t})$ into $w = \dfrac{1}{2}Li^2$ gives

$$w = \frac{1}{2}(2 \times 10^{-3}) \times \left[5(e^{-100t} - e^{-200t})\right]^2$$

$$= (1 \times 10^{-3}) \times 25 \times (e^{-100t} - e^{-200t})^2 \qquad \text{[Taking out } 5^2 = 25\text{]}$$

$$= (25 \times 10^{-3}) \times \underbrace{\left[(e^{-100t})^2 - 2 \cdot (e^{-100t}e^{-200t}) + (e^{-200t})^2\right]}_{\text{by } \boxed{1.4}}$$

$$= (25 \times 10^{-3}) \times \left[\underbrace{e^{-200t}}_{\text{by } \boxed{5.3}} - \underbrace{2e^{-300t}}_{\text{by } \boxed{5.1}} + \underbrace{e^{-400t}}_{\text{by } \boxed{5.3}}\right] \qquad \text{[Simplifying]}$$

$$w = 0.025(e^{-200t} - 2e^{-300t} + e^{-400t})$$

SUMMARY

The exponential function is denoted by e^x where e is 2.71828.... Also

| 5.9 | $e^0 = 1$ |

| 5.10 | $e^{-x} = 1/e^x$ |

e^x is a graph of growth while e^{-x} is a decaying graph.

Exercise 5(b)

Solutions at end of book. Complete solutions available at www.palgrave.com/science/engineering/singh

1 [*mechanics*]   A pulley is driven by a flat belt with an angle of lap $\theta = \dfrac{2\pi}{3}$. Given that T_1, T_2 are tensions on each end of the belt and
$$T_2 = T_1 e^{\mu\theta}$$
Determine T_2 for $T_1 = 1000$ N and $\mu = 0.2$ (coefficient of friction).

2 Complete the following table:

x	-2	-1	0	1	2
$y = e^{-x}$			1.00		

Plot the graph of $y = e^{-x}$.

3 [*electronics*]   The current, i, through a diode is given by

$$i = (0.1 \times 10^{-6})(e^{20v} - 1)$$

where v is the voltage. Complete the following table:

v	0	0.1	0.2	0.3	0.4	0.5	0.6	0.7
i								

Plot the graph of i versus v for the values in the table.

| 1.4 | $(a - b)^2 = a^2 - 2ab + b^2$ | | 5.1 | $a^m a^n = a^{m+n}$ | | 5.3 | $(a^m)^n = a^{m \times n}$ |

Exercise **5(b) continued**

Solutions at end of book. Complete solutions available at www.palgrave.com/science/engineering/singh

4 [aerodynamics]   The pressure, p, of air at an altitude, z, is given by

$$p = (20 \times 10^3)e^{-(1.25 \times 10^{-4})z}$$

Find p to the nearest whole number for
a $z = 5 \times 10^3$ **b** $z = 7 \times 10^3$
c $z = 10 \times 10^3$

5 [heat transfer]   The temperature, θ (°C), at time t (min) of a body is given by

$$\theta = 300 + 100e^{-0.1t}$$

Evaluate θ for $t = 0$, 1, 2 and 5.

6 [electrical principles]   The energy, W, of a capacitor, C, is defined by

$$W = \frac{1}{2}CV^2$$

For $C = 1\ \mu\text{F}$ and $V = e^{-(1 \times 10^3)t}$, find W.

7 [electrical principles]   The instantaneous power, p, of a capacitor is defined as

$$p = vi$$

a For $v = 1 - e^{-t/2 \times 10^{-6}}$ and $i = 0.4e^{-t/2 \times 10^{-6}}$, find p.

b For $v = e^{-(5 \times 10^3)(t-1)}$ and $i = -0.05e^{-(5 \times 10^3)(t-1)}$, find p.

8 [electrical principles]   The voltage, v, and current, i, of an inductor is given by

$$i = 5(e^{-200t} - e^{-800t})\text{ and}$$
$$v = e^{-200t} + 400e^{-800t}$$

Find an expression for the power, $p = vi$, of the inductor.

9 [electrical principles]   The current, i (mA), through a capacitor at time t (s) is given by $i = 100e^{-(5 \times 10^3)t}$. Sketch the graph of i against t for $t \geq 0$.

10 [mechanics]   The velocity, v, of a train along a straight track is given by

$$v = 20(1 - e^{-t})$$

Sketch the graph of v against t for $t \geq 0$. What is the value of v for large t?

SECTION C **The logarithmic function**

By the end of this section you will be able to:
▶ use a calculator to find the logarithm of a number
▶ apply the rules of logarithms to engineering problems
▶ change the base of a logarithm

C1 **Natural logarithms**

The most important logarithm in engineering is the **natural** logarithm and is denoted by ln which stands for logarithm natural. The logarithmic function, $\ln(x)$, is the inverse of the exponential function, e^x; that is,
$y = e^x$ and $\ln(y) = x$ are equivalent.
For example, $e^2 = 7.389$ therefore $\ln(7.389) = 2$. Similarly $e^{3.142} = 23.150$ and so $\ln(23.150) = 3.142$.

> The word logarithm is made up of two Greek words – *logos* (expression) and *arithmos* (number). It was **John Napier** (1550–1617) of Edinburgh who discovered the logarithm.

Figure 5 shows the graphs of $y = e^x$ and $y = \ln(x)$. Note that $\ln(x)$ is not defined for negative x and $x = 0$.

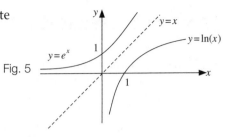

Fig. 5

At one time you had to find the logarithm of a number by using tables, but since the introduction of the hand-held calculator in the 1970s these logarithmic tables, along with the slide rule, have been exiled to history.

You can evaluate the ln function on your calculator.
To evaluate $\ln(3)$
PRESS: | ln | | 3 | | = | , which should show 1.09861.

Example 7 *electronics*

The current gain, G, in neper of a system is defined as

$$G = \ln\left(\frac{I_{out}}{I_{in}}\right)$$

where I_{out} and I_{in} are the output and input current respectively. Evaluate G for $I_{out} = 0.5$ mA and $I_{in} = 0.1$ mA. (Remember that m = milli is 10^{-3}.)

Solution

Substituting the given values into G:

$$G = \ln\left(\frac{0.5 \times 10^{-3}}{0.1 \times 10^{-3}}\right) = \ln(5) = 1.609 \text{ nep}$$

Example 8

Evaluate the following:

a $\ln(2) + \ln(3)$ and $\ln(6)$ **b** $\ln(7) + \ln(5)$ and $\ln(35)$

What do you notice about your results?

Solution

By using a calculator and rounding to 3 decimal places:

a $\ln(2) + \ln(3) = 1.792$ and $\ln(6) = 1.792$

b $\ln(7) + \ln(5) = 3.555$ and $\ln(35) = 3.555$

We have

$$\ln(2) + \ln(3) = \ln(2 \times 3) = \ln(6)$$
$$\ln(7) + \ln(5) = \ln(7 \times 5) = \ln(35)$$

In general we have the following laws of logarithms:

5.12 $\ln(A) + \ln(B) = \ln(A \times B) = \ln(AB)$

5.13 $\ln(A) - \ln(B) = \ln\left(\dfrac{A}{B}\right)$

5.14 $\ln(A^n) = n \ln(A)$

In 5.12 to 5.14 , A and B are positive because the ln function is not defined for negative numbers. We can employ these to simplify our results, as the next example shows.

Example 9 *thermodynamics*

The pressure, P, and volume, V, of a gas are related by

$$\ln(P) = -n \ln(V) + \ln(C)$$

where n and C are constants. Show that $\ln(PV^n) = \ln(C)$.

Solution

Adding $n \ln(V)$ to both sides gives

$$\ln(P) + n \ln(V) = \ln(C)$$

By 5.14 , $n \ln(V) = \ln(V^n)$ so we have

$$\ln(P) + \ln(V^n) = \ln(C)$$

$$\underbrace{\ln(PV^n)}_{\text{by } 5.12} = \ln(C)$$

? **What is ln(e) equal to?**

Since ln and e are inverses of each other:

$$\ln(e) = 1$$

? **What is ln(e^x) equal to?**

$$\ln(e^x) = x \ln(e) \qquad \text{[By } 5.14 \text{]}$$
$$\qquad\quad = x \qquad \text{[Because } \ln(e) = 1\text{]}$$

Thus by noting that $\ln(x)$ is the inverse function of e^x we have the following results:

5.15 $\ln(e) = 1$

5.16 $\ln(e^x) = x$

5.17 $e^{\ln(x)} = x$ (provided x is positive)

We can use these results to solve equations involving the exponential function or show other results such as

$$e^{x \ln(a)} = a^x$$

[See **Exercise 5(c)**, question 15.]

Example 10 *mechanics*

The velocity, v, of a vehicle during the application of brakes according to time t is given by

$$v = 10e^{-kt}$$

where k is a constant. At $t = 30$ s the velocity is reduced to 2.5 m/s.

Determine k.

Solution

Substituting $t = 30$ and $v = 2.5$ into $v = 10e^{-kt}$ gives

$$2.5 = 10e^{-30k}$$
$$\frac{2.5}{10} = e^{-30k} \quad \text{[Dividing by 10]}$$
$$0.25 = e^{-30k}$$

How can we find k?

By taking natural logs, ln, of both sides because ln is the inverse of the exponential function. Applying $\ln(A^n) = n \ln(A)$ to the Right-Hand Side gives

$$\ln(e^{-30k}) = -30k \underbrace{\ln(e)}_{=1} = -30k$$

Therefore we have

$$\ln(0.25) = -30k$$
$$k = \frac{\ln(0.25)}{-30} \quad \text{[Dividing by } -30\text{]}$$
$$= 0.046 \quad (2 \text{ s.f.})$$

C2 Common logarithms

Apart from natural logarithms, ln, there are other logarithms which are also used in engineering. The other important logarithm is called the **common** logarithm and is denoted by log.

The logarithmic function, $\log(x)$, is the inverse function of 10^x. In other words the log of a number is the power of 10 needed to make that number. For example, $1000 = 10^3$ thus $\log(1000) = \log(10^3) = 3$. Similarly $10 = 10^1$ so $\log(10) = \log(10^1) = 1$.

What is log(100) equal to?

$$\log(100) = \log(10^2) = 2$$

What is log(199.53) given that $10^{2.3} = 199.53$?

$$\log(199.53) = 2.3$$

Figure 6 shows the graphs of $y = 10^x$ and $y = \log(x)$. Note that $\log(x)$ is also not defined for negative x and $x = 0$.

Fig. 6

Example 11 *electronics*

The power gain, G, in decibels (dB) of a system is defined as

$$G = 10 \log\left(\frac{P_{out}}{P_{in}}\right)$$

where P_{out} is the output power and P_{in} is the input power. Evaluate G for the following:

a $P_{out} = 5$ W, $\quad P_{in} = 0.5$ W $\qquad$ **b** $P_{out} = 1$ W, $\quad P_{in} = 1 \times 10^{-3}$ W

Solution

Remember that W = watt is the SI unit of power.

a $G = 10 \times \log\left(\dfrac{5}{0.5}\right) = 10 \times \underbrace{\log(10)}_{=1} = 10$ dB

b $G = 10 \times \log\left(\dfrac{1}{1 \times 10^{-3}}\right) = 10 \times \log(1000) = 10 \times 3 = 30$ dB

The same laws apply to the log function.

5.18 $\qquad \log(A) + \log(B) = \log(A \times B) = \log(AB)$

5.19 $\qquad \log(A) - \log(B) = \log\left(\dfrac{A}{B}\right)$

5.20 $\qquad \log(A^n) = n \log(A)$

5.21 $\qquad \log(10) = 1$

5.22 $\qquad \log(10^x) = x$

5.23 $\qquad 10^{\log(x)} = x$

We can use these laws to evaluate G of **Example 11.** For example, if we had $P_{out} = 1$ mW, $P_{in} = 1$ W then

$$G = 10 \log\left(\frac{1 \times 10^{-3}}{1}\right) = 10 \log(10^{-3}) \underset{\text{by } 5.20}{=} -3 \times 10 \log(10) = -30 \underbrace{\log(10)}_{=1 \text{ by } 5.21} = -30 \text{ dB}$$

Of course the easiest way to evaluate log is to use a calculator. For example, to evaluate $\log(3)$ on a calculator PRESS

| log | | 3 | | = | and the screen should show 0.4771213. Try **Example 11** on your calculator.

C3 **Other bases**

In this section we have concentrated on natural, ln, and common, log, logarithms. The logarithm can be **based** on any positive number. For example, we have logarithm to the base a written $\log_a(x)$ which is the inverse of a^x.

Figure 7 shows the graphs of a^x and $\log_a(x)$ where a can be any positive number. Consider the case $a=2$.

? **What is the inverse of 2^x ?**

$\log_2(x)$ [log to the base 2]. This base is used in communication engineering. For example

Fig. 7

$$2^3 = 8$$
$$\log_2(8) = 3$$

Similarly the inverse of 3^x is $\log_3(x)$ [log to the base 3].

? **Given that $3^5 = 243$, what is $\log_3(243)$ equal to?**

$$\log_3(243) = 5$$

In general $a^x = N$ is equivalent to $\log_a(N) = x$.

We can change the base of the logarithm by using the following rule:

5.24
$$\log_a(N) = \frac{\log_b(N)}{\log_b(a)}$$

5.24 changes the logarithm from base a to base b. We can use this change of base rule to evaluate logarithm to any base, as the following example shows.

Example 12

Find the values of

$$\log_2(5), \log_3(7) \text{ and } \log_{8.7}(10.2)$$

Solution

? **Which base should we convert these logs to?**

Since the calculator evaluates the natural and common logarithms, we have to convert to one of these bases. Converting to the natural logarithm by applying 5.24 and rounding to 2 decimal places we have

$$\log_2(5) = \frac{\ln(5)}{\ln(2)} = 2.32$$

$$\log_3(7) = \frac{\ln(7)}{\ln(3)} = 1.77$$

$$\log_{8.7}(10.2) = \frac{\ln(10.2)}{\ln(8.7)} = 1.07$$

Of course you can convert the logarithms of **Example 12** into the common logarithms with exactly the same results.

Log to any base a, $\log_a$, also obey **all** the rules of logarithms stated for the common and natural logarithms.

SUMMARY

The natural logarithm, $\ln(x)$, is the inverse of the exponential function, e^x. The common logarithm, $\log(x)$, is the inverse of 10^x. All logarithms obey the rules:

5.12 $\ln(A) + \ln(B) = \ln(A \times B) = \ln(AB)$

5.13 $\ln(A) - \ln(B) = \ln\left(\dfrac{A}{B}\right)$

5.14 $\ln(A^n) = n \ln(A)$

For any positive number a, the inverse of a^x is $\log_a(x)$. We can change the base of a logarithm by using

5.24 $\log_a(N) = \dfrac{\log_b(N)}{\log_b(a)}$

Exercise 5(c)

Solutions at end of book. Complete solutions available at www.palgrave.com/science/engineering/singh

📺 **Questions 1–11, inclusive, are all in the [*electronics*] discipline.**

1 By using the formula of **Example 7** determine the current gain, G, in neper for
 a $I_{out} = 0.1$ A, $\ I_{in} = 0.01$ A
 b $I_{out} = 10$ mA, $I_{in} = 0.1$ mA
 c $I_{out} = 5$ mA, $I_{in} = 15$ mA

2 The power gain, G, in neper is defined as

$$G = \frac{1}{2}\ln\left(\frac{p_o}{p_i}\right)$$

where p_o = output power and p_i = input power. Find G for
 a $p_o = 1$ W, $p_i = 0.2$ W
 b $p_o = 20$ mW, $p_i = 1$ mW

3 If $V_{out} = 1$ V and $V_{in} = 1$ mV then without using your calculator evaluate G, where

$$G = 20 \log\left(\frac{V_{out}}{V_{in}}\right)$$

4 If $G = 5$ dB, $P_{in} = 0.1$ W, evaluate P_{out} by using the formula of **Example 11**.

5 In a cascaded electronic circuit the power gains at various stages are p_1, p_2 and p_3. Given that the overall power gain, G, is defined as
 $G = 10 \log(p_1) + 10 \log(p_2) + 10 \log(p_3)$
 show that $G = 10 \log(p_1 \times p_2 \times p_3)$.

6 If $p_{out} = I_{out}^2 R$ and $p_{in} = I_{in}^2 R$, show that

$$G = 20 \log\left(\frac{I_{out}}{I_{in}}\right)$$

$\left(p_{out} = \text{output power, } p_{in} = \text{input power,} \right.$
$I_{out} = \text{output current, } I_{in} = \text{input current,}$
$R = \text{resistance and } G = 10 \log\left. \left(\dfrac{p_{out}}{p_{in}}\right).\right)$

7 Given that $G = 20$ dB and $I_{in} = 1$ mA, determine I_{out} by using the formula of question **6**.

Exercise **5(c) continued**

Solutions at end of book. Complete solutions available at www.palgrave.com/science/engineering/singh

8 Given that $p_{out} = \dfrac{V_{out}^2}{R}$ and $p_{in} = \dfrac{V_{in}^2}{R}$,

show that

$$G = 20 \log\left(\frac{V_{out}}{V_{in}}\right)$$

$\left(V_{out} = \text{output voltage, } V_{in} = \text{input}\right.$

voltage, $p_{out} = $ output power,

$p_{in} = $ input power, $R = $ resistance and

$G = 10 \log\left(\dfrac{p_{out}}{p_{in}}\right).\Big)$

9 For the following values of G find the

power ratio $\left(\dfrac{p_o}{p_i}\right)$ and the voltage ratio

$\left(\dfrac{V_o}{V_i}\right)$ where

$$G = 10 \log\left(\frac{p_o}{p_i}\right) = 10 \log\left[\left(\frac{V_o}{V_i}\right)^2\right]:$$

a $G = 3$ dB   **b** $G = 10$ dB
c $G = 20$ dB

10 The voltage, V, across a capacitor is given by

$$V = 9 - 10e^{-t}$$

Find t for $V = 0$.

11 The voltage, V, across a capacitor is given by

$$V = 9(1 - e^{-0.1t})$$

Evaluate the time taken, t, for $V = 3$.

12 Evaluate the following:

a $\log_2(7.6)$ **b** $\log_{7.6}(2)$
c $\log_7(6.3)$ **d** $\log_{6.3}(7)$

13 Evaluate $\ln\big[\ln(N)\big]$ for the following values of N:

a $N = 1000$ **b** $N = 10^{10}$ **c** $N = 10^{50}$

What do you notice about your results?

14 ▨ [*thermodynamics*]   In a thermodynamic system we have the relationship

$$\log(P) + n \log(V) = \log(C)$$

where P represents pressure, V represents volume, C is a constant and n is an index. Show that

$$PV^n = C$$

15 Show that $e^{x \ln(a)} = a^x$.

16 Given that $\ln(1 + y) = x + \dfrac{x^2}{2}$, show that

$$y = e^{x^2/2}e^x - 1$$

17 🌡 [*heat transfer*]   The temperatures θ_1 and θ_2 of a pipe with inner radius r_1 and outer radius r_2 are given by

$$\theta_1 = -\frac{Q}{2\pi kL} \ln(r_1) \text{ and}$$

$$\theta_2 = -\frac{Q}{2\pi kL} \ln(r_2)$$

where Q is the heat transfer rate, L is length of pipe and k is thermal conductivity. Show that

$$Q = \frac{2\pi kL(\theta_1 - \theta_2)}{\ln\left(\dfrac{r_2}{r_1}\right)}$$

18 📺 [*electronics*]   The voltage, v, across a diode and the current, i, through it are related by

$$i = I_s(e^{11600v/\eta T} - 1)$$

where I_s is the reverse saturation current, η is the emission coefficient and T is the temperature. Make v the subject of the formula. Evaluate v for $\eta = 2$, $T = 330$ K, $I_s = 0.1$ μA and $i = 5$ mA.

SECTION D **Applications of logarithms**

By the end of this section you will be able to:
▶ choose appropriate variables for data to obtain a straight line
▶ apply laws of logs to experimental data
▶ plot the graph of $\log(y)$ against $\log(x)$ for data which follows the law $y = kx^n$
▶ find the values of k and n in $y = kx^n$

D1 **Experimental law**

Suppose x and y are related by $y = kx^2 + b$ where k and b are constants. If we plot the graph of y against x the result is a curve. For a straight line graph, $y = mx + c$, we can find the values of m (= gradient) and c (= intercept). **Can we find the values of k and b in $y = kx^2 + b$?**

If we plot y against x^2 instead of y against x then we should obtain a straight line. Compare

$$y = kx^2 + b \qquad (y \text{ against } x^2)$$
and $$y = mx + c \qquad (y \text{ against } x)$$

The gradient of the resulting line gives the value of k and the vertical intercept gives b. Let's do an example.

Example 13

TABLE 6	x	1	2	3	4	5
	y	7.1	10.4	15.9	23.6	33.5

Suppose the experimental data in Table 6 is related by $y = kx^2 + b$. By plotting an appropriate graph find the approximate values of k and b.

Solution

We plot y vertically against x^2 horizontally.

TABLE 7	x^2	1	4	9	16	25
	y	7.1	10.4	15.9	23.6	33.5

We draw the most fitting straight line through the points (Fig. 8).

The gradient of this line gives the value of k:

$$k = 1.08$$

Example 13 *continued*

The vertical intercept
gives the value of b:

$$b = 6$$

By substituting $k = 1.08$
and $b = 6$ into
$y = kx^2 + b$ we obtain
the law relating x and y:

Fig. 8

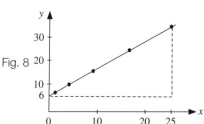

Slope $= \dfrac{33 - 6}{25 - 0} = \dfrac{27}{25} = 1.08$

$$y = 1.08x^2 + 6$$

For other curves such as $y = \dfrac{k}{x} + b$ and $y = k\sqrt{x} + b$ we can obtain straight lines by
plotting y against $\dfrac{1}{x}$ and y against $\sqrt{x}$ respectively. From the resulting straight lines
the gradient gives the value of k and the vertical intercept gives b. Similarly other straight
lines can also be obtained by plotting the appropriate variables.

D2 Determination of law by using logarithms

Consider an equation of the form $y = kx^n$ (where k and n are constants and $n \neq 1$).

? **Is it possible to find values of k and n given certain values of x and y from the graph of $y = kx^n$?**

? This is not a straight line of the form $y = mx + c$ where m is the gradient and c is the
intercept. **How can we find k and n in $y = kx^n$?**

? We need to get $y = kx^n$ into linear form. **How?**

By taking common logs of both sides:

$$\log(y) = \log\left(kx^n\right) \underset{\text{by } 5.18}{=} \log\left(x^n\right) + \log(k)$$

$$\log(y) = \underbrace{n \log(x)}_{\text{by } 5.20} + \log(k)$$

This is similar to the equation of a straight line, $y = mx + c$, except we have $\log(y)$
and $\log(x)$ instead of y and x respectively. If we plot $\log(y)$ against $\log(x)$ we obtain a straight
line with the gradient equal to n and the intercept equal to $\log(k)$. This plot is called
a log−log plot.

Natural logarithms, ln, can also be taken and we would still obtain the same results.
However in this section we restrict ourselves to common logarithms.

5.18 $\log(AB) = \log(A) + \log(B)$ **5.20** $\log(A^n) = n \log(A)$

Example 14 *electronics*

An electronic device gives the values of voltage, v, and current, i, shown in Table 8.

TABLE 8						
v (volts)	1	2	4	6	8	10
i (mA)	0.5	1.0	3.0	4.5	5.9	7.0

Assume that the device obeys the law $i = kv^n$ where k and n are constants. By plotting $\log(v)$ versus $\log(i)$, find the values of k and n.

Solution

We are given $i = kv^n$. Taking logs of $i = kv^n$:

$$\log(i) = \log(kv^n) \underset{\text{by } 5.18}{=} \log(v^n) + \log(k) = \underset{\text{by } 5.20}{n \log(v)} + \log(k)$$

Evaluating $\log(v)$ for each of the v values in Table 8 and $\log(i)$ for the i values gives the results shown in Table 9.

TABLE 9						
$\log(v)$	0	0.30	0.60	0.78	0.90	1.00
$\log(i)$	−0.30	0.00	0.48	0.65	0.77	0.85

Plotting the graph of $\log(v)$ against $\log(i)$ yields Fig. 9.

$$\text{gradient} = \frac{0.9 + 0.3}{1.0 - 0} = \frac{1.2}{1} = 1.2$$

The gradient of the straight line in Fig. 9 gives the value of n. Hence $n = 1.2$. The intercept, where the line cuts the vertical axis, gives $\log(k)$. So $\log(k) = -0.3$. **How can we find k?**

We need to remove the log from the Left-Hand Side of $\log(k) = -0.3$. **How?**

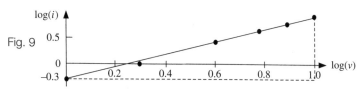

Fig. 9

Taking exponentials to the base 10 because 10^x is the inverse of $\log(x)$:

$$10^{\log(k)} = 10^{-0.3}$$

$$\underset{\text{by } 5.23}{k} = 10^{-0.3} = 0.5$$

Substituting $k = 0.5$ and $n = 1.2$ into $i = kv^n$ gives the law $i = 0.5v^{1.2}$.

5.18 $\log(AB) = \log(A) + \log(B)$ **5.20** $\log(A^n) = n \log(A)$ **5.23** $10^{\log(x)} = x$

Example 15 *thermodynamics*

A gas in a cylinder expands according to the law $PV^n = C$ where P is pressure, V is volume and C, n are constants. An experiment produces the results shown in Table 10.

P (kN/m²)	100	200	300	400	500	600
V (m³)	0.160	0.100	0.076	0.063	0.054	0.048

TABLE 10

Find the relationship between P and V.

Solution

As in **Example 14**, taking common logs of $PV^n = C$ gives

$$\log(PV^n) = \log(C)$$

$$\underbrace{\log(P) + \log(V^n)}_{\text{by } 5.18} = \log(C)$$

$$\log(P) + \underbrace{n \log(V)}_{\text{by } 5.20} = \log(C)$$

Transposing gives

$$\log(P) = -n \log(V) + \log(C)$$

Table 11 shows values of $\log(P)$ against $\log(V)$.

log (P)	2.000	2.301	2.477	2.602	2.699	2.778
log (V)	−0.796	−1.000	−1.119	−1.201	−1.268	−1.319

TABLE 11

What do we plot on the horizontal axis?

Comparing

$$\log(P) = -n \log(V) + \log(C)$$
and $\qquad y = mx + c$

shows that we plot $\log(V)$ on the horizontal axis and $\log(P)$ on the vertical axis (Fig. 10).

$$\text{Slope} = -\frac{2.8 - 0}{1.3 + 0.6} = -1.47$$

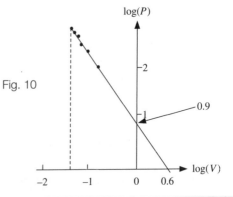

Fig. 10

5.18 $\log(AB) = \log(A) + \log(B)$ **5.20** $\log(A^n) = n \log(A)$

Example 15 *continued*

The gradient (slope) of the straight line is $-n$, so we have

$$-n = -1.47$$
$$n = 1.47$$

How do we find C?

The value of $\log(C)$ is where the line cuts the vertical axis, therefore

$$\log(C) = 0.9$$

Taking exponentials to the base 10:

$$C = 10^{0.9} = 7.94$$

Substituting $n = 1.47$, $C = 7.94$ into $PV^n = C$ gives the relationship

$$PV^{1.47} = 7.94$$

We need to be careful when plotting a graph. If you plot Fig. 10 graph on a grid you should achieve more accurate answers for n and C.

SUMMARY

We need to select suitable variables to obtain a straight line, so that the new equation is of the form $y = mx + c$. If the experimental data follows the law $y = kx^n$ then we can obtain the values of k and n by:
1 plotting the graph of $\log(y)$ against $\log(x)$ results in a straight line
2 the gradient of the straight line = n
3 the vertical intercept = $\log(k)$.
To obtain k, we take exponentials of both sides of **3** to the base 10.

Exercise 5(d)

Solutions at end of book. Complete solutions available at www.palgrave.com/science/engineering/singh

1 The area A for dimension d of a figure is given in the table:

d (cm)	2	3	4	5	6
A (cm²)	12.57	28.27	50.27	78.54	113.10

Assuming the results obey the law $A = kd^2 + b$, find k and b by plotting an appropriate graph. What do you notice about your results? What shape is the figure?

2 [*mechanics*] The force, F, between two objects at a distance d apart is given in the table:

d (m)	2	3	4	5	6	7
F ($\times 10^{-9}$) (N)	2.510	1.110	0.630	0.400	0.280	0.200

Assuming the results obey the rule $F = \dfrac{k}{d^2}$, find k by plotting a suitable graph.

Solutions at end of book. Complete solutions available at www.palgrave.com/science/engineering/singh

Exercise **5(d) continued**

3 [thermodynamics] The pressure, P, and volume, V, of a gas are given in the table:

V (m³)	0.1	0.15	0.2	0.25	0.3	0.35	0.4
P (kPa)	416	277	208	166	139	119	104

  i Plot P vertically against V

  ii Plot P vertically against $\dfrac{1}{V}$

  iii Show that $P = \dfrac{k}{V}$ and find the
     value of k.

4 [aerodynamics] The drag coefficient, C, is related to Reynolds number, R, by
$$C = kR^n$$

where k and n are constants. For the data in the table, find k and n.

R	3	4	5	6	7	8	9	10
C	0.060	0.056	0.054	0.052	0.050	0.049	0.048	0.047

5 [structures] The table shows the deflection, x, of a beam under the corresponding loads.

x (mm)	5.00	7.00	10.00	15.00	20.00
L (kN)	4.39	5.49	6.95	9.12	11.05

Deduce that the results obey $L = kx^n$ and find k and n.

SECTION E **Hyperbolic functions**

By the end of this section you will be able to:
▶ use a calculator to evaluate hyperbolic functions
▶ plot graphs of these functions
▶ investigate properties of hyperbolic functions
▶ examine engineering applications of these functions

This is a particularly demanding section requiring a good grasp of the exponential function and algebra. If you come up against some difficulties and get 'stuck', do not give up too quickly. Try following the steps slowly and go over them several times.

E1 **Hyperbolic functions**

Hyperbolic functions are characterized in terms of the exponential function. We define the hyperbolic sine, written sinh and pronounced 'shine', as

5.25 $$\sinh(x) = \frac{e^x - e^{-x}}{2}$$

We define the hyperbolic cosine, written cosh and pronounced 'cosh', as

> The German mathematician **Lambert** (1728–1777) was the first person who seriously developed hyperbolic functions.

5.26 $$\cosh(x) = \frac{e^x + e^{-x}}{2}$$

These are related by the fundamental identity $\cosh^2(x) - \sinh^2(x) = 1$ which will be shown in **Example 20**.

The hyperbolic tangent, written tanh and pronounced 'than' is defined as

5.27 $$\tanh(x) = \frac{\sinh(x)}{\cosh(x)} = \frac{e^x - e^{-x}}{e^x + e^{-x}}$$

You can use your calculator to find values of these hyperbolic functions.

Example 16

Evaluate, correct to three d.p., $\sinh(2.1)$, $\tanh(1)$, $\cosh(0.3)$, $\cosh(15)$, $\tanh(5000)$, $\tanh(10^6)$, $\sinh(0.01)$ and $\sinh(10^{-6})$.

Solution

To evaluate $\sinh(2.1)$ on a calculator, PRESS $\boxed{\text{hyp}}$ $\boxed{\text{sin}}$ $\boxed{\text{2.1}}$ $\boxed{=}$ which should show 4.021856742.

Correct to 3 d.p., $\sinh(2.1) = 4.022$. Also note that $\sinh(2.1) = 4.022$ means

$$\frac{e^{2.1} - e^{-2.1}}{2} = 4.022$$

(You can check this result on your calculator.)

Evaluating the others on a calculator gives

$$\tanh(1) = 0.762, \cosh(0.3) = 1.045, \cosh(15) = 1634508.686,$$

$$\tanh(5000) = 1, \tanh(10^6) = 1, \sinh(0.01) = 0.01 \text{ and } \sinh(10^{-6}) = 10^{-6}$$

? **What do you notice about the last four results of Example 16?**

It seems that tanh of a large number is approximately 1 and sinh of a small number gives the same small number. These statements are approximately correct, as can be seen by the graphs of these functions.

E2 Graphs of hyperbolic functions

Example 17

Plot the graph of $y = \cosh(x)$ for $x = -4$ to $+4$.

Solution

We select whole numbers of x between -4 and $+4$ and evaluate $\cosh(x)$ on a calculator for each of these values (Table 12).

TABLE 12	x	-4	-3	-2	-1	0	1	2	3	4
	$y = \cosh(x)$	27.31	10.07	3.76	1.54	1.00	1.54	3.76	10.07	27.31

Connecting the points in the table gives the graph shown in Fig. 11.

What do you notice about your graph?

1 $\cosh(-x) = \cosh(x)$, symmetrical about the y axis

2 $\cosh(0) = 1$

3 $\cosh(x)$ is always bigger than or equal to 1.

Fig. 11

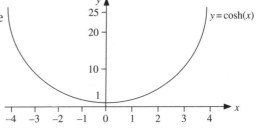

You can confirm the first two statements by using the definition of $\cosh(x)$.

The sinh and tanh graphs can be plotted similarly to the cosh graph (see Figs 12 and 13)

You are asked to plot these in **Exercise 5(e).**

Fig. 12

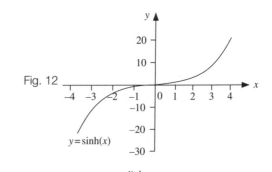

Observe from Fig. 13 that as x gets positively or negatively large then $\tanh(x)$ gets close to 1 or -1 respectively.

(More difficult to observe is the property stated earlier about $\sinh(x) \approx x$ for small x, but if you plot $\sinh(x)$ on a computer algebra system and zoom in close to zero you should be able to read off the x values on the horizontal axis and the corresponding $\sinh(x)$ values.)

Fig. 13

E3 Properties of hyperbolic functions

Example 18

Show that
$$\cosh(-x) = \cosh(x)$$

Solution

This is statement **1** above. **How can we show $\cosh(-x) = \cosh(x)$?**

By substituting $-x$ for x into 5.26 :

$$\cosh(-x) = \frac{e^{-x} + e^{-(-x)}}{2}$$
$$= \frac{e^{-x} + e^{+x}}{2}$$
$$= \frac{e^{x} + e^{-x}}{2} = \cosh(x) \qquad \left[\text{By } 5.26 \right]$$

We have

5.28 $\qquad \cosh(-x) = \cosh(x)$

5.29 $\qquad \sinh(-x) = -\sinh(x)$

You can also use 5.28 and 5.29 to deduce $\tanh(-x) = -\tanh(x)$.

Example 19

Show that

5.30 $\qquad \cosh(x) + \sinh(x) = e^{x}$

5.31 $\qquad \cosh(x) - \sinh(x) = e^{-x}$

Solution

Using 5.25 and 5.26 we have a common denominator:

$$\cosh(x) + \sinh(x) = \frac{e^{x} + e^{-x}}{2} + \frac{e^{x} - e^{-x}}{2}$$
$$= \frac{e^{x} + e^{-x} + e^{x} - e^{-x}}{2}$$
$$= \frac{2e^{x}}{2} \qquad \text{[Adding numerator]}$$
$$= e^{x} \qquad \text{[Cancelling 2's]}$$

5.25 $\quad \sinh(x) = (e^{x} - e^{-x})/2$ $\qquad$ 5.26 $\quad \cosh(x) = (e^{x} + e^{-x})/2$

Example 19 *continued*

Similarly we can show the other result 5.31 :

$$\cosh(x) - \sinh(x) = \frac{e^x + e^{-x}}{2} - \frac{e^x - e^x}{2}$$

$$= \frac{e^x + e^{-x} - (e^x - e^{-x})}{2}$$

$$= \frac{e^x + e^{-x} - e^x + e^{-x}}{2}$$

$$= \frac{2e^{-x}}{2} = e^{-x}$$

Example 20

Show the fundamental identity

5.32 $\cosh^2(x) - \sinh^2(x) = 1$

where $\cosh^2(x) = [\cosh(x)]^2$ and $\sinh^2(x) = [\sinh(x)]^2$.

Solution

How can we show $\cosh^2(x) - \sinh^2(x) = 1$?

We can use the difference of two squares:

1.15 $a^2 - b^2 = (a + b)(a - b)$

You really need to **know** this identity, otherwise it would not occur to us to use this identity in this context:

$$\cosh^2(x) - \sinh^2(x) = [\cosh(x)]^2 - [\sinh(x)]^2$$

$$\equiv [\cosh(x) + \sinh(x)][\cosh(x) - \sinh(x)]$$
by 1.15

$$= e^x \times e^{-x} = e^{x-x} = e^0 = 1$$
by 5.30 by 5.31 by 5.1

E4 Engineering applications of hyperbolic functions

The following example is an increase in the level of sophistication compared to the previous examples.

5.1 $a^m a^n = a^{m+n}$ 5.30 $\cosh(x) + \sinh(x) = e^x$ 5.31 $\cosh(x) - \sinh(x) = e^{-x}$

Example 21 *electrical principles*

A transmission line of length L has voltage V_x at a distance x from the sending end and is given by

$$V_x = \left(\frac{V + IZ_0}{2}\right)e^{-\gamma x} + \left(\frac{V - IZ_0}{2}\right)e^{\gamma x}$$

where V is voltage, I is current, Z_0 is characteristic impedance and γ is propagation coefficient. Show that at $x = L$:

$$V_L = V \cosh(\gamma L) - IZ_0 \sinh(\gamma L)$$

Solution

Don't be put off by all the different symbols used, it is really quite straightforward. Putting $x = L$ into ▨ * gives

$$V_L = \left(\frac{V + IZ_0}{2}\right)e^{-\gamma L} + \left(\frac{V - IZ_0}{2}\right)e^{\gamma L}$$

$$= V\underbrace{\left(\frac{e^{-\gamma L} + e^{\gamma L}}{2}\right)}_{\text{collecting the } V \text{ terms}} + IZ_0\underbrace{\left(\frac{e^{-\gamma L} - e^{\gamma L}}{2}\right)}_{\text{collecting the } IZ_0 \text{ terms}}$$

$$= V\underbrace{\cosh(\gamma L)}_{\text{by } 5.26} - IZ_0\left(\frac{e^{\gamma L} - e^{-\gamma L}}{2}\right)$$

$$V_L = V \cosh(\gamma L) - \underbrace{IZ_0 \sinh(\gamma L)}_{\text{by } 5.25}$$

The next example is particularly difficult. See if you can follow it through. It uses a noteworthy 'trick' which is to write the number 1 as

$$1 = \frac{2}{2} = e^0 = e^{kx - kx}\ldots$$

There are an infinite number of ways you can write any number.

Example 22 *electronics*

In a semiconductor, a force, F, exerted on an electron is given by

$$F = \frac{Qcke^{-kx}}{(1 + e^{-kx})^2}$$

5.25 $\sinh(x) = (e^x - e^{-x})/2$ **5.26** $\cosh(x) = (e^x + e^{-x})/2$

Example 22 *continued*

where c and k are constants, x is the distance from the pn junction and Q is charge. Show that

$$F = \frac{Qck}{2[1 + \cosh(kx)]}$$

Solution

Expanding the denominator of ▮ * ▮ by using

1.13 $\qquad (a + b)^2 = a^2 + 2ab + b^2$

$$(1 + e^{-kx})^2 = 1 + 2e^{-kx} + (e^{-kx})^2$$
$$= 1 + 2e^{-kx} + e^{-kx}e^{-kx}$$

Also we can write 1 as $1 = e^0 = e^{kx-kx} = e^{kx}e^{-kx}$, so in the denominator we have

$$(1 + e^{-kx})^2 = \underbrace{e^{kx}e^{-kx}}_{\text{replacing 1}} \overset{\text{by }5.1}{=} + 2e^{-kx} + e^{-kx}e^{-kx}$$

$$= e^{-kx}(e^{kx} + 2 + e^{-kx}) \qquad \text{[Taking out } e^{-kx}]$$

Substituting this, $(1 + e^{-kx})^2 = e^{-kx}(2 + e^{kx} + e^{-kx})$, into ▮ * ▮ gives

$$F = \frac{Qcke^{-kx}}{e^{-kx}(2 + e^{kx} + e^{-kx})} = \frac{Qck}{2 + e^{kx} + e^{-kx}} \qquad \text{[Cancelling } e^{-kx}]$$

Examining the denominator we have

$$2 + e^{kx} + e^{-kx} = 2 + (e^{kx} + e^{-kx})$$

$$= 2 + 2\left(\frac{e^{kx} + e^{-kx}}{2}\right) \qquad \left[\text{Writing } 1 = \frac{2}{2}\right]$$

$$= 2 + \underbrace{2\cosh(kx)}_{\text{by }5.26} = 2\big[1 + \cosh(kx)\big]$$

Substituting $2 + e^{kx} + e^{-kx} = 2\big[1 + \cosh(kx)\big]$ into the denominator of F gives the required result:

$$F = \frac{Qck}{2\big[1 + \cosh(kx)\big]}$$

SUMMARY

We define the hyperbolic functions in terms of the exponential functions:

5.25 $\qquad \sinh(x) = (e^x - e^{-x})/2$

5.26 $\qquad \cosh(x) = (e^x + e^{-x})/2$

5.27 $\qquad \tanh(x) = \dfrac{\sinh(x)}{\cosh(x)}$

5.1 $\quad a^m\, a^n = a^{m+n}$

Exercise **5(e)**

Solutions at end of book. Complete solutions available at www.palgrave.com/science/engineering/singh

1 Evaluate the following (use your calculator): $\sinh(10)$, $\sinh(-3)$, $\cosh(10)$, $\cosh(1 \times 10^{-3})$, $\tanh(5 \times 10^{7})$ and $\tanh(0.5)$.

2 a Plot the graph of $y = \sinh(x)$ for $x = -4$ to $x = 4$.
   b Plot the graph of $y = \tanh(x)$ for $x = -4$ to $x = 4$.

3 Show that $\sinh(-x) = -\sinh(x)$.

4 Show that $\tanh(-x) = -\tanh(x)$.

5 Show that $\cosh^2(x) - 2\cosh(x)\sinh(x) + \sinh^2(x) = e^{-2x}$.

6 Show that $\cosh^2(x) + 2\cosh(x)\sinh(x) + \sinh^2(x) = e^{2x}$.

7 If $x = a\cosh(t)$ and $y = b\sinh(t)$, show that
$$\frac{x^2}{a^2} - \frac{y^2}{b^2} = 1$$
(Equation of hyperbola, hence the name hyperbolic functions.)

8 [*wave theory*] The speed, v (m/s), of water waves is given by
$$v^2 = 1.8\lambda \tanh\left(\frac{6.3d}{\lambda}\right)$$
where d is depth and λ is wavelength. If $d = 35$ m and $\lambda = 300$ m, evaluate v.

9 [*heat transfer*] The heat transfer rate, Q, of a rod of length L is given by
$$Q = 15\left[\frac{\sinh(2.56L) + (6 \times 10^{-3})\cosh(2.56L)}{\cosh(2.56L) + (6 \times 10^{-3})\sinh(2.56L)}\right]$$
If $L = 30 \times 10^{-3}$ m, find Q.

10 [*electrical principles*] A transmission line of length L has current I_x at a distance x from the sending end given by
$$I_x = \left(\frac{v/z + I}{2}\right)e^{\gamma x} - \left(\frac{v/z - I}{2}\right)e^{-\gamma x}$$

where v is the receiving end voltage, I is the receiving end current, z is the characteristic impedance and γ is the propagation coefficient. Show that at $x = L$
$$I_L = \frac{v}{z}\left[\sinh(\gamma L)\right] + I\left[\cosh(\gamma L)\right]$$

11 [*mechanics*] The tension, T, in a cable is given by
$$T^2 = H^2\left[1 + \sinh^2\left(\frac{wx}{H}\right)\right]$$
where H is horizontal tension, w is weight per unit length and x is horizontal base distance. Show that
$$T = H\cosh\left(\frac{wx}{H}\right).$$

12 [*electrical principles*] The current, I_x, at a distance x along a transmission line of length L is given by
$$I_x = Ie^{-\gamma L}\left[e^{\gamma x} + \left(\frac{Z_0 - Z}{Z_0 + Z}\right)e^{-\gamma x}\right]$$
where I is current, Z_0 is characteristic impedance, Z is load impedance and γ is propagation coefficient. Show that
$$I_x = \frac{2Ie^{-\gamma L}}{Z_0 + Z}\left[Z_0\cosh(\gamma x) + Z\sinh(\gamma x)\right]$$

13 [*electrical principles*] The impedance, Z_x, at a distance x from the sending end of a transmission line is given by
$$Z_x = Z_0\left[\frac{e^{\gamma x} + \left(\dfrac{Z - Z_0}{Z_0 + Z}\right)e^{-\gamma x}}{e^{\gamma x} + \left(\dfrac{Z_0 - Z}{Z_0 + Z}\right)e^{-\gamma x}}\right]$$
where Z_0 is the characteristic impedance, Z is the load impedance and γ is the propagation coefficient. Show that
$$Z_x = Z_0\left[\frac{Z_0\sinh(\gamma x) + Z\cosh(\gamma x)}{Z_0\cosh(\gamma x) + Z\sinh(\gamma x)}\right]$$

Exercise **5(e) continued**

Solutions at end of book. Complete solutions available at www.palgrave.com/science/engineering/singh

14 🔧 [*heat transfer*]   The temperature distribution, $\theta(x)$, along a fin of length L is given by

$$\theta(x) = Ae^{kx} + Be^{-kx}$$

where A, B and k are constants and x is the distance along the fin. A and B can be evaluated from

$$A + B = \theta_0 \quad \text{and} \quad Ae^{kL} + Be^{-kL} = \theta_L$$

(θ_0 is the temperature at the base of the fin and θ_L is the temperature at length L). Show that

i $A = \dfrac{\theta_L - \theta_0\, e^{-kL}}{e^{kL} - e^{-kL}}$

ii $B = \dfrac{\theta_0\, e^{kL} - \theta_L}{e^{kL} - e^{-kL}}$

iii $\theta(x) = \dfrac{\theta_L \sinh(kx) + \theta_0 \sinh\big[k(L - x)\big]}{\sinh(kL)}$

Miscellaneous exercise **5**

Solutions at end of book. Complete solutions available at www.palgrave.com/science/engineering/singh

1 🖥️ [*electronics*]   Evaluate V for $I = 10\ \text{mA}$, $I_0 = 1\ \mu\text{A}$:

$$V = (20 \times 10^{-3}) \ln\!\left(\frac{I}{I_0} + 1\right)$$

2 🖥️ [*electronics*]   The ideal diode equation is given by

$$I = I_0\!\left[\exp\!\left(\frac{qV}{kT}\right) - 1\right]$$

where I, I_0 represent currents, V represents voltage, T is temperature, q is charge and k is a constant. Show that

$$V = \frac{kT}{q} \ln\!\left(\frac{I}{I_0} + 1\right)$$

(Remember that $\exp(x) = e^x$.)

3 〽️ [*vibrations*]   In a vibrational system, the logarithmic decrement, δ, is defined as

$$\delta = \ln\!\left(\frac{ke^{-\zeta\omega t_1}}{ke^{-\zeta\omega t_2}}\right)$$

where ζ is the damping ratio, ω is the angular frequency, t_1, t_2 represent time and k is a constant. Show that

$$\delta = \zeta\omega(t_2 - t_1)$$

4 📊 [*electrical principles*]   During a charging phase of a RC circuit, the voltage, V, across a capacitor, C, is given by

$$V = V_0(1 - e^{-t/\tau})$$

where $V_0 = 100$ volts, t is time (in s) and τ is the time constant (s) [$\tau = RC$].

Plot the graph of V versus t for $0 \le t \le 5\tau$ and state the percentage of voltage V of the initial voltage V_0 at $t = \tau, 2\tau, 3\tau, 4\tau$ and 5τ.

5 🎛️ [*control engineering*]   The output, y, of a system is given by

$$y = 1 - e^{-t/\tau}$$

where $\tau > 0$ is the time constant and t is time. The final value, Y, of the output is defined by

$$Y = \lim_{t \to \infty} (1 - e^{-t/\tau})$$

i Determine Y.

ii Evaluate the values of y for $t = \tau, 2\tau, 3\tau, 4\tau$ and 5τ.

iii What percentage of the final value, Y, is reached at $t = \tau, 2\tau, 3\tau, 4\tau$ and 5τ.

iv Sketch the graph of y versus t.

6 [mechanics]   The velocity, v, of an object falling in air is given by

$$v = \frac{10}{k}(1 - e^{-kt})$$

where k is a positive constant (air resistance).

  i How does v change as k increases if t is large?

  ii What happens to v as $t \to \infty$? (This is called the terminal velocity.)

  iii Plot on the same axes the graphs of v against t for $k = 1$, 2 and 5 and for t between 0 and 10.

  iv Do the graphs satisfy your answers in parts **i** and **ii**?

  v Determine the terminal velocity for $k = 1$, 2 and 5.

7 [aerodynamics]   The friction coefficient, C, is related to Reynolds number, R, by

$$C = kR^n$$

where k and n are constants. Find k and n for the data in the following table.

R	3	4	5	6	7	8
$C\ (\times 10^{-3})$	48	45	43	42	40	39

Use a symbolic manipulator or a graphical calculator for questions 8, 9, 10, 11 and 12.

8 [structures]   A cable is suspended between two level points, $L = 270$ m apart, and has a sag $y = 55$ m. If the load per unit length $w = 0.16$ kN/m, find the horizontal tension H, given that

$$y = \frac{H}{w}\left[\cosh\left(\frac{wL}{2H}\right) - 1\right]$$

9 [heat transfer]   Given that

$\theta = 300 + 100e^{kt}$ where θ is the temperature of a body at time t, plot on the same axes the graphs of θ against t for $t \geq 0$ and $k = -0.01$, -0.1 and -1.

How does the temperature θ change with different values of k?

What is the temperature θ in each case as $t \to \infty$? (This is called the steady-state temperature.)

10 [control engineering]   The output response, y, of a feedback control system is given by

$$y = \frac{k}{1 + k}\left[1 - e^{-(1 + k)t}\right]$$

Plot on the same axes the graphs of y against t (> 0) for $k = 1$, 5, 7 and 10.

11 The normal distribution with zero mean is described by

$$f(x) = \frac{1}{\sigma\sqrt{2\pi}} \exp\left(\frac{-x^2}{2\sigma}\right)$$

Plot the graphs of $f(x)$ against x for $\sigma = 0.1$, 0.2 and 0.3 on the same axes for x between -1 and 1. As σ increases, how does the graph change?

12 [reliability engineering]   The reliability function, $R(t)$, for a particular type of component is given by

$$R(t) = e^{-\sqrt{t}} \qquad t \geq 0$$

  i Plot $R(t)$ and $F(t) = 1 - R(t)$ on the same axes for $t \geq 0$.

  ii What are the values of $R(t)$ and $F(t)$ as $t \to \infty$?

Solutions at end of book. Complete solutions available at www.palgrave.com/science/engineering/singh

13 By using the exponential definition of sinh and cosh, show that

$$\cosh(A)\cosh(B) + \sinh(A)\sinh(B)$$
$$= \cosh(A + B)$$

For questions 14 and 15 use a symbolic manipulator (or a graphical calculator).

14 [mechanics]   The shape of a hanging cable between two fixed points is given by

$$y = k \cosh\left(\frac{x}{k}\right)$$

where $k = T/W$, T is the tension in the cable, W is the load per unit length and x is the horizontal distance. On the same axes, plot the graphs of y against x for x values between -10 and $+10$ and

$k = 10, 20, 30, 40$ and 50. What do you notice about the graph as k increases, that is the tension, T, in the cable increases?

15 [mechanics]   A design engineer wants to design a bridge whose arch is a shape of the form

$$h = c - c\cosh\left(\frac{x}{c}\right)$$

(h gives the height (m) of the arch at a distance x (m) along the base.)

Plot graphs of h, on the same axes, for $-20 \le x \le 20$ and for the following values of c:

i 50   **ii** 100   **iii** 150   **iv** 200
v 250   **vi** 300

CHAPTER 6

Differentiation

SECTION A **The derivative**

By the end of this section you will be able to:
- ▶ evaluate the gradient of a curve at a point
- ▶ understand the notation $\dfrac{dy}{dx}$, $f'(x)$ and $\dot{x}$
- ▶ sketch the derivative graph

A1 Gradient

? **What does the term *gradient* mean?**

Approaching a hill we sometimes see the warning sign

1:4

? **What does this mean?**

The slope of the approaching hill is 4 units across for every
1 unit up (Fig. 1). The fraction $\dfrac{1}{4}$ (or 25%) is called the gradient

Fig. 1

of the hill. Gradient is a measure of the steepness of a hill. Below we examine the gradients of straight lines on graphs. We covered these in Chapter 2 where we defined gradient as

$$\text{Gradient} = \frac{\text{vertical}}{\text{horizontal}}$$

Example 1

Find the gradient of $y = 2x$

Solution

$$\text{Gradient} = \frac{\text{change in } y}{\text{change in } x} = \frac{4-2}{2-1} = 2$$

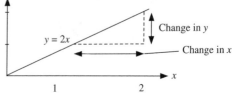
Fig. 2

Of course the gradient in **Example 1** could easily have been evaluated by using 2.1 ,
$y = mx + c$, where m is the gradient and
c is the y-intercept. We have $m = 2$ so the
gradient is 2.

Now let's examine the gradient of the curve
$y = x^2$. Suppose we draw the graph of $y = x^2$
(Fig. 3).

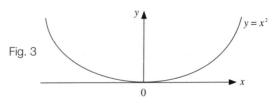
Fig. 3

? **How do we find the gradient of this curve? More precisely the question should be – where do we want to find the gradient of this curve?**

As we move away from the origin, (0, 0), the gradient gets steeper. The gradient is changing as you move along the x axis.

? **Can we find the gradient at an arbitrary point x?**

Consider the graph of $y = x^2$ close to a point x. With graphical calculators or computer algebra packages we can zoom in very close to a point x and see what the graph looks like around x and $x + h$, where h is small (Fig. 4). If this small part is magnified then it will look almost like a straight line (Fig. 5).

Remember the gradient of a straight line:

$$\text{Gradient} = \frac{\text{change in } y}{\text{change in } x}$$

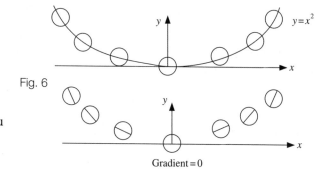

$$= \frac{(x + h)^2 - x^2}{(x + h) - x}$$

$$= \frac{(x^2 + 2hx + h^2) - x^2}{h} = \frac{2hx + h^2}{h} = \frac{h(2x + h)}{h}$$

$$\text{Gradient} = 2x + h \qquad \text{(Cancelling } h\text{'s)}$$

For the graph to be a straight line, h needs to be very small. This can be written as $\lim_{h \to 0}$. Remember $\lim_{h \to 0}$ means we can make h as close as possible to zero. We have

$$\text{Gradient} = \lim_{h \to 0}(2x + h) = 2x$$

When we found the gradient of a straight line it was a real number.

? **But why is the gradient of the curve, $y = x^2$, a function, $2x$?**

Observe the graph in Fig. 6.

? **What do you notice?**

The gradient is getting steeper as you move away from the origin. At the origin, (0, 0), the gradient $= 0$, that is we have **no** steepness (see Fig. 6). As you shift away from the origin, the magnitude of the gradient increases and it depends on the value of x.

$$\text{Gradient function} = 2x$$

The gradient at $x = 0$ is 0.

The gradient at $x = 1$ is 2.

The gradient at $x = -1$ is -2.

The gradient at $x = 10$ is 20.

The **gradient function** is called the **derivative** of the original function and is denoted by $\dfrac{dy}{dx}$. If $y = x^2$ then $\dfrac{dy}{dx} = 2x$. Note that y is the dependent variable and x is the independent variable.

A2 The derivative of a function

If we consider the general function, $y = f(x)$, then we have (Fig. 7):

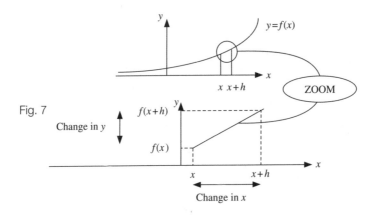

Fig. 7

$$\frac{dy}{dx} = \lim_{h \to 0} \frac{\text{change in } y}{\text{change in } x}$$

$$= \lim_{h \to 0} \frac{f(x + h) - f(x)}{(x + h) - x}$$

$$\frac{dy}{dx} = \lim_{h \to 0} \frac{f(x + h) - f(x)}{h}$$

So for the general function, $y = f(x)$, the derivative is defined to be

This branch of mathematics, along with integration (Chapter 8), is called **calculus**. Two men, **Newton** and **Leibniz**, can claim to have discovered calculus as we know it today. Newton discovered his calculus first in 1655–66 and Leibniz in 1675, although Leibniz's work was published earlier.

6.1

$$\frac{dy}{dx} = \lim_{h \to 0} \frac{f(x + h) - f(x)}{h}$$

The notation, $\dfrac{dy}{dx}$, pronounced 'dee y by dee x' was developed by Leibniz in 1684. This is more popular then Newton's notation, which is $\dot{x}$, pronounced 'x dot' and represents $\dfrac{dx}{dt}$. The $\dot{x}$ symbol is often found in applied mathematics:

$$\frac{dx}{dt} = \dot{x}$$

(Sometimes called the time derivative because t represents time.)

There are also other notations for the derivative. If $y = f(x)$ then the derivative can also be denoted as $f'(x)$, pronounced 'f dashed x':

$$\frac{dy}{dx} = f'(x)$$

Example 2

If $y = 2x$, determine $\dfrac{dy}{dx}$ by using 6.1 .

Solution

Let $f(x) = 2x$ then $f(x + h) = 2(x + h)$. Putting these into 6.1 gives

$$\begin{aligned}
\frac{dy}{dx} &= \lim_{h \to 0} \frac{2(x + h) - 2x}{h} \\
&= \lim_{h \to 0} \frac{2x + 2h - 2x}{h} \qquad \text{[Expanding numerator]} \\
&= \lim_{h \to 0} \frac{2h}{h} = 2 \qquad \text{[Cancelling } h\text{'s]}
\end{aligned}$$

The derivative of $y = 2x$ is 2. Sometimes this is written as

$$\frac{d}{dx}(2x) = 2$$

Note that this is the gradient of the straight line, $y = 2x$, as found in **Example 1**.

Example 3

If $y = x^3$, find $\dfrac{dy}{dx}$ by using 6.1 .

Solution

This is a more challenging example because it uses complicated algebra such as the binomial expansion.

With $f(x) = x^3$ then $f(x + h) = (x + h)^3$. Substituting these into 6.1 gives

† $\qquad \dfrac{dy}{dx} = \lim_{h \to 0} \dfrac{(x + h)^3 - x^3}{h}$

By using the binomial expansion on $(x + h)^3$ we have

$$(x + h)^3 = x^3 + 3x^2 h + 3xh^2 + h^3$$

Substituting this, $(x + h)^3 = x^3 + 3x^2h + 3xh^2 + h^3$, into † gives

$$\frac{dy}{dx} = \lim_{h \to 0} \frac{(x^3 + 3x^2h + 3xh^2 + h^3) - x^3}{h}$$

Example 3 *continued*

$$= \lim_{h \to 0} \frac{3x^2h + 3xh^2 + h^3}{h} \quad \text{[Simplifying numerator]}$$

$$= \lim_{h \to 0} \frac{h(3x^2 + 3xh + h^2)}{h} \quad \text{[Factorizing numerator]}$$

$$= \lim_{h \to 0} (3x^2 + 3xh + h^2) = 3x^2$$

$$\frac{dy}{dx} = 3x^2$$

So the derivative of $y = x^3$ is $3x^2$. It might also be written as $\dfrac{d}{dx}(x^3) = 3x^2$. $\dfrac{dy}{dx}$ means differentiating the **dependent** variable, y, with respect to the **independent** variable, x. Other variables can also be used:

$$\frac{ds}{dt}, \quad \frac{dx}{da}, \quad \frac{d\theta}{d\alpha}, \quad \frac{dM}{dx} \quad \text{etc.}$$

Example 4

Using a Computer algebra system:

i Plot the function $y = x^3$ for $x = 0$ to 6.
ii Plot, on different axes, the function $y = x^3$ for $x = 2.5$ to 3.5, $x = 2.9$ to 3.1 and $x = 2.99$ to 3.01.
   Notice how the graph gets closer and closer to a straight line as x gets close to 3.
iii Evaluate the gradient, $\lim\limits_{h \to 0} \dfrac{(3 + h)^3 - 3^3}{h}$. This value is the derivative, $\dfrac{dy}{dx}$, of $y = x^3$ at $x = 3$.

Solution

The following are the MAPLE commands and output for this example. Even if you don't have access to MAPLE or any other computer algebra system you should be able to follow this example through. It demonstrates how a curve becomes a straight line as two points become closer and closer.

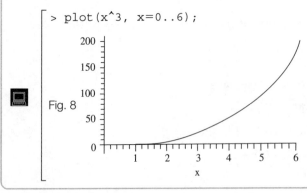

Fig. 8

Example 4 *continued*

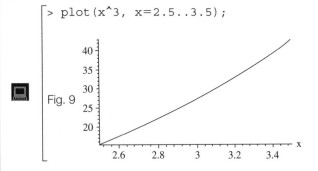

```
> plot(x^3, x=2.5..3.5);
```

Fig. 9

```
> plot(x^3, x=2.9..3.1);
```

Fig. 10

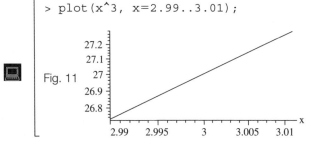

```
> plot(x^3, x=2.99..3.01);
```

Fig. 11

```
> simplify(((3+h)^3 − 3^3)/h);
```
$27 + 9\,h + \bar{}h^2$

```
> grad:=limit(%,h=0);
```
grad: = 27

The graph of $y = x^3$ becomes approximately a straight line as we move closer to 3. Figures 8–11 show the MAPLE output as we have plotted the graph of $y = x^3$ on the stated domains. The gradient is 27 at $x = 3$ for $y = x^3$.

We could also have found the gradient of $y = x^3$ at $x = 3$ by substituting $x = 3$ into the derivative function found in **Example 3.** Since

$$\frac{dy}{dx} = 3x^2$$

at $x = 3$, $\quad \dfrac{dy}{dx} = 3 \times 3^2 = 27$

Example 5 *structures*

The bending moment, M, of a beam of length 10 m is given by

$$M = \begin{cases} x & 0 \le x < 6 \\ 15 - 1.5x & 6 \le x \le 10 \end{cases}$$

where x is the distance along the beam.
Sketch M for $0 \le x \le 10$.

The shear force, F, is given by $F = \dfrac{dM}{dx}$.

Sketch F for $0 \le x \le 10$.

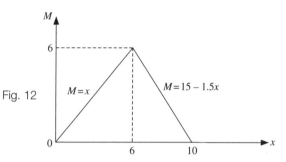

Fig. 12

Solution

The graph of M is shown in Fig. 12.

Since $F = \dfrac{dM}{dx}$, so we need to
evaluate the gradient of M to find F.

For $0 \le x < 6$, the gradient of $M = x$ is 1.
Hence

$$F = 1$$

Fig. 13

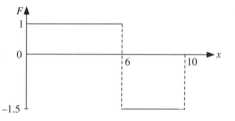

For $6 \le x \le 10$, the gradient of
$M = 15 - 1.5x$ is -1.5. Hence

$$F = -1.5$$

Note that the gradient does not exist at
$x = 6$ (Fig. 13). We say M is **not** differentiable at $x = 6$.

Notice that not **all** functions are differentiable.

SUMMARY

The derivative of $y = f(x)$ is defined as

$$\boxed{6.1} \qquad \frac{dy}{dx} = \lim_{h \to 0} \frac{f(x + h) - f(x)}{h}$$

Also derivatives can be denoted by

$$\frac{dy}{dx} = f'(x) \qquad \dot{x} = \frac{dx}{dt} \quad \text{(time derivative)}$$

Exercise **6(a)**

Solutions at end of book. Complete solutions available
at www.palgrave.com/science/engineering/singh

1 By using the definition $\boxed{6.1}$, find $\dfrac{dy}{dx}$
for the following:

  a $y = x$ **b** $y = 5x$

  c $y = x^2 + x$ **d** $y = \dfrac{1}{x}$

2 *[structures]*   The bending moment,
M (kN m), of a beam of length 8 m is
given by

$$M = \begin{cases} 2.16x & 0 \le x < 5 \\ 28.8 - 3.6x & 5 \le x \le 8 \end{cases}$$

where x is the distance along the beam.

The shear force, V, is given by $V = \dfrac{dM}{dx}$.
Sketch V for $0 \le x \le 8$.

3 [electrical principles]   The current, i, through an inductor is given by

$$i = \begin{cases} t & 0 \le t \le 2 \\ 4 - t & 2 < t \le 4 \end{cases}$$

Sketch the current i.

The voltage, V, applied across the inductor is given by $V = 0.25\dfrac{di}{dt}$. Sketch the voltage, V, on different axes.

4 [electrical principles]   The voltage, V, applied across an inductor of inductance L is

$$V = L\frac{di}{dt}$$

where i represents the current through the inductor. If the current i is as shown in Fig. 14, find V.

Fig. 14

Use a computer algebra system or a graphical calculator for question 5.

5 **i** Plot the function $y = 10x^6$ for $x = 0$ to 3.

  ii Plot on different axes the function $y = 10x^6$ for $x = 1.4$ to 1.6, $x = 1.49$ to 1.51 and $x = 1.499$ to 1.501.

  iii Evaluate $\displaystyle\lim_{h \to 0} \frac{10(1.5 + h)^6 - 10(1.5)^6}{h}$.
   This is $\dfrac{dy}{dx}$ of $y = 10x^6$ at $x = 1.5$.

SECTION B **Derivatives of functions**

By the end of this section you will be able to:
▶ find the derivatives of common engineering functions
▶ apply the chain rule

B1 **General formulae for derivatives**

To find a derivative of a function using 6.1 can be very tedious. For example, if $y = x^{10}$ and you want to obtain an expression for $\dfrac{dy}{dx}$ then you would have to expand $(x + h)^{10}$ and then find $\dfrac{dy}{dx}$:

$$\frac{dy}{dx} = \lim_{h \to 0} \frac{(x + h)^{10} - x^{10}}{h}$$

However by repeating the earlier derivation for the power function, $y = x^n$, we find that there is a general formula for $\dfrac{dy}{dx}$. If $y = x^n$ where n is any real number then

$$\frac{dy}{dx} = nx^{n-1}$$

We will not prove this result but it can be derived by using 6.1 .

If $y = x^{10}$ then $\dfrac{dy}{dx} = 10x^{10-1} = 10x^9$.

There are general formulae for finding the derivative of many functions such as exponential, trigonometric, logarithmic etc. Some of these common formulae are set out in Table 1. Generally it is easier to use this table rather than 6.1 .

TABLE 1

Reference number	$y = f(x)$	$\dfrac{dy}{dx} = f'(x)$
6.2	x^n	nx^{n-1}
6.3	kx^n	nkx^{n-1}
6.4	Constant	0
6.5	e^x	e^x
6.6	$\sin(x)$	$\cos(x)$
6.7	$\cos(x)$	$-\sin(x)$
6.8	$\tan(x)$	$\sec^2(x)$
6.9	$\ln(x)$	$\dfrac{1}{x}$

For 6.6 to 6.8 , x is in radians.

For $y = x^3$ we have

$$\frac{dy}{dx} = 3x^{3-1} = 3x^2 \qquad [\text{By } 6.2 \text{ with } n = 3]$$

The process in differentiating x^3 is to multiply x by the given index, 3, and reduce the index by 1; $3 - 1 = 2$. Thus differentiating x^3 gives $3x^2$.

For $y = 5x^2$ we have

$$\frac{dy}{dx} = (2 \times 5)x^{2-1} \qquad [\text{By } 6.3 \text{ with } n = 2 \text{ and } k = 5]$$

$$= 10x^1 = 10x$$

In Table 1 the variables x and y can be replaced by other variables such as s, t, u, v, α, β, etc., and the same formulae apply. For example if $y = u^n$ then by 6.2 :

$$\frac{dy}{du} = nu^{n-1}$$

In addition to Table 1 we have

$$\frac{d}{dx}(u \pm v) = \frac{du}{dx} \pm \frac{dv}{dx}$$

where u and v are functions of x. This formula means that you can differentiate each part separately and then add or subtract the resulting functions. For example if $y = x^3 + \sin(x)$ then

$$\frac{dy}{dx} = 3x^2 + \cos(x)$$

Also

$$\frac{d}{dx}(ku) = k\frac{du}{dx} \quad \text{(where k is a constant)}$$

which means you can take out the constant term, k, and differentiate the remaining expression. For example if $y = 10\cos(x)$ then

$$\frac{dy}{dx} = 10[-\sin(x)] = -10\sin(x)$$

Example 6

Find $\dfrac{dy}{dx}$ in each of the following cases:

a $y = x^6$ **b** $y = 3x^7 + x^6$ **c** $y = e^x + \dfrac{3}{\sqrt{x}} + \tan(x)$

Solution

a Applying 6.2 on $y = x^6$ we have

$$\frac{dy}{dx} = 6x^{6-1} = 6x^5$$

b $y = 3x^7 + x^6$

$$\frac{dy}{dx} = \underbrace{(7 \times 3)x^{7-1}}_{\text{by }6.3} + \underbrace{6x^5}_{\text{by part a}} = 21x^6 + 6x^5$$

6.2 $\dfrac{d}{dx}(x^n) = nx^{n-1}$ 6.3 $\dfrac{d}{dx}(kx^n) = nkx^{n-1}$

Example 6 *continued*

c $y = e^x + \dfrac{3}{\sqrt{x}} + \tan(x)$

We can find the derivatives of e^x and $\tan(x)$ from Table 1.

? **How do we find the derivative of $\dfrac{3}{\sqrt{x}}$?**

? **Can we write this term so that we can use Table 1?**

We need to convert $3/\sqrt{x}$ to a power:

$$\frac{3}{\sqrt{x}} = \frac{3}{x^{\frac{1}{2}}} = 3x^{-\frac{1}{2}}$$

We have

$$y = e^x + 3x^{-\frac{1}{2}} + \tan(x)$$

$$\frac{dy}{dx} = \underbrace{e^x}_{\text{by } 6.5} + \underbrace{\left(-\frac{1}{2} \times 3\right)x^{-\frac{1}{2}-1}}_{\text{by } 6.3} + \underbrace{\sec^2(x)}_{\text{by } 6.8}$$

$$= e^x - \frac{3}{2}x^{-\frac{3}{2}} + \sec^2(x)$$

Example 7 *mechanics*

The displacement, s, of a particle is given by

a $s = 80t - t^3$ **b** $s = t^3 - 12t^2 + 18t$

In each case find the velocity, v, where $v = \dfrac{ds}{dt}$.

Solution

a $v = \dfrac{ds}{dt} = 80 - 3t^2$

b We have

$$\frac{ds}{dt} = 3t^2 - 24t + 18 \underset{\text{factorizing}}{=} 3(t^2 - 8t + 6) \quad [\text{3 is a common factor}]$$

6.3 $\dfrac{d}{dx}(kx^n) = nkx^{n-1}$ 6.5 $\dfrac{d}{dx}(e^x) = e^x$ 6.8 $\dfrac{d}{dx}[\tan(x)] = \sec^2(x)$

Notice that: $\dfrac{d}{dt}(18t) = \dfrac{d}{dt}(18t^1) \underset{\text{by } \boxed{6.3}}{=} (18 \times 1)t^{1-1} = 18t^0 = 18$. Similarly

$$\frac{d}{dt}(80t) = 80$$

In general, if k is a constant then

$$\frac{d}{dt}(kt) = k$$

Differentiation itself is **not** difficult but you might find the algebra more demanding. If the examples have been difficult to follow then invest some time in the revision of **indices** (Sections **1C** and **5A**).

Example 8

If $y = x^{3/2} + 5\cos(x)$, find $\dfrac{dy}{dx}$.

Solution

Using Table 1 we have:

$$\frac{dy}{dx} = \underbrace{\frac{3}{2}x^{\frac{3}{2}-1}}_{\text{by } \boxed{6.2}} + 5\underbrace{[-\sin(x)]}_{\text{by } \boxed{6.7}}$$

$$= \frac{3}{2}x^{1/2} - 5\sin(x)$$

B2 Chain rule

The **Chain rule** says that if y is a function of u and u is a function of x then

$\boxed{6.10}$
$$\frac{dy}{dx} = \frac{dy}{du} \cdot \frac{du}{dx}$$

The 'chain rule' is also called a 'function of a function rule'. Let's apply the chain rule to an example.

$\boxed{6.2}$ $\dfrac{d}{dx}(x^n) = nx^{n-1}$ $\boxed{6.3}$ $\dfrac{d}{dx}(kx^n) = nkx^{n-1}$ $\boxed{6.7}$ $\dfrac{d}{dx}[\cos(x)] = -\sin(x)$

Example 9

If $y = (x^2 + 2)^3$, find $\dfrac{dy}{dx}$.

Solution

At present we cannot use any of the derivative formulae of Table 1. Let $u = x^2 + 2$ then we have $y = u^3$.

If we differentiate this function, $y = u^3$, we obtain the derivative $\dfrac{dy}{du}$:

$$\frac{dy}{du} = 3u^2 \qquad \left[\text{By } \frac{d}{du}(u^n) = nu^{n-1} \right]$$

? We need to find $\dfrac{dy}{dx}$, not $\dfrac{dy}{du}$. **How can we obtain $\dfrac{dy}{dx}$?**

From the chain rule, 6.10 , we have $\dfrac{dy}{dx} = \dfrac{dy}{du} \cdot \dfrac{du}{dx} = 3u^2 \cdot \dfrac{du}{dx}$ thus we need $\dfrac{du}{dx}$.

Differentiating $u = x^2 + 2$ gives

$$\frac{du}{dx} = 2x$$

Substituting into the chain rule:

$$\frac{dy}{dx} = \frac{dy}{du} \cdot \frac{du}{dx}$$

$$= 3u^2 \cdot 2x = 6xu^2 \quad [\text{Because } 3 \times 2x = 6x]$$

? **Is this our final answer?**

No, because $\dfrac{dy}{dx} = 6xu^2$, but u was not part of the original function. We need to replace u.

? **What is u?**

$u = x^2 + 2$. Hence $\dfrac{dy}{dx} = 6x(x^2 + 2)^2$.

Example 10

If $y = \dfrac{(x^2 + 2)^{3/2}}{3}$, find $\dfrac{dy}{dx}$.

Solution

We cannot use any of the derivative formulae of Table 1. Let $u = x^2 + 2$ then $y = \dfrac{u^{3/2}}{3}$.

Example 10 *continued*

If we differentiate this function we obtain the derivative $\dfrac{dy}{du}$:

$$\frac{dy}{du} = \frac{3}{2}\frac{u^{\frac{3}{2}-1}}{3} \quad \left[\text{By } \frac{d}{du}\left(ku^n\right) = nku^{n-1}\right]$$

$$= \frac{u^{1/2}}{2} \qquad [\text{Cancelling 3's}]$$

Again we need to find $\dfrac{dy}{dx}$, not $\dfrac{dy}{du}$. **How can we obtain** $\dfrac{dy}{dx}$?

From the chain rule, 6.10 , we have $\dfrac{dy}{dx} = \dfrac{dy}{du} \cdot \dfrac{du}{dx} = \dfrac{u^{1/2}}{2} \cdot \dfrac{du}{dx}$, thus we need $\dfrac{du}{dx}$.

Differentiating $u = x^2 + 2$ gives

$$\frac{du}{dx} = 2x$$

Substituting into the chain rule:

$$\frac{dy}{dx} = \frac{dy}{du} \cdot \frac{du}{dx}$$

$$= \frac{u^{1/2}}{2} \cdot 2x = xu^{1/2} \quad [\text{Cancelling 2's}]$$

Since $u = x^2 + 2$ we have $\dfrac{dy}{dx} = x(x^2 + 2)^{1/2}$.

Example 11

If $y = e^{x^2 - 2x}$, find $\dfrac{dy}{dx}$.

Solution

Again in Table 1 we have the derivative of e^x but not e^{x^2-2x}. Therefore let $u = x^2 - 2x$ and we have

$$y = e^u \qquad \frac{dy}{du} = e^u$$

How can we obtain $\dfrac{dy}{dx}$?

Use the chain rule. We need to find $\dfrac{du}{dx}$:

$$u = x^2 - 2x$$

$$\frac{du}{dx} = 2x - 2$$

Example 11 *continued*

Applying

6.10 $\dfrac{dy}{dx} = \dfrac{dy}{du} \cdot \dfrac{du}{dx}$

gives

$$\dfrac{dy}{dx} = e^u \cdot (2x - 2) = (2x - 2)e^{x^2 - 2x} \qquad \text{[Replacing } u = x^2 - 2x]$$

Example 12

If $y = \sin(kx)$ where k is a constant, show that

$$\dfrac{dy}{dx} = k \cos(kx)$$

Solution

Let $u = kx$ then $y = \sin(u)$ and $\dfrac{dy}{du} = \cos(u)$.

It is worth remembering that the derivative of sin is cos. (Derivative of cos is $-\sin$.)

$$u = kx$$

$$\dfrac{du}{dx} = k$$

Applying the chain rule, 6.10 , gives

$$\dfrac{dy}{dx} = k \cos(u) = k \cos(kx) \quad \text{[Replacing } u = kx]$$

Thus by using the chain rule we can easily verify the results shown in Table 2.

Reference number	$y = f(x)$	$\dfrac{dy}{dx} = f'(x)$
6.11	e^{kx}	$k\,e^{kx}$
6.12	$\sin(kx)$	$k \cos(kx)$
6.13	$\cos(kx)$	$-k \sin(kx)$

TABLE 2

Rather than applying the chain rule each time you can use Table 2 to differentiate functions of the type shown there. For example, e^{10x} is like e^{kx} and we have the formula

6.11 $$\frac{d}{dx}(e^{kx}) = k\,e^{kx}$$

Put $k = 10$ and we get

$$\frac{d}{dx}(e^{10x}) = 10e^{10x}$$

Similarly we have

$$\frac{d}{dx}[\sin(5x)] = 5\cos(5x)$$
$$\text{by } 6.12$$

$$\frac{d}{dx}[\cos(3x)] = -3\sin(3x)$$
$$\text{by } 6.13$$

SUMMARY

It is easier to use Tables 1 and 2 to find the derivatives rather than formula 6.1 .
The chain rule says that if y is a function of u and u is a function of x then

6.10 $$\frac{dy}{dx} = \frac{dy}{du} \cdot \frac{du}{dx}$$

Exercise 6(b)

Solutions at end of book. Complete solutions available at www.palgrave.com/science/engineering/singh

1 Find $\dfrac{dy}{dx}$ in each of the following cases:

a $y = x^3$ **b** $y = 3x^5 + x^3$

c $y = \sqrt{x} + \sin(x)$

d $y = \dfrac{1}{\sqrt[3]{x}} + e^x + \cos(x)$

e $y = \dfrac{3}{\sqrt{x}} + \dfrac{5}{\sqrt[3]{x}}$

2 🚗 [*mechanics*]   The displacement, s, of a particle is given by

$$s = 75t - t^3 \quad (t \geq 0)$$

i Find the velocity v where $v = \dfrac{ds}{dt}$.

ii At what value of t is $v = 0$?

3 〰 [*fluid mechanics*]   The density, ρ, is defined as $\rho = \dfrac{1}{V}$ where V represents volume of unit mass (specific volume). Find an expression for $\dfrac{dV}{d\rho}$.

4 Find $\dfrac{dy}{dx}$ in each of the following cases:

a $y = \cos(3x)$ **b** $y = \sin(\pi x)$

c $y = e^{11x}$

Exercise **6(b) continued**

Solutions at end of book. Complete solutions available at www.palgrave.com/science/engineering/singh

5 [electromagnetism] Faraday's law states that the electromotive force, E, induced by N turns of a coil with a flux, ϕ, passing through it, is given by

$$E = -N\frac{d\phi}{dt}$$

If $\phi = K\sin(2\pi ft)$ where K is a constant and f is the frequency, determine E.

6 Find $\dfrac{dy}{dx}$ in each of the following cases:

a $y = \sin(x^2)$ **b** $y = e^{\sin(x)}$

c $y = \ln(x^2)$ **d** $y = \cos(2x^3 - 3x)$

e $y = (x^2 + 1)^{10}$ **f** $y = \tan(5x^7 + 3x^4)$

SECTION C **Chain rule revisited**

By the end of this section you will be able to:
► apply the chain rule to differentiate complicated functions

---

C1 **Applications of the chain rule**

Applying the chain rule to differentiate functions like

$$\sin(x^3 + 2x), \quad e^{\cos(x) + x^2}, \quad \ln[\tan(x) + \sec(x)]$$

can be laborious. Using the rule on the following sine functions we obtain:

$$\frac{d}{dx}\big[\sin(x^2)\big] = 2x\cos(x^2)$$

$$\frac{d}{dx}\big[\sin(x^5)\big] = 5x^4\cos(x^5)$$

$$\frac{d}{dx}\big[\sin(x^3 + 2x)\big] = (3x^2 + 2)\cos(x^3 + 2x)$$

? **What do you notice?**

$$\frac{d}{dx}\big[\sin(u)\big] = (\text{derivative of } u) \times \cos(u)$$

$$\frac{d}{dx}\big[\sin(u)\big] = \frac{du}{dx}\cos(u)$$

where u is a function of x and $\dfrac{du}{dx}$ is its derivative. Note the similarity with the derivative given in Table 1:

$$\frac{d}{dx}\big[\sin(x)\big] = \cos(x)$$

$$\frac{d}{dx}\big[\sin(u)\big] = \cos(u)\frac{du}{dx}$$

We find this similarity with each derivative of Table 1, for example:

Table 1

$$\frac{d}{dx}\big[\cos(u)\big] = -\sin(u)\frac{du}{dx} \qquad \frac{d}{dx}\big[\cos(x)\big] = -\sin(x)$$

$$\frac{d}{dx}(e^u) = e^u\frac{du}{dx} \qquad\qquad \frac{d}{dx}(e^x) = e^x$$

where u is a function of x and $\dfrac{du}{dx}$ is the derivative of u.

Thus we can use Table 1 in a slightly modified form to differentiate complicated functions. The modifications are

1 replace the x in Table 1 by u where u is a function of x

2 multiply your result of **1** by $\dfrac{du}{dx}$.

By implementing these two steps and adding more functions we obtain the results shown in Table 3.

TABLE 3	Reference number	y (u is a function of x)	$\dfrac{dy}{dx}$ (or y')
	6.14	u^n	$nu^{n-1}\dfrac{du}{dx}$
	6.15	ku^n	$nku^{n-1}\dfrac{du}{dx}$
	6.16	e^u	$e^u\dfrac{du}{dx}$
	6.17	a^u	$a^u\ln(a)\dfrac{du}{dx}$
	6.18	$\ln(u)$	$\dfrac{1}{u}\dfrac{du}{dx}$
	6.19	$\sin(u)$	$\cos(u)\dfrac{du}{dx}$
	6.20	$\cos(u)$	$-\sin(u)\dfrac{du}{dx}$
	6.21	$\tan(u)$	$\sec^2(u)\dfrac{du}{dx}$
	6.22	$\sec(u)$	$\sec(u)\tan(u)\dfrac{du}{dx}$
	6.23	$\operatorname{cosec}(u)$	$-\operatorname{cosec}(u)\cot(u)\dfrac{du}{dx}$

TABLE 3 CONTINUED

Reference number	y (u is a function of x)	$\dfrac{dy}{dx}$ (or y')
6.24	$\cot(u)$	$-\operatorname{cosec}^2(u)\dfrac{du}{dx}$
6.25	$\sinh(u)$	$\cosh(u)\dfrac{du}{dx}$
6.26	$\cosh(u)$	$\sinh(u)\dfrac{du}{dx}$
6.27	$\tanh(u)$	$\operatorname{sech}^2(u)\dfrac{du}{dx}$
6.28	$\operatorname{sech}(u)$	$-\operatorname{sech}(u)\tanh(u)\dfrac{du}{dx}$
6.29	$\operatorname{cosech}(u)$	$-\operatorname{cosech}(u)\coth(u)\dfrac{du}{dx}$
6.30	$\coth(u)$	$-\operatorname{cosech}^2(u)\dfrac{du}{dx}$

So in the last column we have the chain rule, $\dfrac{dy}{du} \times \dfrac{du}{dx}$. The first part gives $\dfrac{dy}{du}$ and then we multiply this by $\dfrac{du}{dx}$. For example, 6.14 says

$$\frac{d}{dx}\left[u^n\right] = nu^{n-1}\frac{du}{dx}$$
$$\frac{d}{dx}\left[u^n\right] = nu^{n-1} \times (\text{derivative of } u)$$

where u is any function of x.

This chapter on differentiation is more straightforward if we use Table 3, despite the horrendous appearance of formulae. In practice it is usual to use a table of standard derivatives. This book will use Table 3 as its table of derivatives.

Examples 13–17 use Table 3.

Example 13

Find the derivatives with respect to x of

a $(x^3 + 2x)^5$ **b** $\tan(x^7 - 5x)$ **c** $\sinh(x^2 + 3x)$

Solution

a By

Example 13 *continued*

6.14
$$\frac{d}{dx}(u^n) = nu^{n-1}\frac{du}{dx}$$

with $u = x^3 + 2x$ gives

$$\frac{d}{dx}[(x^3 + 2x)^5] = 5(x^3 + 2x)^4 \times \underbrace{(3x^2 + 2)}_{\text{derivative of } x^3 + 2x}$$

b Similarly by

6.21
$$\frac{d}{dx}[\tan(u)] = \sec^2(u)\frac{du}{dx}$$

with $u = x^7 - 5x$ gives

$$\frac{d}{dx}[\tan(x^7 - 5x)] = \sec^2(x^7 - 5x) \times \underbrace{(7x^6 - 5)}_{\text{derivative of } x^7 - 5x} = (7x^6 - 5)\sec^2(x^7 - 5x)$$

c Also using

6.25
$$\frac{d}{dx}[\sinh(u)] = \cosh(u)\frac{du}{dx}$$

with $u = x^2 + 3x$ gives

$$\frac{d}{dx}[\sinh(x^2 + 3x)] = \cosh(x^2 + 3x) \times \underbrace{(2x + 3)}_{\text{derivative of } x^2 + 3x} = (2x + 3)\cosh(x^2 + 3x)$$

Example 14

Determine $\dfrac{d}{dx}[\sin(x^3 + 2x)]$

Solution

By 6.19, $\dfrac{d}{dx}[\sin(u)] = \cos(u)\dfrac{du}{dx}$ with $u = x^3 + 2x$ we have

$$\frac{du}{dx} = 3x^2 + 2$$

$$\frac{d}{dx}[\sin(x^3 + 2x)] = \cos(x^3 + 2x) \cdot (3x^2 + 2)$$

$$= (3x^2 + 2)\cos(x^3 + 2x)$$

Example 15

Obtain $\dfrac{d}{dx}[e^{x^2 + \cos(x)}]$

Solution

Using 6.16 , $\dfrac{d}{dx}(e^u) = e^u \dfrac{du}{dx}$, with

$$u = x^2 + \cos(x)$$

$$\frac{du}{dx} = 2x - \sin(x)$$

Thus

$$\frac{d}{dx}[e^{x^2 + \cos(x)}] = e^{x^2 + \cos(x)}[2x - \sin(x)]$$

$$= [2x - \sin(x)]e^{x^2 + \cos(x)} \quad \text{[Writing the square bracket term first]}$$

Example 16

Find $\dfrac{dy}{dx}$ given that $y = \ln[\cos(x)]$.

Solution

Applying 6.18 , $\dfrac{d}{dx}[\ln(u)] = \dfrac{1}{u}\dfrac{du}{dx}$, with $u = \cos(x)$ and therefore

$$\frac{du}{dx} = -\sin(x)$$

$$\frac{dy}{dx} = \frac{1}{\cos(x)}[-\sin(x)]$$

$$= -\frac{\sin(x)}{\cos(x)} \underset{\text{by }\ 4.35}{=} -\tan(x)$$

Example 17 *electrical principles*

The voltage, $v = 1 - e^{-t/(2 \times 10^{-6})}$, across a capacitor, $C = 0.2 \times 10^{-6}$ F, has a current i, given by

$$i = C\frac{dv}{dt}$$

Find i.

4.35 $\sin(x)/\cos(x) = \tan(x)$

Example 17 *continued*

Solution

Substituting the given values yields

$$i = (0.2 \times 10^{-6}) \frac{d}{dt} \left[1 - e^{-t/(2 \times 10^{-6})} \right] \qquad \boxed{\ast}$$

Differentiating the 1 inside the square brackets gives 0 so we are left with differentiating the exponential function. Use $\boxed{6.16}$, $\dfrac{d}{dt}[e^u] = e^u \dfrac{du}{dt}$, with

$$u = -\frac{t}{2 \times 10^{-6}}$$

$$\frac{du}{dt} = -\frac{1}{2 \times 10^{-6}} \qquad \left[\text{Using } \frac{d}{dt}(kt) = k \right]$$

$$\frac{d}{dt}\left[1 - e^{-t/(2 \times 10^{-6})} \right] = 0 - e^{-t/(2 \times 10^{-6})} \left(-\frac{1}{2 \times 10^{-6}} \right)$$

$$= \frac{1}{2 \times 10^{-6}} e^{-t/(2 \times 10^{-6})}$$

Substituting this into $\boxed{\ast}$ gives

$$i = (0.2 \times 10^{-6}) \underbrace{\frac{1}{2 \times 10^{-6}}}_{= 0.1} e^{-t/(2 \times 10^{-6})} = 0.1 e^{-t/(2 \times 10^{-6})}$$

SUMMARY

Rather than applying the chain rule to differentiate complicated functions it is easier to use Table 3.

Exercise **6(c)**

Solutions at end of book. Complete solutions available at www.palgrave.com/science/engineering/singh

1 Find $\dfrac{dy}{dx}$ in each of the following cases:

a $y = (x^2 + 2)^3$ **b** $y = \sin(x^2)$

c $y = \cos(x^2 + 2x)$ **d** $y = \ln(x^3 + x)$

e $y = \sinh(3x^5 + x^3)$

2 Differentiate the following with respect to x and simplify your answer:

a $\ln[\sin(x)]$ **b** $\ln[\cosh(x)]$

c $\tan[\cos(x)]$ **d** 10^{x^2}

e $\sin[\ln(x^2)]$

3 [*electrical principles*]   The current, i, in a circuit is given by

$$i = (5 \times 10^{-3})\left[1 - e^{-(1 \times 10^3)t} \right]$$

Exercise **6(c) continued**

Solutions at end of book. Complete solutions available at www.palgrave.com/science/engineering/singh

Find the voltage, v, where

$$v = (3 \times 10^{-3})\frac{di}{dt}$$

4 [electrical principles]   The voltage, v, across a capacitor, C, in series with a resistor, R, is given by

$$v = Ee^{-t/RC}$$

where $E > 0$ is a constant. Determine i where $i = C\frac{dv}{dt}$.

5 [electronics]   The current, i, through a diode is given by

$$i = I_s(e^{11600V/\eta T} - 1)$$

where V is the voltage, η is the emission coefficient, T is the temperature and I_s is the saturation current. The a.c. resistance, r, of a diode is defined as

$$r = \frac{1}{di/dV}$$

Find r by considering η, T and I_s as constants.

6 [electronics]   The flow of charge, q, is given by

$$q = \tau I_s e^{\alpha v/\eta}$$

where τ is the mean lifetime, I_s is the saturation current, v is the voltage, η is the emission coefficient and α is a constant. The diffusion capacitance $C_D = \frac{dq}{dv}$. Find C_D by considering τ, I_s, α and n as constants.

7 [reliability engineering]   The reliability function, $R(t)$, of a component is given by

$$R(t) = e^{-\sqrt{t}} \quad (t \geq 0)$$

Find $\frac{d}{dt}[1 - R(t)]$.

8 [reliability engineering]   The distribution function, F, for a set of components is given by

$$F = 4 - \frac{20}{5 + t/3}$$

where t is in years. Determine the density function $\frac{dF}{dt}$.

SECTION D **Product and quotient rules**

By the end of this section you will be able to:
▶ apply the product rule
▶ apply the quotient rule

D1 **Product rule**

? **How can we differentiate functions like $x^2 \ln(x)$?**

Note that $x^2 \ln(x)$ is a product of two functions, x^2 and $\ln(x)$. We can find the derivative of x^2 and $\ln(x)$ on their own, but

$$\frac{d}{dx}[x^2 \ln(x)] \neq \frac{d}{dx}(x^2) \times \frac{d}{dx}[\ln(x)] \qquad \text{[Not equal]}$$

That is, the derivative of $x^2 \ln(x)$ **does not equal** the derivative of x^2 times the derivative of $\ln(x)$. We use the **product rule** to find $\dfrac{d}{dx}\left[x^2 \ln(x)\right]$.

The product rule is applied to functions of the type

$$y = u \times v = uv$$

where u and v are functions of x. Note that y is a product of u and v. The product rule is

$$\boxed{6.31} \qquad \frac{d}{dx}(uv) = \left(\frac{du}{dx} \cdot v\right) + \left(u \cdot \frac{dv}{dx}\right) \text{ or } (uv)' = u'v + uv'$$

Remember the symbol $'$ means differentiate and we could just as well use $(uv)' = u'v + uv'$. It is the same formula but using different notation.

Example 18

If $y = x \cos(x)$, find $\dfrac{dy}{dx}$.

Solution

Since y is the product of x and $\cos(x)$ we apply the product rule:

$$\boxed{6.31} \qquad \frac{d}{dx}(uv) = \left(\frac{du}{dx} \cdot v\right) + \left(u \cdot \frac{dv}{dx}\right)$$

Let

$$u = x \qquad\qquad v = \cos(x)$$

In order to use the product rule we have to differentiate these:

$$\frac{du}{dx} = 1 \qquad\qquad \frac{dv}{dx} = -\sin(x) \qquad \left[\text{Because } \frac{d}{dx}[\cos(x)] = -\sin(x)\right]$$

Substituting these into $\boxed{6.31}$ we have

$$\frac{d}{dx}\left[x\cos(x)\right] = \left(\frac{du}{dx} \cdot v\right) + \left(u \cdot \frac{dv}{dx}\right)$$

$$= 1 \cdot \cos(x) + x \cdot \left[-\sin(x)\right]$$

$$= \cos(x) - x\sin(x)$$

We could have chosen $u = \cos(x)$ and $v = x$.

You can remember the verbal form of the product rule, 'differentiate the first and leave the second plus differentiate the second and leave the first'.

Example 19

If $y = x^2 \ln(x)$, find $\dfrac{dy}{dx}$.

Solution

Since y is a product of two functions x^2 and $\ln(x)$, we use the product rule, 6.31 . Let

$$u = x^2 \qquad\qquad v = \ln(x)$$

In order to use 6.31 , what else do we need to find?

$$\frac{du}{dx} = 2x \qquad\qquad \underset{\text{by } 6.9}{\frac{dv}{dx} = \frac{1}{x}}$$

Substituting these into $\dfrac{d}{dx}(uv) = \dfrac{du}{dx} \cdot v + u \cdot \dfrac{dv}{dx}$ gives

$$\frac{d}{dx}\left[x^2 \ln(x)\right] = 2x \cdot \ln(x) + x^2 \cdot \frac{1}{x}$$

$$= 2x \ln(x) + x$$

$$= x[2 \ln(x) + 1] \qquad\qquad \text{[Factorizing]}$$

We could select u to be $\ln(x)$ and v to be x^2, it doesn't make any difference. For the product rule u can be the second function and v the first, or vice versa.

Example 20 *electrical principles*

The voltage, E, applied across an inductor with inductance L is given by

$$E = L\frac{di}{dt}$$

where i is the current through the inductor. For $i = te^{-t}$ and $L = 5 \times 10^{-3}$, find E.

Solution

Substituting $L = 5 \times 10^{-3}$ and $i = te^{-t}$ into $E = L\dfrac{di}{dt}$ gives

$$E = (5 \times 10^{-3})\frac{d}{dt}(te^{-t})$$

6.9 $\dfrac{d}{dx}[\ln(x)] = \dfrac{1}{x}$

Example 20 *continued*

How do we differentiate te^{-t}?

Since te^{-t} is a product of t times e^{-t}, we need to use the product rule:

6.31 $(uv)' = u'v + uv'$

Let

$$u = t \qquad v = e^{-t}$$
$$u' = 1 \qquad v' = -e^{-t}$$
$$\text{by } 6.11$$

Hence $E = (5 \times 10^{-3})\big[\underbrace{1 \cdot e^{-t} + t \cdot (-e^{-t})}_{\text{by } u' \cdot v + u \cdot v'}\big]$

$$= (5 \times 10^{-3})(e^{-t} - te^{-t})$$

$$= (5 \times 10^{-3})(1 - t)e^{-t} \qquad [\text{Factorizing out } e^{-t}]$$

D2 Quotient rule

How can we differentiate functions like $\dfrac{\sin(x)}{x^2}$?

There is a formula called the **quotient rule** which gives the derivative of quotients. The quotient rule is applied to functions of the type

$$y = \frac{u}{v} \quad (v \neq 0)$$

where u and v are functions of x. Note that y is the quotient of u by v. The quotient rule is

6.32 $$\frac{d}{dx}\left(\frac{u}{v}\right) = \frac{\left(\dfrac{du}{dx} \cdot v\right) - \left(u \cdot \dfrac{dv}{dx}\right)}{v^2} \quad \text{or} \quad \left(\frac{u}{v}\right)' = \frac{u'v - uv'}{v^2}$$

Example 21

Find $\dfrac{d}{dx}\left(\dfrac{\sin(x)}{x^2}\right)$.

Solution

Since we have a quotient, we apply 6.32 :

$$u = \sin(x) \qquad v = x^2$$
$$\frac{du}{dx} = \cos(x) \qquad \frac{dv}{dx} = 2x$$

6.11 $(e^{kt})' = ke^{kt}$

Example 21 *continued*

Putting these into

$$\frac{d}{dx}\left(\frac{u}{v}\right) = \frac{\frac{du}{dx}\cdot v - u\cdot\frac{dv}{dx}}{v^2}$$

yields

$$\frac{d}{dx}\left(\frac{\sin(x)}{x^2}\right) = \frac{\cos(x)\cdot x^2 - \sin(x)\cdot 2x}{(x^2)^2}$$

$$= \frac{x^2\cos(x) - 2x\sin(x)}{x^4}$$

$$= \frac{x\cos(x) - 2\sin(x)}{x^3} \quad \text{[Cancelling } x\text{]}$$

Example 22

By using $\boxed{6.32}$, show that $\dfrac{d}{dx}[\tan(x)] = \sec^2(x)$.

Solution

From

$\boxed{4.35}$ $\tan(x) = \dfrac{\sin(x)}{\cos(x)}$

we have $\dfrac{d}{dx}[\tan(x)] = \dfrac{d}{dx}\left(\dfrac{\sin(x)}{\cos(x)}\right).$

How do we derive the given result?

We use $\boxed{6.32}$ with $\quad u = \sin(x) \qquad v = \cos(x)$

$$\frac{du}{dx} = \cos(x) \qquad \frac{dv}{dx} = -\sin(x)$$

Substituting these into

$\boxed{6.32}$ $\dfrac{d}{dx}\left(\dfrac{u}{v}\right) = \dfrac{\frac{du}{dx}\cdot v - u\cdot\frac{dv}{dx}}{v^2}$

gives

$$\frac{d}{dx}\left(\frac{\sin(x)}{\cos(x)}\right) = \frac{\cos(x)\cdot\cos(x) - \sin(x)\cdot[-\sin(x)]}{[\cos(x)]^2}$$

$$= \frac{\cos^2(x) + \sin^2(x)}{\cos^2(x)} = \frac{1}{\cos^2(x)} \qquad \text{[By } \boxed{4.64}\text{]}$$

$$= \sec^2(x) \qquad \text{[By } \boxed{4.11}\text{]}$$

$\boxed{4.11}$ $1/\cos(x) = \sec(x)$ $\boxed{4.64}$ $\cos^2(x) + \sin^2(x) = 1$

You often need to use algebra, particularly indices, or trigonometric identities, to simplify your result after differentiating. It might be worth revising **Sections 1C, 4G** and **5A** if you are finding it difficult to follow the above examples.

SUMMARY

Product rule: If $y = uv$ where u and v are functions of x then

6.31 $$\frac{d}{dx}(uv) = \left(\frac{du}{dx} \cdot v\right) + \left(u \cdot \frac{dv}{dx}\right) \text{ or } (uv)' = u'v + uv'$$

Quotient rule: If $y = \dfrac{u}{v}$ where u and v are functions of x then

6.32 $$\frac{d}{dx}\left(\frac{u}{v}\right) = \frac{\left(\frac{du}{dx} \cdot v\right) - \left(u \cdot \frac{dv}{dx}\right)}{v^2} \text{ or } \left(\frac{u}{v}\right)' = \frac{u'v - uv'}{v^2}$$

Exercise **6(d)**

Solutions at end of book. Complete solutions available at www.palgrave.com/science/engineering/singh

1 [electrical principles]   The current, i, through an inductor is given by

$i = 5te^{-5t}$. Find $\dfrac{di}{dt}$.

2 [electrical principles]   The current, i, through a capacitor of capacitance C is given by

$$i = C\frac{dv}{dt}$$

where v is the voltage across the capacitor. For $v = (t + 1)e^{-(1 \times 10^3)t}$ and $C = 1\,\mu F$, find i.

In question 3 don't be put off by the different symbols used for variables; the same differentiation rules apply.

3 Differentiate each of the following expressions with respect to the variables used, simplify your result where possible:

a $t\sin(t)$ **b** $x\ln(x)$ **c** θe^{θ}

d $e^u \sin(u)$ **e** $\sin(\alpha)\cos(\alpha)$

f $\dfrac{q}{1 + q}$ **g** $\dfrac{\cos(a)}{1 + \sin(a)}$

h $\dfrac{e^z}{\cos(z)}$ **i** $M^3 \ln(M)$

j $\dfrac{\beta^2}{\sin(\beta)}$ **k** $\dfrac{h}{\sqrt{1 + h^2}}$

l $J\ln[\sin(J)]$ **m** $e^{\Sigma}\sin(2\Sigma)$

n $\sqrt{\dfrac{Z + 1}{Z - 1}}$ **o** $\Sigma\, e^{\Sigma}\sin(5\Sigma)$

p $\ln\left(\dfrac{\delta^2 + 1}{\delta^2 - 1}\right)$

4 [mechanics]   A damped oscillator has displacement, s, given by

$$s = e^{-kt}\cos(\omega t)$$

where ω is the angular frequency and k is a constant. Find the velocity $v = \dfrac{ds}{dt}$.

SECTION E **Higher derivatives**

By the end of this section you will be able to:

▶ understand second-derivative notation $\dfrac{d^2y}{dx^2}$, $f''(x)$ and $\ddot{x}$

▶ apply these to engineering problems

▶ understand other higher derivatives

E1 **Second derivatives**

Example 23 *mechanics*

A particle moves along a horizontal line according to the equation

$$s = 2t^3 + t^2 - 10t$$

where s is the displacement at time t. Find the velocity, $v = \dfrac{ds}{dt}$. Also determine the acceleration, $a = \dfrac{dv}{dt}$ (differentiate v with respect to t).

Solution

$$v = \frac{d}{dt}\left(2t^3 + t^2 - 10t\right) = 6t^2 + 2t - 10$$

To find the acceleration, $a = \dfrac{dv}{dt}$, we need to differentiate $6t^2 + 2t - 10$ with respect to t:

$$\frac{dv}{dt} = \frac{d}{dt}\left(6t^2 + 2t - 10\right) = 12t + 2$$

$\dfrac{dv}{dt}$ in the above example is called the **second derivative** of s with respect to t and is denoted by $\dfrac{d^2s}{dt^2}$ because we differentiate $s = 2t^3 + t^2 - 10t$ twice.

In general, if $y = f(x)$, a function of x, then $\dfrac{d^2y}{dx^2}$, pronounced 'dee two y by dee x squared', denotes the second derivative of y with respect to x. $f''(x)$, pronounced 'f double dashed x', also denotes the second derivative. Additionally there is Newton's notation, often found in applied mathematics: $\ddot{x} = \dfrac{d^2x}{dt^2}$ where $\ddot{x}$ is pronounced 'x double dot'. In **Example 23**, $\ddot{s} = 12t + 2$ and represents the acceleration of the particle. Differentiating displacement once gives the velocity and twice the acceleration. We will explore differentiation applied to mechanics in Chapter 7.

Example 24 *vibrations*

The displacement, x, of a particle in simple harmonic motion is given by

$$x = A\cos(\omega t) + B\sin(\omega t)$$

where ω represents the angular frequency and A, B are constants. Show that

$$\frac{d^2x}{dt^2} + \omega^2 x = 0$$

Solution

We have to differentiate x twice with respect to t:

$$x = A\cos(\omega t) + B\sin(\omega t)$$

$$\frac{dx}{dt} = \underbrace{-\omega A\,\sin(\omega t)}_{\text{by } 6.13} + \underbrace{\omega B\,\cos(\omega t)}_{\text{by } 6.12} \qquad \text{[Differentiating]}$$

$$\frac{d^2x}{dt^2} = \underbrace{-\omega \cdot \omega A\,\cos(\omega t)}_{\text{by } 6.12} - \underbrace{\omega \cdot \omega B\,\sin(\omega t)}_{\text{by } 6.13} \qquad \text{[Differentiating]}$$

$$= -\omega^2 A \cdot \cos(\omega t) - \omega^2 B\,\sin(\omega t) \qquad \text{[Because } \omega \cdot \omega = \omega^2\text{]}$$

$$= -\omega^2 \underbrace{\left[A\cos(\omega t) + B\sin(\omega t) \right]}_{= x} \qquad \text{[Factorizing]}$$

$$\frac{d^2x}{dt^2} = -\omega^2 x$$

We have

$$\frac{d^2x}{dt^2} + \omega^2 x = -\omega^2 x + \omega^2 x = 0$$

6.12 $[\sin(kt)]' = k\cos(kt)$ 6.13 $[\cos(kt)]' = -k\sin(kt)$

Example 25 *mechanics*

The displacement, s, of a particle is given by

$$s = 2te^{-0.5t}$$

Find the value of t at which the acceleration $\ddot{s} = 0$.

Solution

First we find $\dot{s}$. **How?**

By using the product rule 6.31 on $s = 2te^{-0.5t}$. Let

$$u = 2t \qquad v = e^{-0.5t}$$
$$u' = 2 \qquad v' = -0.5e^{-0.5t}$$

Substituting these into $\dot{s} = u'v + uv'$ gives

$$\dot{s} = 2e^{-0.5t} + 2t(-0.5e^{-0.5t})$$
$$= 2e^{-0.5t} - te^{-0.5t}$$
$$= (2 - t)e^{-0.5t} \quad \text{[Factorizing]}$$

To find the acceleration, $\ddot{s}$, we need to use the product rule again because $\dot{s} = (2 - t)e^{-0.5t}$ is the product of $2 - t$ times $e^{-0.5t}$:

$$u = 2 - t \qquad v = e^{-0.5t}$$
$$u' = -1 \qquad v' = -0.5e^{-0.5t}$$

Putting these into $\ddot{s} = u'v + uv'$ yields

$$\ddot{s} = (-1)e^{-0.5t} + (2 - t)(-0.5)e^{-0.5t}$$
$$= -e^{-0.5t} + (-1 + 0.5t)e^{-0.5t}$$
$$= (-1 - 1 + 0.5t)e^{-0.5t} \qquad \text{[Factorizing]}$$
$$= (-2 + 0.5t)e^{-0.5t}$$

For $\ddot{s} = 0$ we have

$$(-2 + 0.5t)e^{-0.5t} = 0$$

From the properties of the exponential function we know

$$e^{-0.5t} \neq 0 \quad \text{[Cannot equal zero]}$$

Example 25 *continued*

Thus the bracket term, $(-2 + 0.5t)$, must be zero:

$$-2 + 0.5t = 0$$

$$0.5t = 2 \text{ which gives } t = \frac{2}{0.5} = 4$$

The acceleration is zero, $\ddot{s} = 0$, at $t = 4$.

E2 Other higher derivatives

If $y = f(x)$ then differentiating $\dfrac{dy}{dx}$ gives

6.33 $$\frac{d}{dx}\left(\frac{dy}{dx}\right) = \frac{d^2y}{dx^2} \text{ or } f''(x)$$

Other higher derivatives are defined as

6.34 $$\frac{d}{dx}\left(\frac{d^2y}{dx^2}\right) = \frac{d^3y}{dx^3} \text{ or } f'''(x)$$

6.35 $$\frac{d}{dx}\left(\frac{d^3y}{dx^3}\right) = \frac{d^4y}{dx^4} \text{ or } f^{iv}(x)$$

and so on.

Of course these definitions apply to other variables such as i, t, u, v, z etc., and not just to x and y.

Example 26

If $f(x) = \sin(x)$, determine $f'''(0)$.

Solution

$f'''(0)$ means differentiate $\sin(x)$ three times and substitute zero into your result:

$$f(x) = \sin(x)$$
$$f'(x) = \cos(x) \qquad \text{[Differentiating]}$$
$$f''(x) = -\sin(x) \qquad \text{[Differentiating]}$$
$$f'''(x) = -\cos(x) \qquad \text{[Differentiating]}$$

Substituting $x = 0$ into $f'''(x) = -\cos(x)$ gives $f'''(0) = -\cos(0) = -1$

SUMMARY

The second derivative is defined as

6.33
$$\frac{d^2 y}{dx^2} = \frac{d}{dx}\left(\frac{dy}{dx}\right) = f''(x)$$

$\ddot{x}$ also denotes the second derivative. Other higher derivatives are defined in a similar manner.

Exercise 6(e)

Solutions at end of book. Complete solutions available at www.palgrave.com/science/engineering/singh

🚗 **Questions 1–5, inclusive, are in the field of [*mechanics*], with all distances in metres and all times in seconds.**

1 In the following, x represents the displacement of a particle in a horizontal line at time t. In each case find expressions for the velocity, $v = \dfrac{dx}{dt}$, and

acceleration, $a = \dfrac{d^2 x}{dt^2}$ at time t:

a $x = t^3 - 10t^2 + 40t - 30$

b $x = 7t^3 - 50t^2 + 50t$

c $x = 5\cos(2t)$

d $x = 3\cos(10\pi t) + 5\sin(10\pi t)$

2 The displacement, x, of a particle moving in a straight line at time t is given by

$$x = 2t^3 - 15t^2 - 36t \quad (t \geq 0)$$

i Find expressions for the velocity $v = \dot{x}$ and acceleration $a = \ddot{x}$.

ii Find the value of t when the particle comes to rest ($v = 0$ m/s).

3 A projectile is fired vertically upwards and the height, h, is given by

$$h = 10t - 4.9t^2$$

Find the acceleration $a = \ddot{h}$. Comment upon your result.

4 A stone is dropped from a bridge and the distance fallen, h, at time t is given by

$$h = 4.9t^2 \quad 0 \leq t \leq 3$$

Determine the velocity $v = \dot{h}$ and acceleration $a = \ddot{h}$ at $t = 1.5$ s.

5 Find the velocity, $v = \dot{x}$, and acceleration, $a = \ddot{x}$, for the following displacements, x:

a $x = 3t - \sin(3t)$ **b** $x = \dfrac{1 - e^{-t/5}}{5}$

c $x = te^{-t}$ **d** $x = \ln(1 + 3t^2)$

e $x = 0.5\ln\left[\cosh\left(\sqrt{20}t\right)\right]$

6 ⚙ [*vibrations*] The displacement, x, of a particle in simple harmonic motion is given by

$$x = \cos(3t) - \sin(3t)$$

Show that

$$\frac{d^2 x}{dt^2} + 9x = 0$$

7 Determine $f^v(0)$ for $f(x) = \cos(x)$.

8 🌡 [*thermodynamics*] The pressure, P, and volume, V, of a gas in a cylinder are related by

$$PV^n = C \quad \text{(where } C \text{ is a constant)}$$

Show that

$$\frac{d^2 P}{dV^2} = \frac{n(n + 1)P}{V^2}$$

9 ⚙ [*vibrations*] The displacement, x, of a mass in a vibrating system is given by

$$x = (1 + t)e^{-\omega t}$$

Solutions at end of book. Complete solutions available at www.palgrave.com/science/engineering/singh

Exercise **6(e) continued**

where ω is the natural frequency of vibration. Show that

$$\ddot{x} + 2\omega\dot{x} + \omega^2 x = 0$$

10 [vibrations]   The amplitude, u, at displacement x along a bar satisfies

$$u = A\sin\left(\frac{\omega x}{k}\right) + B\cos\left(\frac{\omega x}{k}\right)$$

where ω = natural frequency of vibration and k, A, B are constants. Show that

$$\frac{d^2u}{dx^2} + \frac{\omega^2}{k^2}u = 0$$

SECTION F **Parametric differentiation**

By the end of this section you will be able to:
▶ plot graphs of parametric equations
▶ find $\dfrac{dy}{dx}$ and $\dfrac{d^2y}{dx^2}$ of parametric equations
▶ investigate applications of parametric equations

Parametric differentiation might seem like a huge leap in sophistication from previous sections because of the notation used. Also many of the examples rely on trigonometry, particularly trigonometric identities. Follow each example carefully and try not to skip anything you don't understand.

F1 **Graphs of parametric equations**

Parametric equations arise when the co-ordinate variables x and y are expressed in terms of a common third variable, t, θ, ϕ etc. The third variable is called a *parameter*.

Figure 15 shows a circle of radius r and centre origin. The point (x, y) is defined by

$$x = r \cos(\theta), \quad y = r \sin(\theta)$$

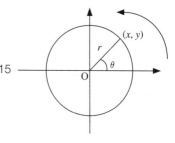
Fig. 15

where r is fixed and θ varies as shown in Fig. 15. (We derived these in Chapter 4.) Hence x and y are expressed in terms of the *parameter* θ.

Parametric equations are commonly used in dynamics, when a particle moves in a plane and the x, y co-ordinates are usually defined in terms of a common third variable, say time t. That is, the x, y co-ordinates give the position of the particle at time t.

Example 27 *mechanics*

A particle moving in the x–y plane has its displacement components, x and y, satisfying

$$x = t - \sin(t), \; y = 1 - \cos(t)$$

By using a computer algebra system or a graphical calculator, plot the path for $0 \le t \le 4\pi$

Solution

By using MAPLE we have Fig. 16.

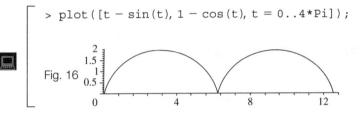

```
> plot([t - sin(t), 1 - cos(t), t = 0..4*Pi]);
```

Fig. 16

You can think of the graph of Fig. 16 as the path traced out by a point on the edge of a bicycle wheel (Fig. 17).

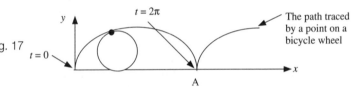

Fig. 17

After t seconds, the position of the point is

$$x = t - \sin(t), \; y = 1 - \cos(t)$$

After 2π seconds, the point on the edge of the wheel reaches A whose co-ordinates are given by substituting $t = 2\pi$ into $x = t - \sin(t)$ and $y = 1 - \cos(t)$:

$$x = 2\pi - \underbrace{\sin(2\pi)}_{= 0} = 2\pi$$

$$y = 1 - \underbrace{\cos(2\pi)}_{= 1} = 1 - 1 = 0$$

As can be seen from the graph, the point A has the co-ordinates $(2\pi, 0)$.

F2 **Parametric differentiation**

For the parametric equations of Example 27:

$$x = t - \sin(t), \; y = 1 - \cos(t)$$

what does $\dfrac{dy}{dx}$ represent?

The gradient at time t. For example, these functions have the gradient graph shown in Fig. 18.

? How can we find $\dfrac{dy}{dx}$?

? First we have $x = t - \sin(t)$. **What derivative can we find from this equation?**

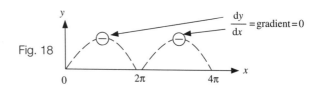

Fig. 18

$$\frac{dx}{dt}$$

$$\frac{dx}{dt} = 1 - \cos(t)$$

Secondly we have $y = 1 - \cos(t)$ [Given equation]

? **What derivative can we find from this equation?**

$$\frac{dy}{dt}$$

$$\frac{dy}{dt} = 0 - [-\sin(t)] = \sin(t)$$

So far we have found $\dfrac{dx}{dt}$ and $\dfrac{dy}{dt}$ but we need to find $\dfrac{dy}{dx}$.

? **How do we determine $\dfrac{dy}{dx}$?**

By the chain rule 6.10 , we have $\dfrac{dy}{dx} = \dfrac{dy}{dt} \cdot \dfrac{dt}{dx}$

We know $\dfrac{dy}{dt}$ but we do not have an expression for $\dfrac{dt}{dx}$. However we do have an expression

? for $\dfrac{dx}{dt}$. **Can we find $\dfrac{dt}{dx}$ from $\dfrac{dx}{dt}$?**

$$\frac{dt}{dx} = 1 \div \frac{dx}{dt} = \frac{1}{dx/dt}$$

Remember that $\dfrac{dt}{dx}$ is a derivative and **not** a fraction.

Substituting $\dfrac{dx}{dt} = 1 - \cos(t)$ into $\dfrac{dt}{dx} = \dfrac{1}{dx/dt}$ yields

$$\frac{dt}{dx} = \frac{1}{1 - \cos(t)}   \quad (\text{provided } 1 - \cos(t) \neq 0)$$

6.10 $\dfrac{dy}{dx} = \dfrac{dy}{du} \cdot \dfrac{du}{dx}$

Putting $\dfrac{dy}{dt} = \sin(t)$ and $\dfrac{dt}{dx} = \dfrac{1}{1 - \cos(t)}$ into $\dfrac{dy}{dx} = \dfrac{dy}{dt} \cdot \dfrac{dt}{dx}$ gives

$$\frac{dy}{dx} = \sin(t) \cdot \frac{1}{1 - \cos(t)} = \frac{\sin(t)}{1 - \cos(t)}$$

In general, if x and y are expressed in terms of a common parameter t then

$$\frac{dy}{dx} = \frac{dy}{dt} \cdot \frac{dt}{dx} = \frac{dy}{dt} \cdot \frac{1}{dx/dt} = \frac{dy/dt}{dx/dt}$$

Thus

6.36 $$\frac{dy}{dx} = \frac{dy/dt}{dx/dt}$$

Example 28

Find $\dfrac{dy}{dx}$ if

$$x = \sin^2(\theta), \ y = \cos^2(\theta)$$

Comment upon your result. Plot the path traced out by x and y.

Solution

We have

$$x = \sin^2(\theta) = [\sin(\theta)]^2$$

$$\frac{dx}{d\theta} = 2\sin(\theta)\cos(\theta) \qquad \left[\text{By } (u^n)' = nu^{n-1}u' \right]$$

$$y = \cos^2(\theta) = [\cos(\theta)]^2$$

$$\frac{dy}{d\theta} = 2\cos(\theta)[-\sin(\theta)] \qquad \left[\text{By } (u^n)' = nu^{n-1}u' \right]$$

$$= -2\cos(\theta)\sin(\theta)$$

Substituting these into 6.36 gives

$$\frac{dy}{dx} = \frac{dy/d\theta}{dx/d\theta} = -\frac{2\cos(\theta)\sin(\theta)}{2\sin(\theta)\cos(\theta)} = -1 \quad \text{[Cancelling]}$$

The gradient at any value of the parameter θ is always -1. The graph of $x = \sin^2(\theta), \ y = \cos^2(\theta)$ is shown in Fig. 19.

Fig. 19

Notice from the graph that the gradient of the line is -1.

The graph in Fig. 19 shows the line $y = 1 - x$ which in terms of θ is written as

$$\cos^2(\theta) = 1 - \sin^2(\theta)$$

[Substituting $y = \cos^2(\theta)$ and $x = \sin^2(\theta)$.]

This is a rearrangement of the well-known trigonometric identity

4.64 $\qquad \sin^2(\theta) + \cos^2(\theta) = 1$

The second derivative, $\dfrac{d^2y}{dx^2}$, of parametric equations x, y written in terms of a parameter t is

6.37 $\qquad \dfrac{d^2y}{dx^2} = \dfrac{\dfrac{d}{dt}\left(\dfrac{dy}{dx}\right)}{\dfrac{dx}{dt}}$

Example 29

Find $\dfrac{dy}{dx}$ and $\dfrac{d^2y}{dx^2}$ for $x = r\cos(\omega t)$, $y = r\sin(\omega t)$

Solution

This is a very challenging example, follow each step very carefully.

Differentiating these:

$$\underset{\text{by } 6.13}{\dfrac{dx}{dt}} = -\omega r\sin(\omega t), \quad \underset{\text{by } 6.12}{\dfrac{dy}{dt}} = \omega r\cos(\omega t)$$

Putting these into 6.36 :

$$\dfrac{dy}{dx} = \dfrac{dy/dt}{dx/dt} = \dfrac{\omega r\cos(\omega t)}{-\omega r\sin(\omega t)}$$

$$= -\dfrac{\cos(\omega t)}{\sin(\omega t)} \qquad [\text{Cancelling } \omega r]$$

$$= -\cot(\omega t) \qquad [\text{Because } \cos/\sin = \cot]$$

To find $\dfrac{d^2y}{dx^2}$ we first determine the numerator of 6.37 :

$$\dfrac{d}{dt}\left(\dfrac{dy}{dx}\right) = \dfrac{d}{dt}\big[-\cot(\omega t)\big] = -\big[\underset{\text{by } 6.24}{-\omega\,\text{cosec}^2(\omega t)}\big] = \omega\,\text{cosec}^2(\omega t)$$

6.12 $\quad [\sin(kt)]' = k\cos(kt)$ $\qquad$ 6.13 $\quad [\cos(kt)]' = -k\sin(kt)$ $\qquad$ 6.24 $\quad [\cot(u)]' = -\text{cosec}^2(u)\,\dfrac{du}{dt}$

Example 29 *continued*

Substituting these into 6.37 gives

$$\frac{d^2y}{dx^2} = \frac{\omega \cosec^2(\omega t)}{-\omega r \sin(\omega t)} = -\frac{1}{r \sin(\omega t)} \cdot \cosec^2(\omega t) \quad [\text{Cancelling } \omega]$$

$$= -\frac{1}{r} \underbrace{\cosec(\omega t) \cdot \cosec^2(\omega t)}_{\text{by } 4.10}$$

$$= -\frac{1}{r} \cosec^3(\omega t)$$

F3 **Applications**

Next we examine parametric equations in mechanics. We find the velocity and acceleration components by determining $\dot{x}$, $\ddot{x}$, $\dot{y}$ and $\ddot{y}$.

 Example 30 *mechanics*

The motion of a projectile is given by

$$x = 3t^2 - 6t, \quad y = 18t^2 - 2t^3$$

The velocity components, v_x and v_y, and acceleration components, a_x and a_y, are defined by

$$v_x = \dot{x}, \; v_y = \dot{y}, \; a_x = \ddot{x}, \; a_y = \ddot{y}$$

Find expressions for v_x, v_y, a_x and a_y.

Solution

Remember that $\dot{x}$ denotes the time derivative, $\dfrac{dx}{dt}$.

$$
\begin{aligned}
x &= 3t^2 - 6t \\
\dot{x} &= 6t - 6 && [\text{Differentiating}] \\
\ddot{x} &= 6 && [\text{Differentiating}]
\end{aligned}
$$

Hence $v_x = 6t - 6$, $a_x = 6$. Similarly

$$
\begin{aligned}
y &= 18t^2 - 2t^3 \\
\dot{y} &= 36t - 6t^2 && [\text{Differentiating}] \\
\ddot{y} &= 36 - 12t && [\text{Differentiating}]
\end{aligned}
$$

Hence $v_y = 36t - 6t^2$, $a_y = 36 - 12t$.

4.10 $\dfrac{1}{\sin(x)} = \cosec(x)$ 6.37 $\dfrac{d^2y}{dx^2} = \dfrac{\dfrac{d}{dt}\left(\dfrac{dy}{dx}\right)}{\dfrac{dx}{dt}}$

SUMMARY

If the parametric equations x, y are expressed in terms of a parameter t then

6.36
$$\frac{dy}{dx} = \frac{dy/dt}{dx/dt}$$

6.37
$$\frac{d^2y}{dx^2} = \frac{\dfrac{d}{dt}\left(\dfrac{dy}{dx}\right)}{\dfrac{dx}{dt}}$$

Velocity and acceleration components are defined by $v_x = \dot{x}$, $a_x = \ddot{x}$, $v_y = \dot{y}$ and $a_y = \ddot{y}$

Exercise **6(f)**

Solutions at end of book. Complete solutions available at www.palgrave.com/science/engineering/singh

1 Find $\dfrac{dy}{dx}$ in each of the following cases:

a $x = \sin(t)$, $y = \cos(t)$

b $x = 2t^3 - t^2$, $y = 10t^2 - t^3$

c $x = (t - 3)^2$, $y = t^3 - 1$

d $x = e^t - 1$, $y = e^{t/2}$

🚗 **Questions 2–7 are in the area of [*mechanics*]. For questions 2–4 use a graphical calculator or a computer algebra system to plot the parametric equations. Note that you can still find the derivatives without plotting the parametric equations.**

2 A particle moves in the plane according to
$$x = t\sin(t), \quad y = t\cos(t)$$

a Plot the path of the particle for $0 \le t \le 50$.

b Find $\dfrac{dy}{dx}$.

3 A particle moving in the plane has its displacement components, x and y, given by
$$x = (t - 2)^2, \quad y = t^3 - 5$$

i Plot the path for $0 \le t \le 3.5$.

ii Determine
$$v_x = \dot{x}, \; v_y = \dot{y} \text{ and } v = \sqrt{v_x^2 + v_y^2}.$$

4 A particle moving in the x–y plane has its displacement components given by

$$x = \frac{(t^2 + 2)^{\frac{3}{2}}}{3}, \quad y = t$$

i Plot the path for $0 \le t \le 5$.

ii Determine the velocity components, $v_x = \dot{x}$ and $v_y = \dot{y}$.

iii Determine the magnitude of the velocity (speed), v, defined by
$$v = \sqrt{v_x^2 + v_y^2}$$

5 A particle moving in the x–y plane has its displacement components, x and y, satisfying

$$x = 8t(2 - t^2), \quad y = t^3$$

i Determine the velocity components, $v_x = \dot{x}$ and $v_y = \dot{y}$.

ii Determine the magnitude of the velocity (speed), v, defined by
$$v = \sqrt{v_x^2 + v_y^2}$$

Exercise **6(f) continued**

Solutions at end of book. Complete solutions available at www.palgrave.com/science/engineering/singh

6 The x, y components of a particle moving in the $x-y$ plane are given by

$$x = r\cos(t), \quad y = r\sin(t)$$

where r is a constant. Determine

$$v_x = \dot{x}, \quad v_y = \dot{y}, \quad a_x = \ddot{x} \text{ and } a_y = \ddot{y}.$$

Show that $v = \sqrt{v_x^2 + v_y^2} = r$

and $a = \sqrt{a_x^2 + a_y^2} = r$.

7 A particle moves with curvilinear motion given by

$$x = t^3 - t^2, \quad y = 5t^2 - t^3$$

i Find

$$v_x = \dot{x}, \quad v_y = \dot{y}, \quad a_x = \ddot{x} \text{ and } a_y = \ddot{y}.$$

ii Find the magnitudes of the velocity, v (m/s) and the acceleration, a (m/s²) where

$$v = \sqrt{v_x^2 + v_y^2}$$

$$a = \sqrt{a_x^2 + a_y^2}$$

at $t = 2$ s.

8 If $x = \cosh(t)$ and $y = \sinh(t)$, show that

$$\frac{d^2y}{dx^2} = -\text{cosech}^3(t)$$

9 The parametric equations of a cycloid are given by

$$x = r[\theta - \sin(\theta)], \quad y = r[1 - \cos(\theta)]$$

where r is a constant. Show that

$$\frac{dy}{dx} = \frac{\sin(\theta)}{1 - \cos(\theta)}$$

and $\dfrac{d^2y}{dx^2} = \dfrac{-1}{r[1 - \cos(\theta)]^2}$

SECTION G Implicit and logarithmic differentiation

By the end of this section you will be able to:

▶ find $\dfrac{dy}{dx}$ of an implicit function

▶ apply logarithmic differentiation

G1 Implicit differentiation

The following are examples of **explicit functions**:

$$y = 1 - x^2$$
$$y = x^3 + e^x$$
$$y = \cos^2(x) + e^x + 3x^2$$

The following are examples of **implicit functions**:

$$x^3 + y^3 + e^{xy} - 2 = 0$$
$$y + \tan(xy) + x^3 = 0$$
$$x^2 + y^2 - 4 = 0$$

? **What is the difference between implicit and explicit functions?**

If y can be explicitly expressed in terms of x then we say that y is an explicit function, as shown by the first three examples; that is, $y = f(x)$. However if the relationship $y = f(x)$ is hidden then we say that the function is an implicit function, as shown by the last three examples. It is sometimes impossible to extract y in terms of x, $y = f(x)$, from an implicit function. An implicit function is denoted by $f(x, y) = 0$; that is, f is a function of x and y.

In this section we are interested in finding the derivative $\dfrac{dy}{dx}$ of implicit functions. We can use the formulae of Table 3.

Example 31

If

$$x^3 + y^2 + x + y = 2$$

then find $\dfrac{dy}{dx}$.

Solution

?
? Differentiating x^3 and x in the above expression is straightforward. **But how do we differentiate y and y^2 with respect to x? If you differentiate y with respect to x, what do you obtain?**

$$\frac{dy}{dx}$$

This derivative is very important. We have done this numerous times before. For example, if $y = x^3$ then $\dfrac{dy}{dx} = 3x^2$.

Differentiating y gives $\dfrac{dy}{dx}$

? **How do we differentiate y^2 with respect to x?**

Using $\dfrac{d}{dx}(u^n) = nu^{n-1}\dfrac{du}{dx}$ with $u = y$ and $n = 2$, we obtain

$$\frac{d}{dx}\left(y^2\right) = 2y\,\frac{dy}{dx}$$

So

$$\frac{d}{dx}\left(x^3 + y^2 + x + y = 2\right) = \frac{d}{dx}(x^3) + \frac{d}{dx}(y^2) + \frac{d}{dx}(x) + \frac{d}{dx}(y) = \frac{d}{dx}(2)$$

★
$$3x^2 + 2y\,\frac{dy}{dx} + 1 + \frac{dy}{dx} = 0$$

Example 31 *continued*

? **How do we find $\dfrac{dy}{dx}$?**

By collecting the $\dfrac{dy}{dx}$ terms on one side of ▢ ★ and the remaining terms on the other side; that is, by transposing to make $\dfrac{dy}{dx}$ the subject of the formula.

$$2y\frac{dy}{dx} + \frac{dy}{dx} = -(3x^2 + 1)$$

$$(2y + 1)\frac{dy}{dx} = -(3x^2 + 1) \qquad \text{[Factorizing]}$$

$$\frac{dy}{dx} = -\frac{(3x^2 + 1)}{2y + 1} \qquad \text{[Dividing by } 2y + 1]$$

Example 32

If

$$x^2 y^2 + 2xy = 3$$

then find $\dfrac{dy}{dx}$.

Solution

? **How do we differentiate the first term, $x^2 y^2$?**

By using the product rule, ▢ 6.31 , because $x^2 y^2 = x^2 \times y^2$:

$$u = x^2 \qquad\qquad v = y^2$$
$$u' = 2x \qquad\qquad v' = 2y\frac{dy}{dx}$$

Substituting into $\dfrac{d}{dx}\left(x^2 y^2\right) = u'v + uv'$ gives

▢ ★
$$\frac{d}{dx}\left(x^2 y^2\right) = 2xy^2 + x^2 \cdot 2y\frac{dy}{dx}$$

Similarly the other term, $2xy$, can be differentiated by using the product rule, ▢ 6.31 , again:

$$u = x \qquad\qquad v = y$$
$$u' = 1 \qquad\qquad v' = \frac{dy}{dx}$$

Example 32 *continued*

$$\frac{d}{dx}(2xy) = 2\frac{d}{dx}(xy)$$

$$= 2[u'v + uv'] = 2\left[(1)y + x\frac{dy}{dx}\right]$$

✱✱ $\dfrac{d}{dx}(2xy) = 2\left(y + x\dfrac{dy}{dx}\right)$

Hence $\dfrac{d}{dx}(x^2y^2 + 2xy = 3)$ gives

$$\underbrace{2xy^2 + 2x^2y\frac{dy}{dx}}_{\text{by } ✱} + \underbrace{2\left(y + x\frac{dy}{dx}\right)}_{\text{by } ✱✱} = 0$$

$$xy^2 + x^2y\frac{dy}{dx} + y + x\frac{dy}{dx} = 0 \qquad \text{[Dividing by 2]}$$

$$(x^2y + x)\frac{dy}{dx} + (xy^2 + y) = 0 \qquad \text{[Collecting like terms]}$$

$$(x^2y + x)\frac{dy}{dx} = -(xy^2 + y) \qquad \text{[Subtracting } xy^2 + y]$$

$$\frac{dy}{dx} = -\frac{xy^2 + y}{x^2y + x} \qquad \text{[Dividing by } x^2y + x]$$

Example 33

Find $\dfrac{d}{dx}\left[\ln(y)\right]$.

Solution
Using

6.18 $\dfrac{d}{dx}\left[\ln(u)\right] = \dfrac{1}{u}\dfrac{du}{dx}$

with $u = y$ we have

$$\frac{d}{dx}\left[\ln(y)\right] = \frac{1}{y}\frac{dy}{dx}$$

G2 **Logarithmic differentiation**

Example 34

Find $\dfrac{dy}{dx}$ if $y = x^2(x^3 - 1)^5 e^x$.

Solution

How do we differentiate functions like $y = x^2(x^3 - 1)^5 e^x$?

Although it is possible to use the product rule (try it), we wish to show another method.
We can apply logs to both sides:

$$\ln(y) = \ln\left[x^2(x^3 - 1)^5 e^x\right]$$

$$\underset{\text{by } 5.12}{= \ln(x^2) + \ln\left[(x^3 - 1)^5\right] + \ln(e^x)} = \underbrace{2\ln(x)}_{} + \underbrace{5\ln(x^3 - 1)}_{\text{by } 5.14} + \underbrace{x}_{\text{by } 5.16}$$

Differentiating this, $\ln(y) = 2\ln(x) + 5\ln(x^3 - 1) + x$, is much easier than applying the product rule:

$$\frac{d}{dx}\left[\ln(y)\right] = 2\frac{d}{dx}\left[\ln(x)\right] + 5\frac{d}{dx}\left[\ln(x^3 - 1)\right] + \frac{d}{dx}(x)$$

$$\frac{1}{y}\frac{dy}{dx} = 2\frac{1}{x} + 5\frac{1}{x^3 - 1}3x^2 + 1 \qquad \left[\text{By } \frac{d}{dx}\left[\ln(u)\right] = \frac{1}{u}\frac{du}{dx}\right]$$

$$\frac{dy}{dx} = y\left(\frac{2}{x} + \frac{15x^2}{x^3 - 1} + 1\right) \qquad \text{[Multiplying by } y\,]$$

We can rewrite the algebraic expression in the brackets on the Right-Hand Side with a common denominator as follows:

$$\frac{2}{x} + \frac{15x^2}{x^3 - 1} + 1 = \frac{2(x^3 - 1) + 15x^2 x + x(x^3 - 1)}{x(x^3 - 1)}$$

$$= \frac{2x^3 - 2 + 15x^3 + x^4 - x}{x(x^3 - 1)} \qquad \text{[Expanding numerator]}$$

$$= \frac{x^4 + 17x^3 - x - 2}{x(x^3 - 1)} \qquad \text{[Simplifying numerator]}$$

Thus we have

$$\frac{dy}{dx} = y\left(\frac{x^4 + 17x^3 - x - 2}{x(x^3 - 1)}\right)$$

5.12 $\ln(AB) = \ln(A) + \ln(B)$ 5.14 $\ln(A^n) = n\ln(A)$ 5.16 $\ln(e^x) = x$

Example 34 *continued*

Substituting our given y, that is $y = x^2(x^3 - 1)^5 e^x$:

$$\frac{dy}{dx} = x^2(x^3 - 1)^5 e^x \left(\frac{x^4 + 17x^3 - x - 2}{x(x^3 - 1)} \right)$$

$$= x(x^3 - 1)^4 e^x \left(x^4 + 17x^3 - x - 2 \right) \qquad \text{[Cancelling like terms]}$$

$$= x(x^3 - 1)^4 \left(x^4 + 17x^3 - x - 2 \right) e^x \qquad \text{[Rearranging]}$$

This method is called **logarithmic differentiation** and turns difficult products and quotients into simple sums and differences. However the algebra is still quite demanding.

? **How would you differentiate x^x?**

This is an arduous and tedious task. It can be made more straightforward by using logs, as the following example shows.

Example 35

Let $y = x^x$, find $\dfrac{dy}{dx}$.

Solution

Taking natural logs, ln, of both sides:

[*] $\ln(y) = \ln(x^x)$

? **Can we rewrite the Right-Hand Side of [*]?**

Using

[5.14] $\ln(A^n) = n \ln(A)$

gives

$$\ln(x^x) = x \ln(x)$$

? **How do we differentiate $x \ln(x)$?**

By using the product rule, [6.31], because we have a product of x and $\ln(x)$:

$$u = x \qquad\qquad v = \ln(x)$$

$$u' = 1 \qquad\qquad v' = \frac{1}{x}$$

Substituting these into $\dfrac{d}{dx}\left[x \ln(x) \right] = u'v + uv'$ gives

$$\frac{d}{dx}\left[x \ln(x) \right] = 1 \cdot \ln(x) + x \cdot \frac{1}{x} = \ln(x) + 1$$

Therefore the derivative of the Right-Hand Side of [*] is $\ln(x) + 1$.

Example 35 *continued*

?

How do we differentiate ln(y)?

From **Example 33** we have $\dfrac{d}{dx}\big[\ln(y)\big] = \dfrac{1}{y}\dfrac{dy}{dx}$.

Hence differentiating $\ln(y) = \ln(x^x)$ gives

$$\frac{1}{y}\frac{dy}{dx} = \ln(x) + 1$$

$$\frac{dy}{dx} = y[\ln(x) + 1] \qquad [\text{Multiplying by } y]$$

$$= x^x\,[\ln(x) + 1] \qquad [\text{Substituting } y = x^x]$$

Logarithmic differentiation is applied to functions of the type:

$$y = u^v \qquad [\text{An example is } x^x]$$

$$y = \frac{u^m\,v^n}{w^k} \qquad \left[\text{An example is } \frac{x^2(1-x)^3}{\ln(x)}\right]$$

$$y = u^m\,v^n\,k^w \qquad [\text{An example is } x^2(x^3-1)^5\,e^x]$$

where u, v and w are functions of x and k, m and n are constants. Generally logarithmic differentiation is applied to cases where logs make differentiation simpler.

SUMMARY

An implicit function is of the form

$$f(x, y) = 0$$

When differentiating y with respect to x we obtain $\dfrac{dy}{dx}$.

Use **Table 3** to differentiate the y terms.
Logarithmic differentiation applies to cases where

$$y = u^v,\; y = \frac{u^m\,v^n}{w^k},\; y = u^m\,v^n\,k^w$$

(u, v and w are functions of x).

Exercise **6(g)**

Solutions at end of book. Complete solutions available at www.palgrave.com/science/engineering/singh

1 Find $\dfrac{dy}{dx}$ for the following:

a $x^2 + y^2 = 4$ **b** $x^3 + y^3 - 2x = 3$

c $\dfrac{x^2}{4} + \dfrac{y^2}{16} = 1$

d $x^2 + y^2 - 4x - 6y = -12$

Exercise **6(g) continued**

2 Find $\dfrac{dy}{dx}$ for each of the following cases:

  a $x^2 + y^2 - xy^2 = 5$

  b $\cos(x^2 + y) - 2xy = 0$

  c $\ln(y) = \ln(x - 1) - \ln(x^2)$

  d $\ln(xy) + 2y = 0$

  e $e^{x+y} + \sin(x) = 0$

3 Show that $\dfrac{d}{dx}(10^x) = 10^x \ln(10)$.

4 Find $\dfrac{dy}{dx}$ for each of the following:

  a $y = x^{\sin(x)}$

  b $y = (x^2 + 1)^{1/2}(3x + 1)^{1/3}$

  c $y = \dfrac{(x^2 + 1)^{1/2}(3x + 1)^{1/3}}{(x^4 - 2)^{1/5}}$

  d $y = e^{(x^x)}$ **e** $y = x^{(x^x)}$

Miscellaneous exercise **6**

1 [electrical principles] The voltage, v, across an inductor with inductance L is defined as

$$v = L\frac{di}{dt}$$

where i represents current. Given that the current i is constant, find v.

2 [mechanics] The velocity, v, of a particle is given by $v = 3e^{2t}$. Show that

$$\frac{dv}{dt} - 2v = 0$$

3 [mechanics] If the velocity of a particle is given by $v = ae^{bt}$, show that

$$\frac{dv}{dt} - bv = 0$$

4 [heat transfer] The heat transfer rate, Q, is given by

$$Q = -kA\frac{dT}{dx}$$

where k is the thermal conductivity of the material, A is the cross-sectional area and $T = T(x)$ is the temperature distribution.

For $T(x) = -10x^3 - 500x + 600$, $k = 0.7$ and $A = 10$, find Q at $x = t$ (t represents thickness of material).

5 [heat transfer] The heat transfer rate, Q, is determined by Fourier's law by

$$Q = -kA\frac{dT}{dx}$$

where k, A and T are defined in question **4**. For $k = 370$, $A = 0.01\pi$ and $T = 100e^{-3x}$, find Q for $x = t$.

6 [vibrations] A system consists of a rod of length l and mass m attached to a spring of stiffness k. The potential energy, V, is given by

$$V = \frac{1}{2}kl^2 \cos^2(\theta) + mgl \sin(\theta)$$

where θ is an angle such that $0 \le \theta \le \pi/2$ and g is acceleration due to gravity. A system is in equilibrium if

$$\frac{dV}{d\theta} = 0$$

Find the value(s) of θ for the system to be in equilibrium.

7 [vibrations] A mass, m, is held by a spring with a stiffness constant k. The potential energy, P, of the system is given by

$$P = \frac{1}{2}kx^2 - mgx$$

where x is displacement and g is acceleration due to gravity. A system is said to be in equilibrium if

$$\frac{dP}{dx} = 0$$

Determine x for which the above system is in equilibrium.

8 [electrical principles] The voltage, v, in a circuit is given by

$$v = -6e^{-(2 \times 10^3)t} + 10e^{-(8 \times 10^3)t}$$

Find the current, i, given that

$$i = C\frac{dv}{dt}$$

where $C = 3\ \mu F$.

9 [reliability engineering] The reliability function, $R(t)$, is given by

$$R(t) = e^{e^t - 1}$$

Show that $\dfrac{d}{dt}[1 - R(t)] = -e^{e^t + t - 1}$

10 [electrical principles] The current, i, through a capacitor C is given by

$$i = C\frac{dv}{dt}$$

Find, i, given that $C = 10\ \mu F$ and $v = 50 \sin(75t)$.

11 [structures] A cantilever beam of length L has a bending moment M given by

$$M = -\frac{\omega_0 L^3}{6} - \frac{kL^6}{24} + \omega_0 Lx + \frac{kL^4 x}{4}$$
$$- \frac{\omega_0 x^2}{2} - \frac{kx^5}{20}$$

where k, ω_0 are constants and x is distance along the beam. The shear force, V, is given by $V = \dfrac{dM}{dx}$. Show that

$$V = (L - x)\left[\omega_0 + \frac{k}{4}(L^3 + L^2 x + Lx^2 + x^3)\right]$$

Hint:

$$a^4 - b^4 = (a - b)(a^3 + a^2 b + ab^2 + b^3)$$

12 [structures] The bending moment, M, of a beam of length L is given by

$$M = \frac{\omega_0 L}{\pi}\left[L + x + \frac{L}{\pi}\sin\left(\frac{\pi x}{L}\right)\right]$$

The shear force, V, is given by $V = \dfrac{dM}{dx}$.

Show that

$$V = \frac{\omega_0 L}{\pi}\left[1 + \cos\left(\frac{\pi x}{L}\right)\right]$$

13 [vibrations] The displacement, x, of a mass from its equilibrium position in a vibrating system is given by

$$x = e^{(-2 + \sqrt{3})\omega t} + e^{-(2 + \sqrt{3})\omega t}$$

where ω is the natural frequency of vibration. Show that

$$\ddot{x} + 4\omega\dot{x} + \omega^2 x = 0$$

14 [electronics] At a pn-junction in a semiconductor the barrier potential, v, is given by

$$v = \frac{A}{1 + e^{-kx}}$$

where x is the distance into the n-type material and A, k are constants. Show that the electric field E where

$$E = -\frac{dv}{dx}$$

is given by $E = \dfrac{-kA}{2[1 + \cosh(kx)]}$

15 [mechanics] The x co-ordinate of a particle in motion is given by

$$x = t^2 \sin(t)$$

Find $v_x = \dot{x}$ and $a_x = \ddot{x}$.

16 [mechanics] A particle moves in a curvilinear motion described by

$$x = t\cos(t), \quad y = t\sin(t)$$

i Determine v_x, v_y, a_x and a_y.

ii The magnitude of the velocity, v, and acceleration, a, are defined by

$$v = \sqrt{v_x^2 + v_y^2} \text{ and } a = \sqrt{a_x^2 + a_y^2}$$

Show that

$$v = \sqrt{1 + t^2} \text{ and } a = \sqrt{4 + t^2}.$$

iii Find $\dfrac{dy}{dx}$.

17 If $y = 10^{(x^v)}$, find $\dfrac{dy}{dx}$.

Engineering Applications of Differentiation

SECTION A **Curve sketching**

By the end of this section you will be able to:
► find stationary points for a given function
► distinguish between a local maximum, a local minimum and a point of inflexion
► sketch curves of some engineering functions

A1 **Stationary points**

Figure 1 shows a graph of a function with stationary points, A, B and C.

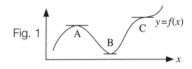

Fig. 1

? **What property do these points have in common?**

The gradient (slope) is zero.

? **What do you think is the definition of a stationary point?**

A point on the curve $y = f(x)$ where the gradient is zero. Remember that $\dfrac{dy}{dx}$ is the gradient function, so we can say that a stationary point is a point on the curve such that

7.1 $$\frac{dy}{dx} = 0 \text{ (or } y' = 0)$$

In Fig. 1, A is called a local maximum of $y = f(x)$.

? **Why do we use the term *local* maximum rather than just maximum?**

Local maximum is a stationary point where the value of y is greater than all the other values of y in the neighbourhood of that point. There may be other greater values of y elsewhere. For example, in Fig. 1 A is a local maximum, but as you can observe there are other values of y greater than A.

Figure 2 shows a local maximum at K.

Similarly B of Fig. 1 is called a **local minimum**.
A stationary point where the value of y is smaller than all the other values of y in the neighbourhood of that point is called a local minimum.

Figure 3 shows a local minimum at K.

Generally we will call local maximum and local minimum just maximum and minimum respectively.

The point C of Fig. 1 is called a (horizontal) point of inflexion. A horizontal point of inflexion is a stationary point which is neither a local maximum nor a local minimum, as shown in Fig. 4.

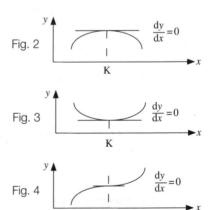

Fig. 2

Fig. 3

Fig. 4

We will return to points of inflexion later in this section.

Example 1

Find the stationary points of $y = x^3 - x^2 - 5x$.

Solution

How do we find the stationary points?

Use 7.1 because at a stationary point the gradient is 0. Differentiating an expression gives us the gradient, so equating that to 0 will give the stationary points.

$$\frac{dy}{dx} = 3x^2 - 2x - 5$$

For stationary points

$$3x^2 - 2x - 5 = 0 \qquad \text{[Quadratic equation]}$$

We need to solve this quadratic equation:

$$(3x - 5)(x + 1) = 0 \quad \text{[Factorizing]}$$

$$x = \frac{5}{3}, x = -1 \qquad \text{[Solving]}$$

What else do we need to find?

The corresponding y values. **How?**

By substituting our x values $\frac{5}{3}$ and -1 into $y = x^3 - x^2 - 5x$.

$$x = \frac{5}{3}, \; y = \left(\frac{5}{3}\right)^3 - \left(\frac{5}{3}\right)^2 - 5\left(\frac{5}{3}\right) = -\frac{175}{27}$$

$$x = -1, y = (-1)^3 - (-1)^2 - (5 \times (-1)) = 3$$

Hence $\left(\frac{5}{3}, -\frac{175}{27}\right)$ and $(-1, 3)$ are the stationary points of $y = x^3 - x^2 - 5x$.

A2 **Maxima and minima**

What are the limitations of 7.1 ?

 7.1 only gives the stationary point(s) but does not reveal whether the stationary point is a local maximum, a local minimum or a point of inflexion.

Example 2

By using a graphical calculator or a computer algebra system, plot the graphs of $y = x^3 - x^2 - 5x$, $\dfrac{dy}{dx}$ and $\dfrac{d^2y}{dx^2}$ against x. Observe what happens at stationary points of y.

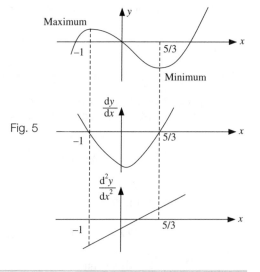

Fig. 5

Solution

Using a technical gadget produces Fig. 5.

Notice that at stationary points, $\dfrac{dy}{dx} = 0$. Also at a maximum, $\dfrac{d^2y}{dx^2} < 0$, and at a minimum, $\dfrac{d^2y}{dx^2} > 0$.

Let's examine the general case. Consider a local maximum at a point $x = C$. If we plot the graphs of y against x, $\dfrac{dy}{dx}$ against x and $\dfrac{d^2y}{dx^2}$ against x close to $x = C$, we obtain Fig. 6.

? What do you notice about the graph of $\dfrac{d^2y}{dx^2}$ close to $x = C$?

$$\dfrac{d^2y}{dx^2} < 0 \quad \text{[Negative]}$$

Note from Fig. 6 that $\dfrac{dy}{dx}$ is decreasing, therefore $\dfrac{d^2y}{dx^2}$ is negative.

Fig. 6

Generally, if at $x = C$ a function, $y = f(x)$, has the property

7.2 $\dfrac{dy}{dx} = 0$ and $\dfrac{d^2y}{dx^2} < 0$ [Negative]

then we say that y has a local maximum at $x = C$.

Similarly, if at $x = C$ a function, $y = f(x)$, has the property

7.3 $\dfrac{dy}{dx} = 0$ and $\dfrac{d^2y}{dx^2} > 0$ [Positive]

then we say that y has a local minimum at $x = C$.

The notation $\dfrac{dy}{dx}$ may be replaced by y' and $\dfrac{d^2y}{dx^2}$ by y''.

To remember these the following mantra might help: 'maximum = frown, minimum = smile'. The frown, being a **negative** feeling, reminds us that the second derivative near a maximum is **negative**, while the smile, showing a **positive** feeling, recalls the **positive** second derivative at a minimum.

Example 3

By using a computer algebra system or a graphical calculator, plot the graph of

$$y = x^4 - 10x^3 + 35x^2 - 50x + 24$$

On the same axes plot the graph of $\dfrac{d^2y}{dx^2}$ against x. **What do you notice about the sign of $\dfrac{d^2y}{dx^2}$ close to the local maxima and minima points of y?**

Solution

At the local maximum the sign of $\dfrac{d^2y}{dx^2}$ is negative, and at both local minima the sign of $\dfrac{d^2y}{dx^2}$ is positive (Fig. 7).

Fig. 7

A3 Curve sketching

7.2 and 7.3 are used as a test for local maxima and local minima. Notice that we first establish a stationary point, $\dfrac{dy}{dx} = 0$, and then use the sign of the second derivative, $\dfrac{d^2y}{dx^2}$, to tell us whether the stationary point is a local maximum or a local minimum.

In this section the purpose of finding local maxima and minima is to sketch graphs of the form $y = f(x)$. In sketching graphs we can also establish where the graph cuts the x and y axes.

How can we find where $y = f(x)$ cuts the x axis?

The graph cuts the x axis at $y = 0$. This means that the x values satisfy $f(x) = 0$.

How can we find where $y = f(x)$ cuts the y axis?

The graph cuts the y axis at $x = 0$. This means that the y value satisfies $y = f(0)$. Thus the graph

7.4   cuts the x axis where $f(x) = 0$, cuts the y axis where $y = f(0)$

Example 4 *mechanics*

A particle moves along a horizontal line according to

$$x = \frac{1}{3}t^3 - 2t^2 + 3t$$

where x is the displacement at time t.

Example 4 *continued*

i Sketch the graph of x against t for $t \geq 0$.
ii At what times do the stationary points of x occur?
iii At what times is the particle at $x = 0$ (at rest)?

Solution

i

$$x = \frac{1}{3} t^3 - 2t^2 + 3t$$

$$\frac{dx}{dt} = t^2 - 4t + 3 \quad \text{[Differentiating]}$$

For stationary points $\dfrac{dx}{dt} = 0$, thus:

$$t^2 - 4t + 3 = 0 \quad \text{[Quadratic equation]}$$

$$(t - 3)(t - 1) = 0 \quad \text{[Factorizing]}$$

$$t = 3, \ t = 1 \quad \text{[Solving]}$$

For maximum or minimum we differentiate again:

$$\frac{d^2x}{dt^2} = 2t - 4$$

Substituting $t = 1$, $\dfrac{d^2x}{dt^2} = (2 \times 1) - 4 < 0$ [Negative]

By 7.2 , $t = 1$ gives local maximum of x. Hence the maximum value is

$$x = \frac{1}{3}(1)^3 - (2 \times 1^2) + (3 \times 1) = \frac{4}{3} \quad \text{[Substituting } t = 1\text{]}$$

Putting $t = 3$, $\dfrac{d^2x}{dt^2} = (2 \times 3) - 4 > 0$ [Positive]

By 7.3 , $t = 3$ gives the local minimum of x. Thus the minimum value is

$$x = \frac{1}{3}(3)^3 - (2 \times 3^2) + (3 \times 3) = 0 \quad \text{[Substituting } t = 3\text{]}$$

The graph cuts the t axis at $x = 0$, that is $t\left(\dfrac{1}{3} t^2 - 2t + 3\right) = 0$ which gives $t = 0$. Also $t = 3$ (twice) gives $x = 0$ because the bracket term can be factorized into

$$\frac{1}{3} t^2 - 2t + 3 = 0$$

$$t^2 - 6t + 9 = 0 \quad \text{[Multiplying by 3]}$$

$$(t - 3)^2 = 0 \quad \text{which gives } t = 3 \text{ (twice)}$$

Example 4 *continued*

Hence we have Fig. 8.

ii Stationary points of x occur when $\dfrac{dx}{dt} = 0$, which is at $t = 1$ and $t = 3$.

iii The particle is at rest, $x = 0$, at $t = 0$ and $t = 3$.

You might ask the question, 'why are we devoting a section to the detailed technique of curve sketching when we have access to computer algebra systems and graphical calculators?'

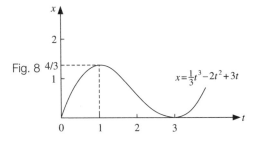

Fig. 8 $x = \frac{1}{3}t^3 - 2t^2 + 3t$

Well generally these technical gadgets **only** give approximations to maxima and minima. Also you cannot visualize asymptotes (described in Chapter 2).

Example 5 *electrical principles*

The current, i, flowing through a circuit is

$$i = te^{-t} \ (t \geq 0)$$

i What is the value of the current i as t gets large, $t \to \infty$?
ii Determine the maximum value of the current.
iii Sketch the graph of i against t for $t \geq 0$.

Solution

i As t gets large, e^{-t} goes to zero. Hence as $t \to \infty$, $e^{-t} \to 0$ and therefore $i \to 0$. This is because e^{-t} decays to zero much faster than t increases. (Try large values of t on your calculator and note how close te^{-t} is to zero.)

ii First we find $\dfrac{di}{dt}$ and then equate this to zero to find the stationary points:

$$i = te^{-t}$$

We need to use the product rule, 6.31 , because i is t times e^{-t}:

$$u = t \qquad v = e^{-t}$$
$$u' = 1 \qquad v' = -e^{-t}$$
$$\frac{di}{dt} = u'v + uv' = (1)e^{-t} + t\,(-e^{-t}) = e^{-t}(1 - t) \quad \text{[Factorizing]}$$

Equating this to zero for stationary points gives $e^{-t}(1 - t) = 0$.

6.31 $(uv)' = u'v + uv'$

Example 5 *continued*

By the properties of the exponential function, $e^{-t} \neq 0$ for any real values of t. Therefore $1 - t = 0$ which yields $t = 1$.

We use the 2nd derivative test to check for the nature of the stationary point. How?

Differentiating, $\dfrac{di}{dt} = e^{-t}(1 - t)$, by using the product rule again:

$$u = 1 - t \qquad\qquad v = e^{-t}$$

$$u' = -1 \qquad\qquad v' = -e^{-t}$$

$$\frac{d^2 i}{dt^2} = u'v + uv' = (-1)e^{-t} + (1 - t)(-e^{-t}) \underset{\text{factorizing}}{=} (-1 - 1 + t)e^{-t} = (t - 2)e^{-t}$$

Substituting $t = 1$ yields

$$\frac{d^2 i}{dt^2} = (1 - 2)e^{-1} = -(1)e^{-1} < 0 \quad \text{[Negative]}$$

Hence i has a local maximum at $t = 1$. Moreover the maximum value is found by putting $t = 1$ into $i = te^{-t}$:

$$i = 1(e^{-1}) = e^{-1}$$

iii At $t = 0$, the current $i = 0$, the graph goes through the origin. Thus combining **i** and **ii** gives Fig. 9.

Fig. 9

A4 Points of inflexion

So far we have described a test for maximum and minimum but we have not established a test for a point of inflexion.

Notice the change in direction of bending at the points of inflexion in Fig. 10.

Fig. 10

A **point of inflexion** at $x = C$ of $y = f(x)$ is where

7.5 $\dfrac{d^2 y}{dx^2} = 0$ and $\dfrac{d^2 y}{dx^2}$ changes sign close to $x = C$

Example 6

Sketch on different axes the graphs of $y = x^3$ and $\dfrac{d^2y}{dx^2}$.

Solution

To plot the graph of $\dfrac{d^2y}{dx^2}$ we need
to differentiate y twice:

$$y = x^3, \quad \frac{dy}{dx} = 3x^2 \text{ and}$$

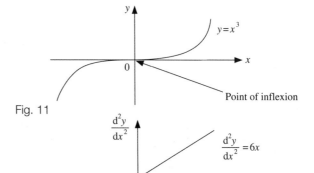

Fig. 11

$$\frac{d^2y}{dx^2} = 6x$$

We have sketched the graph of
$y = x^3$ in Chapter 2 and the graph
of $\dfrac{d^2y}{dx^2} = 6x$ is a straight line going
through the origin (Fig. 11).

Notice that $\dfrac{d^2y}{dx^2}$ changes sign close to the point of inflexion (0) and this point, shown in
Fig. 11, is an example of a **horizontal point of inflexion**.

If at $x = C$ a function, $y = f(x)$, has the property

7.6 $\qquad \dfrac{dy}{dx} = 0, \; \dfrac{d^2y}{dx^2} = 0$ and also $\dfrac{d^2y}{dx^2}$ changes sign close to $x = C$

then the function y has a horizontal point of inflexion at $x = C$.

If $\dfrac{d^2y}{dx^2}$ goes from negative to zero
to positive as x increases through C,
then the y graph has a horizontal
point of inflexion of the form shown
in Fig. 12.

Fig. 12

If $\dfrac{d^2y}{dx^2}$ goes from positive to zero
to negative as x increases through C,
then the y graph has a horizontal
point of inflexion of the form shown
in Fig. 13.

Fig. 13

We can confirm that $y = x^3$ has a horizontal point of inflexion at $x = 0$ by checking the
above. Thus differentiating and equating to zero we have

$$\frac{dy}{dx} = 3x^2 = 0 \text{ gives } x = 0$$

At $x = 0$ we have a stationary point. Differentiating again:

$$\frac{d^2y}{dx^2} = 6x$$

At $x = 0$, $\frac{d^2y}{dx^2} = 6 \times 0 = 0$. It's **not** enough to say that since both $\frac{dy}{dx} = 0$ and $\frac{d^2y}{dx^2} = 0$ we have a horizontal point of inflexion at $x = 0$. We need to check the change of signs of $\frac{d^2y}{dx^2}$ for $x < 0$ and $x > 0$.

If $x < 0$ then $\frac{d^2y}{dx^2} = 6x < 0$ and if $x > 0$ then $\frac{d^2y}{dx^2} = 6x > 0$. Thus $\frac{d^2y}{dx^2}$ changes sign at $x = 0$ and therefore we have a horizontal point of inflexion.

The next example is particularly challenging because it involves a number of different concepts such as differentiation, solving trigonometric equations, substituting and sketching graphs.

Example 7 *electrical principles*

The energy, w, absorbed by a particular resistor at time t is given by

$$w = 5\left(t - \frac{\sin(2t)}{2}\right)$$

Sketch the graph of w against t for $0 \le t \le 2\pi$.

Solution

First are there any obvious points we can plot on the graph?

When $t = 0$, $w = 5\left(0 - \frac{\sin(0)}{2}\right) = 0$. So our first point on the graph is $(0, 0)$.

Next we check for any stationary points. **How do we do that?**

By differentiating and finding values of t which satisfy $\frac{dw}{dt} = 0$:

$$\frac{dw}{dt} = 5\left(1 - \frac{2[\cos(2t)]}{2}\right) = 5[1 - \cos(2t)] \quad \text{[Cancelling 2's]}$$

For stationary points:

$$5[1 - \cos(2t)] = 0 \text{ gives } 1 - \cos(2t) = 0$$

Example 7 *continued*

Therefore $\cos(2t) = 1$ which gives $2t = 0$, 2π, 4π (by examining the cos graph, Fig. 53 of Chapter 4) and it follows that

$$t = 0, \pi, 2\pi \quad \text{[Dividing by 2]}$$

Differentiating again yields

$$\frac{d^2w}{dt^2} = 5\big(0 - [-2\sin(2t)]\big) = 10\sin(2t)$$

Substituting $t = 0$, $\dfrac{d^2w}{dt^2} = 10\sin(2 \times 0) = 0$, unlike previous examples which have given us a positive or a negative answer.

We therefore need to check for a point of inflexion:

$$\text{For } t \text{ close to zero and } t < 0, \frac{d^2w}{dt^2} = 10\sin(2t) < 0 \quad \text{[Negative]}$$

$$\text{For } t \text{ close to zero and } t > 0, \frac{d^2w}{dt^2} = 10\sin(2t) > 0 \quad \text{[Positive]}$$

$$\left[\text{In each case you can try values of } t \text{ close to zero to check the sign of } \frac{d^2w}{dt^2}. \right]$$

By 7.6 the graph has a horizontal point of inflexion at $t = 0$ of the form ╱—╱.

Substituting $t = \pi$:

$$\frac{d^2w}{dt^2} = 10\sin(2\pi) = 0$$

$$\text{For } t \text{ close to } \pi \text{ and } t < \pi: \frac{d^2w}{dt^2} = 10\sin(2t) < 0 \quad \text{[Negative]}$$

$$\text{For } t \text{ close to } \pi \text{ and } t > \pi: \frac{d^2w}{dt^2} = 10\sin(2t) > 0 \quad \text{[Positive]}$$

By 7.6 , the graph has a horizontal point of inflexion at $t = \pi$ of the form ╱—╱.

Substituting $t = 2\pi$:

$$\frac{d^2w}{dt^2} = 10\sin(4\pi) = 0$$

$$\text{For } t \text{ close to } 2\pi \text{ and } t < 2\pi: \frac{d^2w}{dt^2} < 0 \quad \text{[Negative]}$$

$$\text{For } t \text{ close to } 2\pi \text{ and } t > 2\pi: \frac{d^2w}{dt^2} > 0 \quad \text{[Positive]}$$

Example 7 continued

We have a horizontal point of inflexion at $t = 2\pi$ of the form ⁄—⁄ .

We can find the w values at the three points of inflexion, at $t = 0$, π and 2π, by substituting these into $w = 5\left[t - \dfrac{\sin(2t)}{2}\right]$.

When $t = 0$, $w = 0$

$$t = \pi, \qquad w = 5\left[\pi - \frac{\sin(2\pi)}{2}\right] = 5\pi$$

$$t = 2\pi, \qquad w = 5\left[2\pi - \frac{\sin(4\pi)}{2}\right] = 10\pi$$

Fig. 14

$(0, 0)$, $(\pi, 5\pi)$ and $(2\pi, 10\pi)$ are all horizontal points of inflexion of the form ⁄—⁄ , giving Fig. 14.

Remember that as well as horizontal points of inflexion we have other points of inflexion, see Fig. 10.

SUMMARY

A stationary point is a point on the curve $y = f(x)$ such that

7.1 $\dfrac{dy}{dx} = 0$

If at $x = C$

7.2 $\dfrac{dy}{dx} = 0$ and $\dfrac{d^2y}{dx^2} < 0$ [Negative]

then y has a local maximum at $x = C$ and if

7.3 $\dfrac{dy}{dx} = 0$ and $\dfrac{d^2y}{dx^2} > 0$ [Positive]

then y has a local minimum at $x = C$.
However if at $x = C$

7.6 $\dfrac{dy}{dx} = 0$, $\dfrac{d^2y}{dx^2} = 0$ and also $\dfrac{d^2y}{dx^2}$ changes sign close to C then y has a

horizontal point of inflexion at $x = C$.

Exercise **7(a)**

Solutions at end of book. Complete solutions available at www.palgrave.com/science/engineering/singh

1 Obtain the stationary points for the following functions:

a $y = x^2 + 3x + 1$

b $y = x^3 - 3x$

c $y = \dfrac{x^3}{3} - x$

d $y = x^2(50 - x^2)$

2 🚗 [*mechanics*]   The trajectory of a projectile is related by

$$y = \frac{10x - x^2}{10}$$

where x and y are the horizontal and vertical displacement components respectively. Sketch the graph of y against x.

3 Find the maximum and minimum points of $y = x^3 + x^2 - 5x$.

4 🚗 [*mechanics*]   The displacement, s, of a particle is given by

$$s = t^3 - 6t^2 + 12t$$

Find the maximum, minimum and points of inflexion of s.

5 🚗 [*mechanics*]   The displacement, s, of a particle moving along a horizontal line at time t is given by

$$s = t^2(8 - t^2)$$

Sketch the graph of s against t.

🔲 **The remaining questions in this exercise are in the field of [*electrical principles*].**

6 The energy, w, of an inductor at time t is given by

$$w = 1 - \cos(50t)$$

Sketch the graph of w against t for

$$0 \le t \le \frac{\pi}{25}.$$

7 The power, p, in a resistor is given by

$$p = 10\sin^2(t) \text{ where } 0 \le t \le 2\pi$$

Sketch the graph of p against t, marking all the points of maximum, minimum and inflexion.

8 The energy, w, absorbed by a resistor at time t is given by

$$w = 5\sin(2t) + 10t$$

Sketch the graph of w against t for $0 \le t \le 2\pi$.

9 The power, p, of an inductor is given by

$$p = (t^2 - 2t)e^{-2t} \qquad (t \ge 0)$$

i What is the value of p as t gets large, $t \to \infty$?

ii At what values of t is $p = 0$?

iii Find the minimum value of p.

iv Find the maximum value of p.

v Sketch the graph of p against t for $t \ge 0$.

10 The current, i, through a circuit is given by

$$i = 5te^{-5t} \qquad t \ge 0$$

i What is the value of i for large t, $t \to \infty$?

ii Find the maximum current through the circuit.

iii Sketch the graph of i against t for $t \ge 0$.

SECTION B **Optimization problems**

By the end of this section you will be able to:
- ► formulate equations from given data
- ► use the derivative test technique for solving engineering problems which require maximizing or minimizing

B1 **General optimization**

In the previous section we used derivative tests to find the local maximum or local minimum in order to sketch graphs of various functions. In this section we use the same techniques to maximize or minimize a given engineering problem.

Example 8

Figure 15 shows a rectangular sheet of metal. A square of edge x (m) is cut from each corner and the remaining piece is folded along the broken line to make an open box of volume V. Show that

$$V = 4x^3 - 14x^2 + 10x$$

Find the value of x to 2 d.p. for which the volume V is a maximum.

Fig. 15

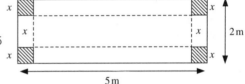

Solution

The base of the box after cutting away the corners is shown in Fig. 16.

Fig. 16

The height of the box is x, so the volume, V, is given by

$$V = x \times [\text{base area}]$$
$$= x(5 - 2x)(2 - 2x)$$
$$= x(10 - 14x + 4x^2) = 4x^3 - 14x^2 + 10x$$

We have seen expressions similar to this: $V = 4x^3 - 14x^2 + 10x$, in **Section A**.

So how do we find when the volume is a maximum?

We need to find $\dfrac{dV}{dx}$ and to solve the equation $\dfrac{dV}{dx} = 0$ for x.

$$V = 4x^3 - 14x^2 + 10x$$
$$\frac{dV}{dx} = 12x^2 - 28x + 10 = 0$$

Dividing by 2 gives

$$6x^2 - 14x + 5 = 0$$

Example 8 *continued*

? **How do we solve this quadratic equation?**

By using the quadratic formula

$$1.16 \qquad x = \frac{-b \pm \sqrt{b^2 - 4ac}}{2a}$$

with $a = 6$, $b = -14$ and $c = 5$:

$$x = \frac{14 \pm \sqrt{(-14)^2 - (4 \times 6 \times 5)}}{(2 \times 6)} = 0.44, \ 1.89 \ (2 \ \text{d.p.})$$

? **x cannot be 1.89 m. Why not?**

Because the length of the base is $2 - 2x$. If $x = 1.89$ m then $2 - 2x$ will be negative. So we only need to check if $x = 0.44$ m gives a maximum volume.

? **How?**

Use the second derivative test:

$$\frac{dV}{dx} = 12x^2 - 28x + 10$$

$$\frac{d^2V}{dx^2} = 24x - 28$$

Putting $x = 0.44$, $\dfrac{d^2V}{dx^2} = (24 \times 0.44) - 28 < 0$ [Negative]

By 7.2 , a square piece of edge 0.44 m gives maximum volume of box.

Example 9

Figure 17 shows a cylindrical container which holds a volume of 1 litre (1000 cm³). Show that the surface area, A, is given by
$A = 2\pi r^2 + 2000r^{-1}$

Fig. 17

? Find the dimensions so that the surface area is a minimum (minimizing the material used). **What do you notice about your results?**

(Volume $= \pi r^2 h$ and surface area $= 2\pi r^2 + 2\pi rh$)

Solution

Since the volume is 1000 cm³ we have

$$\pi r^2 h = 1000$$

$$h = \frac{1000}{\pi r^2} \qquad \text{[Dividing by } \pi r^2 \text{]}$$

Example 9 *continued*

Substituting $h = \dfrac{1000}{\pi r^2}$ into the surface area formula gives

$$A = 2\pi r^2 + 2\pi rh$$

$$= 2\pi r^2 + 2\pi r\left(\dfrac{1000}{\pi r^2}\right)$$

$$A = 2\pi r^2 + 2000r^{-1} \quad [\text{Cancelling}]$$

For stationary points we differentiate A with respect to r and equate the result to zero:

$$\dfrac{dA}{dr} = 4\pi r - 2000r^{-2} = 0$$

$$4\pi r = 2000r^{-2} = \dfrac{2000}{r^2} \qquad \left[\text{Adding }\dfrac{2000}{r^2}\right]$$

$$r^3 = \dfrac{2000}{4\pi} = \dfrac{500}{\pi}$$

? **How do we find r?**

Take the cube root of both sides:

$$r = \sqrt[3]{\dfrac{500}{\pi}} = 5.42 \text{ cm} \quad (2 \text{ d.p.})$$

Using the second derivative test to ascertain that this value of r, $r = 5.42$, does indeed give minimum surface area:

$$\dfrac{dA}{dr} = 4\pi r - 2000r^{-2}$$

$$\dfrac{d^2A}{dr^2} = 4\pi + 4000r^{-3}$$

Putting $r = 5.42$, $\quad \dfrac{d^2A}{dr^2} = 4\pi + \dfrac{4000}{(5.42)^3} > 0 \quad [\text{Positive}]$

By $\boxed{7.3}$, $r = 5.42$ cm gives minimum A.

? **How do we find the other dimension, h?**

Substituting $r = 5.42$ into $h = \dfrac{1000}{\pi r^2}$ gives

$$h = \dfrac{1000}{\pi(5.42)^2} = 10.84 \text{ cm} \quad (2 \text{ d.p.})$$

Note that the height is twice the radius.

$\boxed{7.3}$ $\quad A' = 0$, $A'' > 0$ min

B2 Engineering problems

 Example 10 *electrical principles*

Consider an electrical circuit with resistors of resistance R_1, R_2 and R_3. In the following, treat R_1, R_2, R_3 and i as constants.

i The power, p, dissipated by R_1, R_2 and R_3 is given by

$$p = i_1^2 R_1 + i_1^2 R_2 + (i - i_1)^2 R_3$$

where i_1 and i represent current. Show that p is a minimum at

$$i_1 = \frac{R_3}{R_1 + R_2 + R_3} i$$

ii The voltage, V, is given by $V = iR = i_1(R_1 + R_2)$ where R is the equivalent resistance of R_1, R_2 and R_3.

For minimum power dissipation, minimum p, show that

$$R = \frac{(R_1 + R_2)R_3}{R_1 + R_2 + R_3}$$

Solution

Don't be confused by the R notation; we apply the same rules.

i Copying the equation for p we have

$$p = i_1^2 R_1 + i_1^2 R_2 + (i - i_1)^2 R_3$$

For stationary points we differentiate p with respect to i_1 and equate the result to zero:

$$\frac{dP}{di_1} = 2i_1 R_1 + 2i_1 R_2 + 2(i - i_1)R_3(-1) = 0$$

$$2i_1 R_1 + 2i_1 R_2 - 2iR_3 + 2i_1 R_3 = 0$$

$$i_1 R_1 + i_1 R_2 - iR_3 + i_1 R_3 = 0 \qquad \text{[Dividing by 2]}$$

$$i_1(R_1 + R_2 + R_3) = iR_3 \qquad \text{[Adding } iR_3 \text{ and factorizing]}$$

$$i_1 = \frac{iR_3}{R_1 + R_2 + R_3} = \frac{R_3}{R_1 + R_2 + R_3} i$$

For minimum we need to differentiate again:

$$\frac{dp}{di_1} = 2i_1 R_1 + 2i_1 R_2 - 2iR_3 + 2i_1 R_3$$

$$\frac{d^2 p}{di_1^2} = 2R_1 + 2R_2 - 0 + 2R_3$$

Example 10 *continued*

(Remember that $2iR_3$ is a constant, so differentiating this gives 0.)

Since resistance values are positive, therefore $\dfrac{d^2p}{di_1^{\,2}} = 2R_1 + 2R_2 + 2R_3 > 0$.

By $\boxed{7.3}$, $i_1 = \dfrac{iR_3}{R_1 + R_2 + R_3}$ gives minimum p.

ii We are given $iR = i_1(R_1 + R_2)$

Substituting $i_1 = \dfrac{iR_3}{R_1 + R_2 + R_3}$ into the given equation, $iR = i_1(R_1 + R_2)$, yields

$$iR = \dfrac{iR_3}{R_1 + R_2 + R_3}(R_1 + R_2)$$

Dividing both sides by i gives the required result:

$$R = \dfrac{(R_1 + R_2)R_3}{R_1 + R_2 + R_3}$$

SUMMARY

We use the derivative tests from **Section A** to find which values maximize or minimize a given problem.

Exercise 7(b)

Solutions at end of book. Complete solutions available at www.palgrave.com/science/engineering/singh

1

Fig. 18

Figure 18 shows a fence of dimensions x and y. The perimeter of the fence is 100 m. Show that the area, A, constrained by the fence is

$$A = 50x - x^2$$

Find the values of x and y for which the area A is a maximum. Is there a relationship between x and y which gives maximum area?

2 Figure 19 shows a field surrounded by 240 m of fencing on three sides.

Fig. 19

The shaded part is a hedge where no fencing is required. Find the dimensions, x and y, that command maximum area of the field. What do you notice about your results?

$\boxed{7.3}$ $p' = 0$, $p'' > 0$ min

Exercise **7(b) continued**

Solutions at end of book. Complete solutions available at www.palgrave.com/science/engineering/singh

3 [aerodynamics]   Minimum drag occurs at the minimum value of

$$\frac{C_D + kC_L^2}{C_L}$$

where C_L = lift coefficient, C_D = zero lift drag coefficient and k = non-zero constant. For what value of C_L do we have minimum drag? (Treat C_D and k as constants.)

4 Consider a cylindrical container without a top. If the volume of the container is 8 m³ then find the dimensions, radius (r in m) and height (h in m), so that the surface area required is a minimum. What do you notice about your results?

5 Show that for any given volume, V, the minimum surface area required for a closed cylindrical can is when the height, h, is twice the radius, r.

6 [materials]   The bending moment, M, at a distance x of a beam of length L is given by

$$M = \frac{WLx}{2} - \frac{Wx^2}{2}$$

where W is the weight per unit length. Find the value of x which gives the maximum bending moment and evaluate this maximum.

7 [materials]   The downward deflection, y, of a cantilever of length L with a load W is given by

$$y = \frac{W}{6EI}(3L^2x - x^3) \quad \left(\frac{W}{EI} > 0\right)$$

where x is the horizontal distance from the fixed end. Find the maximum deflection and the value of x to achieve this.

8 [materials]

Fig. 20

Figure 20 shows a cantilever of length L with uniform distributed loading of W per unit length. The downward deflection of the cantilever at a distance x is denoted by y and satisfies

$$y = \frac{W}{24EI}[x^4 - 4Lx^3 + 4L^2x^2]$$

where $EI > 0$ is the constant flexural rigidity of the cantilever. Show that the maximum deflection occurs at $x = L$ and determine this maximum deflection.

9 [mechanics]   A belt of mass m per unit length is wound partly around a pulley. The power, p, transmitted is given by

$$p = Tv - mv^3$$

where T is the tension and v is the velocity of the belt. Show that the maximum power transmitted occurs

when $v = \sqrt{\dfrac{T}{3m}}$.

10 [electrical principles]   The torque, T, of an electric machine is given by

$$T = 3\sin(2\theta) + 6\sin(\theta)$$

where θ is the torque position and $0 \le \theta < \pi$.

  i Find the value of θ for which maximum torque occurs.
  ii Determine the maximum torque.

11 [electrical principles]   Consider an electrical circuit with resistors of resistance R_1 and R_2.
  i The power, P, dissipated by R_1 and R_2 is given by

$$P = i_1^2 R_1 + (i - i_1)^2 R_2$$

where i_1 and i represent current.

Solutions at end of book. Complete solutions available at www.palgrave.com/science/engineering/singh

Exercise **7(b) continued**

Show that P is a minimum when

$$i_1 = \frac{R_2}{R_1 + R_2} i$$

ii The voltage, V, is given by

$$V = i_1 R_1 = iR$$

where R is the equivalent resistance of R_1 and R_2. For minimum power dissipation, show that

$$R = \frac{R_1 R_2}{R_1 + R_2}$$

SECTION C **First derivative test**

By the end of this section you will be able to:
▶ apply the first derivative test for maxima and minima
▶ investigate examples using the first derivative test

C1 **Maxima and minima problems**

We can use a first derivative test to establish whether a stationary point is a maximum or a minimum. This test is particularly beneficial if the second differentiation is lengthy or laborious.

If we plot the graph of y against x for a **local maximum** at $x = C$, we obtain Fig. 21.

Fig. 21

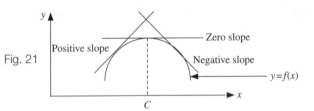

? **What do you notice about the gradient, $\dfrac{dy}{dx}$, of y for x close to C?**

It changes sign from left to right – positive to zero to negative. In general a function, $y = f(x)$, with local maximum at $x = C$ means that

7.7

$$
\left\{
\begin{array}{lll}
\mathbf{1} & \text{at } x = C, & \dfrac{dy}{dx} = 0 \quad \text{[Stationary point]} \\[2mm]
\mathbf{2} & \text{for } x < C, & \dfrac{dy}{dx} > 0 \quad \text{[Positive]} \\[2mm]
\mathbf{3} & \text{for } x > C, & \dfrac{dy}{dx} < 0 \quad \text{[Negative]}
\end{array}
\right.
$$

$\dfrac{dy}{dx}$ changes sign as it passes through the stationary point, $x = C$. Also note that when we say $x < C$ or $x > C$ we mean x values close to C.

If we carry out the same procedure for a **local minimum** at $x = C$ we obtain Fig. 22.

Again $\dfrac{dy}{dx}$ changes sign as it passes through the stationary point, $x = C$.

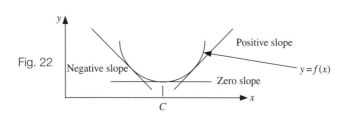

Fig. 22

It changes sign from left to right – negative to zero to positive.

In general a function, $y = f(x)$, with local minimum at $x = C$ means that

7.8

$$\begin{cases} \textbf{1} \quad \text{at } x = C, \quad \dfrac{dy}{dx} = 0 \quad \text{[Stationary point]} \\[2mm] \textbf{2} \quad \text{for } x < C, \quad \dfrac{dy}{dx} < 0 \quad \text{[Negative]} \\[2mm] \textbf{3} \quad \text{for } x > C, \quad \dfrac{dy}{dx} > 0 \quad \text{[Positive]} \end{cases}$$

Let's do an example.

Example 11

Find the points of maxima and minima of $y = \dfrac{x}{1 + x^2}$.

Solution

For stationary points we need to differentiate y. **How?**

We need to apply the quotient rule, 6.32 , with

$$u = x \qquad v = 1 + x^2$$
$$u' = 1 \qquad v' = 2x$$

Substituting these gives

$$\frac{dy}{dx} = \frac{u'v - uv'}{v^2}$$
$$= \frac{1(1 + x^2) - 2x(x)}{(1 + x^2)^2}$$

6.32 $\quad (u/v)' = (u'\,v - uv')/v^2$

Example 11 *continued*

$$= \frac{1 + x^2 - 2x^2}{(1 + x^2)^2} = \frac{1 - x^2}{(1 + x^2)^2} \qquad \text{[Simplifying numerator]}$$

Equating this to zero:

$$\frac{dy}{dx} = \frac{1 - x^2}{(1 + x^2)^2} = 0$$

The derivative $\dfrac{dy}{dx} = 0$ only when the numerator $1 - x^2 = 0$. Solving gives

$$x^2 = 1$$
$$x = \pm 1 \qquad \text{[Square root]}$$

Hence we have two stationary points, $x = 1$ and $x = -1$. The denominator $(1 + x^2)^2$ is positive so we only need to check the sign of the numerator, $1 - x^2$:

If $x < 1$, try $x = 0.9$, then $1 - x^2 = 1 - 0.9^2 > 0$ so $\dfrac{dy}{dx} > 0$ [Positive]

If $x > 1$, try $x = 1.1$, then $1 - x^2 = 1 - 1.1^2 < 0$ so $\dfrac{dy}{dx} < 0$ [Negative]

Thus we have

$$\frac{dy}{dx} > 0 \qquad \frac{dy}{dx} < 0$$
$$x = 1$$

By 7.7 , $x = 1$ gives a maximum.

Now we test the other point, $x = -1$:

If $x < -1$, try $x = -1.1$, then $1 - x^2 = 1 - (-1.1)^2 < 0$ so $\dfrac{dy}{dx} < 0$ [Negative]

If $x > -1$, try $x = -0.9$, then $1 - x^2 = 1 - (-0.9)^2 > 0$ so $\dfrac{dy}{dx} > 0$ [Positive]

Hence we have

$$\frac{dy}{dx} < 0 \qquad \frac{dy}{dx} > 0$$
$$x = -1$$

By 7.8 , $x = -1$ gives a minimum.

? **Why would you *not* use the second derivative test for the above example?**

We have
$$\frac{dy}{dx} = \frac{1 - x^2}{(1 + x^2)^2}$$

Differentiating this to find $\dfrac{d^2y}{dx^2}$ by using the quotient rule is a horrendous task. The only other way to differentiate the original function, $y = \dfrac{x}{1 + x^2}$, twice is to use implicit differentiation but then we have to remember the rules connected with implicit differentiation.

SUMMARY

We can use a first derivative test for maxima and minima by finding the stationary points of a given function and then checking the sign of $\dfrac{dy}{dx}$ on either side of the stationary point.

Remember that care needs to taken as we test x values close to the stationary point.

Exercise **7(c)**

Solutions at end of book. Complete solutions available at www.palgrave.com/science/engineering/singh

1 Find the stationary points of

$$y = x^2(8 - x^2)$$

and distinguish between them using the first derivative test.

2 🚗 [*mechanics*] The acceleration, a, of a machine is given by

$$a = \frac{10r + 1}{5r^2 + 3150}$$

where r is the gear ratio. Determine the value of r for maximum acceleration to occur.

3 🔧 [*electrical principles*]

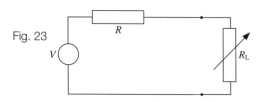

Fig. 23

Figure 23 shows a circuit consisting of a source voltage V and resistor R applied across a variable load R_L. The power, P, delivered to the load R_L is given by

$$P = \frac{V^2 R_L}{(R + R_L)^2}$$

Show that maximum power occurs when $R_L = R$.

4 🔧 [*electrical principles*] A capacitor is formed by two concentric spheres of radii a and b respectively. The electric stress, E, is given by

$$E = \frac{Vb}{ba - a^2} \qquad (ba - a^2 \neq 0)$$

where $V\,(>0)$ is the potential difference between the spheres. Show that the minimum electric stress occurs at $a = \dfrac{b}{2}$.

5 🧲 [*electromagnetism*] The induced e.m.f., E, in a coil is given by

$$E = 2\pi f k \cos(2\pi f t)$$

Find the maximum value of E.

SECTION D **Applications to kinematics**

By the end of this section you will be able to:
► determine the velocity and acceleration of particles in rectangular (straight line) motion
► determine the angular velocity and angular acceleration of particles in circular motion

D1 **Rectangular motion**

Figure 24 shows a particle moving along a straight line. The displacement, s, from the origin is a function of time, t. The units of

Fig. 24

displacement are metres (m). The velocity, v, of a particle at time t is defined as

7.9
$$v = \frac{ds}{dt} \quad \text{[1st derivative]}$$

Velocity is the rate of change of displacement with respect to time. The units of velocity, v, are metres per second (m/s). A particle with $v = 10$ m/s means that the particle covers 10 m for every second.

The acceleration, a, of a particle at time t is defined as

7.10
$$a = \frac{dv}{dt} = \frac{d}{dt}\left(\frac{ds}{dt}\right) = \frac{d^2s}{dt^2} \quad \text{[2nd derivative]}$$

Acceleration is the rate of change of velocity with respect to time t. The units of acceleration, a, are metres per second^2 (m/s^2). A particle with an acceleration of 9.81 m/s^2 means that the velocity of the particle increases by 9.81 m/s for every second.

We have done many examples of finding the velocity and acceleration of particles in Chapter 6.

Example 12 *mechanics*

A particle moves along a horizontal line according to the equation
$$x = 2t^3 + t^2 - 10t + 10$$
where x is the displacement at time t. Find the displacement, velocity and acceleration at $t = 1$.

Solution

Differentiating x once gives velocity, and differentiating x twice gives acceleration:

$$x = 2t^3 + t^2 - 10t + 10$$
$$v = \frac{dx}{dt} = 6t^2 + 2t - 10$$
$$a = \frac{d^2x}{dt^2} = 12t + 2$$

To find displacement, velocity and acceleration at $t = 1$, we substitute $t = 1$ into the above equations:

⊕ Example 13 *mechanics*

$$x = 2 + 1 - 10 + 10 = 3 \text{ m}$$
$$v = 6 + 2 - 10 = -2 \text{ m/s}$$
$$a = 12 + 2 = 14 \text{ m/s}^2$$

The displacement, s, of a particle is given by

$$s = (1 - t)e^{-t}$$

i Find the time t for the velocity $v = 0$.
ii Find the time t for the acceleration $a = 0$.
iii Sketch the graph of v against t showing all the points of maxima and minima.

Solution

? **i** We need to differentiate $s = (1 - t)e^{-t}$ to find the velocity. **How?**

Apply the product rule, 6.31 , because s is $1 - t$ times e^{-t}. Let

$$p = 1 - t \qquad q = e^{-t}$$
$$p' = -1 \qquad q' = -e^{-t}$$

We use p and q because v is velocity in this example. Thus

$$v = \frac{ds}{dt} = p'q + pq'$$
$$= (-1)e^{-t} + (1 - t)(-e^{-t})$$
$$= e^{-t}(-1 - 1 + t) = e^{-t}(t - 2)$$
$$v = e^{-t}(t - 2)$$

For $v = 0$ we have

$$e^{-t}(t - 2) = 0$$

Since the exponential function is never zero we have

$$t - 2 = 0 \text{ which gives } t = 2 \text{ s}$$

ii To find acceleration, a, we have to differentiate v. By applying the product rule, 6.31 , on $v = (t - 2)e^{-t}$ we have

$$p = t - 2 \qquad q = e^{-t}$$
$$p' = 1 \qquad q' = -e^{-t}$$

Thus

$$a = \frac{d^2s}{dt^2} = p'q + pq'$$
$$= (1)e^{-t} + (t - 2)(-e^{-t}) = e^{-t}(1 - t + 2) = e^{-t}(3 - t)$$
$$a = e^{-t}(3 - t)$$

We want to find t which satisfies $a = 0$:

$$(3 - t)e^{-t} = 0 \text{ gives } 3 - t = 0, \text{ therefore } t = 3 \text{ s}$$

iii From part **i** we know that $v = (t - 2)e^{-t}$.

6.31 $(pq)' = p'q + pq'$

Example 13 *continued*

Where does the *v* graph cut the *t* axis?

This is when $v = 0$ therefore $t = 2$. Also when $t = 0$, $v = (0 - 2)e^0 = -2$. Hence the graph goes through the points $(2, 0)$ and $(0, -2)$.

For sketching the graph we must find stationary points of *v* and their nature. **How can we find the stationary points of *v*?**

We covered this in part **ii**:

$$v' = \frac{dv}{dt} = (3 - t)e^{-t} \left(= \frac{d^2s}{dt^2} \right)$$

For stationary points, v' needs to be zero, that is when $a = 0$:

$$(3 - t)e^{-t} = 0 \text{ gives } t = 3$$

The exponential function e^{-t} is positive so we just need to check the sign of $(3 - t)$:

If $t < 3$ then $3 - t > 0$, therefore $v' > 0$ [Positive]
If $t > 3$ then $3 - t < 0$, therefore $v' < 0$ [Negative]

Thus we have $v'>0$ ⟋ ⟍ $v'<0$
                      $t=3$

Hence *v* has a maximum at $t = 3$. The maximum value is found by substituting $t = 3$ into $v = (t - 2)e^{-t}$:

$$v = (3 - 2)e^{-3} = e^{-3}$$

Also what can you say about *v* as $t \to \infty$?

Since $v = (t - 2)e^{-t}$ and the e^{-t} term goes to zero as *t* approaches infinity, we can say as $t \to \infty$, $v \to 0$. Collecting all the above gives the graph shown in Fig. 25.

Fig. 25

D2   Angular motion

So far we have dealt with a particle moving in a straight line. Now we discuss a particle having a circular motion.

Consider a particle moving around a circle so that the angular displacement, θ, is a function of time *t*, $\theta = f(t)$. [Fig. 26]

θ is the angle swept in time *t* and the units of θ are radians.

The angular velocity, ω, of a particle at time *t* is defined as

Fig. 26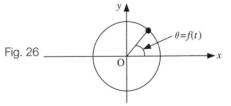

7.11 $\omega = \dfrac{d\theta}{dt}$ [1st derivative]

ω measures the rate at which the particle is going around the origin. The units of ω are radians per second, denoted by rad/s. A particle with angular velocity $\omega = \pi$ rad/s means that the particle goes through 180° (π radians) in every second.

The angular acceleration, α, of the particle at time t is defined as

7.12 $$\alpha = \frac{d\omega}{dt} = \frac{d}{dt}\left(\frac{d\theta}{dt}\right) = \frac{d^2\theta}{dt^2} \quad \text{[2nd derivative]}$$

The units of angular acceleration, α, are radians per second², denoted by rad/s². A particle with angular acceleration $\alpha = 3$ rad/s² means that the angular velocity increases by 3 rad/s for every second.

Example 14 *mechanics*

The angular displacement, θ, of a particle in circular motion is given by

$$\theta = \frac{1}{2}\cos\left(2t + \frac{\pi}{4}\right) \text{ where } 0 \leq t \leq \pi$$

i Obtain an expression for angular acceleration, α.
ii Find the first time t for $\alpha = 0$.

Solution

i By 7.12 we need to find $\dfrac{d^2\theta}{dt^2}$. First we obtain $\dfrac{d\theta}{dt}$:

$$\theta = \frac{1}{2}\cos\left(2t + \frac{\pi}{4}\right)$$

$$\frac{d\theta}{dt} = \frac{1}{2}(-2)\sin\left(2t + \frac{\pi}{4}\right) = -\sin\left(2t + \frac{\pi}{4}\right)$$

Differentiating again:

$$\frac{d^2\theta}{dt^2} = -2\cos\left(2t + \frac{\pi}{4}\right) = \alpha$$

ii For $\alpha = 0$ we have

$$-2\cos\left(2t + \frac{\pi}{4}\right) = 0$$

Dividing both sides by -2:

$$\cos\left(2t + \frac{\pi}{4}\right) = 0$$

How can we find t?
We need to remove the cos by taking inverse cos, $\cos^{-1}$, of both sides:

$$2t + \frac{\pi}{4} = \cos^{-1}(0) = \frac{\pi}{2}$$

$$2t = \frac{\pi}{2} - \frac{\pi}{4} = \frac{\pi}{4}$$

Divide both sides by 2: $t = \dfrac{\pi}{8}$ gives $\alpha = 0$.

SUMMARY

The velocity, v, and acceleration, a, of a particle in a straight line motion with displacement s at time t are defined as

7.9 $v = \dfrac{ds}{dt}$

7.10 $a = \dfrac{d^2s}{dt^2}$

The angular velocity, ω, and angular acceleration, α, of a particle in a circular motion with angle θ swept in time t are defined as

7.11 $\omega = \dfrac{d\theta}{dt}$

7.12 $\alpha = \dfrac{d^2\theta}{dt^2}$

Exercise **7(d)**

Solutions at end of book. Complete solutions available at www.palgrave.com/science/engineering/singh

🚗 **All questions in this exercise belong to the field of [*mechanics*] with units as defined in Sections D1 and D2.**

1 A particle moves along a horizontal line with displacement, s, given by
$$s = t^2 - 4t + 3$$
i Find the time(s) t for $s = 0$.
ii Determine expressions for the velocity and acceleration.

2 A projectile is fired vertically upwards with height h given by
$$h = 25t - 4.9t^2$$
$$\text{where} \quad 0 \le t \le 5$$
Find expressions for the velocity and acceleration. Determine the maximum height reached by the projectile.

3 The displacement, s, of a particle is given by
$$s = 3t^3 - 9t^2 + 10$$
Find the time(s) t for the velocity $v = 0$ and acceleration $a = 0$.

4 The displacement, s, of an object is given by
$$s = 3t^3 - 10t^2 + t - 10$$
Find expressions for the velocity and acceleration.

5 The displacement, s, of a particle is given by
$$s = t^3 - 6t^2 + 12t$$
Find the minimum velocity and sketch the graph of v against t for $t \ge 0$.

6 The displacement, s, of a particle is given by
$$s = t^3 - 12t^2 + 18t$$
Find the velocity and sketch the graph of v against t for $t \ge 0$.

7 The displacement, s, of a particle is given by
$$s = \frac{1}{v} \quad (v > 0)$$
where v is the velocity of the particle. Find $\dfrac{ds}{dv}$ and sketch the graphs, on the same axes, of s and $\dfrac{ds}{dv}$ against v for $v > 0$.

8 The angular displacement, θ, of a flywheel at time t is given by
$$\theta = t^3 - 80\pi t$$
Determine the angular velocity, ω, and angular acceleration, α, at $t = 1$.

Exercise **7(d) continued**

Solutions at end of book. Complete solutions available at www.palgrave.com/science/engineering/singh

9 The angular displacement, θ, of a particle in circular motion is given by

$$\theta = \frac{1}{4}\sin\left(2t + \frac{\pi}{6}\right) \quad (0 \le t \le \pi)$$

  i Find an expression for angular acceleration, α.

  ii Determine the first time t for $\alpha = 0$.

10 The displacement, s, of a particle is given by

$$s = (1 + t)e^{-0.25t}$$

By using a computer algebra system or a graphical calculator, plot, on different axes, the displacement, velocity and acceleration graphs as functions of time t ($0 \le t \le 10$).

SECTION E **Tangents and normals**

By the end of this section you will be able to:
- ▶ obtain the equation of a tangent at a given point on a curve
- ▶ find the equation of the normal at a given point on a curve

E1 **Tangents and normals**

Figure 27 shows a curve $y = f(x)$ with a tangent and normal at the point P.

A **tangent** at point P is defined as a line which touches the curve at **only** P.

A **normal** at point P is defined as a line which is at right angles (90°) to the tangent at P.

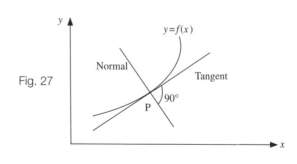

Fig. 27

The equation of a tangent can be determined by using the general straight line formula $y = mx + c$, where m is the gradient of the tangent and c is the y-intercept.

How can we determine the gradient of a tangent?

The derivative $\dfrac{dy}{dx}$ gives an expression for the gradient of a tangent.

For points where the gradient (m) of the tangent is not zero the gradient of the normal can be shown to be

7.13 $\text{Gradient of normal} = -\dfrac{1}{m}$

Example 15

Find the equation of the tangent and normal to the curve $y = \dfrac{1}{x}$ at $x = 2$.

Solution

Let $y = mx + c$ be the equation of the tangent.

? **What do we need to find?**

The gradient $m \left(= \dfrac{dy}{dx} \right)$ and the y-intercept c.

$$y = \frac{1}{x} = x^{-1}$$

$$\frac{dy}{dx} = -x^{-2} = -\frac{1}{x^2}$$

Substituting $x = 2$, $\dfrac{dy}{dx} = -\dfrac{1}{2^2} = -\dfrac{1}{4}$. Therefore the gradient $m = -\dfrac{1}{4}$.

? **How can we find c?**

We know the tangent goes through the point $\left(2, \dfrac{1}{2} \right)$ because when $x = 2$, $y = \dfrac{1}{x} = \dfrac{1}{2}$.

By substituting $x = 2$, $y = \dfrac{1}{2}$ and $m = -\dfrac{1}{4}$ into $y = mx + c$ we can find c:

$$\frac{1}{2} = \left(-\frac{1}{4} \times 2 \right) + c = -\frac{1}{2} + c$$

Hence $c = \dfrac{1}{2} + \dfrac{1}{2} = 1$.

The equation of the tangent is $y = -\dfrac{1}{4}x + 1$ or $y = 1 - \dfrac{1}{4}x$.

? **How can we find the equation of the normal at the point $\left(2, \dfrac{1}{2} \right)$?**

By $\boxed{7.13}$, the gradient of the normal $= -\dfrac{1}{-1/4} = 4$. The equation of the normal has the form

$$y = 4x + c_1$$

where c_1 is the y-intercept. Since the normal also goes through the **same** point, $\left(2, \dfrac{1}{2} \right)$,

we can substitute $x = 2$ and $y = \dfrac{1}{2}$ into $y = 4x + c_1$:

$$\frac{1}{2} = (4 \times 2) + c_1, \quad \frac{1}{2} = 8 + c_1$$

$$c_1 = \frac{1}{2} - 8 = -7\frac{1}{2} = -\frac{15}{2}$$

Hence the equation of the normal is $y = 4x - \dfrac{15}{2}$.

$\boxed{7.13}$ Gradient of normal $= -1/m$

Example 15 *continued*

Be careful not to be confused by the three different x and y relationships:

$$y = 1/x \qquad \text{[Given equation]}$$

$$y = 1 - \frac{1}{4}x \qquad \left[\text{Tangent equation at } \left(2, \frac{1}{2}\right)\right]$$

$$y = 4x - \frac{15}{2} \qquad \left[\text{Normal equation at } \left(2, \frac{1}{2}\right)\right]$$

Example 16 *electrical principles*

An electric circuit obeys the law

$$P = (10 \times 10^3)i^2$$

where P represents power in W and i represents current in A. Find the equation of the tangent relating P and i for a current $i = 5$ mA.

Solution

Fig. 28

Let $P = mi + c$ be the equation of the tangent at

$$i = 5 \text{ mA} = (5 \times 10^{-3}) \text{ A}$$

The gradient is obtained by differentiating $P = (10 \times 10^3)i^2$:

$$m = \frac{dP}{di} = (20 \times 10^3)i$$

Substituting $i = 5 \times 10^{-3}$ into $m = (20 \times 10^3)i$ yields

$$m = (20 \times 10^3) \times (5 \times 10^{-3}) = 100 \quad \text{[Gradient]}$$

We have

 $\qquad P = 100i + c$

We need to find c. How?

We know the tangent goes through the value of $i = 5 \times 10^{-3}$. Substituting $i = 5 \times 10^{-3}$ into the given equation $P = (10 \times 10^3)i^2$:

$$P = (10 \times 10^3) \times (5 \times 10^{-3})^2 = 0.25$$

Hence the tangent goes through the point $(5 \times 10^{-3}, 0.25)$. Substituting $i = 5 \times 10^{-3}$ and $P = 0.25$ into ⬛* yields

$$0.25 = (100 \times 5 \times 10^{-3}) + c \text{ gives } c = -0.25$$

Putting $c = -0.25$ into ⬛* :

$$P = 100i - 0.25$$

This is the equation of the tangent relating P and i at $i = 5 \times 10^{-3}$ A.

Sometimes it is known as the linear (straight line) relationship between P and i at $i = 5 \times 10^{-3}$ A.

The concept of finding an equation of a tangent is **important** because it gives a linear relationship between the variables at a given point, that is it simplifies functions in a given neighbourhood.

SUMMARY

The equation of the tangent line to a curve at a given point on that curve is found from the straight line equation

$$y = mx + c$$

where m is the gradient and c is the vertical intercept.
The gradient of the normal is given by

> 7.13 $\text{gradient} = -\dfrac{1}{m}$

where $m \neq 0$ is the gradient of the tangent.

Exercise 7(e)

Solutions at end of book. Complete solutions available at www.palgrave.com/science/engineering/singh

1 Obtain the equations of the tangent and normal of the following curves at the corresponding point:

a $y = x^2 - 3$, $x = 2$

b $y = \cos(x)$, $x = \dfrac{\pi}{2}$

c $y = e^x$, $x = 1$

d $y = \ln(x)$, $x = 1$

2 [electrical principles]   A non-linear resistor obeys the law
$$i = v^3$$

where v and i represent voltage and current respectively. Find the equation of the tangent relating v and i at $v = 1.6$ volts.

3 [electrical principles]   The current, i, through an inductor is given by

$$i = 2(1 - e^{-2000t})$$

Find the equation of the tangent relating i and t at $t = 1 \times 10^{-3}$ s.

SECTION F Series expansion

By the end of this section you will be able to:
▶ obtain the Maclaurin series of a given function
▶ understand the Maclaurin series expansion of some common engineering functions
▶ use these established series to evaluate limits of functions

This section might seem like a huge jump from previous sections. One of the difficulties in this section is understanding the notation: we use $f'(x)$ to denote differentiation rather than $\dfrac{dy}{dx}$. Moreover we differentiate a given function a number of times and then substitute $x = 0$ into the resulting derivatives.

F1 General Maclaurin series

Example 17

Find the equation of the tangent at $x = 0$ of $f(x) = e^x$.

Solution

Let the equation of the tangent be $y = mx + c$. We need to find the gradient, m, and y-intercept, c. Differentiating $f(x) = e^x$ gives

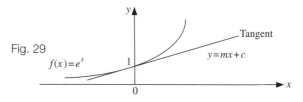

Fig. 29

$$m = f'(x) = e^x$$

Since we want to find the gradient at $x = 0$, we substitute this value

$$m = f'(0) = e^0 = 1$$

By looking at Fig. 29 we have

$$c = e^0 = 1$$

Thus the equation of the tangent is given by substituting $m = 1$ and $c = 1$:

$$y = mx + c = x + 1$$

? The tangent, $y = x + 1$, is **not** a good approximation to $f(x) = e^x$ away from zero. **Why not?**
Because it is a straight line and our function $f(x)$ is a curve.

? **How can we approximate by a curve?**

Add a quadratic term (x^2) to $x + 1$ (Fig. 30). Moreover we can achieve better and better approximations to a curve by adding more and more terms such as a cubic term (x^3), a quartic term (x^4), etc.

Fig. 30

We can write down the exact function if we consider an infinite number of terms and add these to make what is called an **infinite series**.

In this section we express general engineering functions by an infinite series. For example, the exponential function is given by

$$e^x = \underbrace{1 + x}_{\text{tangent}} + \underbrace{\frac{x^2}{2!}}_{\text{quadratic}} + \underbrace{\frac{x^3}{3!}}_{\text{cubic}} + \underbrace{\frac{x^4}{4!}}_{\text{quartic}} + \cdots$$

The sine function is

$$\sin(x) = x - \frac{x^3}{3!} + \frac{x^5}{5!} - \cdots$$

The ! sign in the series is called factorial. For example, '3!', pronounced '3 factorial', is

$$1 \times 2 \times 3 = 6$$

That is 3! = 6. Similarly

$$4! = 1 \times 2 \times 3 \times 4 = 24$$

We define 0! = 1 and 1! = 1. You can evaluate n! on a calculator; look for the ! button on your calculator.

? In this section we aim to determine the infinite series on the Right-Hand Side. **How can we discover this series?**

Let's first try to identify the series for a general function $f(x)$ and then find the series for e^x and $\sin(x)$ in the examples.

We write down the general expression of the infinite series as

> ⬛ * $\qquad f(x) = A + Bx + Cx^2 + Dx^3 + Ex^4 + \cdots$

? where A, B, C, D, E, ... are constants and $f(x)$ is a general function such that $f(x)$ and its derivatives are defined at the origin and the series sums to a finite value. **What do we need to find?**

? The constants A, B, C, D, E, ... **How can we find A?**

? By putting $x = 0$ into ⬛ * . **Why?**

Because this will remove all the other terms on the Right-Hand Side of ⬛ * except A. We have

$$f(0) = A$$

? **How can we find B?**

? We have to remove all the terms of ⬛ * except B. **How?**

First differentiate ⬛ * with respect to x:

> ⬛ ** $\qquad f'(x) = B + 2Cx + 3Dx^2 + 4Ex^3 + \cdots$

Now substitute $x = 0$ into ⬛ ** :

$$f'(0) = B + 0$$

? Therefore $B = f'(0)$. **How can we find C?**

Differentiate ⬛ ** with respect to x:

> ⬛ *** $\qquad f''(x) = 0 + 2C + (3 \times 2)Dx + (4 \times 3)Ex^2 + \cdots$

Substitute $x = 0$ into ▮***▮ :

$$f''(0) = 2C + 0$$

$$C = \frac{f''(0)}{2} = \frac{f''(0)}{2!} \quad \text{[Remember that 2! = 2]}$$

? **How can we find D?**

Differentiate ▮***▮ with respect to x:

$$f'''(x) = 0 + (3 \times 2)D + (4 \times 3 \times 2)Ex + \cdots$$

Substitute $x = 0$:

$$f'''(0) = (3 \times 2)D + 0$$

$$D = \frac{f'''(0)}{3 \times 2} = \frac{f'''(0)}{3!} \quad \text{[Remember that 3! = 1 \times 2 \times 3]}$$

Substituting $A = f(0)$, $B = f'(0)$, $C = \dfrac{f''(0)}{2!}$, $D = \dfrac{f'''(0)}{3!}$ into ▮*▮ gives

$$f(x) = f(0) + f'(0) + x\frac{f''(0)}{2!}x^2 + \frac{f'''(0)}{3!}x^3 + Ex^4 + \cdots$$

? **By observing the pattern of the series, what do you think is the value of the constant E?**

$$E = \frac{f^{(4)}(0)}{4!}$$

In the general case we have

▮7.14▮ $\qquad f(x) = f(0) + f'(0)x + \dfrac{f''(0)}{2!}x^2 + \dfrac{f'''(0)}{3!}x^3 + \dfrac{f^{(4)}(0)}{4!}x^4 + \cdots$

▮7.14▮ is called the Maclaurin series for the function $f(x)$. This is an infinite series with the $(n + 1)$th term given by

$$\frac{f^{(n)}(0)}{n!}x^n$$

where $f^{(n)}(0)$ represents the value of $\dfrac{d^n y}{dx^n}$ at $x = 0$.

We can only find the Maclaurin series if $f(x)$ and its derivatives are defined at $x = 0$ and the series approaches a finite value.

Colin Maclaurin was born in February 1698 in Argyllshire, Scotland and graduated from Glasgow University. In 1717 he became Professor of Mathematics at the University of Aberdeen and in 1725 he moved to Edinburgh University where he spent the remaining part of his career. The Maclaurin series is a special case, the case of approximating close to zero, of the general Taylor series. (We do not cover Taylor series in this book.)

F2 **Maclaurin series of functions**

Example 18

Obtain the Maclaurin series expansion of e^x.

Solution

Let $f(x) = e^x$. We need to differentiate this function a number of times to obtain the Maclaurin series and then substitute $x = 0$ into the resulting derivatives. Don't be confused by the f dashed notation.

$$f(x) = e^x \qquad\qquad\qquad\qquad f(0) = e^0 = 1$$

$$f'(x) = e^x \quad [\text{Differentiating } f] \qquad f'(0) = 1$$

$$f''(x) = e^x \quad [\text{Differentiating } f'] \qquad f''(0) = 1$$

$$f'''(x) = e^x \quad [\text{Differentiating } f''] \qquad f'''(0) = 1$$

$$f^{(4)}(x) = e^x \quad [\text{Differentiating } f'''] \qquad f^{(4)}(0) = 1$$

Substituting 1 for $f(0)$, $f'(0)$, $f''(0)$, $f'''(0)$, $f^{(4)}(0)$ into $\boxed{7.14}$ gives

$$e^x = f(0) + f'(0)x + \frac{f''(0)}{2!}x^2 + \frac{f'''(0)}{3!}x^3 + \frac{f^{(4)}(0)}{4!}x^4 + \cdots$$

$$= 1 + (1)x + \frac{1}{2!}x^2 + \frac{1}{3!}x^3 + \frac{1}{4!}x^4 + \cdots \quad [\text{Substituting}]$$

$$= 1 + x + \frac{x^2}{2!} + \frac{x^3}{3!} + \frac{x^4}{4!} + \cdots \quad [\text{Simplifying}]$$

Use a computer algebra package, or a graphical calculator, to plot e^x, and $1 + x$, $1 + x + \dfrac{x^2}{2!}$, $1 + x + \dfrac{x^2}{2!} + \dfrac{x^3}{3!}$ and $1 + x + \dfrac{x^2}{2!} + \dfrac{x^3}{3!} + \dfrac{x^4}{4!}$ on the same axes, and observe how the graphs approach e^x (Fig. 31).

By using MAPLE we have

```
> f := 1 + x;
```
$$f := 1 + x$$

```
> g := 1 + x + (x^2)/2!;
```
$$g := 1 + x + \frac{1}{2}x^2$$

```
> h := 1 + x + (x^2)/2! + (x^3)/3!;
```
$$h := 1 + x + \frac{1}{2}x^2 + \frac{1}{6}x^3$$

```
> k:=1 + x + (x^2)/2! + (x^3)/3! + (x^4)/4!;
```

$$k := 1 + x + \frac{1}{2}x^2 + \frac{1}{6}x^3 + \frac{1}{24}x^4$$

```
> plot([f,g,h,k,exp(x)], x = -2..2,linestyle = [2,6,10,14,26],color = black);
```

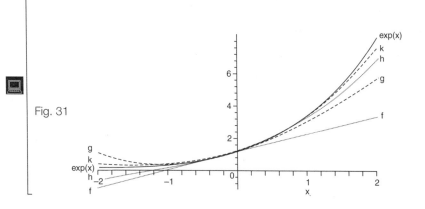

Fig. 31

Example 19

Obtain the Maclaurin series expansion of $\sin(x)$.

Solution

Let $f(x) = \sin(x)$, then differentiating and substituting $x = 0$ into the resulting derivatives gives

$$f(x) = \sin(x) \qquad\qquad f(0) = \sin(0) = 0$$
$$f'(x) = \cos(x) \qquad\qquad f'(0) = \cos(0) = 1$$
$$f''(x) = -\sin(x) \qquad\qquad f''(0) = -\sin(0) = 0$$
$$f'''(x) = -\cos(x) \qquad\qquad f'''(0) = -\cos(0) = -1$$
$$f^{(4)}(x) = \sin(x) \qquad\qquad f^{(4)}(0) = \sin(0) = 0$$
$$f^{(5)}(x) = \cos(x) \qquad\qquad f^{(5)}(0) = \cos(0) = 1$$

Substituting the above values into 7.14 gives

$$\sin(x) = f(0) + f'(0)x + \frac{f''(0)}{2!}x^2 + \frac{f'''(0)}{3!}x^3 + \frac{f^{(4)}(0)}{4!}x^4 + \frac{f^{(5)}(0)}{5!}x^5 + \cdots$$

$$= 0 + (1)x + \frac{0}{2!}x^2 + \frac{(-1)}{3!}x^3 + \frac{0}{4!}x^4 + \frac{1}{5!}x^5 + \cdots \qquad \text{[Substituting]}$$

$$= x - \frac{x^3}{3!} + \frac{x^5}{5!} - \cdots \qquad \text{[Simplifying]}$$

The infinite series for e^x and $\sin(x)$ are evaluated in the above examples and are valid for all values of x. We have placed no restriction on x. However in some cases there is a restriction on the x values, for example

$$\ln(1 + x) = x - \frac{x^2}{2} + \frac{x^3}{3} - \frac{x^4}{4} + \cdots \text{ for } -1 < x \le 1$$

This is because the series would **not** sum to a finite value but infinity if x was outside the stated domain, $x \le -1$ or $x > 1$.

In the formulae below there are some common engineering functions and their Maclaurin series expansion with the stated domains:

| 7.14 | $$f(x) = f(0) + f'(0)x + \frac{f''(0)}{2!}x^2 + \frac{f'''(0)}{3!}x^3 + \frac{f^{(4)}(0)}{4!}x^4 + \cdots$$ |

| 7.15 | $$e^x = 1 + x + \frac{x^2}{2!} + \frac{x^3}{3!} + \cdots + \frac{x^n}{n!} + \cdots \quad \text{all } x$$ |

| 7.16 | $$\sin(x) = x - \frac{x^3}{3!} + \frac{x^5}{5!} - \cdots + \frac{(-1)^n x^{2n+1}}{(2n+1)!} + \cdots \quad \text{all } x$$ |

| 7.17 | $$\cos(x) = 1 - \frac{x^2}{2!} + \frac{x^4}{4!} - \cdots + \frac{(-1)^n x^{2n}}{(2n)!} + \cdots \quad \text{all } x$$ |

| 7.18 | $$\tan(x) = x + \frac{x^3}{3} + \frac{2x^5}{15} + \frac{17x^7}{315} + \cdots \quad -\pi/2 < x < \pi/2$$ |

| 7.19 | $$\sinh(x) = x + \frac{x^3}{3!} + \frac{x^5}{5!} + \cdots + \frac{x^{2n+1}}{(2n+1)!} + \cdots \quad \text{all } x$$ |

| 7.20 | $$\cosh(x) = 1 + \frac{x^2}{2!} + \frac{x^4}{4!} + \cdots + \frac{x^{2n}}{(2n)!} + \cdots \quad \text{all } x$$ |

| 7.21 | $$\ln(1 + x) = x - \frac{x^2}{2} + \frac{x^3}{3} - \frac{x^4}{4} + \cdots + \frac{(-1)^{n+1} x^n}{n} \cdots \quad -1 < x \le 1$$ |

| 7.22 | $$\ln(1 - x) = -x - \frac{x^2}{2} - \frac{x^3}{3} - \frac{x^4}{4} - \cdots - \frac{x^n}{n} \cdots \quad -1 \le x < 1$$ |

(The x for the trigonometric functions is in radians.)

We can use these series to evaluate limits of functions, as the following example shows.

Example 20

Determine $\lim\limits_{x \to 0} \dfrac{\sin(x)}{x}$.

Solution

Substituting $x = 0$, the function $\dfrac{\sin(x)}{x}$ gives $\dfrac{0}{0}$ which is not defined. **How can we find the value of** $\dfrac{\sin(x)}{x}$ **close to 0?**

Example 20 *continued*

Using 7.16 we have

$$\frac{\sin(x)}{x} = \frac{x - \frac{x^3}{3!} + \frac{x^5}{5!} - \cdots}{x}$$

$$= \frac{x\left(1 - \frac{x^2}{3!} + \frac{x^4}{5!} - \cdots\right)}{x} \qquad \text{[Factorizing numerator]}$$

$$= 1 - \frac{x^2}{3!} + \frac{x^4}{5!} - \cdots \qquad \text{[Cancelling } x\text{'s]}$$

Hence

$$\lim_{x \to 0} \frac{\sin(x)}{x} = \lim_{x \to 0} \left(1 - \frac{x^2}{3!} + \frac{x^4}{5!} - \cdots\right) = 1$$

SUMMARY

We can find the Maclaurin series of a function, $f(x)$, which has its derivatives defined at $x = 0$ and the series sums to a finite value by

7.14 $\qquad f(x) = f(0) + f'(0)x + \frac{f''(0)}{2!}x^2 + \frac{f'''(0)}{3!}x^3 + \frac{f^{(4)}(0)}{4!}x^4 + \cdots$

We can use the established series to evaluate limits of functions.

Exercise **7(f)**

Solutions at end of book. Complete solutions available at www.palgrave.com/science/engineering/singh

1 Obtain the Maclaurin series expansion of $\cos(x)$ by using 7.14 .

2 Obtain the Maclaurin series expansion of $\ln(1 + x)$ by applying 7.14 and where $-1 < x \le 1$.

3 By using the Maclaurin series, show that

$$(1 + x)^n = 1 + nx + \left[\frac{n(n - 1)}{2!}\right] x^2$$
$$+ \left[\frac{n(n - 1)(n - 2)}{3!}\right] x^3 + \cdots$$
$$(-1 < x < 1)$$

where n is any fraction or a negative number.

4 Find the Maclaurin series expansion of $\sin^2(x)$ and $\cos^2(x)$.
(*Hint*: Use a trigonometrical relation for $\cos^2(x)$ and substitute $2x$ into a known series.)

5 Determine

  a $\displaystyle\lim_{x \to 0} \frac{\ln(1 + x)}{x}$ **b** $\displaystyle\lim_{x \to 0} \frac{1 - \cos(x)}{x^2}$

6 Explain why the Maclaurin series for $\ln(x)$ does not exist.

Exercise **7(f) continued**

Solutions at end of book. Complete solutions available at www.palgrave.com/science/engineering/singh

7 Find the Maclaurin series expansion of

$$e^{-\frac{1}{2}x^2}$$

8 Determine the Maclaurin series of

$$\frac{1}{2}\ln\left(\frac{1+x}{1-x}\right)$$

SECTION G **Binomial revisited**

By the end of this section you will be able to:
- ► expand $(a + b)^n$ where n is a positive whole number
- ► expand $(1 + x)^n$ where n is a fraction or a negative whole number
- ► apply the binomial expansion to engineering examples

G1 **Expanding $(a + b)^n$**

We have covered the binomial expansion (**Section 2F**) in Chapter 2, where we expanded terms like $(1 + x)^5$ by using Pascal's triangle.

However you would *not* expand $(1 + x)^{15}$ by using Pascal's triangle. Why not?

Evaluating Pascal's triangle for $n = 15$ is a tedious task, so we need to examine some other technique.

Expansion of the general term $(a + b)^n$ is given by

7.23
$$(a + b)^n = a^n + na^{n-1}b + \left[\frac{n(n-1)}{2!}\right]a^{n-2}b^2$$

$$+ \left[\frac{n(n-1)(n-2)}{3!}\right]a^{n-3}b^3 + \cdots + b^n$$

7.23 looks like an horrendous formula but once it has been applied a number of times you will become confident in using it. Also it might be easier to apply the formula if you notice the patterns that it contains.

The index of a starts with n and decreases by 1 in each term to the right:

$$a^n, \; a^{n-1}, \; a^{n-2}, \; \cdots, \; a^1 = a, \; a^0 = 1$$

Similarly the index of b increases by 1 with each term to the right:

$$b^0 = 1, \; b^1 = b, \; b^2, b^3, \cdots, \; b^n$$

Moreover notice the coefficients of a and b in the middle terms:

$$n, \; \frac{n(n-1)}{2!}, \; \frac{n(n-1)(n-2)}{3!}, \; \frac{n(n-1)(n-2)(n-3)}{4!}, \; \cdots$$

Example 21

Find the first four terms in the expansion of $(1 + x)^{15}$.

Solution

Substituting $a = 1$, $b = x$ and $n = 15$ into 7.23 gives

$$(1 + x)^{15} = 1^{15} + 15(1)^{14}\, x + \left[\frac{15(15 - 1)}{2!} \right](1)^{13}\, x^2$$

$$+ \left[\frac{15(15 - 1)(15 - 2)}{3!} \right](1)^{12}\, x^3 + \cdots$$

$$= 1 + 15x + \left[\frac{15(14)}{2!} \right]x^2 + \left[\frac{15(14)(13)}{3!} \right]x^3 + \cdots \qquad [\text{Remember that } 1^{\text{index}} = 1]$$

$$= 1 + 15x + 105x^2 + 455x^3 + \cdots \qquad [\text{Evaluating brackets}]$$

Example 22

Determine up to and including the x^3 term in the expansion of $(5 - x)^{10}$.

Solution

Putting $a = 5$, $b = -x$ and $n = 10$ into 7.23 gives

$$(5 - x)^{10} = [5 + (-x)]^{10} = 5^{10} + [10 \times 5^9 \times (-x)] + \left[\frac{10(10-1)}{2!} \right]5^8(-x)^2$$

$$+ \left[\frac{10(10 - 1)(10 - 2)}{3!} \right]5^7(-x)^3 + \cdots$$

$$= 5^{10} + [10 \times 5^9](-x) + \left[\frac{10(9)}{2!} \times 5^8 \right](-x)^2$$

$$+ \left[\frac{10(9)(8)}{3!} \times 5^7 \right](-x)^3 + \cdots$$

$$\underset{\text{by calculator}}{=} 9765625 - 19531250x + 17578125x^2 - 9375000x^3 + \cdots$$

Notice how the numbers go down one step at a time on the numerator: 10, 9, 8 etc.

G2 Binomial series

In question **3** of **Exercise 7(f)** you were asked to show the following result by using the Maclaurin series:

7.24 $$(1 + x)^n = 1 + nx + \left[\frac{n(n-1)}{2!}\right]x^2 + \left[\frac{n(n-1)(n-2)}{3!}\right]x^3 + \cdots \qquad -1 < x < 1$$

where n is not a positive whole number. If n is a positive whole number then the series terminates after $n + 1$ terms, as can be seen in 7.23 .

? **How many terms does $(1 + x)^{15}$ have?**

16

? **How many terms does $(5 - x)^{10}$ have?**

11

Example 23

Show that $\dfrac{1}{1 + x} = 1 - x + x^2 - x^3 + x^4 - \cdots$ where $-1 < x < 1$.

Solution

$\dfrac{1}{1 + x} = (1 + x)^{-1}$. Substituting $n = -1$ into 7.24 gives

$$(1 + x)^{-1} = 1 + (-1)x + \left[\frac{(-1)(-2)}{2!}\right]x^2 + \left[\frac{(-1)(-2)(-3)}{3!}\right]x^3$$

$$+ \left[\frac{(-1)(-2)(-3)(-4)}{4!}\right]x^4 \cdots$$

$$= 1 - x + x^2 - x^3 + x^4 \cdots \qquad \text{[Cancelling]}$$

Note that this is an infinite series.

Example 24

Expand $(1 + x)^{1/2}$.

Solution

Substituting $n = \dfrac{1}{2}$ into 7.24 yields

Example 24 *continued*

$$(1 + x)^{1/2} = 1 + \frac{1}{2}x + \left[\frac{\frac{1}{2}\left(\frac{1}{2} - 1\right)}{2!}\right]x^2 + \left[\frac{\frac{1}{2}\left(\frac{1}{2} - 1\right)\left(\frac{1}{2} - 2\right)}{3!}\right]x^3 + \cdots$$

$$+ \left[\frac{\frac{1}{2}\left(-\frac{1}{2}\right)}{2}\right]x^2 + \left[\frac{\frac{1}{2}\left(-\frac{1}{2}\right)\left(-\frac{3}{2}\right)}{6}\right]x^3 + \cdots$$

$$- \frac{1}{8}x^2 + \frac{3}{48}x^3 + \cdots \qquad \text{[Evaluating brackets]}$$

tions

n the expansion of $(1 + x)^n$ then we have a finite series,
whole number then we have an infinite series.

ing. It involves a number of algebraic techniques which
pplying 7.24 .

n L and sag h. The length of the cable can be determined

Fig. 32

where x is the horizontal distance from the
lowest point of the cable. Use the binomial expansion to expand the above
expression up to and including the x^6 term, and state the range for which the expression
is valid.

Solution

Squaring the bracket term:

$$\left(\frac{8hx}{L^2}\right)^2 = \frac{8^2h^2x^2}{(L^2)^2} = \frac{64h^2x^2}{L^4}$$

Example 25 *continued*

Substituting $n = \dfrac{1}{2}$ into 7.24 yields

$$\left[1 + \frac{64h^2x^2}{L^4}\right]^{\frac{1}{2}} = 1 + \frac{1}{2}\left(\frac{64h^2x^2}{L^4}\right) + \left[\frac{\frac{1}{2}\left(\frac{1}{2} - 1\right)}{2!}\right]\left(\frac{64h^2x^2}{L^4}\right)^2$$

$$+ \left[\frac{\frac{1}{2}\left(\frac{1}{2} - 1\right)\left(\frac{1}{2} - 2\right)}{3!}\right]\left(\frac{64h^2x^2}{L^4}\right)^3 \cdots$$

$$= 1 + \frac{1}{2}\left(\frac{64h^2x^2}{L^4}\right) + \left[\frac{\frac{1}{2}\left(-\frac{1}{2}\right)}{2}\right]\left(\frac{64^2(h^2)^2(x^2)^2}{(L^4)^2}\right)$$

$$+ \left[\frac{\frac{1}{2}\left(-\frac{1}{2}\right)\left(-\frac{3}{2}\right)}{6}\right]\left(\frac{64^3(h^2)^3(x^2)^3}{(L^4)^3}\right) \cdots$$

$$= 1 + \frac{64}{2}\frac{h^2x^2}{L^4} + \left[-\frac{1}{8}\right]64^2\frac{h^4x^4}{L^8} + \left[\frac{3}{48}\right]64^3\frac{h^6x^6}{L^{12}} \cdots$$
$$\text{[Using indices]}$$

$$= 1 + 32\frac{h^2x^2}{L^4} - 512\frac{h^4x^4}{L^8} + 16384\frac{h^6x^6}{L^{12}} \cdots \quad \text{[Simplifying]}$$

where $-1 < \dfrac{64h^2x^2}{L^4} < 1$

SUMMARY

For a positive whole number, n, we have a finite series:

7.23
$$(a + b)^n = a^n + na^{n-1}b + \left[\frac{n(n-1)}{2!}\right]a^{n-2}b^2$$
$$+ \left[\frac{n(n-1)(n-2)}{3!}\right]a^{n-3}b^3 + \cdots + b^n$$

If n is a fraction or a negative whole number we have an infinite series:

7.24
$$(1 + x)^n = 1 + nx + \left[\frac{n(n-1)}{2!}\right]x^2 + \left[\frac{n(n-1)(n-2)}{3!}\right]x^3 + \cdots \quad -1 < x < 1$$

Exercise 7(g)

Solutions at end of book. Complete solutions available at www.palgrave.com/science/engineering/singh

1 Find the first four terms in the binomial expansion of

a $(1 + x)^{20}$ **b** $(x + y)^{15}$

c $(2 - Z)^{10}$

2 Expand $\dfrac{1}{1 - Z}$ where $-1 < Z < 1$.

Exercise **7(g) continued**

3 [structures]   The length of a part of a cable of a suspension bridge can be found by integrating the term

$$\left[1 + \left(\frac{Wx}{T} \right)^2 \right]^{\frac{1}{2}}$$

where W is load per unit length, T is tension and x is distance along the bridge. Use the binomial expansion to expand the above expression up to and including the x^6 term, and state the range for which the expansion is valid.

4 [structures]   Figure 33 shows part of a suspension bridge cable where ℓ = total span and s = sag. The maximum tension, T, in the cable is given by

Fig. 33

$$T = \frac{W\ell}{2} \sqrt{1 + \left(\frac{\ell}{4s} \right)^2}$$

where W is the load per unit length. Show that

$$T = \frac{W\ell}{2} + \frac{W\ell^3}{64s^2} - \frac{W\ell^5}{4096s^4} + \frac{W\ell^7}{131072s^6} \cdots$$

5 [fluid mechanics]   In fluid flow the pressure ratio P (between a stagnation point and free stream) is given by

$$P = \left(1 + \frac{1}{2}(\gamma - 1)M^2 \right)^{\frac{\gamma}{\gamma - 1}}$$

where γ is the specific heat ratio and M is the Mach number. Show that

$$P = 1 + \gamma \left(\frac{M^2}{2} + \frac{M^4}{8} + \frac{2 - \gamma}{48} M^6 + \cdots \right)$$

SECTION H **Numerical solution of equations**

By the end of this section you will be able to:
► derive the Newton–Raphson formula
► apply the Newton–Raphson formula

Linear equations can be easily solved by using basic algebra. Non-linear equations such as quadratic equations can be solved by factorizing or by using the formula

$$x = \frac{-b \pm \sqrt{b^2 - 4ac}}{2a}$$

There is **no simple** formula for solving a cubic or equations of higher order.

Most of today's society have heard of Newton but very little is known of **Joseph Raphson**. He was born in Middlesex, England in 1648 and graduated from Cambridge. He wrote a book 'Analysis Aequationum Universalis' which contained Newton's method for approximating roots of an equation. Thus the name Newton–Raphson for solving equations.

We can solve these using a computer algebra system, a spreadsheet or a graphical calculator. Another approach is to solve these equations using numerical methods, which involves locating a root of an equation by successive iterations. Computers and calculators use such methods to solve these equations. One such method is called the **Newton–Raphson**.

H1 Newton–Raphson method

We consider the non-linear equation $f(x) = 0$. The root(s) of this equation lie where the graph of $f(x)$ cuts the horizontal axis. Suppose the graph of $f(x)$ follows the path shown in Fig. 34.

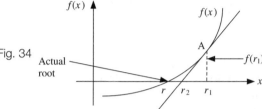

Fig. 34

Let r be the actual root of the equation $f(x) = 0$. The problem is to find the root of the equation, r. If we first estimate the root of $f(x) = 0$ to be at r_1 (say) and we draw a tangent at point A as shown, then

$$f'(r_1) = \frac{f(r_1)}{r_1 - r_2} \quad \text{[Gradient]}$$

Rearranging:

$$r_1 - r_2 = \frac{f(r_1)}{f'(r_1)} \text{ gives } r_2 = r_1 - \frac{f(r_1)}{f'(r_1)}$$

In Fig. 34 we can see that r_2 is closer to the actual root r than it is to r_1.

We can now magnify between r and r_2 and get a closer approximation to the root r (Fig. 35).

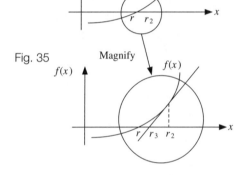

Fig. 35   Magnify

By the same procedure as above we get r_3. Now r_3 is closer to the actual root r than it is to r_2. So as above, we have

$$r_3 = r_2 - \frac{f(r_2)}{f'(r_2)}$$

By repeating we get

$$r_4 = r_3 - \frac{f(r_3)}{f'(r_3)}$$

$$r_5 = r_4 - \frac{f(r_4)}{f'(r_4)}$$

? **What do you think r_{n+1} is going to be?**

7.25 $$r_{n+1} = r_n - \frac{f(r_n)}{f'(r_n)}$$

The above method is known as the Newton–Raphson procedure. The number of iterations $n + 1$ is determined by the accuracy required.

Example 26

Find the root of the equation

$$x^3 - 2x - 5 = 0$$

correct to two decimal places using the Newton–Raphson procedure with your first approximation, $r_1 = 2$.

Solution

Let

$$f(x) = x^3 - 2x - 5$$

$$f'(x) = 3x^2 - 2 \quad \text{[Differentiating]}$$

We are given $r_1 = 2$ and using $\boxed{7.25}$ with $n = 1$ we have

$$r_2 = r_1 - \frac{f(r_1)}{f'(r_1)}$$

† $\qquad r_2 = 2 - \dfrac{f(2)}{f'(2)}$

$$f(2) = 2^3 - (2 \times 2) - 5 = -1 \quad \text{[Substituting } r_1 = 2\text{]}$$

$$f'(2) = 3(2)^2 - 2 = 10 \qquad\qquad \text{[Substituting } r_1 = 2\text{]}$$

Putting these values, $f(2) = -1$ and $f'(2) = 10$, into ┆ † ┆ :

$$r_2 = 2 - \left(\frac{-1}{10}\right) = 2.10$$

To find r_3, use $\boxed{7.25}$ again:

$$r_3 = r_2 - \frac{f(r_2)}{f'(r_2)}$$

with $r_2 = 2.1$:

†† $\qquad r_3 = 2.1 - \dfrac{f(2.1)}{f'(2.1)}$

$$f(2.1) = 2.1^3 - (2 \times 2.1) - 5 = 0.061 \quad \text{[Substituting } r_2 = 2.1\text{]}$$

$$f'(2.1) = 3 \times (2.1)^2 - 2 = 11.230 \qquad \text{[Substituting } r_2 = 2.1\text{]}$$

Putting these, $f(2.1) = 0.061$ and $f'(2.1) = 11.230$, into ┆ †† ┆ gives

$$r_3 = 2.1 - \frac{0.061}{11.23} = 2.095 = 2.10 \quad \text{(2 d.p.)}$$

Since the r_2 and r_3 approximations agree to 2 d.p., the root of $x^3 - 2x - 5 = 0$ is 2.10 (2 d.p.)

Sometimes we are not given the initial value, r_1, to start the Newton–Raphson procedure. To find r_1, we can either

i plot a graph of the function and see approximately where the roots lie, or
ii evaluate the function at some obvious values.

Example 27 *mechanics*

The velocity, v, of an object is given by
$$v = t^3 - 3t^2 - 2 \quad \text{where} \quad 2 \le t \le 4$$
By using the Newton–Raphson procedure, find t correct to four significant figures for the velocity to be zero.

Solution

Here we are not given our first approximation r_1.

So what should we take r_1 to be?

Find v for obvious values of t, that is $t = 2$, 3 and 4 since $2 \le t \le 4$:

Substituting $t = 2$; $v = 2^3 - 3(2)^2 - 2 = -6$
Substituting $t = 3$; $v = 3^3 - 3(3)^2 - 2 = -2$
Substituting $t = 4$; $v = 4^3 - 3(4)^2 - 2 = 14$

Clearly by observing the last two values of v, there is a root of $t^3 - 3t^2 - 2 = 0$ between $t = 3$ and $t = 4$ because v goes from being negative to positive. Since -2 is closer to zero, take $r_1 = 3$. By 7.25 :

$$r_2 = r_1 - \frac{f(r_1)}{f'(r_1)}$$

With $r_1 = 3$, this gives

$$r_2 = 3 - \frac{f(3)}{f'(3)} \quad \boxed{*}$$

Let $v = f(t)$:

$$f(t) = t^3 - 3t^2 - 2, \qquad\qquad f'(t) = 3t^2 - 6t$$
$$f(3) = 3^3 - 3(3)^2 - 2 = -2, \quad f'(3) = 3(3)^2 - (6 \times 3) = 9$$

Substituting $f(3) = -2$ and $f'(3) = 9$ into $\boxed{*}$:

$$r_2 = 3 - \left(-\frac{2}{9}\right) = 3.222$$

Using 7.25 again:

$$r_3 = 3.222 - \frac{f(3.222)}{f'(3.222)} \quad \boxed{**}$$

$$f(3.222) = 3.222^3 - [3 \times (3.222)^2] - 2 = 0.3046 \quad [\text{Substituting } r_2 = 3.222]$$
$$f'(3.222) = [3 \times (3.222)^2] - (6 \times 3.222) = 11.811 \quad [\text{Substituting } r_2 = 3.222]$$

7.25 $r_{n-1} = r_n - \dfrac{f(r_n)}{f'(r_n)}$

Example 27 *continued*

Putting these into ** :

$$r_3 = 3.222 - \frac{0.3046}{11.811} = 3.196$$

Applying 7.25 with $r_3 = 3.196$:

*** $$r_4 = 3.196 - \frac{f(3.196)}{f'(3.196)}$$

$$f(3.196) = 3.196^3 - [\,3 \times (3.196)^2\,] - 2 = 0.002 \quad \text{[Substituting } r_3 = 3.196]$$
$$f'(3.196) = [3 \times (3.196)^2] - [6 \times 3.196] = 11.467 \quad \text{[Substituting } r_3 = 3.196]$$

Putting these into *** :

$$r_4 = 3.196 - \frac{0.002}{11.467} = 3.1958 = 3.196 \ (4 \ \text{s.f.})$$

Since r_3 and r_4 are identical, the velocity of the object is zero when $t = 3.196$ s.

The Newton–Raphson method normally requires a close initial estimate to the actual root, otherwise it may fail.

SUMMARY

We can use the Newton–Raphson procedure to solve non-linear equations. With an initial guess, r_1, we obtain $r_2, r_3, \ldots$ by using

7.25 $$r_{n+1} = r_n - \frac{f(r_n)}{f'(r_n)}$$

The number of iterations depends on the accuracy required.

Exercise **7(h)**

Solutions at end of book. Complete solutions available at www.palgrave.com/science/engineering/singh

1 Apply the Newton–Raphson procedure, with the first approximation $r_1 = -2$, to find a root of the following equations to two decimal places:

a $x^3 - 2x + 7 = 0$
b $x^4 - 3x^2 - 2 = 0$
c $e^x - 2x - 5 = 0$

2 [vibrations] The natural frequencies of a three-storey building are related to λ where λ satisfies

$$\lambda^3 + 28\lambda^2 + 231\lambda + 541 = 0$$

By applying the Newton–Raphson procedure, determine all three roots of

7.25 $$r_{n-1} = r_n - \frac{f(r_n)}{f'(r_n)}$$

Exercise **7(h) continued**

Solutions at end of book. Complete solutions available at www.palgrave.com/science/engineering/singh

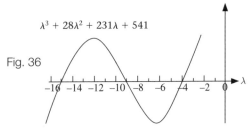

$\lambda^3 + 28\lambda^2 + 231\lambda + 541$

Fig. 36

the above equation correct to three decimal places.

3 [fluid mechanics]   The velocity flow, v, of a liquid along a channel satisfies

$$v^3 - 6v^2 - 348v + 3112 = 0$$

Given that there is a root of this equation between $v = 10$ and $v = 11$,

find this root correct to three decimal places.

4 [materials]   The deflection, y, at a distance x along a beam of length L is given by

$$y = \frac{WL}{EI}(x^3 - 3x^2 + 2x - 1)$$

where W = weight, E = modulus of elasticity and I = second moment of area. These are all non-zero constants.

By applying the Newton–Raphson procedure, find the distance (correct to three decimal places), x, where the deflection y is zero.

Miscellaneous exercise 7

Solutions at end of book. Complete solutions available at www.palgrave.com/science/engineering/singh

1 A rectangular tank with no top and a volume of 100 m³ is to be made from a sheet of metal. If one side has a length of 2 m, find the dimensions of the other sides so that the minimum material is used (minimum surface area).

2 A parcel with a square cross-section is limited by the sum of its length and girth (perimeter of cross-section) to 2 m. Find the dimensions of the parcel which maximizes the volume.

3 Repeat the above question, showing that whatever may be the sum of length and girth of a square-based parcel, the maximum volume will always be when the length is twice the side of the square.

4 [materials]   The loading distribution, w, on a beam is given by

$$w = -36x^2 + 50x$$

where x is the distance in metres from one end of the beam. Find the value of x at which maximum loading occurs.

5 [materials]   A beam of length 8 m has a deflection y (m) at a distance x (m) from one end given by

$$y = \frac{1}{12 \times 10^3}(x^4 - 14x^3 + 36x^2)$$

Determine the maximum deflection, and the value of x to achieve this.

6 [mechanics]   A projectile is fired at an angle θ with the horizontal distance, x, given by

$$x = 2.5\sin(2\theta)   \text{where } 0 < \theta < \frac{\pi}{2}$$

Find the value of θ which gives maximum x.

7 [mechanics]   The horizontal distance, x, travelled by a projectile is given by

$$x = \frac{u^2}{25}\sin(\theta)\cos(\theta)   0 < \theta < \frac{\pi}{2}$$

where u is the initial velocity and θ is the angle that the projectile makes with the horizontal. Determine the value of θ which maximizes x.

8 🚗 [*mechanics*]   A particle moves along a horizontal line with displacement, s, given by

$$s = 2 - te^{-t}$$

Find expressions for the velocity and acceleration of the particle. Sketch v against t for $t \geq 0$.

9 🚗 [*mechanics*]   The velocity, v, of a particle moving in a straight line at time t is given by

$$v = 4e^{-50t^2}$$

Sketch the graph of v against t, marking all points of maximum, minimum and inflexion.

(*Note*: The inflexion points are **not horizontal**, but are general points of inflexion as defined in the text: the point $x = c$ of $y = f(x)$ where $\dfrac{d^2y}{dx^2} = 0$ and $\dfrac{d^2y}{dx^2}$ changes sign close to $x = c$.)

10 🌡 [*heat transfer*]   The critical insulation thickness t of a tube is defined to be the value of t which gives minimum resistance R where

$$R = \frac{\ln(t/t_1)}{2\pi k} + \frac{1}{2\pi t h}$$

where $k > 0$ is thermal conductivity, h is heat transfer coefficient and t is inner thickness of insulation tube. Show that the critical thickness is $t = \dfrac{k}{h}$.
(Treat t_1, k and h as constants.)

11 🚗 [*mechanics*]   The angular acceleration, α, of a machine is given by

$$\alpha = \frac{n^2 + 12}{3 - n}$$

where n is the gear ratio. Determine the value of n which gives the maximum acceleration α.

12 Evaluate $\displaystyle\lim_{x \to 0} \frac{e^x - 1}{x}$.

13 Find the equations of the tangent and normal for $y = \sin^2(x)$ at $x = \pi/4$.

14 📱 [*signal processing*]   The velocity, v, of a signal in a cable at a distance x is given by

$$v = kx \ln\left(\frac{1}{x}\right) \quad (0 < x < 1)$$

where k is a positive constant. Find the value of x which gives maximum velocity.

15 🔌 [*electrical principles*]   The voltage, v, across an inductor with inductance L, is given by

$$v = L\frac{di}{dt}$$

For $i = 5e^{-500t}$ and $L = 2$ mH, determine v. Sketch, on different axes, the graphs of v and i against t for $t \geq 0$, marking any maxima, minima and horizontal points of inflexion.

16 🧲 [*electromagnetism*]   The magnetising force, F, at a point half way along the line joining the centres of two coils of equal radius is given by

$$F = \frac{Ir^2}{2(x^2 + r^2)^{\frac{3}{2}}}$$

where r is the radius of each coil, I is the current carried in each coil and x is the distance between the coils.

The distance x is varied and the graph of F against x has a general point of inflexion. Determine the point x in terms of r.

17 🧲 [*electromagnetism*]   The efficiency, η, of a transformer with a voltage–ampere rating of s is given by

$$\eta = \frac{xs\cos(\phi)}{L_i + xs\cos(\phi) + x^2 L_c}$$

$$(s\cos(\phi) > 0)$$

where x is the fraction of output power of full load, $\cos(\phi)$ is the power factor, L_c is the copper loss and L_i is the iron loss. Show that the maximum efficiency occurs at

$$x = \sqrt{\frac{L_i}{L_c}}$$

(Treat s, ϕ, L_i and L_c as constants.)

18 i Obtain the first three non-zero terms of the Maclaurin series for $\sinh(x)$.

ii Use a computer algebra system or a graphical calculator to plot $y = \sinh(x)$. On the same axes, plot the first three non-zero terms obtained in part **i**.

19 i Show that the first three non-zero terms of the Maclaurin series for $\tan^{-1}(x)$ are given by

$$x - \frac{x^3}{3} + \frac{x^5}{5}$$

Hint: $[\tan^{-1}(x)]' = \dfrac{1}{1 + x^2}$. Also the series is only valid for $-1 \le x \le 1$.

ii Show that

$$\frac{\pi}{4} = 1 - \frac{1}{3} + \frac{1}{5} \cdots$$

iii By using your calculator, evaluate the following:

$$\tan^{-1}\left(\frac{1}{2}\right) + \tan^{-1}\left(\frac{1}{3}\right),$$

$$\tan^{-1}\left(\frac{1}{2}\right) + \tan^{-1}\left(\frac{1}{5}\right) + \tan^{-1}\left(\frac{1}{8}\right),$$

$$4\tan^{-1}\left(\frac{1}{5}\right) - \tan^{-1}\left(\frac{1}{239}\right)$$

What do you notice about your results?

20 i Sketch the graph

$$y = \frac{1}{3}x^3 - 3x^2 + 8x - 3$$

ii By using the Newton–Raphson procedure, solve

$$\frac{1}{3}x^3 - 3x^2 + 8x - 3 = 0$$

correct to three decimal places.

21 Expand the following:

$$\left(1 - \frac{v^2}{c^2}\right)^{\frac{1}{2}}$$

where v is velocity and c is speed of light.

CHAPTER 8
Integration

SECTION A **Integrals**

By the end of this section you will be able to:
▶ understand an indefinite integral as the anti-derivative
▶ use formulae to determine indefinite integrals

A1 **Integration – inverse of differentiation**

Integration is the **reverse** process of **differentiation**. It is sometimes called *anti-differentiation*. Consider a function $y = f(x)$. The anti-derivative of y is called the **indefinite integral** and is denoted by $\int y\,dx$. The integral notation, $\int y\,dx$, means integrate, $\int$, y with respect to x, dx. The function y is called the **integrand**.

For example, if we differentiate x^2 we obtain $2x$:

$$\frac{d}{dx}(x^2) = 2x$$

? **So does that mean the integral of $2x$ should be x^2?**

No.

$$\int 2x\,dx = x^2 + C \text{ where } C \text{ is a constant}$$

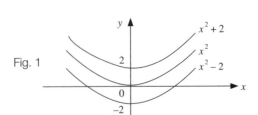

Differentiate

$x^2 + c$ ⟶ $2x$

Integrate

? **Why do we need to add a constant C?**

When we differentiate $x^2 + C$ we obtain $2x$ because differentiating a constant gives zero. It follows that

$$\int 2x\,dx = x^2 + C \text{ holds for any constant } C$$

There are an infinite number of different functions which are the integral of $2x$, that is why $\int 2x\,dx$ is called an indefinite integral.

Figure 1 shows some of the solutions to $\int 2x\,dx$.

All these function, $x^2 + C$, are similar in nature apart from the constant C. This C is called the *constant of integration*. Of course other symbols can also represent the constant of integration. In general, if $y = f(x)$ and

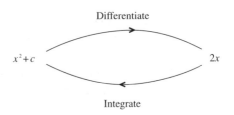

Fig. 1

$$\frac{dy}{dx} = f'(x) \text{ then integrating this gives } y = f(x) + C$$

For example, if $y = x^4$ then

$$\frac{dy}{dx} = 4x^3 \text{ and } \int 4x^3 \, dx = x^4 + C$$

When we differentiate x^n we multiply by n and reduce the index of x by 1, thus

$$\frac{d}{dx}(x^n) = nx^{n-1}$$

? **What is the reverse process of this?**

We add 1 to the index of x and divide by the new index, $n + 1$, thus

$$\int x^n dx = \frac{x^{n+1}}{n+1} + C \text{ (provided } n \neq -1)$$

Example 1

Determine

a $\int x^3 dx$ **b** $\int x^7 dx$ **c** $\int x^{1/2} dx$

Solution

To integrate, we add 1 to the index of x and divide by 'index plus 1'.

a $\int x^3 dx = \dfrac{x^{3+1}}{3+1} + C = \dfrac{x^4}{4} + C$

b Similarly we have

$$\int x^7 dx = \frac{x^{7+1}}{7+1} + C = \frac{x^8}{8} + C$$

c Even if the index of x is not a whole number, we still apply the same rule:

$$\int x^{1/2} dx = \frac{x^{1/2+1}}{1/2+1} + C = \frac{x^{3/2}}{3/2} + C$$

$$= \frac{2x^{3/2}}{3} + C \quad \left[\text{Because } \frac{1}{3/2} = \frac{2}{3} \right]$$

? Let's consider other functions. **What is the derivative $\dfrac{d}{du}[\sin(u)]$ equal to?**

$$\frac{d}{du}[\sin(u)] = \cos(u)$$

? **What is $\int \cos(u) \, du$ equal to?**

Since integration is the reverse of differentiation we have

$$\int \cos(u) \, du = \sin(u) + C$$

? Hence the integral of cos is sin. **What is** $\int e^u\, du$ **equal to?**

Well $\dfrac{d}{du}(e^u) = e^u$, therefore

$$\int e^u du = e^u + C \quad \text{[Anti-derivative]}$$

Also from Chapter 6 on differentiation, the derivative of a natural logarithm, ln, is

$$\frac{d}{du}[\ln(u)] = \frac{1}{u}$$

? **What is** $\int \dfrac{du}{u}$ **equal to?**

$$\int \frac{du}{u} = \ln|u| + C$$

We place u with a modulus sign, | |, because the logarithmic function is only defined for the positive real numbers.

Following in the above manner we can establish an integration table of general engineering functions by reversing the differentiation process. There are many more functions in the integration table (Table 1) than there were in the differentiation table (Table 3 of Chapter 6) because integration is **more difficult** than differentiation. Don't be put off by the many complicated functions. We use the formulae in this table to find the integrals of various functions.

TABLE 1

Add a constant (C) to all of these

8.1 $\int u^n du = \dfrac{u^{n+1}}{n+1} \quad (n \neq -1)$

8.2 $\int \dfrac{du}{u} = \ln|u|$

8.3 $\int e^u du = e^u$

8.4 $\int a^u du = a^u/\ln(a)$

8.5 $\int \ln(u)du = u[\ln(u) - 1]$

8.6 $\int \log(u)du = u[\log(u) - \log(e)]$

8.7 $\int \sin(u)du = -\cos(u)$

8.8 $\int \cos(u)du = \sin(u)$

TABLE 1 CONTINUED

8.9 $\displaystyle\int \tan(u)du = \ln|\sec(u)|$

8.10 $\displaystyle\int \sec(u)du = \ln|\sec(u) + \tan(u)|$

8.11 $\displaystyle\int \operatorname{cosec}(u)du = \ln|\operatorname{cosec}(u) - \cot(u)|$

8.12 $\displaystyle\int \cot(u)du = \ln|\sin(u)|$

8.13 $\displaystyle\int \sin^{-1}(u)du = u\sin^{-1}(u) + \sqrt{1 - u^2}$

8.14 $\displaystyle\int \cos^{-1}(u)du = u\cos^{-1}(u) - \sqrt{1 - u^2}$

8.15 $\displaystyle\int \tan^{-1}(u)du = u\tan^{-1}(u) - \frac{1}{2}\ln(1 + u^2)$

8.16 $\displaystyle\int \sinh(u)du = \cosh(u)$

8.17 $\displaystyle\int \cosh(u)du = \sinh(u)$

8.18 $\displaystyle\int \tanh(u)du = \ln[\cosh(u)]$

8.19 $\displaystyle\int \operatorname{sech}(u)du = \tan^{-1}[\sinh(u)]$

8.20 $\displaystyle\int \operatorname{cosech}(u)du = \ln|\tanh(u/2)|$

8.21 $\displaystyle\int \coth(u)du = \ln|\sinh(u)|$

8.22 $\displaystyle\int \sinh^{-1}(u)du = u\sinh^{-1}(u) - \sqrt{1 + u^2}$

8.23 $\displaystyle\int \cosh^{-1}(u)du = u\cosh^{-1}(u) - \sqrt{u^2 - 1} \qquad (\cosh^{-1}(u) \geq 0)$

8.24 $\displaystyle\int \tanh^{-1}(u)du = u\tanh^{-1}(u) + \frac{1}{2}\ln(1 - u^2)$

8.25 $\displaystyle\int \frac{du}{\sqrt{a^2 - u^2}} = \sin^{-1}\left(\frac{u}{a}\right)$

8.26 $\displaystyle\int \frac{du}{u^2 + a^2} = \frac{1}{a}\tan^{-1}\left(\frac{u}{a}\right)$

8.27 $\displaystyle\int \frac{du}{u\sqrt{u^2 - a^2}} = \frac{1}{a}\sec^{-1}\left(\frac{u}{a}\right)$

8.28 $\displaystyle\int \frac{du}{\sqrt{a^2 + u^2}} = \sinh^{-1}\left(\frac{u}{a}\right) = \ln\left|u + \sqrt{u^2 + a^2}\right|$

TABLE 1 CONTINUED

8.29	$\displaystyle \int \frac{du}{\sqrt{u^2 - a^2}} = \cosh^{-1}\left(\frac{u}{a}\right) = \ln\left| u + \sqrt{u^2 - a^2} \right|$
8.30	$\displaystyle \int \frac{du}{a^2 - u^2} = \frac{1}{a}\tanh^{-1}\left(\frac{u}{a}\right) = \frac{1}{2a}\ln\left|\frac{a + u}{a - u}\right|$
8.31	$\displaystyle \int \sqrt{u^2 \pm a^2}\, du = \frac{u}{2}\sqrt{u^2 \pm a^2} \pm \frac{a^2}{2}\ln\left| u + \sqrt{u^2 \pm a^2}\right|$
8.32	$\displaystyle \int \sqrt{a^2 - u^2}\, du = \frac{u}{2}\sqrt{a^2 - u^2} + \frac{a^2}{2}\sin^{-1}\left(\frac{u}{a}\right)$
8.33	$\displaystyle \int e^{au}\cos(bu)\, du = \frac{e^{au}}{a^2 + b^2}[a\cos(bu) + b\sin(bu)]$
8.34	$\displaystyle \int e^{au}\sin(bu)\, du = \frac{e^{au}}{a^2 + b^2}[a\sin(bu) - b\cos(bu)]$

Remember that u in this table is a dummy variable and can be replaced by any letter. For example, we could have replaced the u in this table by x.

Let p and q be functions of x, then we have

8.35
$$\int (p + q)\,dx = \int p\,dx + \int q\,dx$$

8.36
$$\int (p - q)\,dx = \int p\,dx - \int q\,dx$$

8.37
$$\int kp\,dx = k\int p\,dx \text{ where } k \text{ is a constant}$$

8.35 and 8.36 mean that we can take the integral of each function separately and then add or subtract the resulting integrals. 8.37 means that we can take a constant outside the integral and then multiply the result by the constant.

You can also use the differentiation table in reverse to find the integrals of functions not listed in Table 1 above.

Example 2

Determine

a $\displaystyle \int x^6\,dx$ **b** $\displaystyle \int (x^3 - 2x^2 + 5x)\,dx$ **c** $\displaystyle \int \frac{2}{\sqrt{x}}\,dx$

Solution

a $\displaystyle \int x^6\,dx = \frac{x^{6+1}}{6 + 1} + C = \frac{x^7}{7} + C$
by 8.1 with
$n = 6$

8.1	$\displaystyle \int u^n\,du = \frac{u^{n+1}}{n + 1} \quad (n \neq -1)$

Example 2 *continued*

b Splitting the integrand and taking out the constants we have

$$\int (x^3 - 2x^2 + 5x)dx = \int x^3 dx - 2\int x^2 dx + 5\int x\, dx$$

$$\underset{\text{by } 8.1}{=} \frac{x^{3+1}}{3+1} - 2\left[\frac{x^{2+1}}{2+1}\right] + 5\left[\frac{x^{1+1}}{1+1}\right] + C$$

$$= \frac{x^4}{4} - \frac{2x^3}{3} + \frac{5x^2}{2} + C$$

c Which formula should we use to find $\int \frac{2}{\sqrt{x}}dx$?

We need to rewrite $\frac{2}{\sqrt{x}}$: $\frac{2}{\sqrt{x}} = \frac{2}{x^{1/2}} = 2x^{-1/2}$

We have

$$\int \frac{2}{\sqrt{x}}dx = \int 2x^{-1/2}dx = 2\int x^{-1/2}dx \quad \text{[Taking out 2]}$$

$$= 2\left[\frac{x^{-\frac{1}{2}+1}}{-1/2 + 1}\right] + C$$

$$= \frac{2x^{\frac{1}{2}}}{1/2} + C = 4x^{1/2} + C \quad \left[\text{Because } \frac{2}{1/2} = 4\right]$$

Example 3 *thermodynamics*

Find **a** $\int PV^{1.3}dV$ **b** $\int \frac{P}{V}dV$

where P and V represent pressure and volume respectively. P is constant and V varies.

Solution

a $\int PV^{1.3}dV = P\int V^{1.3}dV$ [Taking out P]

$$= P\left[\frac{V^{1.3+1}}{1.3+1}\right] + C = \frac{PV^{2.3}}{2.3} + C$$

by 8.1 with $n = 1.3$

8.1 $\int x^n dx = \frac{x^{n+1}}{n+1}$

Example 3 *continued*

b How do we determine $\int \dfrac{P}{V} \, dV$?

$$\int \frac{P}{V} \, dV = P \int \frac{dV}{V} = P \underbrace{\ln(V)}_{\text{by } 8.2} + C$$

We can write $\ln|V| = \ln(V)$ because V is volume and therefore is positive.

Example 4

Determine

a $\displaystyle\int \sin(\omega t) \, d(\omega t)$ **b** $\displaystyle\int [3e^t + 2\cos(t)] \, dt$ **c** $\displaystyle\int \left(\frac{1}{x} - x^{-3} + 5x^{3/5} \right) dx$

Solution

a From Table 1, using $\displaystyle\int \sin(u) \, du = -\cos(u) + C$ with $u = \omega t$ we have

$$\int \sin(\omega t) \, d(\omega t) = -\cos(\omega t) + C$$

b We can integrate each part and take the constants out, thus

$$\int [3e^t + 2\cos(t)] \, dt = 3 \int e^t dt + 2 \int \cos(t) dt$$

$$= \underbrace{3e^t}_{\text{by } 8.3} + \underbrace{2\sin(t)}_{\text{by } 8.8} + C$$

c Again splitting the integrand we have

$$\int \left(\frac{1}{x} - x^{-3} + 5x^{3/5} \right) dx = \int \frac{dx}{x} - \int x^{-3} dx + 5 \int x^{3/5} dx$$

$$= \underbrace{\ln|x|}_{\text{by } 8.2} - \underbrace{\frac{x^{-3+1}}{-3+1}}_{\text{by } 8.1} + \underbrace{5\frac{x^{3/5+1}}{3/5+1}}_{\text{by } 8.1} + C$$

$$= \ln|x| - \frac{x^{-2}}{-2} + 5\frac{x^{8/5}}{8/5} + C \quad \text{[Simplifying]}$$

$$= \ln|x| + \frac{x^{-2}}{2} + 25\frac{x^{8/5}}{8} + C \quad \left[\text{Because } \frac{5}{8/5} = \frac{25}{8} \right]$$

8.1 $\displaystyle\int x^n dx = \frac{x^{n+1}}{n+1}$ **8.2** $\displaystyle\int du/u = \ln|u|$ **8.3** $\displaystyle\int e^t dt = e^t$

8.8 $\displaystyle\int \cos(t) dt = \sin(t)$

SUMMARY

The **indefinite integral** of $y = f(x)$ is defined as the **anti-derivative** of y and is denoted by $\int y \, dx$.

Let p and q be functions of x then

8.35 $\qquad \displaystyle\int (p + q)dx = \int p\,dx + \int q\,dx$

8.36 $\qquad \displaystyle\int (p - q)dx = \int p\,dx - \int q\,dx$

8.37 $\qquad \displaystyle\int kp\,dx = k\int p\,dx$ where k is a constant

Exercise 8(a)

Solutions at end of book. Complete solutions available at www.palgrave.com/science/engineering/singh

1 Determine the following integrals:

a $\displaystyle\int x\,dx$ **b** $\displaystyle\int u^2\,du$

c $\displaystyle\int z^3\,dz$ **d** $\displaystyle\int t^{1/2}\,dt$

e $\displaystyle\int (\omega t)\,d(\omega t)$ **f** $\displaystyle\int 3\,dt$

g $\displaystyle\int t^{-2}\,dt$ **h** $\displaystyle\int x^{-1/2}\,dx$

2 Find

a $\displaystyle\int \cos(\omega t)\,d(\omega t)$ **b** $\displaystyle\int 10^x\,dx$

c $\displaystyle\int \sinh(t)\,dt$ **d** $\displaystyle\int \sec(x)\,dx$

3 🔲 [*thermodynamics*] Find

a $\displaystyle\int PV^{1.35}\,dV$

b $\displaystyle\int PV^{1.61}\,dV$

(P is constant and V varies.)

4 Find

a $\displaystyle\int (x^2 + 2x)\,dx$

b $\displaystyle\int \left(\frac{1}{\sqrt{x}} + \tan(x) \right)\,dx$

c $\displaystyle\int (\sec^2(x) - 5e^x + 1)\,dx$

Hint: Use the differentiation table in reverse to find the integral of $\sec^2(x)$.

d $\displaystyle\int (\sin(x) + 2\sqrt{x} - 1)\,dx$

5 i Show that

$$\frac{d}{dx}\left(\frac{3 - 4x}{1 + x^2} \right) = \frac{4x^2 - 6x - 4}{\left(1 + x^2\right)^2}$$

ii Determine $\displaystyle\int \frac{4x^2 - 6x - 4}{\left(1 + x^2\right)^2}\,dx$

SECTION B **Integration by substitution**

By the end of this section you will be able to:
▶ use substitution to find indefinite integrals
▶ apply integration to engineering problems

B1 **Using substitution**

Example 5

Find $\int \cos(2x + 5)\mathrm{d}x$

Solution

? **What formula of Table 1 can we use to determine**

$$\int \cos(2x + 5)\mathbf{dx?}$$

Using 8.8 , $\int \cos(u)\mathrm{d}u = \sin(u) + C$ with $u = 2x + 5$, we have

* $\qquad \int \cos(2x + 5)\mathrm{d}x = \int \cos(u)\mathrm{d}x = ?$

? We **cannot** use 8.8 to determine ***** because we have a dx in place of a du. **What can we substitute for dx?**

We know $u = 2x + 5$ and we can differentiate this function to obtain

$$\frac{\mathrm{d}u}{\mathrm{d}x} = 2 \text{ which gives } \mathrm{d}u = 2\mathrm{d}x$$

because $\mathrm{d}u = \left(\dfrac{\mathrm{d}u}{\mathrm{d}x}\right)\mathrm{d}x$, where $\dfrac{\mathrm{d}u}{\mathrm{d}x}$ is the 'differential coefficient' and du is the 'differential'.

Therefore we have $\mathrm{d}x = \dfrac{\mathrm{d}u}{2}$. Putting this into ***** :

$$\int \cos(u)\frac{\mathrm{d}u}{2} = \frac{1}{2}\int \cos(u)\mathrm{d}u = \frac{1}{2}\underbrace{\sin(u)}_{\text{by } 8.8} + C$$

It follows that

$$\int \cos(2x + 5)\mathrm{d}x = \frac{1}{2}\sin(u) + C = \frac{1}{2}\underbrace{\sin(2x + 5)}_{\substack{\text{substituting}\\ u = 2x+5}} + C$$

Example 6

Determine $\int \sin(2x)dx$

Solution

We can use 8.7 , $\int \sin(u)du = -\cos(u) + C$. Let $u = 2x$ then

$$\frac{du}{dx} = 2 \qquad dx = \frac{du}{2}$$

Substituting $u = 2x$ and $dx = \frac{du}{2}$ gives

$$\int \sin(2x)dx = \int \sin(u)\frac{du}{2}$$

$$= \frac{1}{2}\int \sin(u)du \quad \left[\text{Taking out } \frac{1}{2}\right]$$

$$= \frac{1}{2}[-\cos(u)] + C = -\frac{\cos(2x)}{2} + C$$

In general we have (k and m are constants):

8.38 $$\int \cos(kx + m)dx = \frac{\sin(kx + m)}{k} + C$$

8.39 $$\int \sin(kx + m)dx = -\frac{\cos(kx + m)}{k} + C$$

8.40 $$\int \sec^2(kx + m)dx = \frac{\tan(kx + m)}{k} + C$$

Example 7

Obtain $\int e^{7x+3}dx$

Solution

Which formula can we use to find $\int e^{7x+3}dx$?

We use 8.3 , $\int e^u du = e^u + C$. Let $u = 7x + 3$ then we have

$$\int e^{7x+3}dx = \int e^u dx$$

We need to replace the dx. How?

Differentiating $u = 7x + 3$ gives

$$\frac{du}{dx} = 7 \qquad\qquad dx = \frac{du}{7}$$

Substituting these: $\int e^{7x+3}dx = \int e^u\frac{du}{7} = \frac{1}{7}\int e^u du = \frac{1}{7}e^u + C = \frac{1}{7}e^{7x+3} + C$

In general

8.41
$$\int e^{kx+m}dx = \frac{e^{kx+m}}{k} + C$$

B2 Engineering application

Let's investigate an engineering example. The next example is particularly challenging. It is sophisticated compared to the above examples. Follow it through carefully.

Example 8 *structures*

The equation, y, of a catenary formed by a cable under its own weight is given by

$$y = \int \sinh\left(\frac{wx}{T}\right)dx$$

where x is horizontal distance, T is horizontal tension and w is weight per unit length of cable. (T and w are constants.) Given that when $x = 0$, $y = 0$, show that

$$y = \frac{T}{w}\left[\cosh\left(\frac{wx}{T}\right) - 1\right]$$

Solution

We use 8.16, $\int \sinh(u)du = \cosh(u) + C$ with $u = \frac{wx}{T}$. Differentiating:

$$\frac{du}{dx} = \frac{w}{T} \qquad dx = \frac{T}{w}du$$

Putting these into the given integral:

$$y = \int \sinh\left(\frac{wx}{T}\right)dx = \frac{T}{w}\int \sinh(u)du$$

$$= \frac{T}{w}\cosh(u) + C = \frac{T}{w}\cosh\left(\frac{wx}{T}\right) + C$$

Substituting $x = 0$, $y = 0$ into $y = \frac{T}{w}\cosh\left(\frac{wx}{T}\right) + C$ and using $\cosh(0) = 1$:

$$0 = \frac{T}{w} + C \text{ gives } C = -\frac{T}{w}$$

Hence

$$y = \frac{T}{w}\cosh\left(\frac{wx}{T}\right) - \frac{T}{w} = \frac{T}{w}\left[\cosh\left(\frac{wx}{T}\right) - 1\right] \quad \text{[Factorizing]}$$

B3 **An important integral**

Example 9

Obtain $\displaystyle\int \frac{3x^2}{x^3 - 5}\, dx$

Solution

What can we use to find $\displaystyle\int \frac{3x^2}{x^3 - 5}\, dx?$

We can try using $\displaystyle\int \frac{du}{u}$. Let $u = x^3 - 5$. We need to replace dx. **How can we find dx?**

Differentiating $u = x^3 - 5$ with respect to x gives

$$\frac{du}{dx} = 3x^2 \qquad\qquad dx = \frac{du}{3x^2}$$

$$\int \frac{3x^2}{x^3 - 5}\, dx = \int \left(\frac{3x^2}{u}\right)\frac{du}{3x^2} = \int \frac{du}{u} \quad [\text{Cancelling } 3x^2]$$

$$= \ln|u| + C = \ln|x^3 - 5| + C$$
$$\text{by } \boxed{8.2}$$

Take another look at $\displaystyle\int \frac{3x^2}{x^3 - 5}\, dx = \ln|x^3 - 5| + C.$

What do you notice about this integrand, $\dfrac{3x^2}{x^3 - 5}?$

Similarly we have

$$\int \frac{2x}{x^2 - 1}\, dx = \ln|x^2 - 1| + C$$

$$\int \frac{2x - 2}{x^2 - 2x}\, dx = \ln|x^2 - 2x| + C$$

$$\int \frac{5x^4 - 6x}{x^5 - 3x^2}\, dx = \ln|x^5 - 3x^2| + C$$

What do you notice about all these integrands and associated solutions to the integral?

In each case the numerator is the derivative of the denominator. That is, if you differentiate the denominator you get the numerator. In **Example 9** notice that the derivative, $3x^2$, of the denominator cancels out with the numerator and that is why we can use the $\displaystyle\int \frac{du}{u}$ formula.

Also observe that the solution of these integrals is the natural log, $\ln$, of the denominator.

$\boxed{8.2}$ $\displaystyle\int du/u = \ln|u|$

Example 10

Show that

$$\int \frac{f'(x)}{f(x)}\, dx = \ln|f(x)| + C$$

Solution

Let $u = f(x)$, then differentiating yields

$$\frac{du}{dx} = f'(x) \text{ therefore } dx = \frac{du}{f'(x)}$$

Substituting $u = f(x)$ and $dx = \dfrac{du}{f'(x)}$ into the given integral:

$$\int \frac{f'(x)}{f(x)}\, dx = \int \left(\frac{f'(x)}{u}\right)\frac{du}{f'(x)}$$

$$= \int \frac{du}{u} \quad \text{[Cancelling } f'(x)\text{]}$$

$$= \ln|u| + C = \ln|f(x)| + C \quad \text{[Replacing } u = f(x)\text{]}$$

In general if $f(x)$ is a function of x and its derivative is $f'(x)$ then

8.42

$$\int \frac{f'(x)}{f(x)}\, dx = \ln|f(x)| + C$$

8.42 is a **very important** integral because it can be used to integrate a wide range of functions such as

$$\int \frac{2x}{x^2 + 1}\, dx = \ln|x^2 + 1| + C \qquad \left[\text{Because } \frac{d}{dx}[x^2 + 1] = 2x\right]$$

$$\int \frac{\cos(x)}{\sin(x)}\, dx = \ln|\sin(x)| + C \qquad \left[\text{Because } \frac{d}{dx}[\sin(x)] = \cos(x)\right]$$

$$\int \frac{2}{2x + 1}\, dx = \ln|2x + 1| + C \qquad \left[\text{Because } \frac{d}{dx}[2x + 1] = 2\right]$$

Say we had 3 in place of 2 in the above example, that is $\displaystyle\int \frac{3}{2x + 1}\, dx$.

Then how do we integrate this?

Differentiating $2x + 1$ with respect to x gives 2, so we have to rewrite the integrand as follows:

$$\frac{3}{2x + 1} = \frac{3}{2}\, \frac{2}{2x + 1}$$

Thus

$$\int \frac{3}{2x + 1}\, dx = \frac{3}{2} \int \frac{2}{2x + 1}\, dx$$

$$= \frac{3}{2} \ln|2x + 1| + C \qquad \text{[Integrating]}$$

This is normal procedure in these types of examples. Normally you have to rewrite the integrand so that you can apply formula 8.42 . Similarly we have

$$\int \frac{5}{3x + 2}\, dx = \frac{5}{3} \ln|3x + 2| + C$$

$$\int \frac{x + 1}{x^2 + 2x + 5}\, dx = \frac{1}{2} \ln|x^2 + 2x + 5| + C$$

The last result might be more difficult to follow. Differentiating $x^2 + 2x + 5$ with respect to x gives $2x + 2$ and so rewriting the integrand we have

$$\frac{x + 1}{x^2 + 2x + 5} = \frac{1}{2}\, \frac{2x + 2}{x^2 + 2x + 5}$$

Thus

$$\int \frac{x + 1}{x^2 + 2x + 5}\, dx = \frac{1}{2} \int \frac{2x + 2}{x^2 + 2x + 5}\, dx$$

$$= \frac{1}{2} \ln|x^2 + 2x + 5| + C$$

Example 11

Determine $\displaystyle\int \frac{t^2}{t^3 - 1}\, dt$

Solution

Differentiating $t^3 - 1$ gives us $3t^2$ and we are missing the 3 in the numerator. But it is close so let's try using the $\displaystyle\int \frac{du}{u}$ formula. Let $u = t^3 - 1$ and differentiating:

$$\frac{du}{dt} = 3t^2 \text{ which gives } dt = \frac{du}{3t^2}$$

Substituting these gives

$$\int \frac{t^2}{t^3 - 1}\, dt = \int \left(\frac{t^2}{u}\right) \underbrace{\frac{du}{3t^2}}_{\text{cancelling } t^2} = \frac{1}{3} \int \frac{du}{u} = \frac{1}{3} \ln|u| + C = \underbrace{\frac{1}{3} \ln|t^3 - 1| + C}_{\text{replacing } u = t^3 - 1}$$

If you know that the derivative of the denominator is 3 times the numerator then take out 1/3 and integrate the remaining function of the form $\dfrac{f'(x)}{f(x)}$.

SUMMARY

An important integral well worth remembering is

8.42 $\qquad \displaystyle\int \frac{f'(x)}{f(x)}\, dx = \ln|f(x)| + C$

Solutions at end of book. Complete solutions available at www.palgrave.com/science/engineering/singh

Exercise 8(b)

1 Obtain **a** $\displaystyle\int \sin(7x + 1)dx$

  b $\displaystyle\int \cos(7x + 1)dx$

2 Find **a** $\displaystyle\int \cos(\omega t)dt$ **b** $\displaystyle\int \cos(\omega t + \theta)dt$

3 Obtain **a** $\displaystyle\int \sin(\omega t)dt$ **b** $\displaystyle\int \sin(\omega t)\, d(\omega t)$

4 Determine **a** $\displaystyle\int \frac{2x}{x^2 - 1}dx$

  b $\displaystyle\int \frac{3x^2 - 6x}{x^3 - 3x^2 + 1}dx$

5 Show that $\displaystyle\int \cot(x)dx = \ln|\sin(x)| + C$

6 Show that

  a $\displaystyle\int \tanh(x)\, dx = \ln|\cosh(x)| + C$

  b $\displaystyle\int \coth(x)\, dx = \ln|\sinh(x)| + C$

7 Find **a** $\displaystyle\int \frac{dt}{7t - 1}$ **b** $\displaystyle\int \frac{t^3}{t^4 - 1}\, dt$

  c $\displaystyle\int \frac{t^2}{5 - t^3}\, dt$

8 Determine **a** $\displaystyle\int e^{11x + 5}\, dx$ **b** $\displaystyle\int e^{-2x + 1000}\, dx$

🚗 **Questions 9 to 12 belong to the field of [mechanics].**

9 The vertical velocity component, v, of a projectile is defined as

$$v = \int (-g)dt$$

where g is the constant acceleration due to gravity. Given that at $t = 0$, $v = v_0$, show that

$$v = v_0 - gt$$

10 The velocity, v, of a projectile is given by

$$v = \int (-g)dt$$

Given that when $t = 0$, $v = u$, find an expression for v in terms of t.

11 The position, s, of a particle moving along a straight line is given by

$$s = \int 10(30t + 1)^{-1/2}dt$$

Given that when $t = 0$, $s = \dfrac{2}{3}$, show that

$$s = \frac{2}{3}(30t + 1)^{1/2}$$

12 The velocity, v, of a particle is given by

$$v = \int (-6t)dt$$

For the initial condition that when $t = 0$, $v = 48$ m/s, find

  i an expression for the velocity, v

  ii the time taken, t, for the velocity to become $v = 0$

13 ≈ [fluid mechanics]   The pressure, P, and density, ρ, of an airstream in an isentropic flow are related by

$$\frac{P}{\rho^\gamma} = k$$

where k is a constant and γ is the specific heat constant. Find $\displaystyle\int \frac{dP}{\rho}$.

SECTION C **Definite integrals**

By the end of this section you will be able to:
▶ understand the definite integral
▶ evaluate definite integrals for engineering examples

C1 **Definite integrals**

Say we want to find the area A under the curve $y = f(x)$ between a and b of Fig. 2.
❓ How can we obtain this area?

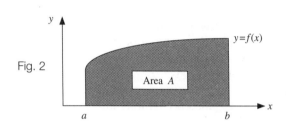

Fig. 2

It is not a regular shape like a circle or a rectangle so we cannot use any of the established formulae such as πr^2 or length $\times$ height.

We can cut the area A into smaller rectangular blocks as shown in Fig. 3.

❓ So does the area A equal $B + C + D + E$?

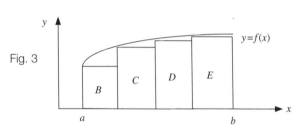

Fig. 3

No, because $B + C + D + E$ does **not** cover all the area A.

Some of the area under the curve is **not** covered by the rectangular blocks.

❓ How can we obtain a more accurate answer for the area A?

The area A can be chopped into even smaller pieces of width Δx as shown in Fig. 4.

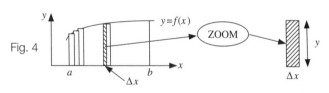

Fig. 4

If the width Δx is small enough then we can obtain an approximation for the entire area A.

❓ Consider the shaded area of Fig. 4. How can we find this area?

The shaded area is a rectangle of width Δx and height y:

$$\text{Rectangle area} = y\Delta x$$

❓ How can we obtain an approximation to the original area A under the curve?

By adding together all the rectangles of height y and width Δx.

Observe that the height y changes as you move along the curve. Hence the area A is approximately the sum from a to b of $y\Delta x$ and can be written as

$$\text{Area } A \simeq S_a^b\, y\Delta x$$

where S_a^b denotes the sum from a to b.

The approximation becomes closer and closer to the exact area A as Δx gets smaller and smaller. To obtain the exact area A we need Δx to be very close to zero, $\Delta x \to 0$. We have

$$\text{Area } A = \lim_{\Delta x \to 0} S_a^b \, y\Delta x = S_a^b y \, dx$$

where dx has replaced Δx as $\Delta x \to 0$. [In the notation, over the years, the letter S was replaced by an elongated $S, \int .$]

$$\text{Area } A = \int_a^b y \, dx$$

Some of this notation might seem baffling but after a few examples it will become pretty straightforward. The notation was introduced by the German mathematician Liebniz who thought of obtaining the area under the curve by summing areas of rectangles. Let's define some of the terms.

$\int_a^b y \, dx$ is called the **definite integral** of y. The process of determining the area is called integration and $x = a$, $x = b$ are called the **limits of integration.** Remember that y is a function of x, $y = f(x)$.

We can witness how the area under the curve can be determined by a definite integral by displaying graphs in a computer algebra system such as MAPLE.

If you don't have access to MAPLE or a similar package such as MATLAB, MATHEMETICA, DERIVE etc., read through the example carefully and see if you understand the results.

Example 12

Plot the graph of $y = x^2$ from 0 to 3.

i Apply the leftbox command in MAPLE to draw 5 boxes on the graph of $y = x^2$ and find the total area of these boxes by using the command leftsum.

ii Repeat **i** for 10 boxes.

iii Repeat **i** for 50 boxes.

iv Repeat **i** for 100 boxes.

v Evaluate $\int_0^3 x^2 \, dx$ by using the evalf (int(x^2, $x = 0..3$)); command

What do you notice about your graphs and their corresponding total area evaluation?

Solution

```
> with (student):
Warning, new definition for D
```

Example 12 *continued*

```
> leftbox (x^2,x=0..3,5);
```

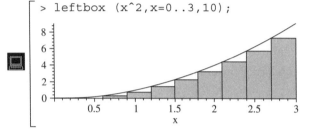

```
> evalf (leftsum(x^2,x=0..3,5));
6.480000000
```

```
> leftbox (x^2,x=0..3,10);
```

```
> evalf (leftsum(x^2,x=0..3,10));
7.695000000
```

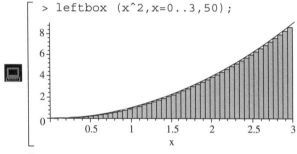

```
> leftbox (x^2,x=0..3,50);
```

```
> evalf (leftsum(x^2,x=0..3,50));
8.731800000
```

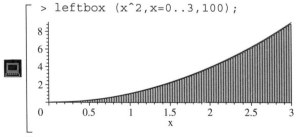

```
> leftbox (x^2,x=0..3,100);
```

```
> evalf (leftsum(x^2,x=0..3,100));
8.865450000
```

```
> evalf (int(x^2,x=0..3));
9.
```

Example 12 *continued*

Notice as the number of boxes increases, more of the area under the curve becomes covered. Using more boxes in MAPLE and summing the corresponding area (leftsum) gives the results shown in Table 2.

TABLE 2	*Number of boxes*	100	200	500	1000	100 000
	Area (5 d.p.)	8.86545	8.93261	8.97302	8.98650	8.99987

If we consider more and more boxes, the limiting value of the sum will be 9. Thus the total area under the curve $y = x^2$ between 0 to 3 is 9. We have

$$\int_0^3 x^2 \, dx = 9$$

As the number of boxes increases, the area gets closer and closer to the integration of x^2 from 0 to 3.

C2 Engineering applications of the definite integral

We defined the area under the curve $y = f(x)$, between $x = a$ to $x = b$ (Fig. 5) as

8.43 $$\int_a^b f(x)dx \left[= \int_{x=a}^{x=b} f(x)dx \right]$$

8.43 is called the definite integral.

If $\int f(x)dx = F(x)$ then

Fig. 5

Shaded area = $\int_a^b f(x)dx$

8.44 $$\int_a^b f(x)dx = F(b) - F(a)$$

In other words

$$\int_a^b f(x)dx = [\text{the value of the integral at } x = b] - [\text{the value of the integral at } x = a]$$

Example 13

Evaluate $\int_2^4 x^2 dx$.

Solution

We first integrate x^2 with respect to x. Then we evaluate the integral at $x = 4$ and $x = 2$ and subtract the two values. Notice how the 4 and 2 are transferred to the top right and bottom right of the square brackets after the integration is carried out.

Example 13 *continued*

$$\int_2^4 x^2 dx = \left[\frac{x^3}{3} + C \right]_2^4 \qquad [\text{Integrating by } 8.1]$$

$$= \underbrace{\left(\frac{4^3}{3} + C \right)}_{\text{substituting } x = 4} - \underbrace{\left(\frac{2^3}{3} + C \right)}_{\text{substituting } x = 2}$$

$$= \frac{64}{3} + C - \frac{8}{3} - C$$

$$\int_2^4 x^2 dx = \frac{56}{3}$$

? Note that the definite integral $\int_2^4 x^2 dx$ is a number, $\frac{56}{3}$, and does not contain x. **Moreover there is *no* constant of integration (C). Why?**

Because the C's cancel each other out. When there is a definite integral you do not need to include the constant C. Let's examine some engineering examples.

Example 14 *mechanics*

The displacement, s, of an object is given by

$$s = \int_0^3 (t^4 + t) dt$$

Evaluate s.

Solution

We integrate and then substitute $t = 3$ and $t = 0$:

$$s = \int_0^3 (t^4 + t) dt$$

$$= \left[\frac{t^5}{5} + \frac{t^2}{2} \right]_0^3 \qquad [\text{Integrating by } 8.1]$$

$$= \underbrace{\left(\frac{3^5}{5} + \frac{3^2}{2} \right)}_{\text{substituting } t = 3} - \underbrace{\left(\frac{0^5}{5} + \frac{0^2}{2} \right)}_{\text{substituting } t = 0} = 53.1$$

$$s = 53.1 \text{ m}$$

8.1 $\displaystyle\int u^n du = \frac{u^{n+1}}{n+1}$

Example 15 *thermodynamics*

The change in specific enthalpy, Δh, in J/kg, is given by

$$\Delta h = \int_{150}^{300} (2.1 + (7 \times 10^{-3})T)\,\mathrm{d}T$$

Evaluate Δh.

Solution

Our limits of integration are $T = 300$ and $T = 150$:

$$\Delta h = \int_{150}^{300} (2.1 + (7 \times 10^{-3})T)\,\mathrm{d}T$$

$$= \left[2.1T + \frac{(7 \times 10^{-3})T^2}{2} \right]_{150}^{300} \qquad \left[\text{Integrating by } \boxed{8.1}\,\right]$$

$$= \underbrace{\left((2.1 \times 300) + \frac{(7 \times 10^{-3})300^2}{2} \right)}_{\text{substituting } T = 300} - \underbrace{\left((2.1 \times 150) + \frac{(7 \times 10^{-3})150^2}{2} \right)}_{\text{substituting } T = 150}$$

$$= 945 - 393.75 = 551.25$$

$$\Delta h = 551.25 \text{ J/kg}$$

Sometimes in a definite integral we substitute a letter or a symbol rather than a number in the limits of integration. This results in an expression.

Example 16 *mechanics*

The moment of inertia, I, of a rod of length $2r$ is given by

$$I = \int_{-r}^{r} \frac{mx^2}{2r}\,\mathrm{d}x$$

where m is the mass of the rod and x is the distance from the axis. Show that $I = \dfrac{mr^2}{3}$. (The mass, m, and length, r, are constant.)

$\boxed{8.1}$ $\displaystyle\int u^n\,\mathrm{d}u = u^{n+1}/n + 1$

Example 16 *continued*

Solution .

Taking out the $m/2r$ gives

$$I = \frac{m}{2r}\int_{-r}^{r} x^2 dx = \frac{m}{2r}\left[\frac{x^3}{3}\right]_{x=-r}^{x=r} = \frac{m}{2r}\left(\frac{r^3 - (-r)^3}{3}\right) = \frac{m}{2r}\left(\frac{2r^3}{3}\right) = \frac{mr^2}{3} \quad \text{[Cancelling]}$$

The next example is difficult because it involves trigonometric identities, integration, substitution, unfamiliar symbols etc. It is a lot more complicated than previous examples.

Example 17 *electrical principles*

The power, P, in a circuit is given by

$$P = \frac{\omega R}{2\pi}\int_{0}^{2\pi/\omega} (i^2) dt$$

where $i = I\sin(\omega t)$, R is resistance and ω is angular frequency. Show that

$$P = \frac{I^2 R}{2}$$

Solution

We have $i = I\sin(\omega t)$, therefore $i^2 = [I\sin(\omega t)]^2 = I^2\sin^2(\omega t)$. Substituting $i^2 = I^2\sin^2(\omega t)$

into $P = \frac{\omega R}{2\pi}\int_{0}^{2\pi/\omega} (i^2) dt$ gives

$$P = \frac{\omega R}{2\pi}\int_{0}^{2\pi/\omega} I^2\sin^2(\omega t)\, dt = \frac{\omega R I^2}{2\pi}\int_{0}^{2\pi/\omega} \sin^2(\omega t) dt \quad \text{[Taking out } I^2\text{]}$$

The difficulty in this example is to find $\int_{0}^{2\pi/\omega} \sin^2(\omega t) dt$.

 How do we integrate the sin² term?

We need to avoid $\sin^2$ by finding another way of expressing it.

We use the trigonometric identity

4.68 $$\sin^2(x) = \frac{1}{2}\left[1 - \cos(2x)\right]$$

$$\sin^2(\omega t) = \frac{1}{2}\left[1 - \cos(2\omega t)\right]$$

Example 17 *continued*

Therefore we have

$$P = \frac{\omega R I^2}{2\pi} \int_0^{2\pi/\omega} \sin^2(\omega t)\, dt = \frac{\omega R I^2}{2\pi} \int_0^{2\pi/\omega} \frac{1}{2}\left[1 - \cos(2\omega t)\right] dt$$

$$= \frac{\omega R I^2}{4\pi} \int_0^{2\pi/\omega} \left[1 - \cos(2\omega t)\right] dt \quad \left[\text{Taking out } \frac{1}{2}\right]$$

$$= \frac{\omega R I^2}{4\pi} \left[t - \frac{\sin(2\omega t)}{2\omega}\right]_0^{2\pi/\omega} \quad \left[\text{Integrating using } \int \cos(kt)dt = \frac{\sin(kt)}{k}\right]$$

$$= \frac{\omega R I^2}{4\pi} \left\{ \underbrace{\left(\frac{2\pi}{\omega} - \frac{\sin[2\omega\,(2\pi/\omega)]}{2\omega}\right)}_{\text{substituting } t = 2\pi/\omega} - \underbrace{\left[0 - \frac{\sin(0)}{2\omega}\right]}_{\text{substituting } t = 0} \right\}$$

$$= \frac{\omega R I^2}{4\pi} \left[\left(\frac{2\pi}{\omega} - \underbrace{\frac{\sin(4\pi)}{2\omega}}_{\substack{= 0 \text{ because} \\ \sin(4\pi)\,=\,0}}\right) - 0\right]$$

$$= \frac{\omega R I^2}{4\pi}\left[\frac{2\pi}{\omega}\right] = \frac{I^2 R}{2} \qquad \text{[Cancelling]}$$

$$P = \frac{I^2 R}{2}$$

Example 18 *reliability engineering*

The failure density function, $f(t)$, for a class of electronic components is given by

$$f(t) = 0.2e^{-0.2t}$$

The hazard function, $h(t)$, is defined as

$$h(t) = \frac{f(t)}{1 - \displaystyle\int_0^t f(x)dx}$$

Determine $h(t)$.

Example 18 *continued*

Solution

We are given $f(t)$, but what is $f(x)$ equal to?

Since $f(x)$ is a function of x, we replace the t with x, thus $f(x) = 0.2e^{-0.2x}$.

$$\int_0^t f(x)dx = 0.2\int_0^t e^{-0.2x}dx \quad \text{[Taking out 0.2]}$$

$$= 0.2\left[\frac{e^{-0.2x}}{-0.2}\right]_0^t \quad \left[\text{Integrating using}\int e^{kx}dx = \frac{e^{kx}}{k}\right]$$

$$= -\left[e^{-0.2x}\right]_0^t \quad \text{[Cancelling 0.2's]}$$

$$= -\left(\underbrace{e^{-0.2t}}_{\text{substituting } x=t} - \underbrace{e^{-(0.2\times 0)}}_{\text{substituting } x=0}\right)$$

$$= -(e^{-0.2t} - 1) = -e^{-0.2t} + 1$$

$$\int_0^t f(x)dx = 1 - e^{-0.2t}$$

Simplifying the denominator in $h(t)$ gives

$$1 - \int_0^t f(x)dx = 1 - (1 - e^{-0.2t})$$

$$= 1 - 1 + e^{-0.2t}$$

$$= e^{-0.2t}$$

Replacing $1 - \int_0^t f(x)dx$ with $e^{-0.2t}$ in the denominator of the hazard function, $h(t)$, gives

$$h(t) = \frac{0.2\, e^{-0.2t}}{e^{-0.2t}} = 0.2 \quad \text{[Cancelling } e^{-0.2t}\text{]}$$

Remember the variable x in $\int_a^b f(x)dx$ is called the **dummy variable** since x can be replaced by any letter because the value of the integral is the same. That is

$$\int_a^b f(x)dx = \int_a^b f(t)dt = \int_a^b f(\xi)d\xi = \cdots \quad \text{(A definite integral is a function of the limits \textbf{only}.)}$$

SUMMARY

The **definite integral** is defined as **the area under the curve**, $y = f(x)$, between $x = a$ and $x = b$. Area $= \int_a^b y\, dx$.

If $\int_a^b f(x)dx = F(x)$ then

$$8.44 \qquad \int_a^b f(x)dx = F(b) - F(a)$$

There is no constant (C) of integration in the evaluation of a definite integral.

Solutions at end of book. Complete solutions available at www.palgrave.com/science/engineering/singh

1 Plot the graph of $y = x^3$ from $x = 0$ to 1.

　i Use the leftbox command to draw four boxes on the graph of $y = x^3$. Evaluate the total area of the four boxes by using the leftsum command.

　ii Repeat **i** for 8 boxes.

　iii Repeat **i** for 16 boxes.

　iv Repeat **i** for 32 boxes.

　v Repeat **i** for 64 boxes.

　vi Repeat **i** for 128 boxes.

　vii By applying the command evalf(int($x^{3,}x$ = 0..1)), evaluate

$$\int_0^1 x^3 dx.$$ What do you notice about your results?

🚗 **Questions 2 to 6, inclusive, belong to the field of [*mechanics*].**

2 A particle moves along a horizontal line with displacement, s, given by

$$s = \int_0^2 (t^2 - 2t)dt$$

Determine s.

3 The work done, W, to stretch a spring from its natural length to an extension of 0.5 m is given by

$$W = \int_0^{0.5} 100x\, dx$$

Evaluate W.

4 The force, F, required to compress a spring is given by

$$F = 1000x + 50x^3$$

where x is the displacement from its unstretched length. The work done, W, to compress a spring by 0.3 m is given by

$$W = \int_0^{0.3} F\, dx$$

Determine W.

5 The distance, s, travelled by a train on a straight track in the first two seconds is given by

$$s = \int_0^2 20(1 - e^{-t})dt$$

Find s.

6 The distance, s, covered by a vehicle between the 5th and 6th second is given by

$$s = \int_5^6 10e^{0.4t}dt$$

Evaluate s.

Exercise **8(c) continued**

Solutions at end of book. Complete solutions available at www.palgrave.com/science/engineering/singh

7 [structures]   A beam of length 6 m has a uniform distributed load, w, given by

$$w = 800 + \frac{1}{2}x^3$$

where x is the distance along the beam. The total load P (N), and the moment about the origin, R (N m), are given by

$$P = \int_0^6 w\,dx \text{ and } R = \int_0^6 (wx)\,dx$$

Determine P and R.

8 [thermodynamics]   The specific heat, c, of steam is given by

$$c = \frac{3000 + T - 100}{1300}$$

The enthalpy change, Δh, is given by

$$\Delta h = \int_{150}^{500} 1.5c\,dT$$

Evaluate Δh.

9 [mechanics]   The time taken, t (hours), for a vehicle to reach a speed of 120 km/h with an initial speed of 80 km/h is given by

$$t = \int_{80}^{120} \frac{dv}{600 - 3v}$$

where v is velocity (km/h). Determine t.

10 [thermodynamics]   A gas in a cylinder obeys the law $PV^{1.25} = 1789$. The work done, W, is given by

$$W = \int_{0.01}^{0.1} P\,dV$$

Determine W.

  Questions 11 to 15, inclusive, are in the field of [mechanics].

11 The moment of inertia, I, of a disc of radius r and mass m is given by

$$I = \int_0^r \left(\frac{2mx^3}{r^2}\right)dx$$

where x is the distance from an axis of rotation. Show that $I = \frac{mr^2}{2}$.

12 The moment of inertia, I, of an annulus of inner radius a, outer radius b and mass m is given by

$$I = \int_a^b \frac{2mx^3}{b^2 - a^2}\,dx$$

where x is the distance from the axis of rotation. Show that $I = \frac{1}{2}m(a^2 + b^2)$.

13 The moment of inertia, I, of a rod of mass m and length $2r$ is given by

$$I = \int_0^{2r} \frac{mx^2}{2r}\,dx$$

where x is the distance from the axis of rotation. Show that $I = \frac{4mr^2}{3}$.

14 The moment of inertia, I, of a rod of length l and mass m is given by

$$I = \int_0^l \frac{mx^2\sin^2(\theta)}{l}\,dx$$

where x is the distance along the rod and θ is the angle made between the rod and axis of rotation. Show that

$$I = \frac{ml^2\sin^2(\theta)}{3}$$

15 The moment of inertia, J, of a disc of radius r and mass m is defined as

$$J = \int_0^r \frac{2m}{r^2}x^3\,dx$$

Evaluate J.

16 [fluid mechanics]   The momentum per unit time, M, for a fluid flow through a pipe of radius r is given by

$$M = k\int_0^r x^{2/7}(r - x)\,dx$$

where x is the distance from the pipe wall and k is a constant. Evaluate M.

Exercise **8(c) continued**

Solutions at end of book. Complete solutions available at www.palgrave.com/science/engineering/singh

17 🌊 [*fluid mechanics*]   The rate of flow, Q, of a fluid at a radius r in a pipe of radius R is given by

$$Q = \int_0^R (2\pi ur)\,dr$$

where u is the mean velocity, which is constant. Evaluate Q.

18 📶 [*signal processing*]   The average voltage, v_{avg}, of a waveform with period T is given by

$$v_{avg} = \frac{1}{T}\int_0^T V\cos(\omega t + \theta)\,dt$$

where V is the peak voltage, ω is the angular frequency and θ is the phase. Evaluate v_{avg}.

19 📶 [*signal processing*]   The mean, $\bar{x}$, of a signal over a period T is given by

$$\bar{x} = \frac{1}{T}\int_0^T a\cos(\omega t + \phi)\,dt$$

where a is the amplitude, ω is the angular frequency and ϕ is the phase. Determine $\bar{x}$.

🔧 **Questions 20 to 26, inclusive, are in the field of [*electrical principles*].**

20 The average value of current, I_{AV}, for a rectified alternating current is given by

$$I_{AV} = \frac{\omega}{\pi}\int_0^{\pi/\omega} i\,dt$$

where $i = I\sin(\omega t)$, ω is angular frequency, i is instantaneous current, I is peak current and t is time. Show that

$$I_{AV} = \frac{2}{\pi}I$$

21 The energy, W, stored in a capacitor of capacitance C charged with voltage V, is defined by

$$W = \int_0^V CV\,dV$$

Show that $W = \dfrac{1}{2}CV^2$.

22 A resistor shaped in the form of an annulus has an outer radius b and inner radius a and its resistance R is given by

$$R = \int_a^b \frac{\rho\,dr}{2\pi r}$$

where $a < r < b$ and ρ is the resistance per unit area. Evaluate R.

23 The voltage, v, across a capacitor is given by

$$v = \frac{1}{10 \times 10^{-6}}\int_0^t 100e^{-(5\times 10^3)t}\,dt$$

Evaluate v.

24 The magnetising force, F, of a circular coil of radius r and at a distance d from the centre is given by

$$F = \int_0^{2\pi} \frac{I r\sin(\beta)}{4\pi(d^2 + r^2)}\,d\phi \quad \left[\frac{I r\sin(\beta)}{4\pi(d^2 + r^2)}\text{ is a constant}\right]$$

where I is the current carried by the coil, ϕ and β are angles. Evaluate F.

25 The magnetising force, F, of an infinite wire is given by

$$F = \int_0^{\pi} \frac{I\sin(\theta)}{4\pi r}\,d\theta$$

where r is perpendicular distance, θ is angle and I is current. Evaluate F.

26 The potential difference, V, between two concentric spheres of radii a and b is given by

$$V = \int_a^b \frac{Q}{4\pi\varepsilon_0 r^2}\,dr \text{ where } a < r < b \ (r \neq 0)$$

ε_0 is the permittivity constant and Q is charge. Show that

$$V = \frac{Q}{4\pi\varepsilon_0}\left(\frac{1}{a} - \frac{1}{b}\right) \text{ (where } a \neq 0, b \neq 0)$$

Exercise **8(c) continued**

Solutions at end of book. Complete solutions available at www.palgrave.com/science/engineering/singh

Questions 27 to 30, inclusive, are in the field of [*reliability engineering*].

27 The hazard function, $h(t)$, for a component is given by

$$h(t) = t^2 - 4t + 9 \quad (0 < t \le 1)$$

The failure density function, $f(t)$, is defined as

$$f(t) = h(t)\exp\left[-\int_0^t h(x)dx\right]$$

where exp is the exponential function. Determine $f(t)$.

28 The hazard function, $h(t)$, for a component is given by

$$h(t) = 3 - t \quad (0 < t \le 2)$$

Determine $f(t)$ where $f(t)$ is as defined in question **27**.

29 The failure density function, $f(t)$, for a set of components is given by

$$f(t) = 0.02(10 - t) \quad (0 < t \le 10)$$

The reliability function, $R(t)$, is defined as

$$R(t) = 1 - \int_0^t f(x)dx$$

and the hazard function, $h(t)$, is defined as

$$h(t) = \frac{f(t)}{R(t)}$$

Determine $R(t)$ and $h(t)$.

30 The hazard function, $h(t)$, is given by

$$h(t) = (2 \times 10^{-3})t^{-1/2}$$

The reliability function, $R(t)$, is defined as

$$R(t) = \exp\left[-\int_0^t h(x)dx\right]$$

Find $R(t)$.

Questions 31 to 34, inclusive, are in the field of [*thermodynamics*].

31 The work done, W, by a gas on a piston is given by

$$W = \int_{V_1}^{V_2} \frac{C}{V}\,dV$$

where C is a constant and V is volume. Show that $W = C\ln\left(\dfrac{V_2}{V_1}\right)$.

32 The work done, W, by a gas on a piston is given by

$$W = \int_{P_1}^{P_2} CP^{-1/n}dP$$

where P is pressure and C is constant. Evaluate W.

33 A gas expands according to the law

$$P_1V_1^k = P_2V_2^k = C$$

where P is pressure, V is volume, k and C are constants. The work done, W, by the gas is given by

$$W = \int_{V_1}^{V_2} CV^{-k}dV$$

Show that $W = \dfrac{P_1V_1 - P_2V_2}{k - 1}$.

34 A gas obeys the law

$$PV^{1.5} = C \quad \text{(constant)}$$

If the work done, $W = \displaystyle\int_{V_1}^{V_2} P\,dV$ then show that

$$W = 2C\left[\frac{1}{\sqrt{V_1}} - \frac{1}{\sqrt{V_2}}\right]$$

35 [*aerodynamics*]   The drag coefficient, C_F, is defined as

$$C_F = k\int_0^1 \left(\frac{x}{L}\right)^{-\frac{1}{2}}d\left(\frac{x}{L}\right)$$

where k is a constant, L and x are lengths. Show that $C_F = 2k$.

SECTION D **Integration by parts**

By the end of this section you will be able to:
▶ understand the integration by parts formula
▶ apply the integration by parts formula
▶ apply the formula to engineering applications

This section might seem like a huge leap in sophistication from previous sections. The difficulty is in understanding the notation of the integration by parts formula.

D1 **Integration by parts formula**

The product rule for differentiation is given by

6.31
$$\frac{d}{dx}(uv) = u'v + uv'$$

where u and v are functions of x. The dash notation, $'$, represents derivatives. Integrating both sides of 6.31:

$$uv = \int u'v\,dx + \int uv'\,dx$$

Rearranging yields

8.45
$$\int uv'\,dx = uv - \int u'v\,dx$$

8.45 is called the **integration by parts** formula. It is used to integrate **some** product of functions. If you have never seen this formula you might be confused with the symbols and notation. The v on the Right-Hand Side is the integral of v', that is

$$v = \int v'\,dx$$

In general the verbal form for integration by parts formula is 'differentiate one (u) and integrate the other, (v')'.

Example 19

Find $\int xe^x dx$.

Solution

Let $u = x$ and $v' = e^x$. Then to use the integration by parts formula, 8.45 , we need to differentiate u and integrate v':

$$u' = 1 \qquad v = \int e^x dx = e^x$$

Example 19 *continued*

Don't worry about the constant (C) of integration, we add it on at the end. Substituting these into 8.45 :

$$\int xe^x dx = uv - \int vu' dx$$

$$= xe^x - \int (e^x)(1)dx$$

$$= xe^x - \int e^x dx$$

We have $\int xe^x dx = xe^x - \int e^x dx$. Notice that we still need to integrate e^x, the last term on the Right-Hand Side. Hence

$$\int xe^x dx = xe^x - \underbrace{e^x}_{=\int e^x dx} + C$$

$$= e^x(x - 1) + C \qquad \text{[Taking out } e^x\text{]}$$

Observe that in the application of 8.45 , we still have an integral that we need to determine, $\int vu' dx$.

? **What is** $\dfrac{d}{dx}\left[e^x(x - 1) + C\right]$**?**

By inspecting **Example 19** we have

$$\frac{d}{dx}[e^x(x - 1) + C] = xe^x$$

because when we integrate xe^x we obtain $e^x(x - 1) + C$. Remember that differentiation is the reverse process of integration.

? **Why did we choose $u = x$ and $v' = e^x$ in Example 19?**

If you selected u and v' the other way round, that is $u = e^x$ and $v' = x$ then the integration of xe^x becomes very complicated. Applying 8.45 gives

$$\int xe^x dx = e^x \cdot \frac{x^2}{2} - \int \frac{x^2}{2} e^x dx$$

The integral on the Right-Hand Side is more complicated then the integral of the given function, xe^x.

8.45 $\int uv' dx = uv - \int vu' dx$

? **In general, how do we know what to take *u* and *v'* to be?**

From previous experience we can establish some sort of ranking. Take *u* to be the function in the following order:

1 $\ln(x)$

2 x^n

3 e^{kx} where *k* is a constant

? **So if we want to find $\int x\ln(x)dx$, what is our *u* and *v'*?**

We take $u = \ln(x)$ and $v' = x$ because $\ln(x)$ has priority over *x*. You can use this list to determine your *u* and *v'* of the given function. In general, you choose your *u* and *v'* such that the Right-Hand Side integral, $\int vu'dx$ of $\boxed{8.45}$, should be simpler than the original integral, $\int uv'dx$.

Example 20

Determine $\int x\ln(x)dx$

Solution

Let

$$u = \ln(x) \qquad\qquad v' = x$$

$$u' = \frac{1}{x} \quad \text{[Differentiating]} \quad v = \int xdx = \frac{x^2}{2} \qquad \text{[Integrating]}$$

Substituting these into the integration by parts formula:

$$\int x\ln(x)dx = uv - \int u'vdx$$

$$= \ln(x)\left(\frac{x^2}{2}\right) - \int\left(\frac{1}{x}\right)\left(\frac{x^2}{2}\right)dx$$

$$= \left(\frac{x^2}{2}\right)\ln(x) - \frac{1}{2}\int xdx \quad \text{[Simplifying]}$$

$$= \left(\frac{x^2}{2}\right)\ln(x) - \frac{1}{2}\left(\frac{x^2}{2}\right) + C$$

$$= \frac{x^2}{2}\left(\ln(x) - \frac{1}{2}\right) + C \quad \left[\text{Taking out } \frac{x^2}{2}\right]$$

$\boxed{8.45}$ $\int uv'dx = uv - \int u'vdx$

D2 **Engineering applications of integration by parts**

Example 21 *electrical principles*

The energy, W, of an inductor is given by

$$★ \qquad W = \int t \cos(t) \, dt$$

Determine W.

Solution

How do we integrate $t \cos(t)$?

Since we have a product, $t \times \cos(t)$, we try using the integration by parts formula 8.45.
What is u and v'?

By using our list we see that t is on the list whilst $\cos(t)$ is **not**. Hence

$$u = t \quad \text{and} \quad v' = \cos(t)$$

Differentiating u and integrating v' gives

$$u' = 1 \qquad v = \int \cos(t) \, dt = \sin(t)$$

Putting these into 8.45 gives

$$\int t \cos(t) dt = uv - \int u' v \, dt$$

$$= t \sin(t) - \int (1)\sin(t) dt$$

$$= t \sin(t) - \int \sin(t) \, dt$$

$$= t \sin(t) - \underbrace{\left[-\cos(t) \right]}_{\text{by } 8.7} + C$$

$$W = \int t \cos(t) \, dt = t \sin(t) + \cos(t) + C$$

8.7 $\displaystyle \int \sin(t) dt = -\cos(t)$

🚗 Example 22 *mechanics*

The acceleration, $\ddot{x}$, of a particle is given by

$$\ddot{x} = te^{-t}$$

Find the velocity, $v = \dot{x}$, for the initial condition $t = 0$, $v = 0$.

Solution

$\dot{x}$ is obtained from $\ddot{x}$ by integrating $\ddot{x}$. Hence

$$v = \dot{x} = \int te^{-t}dt$$

? **How do we integrate this function te^{-t}?**

Use the integration by parts formula 8.45 :

$$u = t \qquad\qquad\qquad w' = e^{-t}$$

$$u' = 1 \quad \text{[Differentiating]} \qquad w = \int e^{-t}dt = -e^{-t} \quad \text{[Integrating]}$$
$$\text{by } 8.41$$

We use w because v represents velocity in this example.

Applying 8.45 :

$$v = \int te^{-t}dt = uw - \int u'w \, dt$$

$$= (t)(-e^{-t}) - \int (1)(-e^{-t})dt$$

$$= -te^{-t} + \int e^{-t}dt \qquad \text{[Simplifying]}$$

$$v = -te^{-t} - e^{-t} + C$$

Substituting the given conditions $t = 0$, $v = 0$ into $v = -te^{-t} - e^{-t} + C$ yields

$$0 = 0 - 1 + C \qquad \text{[Remember that } e^0 = 1]$$
$$C = 1$$

Hence putting $C = 1$ and taking out the common factor $-e^{-t}$ we have

$$v = -e^{-t}(t + 1) + 1 \text{ or } v = 1 - e^{-t}(t + 1)$$

For definite integrals the integration by parts formula is given by

8.46 $$\int_a^b uv'dx = [uv]_a^b - \int_a^b u'vdx$$

8.41 $$\int e^{kt+m} dt = e^{kt+m}/k$$

⌗ | Example 23 *electrical principles*

The energy stored, W, in an inductor is given by

$$W = \int_0^1 (10 \times 10^{-3})te^{-2t}dt$$

Find the exact value of W.

Solution

By taking out the constant, (10×10^{-3}), we have

† $$W = (10 \times 10^{-3})\int_0^1 te^{-2t}dt$$

? | **How do we find the definite integral $\int_0^1 te^{-2t}dt$?**

Since the function is a product, we try the integration by parts formula for definite integrals, 8.46 . Differentiate u and integrate v':

$$u = t \qquad v' = e^{-2t}$$

$$u' = 1 \qquad v = \int e^{-2t}dt = -\frac{e^{-2t}}{2} \qquad [\text{By } 8.41 \ ]$$

Thus, applying 8.46 with $a = 0$ and $b = 1$:

$$\int_0^1 te^{-2t}dt = [\ uv\ ]_0^1 - \int_0^1 u'\ v\ dt$$

$$= \left[-\frac{te^{-2t}}{2}\right]_0^1 - \int_0^1 \left(-\frac{e^{-2t}}{2}\right)dt$$

∗ $$= -\frac{1}{2}\left(\underbrace{1e^{-2}}_{\substack{\text{substituting}\\t=1}} - \underbrace{0}_{\substack{\text{substituting}\\t=0}}\right) + \frac{1}{2}\int_0^1 e^{-2t}dt$$

We still need to integrate the last term on the Right-Hand Side of ∗ :

$$\underbrace{\int_0^1 e^{-2t}dt}_{\text{by } 8.41} = \left[\frac{e^{-2t}}{-2}\right]_0^1 = -\frac{1}{2}\left(e^{-2} - e^0\right) = -\frac{1}{2}\left(e^{-2} - 1\right)$$

8.41 $$\int e^{(kt+m)}dt = e^{kt+m}/k$$ 8.46 $$\int_a^b uv'dt = [\ uv\ ]_a^b - \int_a^b u'vdt$$

Example 23 *continued*

Substituting this into gives:

$$\int_0^1 te^{-2t}dt = -\frac{1}{2}e^{-2} + \frac{1}{2}\left[-\frac{1}{2}\left(e^{-2} - 1\right)\right]$$

$$= \frac{1}{2}\left[-e^{-2} - \frac{e^{-2}}{2} + \frac{1}{2}\right] \quad \left[\text{Taking out } \frac{1}{2}\right]$$

$$= \frac{1}{2}\left[-\frac{3e^{-2}}{2} + \frac{1}{2}\right] \quad \text{[Simplifying]}$$

$$= \frac{1}{4}\left[1 - 3e^{-2}\right] \quad \left[\text{Taking out } \frac{1}{2}\right]$$

The exact value of W is evaluated by putting this into †:

$$W = \frac{10 \times 10^{-3}}{4}\left[1 - 3e^{-2}\right] = 2.5 \times 10^{-3}[1 - 3e^{-2}] \text{ (joule)}$$

Sometimes we need to apply the integration by parts formula, 8.46 , twice as the following example shows.

Example 24 *electrical principles*

The energy stored, W, in an inductor is given by

$$W = \int_0^2 (t^2 - 2t)e^{-2t}dt$$

Evaluate W.

Solution

We need to use 8.46 . **What is *u* and *v'* in this formula?**

By examining the priority list we let

$$u = t^2 - 2t \qquad v' = e^{-2t}$$

$$u' = 2t - 2 \qquad v = \int e^{-2t}dt = \frac{e^{-2t}}{-2} \quad \left[\text{Using } \int e^{kt}\,dt = \frac{e^{kt}}{k}\right]$$

Substituting these into 8.46 with $a = 0$ and $b = 2$ gives

$$W = \int_0^2 (t^2 - 2t)e^{-2t}dt = [uv]_0^2 - \int_0^2 u'v\,dt$$

$$= \left[(t^2 - 2t)\left(\frac{e^{-2t}}{-2}\right)\right]_0^2 - \int_0^2 (2t - 2)\left(\frac{e^{-2t}}{-2}\right)dt$$

$$= \underbrace{\left(\left(2^2 - (2 \times 2)\right)\frac{e^{-4}}{-2} - 0\right)}_{= 0} + \frac{1}{2}\int_0^2 2(t - 1)e^{-2t}dt$$

$$W = \int_0^2 (t - 1)e^{-2t}dt \qquad \text{[Cancelling 2's]}$$

Example 24 *continued*

How can we find this integral?

Use 8.46 again since we have a product of $t - 1$ and e^{-2t}. Let

$$u = t - 1 \qquad\qquad v' = e^{-2t}$$

$$u' = 1 \qquad\qquad v = \int e^{-2t}\,dt = \frac{e^{-2t}}{-2}$$

Note that these u's and v's are different from above. Putting these into the formula:

$$W = [\,uv\,]_0^2 - \int_0^2 u'\,v\,dt$$

$$= \left[(t - 1)\frac{e^{-2t}}{-2}\right]_0^2 - \int_0^2 (1)\frac{e^{-2t}}{-2}\,dt$$

$$= -\frac{1}{2}\left[(t - 1)e^{-2t}\right]_0^2 + \frac{1}{2}\int_0^2 e^{-2t}\,dt \qquad \left[\text{Taking out } -\frac{1}{2}\right]$$

$$= -\frac{1}{2}\Big(\underbrace{(2 - 1)e^{-4}}_{\text{substituting } t\,=\,2} - \underbrace{(0 - 1)e^0}_{\text{substituting } t\,=\,0}\Big) + \frac{1}{2}\int_0^2 e^{-2t}\,dt$$

$$= -\frac{1}{2}(e^{-4} + 1) + \frac{1}{2}\left[\frac{e^{-2t}}{-2}\right]_0^2$$

$$= -\frac{1}{2}(e^{-4} + 1) - \frac{1}{4}\Big(\underbrace{e^{-4}}_{\text{substituting } t=2} - \underbrace{e^0}_{\text{substituting } t=0}\Big)$$

$$= -\frac{1}{2}(e^{-4} + 1) - \frac{1}{4}(e^{-4} - 1) = -0.264 \qquad \text{[Evaluating]}$$

$$W = -0.264\,\text{J} \quad \text{(3 d.p.)}$$

The integration by parts formula **cannot** be applied to all products of functions. For example, to find $\int \cos(2x)\sin(3x)\,dx$ we would **not** use the integration by parts formula.

SUMMARY

Let u and v be functions of x, then

8.45 $\qquad \displaystyle\int uv'\,dx = uv - \int vu'\,dx$

8.45 is called the integration by parts formula and gives the integral of a product of functions.

Take u to be the function in the following order:
1 $\ln(x)$
2 x^n
3 e^{kx} where k is a constant

Exercise 8(d)

1 Determine the following integrals:

a $\int 2xe^x dx$ **b** $\int t \sin(t)\, dt$

2 Find **a** $\int q \cos(3q)\, dq$ **b** $\int s^2 \ln(s)\, ds$

3 Obtain $\int te^{2t}\, dt$.

4 Show that

$$\int p\sqrt{1+p}\, dp = \frac{2}{15}(1+p)^{3/2}[3p-2] + C$$

5 By writing $\ln(x) = 1 \times \ln(x)$, show that

$$\int \ln(x)\, dx = x[\ln(x) - 1] + C$$

6 [electrical principles]   The current, i, through an inductor with inductance L is given by

$$i = \frac{1}{L}\int_0^t v\, dt$$

where v is the voltage across the inductor. For a circuit with $L = 10 \times 10^{-3}$ H and $v = 5te^{-t}$, find i.

7 [electrical principles]   The energy, w, of an inductor with inductance L is given by

$$w = \frac{1}{2L}\left(\int_0^t v\, dt\right)^2$$

where v is the voltage across the inductor. For $L = 10$ mH and $v = t\cos(t)$, find w.

8 [electrical principles]   The current, i, through an inductor with inductance L is given by

$$i = \frac{1}{L}\int_0^1 v\, dt$$

For $L = (1 \times 10^{-3})$ H and $v = t^2 e^{-t}$, find i.

SECTION E **Algebraic fractions**

By the end of this section you will be able to:
▶ understand what is meant by the term partial fraction
▶ find partial fractions of various expressions
▶ understand the procedure for finding partial fractions
▶ find partial fractions of improper fractions

Partial fractions are an algebraic topic which can be used to integrate algebraic fractions.
For example, how do we integrate the following:

$$\int \frac{x}{(x+2)(x+1)}\, dx?$$

We can't use any of the techniques discussed earlier. We are compelled to express $\frac{x}{(x+2)(x+1)}$ as partial fractions and then integrate the result. This section could have been placed in the earlier Chapters 1 and 2 on algebra, however partial fractions is a difficult topic and requires greater algebraic skills.

E1 **Partial fractions**

Consider the following example:

$$\frac{7}{12} = \frac{1}{4} + \frac{1}{3}$$

We say that $\frac{1}{4}$ and $\frac{1}{3}$ are partial fractions of $\frac{7}{12}$. The two or more fractions which create a single fraction are called **partial fractions**.

Consider the following algebraic expression:

$$\frac{x}{(x + 2)(x + 1)} = \frac{2}{x + 2} - \frac{1}{x + 1}$$

[?] **What are the partial fractions of $\dfrac{x}{(x + 2)(x + 1)}$?**

Clearly they are $\dfrac{2}{x + 2}$ and $\dfrac{-1}{x + 1}$.

In this section we study the processes of splitting a fraction such as $\dfrac{x}{(x + 2)(x + 1)}$ into its partial fractions.

[?] **How do we split $\dfrac{x}{(x + 2)(x + 1)}$ into its partial fractions?**

First we write this as

[†] $$\frac{x}{(x + 2)(x + 1)} = \frac{A}{x + 2} + \frac{B}{x + 1}$$

where A and B are constants which we need to find.

[?] **How do we know that $\dfrac{x}{(x + 2)(x + 1)}$ breaks into $\dfrac{A}{x + 2} + \dfrac{B}{x + 1}$?**

Well, if we add the fractions

$$\frac{A}{x + 2} + \frac{B}{x + 1}$$

then we obtain a common denominator of $(x + 2)(x + 1)$. Thus

$$\frac{A}{x + 2} + \frac{B}{x + 1} = \frac{A(x + 1) + B(x + 2)}{(x + 2)(x + 1)}$$

So all we are left to achieve is to find the constants A and B so that the numerator gives an x.

[††] $$\frac{x}{(x + 2)(x + 1)} = \frac{A(x + 1) + B(x + 2)}{(x + 2)\,(x + 1)}$$

Since on both sides of the equal sign we have a common denominator, the numerator must be the same, that is

[*] $$x = A(x + 1) + B(x + 2)$$

This can also be attained by multiplying both sides of [††] by $(x + 2)(x + 1)$.

[?] **But how can we find A and B?**

We can choose values of x and substitute these values into ▮*▮. The evaluations of A and B are simpler if we select our x values such that some of the terms on the Right-Hand Side of ▮*▮ vanish. For example, choosing $x = -1$ gives

$$-1 = A(-1 + 1) + B(-1 + 2)$$
$$-1 = 0 + B$$

Hence $B = -1$.

[?] **How can we find A?**

By choosing $x = -2$ and substituting into ▮*▮:

$$-2 = A(-2 + 1) + B(-2 + 2)$$
$$-2 = A(-1) + 0$$

Hence $A = 2$.

[?] **Why did we choose $x = -2$?**

Because this removes the B term of ▮*▮.

Putting $A = 2$ and $B = -1$ into ▮†▮ gives the identity

$$\frac{x}{(x + 2)(x + 1)} = \frac{2}{x + 2} - \frac{1}{x + 1}$$

We have broken the single fraction, $\dfrac{x}{(x + 2)(x + 1)}$, into the difference of two fractions, $\dfrac{2}{x + 2} - \dfrac{1}{x + 1}$. This is what we started with at the beginning of this discussion.

For the appropriate values of the constants, the partial fractions become an identity; that is, both sides are equal for all values of x.

Let's investigate another example.

Example 25

Resolve

$$\frac{2x}{x^2 - x - 2}$$

into partial fractions.

Solution

We first try to place the denominator in the form of two bracketed terms, that is to factorize the denominator $x^2 - x - 2$. This factorizes into

$$x^2 - x - 2 = (x + 1)(x - 2)$$

Therefore

$$\frac{2x}{x^2 - x - 2} = \frac{2x}{(x + 1)(x - 2)} = \frac{A}{x + 1} + \frac{B}{x - 2}$$

Multiply both sides by $(x + 1)(x - 2)$:

$$2x = \frac{A(x + 1)(x - 2)}{(x + 1)} + \frac{B(x + 1)(x - 2)}{(x - 2)}$$
$$= A(x - 2) + B(x + 1) \quad \text{[Cancelling]}$$

We have

▮*▮ $\qquad 2x = A(x - 2) + B(x + 1)$

Example 25 *continued*

Which values of x should we select to find A and B?

Putting $x = 2$ into ▮*▮ removes the A term:

$$2 \times 2 = A(2 - 2) + B(2 + 1)$$
$$4 = 3B$$
$$B = \frac{4}{3}$$

To find A, put $x = -1$ into ▮*▮ which removes the B term:

$$2 \times (-1) = A(-1 - 2) + B(0)$$
$$-2 = -3A$$
$$A = \frac{2}{3}$$

Putting $A = \frac{2}{3}$ and $B = \frac{4}{3}$ into

$$\frac{2x}{x^2 - x - 2} = \frac{A}{x + 1} + \frac{B}{x - 2}$$

gives

▮**▮
$$\frac{2x}{x^2 - x - 2} = \frac{\frac{2}{3}}{x + 1} + \frac{\frac{4}{3}}{x - 2}$$

Consider the first term on the Right-Hand Side of ▮**▮ :

$$\frac{\frac{2}{3}}{x + 1} = \frac{2}{3} \div (x + 1) = \frac{2}{3} \div \frac{(x + 1)}{1} = \frac{2}{3} \times \frac{1}{(x + 1)}$$
$$= \frac{2 \times 1}{3 \times (x + 1)} = \frac{2}{3(x + 1)}$$

Similarly the second term on the Right-Hand Side of ▮**▮ becomes

$$\frac{\frac{4}{3}}{x - 2} = \frac{4}{3(x - 2)}$$

Combining these two results we have the identity

$$\frac{2x}{x^2 - x - 2} = \frac{2}{3(x + 1)} + \frac{4}{3(x - 2)}$$

$$= \frac{2}{3}\left[\frac{1}{x + 1} + \frac{2}{x - 2}\right] \qquad \left[\text{Taking out } \frac{2}{3}\right]$$

Consider a single algebraic fraction

8.47 $$\frac{f(x)}{g(x)}$$

? where $g(x)$ factorizes and is of a higher degree polynomial than $f(x)$. **What do we mean by a higher degree polynomial?**

For example, if $g(x)$ is a cubic (containing x^3) then $f(x)$ is at most a quadratic (containing x^2). See Table 3.

Polynomial	Degree	Name
$x - 5$	1	linear
$x^2 + x + 1$	2	quadratic
$x^3 - x^2$	3	cubic
$x^4 + x - 3$	4	quartic

TABLE 3

We need to write 8.47, $\frac{f(x)}{g(x)}$, as a sum of partial fractions. This can be obtained by factorizing the denominator, $g(x)$, and then using one of the following rules.

Each linear factor of the denominator has partial fractions of the form $\frac{A}{ax + b}$. Thus

8.48 $$\frac{f(x)}{(ax + b)(cx + d)} = \frac{A}{ax + b} + \frac{B}{cx + d}$$

Each repeated linear factor of the denominator has partial fractions of the form
$\frac{A}{ax + b} + \frac{B}{(ax + b)^2}$. Thus

8.49 $$\frac{f(x)}{(ax + b)^2} = \frac{A}{ax + b} + \frac{B}{(ax + b)^2}$$

Each quadratic factor of the denominator has the partial fractions of the form $\frac{Ax + B}{ax^2 + bx + c}$.
Hence

8.50 $$\frac{f(x)}{ax^2 + bx + c} = \frac{Ax + B}{ax^2 + bx + c}$$

A combination of linear and quadratic factors gives

8.51 $$\frac{f(x)}{(ax^2 + bx + c)(dx + e)} = \frac{Ax + B}{ax^2 + bx + c} + \frac{C}{dx + e}$$

For the appropriate values of A, B and C, 8.48 to 8.51 are identities.

These identities look horrendous but once we do a few examples then you will become familiar with these 'horrendous' identities. However the procedure in finding the partial fractions of

8.47 $$\frac{f(x)}{g(x)}$$

is

1 Factorize the denominator, $g(x)$, as far as possible.
2 Write 8.47 as one of the general partial fractions by choosing the appropriate identity from 8.48 to 8.51 .
3 Find the values of the unknown constants A, B, C, etc.

Let's do an example.

Example 26

Express

$$\frac{3t^2 - t + 1}{(t^2 - t + 3)(t - 1)}$$

as partial fractions.

Solution

The denominator does not factorize further into simpler factors. Each factor of the denominator, $t^2 - t + 3$ and $t - 1$, produces a partial fraction. **Which identity, 8.48 to 8.51 , is appropriate for**

$$\frac{3t^2 - t + 1}{(t^2 - t + 3)(t - 1)}?$$

Since we have quadratic and linear factors we use 8.51 :

† $$\frac{3t^2 - t + 1}{(t^2 - t + 3)(t - 1)} = \frac{At + B}{t^2 - t + 3} + \frac{C}{t - 1}$$

What do we need to find?

The constants A, B and C. **How can we find A, B and C?**

Multiplying both sides of † by $(t^2 - t + 3)(t - 1)$ we have

†† $$3t^2 - t + 1 = (At + B)(t - 1) + C(t^2 - t + 3)$$

To remove the first term on the Right-Hand Side of †† we substitute $t = 1$ into †† :

$$3 - 1 + 1 = 0 + C(1^2 - 1 + 3)$$
$$3 = 3C \text{ gives } C = 1$$

Putting $C = 1$ into †† yields

* $$3t^2 - t + 1 = (At + B)(t - 1) + C(t^2 - t + 3)$$

How can we find A and B?

We can now substitute other values of t such as $t = 0$, but it is generally easier to **equate coefficients.** Conventionally we start to equate coefficients of the highest power first. So we first equate the number of t^2 on the left of * to the number of t^2 on the right of * .

Example 26 *continued*

We need to expand the Right-Hand Side of ▨ because $At \times t$ gives At^2 and so the number of t^2 on the right is $A + 1$.

How many t^2 are on the Left-Hand Side of ▨?

3. Thus equating coefficients of t^2 we have

$$3 = A + 1 \text{ therefore } A = 2$$

Equating coefficients of t in ▨ by expanding the Right-Hand Side of ▨ gives

$$At \times (-1) = -At \text{ and } B \times t = Bt$$
$$-1 = -A + B - 1 \qquad\qquad [\text{Equating } t\text{'s of } ▨]$$

Substituting $A = 2$ yields

$$-1 = -2 + B - 1$$
$$-1 = -3 + B \text{ gives } B = 2$$

So we have $A = 2$, $B = 2$ and $C = 1$. Substituting these into ▨ displays the partial fractions

$$\frac{3t^2 - t + 1}{(t^2 - t + 3)(t - 1)} = \frac{2t + 2}{t^2 - t + 3} + \frac{1}{t - 1}$$

E2 Improper fractions

If in the algebraic fraction

$$\frac{f(x)}{g(x)}$$

$f(x)$ is a polynomial of higher, or same, degree compared with $g(x)$ then $\dfrac{f(x)}{g(x)}$ is called an **improper fraction.** The degree of a polynomial, $f(x)$, is the highest power of x contained in the polynomial.

To find the partial fractions of an improper fraction, $\dfrac{f(x)}{g(x)}$, we first divide $f(x)$ by $g(x)$ by applying long division. If this division results in a polynomial $q(x)$ with remainder $R(x)$ then we can write

8.52 $$\frac{f(x)}{g(x)} = q(x) + \frac{R(x)}{g(x)}$$

where the remainder $R(x)$ is a polynomial of lower degree than $g(x)$. We can now treat $\dfrac{R(x)}{g(x)}$ in the usual way for partial fractions.

Let's first do an example on arithmetic long division.

Applying long division to $200 \div 15$ we have

$$
\begin{array}{r}
13 \\
15\overline{)200} \\
195 \\
\hline
5
\end{array}
\quad \text{(Subtracting 195 from 200)}
$$

This says that 15 goes 13 whole times into 200 with remainder 5:

$$\frac{200}{15} = 13 + \frac{5}{15}$$

Example 27

Express $\dfrac{x^3}{x^2 - 4}$ in partial fractions.

Solution

Since x^3 is of degree 3 and $x^2 - 4$ is of degree 2 we first apply long division.

Long division of algebraic expressions is the same procedure as long division of numbers. To obtain an x^3 from $x^2 - 4$ we multiply by x.

$x(x^2 - 4) = x^3 - 4x$, thus

$$
\begin{array}{r}
x \\
x^2 - 4 \overline{)\ x^3 } \\
\underline{x^3 - 4x} \\
0 + 4x
\end{array}
$$
 (Subtracting $x^3 - 4x$ from x^3)

Using ▮8.52▮ with $q(x) = x$ and $R(x) = 4x$ we have

▮†▮ $\qquad \dfrac{x^3}{x^2 - 4} = x + \dfrac{4x}{x^2 - 4}$

▮8.52▮ puts the expression $\dfrac{x^3}{x^2 - 4}$ into partial fractions but we can cut it into even more partial fractions because the denominator, $x^2 - 4$, factorizes. Thus

$$x^2 - 4 = x^2 - 2^2 = (x - 2)(x + 2)$$

▮?▮ **Which identity, ▮8.48▮ to ▮8.51▮, can we use to place $\dfrac{4x}{(x - 2)(x + 2)}$ into partial fractions?**

Use

▮8.48▮ $\qquad \dfrac{f(x)}{(ax + b)(cx + d)} = \dfrac{A}{ax + b} + \dfrac{B}{cx + d}$

▮††▮ $\qquad \dfrac{4x}{(x - 2)(x + 2)} = \dfrac{A}{x - 2} + \dfrac{B}{x + 2}$

Multiplying both sides by $(x - 2)(x + 2)$ gives

▮*▮ $\qquad 4x = A(x + 2) + B(x - 2)$

▮?▮ **What are we trying to find?**

The values of the constants A and B.

▮?▮ **What should we substitute for x into ▮*▮ to obtain A and B?**

Putting $x = 2$ into ▮*▮ gives

$$(4 \times 2) = A(2 + 2) + 0$$
$$8 = 4A \text{ which gives } A = 2$$

Example 27 *continued*

Putting $x = -2$ into ▉*▉ gives

$$[4 \times (-2)] = 0 + B(-2 - 2)$$
$$- 8 = -4B \text{ which gives } B = 2$$

Substituting $A = 2$, $B = 2$ into ▉††▉ yields

$$\frac{4x}{(x - 2)(x + 2)} = \frac{2}{x - 2} + \frac{2}{x + 2} \underset{\substack{\text{taking out the}\\\text{common factor 2}}}{=\!=} 2\left[\frac{1}{x - 2} + \frac{1}{x + 2}\right]$$

Putting $\dfrac{4x}{x^2 - 4} = 2\left[\dfrac{1}{x - 2} + \dfrac{1}{x + 2}\right]$ into ▉†▉ gives the identity

$$\frac{x^3}{x^2 - 4} = x + 2\left[\frac{1}{x - 2} + \frac{1}{x + 2}\right]$$

SUMMARY

The procedure for finding partial fractions of

▉8.47▉ $\dfrac{f(x)}{g(x)}$

where $g(x)$ factorizes and is of a higher degree than $f(x)$ is

1 factorize the denominator
2 write ▉8.47▉ as one of the general partial fractions determined by ▉8.48▉ to ▉8.51▉
3 find the values of the unknown constants A, B, C, etc.

If $f(x)$ is of the same or higher degree than $g(x)$ in ▉8.47▉ then this is called an improper fraction and we need to apply long division.

Exercise 8(e)

Solutions at end of book. Complete solutions available at www.palgrave.com/science/engineering/singh

1 Express the following in partial fractions:

a $\dfrac{3x + 4}{(x + 1)(x + 2)}$ **b** $\dfrac{2t}{t^2 - 1}$

c $\dfrac{2s + 7}{s^2 + s - 2}$ **d** $\dfrac{-12u - 13}{(2u + 1)(u - 3)}$

2 Resolve the following into partial fractions:

a $\dfrac{x^2}{x + 1}$ **b** $\dfrac{x^5 - 2x^2}{x^2 - 1}$

3 Resolve the following into partial fractions:

a $\dfrac{4t^2 + t - 3}{(t^2 + t - 1)(t - 1)}$

b $\dfrac{z + 1}{(z - 1)^2}$

c $\dfrac{2x^3 + 3x^2 + 5x + 2}{(x^2 + x + 1)^2}$

SECTION F **Integration of algebraic fractions**

By the end of this section you will be able to:

▶ integrate $\dfrac{f(x)}{g(x)}$ by using partial fractions

F1 **Integration by partial fractions**

In this section we integrate $\dfrac{f(x)}{g(x)}$ by expressing this in its partial fractions and then integrating. Generally you will discover that once we have found the partial fractions then to integrate we employ formula 8.42 :

8.42
$$\int \frac{f'(x)}{f(x)}\,dx = \ln|f(x)| + C$$

Example 28

Determine $\displaystyle\int \frac{s}{(s+2)(s+1)}\,ds$

Solution

First we express $\dfrac{s}{(s+2)(s+1)}$ in partial fractions. This was obtained in **Section E1** with x in place of s:

$$\frac{s}{(s+2)(s+1)} = \frac{2}{s+2} - \frac{1}{s+1}$$

How do we integrate $\dfrac{s}{(s+2)(s+1)}$ **with respect to** s?

$$\int \frac{s}{(s+2)(s+1)}\,ds = \int \left(\frac{2}{s+2} - \frac{1}{s+1} \right) ds$$

$$= \int \frac{2ds}{s+2} - \int \frac{ds}{s+1} \qquad \text{[Splitting integrand]}$$

$$= 2\int \frac{ds}{s+2} - \int \frac{ds}{s+1} \qquad \text{[Taking out 2]}$$

$$= 2\ln|s+2| - \ln|s+1| + C \quad \text{[Integrating]}$$
$$\text{by } 8.42$$

$$= \ln(|s+2|^2) - \ln|s+1| + C$$
$$\text{by } 5.14$$

$$= \ln\left| \frac{(s+2)^2}{s+1} \right| + C \qquad \left[\text{By } 5.13 \right]$$

5.13 $\ln(A) - \ln(B) = \ln(A/B)$ 5.14 $n\ln(A) = \ln(A^n)$

Note that to simplify the result of integration we apply the laws of logarithms. This is generally the case for integration by partial fractions and so make sure that the laws of logarithms, 5.12 to 5.14 , are second nature to you.

Example 29

Find $\displaystyle\int \frac{2x}{x^2 - x - 2}\, dx$

Solution

We have already placed the integrand, $\dfrac{2x}{x^2 - x - 2}$, in partial fractions in **Example 25**, thus

$$\frac{2x}{x^2 - x - 2} = \frac{2}{3}\left[\frac{1}{x + 1} + \frac{2}{x - 2}\right]$$

Carrying out the integration by splitting the integrand and taking out the constant, 2/3, we have

$$\int \frac{2x}{x^2 - x - 2}\,dx = \frac{2}{3}\int \left[\frac{1}{x + 1} + \frac{2}{x - 2}\right]dx$$

$$= \frac{2}{3}\left[\int \frac{1}{x + 1}\,dx + \int \frac{2}{x - 2}\,dx\right]$$

$$= \frac{2}{3}\left[\ln|x + 1| + 2\ln|x - 2|\right] + C$$

$$= \frac{2}{3}\left[\ln|x + 1| + \underline{\ln|x - 2|^2}\right] + C$$

$$\text{by } 5.14$$

The next example is long but not difficult. First we need to find the partial fractions and then integrate.

🚗 Example 30 *mechanics*

The velocity, v, of an object falling in air at time t is given by

$$t = \int_0^v \frac{dv}{9 - 0.25v^2}$$

By using partial fractions, show that

$$t = \frac{1}{3}\ln\left|\frac{3 + 0.5v}{3 - 0.5v}\right|$$

5.14 $n \ln(A) = \ln(A^n)$

Example 30 *continued*

Is there a more straightforward way to obtain this result?

Solution

First we need to factorize the denominator, $9 - 0.25\,v^2$:

$$9 - 0.25v^2 = 3^2 - (0.5v)^2$$
$$= (3 - 0.5v)(3 + 0.5v)$$

By placing into partial fractions we have

†
$$\frac{1}{9 - 0.25v^2} = \frac{1}{(3 - 0.5v)(3 + 0.5v)} = \frac{A}{3 - 0.5v} + \frac{B}{3 + 0.5v}$$

by **8.48**

Multiplying both sides by $(3 - 0.5v)(3 + 0.5v)$ gives

∗
$$1 = A(3 + 0.5v) + B(3 - 0.5v)$$

How can we find A and B?

Put $v = -6$ into ∗ because this will remove the A term and therefore we can find B:

$$1 = A[3 + 0.5(-6)] + B[3 - 0.5(-6)]$$
$$1 = 0 + 6B$$

Thus $B = \dfrac{1}{6}$. Putting $v = 6$ into ∗ gives

$$1 = A[3 + (0.5 \times 6)] + 0$$
$$1 = 6A$$

So $A = \dfrac{1}{6}$. Substituting $A = \dfrac{1}{6}$ and $B = \dfrac{1}{6}$ into † gives the partial fractions

$$\frac{1}{9 - 0.25v^2} = \frac{\dfrac{1}{6}}{3 - 0.5v} + \frac{\dfrac{1}{6}}{3 + 0.5v}$$

$$= \frac{1}{6}\left[\frac{1}{3 - 0.5v} + \frac{1}{3 + 0.5v}\right] \quad \left[\text{Taking out } \frac{1}{6}\right]$$

Considering the given integral we have

$$t = \int_0^v \frac{dv}{9 - 0.25v^2} = \frac{1}{6}\int_0^v \left(\frac{1}{3 - 0.5v} + \frac{1}{3 + 0.5v}\right) dv$$

††
$$t = \frac{1}{6}\left[\int_0^v \frac{dv}{3 - 0.5v} + \int_0^v \frac{dv}{3 + 0.5v}\right]$$

8.48 $\dfrac{f(v)}{(av + b)(cv + d)} = \dfrac{A}{av + b} + \dfrac{B}{cv + d}$

Example 30 *continued*

How do we find the first integral on the Right-Hand Side, $\int_0^v \dfrac{dv}{3 - 0.5v}$?

We can spot that differentiating the denominator gives -0.5. If we want to use

8.42 $\qquad \int \dfrac{f'(v)}{f(v)}\, dv = \ln|f(v)| + C$

then we need to have the derivative of the denominator on the numerator.
We can rewrite the integrand as follows:

$$\frac{1}{3 - 0.5v} = \frac{1}{-0.5}\left(\frac{-0.5}{3 - 0.5v}\right) = -2\left(\frac{-0.5}{3 - 0.5v}\right)$$

Evaluating the integral:

$$\int_0^v \frac{dv}{3 - 0.5v} = -2\int_0^v \frac{-0.5}{3 - 0.5v}\, dv$$

$$= -2\Big[\underbrace{\ln|3 - 0.5v|}_{\text{by }8.42}\Big]_0^v$$

$$= -2\big[\ln|3 - 0.5v| - \ln(3)\big] \quad \text{[Substituting]}$$

Similarly

$$\int_0^v \frac{dv}{3 + 0.5v} = \frac{1}{0.5}\int_0^v \frac{0.5\,dv}{3 + 0.5v}$$

$$= \frac{1}{0.5}\Big[\ln|3 + 0.5v|\Big]_0^v \quad \text{[Integrating]}$$

$$= 2\big[\ln|3 + 0.5v| - \ln(3)\big] \quad \text{[Substituting]}$$

Substituting these results into †† produces

$$t = \frac{1}{6}\Big[-2\big[\ln|3 - 0.5v| - \ln(3)\big] + 2\big[\ln|3 + 0.5v| - \ln(3)\big]\Big]$$

$$= \frac{2}{6}\Big[\ln(3) - \ln|3 - 0.5v| + \ln|3 + 0.5v| - \ln(3)\Big] \quad \text{[Taking out 2]}$$

$$= \frac{1}{3}\Big[\ln|3 + 0.5v| - \ln|3 - 0.5v|\Big]$$

$$= \frac{1}{3}\underbrace{\ln\left|\frac{3 + 0.5v}{3 - 0.5v}\right|}_{\text{by }5.13} \quad \text{(Simplifying)}$$

However, there is an effortless way of obtaining the above result. We can use

8.30 $\qquad \int \dfrac{du}{a^2 - u^2} = \dfrac{1}{2a}\ln\left|\dfrac{a + u}{a - u}\right|$

(this is given in Table 1). Thus

$$\int_0^v \frac{dv}{9 - 0.25v^2} = \int_0^v \frac{dv}{3^2 - (0.5v)^2}$$

5.13 $\ln(A) - \ln(B) = \ln(A/B)$ **8.42** $\int \dfrac{f'(v)}{f(v)}\, dv = \ln|f(v)|$

⊞ Example 30 *continued*

Let $u = 0.5v$, then

$$\frac{du}{dv} = 0.5 \text{ gives } dv = \frac{du}{0.5} = 2du$$

We can first evaluate the integral without the limits:

$$\int \frac{dv}{3^2 - (0.5v)^2} = \int \frac{2du}{3^2 - u^2} = 2\int \frac{du}{3^2 - u^2}$$

$$= 2\left[\frac{1}{2 \times 3}\ln\left|\frac{3 + u}{3 - u}\right|\right] \quad [\text{By } \boxed{8.30}]$$

$$= \frac{1}{3}\ln\left|\frac{3 + u}{3 - u}\right| \quad [\text{Cancelling 2's}]$$

$$= \frac{1}{3}\ln\left|\frac{3 + 0.5v}{3 - 0.5v}\right| \quad [\text{Remember that } u = 0.5v]$$

Using the limits of integration:

$$\int_0^v \frac{dv}{9 - 0.25v^2} = \frac{1}{3}\left[\ln\left|\frac{3 + 0.5v}{3 - 0.5v}\right|\right]_0^v$$

$$= \frac{1}{3}\left[\ln\left|\frac{3 + 0.5v}{3 - 0.5v}\right| - \ln\left|\frac{3}{3}\right|\right] \quad [\text{Substituting}]$$

$$= \frac{1}{3}\ln\left|\frac{3 + 0.5v}{3 - 0.5v}\right| \quad \left[\text{Because } \ln\left|\frac{3}{3}\right| = \ln|1| = 0\right]$$

SUMMARY

Integration by partial fractions involves first putting the function into partial fractions and then integrating. The most important integral formula for these fractions is $\boxed{8.42}$. Generally we need to use the laws of logarithms to simplify.

Exercise 8(f)

Solutions at end of book. Complete solutions available at www.palgrave.com/science/engineering/singh

1 Determine

a $\displaystyle\int \frac{3c + 4}{(c + 1)(c + 2)}\, dc$ **b** $\displaystyle\int \frac{2\lambda}{\lambda^2 - 1}\, d\lambda$

c $\displaystyle\int \frac{2a + 7}{a^2 + a - 2}\, da$

d $\displaystyle\int \frac{-12y - 13}{(2y + 1)(y - 3)}\, dy$

2 Determine

a $\displaystyle\int \frac{4p^2 + p - 3}{(p^2 + p - 1)(p - 1)}\, dp$

b $\displaystyle\int \frac{z + 1}{(z - 1)^2}\, dz$

3 Evaluate $\displaystyle\int_1^2 \frac{5z^2}{(z^2 + 1)(2z - 1)}\, dz$

| 8.30 | $\displaystyle\int \frac{du}{a^2 - u^2} = \frac{1}{2a}\ln\left|\frac{a + u}{a - u}\right|$ |

Exercise 8(f) continued

4 [mechanics] The velocity, v, of an object moving in a medium at time t is given by

$$t = \int_{10}^{100} \frac{dv}{v(2v + 1)}$$

Evaluate t.

5 [mechanics] The velocity, v, of an object falling in air is given by

$$v = \int \frac{dv}{1 - (kv)^2}$$

where k is a non-zero constant. By using partial fractions, show that

$$v = \frac{1}{2k}\ln\left|\frac{1 + kv}{1 - kv}\right| + C$$

6 Show that $\displaystyle\int_0^1 \frac{5 + 2x - x^2}{(x^2 + 1)(x + 1)} \, dx = \pi$.

7 i Express $\dfrac{x^3 + 1}{x^2 + 3x + 2}$ in partial fractions.

ii Determine $\displaystyle\int \frac{x^3 + 1}{x^2 + 3x + 2} \, dx$

SECTION G Integration by substitution revisited

By the end of this section you will be able to:
▶ evaluate integrals by substitution
▶ integrate $k\left[f(t)\right]^n f'(t)$

G1 Integration by substitution

In **Section B** of this chapter we found indefinite integrals (integrals without limits) by using a substitution. **Why did we use a substitution?**

The integral of interest was not among the list of standard integrals of Table 1. Additionally, the evaluation of the integral became elementary once we used the substitution.

Since then we have developed other methods of integration such as integration by parts and by partial fractions. In this section we examine integrals with limits which demand a substitution. Let's do an example.

Example 31

Evaluate $\displaystyle\int_1^4 \frac{t}{\sqrt{3t^2 + 1}} \, dt$.

Solution

Can you identify a substitution that we can use to evaluate this integral?

One choice might be to let $u = 3t^2 + 1$ because differentiating this we have

$$\frac{du}{dt} = 6t$$

Example 31 *continued*

which gives

$$\frac{du}{6t} = dt$$

Therefore we might be able to cancel the t's once substitution has taken place.

We also need to substitute new values for the limits $t = 1$ and $t = 4$. Why?

Because we are integrating with respect to a new variable, $u = 3t^2 + 1$.

When $t = 1$, $u = (3 \times 1^2) + 1 = 4$

When $t = 4$, $u = (3 \times 4^2) + 1 = 49$

Substituting the new limits and $u = 3t^2 + 1$, $dt = \dfrac{du}{6t}$, we have

$$\int_1^4 \left(\frac{t}{\sqrt{3t^2 + 1}} \right) dt = \int_{u=4}^{u=49} \left(\frac{t}{\sqrt{u}} \right) \frac{du}{6t}$$

$$= \frac{1}{6} \int_4^{49} \frac{du}{u^{1/2}} \quad \text{[Cancelling } t\text{'s]}$$

$$= \frac{1}{6} \int_4^{49} u^{-1/2} du$$

$$= \frac{1}{6} \left[\frac{u^{1/2}}{1/2} \right]_4^{49} \underset{\substack{\text{substituting} \\ \text{the limits}}}{=} \frac{1}{3} \left[49^{1/2} - 4^{1/2} \right] = \frac{5}{3}$$

Notice that the t's in the above example cancel, and for this reason the integration becomes straightforward. In general, to integrate a function of the type

8.53 $k\,[f(t)]^n\, f'(t)$ (where $n \neq -1$ and k is a constant)

with respect to t, we use the substitution $u = f(t)$.

8.53 might look terrifying but it is only asserting that if you have a function, $f(t)$, to the power n and its derivative multiplied by a constant in the form of 8.53, then use the substitution $u = f(t)$. For **Example 31**:

$$f(t) = 3t^2 + 1, \; f'(t) = 6t, \; k = \frac{1}{6} \text{ and } n = -\frac{1}{2}$$

$$\frac{1}{6} (3t^2 + 1)^{-1/2}\, 6t = \frac{t}{\sqrt{3t^2 + 1}}$$

Let's explain this further by trying another example.

> ### Example 32
>
> Determine
>
> $$\int \frac{t}{(t^2 + 1)^5} \, dt$$
>
> Solution
>
> If we let $u = t^2 + 1$, differentiating gives
>
> $$\frac{du}{dt} = 2t$$
>
> (In relation to 8.53 , we have $f(t) = t^2 + 1$ and $f'(t) = 2t$.)
>
> Rearranging $\dfrac{du}{dt} = 2t$ gives
>
> $$dt = \frac{du}{du/dt} = \frac{du}{2t}$$
>
> Remember that we need to replace the dt in the original integral because we are integrating with respect to u. Thus, substituting $u = t^2 + 1$ and $dt = \dfrac{du}{2t}$ gives
>
> $$\int \left(\frac{t}{(t^2 + 1)^5} \right) dt = \int \left(\frac{t}{u^5} \right) \frac{du}{2t}$$
>
> $$= \frac{1}{2} \int \frac{du}{u^5} \quad \text{[Cancelling } t\text{'s]}$$
>
> $$= \frac{1}{2} \int u^{-5} du$$
>
> $$= \frac{1}{2} \left(\frac{u^{-4}}{-4} \right) + C = -\frac{u^{-4}}{8} + C = C - \frac{1}{8u^4}$$
>
> We have
>
> $$\int \frac{t}{(t^2 + 1)^5} \, dt = C - \frac{1}{8(t^2 + 1)^4} \qquad \text{[Substituting } u = t^2 + 1]$$

G2 Some trigonometric substitutions

The method of substitution is important when integrating trigonometric functions.

> ### Example 33
>
> Evaluate $\displaystyle\int_0^{\frac{\pi}{2}} \left(\cos(\alpha)\sqrt{\sin(\alpha)} \right) d\alpha$

Example 33 *continued*

Solution

When we differentiate $\sin(\alpha)$ we obtain $\cos(\alpha)$.

Thus the function $\cos(\alpha)\sqrt{\sin(\alpha)}$ seems to be of the form of 8.53. So use the substitution $u = \sin(\alpha)$. Differentiating gives

$$\frac{du}{d\alpha} = \cos(\alpha)$$

$$d\alpha = \frac{du}{\cos(\alpha)}$$

We also need to change the limits of integration:

When $\alpha = 0$, $u = \sin(0) = 0$

When $\alpha = \dfrac{\pi}{2}$, $u = \sin\left(\dfrac{\pi}{2}\right) = 1$

Replacing the limits and $u = \sin(\alpha)$, $d\alpha = \dfrac{du}{\cos(\alpha)}$, into the original integral we have

$$\int_{\alpha=0}^{\alpha=\frac{\pi}{2}}\left(\cos(\alpha)\sqrt{\sin(\alpha)}\,\right)d\alpha = \int_{u=0}^{u=1}\left(\cos(\alpha)\sqrt{u}\,\right)\frac{du}{\cos(\alpha)}$$

$$= \int_0^1 \sqrt{u}\,du \quad [\text{Cancelling } \cos(\alpha)]$$

$$= \int_0^1 u^{1/2}du$$

$$= \left[\frac{u^{3/2}}{3/2}\right]_0^1 = \frac{2}{3}\left[u^{3/2}\right]_0^1 = \frac{2}{3}\left(1^{3/2} - 0^{3/2}\right) = \frac{2}{3}$$

$$\int_0^{\frac{\pi}{2}}\left(\cos(\alpha)\sqrt{\sin(\alpha)}\,\right)d\alpha = \frac{2}{3}$$

SUMMARY

To integrate a function of the type

8.53 $k[f(t)]^n\, f'(t)$ (where $n \neq -1$ and k is a constant)

with respect to t, use the substitution $u = f(t)$.

For a definite integral we also need to replace the limits of integration:

$$\int_{x=a}^{x=b} f(x)dx = \int_{u=c}^{u=d} g(u)du$$

Solutions at end of book. Complete solutions available at www.palgrave.com/science/engineering/singh

Exercise 8(g)

1 Determine the following integrals:

a $\displaystyle\int b(b^2 - 3)^7 db$

b $\displaystyle\int (5s - 1)^9 ds$

c $\displaystyle\int (3a^2 - 4a)(a^3 - 2a^2 + 6)^4 \, da$

d $\displaystyle\int 21q^2 \sqrt{7q^3 - 5} \, dq$

e $\displaystyle\int \frac{p^2 - 1}{\sqrt{p^3 - 3p}} \, dp$

f $\displaystyle\int \frac{\alpha - 1}{(\alpha^2 - 2\alpha + 10)^2} \, d\alpha$

2 🔧 [*reliability engineering*]   The mean time to failure, MTTF, in years, for a set of components is given by

$$\text{MTTF} = \int_0^5 (1 - 0.2t)^{1.5} dt$$

Evaluate MTTF.

3 Determine $\displaystyle\int_0^1 xe^{-x^2} dx$. [Consider $u = x^2$]

4 Evaluate the following integrals:

a $\displaystyle\int_0^{\frac{\pi}{2}} \sin(\theta)\sqrt{\cos(\theta)} \, d\theta$

b $\displaystyle\int_0^{\pi} \sin(\theta) \cos^5(\theta) d\theta$

What do you notice about your result for **a**?

5 Determine the following indefinite integrals:

a $\displaystyle\int \sec^7(\beta)\tan(\beta) d\beta$

b $\displaystyle\int \tan^5(A)\sec^2(A) dA$

c $\displaystyle\int \cot^3(A)\cosec(A) dA$

[Consider $u = \cosec(A)$]

SECTION H **Trigonometric techniques for integration**

By the end of this section you will be able to:
▶ use trigonometric substitutions to integrate trigonometric functions
▶ show some integral results of Table 1

This is a challenging section which requires knowledge of trigonometric identities, integration and substitution. Go through each example very carefully.

H1 **Trigonometric substitutions**

? How do we integrate $\cos^2(t)$ with respect to t?

? Can we use integration by parts because $\cos^2(t) = \cos(t) \times \cos(t)$?

No, because if we use integration by parts then we will end up in a vicious circle.

We need to substitute something for $\cos^2(t)$. In Chapter 4 we had

4.67 $$\cos^2(t) = \frac{1}{2}\left[1 + \cos(2t)\right]$$

Of course this identity might **not** have occurred to you since it is among many trigonometric identities in Chapter 4. However you can derive this identity from the fundamental identity:

$$\cos(2t) = \cos^2(t) - \sin^2(t) = 2\cos^2(t) - 1$$

Example 34

Determine $\displaystyle\int \cos^2(t)\, dt$.

Solution

Using 4.67 we have

$$\int \cos^2(t)\, dt = \int \frac{1}{2}\left[1 + \cos(2t)\right] dt$$

$$= \frac{1}{2}\int \left[1 + \cos(2t)\right] dt \quad \left[\text{Taking out } \frac{1}{2}\right]$$

$$= \frac{1}{2}\left[t + \frac{\sin(2t)}{2}\right] + C \quad \left[\text{By } \int \cos(kt)\, dt = \frac{\sin(kt)}{k}\right]$$

For these types of trigonometric functions, we need to use the appropriate identity from Chapter 4. Let's investigate an engineering example.

Example 35 *electrical principles*

The voltage, v, and current, i, across a pure capacitance is given by

$$v = V\sin(\omega t) \text{ and } i = I\cos(\omega t)$$

where V is the peak voltage, I is the peak current, ω is the angular frequency and t is time. The average power, p, is given by

$$p = \frac{\omega}{2\pi}\int_0^{2\pi/\omega} (vi)\, dt$$

Show that $p = 0$.

Solution

Substituting $v = V\sin(\omega t)$ and $i = I\cos(\omega t)$ into p produces:

$$p = \frac{\omega}{2\pi}\int_0^{2\pi/\omega} \left[V\sin(\omega t)\, I\cos(\omega t)\right] dt$$

  * $$= \frac{\omega VI}{2\pi}\int_0^{2\pi/\omega} \left[\sin(\omega t)\cos(\omega t)\right] dt$$

Example 35 *continued*

The problem is how do we integrate sin(ωt)cos(ωt)?

Using

4.53

$$2\sin(A)\cos(A) = \sin(2A)$$

$$2\sin(\omega t)\cos(\omega t) = \sin(2\omega t)$$

$$\sin(\omega t)\cos(\omega t) = \frac{1}{2}\sin(2\omega t) \quad \text{[Dividing by 2]}$$

Substituting this into ⬛ ✱ ⬛ :

$$p = \frac{\omega VI}{2\pi}\int_0^{2\pi/\omega}\frac{1}{2}\sin(2\omega t)dt$$

$$= \frac{\omega VI}{4\pi}\int_0^{2\pi/\omega}\sin(2\omega t)dt \qquad \left[\text{Taking out }\frac{1}{2}\right]$$

$$= \frac{\omega VI}{4\pi}\left[-\frac{\cos(2\omega t)}{2\omega}\right]_0^{2\pi/\omega} \qquad \left[\text{Using }\int\sin(kt)dt = -\frac{\cos(kt)}{k}\right]$$

$$= -\frac{\omega VI}{4\pi 2\omega}\left(\underbrace{\cos\left[2\omega\left(\frac{2\pi}{\omega}\right)\right]}_{\text{substituting } t = 2\pi/\omega} - \underbrace{\cos(0)}_{\substack{\text{substituting} \\ t=0}}\right)$$

$$= -\frac{VI}{8\pi}\left(\underbrace{1}_{=\cos(4\pi)} - 1\right) = 0$$

$$p = 0$$

You could also have done Example 35 by using the substitution $u = \sin(\omega t)$. Try it!

There are more problems on trigonometric substitutions which are given in **Exercise 8(h).** You need to search for the appropriate trigonometric identity.

H2 **Further trigonometric substitutions**

Sometimes the substitution is not as clear-cut as in the previous examples. For some standard integrals there are suggested substitutions given in Table 4.

TABLE 4	Formulae number	The function that needs integrating contains	Try
	8.54	$\sqrt{a^2 - u^2}$	$u = a\sin(\theta)$
	8.55	$\sqrt{u^2 - a^2}$	$u = a\sec(\theta)$
	8.56	$a^2 + u^2$	$u = a\tan(\theta)$

Let's do an example

Example 36

Show that

$$\int \frac{du}{a^2 + u^2} = \frac{1}{a}\tan^{-1}\left(\frac{u}{a}\right) + C$$

This is result 8.26 of Table 1.

Solution

? **Which substitution should we use?**

By 8.56, let $u = a\tan(\theta)$.

? We also need to replace the du. **What is du?**

Differentiating $u = a\tan(\theta)$ gives

$$\frac{du}{d\theta}_{\text{by } 6.8} = a\sec^2(\theta)$$

$$du = a\sec^2(\theta)d\theta$$

By substituting these into the given integral we have

$$\int \frac{du}{a^2 + u^2} = \int \frac{a\sec^2(\theta)d\theta}{a^2 + a^2\tan^2(\theta)}$$

$$= \int \frac{a\sec^2(\theta)d\theta}{a^2\underbrace{(1 + \tan^2(\theta))}_{= \sec^2(\theta)}}$$

$$= \frac{a}{a^2}\int \frac{\sec^2(\theta)}{\sec^2(\theta)}d\theta \quad \left[\text{Taking out } \frac{a}{a^2}\right]$$

$$= \frac{1}{a}\int d\theta \qquad \text{[Cancelling]}$$

Integrating dθ gives θ, hence

$$\boxed{*} \qquad \int \frac{du}{a^2 + u^2} = \frac{1}{a}\theta + C$$

? **What is θ?**

From above we have

$$a\tan(\theta) = u$$

$$\tan(\theta) = \frac{u}{a}$$

6.8 $[\tan(\theta)]' = \sec^2(\theta)$

Example 36 *continued*

How do we find θ?

Take inverse tan, $\tan^{-1}$, of both sides:

$$\theta = \tan^{-1}\left(\frac{u}{a}\right)$$

Replacing $\theta = \tan^{-1}\left(\dfrac{u}{a}\right)$ in [⋆] displays the required result:

$$\int \frac{du}{a^2 + u^2} = \frac{1}{a}\tan^{-1}\left(\frac{u}{a}\right) + C$$

Similarly we can show some of the other standard integrals of Table 1 by using the suggested substitution of Table 4.

In some cases you will be given the substitution to use, as the following example shows.

Example 37

By using the substitution $u = a\sinh(\theta)$, show that

$$\int \frac{du}{\sqrt{a^2 + u^2}} = \sinh^{-1}\left(\frac{u}{a}\right) + C$$

This is result 8.28 of Table 1.

Solution

Differentiating $u = a\sinh(\theta)$ gives

$$\frac{du}{d\theta} = a\cosh(\theta) \quad \left[\text{By } 6.25 \right]$$

$$du = a\cosh(\theta)d\theta$$

Substituting $u = a\sinh(\theta)$ and $du = a\cosh(\theta)d\theta$ into the integral gives

$$\int \frac{du}{\sqrt{a^2 + u^2}} = \int \frac{a\cosh(\theta)d\theta}{\sqrt{a^2 + (a\sinh(\theta))^2}}$$

[†]
$$= \int \frac{a\cosh(\theta)d\theta}{\sqrt{a^2 + a^2\sinh^2(\theta)}}$$

The denominator $\sqrt{a^2 + a^2\sinh^2(\theta)}$ simplifies to

$$\sqrt{a^2 + a^2\sinh^2(\theta)} = \sqrt{a^2\underbrace{[1 + \sinh^2(\theta)]}_{=\,\cosh^2(\theta)}} = \sqrt{a^2\cosh^2(\theta)} = a\cosh(\theta)$$

6.25 $[\sinh(\theta)]' = \cosh(\theta)$

Example 37 *continued*

Replacing this in ⬛ † ⬛ gives

$$\int \frac{a\cosh(\theta)\mathrm{d}\theta}{a\cosh(\theta)} = \int \mathrm{d}\theta \quad [\text{Cancelling } a\cosh(\theta)]$$

$$= \theta + C \quad [\text{Integrating}]$$

Thus we have

⬛ * ⬛ $$\int \frac{\mathrm{d}u}{\sqrt{a^2 + u^2}} = \theta + C$$

We need to replace θ. From the given substitution

$$a\sinh(\theta) = u$$

we have

$$\sinh(\theta) = \frac{u}{a}$$

Take inverse sinh, $\sinh^{-1}$, of both sides:

$$\theta = \sinh^{-1}\left(\frac{u}{a}\right)$$

We have our result by substituting this into ⬛ * ⬛:

$$\int \frac{\mathrm{d}u}{\sqrt{a^2 + u^2}} = \sinh^{-1}\left(\frac{u}{a}\right) + C$$

SUMMARY

For integrating some trigonometric functions we select a relevant trigonometric identity and then integrate.
Some of the standard integrals of Table 1 can be established by using an appropriate substitution.

Exercise 8(h)

Solutions at end of book. Complete solutions available at www.palgrave.com/science/engineering/singh

⬛ **Questions 1 to 5 are in the field of [*electrical principles*].**

1 The average power, P, of an a.c. circuit is given by

$$P = \frac{\omega}{2\pi} \int_0^{\frac{2\pi}{\omega}} \left(i^2 R\right)\mathrm{d}t$$

where $i = I\sin(\omega t)$, ω is angular frequency, I is peak current, R is resistance and t is time. Show that

$$P = \frac{I^2 R}{2}$$

2 The voltage, v, produced by an electronic circuit is given by

$$v = 10\sin(\omega t)$$

The root mean square value, V_{RMS}, is defined as

$$V_{\text{RMS}} = \sqrt{\frac{\omega}{2\pi} \int_0^{\frac{2\pi}{\omega}} v^2\,\mathrm{d}t}$$

Evaluate V_{RMS}.

Solutions at end of book. Complete solutions available at www.palgrave.com/science/engineering/singh

3 The average power, P, in an a.c. circuit is given by

$$P = \frac{\omega}{2\pi} \int_0^{2\pi/\omega} (vi)dt$$

where $i = I\sin(\omega t)$ and $v = V\sin(\omega t)$.

(V and I are peak voltage and peak current respectively, ω is angular frequency and t is time.)

Show that $P = \dfrac{VI}{2}$.

4 The energy, W, of an inductance, L, is defined as

$$W = \frac{1}{2L}\left(\int_0^t Vdt + 1\right)^2$$

where V is the voltage across the inductor. For $L = 1 \times 10^{-3}$ henry and $V = \cos(t) - \sin(t)$, show that

$$W = 500[1 + \sin(2t)]$$

5 For $v = V\sin(\omega t)$ and $i = I\sin(\omega t + \phi)$, show that the power, P, in an a.c. circuit is given by

$$P = \frac{VI}{2}\cos(\phi)$$

where ϕ is the phase and P is as defined in question **3**.

6 Show that

$$\int \frac{du}{u\sqrt{u^2 - a^2}} = \frac{1}{a}\sec^{-1}\left(\frac{u}{a}\right) + C.$$

7 Show that $\displaystyle\int \frac{du}{\sqrt{a^2 - u^2}} = \sin^{-1}\left(\frac{u}{a}\right) + C.$

Solutions at end of book. Complete solutions available at www.palgrave.com/science/engineering/singh

1 [mechanics]   The kinetic energy, KE, of a body is given by

$$KE = \int_0^v (mv)dv$$

where m is the mass of the body and v is the velocity. Evaluate KE.

2 [mechanics]   The work done, W, to stretch a spring by length l is given by

$$W = \int_0^l \frac{EAx}{L}dx \quad \left(\frac{EA}{L} \text{ is constant}\right)$$

where E is the modulus of elasticity, A is the cross-sectional area, L is the unstretched length and x is the displacement. Determine W.

3 [mechanics]   The work done, W, to stretch a spring from length x_1 to length x_2 is given by

$$W = \int_{x_1}^{x_2} (kx)dx$$

where k is the stiffness constant of the spring. Evaluate W.

4 [mechanics]   Figure 6 shows a belt wound partly around a pulley. The angle θ is given by

$$\theta = \int_{T_1}^{T_2} \frac{dT}{\mu T}$$

Fig. 6

where μ is the coefficient of friction between the pulley and the belt and T_1, T_2 are tensions as shown. Show that

$$T_2 = T_1 e^{\mu\theta}$$

5 [structures]   The work done, W, (by external forces) on a cantilever beam of length L with uniformly distributed load q and a concentrated load p at the free end is given by

$$W = \int_0^L \frac{pq}{6EI}(2L^3 x - 3L^2 x^2 + x^4)dx$$

(EI is the flexural rigidity and x is the distance along the beam).
Show that

$$W = \frac{pqL^5}{30EI}$$

6 [materials]   The polar second moment of area, J, of a hollow circular shaft of outer diameter D_o and inner diameter D_i is given by

$$J = \int_{D_i/2}^{D_o/2} 2\pi r^3 dr$$

Show that

$$J = \frac{\pi}{32}\left[D_o^4 - D_i^4 \right]$$

7 Find $\displaystyle\int \frac{x^3}{x^2 - 4}\, dx$.

8 [thermodynamics]   The change in specific enthalpy, Δh, of a gas is given by

$$\Delta h = \int_{200}^{1000}\left(1.8 + (12 \times 10^{-3})T \right)dT$$

Evaluate Δh.

9 [thermodynamics]   The enthalpy change, Δh, is given by

$$\Delta h = \int_{T_1}^{T_2} C_p dT$$

where $C_p = R(a + bT + cT^2 + dT^3 + eT^4)$. ($R$, a, b, c, d and e are constants.) Find an expression for Δh.

10 Find $\displaystyle\int \frac{\cos(2x)}{\cos^2(x)\sin^2(x)}\, dx$.

11 [electronics]   The potential, V, at the surface of a conductor at distance r from zero potential at distance z, is given by

$$V = -\int_z^r \frac{q}{2\pi \varepsilon_0 r}\, dr \quad (r \neq 0)$$

where q is the charge and ε_0 is the permittivity constant. Show that

$$V = \frac{q}{2\pi \varepsilon_0}\, \ln\!\left(\frac{z}{r}\right)$$

12 [thermodynamics]   The work done, W, by a gas is given by

$$W = \int_{V_1}^{V_2} P dV$$

where P is the pressure, V is the volume, V_1 and V_2 are the initial and final volumes of the gas respectively. If the gas obeys the law $PV = C$, where C is constant, show that $W = PV \ln\!\left(\dfrac{V_2}{V_1}\right)$.

13 [thermodynamics]   The work done, W, on a piston by a gas in a cylinder is given by $W = \displaystyle\int_{V_1}^{V_2} P dV$ where P is pressure and V is volume. If $PV^{1.32} = C$ (constant), find an expression for W.

14 In Fig. 7:

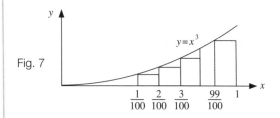

Fig. 7

i Evaluate the total area of the 100 rectangles under the graph $y = x^3$.

ii Determine $\int_0^1 x^3 dx$.

iii Find the difference between the area of part **i** and $\int_0^1 x^3 dx$. How can this difference be made smaller?

(*Hint*:

$$1^3 + 2^3 + 3^3 + \ldots + n^3 = \frac{n^2}{4}(n + 1)^2$$

where n is a positive whole number.)

15 〰 [*fluid mechanics*]   The pressure, P, at a distance x of a fluid flowing around a sphere of radius r is given by

$$P = \int_{-\infty}^x k\left(\frac{1 + r^3/x^3}{x^4}\right) dx$$

where k is a constant. Evaluate P.

16 🏛 [*structures*]   A beam of length L has a uniform load w given by

$$w = w_0 \sin\left(\frac{\pi x}{L}\right)$$

where w_0 is a constant and x is the distance along the beam. The total load P and reaction R are given by

$$P = \int_0^L w dx \text{ and } R = \int_0^L (wx)\, dx$$

Show that **i** $P = \dfrac{2w_0 L}{\pi}$ and

ii $R = \dfrac{w_0 L^2}{\pi}$.

17 📺 [*electronics*]   In a RC (resistor capacitor) network the charging voltage w and the instantaneous voltage v are related by $\dfrac{t}{RC} = \displaystyle\int_{-w}^v \dfrac{dv}{w - v}$.

Show that $v = w(1 - 2e^{-t/RC})$.

18 🔲 [*electrical principles*]   The current, i, through an inductor of inductance $L = 10$ mH is given by

$$i = \frac{1}{L} \int_0^t v dx$$

where v represents voltage. The voltage v across the inductor is given by

$$v = -6e^{-(2 \times 10^3)x} + 10e^{-(8 \times 10^3)x}$$

Determine the current i.

19 🔲 [*electrical principles*]   The current, i, through an inductor of inductance L is given by

$$i = \frac{1}{L} \int_0^t v dt$$

where v is the voltage across the inductor. The energy, w, of an inductance L, is defined to be

$$w = \frac{1}{2}Li^2$$

For $L = 5$ mH and $v = 10\sin(100\pi t)$, find an expression for

i i **ii** w

20 🔳 [*electromagnetism*]   The flux, Φ, for a toroid of height, h, inner radius, a, and outer radius, b, is given by

$$\Phi = \int_a^b \frac{\mu i N h}{2\pi r} dr \quad (r \neq 0)$$

where $a < r < b$, N is the number of turns, μ is the permeability constant and i is the current carried by the toroid coil.

If $L = N\dfrac{d\Phi}{di}$,

show that $L = \dfrac{\mu h N^2}{2\pi}\ln\left(\dfrac{b}{a}\right)$.

21 Evaluate $\displaystyle\int_0^1 \frac{4}{1 + x^2} dx$.

Solutions at end of book. Complete solutions available at www.palgrave.com/science/engineering/singh

22 [mechanics] The moment of inertia, I, of a triangular lamina of height h, base b and mass m is defined as

$$I = \int_0^h \frac{2m(h - y)y^2}{h^2}\,dy$$

where y is the distance from the axis of rotation. Show that $I = \dfrac{mh^2}{6}$.

23 [structures] A cantilever beam of length L, fixed at one end and deflected by a distance D at the free end has strain energy V given by

$$V = \frac{EI}{2}\int_0^L \left(\frac{d^2 y}{dx^2}\right)^2 dx$$

where EI is the flexural rigidity. The deflection y at a distance x from the fixed end is given by

$$y = D\left[1 - \cos\left(\frac{\pi x}{2L}\right)\right]$$

Find V.

24 [communication systems] The total power, p, of an antenna of length L, carrying a peak current I at a distance r, is given by

$$p = \int_0^\pi \frac{\eta I^2\, L^2\, \pi \sin^3(\theta)d\theta}{4\lambda^2}$$

(λ = wavelength, η = space impedance and θ = angle).

Show that $p = \dfrac{\eta I^2\, L^2\pi}{3\lambda^2}$.

25 Sketch the area represented by $\displaystyle\int_1^e \frac{1}{x}\,dx$ and show that

$$\int_1^e \frac{1}{x}\,dx = 1$$

26 i By using a computer algebra system, or a graphical calculator, plot $y = (1 - x^2)^{1/2}$.

ii Evaluate $\displaystyle\int_{-1}^1 (1 - x^2)^{1/2}dx$.

27 Show that

$$\int_0^2 (4 - x^2)^{1/2}dx = \pi$$

28 By using the substitution $u = a\cosh(\theta)$, show that

$$\int \frac{du}{\sqrt{u^2 - a^2}} = \cosh^{-1}\left(\frac{u}{a}\right) + C$$

29 By substituting $u = a\tanh(\theta)$, show that

$$\int \frac{du}{a^2 - u^2} = \frac{1}{a}\tanh^{-1}\left(\frac{u}{a}\right) + C$$

30 The Fourier coefficient, a_n, for a triangular waveform is given by

$$a_n = \frac{1}{\pi^2\, n^2}\int_0^{2\pi} \omega t\cos(n\omega t)d(\omega t)$$

where n is a positive whole number. Show that $a_n = 0$.

For questions 31 and 32 use a computer algebra system or a graphical calculator.

31 [thermodynamics] The rate of heat flow, q, is given by

$$q = 300\int_{306}^{853} [3.7 - (1.9 \times 10^{-3})T$$
$$+ (4.7 \times 10^{-6})T^2 - (3.45 \times 10^{-11})T^3$$
$$+ (8.5 \times 10^{-13})T^4]dT$$

Evaluate q.

Solutions at end of book. Complete solutions available at www.palgrave.com/science/engineering/singh

32 [*thermodynamics*]   Determine the specific enthalpy change, Δh, for the following:

a $\Delta h = \displaystyle\int_{400}^{1000}\left[56 - \left(\dfrac{22 \times 10^3}{T^{0.75}}\right)\right.$

$\left. + \left(\dfrac{116 \times 10^3}{T}\right) - \left(\dfrac{561 \times 10^3}{T^{1.5}}\right)\right]\mathrm{d}T$

b $\Delta h = \displaystyle\int_{400}^{1000}\left[81 - 18.66T^{0.25}\right.$

$\left. + 0.54T^{0.75} - 0.04T\right]\mathrm{d}T$

c $\Delta h = \displaystyle\int_{400}^{1000}\left[143 - 58T^{0.25}\right.$

$\left. + 8.3T^{0.5} - (37 \times 10^{-3})T\right]\mathrm{d}T$

d $\Delta h = \displaystyle\int_{400}^{1000}\left[(2.42 \times 10^{-6})T^2\right.$

$\left. - (41 \times 10^{-3})T + 3.05T^{0.5} - 3.7\right]\mathrm{d}T$

(In each of these examples, T represents temperature and Δh is the specific enthalpy change of the gas being heated from 400 K to 1000 K. Also the units of Δh are kJ/kmol.)

CHAPTER 9

Engineering Applications of Integration

SECTION A **Trapezium rule**

By the end of this section you will be able to:
▶ understand the need for numerical methods
▶ apply the trapezium rule
▶ determine the discrepancy between the exact and approximate solutions

A1 **Introduction to numerical integration**

The advancement of technology has had a profound effect in engineering mathematics. The emphasis of many university engineering mathematics syllabi is growing towards numerical methods in solving problems rather than analytical techniques. The reason for this is that most realistic engineering problems require a numerical method because the problem cannot be solved analytically. **What do we mean by analytical techniques?** Analytical techniques are based on algebraic methods. All the integration examples we have done so far have involved analytical techniques such as integration by parts, substitution, partial fractions, etc. With these techniques you obtain an **exact** answer.

What does a numerical method involve?

Numerical method is based on arithmetic operations. This normally involves tedious calculations, so the use of technology is essential. Generally with numerical methods your final result is an **approximation**.

Why do we bother with an approximation when you can achieve an exact answer?

Most pragmatic engineering problems cannot be solved by analytical methods. For example, the analysis of experimental data normally requires a numerical approach because a formula for the relationship between input and results in experiments simply isn't known – we don't have an $f(x)$ rule. Additionally in many engineering problems an approximate solution is adequate. With the advancement of technology these numerical methods are becoming more important than the algebraic techniques.

Can you think of any examples of numerical methods?

The Newton–Raphson method for solving equations is an example of a numerical method. In sections A and B we evaluate integrals with limits by using numerical methods such as the trapezium rule and Simpson's rule.

A2 **Trapezium rule**

What is the area represented by $\int_a^b y\,\mathrm{d}x$**, where** $y = f(x)$**?**

It is the area under the curve $y = f(x)$ and the x axis between a and b.

We can approximate this area, $\int_a^b y \, dx$, by blocks of trapezia. Consider the shaded block of Fig. 1.

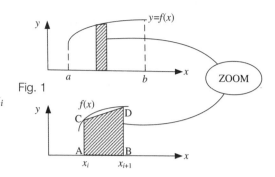

Fig. 1

The area under the curve $y = f(x)$ between x_i and x_{i+1} can be approximated by the area

? of the trapezium ABCD. **What is the area of the trapezium?**

It is equal to

$$\frac{1}{2}AB \times (AC + BD)$$

Let's rewrite $\frac{1}{2}AB \times (AC + BD)$. If we use h as the distance AB and y_i as the value of $f(x_i)$, that is height AC, and y_{i+1} as the value of $f(x_{i+1})$, that is height BD (Fig. 2),

? *then* **what is the formula for the area of the trapezium ABCD?**

Fig. 2

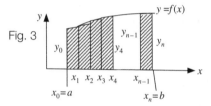

$$\text{Area ABCD} = \frac{1}{2}h(y_i + y_{i+1})$$

Let's consider the whole area under $y = f(x)$ between a and b. Cut this area into n blocks of equal width h and height y_i at the point x_i (Fig. 3).

Fig. 3

? **What is the area of the first block, between x_0 and x_1?**

The heights are y_0 and y_1, therefore

$$\text{area of first block} = \frac{1}{2} h(y_0 + y_1)$$

? **What is the area of the second block between x_1 and x_2?**

$$\frac{1}{2}h(y_1 + y_2)$$

The area of the nth block, between x_{n-1} and x_n is

$$\frac{1}{2}h(y_{n-1} + y_n)$$

? **How do we find the total area of these trapezia under the curve $y = f(x)$?**

Add all these blocks together.

$$\text{Total area} = \frac{1}{2} h(y_0 + y_1) + \frac{1}{2} h(y_1 + y_2) + \frac{1}{2} h(y_2 + y_3) + \ldots + \frac{1}{2}h(y_{n-1} + y_n)$$

$$= \frac{1}{2}h\left[\underbrace{(y_0 + y_1)}_{} + \underbrace{(y_1 + y_2)}_{= 2y_1} + \underbrace{(y_2 + y_3)}_{= 2y_2} + \ldots + (y_{n-1} + y_n)\right]$$

taking out the
common factor of
$\frac{1}{2}h$

$$= \frac{1}{2}h\,[y_0 + 2y_1 + 2y_2 + \ldots + 2y_{n-1} + y_n]$$

Remember the area under the curve is represented by $\int_a^b y\,dx$. So we have

9.1
$$\int_a^b y\,dx \approx \frac{h}{2}[y_0 + 2(y_1 + y_2 + \ldots + y_{n-1}) + y_n]$$

? **Why is there an approximation sign, ≈, in formula 9.1 ?**

Because near the edge of the curve we have the
situation shown in Fig. 4.

Fig. 4 $f(x)$ —————→   ←——— Trapezium

Generally the area under the curve, $\int_a^b y\,dx,$

cannot be evaluated exactly by the trapezia. Remember that y_i represents the height of the
given function at the point x_i, this y_i is called the **ordinate**. Formula 9.1 can also be
written as:

9.2   Area $\approx \dfrac{\text{width of block}}{2}$ [(first ordinate) + (last ordinate) + 2(remaining ordinates)]

These formulae, 9.1 and 9.2 , are known as the trapezium rule.

In Examples 1 and 2 below, the analytical integration is a lot easier than it would be by
applying the trapezium rule. However in these examples we can compare exact and
approximate evaluations of the integral.

Example 1

i Evaluate $\int_0^1 x^2\,dx$ by using the trapezium rule with four intervals.

ii Find the exact value of $\int_0^1 x^2\,dx$

iii Determine the discrepancy between the exact and the value obtained by the
trapezium rule, **i**.

Solution

i We need to apply

9.1
$$\int_b^a y\,dx \approx \frac{h}{2}[y_0 + 2(y_1 + y_2 + \ldots + y_{n-1}) + y_n]$$

Example 1 *continued*

? **What are the values of a, b and h for this example?**

The lower limit $a = 0$ and therefore $b = 1$.
Remember that h is the uniform width between 0 and 1. Since this is split into four equal intervals, we have the situation shown in Fig. 5.

Fig. 5

$$h = \frac{1-0}{4} = \frac{1}{4}$$

$x_0 = 0 \quad x_1 = \frac{1}{4} \quad x_2 = \frac{1}{2} \quad x_3 = \frac{3}{4} \quad x_4 = 1$

? **What is y?**

$y = x^2$ (the function that needs to be integrated)

? **What else do we require in order to use 9.1 ?**

We need to obtain the values of $y_0, y_1, y_2, \ldots$ These are determined by y at $x_0, x_1, x_2, \ldots$ respectively. Thus

$$x_0 = 0 \qquad\qquad y_0 = 0^2 = 0$$

$$x_1 = \frac{1}{4} \qquad\qquad y_1 = \left(\frac{1}{4}\right)^2 = \frac{1}{16}$$

$$x_2 = \frac{1}{2} \qquad\qquad y_2 = \left(\frac{1}{2}\right)^2 = \frac{1}{4}$$

$$x_3 = \frac{3}{4} \qquad\qquad y_3 = \left(\frac{3}{4}\right)^2 = \frac{9}{16}$$

$$x_4 = 1 \qquad\qquad y_4 = 1^2 = 1$$

Putting all these into the trapezium rule gives

$$\int_0^1 x^2\, dx \approx \frac{h}{2}\, [y_0 + 2(y_1 + y_2 + y_3) + y_4]$$

$$= \frac{1/4}{2}\left[0 + 2\left(\frac{1}{16} + \frac{1}{4} + \frac{9}{16}\right) + 1\right] = \frac{11}{32}$$

With the trapezium rule, $\int_0^1 x^2\, dx$ is $\dfrac{11}{32}$.

? **ii What is the exact value of $\int_0^1 x^2\, dx$?**

$$\int_0^1 x^2\, dx \underset{\text{by } 8.1}{=} \left[\frac{x^3}{3}\right]_0^1 = \left(\frac{1}{3} - 0\right) = \frac{1}{3}$$

The exact value is 1/3.

iii The discrepancy between the exact and approximate (trapezium) value is

$$\frac{1}{3} - \frac{11}{32} = -\frac{1}{96}$$

8.1 $\int x^n\, dx = x^{n+1}/n + 1$

Figure 6 is the graphical representation of the trapezium rule with four intervals applied to $\int_0^1 x^2 \, dx$.

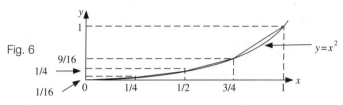

Fig. 6

We can compare exact and approximate evaluations of the integral by determining the percentage error. The percentage error is given by

9.3 $\%\ error = \left(\dfrac{exact - approximation}{exact}\right) \times 100$

$\left(= \dfrac{discrepancy}{exact} \times 100\right)$

In **Example 1**:

$$\%\ error = \frac{-1/96}{1/3} \times 100 = -3.125\%$$

The negative sign indicates that the exact value is smaller than the approximate value as seen in Fig. 6. In some engineering cases an error of more than 3% may be too large. **How can we achieve a better approximation with the trapezium rule?**

You could consider more intervals such as 8, 16, 32 etc.

Yes, but what penalty do we pay if we make an allowance for a large number of intervals?

More calculations are involved and of course in many cases it will not be feasible by hand. You either need to use a graphical calculator or a computer algebra system such as MAPLE. Let's repeat **Example 1** using eight intervals, without a computer algebra system.

Example 2

Evaluate $\int_0^1 x^2 \, dx$ using the trapezium rule with eight intervals, and find the discrepancy from the exact value.

Solution

Since the interval 0 to 1 is sliced into eight equal blocks, the uniform width, h, is given by

$$h = \frac{1-0}{8} = \frac{1}{8}$$

In order to use the trapezium formula:

$$\int_a^b y \, dx \approx \frac{h}{2}\left[y_0 + 2(y_1 + y_2 + \ldots + y_{n-1}) + y_n\right]$$

Example 2 continued

we need to determine

$$x_0 = 0 \qquad\qquad y_0 = 0^2$$

$$x_1 = \frac{1}{8} \qquad\qquad y_1 = \left(\frac{1}{8}\right)^2$$

$$x_2 = \frac{2}{8} \qquad\qquad y_2 = \left(\frac{2}{8}\right)^2$$

$$\vdots \qquad\qquad\qquad \vdots$$

Thus we have

$$\int_0^1 x^2 \, dx \approx \frac{1/8}{2}\left\{0 + 2\left[\left(\frac{1}{8}\right)^2 + \left(\frac{2}{8}\right)^2 + \left(\frac{3}{8}\right)^2 + \left(\frac{4}{8}\right)^2 + \left(\frac{5}{8}\right)^2 + \left(\frac{6}{8}\right)^2 + \left(\frac{7}{8}\right)^2\right] + 1\right\}$$

$$= \frac{43}{128}$$

The exact value is $\frac{1}{3}$, therefore

$$\text{discrepancy} = \frac{1}{3} - \frac{43}{128} = -\frac{1}{384}$$

and

$$\% \text{ error} = -\frac{1/384}{1/3} \times 100 = -0.781\%$$

compared with a percentage error of more than -3% for four intervals. Table 1 shows how the trapezium rule evaluation gets closer and closer to the exact value, $\frac{1}{3} = 0.33333\ldots$, as we consider more and more intervals.

TABLE 1	Number of intervals	4	8	16	32	64
	$\int_0^1 x^2 \, dx$ (trapezium rule (5 d.p.))	0.34375	0.33594	0.33398	0.33349	0.33337
	Percentage error	−3.125	−0.782	−0.194	−0.047	−0.011

The integral $\int_0^1 x^2 \, dx$ was comfortably evaluated by using an analytical method. However the trapezium rule (and other numerical methods) are effective when the solution cannot be evaluated, or is difficult to evaluate, analytically. Let's consider an example which involves experimental data.

A3 **Engineering applications**

Example 3 *structures*

Fig. 7

		0	1.2	2.4	3.6	4.8	6.0	7.2	8.4	9.6	10.8	12.0
TABLE 2	x (m)	0	1.2	2.4	3.6	4.8	6.0	7.2	8.4	9.6	10.8	12.0
	w (kN/m)	2.6	3.8	5.6	7.1	9.3	13.7	15.0	10.7	8.1	6.2	3.5

Figure 7 and Table 2 show the loading on a beam of length 12 m and the corresponding figures for the loading in kN/m. The total force, F, exerted by the load is given by

$$F = \int_0^{12} w \, dx$$

Estimate F.

Solution

We cannot use analytical methods. Why not?

We are not given an algebraic function but a set of data values, so we are compelled to use a numerical method. For the trapezium rule:

9.1
$$\int_a^b y \, dx \approx \frac{h}{2}\left[y_0 + 2\left(y_1 + y_2 + \ldots + y_{n-1}\right) + y_n\right]$$

We have to find a, b, h, y_0, y_1 etc.

We know $a = 0$, $b = 12$ and $h = 1.2$, and remember that h is the uniform width.

But what are the values of $y_0, y_1, y_2, \ldots$?

These can be read off from the second row of Table 2:

$$y_0 = 2.6, \quad y_1 = 3.8, \quad y_2 = 5.6, \text{ etc.}$$

Thus we have

$$\int_0^{12} w \, dx \approx \frac{1.2}{2} \, [2.6 + 2\{3.8 + 5.6 + 7.1 + 9.3 + 13.7 + 15.0 + 10.7 + 8.1 + 6.2\} + 3.5]$$

$$= 99.06$$

The total force, $F \approx 99.06$ kN

SUMMARY

Generally numerical methods give an approximate solution. One numerical method to determine integrals with limits is the trapezium rule:

9.1 $$\int_a^b y \, dx \approx \frac{h}{2}\left[y_0 + 2(y_1 + y_2 + y_3 + \ldots) + y_n\right]$$

This can also be written as

9.2 $$\text{Area} \approx \frac{\text{width of block}}{2}\left[(\text{first ordinate}) + (\text{last ordinate}) + 2(\text{remaining ordinates})\right]$$

Exercise 9(a)

Solutions at end of book. Complete solutions available at www.palgrave.com/science/engineering/singh

1 🚗 [*mechanics*] A force, F, acting on a particle varies with time, t, according to the table below.

t (s)	0	0.5	1.0	1.5	2.0	2.5	3.0
F (N)	3.2	5.6	7.0	7.7	8.4	9.9	11.6

The impulse of this force is given by

$$\int_0^3 F \, dt$$

Find an approximate value for the impulse.

2 Use the trapezium rule with four equal intervals to find the approximate values of

a $\int_0^1 e^{-x^2} \, dx$ **b** $\int_0^{\pi/2} \sqrt{\cos(x)} \, dx$

3 **i** Apply the trapezium rule to find the approximate value of

$$\int_0^1 x^3 \, dx$$

with **a** four equal intervals
b eight equal intervals

ii Find the exact value of $\int_0^1 x^3 \, dx$

iii Determine the percentage error between the exact value and the estimated values found in **i**.

iv Comment upon your results to **iii**.

4 🚗 [*mechanics*] The velocity, v, of a model for a new ship is tested in an experiment for 5 seconds and has the values shown in the table below.

t (s)	0	1	2	3	4	5
v (m/s)	2.10	9.56	11.36	12.08	12.98	13.76

The distance travelled is given by

$$\int_0^5 v \, dt$$

Estimate the distance.

5 〰 [*fluid mechanics*] A river is 15 m wide. The depth of river is found in metres from one side of the embankment to the other with the results shown in the table below.

Dist. (m)	0	1.5	3.0	4.5	6.0	7.5
Depth (m)	0	1.04	1.65	3.10	4.66	4.12

Dist. (m)	9.0	10.5	12.0	13.5	15.0
Depth (m)	3.21	2.33	1.78	0.76	0

Given that the velocity of water is 2.05 m/s, find the approximate number of cubic metres of water flowing down the river per second.

(*Hint*: Volume of flow per second = cross-sectional area × velocity.)

Solutions at end of book. Complete solutions available at www.palgrave.com/science/engineering/singh

Exercise **9(a) continued**

6 🔲 [*electrical principles*]   A voltage, v, has the values at intervals of 0.1 s shown in the table below.

t (s)	0	0.1	0.2	0.3	0.4	0.5	0.6
v (volts)	4	3.92	3.86	3.77	3.61	3.52	3.41

The mean voltage, $\bar{v}$, is given by

$$\bar{v} = \frac{1}{0.6}\int_0^{0.6} v\, dt.$$ Find an approximate value for $\bar{v}$.

SECTION B **Further numerical integration**

By the end of this section you will be able to:
▶ apply Simpson's rule
▶ use a computer algebra system to determine definite integrals

B1 **Simpson's rule**

In this section we state the formula of another numerical method used to find a value of an integral with limits – Simpson's rule.

In Simpson's rule the curve $y = f(x)$ is approximated by a number of quadratic curves. The formula is defined as

Simpson's rule is named after **Thomas Simpson** (1710–61) who was born in Leicestershire, England and was a private tutor in mathematics. He is primarily known for his work in numerical integration.

$$\boxed{9.4}\qquad \int_a^b y\, dx \approx \frac{h}{3}\left[y_0 + 4(y_1 + y_3 + y_5 + \cdots) + 2(y_2 + y_4 + y_6 + \cdots) + y_n\right]$$

where h is the uniform width, $y_0, y_1, y_2, \cdots, y_n$ are the ordinates shown in Fig. 8 and n is the number of blocks, which must be **even**.

❓ **What do you notice about the ordinates $y_0, y_1, \cdots, y_n$ of formula $\boxed{9.4}$?**

Formula $\boxed{9.4}$ can also be written as

Fig. 8

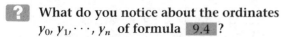

$$\boxed{9.5}\qquad \text{Area} \approx \frac{h}{3}\left[\begin{array}{l}\text{(first ordinate)} + 4(\text{sum of odd ordinates}) \\ + 2(\text{sum of even ordinates}) + (\text{last ordinate})\end{array}\right]$$

Note that to use formulae 9.4 or 9.5 , the number of blocks, n, must be even.

Let's repeat **Example 1** with Simpson's rule.

Example 4

Evaluate $\int_0^1 x^2 \, dx$ using Simpson's rule with four intervals and find the percentage error with the exact value. Comment on your result for percentage error.

Solution

? **Can we use Simpson's rule?**

Yes, since we have an even number (4) of intervals. Putting $n = 4$ into Simpson's rule gives

$$\int_0^1 x^2 \, dx \approx \frac{h}{3}\left[y_0 + 4(y_1 + y_3) + 2y_2 + y_4 \right]$$

? **What is the value of the uniform width h?**

We require four equal intervals, therefore

$$h = \frac{1 - 0}{4} = \frac{1}{4}$$

? **What else do we need to find in order to use ★ ?**

The ordinates y_0, y_1, y_2, y_3 and y_4. With $y = x^2$ we have the same y values as those in **Example 1**:

$$x_0 = 0 \qquad\qquad y_0 = 0^2 = 0$$
$$x_1 = \frac{1}{4} \qquad\qquad y_1 = \left(\frac{1}{4}\right)^2 = \frac{1}{16}$$
$$x_2 = \frac{2}{4} = \frac{1}{2} \qquad\qquad y_2 = \left(\frac{1}{2}\right)^2 = \frac{1}{4}$$
$$x_3 = \frac{3}{4} \qquad\qquad y_3 = \left(\frac{3}{4}\right)^2 = \frac{9}{16}$$
$$x_4 = 1 \qquad\qquad y_4 = 1^2 = 1$$

Putting all these into ★ gives

$$\int_0^1 x^2 \, dx \approx \frac{h}{3}\left[y_0 + 4(y_1 + y_3) + 2y_2 + y_4 \right]$$
$$= \frac{1/4}{3}\left[0 + 4\left(\frac{1}{16} + \frac{9}{16}\right) + 2\left(\frac{1}{4}\right) + 1 \right] = \frac{1}{3}$$

The exact value (evaluated in **Example 1**) of $\int_0^1 x^2 \, dx = \frac{1}{3}$. So

$$\text{percentage error} = 0\%$$

We arrive at the exact answer because Simpson's rule approximates the area by considering **quadratic curves**. Our function, x^2, that needs integrating is also a quadratic, hence there is no error.

Example 5

Find the approximate value of

$$\int_0^{\pi/2} \sqrt{\cos(x)}\, dx$$

by using Simpson's rule with eight equal intervals and working to four decimal places.

Solution

Applying Simpson with $n = 8$ we have

$$\boxed{\!\!*\!\!}\qquad \int_0^{\pi/2} \sqrt{\cos(x)}\, dx \approx \frac{h}{3}\left[y_0 + 4\left(y_1 + y_3 + y_5 + y_7\right) + 2\left(y_2 + y_4 + y_6\right) + y_8\right]$$

What is the value of h?

$$h = \frac{\pi/2 - 0}{8} = \frac{\pi}{16}$$

We can find the y values, as shown in Table 3.

TABLE 3	x	$\sqrt{\cos(x)}$
	0	$\sqrt{\cos(0)} = 1 = y_0$
	$\dfrac{\pi}{16}$	$\sqrt{\cos\left(\dfrac{\pi}{16}\right)} = 0.9903 = y_1$
	$\dfrac{2\pi}{16} = \dfrac{\pi}{8}$	$\sqrt{\cos\left(\dfrac{\pi}{8}\right)} = 0.9612 = y_2$
	$\dfrac{3\pi}{16}$	$\sqrt{\cos\left(\dfrac{3\pi}{16}\right)} = 0.9118 = y_3$
	$\dfrac{4\pi}{16} = \dfrac{\pi}{4}$	$\sqrt{\cos\left(\dfrac{\pi}{4}\right)} = 0.8409 = y_4$
	$\dfrac{5\pi}{16}$	$\sqrt{\cos\left(\dfrac{5\pi}{16}\right)} = 0.7454 = y_5$
	$\dfrac{6\pi}{16} = \dfrac{3\pi}{8}$	$\sqrt{\cos\left(\dfrac{3\pi}{8}\right)} = 0.6186 = y_6$
	$\dfrac{7\pi}{16}$	$\sqrt{\cos\left(\dfrac{7\pi}{16}\right)} = 0.4417 = y_7$
	$\dfrac{8\pi}{16} = \dfrac{\pi}{2}$	$\sqrt{\cos\left(\dfrac{\pi}{2}\right)} = 0 = y_8$

Substituting $h = \dfrac{\pi}{16}$ and the above y values into $\boxed{\!\!*\!\!}$ yields

Example 5 continued

$$\int_0^{\pi/2} \sqrt{\cos(x)}dx \approx \frac{h}{3}\left[y_0 + 4(y_1 + y_3 + y_5 + y_7) + 2(y_2 + y_4 + y_6) + y_8\right]$$

$$= \frac{\pi/16}{3}\left[1 + 4(0.9903 + 0.9118 + 0.7454 + 0.4417) \atop + 2(0.9612 + 0.8409 + 0.6186) + 0\right]$$

$$= \frac{\pi}{48}[18.1982] = 1.1911$$

Example 6 *mechanics*

The height, s, of a rocket in the first 6 seconds is given by

$$s = (5 \times 10^3)\int_0^6 \ln\left(\frac{100}{100 - 4t}\right)dt - 180$$

i Evaluate $\int_0^6 \ln\left(\frac{100}{100 - 4t}\right)dt$ by using Simpson's rule with six equal intervals.

ii Determine s

Solution

Let $y = \ln\left(\frac{100}{100 - 4t}\right)$

i Using

9.4 $$\int_a^b y\, dx \approx \frac{h}{3}\left[y_0 + 4(y_1 + y_3 + \cdots) + 2(y_2 + y_4 + \cdots) + y_n\right]$$

with six equal intervals, $n = 6$, and $dx = dt$ gives

* $$\int_0^6 y\, dt \approx \frac{h}{3}\left[y_0 + 4(y_1 + y_3 + y_5) + 2(y_2 + y_4) + y_6\right]$$

What is the value of the uniform width h?

$$h = \frac{6 - 0}{6} = 1$$

Thus $t_0 = 0, t_1 = 1, \cdots, t_6 = 6$.

What else do we require in order to use Simpson's rule?

We need to find $y_0, y_1, \cdots$ and y_6.

$$t_0 = 0 \qquad\qquad y_0 = \ln\left[\frac{100}{100 - (4 \times 0)}\right] = 0$$

$$t_1 = 1 \qquad\qquad y_1 = \ln\left[\frac{100}{100 - (4 \times 1)}\right] = 0.041$$

$$t_2 = 2 \qquad\qquad y_2 = \ln\left[\frac{100}{100 - (4 \times 2)}\right] = 0.083$$

Example 6 *continued*

$$t_3 = 3 \qquad y_3 = \ln\left[\frac{100}{100 - (4 \times 3)}\right] = 0.128$$

$$t_4 = 4 \qquad y_4 = \ln\left[\frac{100}{100 - (4 \times 4)}\right] = 0.174$$

$$t_5 = 5 \qquad y_5 = \ln\left[\frac{100}{100 - (4 \times 5)}\right] = 0.223$$

$$t_6 = 6 \qquad y_6 = \ln\left[\frac{100}{100 - (4 \times 6)}\right] = 0.274$$

Putting all this into ▨ * gives

$$\int_0^6 y\,dt \approx \frac{1}{3}\left[0 + 4(0.041 + 0.128 + 0.223) + 2(0.083 + 0.174) + 0.274\right] = 0.785$$

ii $s \approx [(5 \times 10^3) \times (0.785)] - 180 = 3745 \text{ m} = 3.75 \text{ km (3 s.f.)}$

Numerical methods are particularly useful if we have to integrate a function which is a set of experimental data.

Example 7

The shape of a piece of land is shown in Fig. 9. All dimensions are in metres.

By using Simpson's rule, estimate the area.

Solution

Can we use Simpson's rule? What do we need to check first?

Fig. 9

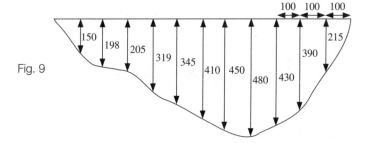

Check that there are an even number of blocks. In fact there are 12.

9.5 $\text{Area} \approx \dfrac{h}{3}\Big[(\text{first ordinate}) + 4(\text{sum of odd ordinates}) + 2(\text{sum of even ordinates}) + (\text{last ordinate})\Big]$

What is the value of the uniform width h?

From Fig. 9, $h = 100$

In order to use Simpson's rule, what else do we need to find?

First ordinate = 0. Last ordinate = 0.

Even ordinates are 198, 319, 410, 480 and 390.

Example 7 *continued*

Odd ordinates are 150, 205, 345, 450, 430 and 215.

Substituting these into ⬛9.5⬛ gives

$$\text{Area} \approx \frac{100}{3}\left[\begin{array}{l} 0 + 4(150 + 205 + 345 + 450 + 430 + 215) \\ \qquad + 2(198 + 319 + 410 + 480 + 390) + 0 \end{array}\right]$$

$$= 359\ 133.33\ \text{m}^2 = 3.59 \times 10^5\ \text{m}^2\ (3\ \text{s.f.})$$

B2 Integration using a computer algebra system

We can use a computer algebra system to apply Simpson's rule or the trapezium rule with a large number of intervals. **Example 8** uses MAPLE to evaluate a definite integral with 100 intervals. Even if you don't have access to a computer algebra system, try following the example.

Example 8 *electrical principles*

The average value of an a.c. current waveform, i_{avg}, is given by

$$i_{\text{avg}} = \frac{1}{2\pi}\int_0^{2\pi} \sqrt{|\sin(t)|}\,dt$$

By using MAPLE, determine an estimate for i_{avg} with 100 intervals using

a Simpson's rule   **b** the trapezium rule

Solution

```
> f:=abs (sqrt(sin(t)));
```
$f: = \sqrt{|\sin\,(t)|}$

```
> with (student):
```

```
> i[avg]:=evalf((1/(2*Pi))*simpson(f,t=0 .. 2*Pi,100));
```
$i_{\text{avg}}: = 0.7619458331$

```
> i[avg]:=evalf((1/(2*Pi))*trapezoid(f,t=0 .. 2*Pi,100));
```
$i_{\text{avg}}: = 0.7606753615$

The next example shows how we can evaluate a difficult definite integral on a computer algebra system (MAPLE).

Example 9 *signal processing*

The probability, P, that a signal is greater than 0.1 volt is given by

$$P = \frac{5}{\sqrt{2\pi}} \int_{0.1}^{\infty} \exp\left(-\frac{25x^2}{2}\right) dx$$

Determine P.

Solution

Remember that exp is the exponential, e. Using MAPLE:

```
> g:=exp(-25*x^2/2);
        g := e^(-25/2x²)
```

```
> f:=int(g, x=0.1 .. infinity);
        f: = 0.1546777837
```

```
> evalf(5/sqrt(2*Pi)*f);
        > 0.3085375386
```

SUMMARY

Simpson's rule is

9.4 $$\int_a^b y\,dx \approx \frac{h}{3}\left[y_0 + 4(y_1 + y_3 + y_5 + \cdots) + 2(y_2 + y_4 + \cdots) + y_n\right]$$

9.5 $$\text{Area} \approx \frac{h}{3}\left[(\text{first ordinate}) + 4(\text{sum of odd ordinates}) + 2(\text{sum of even ordinates}) + (\text{last ordinate})\right]$$

The above rules are particularly useful for functions which are given in experimental data form.

Normally it is more appropriate to use a computer algebra system to integrate difficult algebraic functions.

Exercise **9(b)**

Solutions at end of book. Complete solutions available at www.palgrave.com/science/engineering/singh

1 Evaluate $\int_0^{\pi/2} \sin(t)\,dt$ by applying Simpson's rule with ten equal intervals. (Work to 3 d.p. but give the final answer to 4 d.p.) Determine the exact value of $\int_0^{\pi/2} \sin(t)\,dt$. Find the percentage error from the exact value for the above case.

Exercise **9(b) continued**

Solutions at end of book. Complete solutions available at www.palgrave.com/science/engineering/singh

2 Evaluate $\displaystyle\int_0^{\pi/3} \sqrt{\sin(x)}\,dx$ to 3 d.p., using six equal intervals.

Use a computer algebra system to find $\displaystyle\int_0^{\pi/3} \sqrt{\sin(x)}\,dx$ by applying

a the trapezium rule,   **b** Simpson's rule

with **i** 10 intervals, **ii** 50 intervals,

iii 100 intervals, **iv** 1000 intervals.

3 🚗 [*mechanics*] A force, F, acting on a particle varies with time according to the following experimental data:

t (s)	2.0	2.1	2.2	2.3
F (N)	230.0	202.6	188.3	178.7

t (s)	2.4	2.5	2.6	2.7
F (N)	172.4	168.0	165.6	163.4

t (s)	2.8	2.9	3.0
F (N)	162.0	161.1	160.5

The impulse of this force from 2 to 3 s is given by

$$\int_2^3 F\,dt$$

Estimate this impulse.

x (m)	0	0.5	1.0	1.5	2.0
w (kN/m)	0	1.7	2.3	5.6	8.8

x (m)	2.5	3.0	3.5	4.0
w (kN/m)	7.3	4.9	3.2	2.5

4 ⛰️ [*structures*] The above table shows the loading, w, on a beam of length 4 m. The total force, F, exerted by the load is given by

$$F = \int_0^4 w\,dx$$

Estimate F.

5 🔌 [*electrical principles*] An a.c. current, i, varies with time, t, according to the following table:

t (s)	0	0.1	0.2	0.3	0.4	0.5	0.6
i (A)	0.0	0.25	0.31	0.43	0.37	0.20	0.0

The RMS current, i_{RMS}, is given by

$$i_{\text{RMS}} = \sqrt{\frac{1}{0.6}\int_0^{0.6} i^2\,dt}$$

Find an approximate value for i_{RMS}.

6 🌡️ [*thermodynamics*] The volume, V, and pressure, P, of a gas in a cylinder is measured in an experiment and the results are shown in the table below.

V (m³)	0.01	0.02	0.03
P (kPa)	560.73	237.86	140.33

V (m³)	0.04	0.05	0.06
P (kPa)	98.67	73.66	59.24

V (m³)	0.07	0.08	0.09
P (kPa)	48.73	42.31	35.20

The work done, W, by the gas on the face of the piston is given by

$$W = \int_{0.01}^{0.09} P\,dV$$

Find an approximate value for W.

Solutions at end of book. Complete solutions available at www.palgrave.com/science/engineering/singh

Dist. (m)	0	2.5	5.0	7.5
Depth (m)	0.3	1.78	2.05	2.96

Dist. (m)	10.0	12.5	15.0	17.5
Depth (m)	4.01	5.30	6.78	6.11

Dist. (m)	20.0	22.5	25.0
Depth (m)	4.35	1.76	0.4

7 [fluid mechanics]   A river is 25 m wide. The depth of river is measured at eleven equidistant points from one end, with the following results:
Given that the velocity of water is 1.67 m/s, find the approximate number of cubic metres of water flowing down the river per second.

(*Hint*: Volume of flow per second = cross-sectional area × velocity.)

Use a computer algebra system for the remaining questions in this exercise.

8 [structures]   The force, F, on a beam of length 5 m is given by

$$F = \int_0^5 \frac{e^x}{1 + \sqrt{x}}\, dx \quad \text{(where } x \text{ is distance)}$$

Evaluate F.

9 [signal processing]   The probability, P, that a signal is greater than 3 V is given by

$$P = \sqrt{\frac{2}{\pi}} \int_3^\infty e^{-2x^2}\, dx$$

Evaluate P.

10 [signal processing]   The Q function, $Q(x)$, in signal processing is defined as

$$Q(x) = \frac{1}{\sqrt{2\pi}} \int_x^\infty \exp\left(-\frac{t^2}{2}\right) dt$$

Evaluate $Q(1)$, $Q(2)$, $Q(3)$ and $Q(4)$.

11 [mechanics]   The time taken, t, for an object falling in air to reach a velocity of 20 m/s is given by

$$t = \int_0^{20} \frac{dv}{9.8 - 0.1v^{1.23}}$$

Evaluate t with 100 intervals by using
a Simpson's rule **b** the trapezium rule

SECTION C **Engineering applications**

By the end of this section you will be able to:
▶ determine the area under a curve
▶ find the area between curves
▶ find the average value of a function
▶ find the root mean square (RMS) value of a function

C1 **Area under a curve**

? How can we write the shaded area of Fig. 10 in terms of an integral?

9.6   Shaded area $= \displaystyle\int_a^b y\, dx$

Let's examine this formula by doing an example.

Fig. 10

Example 10

Find the area bounded by sin(x) and the x axis between $-\pi$ and π.

Solution

Using **9.6** we have:

$$\int_{-\pi}^{\pi} \sin(x)\ \mathrm{d}x \underset{\text{by }\boxed{8.7}}{=} \Big[-\cos(x)\Big]_{-\pi}^{\pi}$$

$$= -[\cos(\pi)-\cos(-\pi)] \qquad \text{[Substituting]}$$

$$= -[-1-(-1)] = 0$$

$$\int_{-\pi}^{\pi} \sin(x)\ \mathrm{d}x = 0$$

This says that the area between sin(x) and the x axis ($-\pi$ to π) is zero.

However Fig. 11 shows that there is an area $(A + B)$ between sin(x) and the x axis.

Fig. 11

? **What's gone wrong with our calculation which evaluates the area to zero?**

Consider the following:

For area A we have

$$\int_{-\pi}^{0} \sin(x)\ \mathrm{d}x = \Big[-\cos(x)\Big]_{-\pi}^{0}$$

$$= -[\cos(0) - \cos(-\pi)] = -[1 - (-1)] = -2$$

? **What does the negative sign in -2 for area A signify?**

The -2 for area A means that the area is two units below the x axis. For area B we have

$$\int_{0}^{\pi} \sin(x)\ \mathrm{d}x = \Big[-\cos(x)\Big]_{0}^{\pi}$$

8.7 $\displaystyle\int \sin(x)\mathrm{d}x = -\cos(x)$

Example 10 *continued*

$$= -\left[\cos(\pi) - \cos(0)\right] = -\left[-1 - 1\right] = 2$$

So the areas cancel each other out: $-2 + 2 = 0$

Thus the total area $= 2 + 2 = 4$

The purpose of **Example 10** was to show the need to sketch a graph before finding the area. First we draw the graph and then we integrate each positive and negative area separately. If you get a negative answer then the area is below the x axis. Consider each area to be positive and then add to find the total area.

Example 11

Calculate the area between the curve $y = \dfrac{2}{x}$ and the line $y = 3 - x$.

Solution

The graph of $y = 3 - x$ is a straight line crossing the y axis at 3 and the x axis at 3.

The graph of $y = \dfrac{2}{x}$ is the normal reciprocal graph which is asymptotic to the x and y axes. (Refer back to Chapter 2 if you need to remind yourself what these terms mean.)

Where do these graphs intersect?

Intersection occurs where $y = \dfrac{2}{x}$ and $y = 3 - x$ are equal. Thus

$$3 - x = \frac{2}{x}$$

$$3x - x^2 = 2 \qquad \text{[Multiplying by } x]$$

Collecting on one side:

$$x^2 - 3x + 2 = 0 \qquad \text{[Quadratic equation]}$$

$$(x - 1)(x - 2) = 0 \qquad \text{[Factorizing]}$$

$$x = 1, \quad x = 2$$

What are the y values at $x = 1$ and $x = 2$?

At $x = 1$, $y = \dfrac{2}{1} = 2$, the point $(1, 2)$. At $x = 2$, $y = \dfrac{2}{2} = 1$, the point $(2, 1)$.

Hence the points of intersection are $(1, 2)$ and $(2, 1)$. Thus we have the situation shown in Fig. 12.

Example 11 *continued*

? **How do we find the shaded area?**

We can calculate the area below the
curve by integration and the area below
the line is a trapezium, ABCD. The
difference between the two is therefore
the shaded area.

Fig. 12

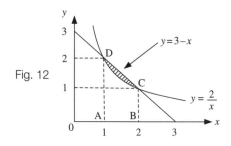

Area below $y = \dfrac{2}{x}$ and the x axis

between 1 and 2 is

$$= \int_1^2 \frac{2}{x} \, dx \underset{\text{by } \boxed{8.2}}{=} \big[2 \ln|x| \big]_1^2$$

$$= 2 \, [\ln(2) - \ln(1)] = 2 \ln(2) \ (\text{units})^2$$

Area below $y = 3 - x$ and the x axis is the area of trapezium
ABCD (Fig. 13).

$$\text{Area ABCD} = \frac{1}{2}1(2 + 1) = \frac{3}{2} \ (\text{units})^2$$

Fig. 13

To find the required area we subtract the area below $y = \dfrac{2}{x}$
from the area of the trapezium.

So the required area $= \dfrac{3}{2} - 2 \ln(2) = 0.114$.

The area between the line $y = 3 - x$ and $y = 2/x$ is 0.114 (units)^2

C2 **Mean and root mean square (RMS) values**

? **What does the term 'mean' represent?**

Average. Average values of current, force and
other engineering quantities are important.
In this section we are interested in the average
value of a function. We can use what we know
about areas and curves to discover what the
mean value of a function is. Consider the function $y = f(x)$ between a and b (Fig. 14).

Fig. 14

Let M be the mean value of the function, $y = f(x)$, between a and b. Then M is defined as the
y value such that

$$M(b - a) = \text{shaded area } A$$

$M(b - a)$ is equivalent in size to an area of a rectangle of width $b - a$ and height M (Fig. 15).

8.2 $\displaystyle \int dx/x = \ln|x|$

? **How can we determine M?**

We have

$$M(b - a) = \text{shaded area } A$$

Fig. 15

Same as the shaded area of Fig. 14

$b - a$

Therefore transposing gives

† $$M = \frac{\text{shaded area } A}{b - a}$$

We can define the mean value, M, as

$$M = \frac{\text{area}}{\text{interval}}$$

? **How can we write the shaded area A of Fig. 14 in terms of an integral?**

It is given by

$$\text{shaded area } A = \int_a^b y \, dx$$

Putting this into † gives

$$M = \frac{\int_a^b y \, dx}{b - a} = \frac{1}{b - a}\int_a^b y \, dx \qquad \left[\text{Taking out } \frac{1}{b - a}\right]$$

So we have the mean value of the function $y = f(x)$ between $x = a$ and $x = b$ given by

9.7 $$\text{Mean value of } y = \frac{1}{b - a}\int_a^b y \, dx$$

Similarly we can define the root mean square, RMS, value of $y = f(x)$ between $x = a$ and $x = b$ as

$$\text{RMS} = \sqrt{\text{Mean value of } y^2}$$

$$(\text{RMS})^2 = \text{Mean value of } y^2$$

? **What is the mean value of y^2 between $x = a$ and $x = b$?**

Using 9.7 :

$$\text{Mean value of } y^2 = \frac{1}{b - a}\int_a^b y^2 \, dx$$

Since $(\text{RMS})^2 = \text{Mean value of } y^2$:

9.8 $$(\text{RMS})^2 = \frac{1}{b - a}\int_a^b y^2 \, dx$$

and hence taking the square root gives

9.9 $$\text{RMS} = \sqrt{\frac{1}{b - a}\int_a^b y^2 \, dx}$$

Generally it is more appropriate to use 9.8 , evaluate $(\text{RMS})^2$ and then take the square root of the result to give the RMS value.

The root mean square (RMS) value is important in electrical principles.

Example 12 *electrical principles*

Evaluate the RMS value of the current $i = 10\sin(t)$ over the interval 0 to 2π.

Solution

Which formula do we use?

9.8 $$(\text{RMS})^2 = \frac{1}{b-a} \int_a^b y^2 \, dx$$

and then take the square root of the result.

What are the values of a and b?

$$a = 0 \text{ and } b = 2\pi$$

What is y?

In this case $y = i = 10\sin(t)$ and we integrate with respect to t. Using **9.8** :

$$(\text{RMS})^2 = \frac{1}{2\pi - 0} \int_0^{2\pi} \left[10\sin(t) \right]^2 dt$$

$$= \frac{1}{2\pi} \int_0^{2\pi} 100\sin^2(t) \, dt$$

$$(\text{RMS})^2 \underset{\substack{\text{taking out the}\\ \text{constant } 100}}{=} \frac{100}{2\pi} \int_0^{2\pi} \sin^2(t) \, dt$$

How do we evaluate $\int_0^{2\pi} \sin^2(t) \, dt$?

We need to use the trigonometric identity

4.68 $$\sin^2(t) = \frac{1}{2} [1 - \cos(2t)]$$

Therefore

$$\int_0^{2\pi} \sin^2(t) \, dt = \int_0^{2\pi} \frac{1}{2} \left[1 - \cos(2t) \right] dt$$

$$= \frac{1}{2} \int_0^{2\pi} \left[1 - \cos(2t) \right] dt \quad \left[\text{Taking out } \frac{1}{2} \right]$$

$$= \frac{1}{2} \left[t - \frac{\sin(2t)}{2} \right]_0^{2\pi} \quad \left(\text{By} \int \cos(kt) \, dt = \frac{\sin(kt)}{k} \right)$$

$$= \frac{1}{2} \left\{ \left[2\pi - \frac{\sin[2(2\pi)]}{2} \right] - 0 \right\}$$

$$= \frac{1}{2} \left\{ [2\pi - 0] - 0 \right\} = \pi \quad \text{[Cancelling 2's]}$$

Example 12 *continued*

$$\int_0^{2\pi} \sin^2(t)\, dt = \pi$$

Substituting $\int_0^{2\pi} \sin^2(t)\, dt = \pi$ into ⬛ * ⬛ gives

$$(\text{RMS})^2 = \frac{100}{2\pi}(\pi) = 50 \quad [\text{Cancelling } \pi's]$$

?   **Have we found the RMS value of $i = 10\sin(t)$?**

No, but we have established the $(\text{RMS})^2$ value of $i = 10\sin(t)$.

?   **How do we find the RMS value?**

Take the square root. Hence

$$\text{RMS} = \sqrt{50} = 7.07 \text{ A} \qquad (2 \text{ d.p.})$$

Example 13 *electrical principles*

Find the RMS value, i_{RMS}, of the current, i, shown in Fig. 16.

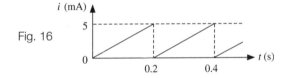

Fig. 16

Solution

To find the RMS value we first obtain $(i_{\text{RMS}})^2$ and then take the square root of our result.

Since the waveform repeats itself we only need to consider ONE cycle, say between $t = 0$ and $t = 0.2$.

?   **What is the equation of the function, i, for t between 0 and 0.2?**

The waveform is a straight line going through 0 and having a gradient given by

$$= \frac{5}{0.2} = 25$$

Therefore using the straight line equation, $i = mt + c$, we have

$$i = 25t \qquad [\text{Because } c = 0]$$

Substituting $a = 0$, $b = 0.2$, $y = i = 25t$ and $dx = dt$ into

⬛ 9.8 $$(\text{RMS})^2 = \frac{1}{b-a}\int_a^b y^2\, dx$$

gives

$$(i_{\text{RMS}})^2 = \frac{1}{0.2}\int_0^{0.2} (25t)^2\, dt$$

Example 13 *continued*

$$= \frac{25^2}{0.2} \int_0^{0.2} t^2 \, dt \qquad \text{[Taking out } 25^2\text{]}$$

$$= 3125 \left[\frac{t^3}{3} \right]_0^{0.2} \qquad \text{[Integrating]}$$

$$= \frac{3125}{3} (0.2^3 - 0^3) = 8.33$$

Hence

$$i_{RMS} = \sqrt{8.33} = 2.89 \text{ mA} \qquad \text{(2 d.p.)}$$

Example 14 *mechanics*

The force, F, acting on a particle for the first 10 seconds is shown in Table 4.

TABLE 4	t (s)	0	1	2	3	4	5	6	7	8	9	10
	F (kN)	0	3.4	5.1	7.6	9.1	8.4	6.2	3.1	2.7	1.3	0.9

Estimate the mean value of the applied force, F.

Solution

What do we use to find the mean value of F?

Use 9.7 with $a = 0$, $b = 10$, $y = F$ and $dx = dt$:

9.7   Mean value of $y = \dfrac{1}{b - a} \displaystyle\int_a^b y \, dx$

★   Mean value of $F = \dfrac{1}{10} \displaystyle\int_0^{10} F \, dt$

How can we find $\displaystyle\int_0^{10} F \, dt$?

Since our F is in tabular form and goes up/down in steps we can only use a numerical method to evaluate $\displaystyle\int_0^{10} F \, dt$. Let's use Simpson's rule with $n = 10$, therefore

† $\displaystyle\int_0^{10} F \, dt \approx \dfrac{h}{3} [F_0 + 4(F_1 + F_3 + F_5 + F_7 + F_9) + 2(F_2 + F_4 + F_6 + F_8) + F_{10}]$

What are the values of h, F_0, F_1, $\cdots$, F_{10}?

From Table 4 the uniform width $h = 1$. Using the second row of Table 4 we have
$F_0 = 0$, $F_1 = 3.4$, $F_2 = 5.1$, $F_3 = 7.6$, $F_4 = 9.1$, $F_5 = 8.4$, $F_6 = 6.2$, $F_7 = 3.1$, $F_8 = 2.7$,
$F_9 = 1.3$ and $F_{10} = 0.9$.

Example 14 *continued*

Substituting all this into 　†　 gives

$$\int_0^{10} F\, dt \approx \frac{1}{3} [0 + 4(3.4 + 7.6 + 8.4 + 3.1 + 1.3) + 2(5.1 + 9.1 + 6.2 + 2.7) + 0.9]$$

$$= 47.43$$

Is 47.43 the mean value of the force F?

No, we need to substitute this into 　*　 :

$$\text{Mean value of } F = \frac{1}{10}(47.43) = 4.74 \text{ kN} \quad (2 \text{ d.p.})$$

SUMMARY

The area between the graph and x axis is given by

9.6　　　　$$\text{area} = \int_a^b y\, dx$$

Remember, before finding areas under curves we must make a sketch of the graph(s) because areas under the x axis will otherwise be subtracted from areas above it and give the wrong answer.
Mean value of the function $y = f(x)$ between $x = a$ and $x = b$ is

9.7　　　　$$\text{Mean value of } y = \frac{1}{b-a}\int_a^b y\, dx$$

Also root mean square, RMS, is

9.8　　　　$$(\text{RMS})^2 = \frac{1}{b-a}\int_a^b y^2\, dx$$

9.9　　　　$$\text{RMS} = \sqrt{\frac{1}{b-a}\int_a^b y^2\, dx}$$

Exercise 9(c)

Solutions at end of book. Complete solutions available at www.palgrave.com/science/engineering/singh

1 Obtain the area bounded by $y = \sin(x)$ and the x axis between 0 and 2π.

▦ **Questions 2 to 10, inclusive, belong to the field of [*electrical principles*].**

2

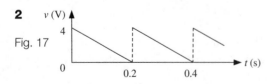

Fig. 17

Determine the mean and RMS values of the voltage, v, shown in Fig. 17. (Consider ONE cycle.)

3 Show that the mean value of $i = I \sin(t)$ over the interval 0 to π is $2I/\pi$ where the constant I is the maximum current. Determine the mean value of $i = \sin(t)$ over this period.

Exercise **9(c) continued**

Solutions at end of book. Complete solutions available at www.palgrave.com/science/engineering/singh

4 Determine the mean value, over the interval 0 to 2π, of the following voltages:
i $v = 10\sin(t)$ **ii** $v = 10\cos(t)$

5 By using

$\boxed{9.7}$ Mean value of $y = \dfrac{1}{b-a}\displaystyle\int_a^b y\,dx$

show that the mean value of the voltage $v = V\sin(\omega t)$ over the interval $\omega t = 0$ to $\omega t = 2\pi$ is 0 V.

6 Show that the RMS value, i_{RMS}, of the current $i = I\cos(t)$ over the interval 0 to 2π is given by

$$i_{RMS} = \frac{I}{\sqrt{2}}$$

where the constant I is the maximum current.

7 Evaluate the mean and RMS values of the voltage, v, over an interval 0 to 10 where

$$v = 10(1 - e^{-0.1t})$$

8

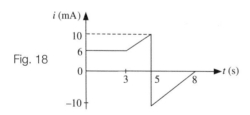

Fig. 18

Determine the mean value of the waveform in Fig. 18. Consider the waveform between 0 and 8 seconds.

9 Determine the mean and RMS values of the voltage waveform, $v = \omega t \sin(\omega t)$ across the interval $\omega t = 0$ to $\omega t = \pi$. The form factor, f, is defined as

$$f = \frac{\text{RMS value}}{\text{Mean value}}$$

Determine f for the waveform v.

10 Find the mean value of the current, $i = \sqrt{\cos(\omega t)}$ for $\omega t = 0$ to $\omega t = \pi/2$. Note that we cannot integrate $\sqrt{\cos(\omega t)}$ analytically, so use Simpson's rule with four equal intervals.

11 [mechanics]   The force, F, acting on a particle for the first 10 seconds is given by:

t (s)	0	1	2	3	4	5
F (kN)	0	2.6	3.1	3.9	4.5	6.7

t (s)	6	7	8	9	10
F (kN)	7.1	4.5	3.1	2.0	1.5

Estimate the mean value of the applied force, F.

12

Fig. 19

i Evaluate the shaded area.

ii Evaluate $\displaystyle\int_0^1 x^{1/2}\,dx$.

iii Draw the diagram of Fig. 19 and sketch the graph of $y = x^{1/2}$. Shade in the area represented by

$$\int_0^1 x^{1/2}\,dx$$

iv What do you notice about the results of **i** and **ii**?

SECTION D **Applications in mechanics**

By the end of this section you will be able to:
► use integration to find the displacement and velocity of a particle

D1 **Displacement, velocity and acceleration**

In Chapter 7, **Section D** (page 330) we stated that for a particle in straight line motion the displacement, s, velocity, v, and acceleration, a, are related as follows:

$$v = \frac{ds}{dt}$$

$$a = \frac{dv}{dt}\left(=\frac{d^2s}{dt^2}\right)$$

In this section we carry out the inverse process, that is we integrate both sides of these equations to give:

9.10 $s = \displaystyle\int v \, dt$

9.11 $v = \displaystyle\int a \, dt$

Remember that, when integrating, we obtain a constant of integration, C, and we can usually find the value of C if we are given some conditions associated with the problem, as the following example shows.

Example 15 *mechanics*

The acceleration, a, of a particle is given by

$$a = \cos(3t)$$

At $t = 0$, $s = \dfrac{1}{3}$ and $\dfrac{ds}{dt} = 0$. Obtain an expression for the displacement, s.

Solution

We first obtain the velocity by using $v = \displaystyle\int a \, dt$ with $a = \cos(3t)$:

$$v = \int \cos(3t) \, dt$$

$$= \frac{\sin(3t)}{3} + C$$

Example 15 *continued*

To find C we use the initial condition, $t = 0$, $\dfrac{ds}{dt} = 0 = v$. Substituting these into :

$$0 = \frac{\sin(0)}{3} + C \text{ gives } C = 0$$

Hence putting $C = 0$ into ⬛ * :

$$v = \frac{\sin(3t)}{3}$$

We integrate v to obtain the displacement, s:

$$s = \int \frac{\sin(3t)}{3}\, dt$$

⬛ ** $$= -\frac{\cos(3t)}{9} + D$$

The new constant of integration is called D because we have already used C for the first constant of integration in ⬛ * .

[?] How do we find the value of D?

Substituting the other condition $t = 0$, $s = 1/3$ into ⬛ ** :

$$\frac{1}{3} = -\frac{\cos(0)}{9} + D$$

$$= -\frac{1}{9} + D \qquad\qquad [\text{Because } \cos(0) = 1]$$

$$\frac{1}{3} + \frac{1}{9} = D \text{ which gives } D = \frac{4}{9}$$

Substituting $D = 4/9$ into ⬛ ** gives us the expression for s:

$$s = -\frac{\cos(3t)}{9} + \frac{4}{9} = \frac{1}{9}\left[4 - \cos(3t)\right]$$

The given conditions in the above are called the initial conditions. They give the values of displacement and velocity at $t = 0$.

 Example 16 *mechanics*

A projectile is fired vertically upwards from the ground with an initial velocity of 50 m/s. Show that if the acceleration

$$a = -9.81$$

then the height, h, of the projectile is given by

$$h = 50t - 4.9t^2$$

Find the maximum height reached by the projectile and sketch the graph of h against t.

Example 16 *continued*

Solution

The acceleration
$$a = -9.81$$

(Negative sign because the projectile is fired against gravity.) Hence by $v = \int a\, dt$ where v is velocity we have

$$v = -\int 9.81\, dt = -9.81t + C \quad \text{[Integrating]}$$

When $t = 0$, $v = 50$ because the initial velocity is 50 m/s. Substituting these into $v = -9.81t + C$:

$$50 = 0 + C \text{ gives } C = 50$$
$$v = -9.81t + 50$$

The height, h, is obtained by using $h = \int v\, dt$

$$h = \int (-9.81t + 50)dt$$

$$h = -4.9t^2 + 50t + D \qquad \text{[Integrating]}$$

When $t = 0$, $h = 0$ (the projectile is fired from the ground, $h = 0$) gives $D = 0$:

   $h = -4.9t^2 + 50t$

How do we find the maximum height?

We use differentiation. The stationary point occurs at $\dfrac{dh}{dt} = 0$:

$$\frac{dh}{dt} = v = -9.81t + 50 = 0$$

$$9.81t = 50$$

$$t = \frac{50}{9.81} = 5.10$$

To check whether we have a maximum or minimum at $t = 5.10$ we need to differentiate again:

$$\frac{d^2h}{dt^2} = a = -9.81 < 0 \quad \text{[Negative]}$$

At $t = 5.1$, h is maximum. The maximum height is obtained by substituting $t = 5.1$ into :

$$h = -[4.9 \times 5.1^2] + (50 \times 5.1) = 127.55 \text{ m}$$

For a sketch we also need to find the values of t where $h = 0$

$$50t - 4.9t^2 = 0$$
$$t(50 - 4.9t) = 0$$
$$t = 0, t = 10.2$$

Thus we have the situation shown in Fig. 20.

Fig. 20

SUMMARY

The displacement, s, velocity, v, and acceleration, a, are related by

9.10 $s = \int v \, dt$

9.11 $v = \int a \, dt$

Moreover if we are given initial conditions we can find the value of the integration constant C.

Exercise **9(d)**   Solutions at end of book. Complete solutions available at www.palgrave.com/science/engineering/singh

🚗 **All questions belong to the field of [*mechanics*].**

1 The acceleration, a, of a mass vibrating on a spring in simple harmonic motion is given by

$$a = -2.5 \cos(2\pi t)$$

Given that at $t = 0$, $v = 0$ and displacement $x = 0$, find an expression for displacement x.

2 The horizontal component of acceleration, a_x, of a particle moving with curvilinear motion is given by

$$a_x = t$$

Given that when $t = 0$, the velocity $v_x = 2$ and displacement $x = 3$, find an expression for displacement x.

3 The velocity, v, of an object is given by

$$v = 0.6t^2 + t$$

Determine the acceleration and displacement from rest after the first 2 s.

4 The acceleration, a, of a particle is given by

$$a = 4t^3$$

and has an initial velocity of 3 m/s. Find the velocity after 1.5 s.

5 A rocket is fired vertically upwards with an acceleration of 2.9 m/s². Find the height reached by the projectile after 2 minutes.

6 The table below shows the velocity, v, for the corresponding time, t, of a vehicle.

t (s)	0	25	50	75	100
v (m/s)	0	33	45	55	64

t (s)	125	150	175	200
v (m/s)	72	77	80	82

Find the approximate distance covered by the vehicle.

7 A particle has an acceleration, a, and at $t = 0$ the velocity $v = u$ and displacement $s = 0$. Show that

$$s = ut + \frac{1}{2}at^2$$

8 A stone is dropped vertically from a building of height 20 m with an acceleration, a, given by

$$a = 9.81$$

Calculate the time taken to reach the ground.

Exercise 9(d) continued

Solutions at end of book. Complete solutions available at www.palgrave.com/science/engineering/singh

9 The acceleration, a, of a particle in simple harmonic motion is given by

$$a = -\cos(\omega t) + \sin(\omega t)$$

where ω is the angular frequency. Show that

$$\omega^2 x = \omega t + \cos(\omega t) - \sin(\omega t) - 1$$

given that x is the displacement and at $t = 0$ both $v = 0$ and $x = 0$.

SECTION E **Miscellaneous applications of integration**

By the end of this section you will be able to:
▶ use integration to find the work done in thermodynamics
▶ use integration to find the deflection of a beam

This section is quite sophisticated compared to the previous ones. It applies integration to two fields of engineering – thermodynamics and structural mechanics. You should be able to follow the examples because they do **not** rely on previous knowledge of these two engineering disciplines.

E1 **Thermodynamics**

Figure 21 shows a gas expanding in a cylinder according to the relation

$$PV^n = C \text{ (constant)}$$

The initial volume of gas is V_1 and the final volume is V_2. The work done, W, by the gas on the face of the piston is given by

Fig. 21

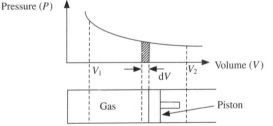

9.12 $$W = \int_{V_1}^{V_2} P \, dV$$

(This is the area under the curve shown in Fig. 21.)

📖 Example 17 *thermodynamics*

A gas expands in a cylinder according to the relation

$$PV^{1.3} = 1560$$

The initial volume of gas is 0.025 m^3 and the final volume is 0.04 m^3.

Evaluate the work done by the gas on the face of the piston.

Example 17 *continued*

Solution

Rearranging $PV^{1.3} = 1560$ to make P the subject gives

$$P = \frac{1560}{V^{1.3}} = 1560V^{-1.3}$$

Using

9.12 $W = \int_{V_1}^{V_2} P \, dV$

with $V_1 = 0.025$, $V_2 = 0.04$, $P = 1560V^{-1.3}$ and $W = $ work done gives

$$W = \int_{0.025}^{0.04} 1560V^{-1.3} \, dV$$

$$= 1560 \int_{0.025}^{0.04} V^{-1.3} \, dV \qquad \text{[Taking out 1560]}$$

$$= 1560 \left[\frac{V^{-0.3}}{-0.3} \right]_{0.025}^{0.04} \qquad \left[\text{By} \int V^n dV = \frac{V^{n+1}}{n+1} \right]$$

$$= -\frac{1560}{0.3} \left[(0.04)^{-0.3} - (0.025)^{-0.3} \right]$$

$$W = 2068.17 \text{ J} = 2.07 \text{ kJ} \ (\text{J} = \text{Joule})$$

If W is positive then this indicates that the gas is expanding, and a negative W signifies that the gas is being compressed.

E2 **Structural mechanics**

Figure 22 shows a beam of length L with a uniform distributed load of w per unit length.

In the general case, the deflection, y, of the beam at a distance x along the beam is related by

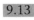

Fig. 22

9.13 $\dfrac{d^2 y}{dx^2} = \dfrac{M}{EI}$

where M is the bending moment and EI is the flexural rigidity.

In this section we are interested in finding an expression for y, so we have to integrate both sides of 9.13 twice with respect to x. Generally the values of the (two) constants of integration are obtained by using the given conditions, as the following example shows.

Example 18 *structures*

The bending moment, M, of the beam shown in Fig. 22 is

$$M = \frac{w}{2}(x^2 - Lx)$$

Given the conditions $x = 0$, $y = 0$ and $x = L$, $y = 0$ show that

$$y = \frac{w}{24EI}(x^4 - 2Lx^3 + L^3x)$$

Solution

Substituting $M = \dfrac{w}{2}(x^2 - Lx)$ into

9.13 $$\frac{d^2y}{dx^2} = \frac{M}{EI}$$

gives

$$\frac{d^2y}{dx^2} = \frac{w}{2EI}(x^2 - Lx)$$

Integrating this:

$$\frac{dy}{dx} = \frac{w}{2EI}\int (x^2 - Lx)\, dx \quad \left[\text{Taking out } \frac{w}{2EI}\right]$$

$$= \frac{w}{2EI}\left(\frac{x^3}{3} - \frac{Lx^2}{2}\right) + C \quad [\text{Integrating}]$$

To find y we integrate again:

$$y = \frac{w}{2EI}\left(\frac{x^4}{12} - \frac{Lx^3}{6}\right) + Cx + D$$

$$= \frac{w}{2EI}\left(\frac{x^4}{12} - \frac{2Lx^3}{12}\right) + Cx + D \quad \left[\text{Writing } \frac{1}{6} = \frac{2}{12}\right]$$

Taking out a common factor of $\dfrac{1}{12}$ from the brackets gives

★ $$y = \frac{w}{24EI}(x^4 - 2Lx^3) + Cx + D$$

Substituting the condition $x = 0$, $y = 0$ into ★ :

$$0 = \frac{w}{24EI}[0] + C(0) + D$$

gives $D = 0$.
Substituting the other condition $x = L$, $y = 0$ into ★ :

$$0 = \frac{w}{24EI}[L^4 - 2LL^3] + CL + D$$

The square brackets become $-L^4$ because

$$L^4 - 2LL^3 = L^4 - 2L^4 = -L^4$$

Example 18 *continued*

Also $D = 0$ therefore

$$0 = \frac{w}{24EI}[-L^4] + CL$$

Transposing for C:

$$C = \frac{w(L^4)}{24EI \, L} = \frac{wL^3}{24EI} \quad \text{[Cancelling]}$$

Putting $C = \dfrac{wL^3}{24EI}$ and $D = 0$ into [★] gives the required result:

$$y = \frac{w}{24EI}\left(x^4 - 2Lx^3\right) + \frac{wL^3x}{24EI}$$

$$= \frac{w}{24EI}\left(x^4 - 2Lx^3 + L^3x\right) \quad \text{[Factorizing]}$$

There are many other applications of integration in engineering, such as finding the moment of inertia, centroids, arc length etc. ... , which are not covered in this book.

SUMMARY

In this section we covered two areas of applications of integration, thermodynamics and structural mechanics.
In thermodynamics the work done, W, by a gas on the face of a piston from an initial volume V_1 to a new volume V_2 is given by

9.12 $$W = \int_{V_1}^{V_2} P \, dV \qquad (P \text{ is pressure})$$

In structural mechanics the deflection, y, of a beam at a distance x is given by

9.13 $$\frac{d^2y}{dx^2} = \frac{M}{EI}$$

where M = bending moment and EI = flexural rigidity.

Exercise **9(e)**

Solutions at end of book. Complete solutions available at www.palgrave.com/science/engineering/singh

Questions 1 to 4, inclusive, are in the field of [*thermodynamics*].

1 Find the work done on the face of a piston by a gas expanding from 0.01 m³ to 0.1 m³ according to $PV^{1.35} = 1500$.

2 The initial volume of a gas in a cylinder is 0.6 m³. Find the work done by the gas if the final volume is 0.023 m³ and the gas obeys the law

$$PV^{1.55} = 1789$$

3 A gas is compressed according to the relation

$$PV^{1.6} = 1675$$

If the volume changes from 0.15 m³ to 0.037 m³, find the work done by the gas.

Solutions at end of book. Complete solutions available at www.palgrave.com/science/engineering/singh

Exercise **9(e) continued**

4 A gas in a cylinder obeys the law

$$PV = C$$

Show that the work done, W, by the gas is given by

$$W = C \ln\left(\frac{V_2}{V_1}\right)$$

where V_1 and V_2 are the initial and final volumes of the gas respectively.

🌉 **Questions 5 to 8, inclusive, are in the field of [structures].**

5

Fig. 23

Figure 23 shows a cantilever with a concentrated load P at a distance l from the fixed end. The bending moment, M, is given by

$$M = P(l - x)$$

At $x = 0$, $y = 0$ and $\dfrac{dy}{dx} = 0$. Determine y.

6 Consider the cantilever beam of length L (Fig. 24).

Fig. 24

Find the deflection y given $M = -\dfrac{wx^2}{2}$ (where w is the load per unit length) and at $x = L$, $y = 0$ and $\dfrac{dy}{dx} = 0$.

7 A cantilever beam of length L is subject to a uniform distributed load, w per unit length, and the bending moment M is given by

$$M = -\frac{w}{2}(L-x)^2$$

Obtain an expression for y for the initial conditions at $x = 0$, $y = 0$ and $\dfrac{dy}{dx} = 0$.

8

Fig. 25

Figure 25 shows a simply supported beam of length L carrying a load, wx, which increases uniformly. The bending moment, M, is given by

$$M = \frac{w}{6}(x^3 - L^2x)$$

For the conditions $x = 0$, $y = 0$ and $x = L$, $y = 0$ show that the deflection y is given by

$$y = \frac{w}{360EI}[3x^5 - 10L^2x^3 + 7L^4x]$$

Miscellaneous exercise **9**

Solutions at end of book. Complete solutions available at www.palgrave.com/science/engineering/singh

1 🖼 [*thermodynamics*] The volume and pressure of a gas are measured in an experiment and give the results shown in the table to the right.

Find the approximate work done on the face of the piston by the gas to produce a volume change from 0.01 m³ to 0.09 m³.

V (m³)	0.01	0.02	0.03	0.04	0.05
P (kPa)	946	358	203	136	99

V (m³)	0.06	0.07	0.08	0.09
P (kPa)	77	62	51	44

2 [mechanics]   The following table shows the experimental results of a force, F, on a piston with displacement, x:

x (m)	0.025	0.05	0.075	0.1	0.125
F (kN)	10.06	4.23	2.55	1.78	1.34

x (m)	0.15	0.175	0.2	0.225
F (kN)	1.07	0.88	0.75	0.64

The work done, W, in moving the piston from 0.025 m to 0.225 m is given by

$$W = \int_{0.025}^{0.225} F \, dx$$

Find an approximate value for W.

3 [thermodynamics]   The specific heat, c, of steam at temperature T, is given in the following table:

T (K)	200	250	300
c (kJ/kg K)	3.57	3.63	3.70

T (K)	350	400	450
c (kJ/kg K)	3.75	3.80	3.86

The enthalpy change, Δh, is given by

$$\Delta h = \int_{200}^{450} c \, dT$$

Find an approximation for Δh.

4 [mechanics]   The height, h, of a projectile is given by

$$h = \int_{0}^{10} \left[5 \ln \left| \frac{2}{2 + 3t} \right| + 9.81t \right] dt$$

By using Simpson's rule with four equal intervals, find an approximation for the height h.

5 [mechanics]   The time taken, t, for a particle to reach a velocity of 90 m/s with an initial velocity of 50 m/s is given by

$$t = \int_{50}^{90} \frac{dv}{5 - 2v^{0.11}}$$

where v is the velocity. Find an approximation for t.

6 [electrical principles]   An a.c. voltage waveform, v, is given by

$$v = \frac{\sin(t)}{t}$$

Find an approximation for the mean value of v over the interval 0.4 to 1. (You cannot find an analytical solution to the integral of $\sin(t)/t$, so use Simpson's rule with six equal intervals.)

7 [electrical principles]   An a.c. current, i, is given by

$$i = \sqrt{1 - 2\sin(t)}$$

Find an approximation for the mean value of i over the interval π to $3\pi/2$.

8 [aerodynamics]   The area of an aerofoil is shown shaded in Fig. 26. Determine the shaded area.

Fig. 26

9 [electrical principles]   A voltage waveform, v, is given by

$$v = 5(1 - e^{-t/4}) \quad 0 < t < 20$$

Find the mean and RMS values of v.

10 [mechanics]   The acceleration, a, of a projectile fired vertically upwards is given by

$$a = \frac{10}{10 - t} + 9.8$$
$$\text{where } 0 \le t < 10$$

a Show that the velocity, v, is given by

$$v = 10 \ln \left(\frac{10}{10 - t} \right) + 9.8t$$

b Find an approximation for the height reached by the projectile after 6 seconds.

11 [thermodynamics]   Find the work done on the face of a piston by a gas expanding from 0.013 m³ to 0.09 m³ according to $PV^{1.41} = 1961$.

12 [structures]   The bending moment, M, of a beam is given by

$$M = 2.5(x - 5)^2 + 8(x - 3) - 5x$$

where x is the distance along the beam. Given that at $x = 0$, $y = 0$ and at $x = 5$, $y = 0$, find an expression for the deflection of the beam, y.

13 Use a computer algebra system or a graphical calculator to determine

$$\int_0^1 \frac{x^4(1-x)^4}{1 + x^2} \, dx$$

14 [mechanics]

Fig. 27

Figure 27 shows the motion of a projectile.

i The horizontal acceleration, $\dfrac{d^2x}{dt^2} = 0$.

For the initial conditions when

$t = 0$, $\dfrac{dx}{dt} = v \cos(\theta)$ and $x = 0$,

show that

$$x = [v \cos(\theta)]t$$

ii The vertical acceleration, $\dfrac{d^2y}{dt^2} = -g$.

For the initial conditions when

$t = 0$, both $\dfrac{dy}{dt} = v \sin(\theta)$ and $y = 0$,

show that

$$y = [v \sin(\theta)]t - \frac{gt^2}{2}$$

15 [electrical principles]   A voltage signal is given by

$$v = V_1 \sin(\omega t) + V_3 \sin(3\omega t)$$

where V_1, V_3 are the constant maximum voltages and ω is angular frequency. Show that for the interval $t = 0$ to $t = \pi/\omega$:

$$(v_{RMS})^2 = \frac{V_1^2 + V_3^2}{2}$$

where v_{RMS} is the RMS value of v.

CHAPTER 10

Complex Numbers

SECTION A **Arithmetic of complex numbers**

By the end of this section you will be able to:
▶ understand the term *complex number*
▶ add and subtract complex numbers
▶ evaluate and simplify powers of *j*
▶ multiply and divide two complex numbers
▶ use complex numbers to solve engineering problems

A1 Introduction

Complex numbers and complex functions are used in many engineering fields such as alternating current (a.c.) theory, vibration problems, fluid flow, elasticity etc. The word 'complex' in this chapter does not mean complicated.

To understand complex numbers we use the following rules of indices:

10.1
$$\sqrt{a \times b} = \sqrt{a} \times \sqrt{b} = a^{1/2} \times b^{1/2}$$

10.2
$$(a^m)^n = a^{m \times n}$$

10.3
$$a^m a^n = a^{m+n}$$

10.4
$$\frac{a^m}{a^n} = a^{m-n}$$

Example 1

What is $\sqrt{-25}$ equal to?

Solution

If we try to evaluate this on a simple calculator we get an error message. **So if, in the course of trying to solve a mathematical problem, we needed to evaluate $\sqrt{-25}$, how would we go about it?**

There is **no** real number which when multiplied by itself gives -25:

$$5^2 = +25 \qquad\qquad (-5)^2 = +25$$

We can rewrite $-25 = 25 \times (-1)$ and then we have

$$\sqrt{-25} = \sqrt{25 \times (-1)} \underset{\text{by } 10.1}{=} \sqrt{25} \times \sqrt{(-1)}$$

$$= 5 \times \sqrt{(-1)}$$

$$= 5\sqrt{(-1)}$$

$$= 5\,j \text{ where } j = \sqrt{(-1)}$$

Example 1 *continued*

We introduce the number j $\left(=\sqrt{(-1)}\right)$ so that we can find the square root of a negative number.

Of course we also have $\sqrt{-25} = -5j$ because $\sqrt{25} = -5$ or 5.

How do we solve the equation $x^2 + 1 = 0$?

The graph of $x^2 + 1$ is shown in Fig. 1.

Since $x^2 + 1$ does not cut the x axis there cannot be a real number which satisfies

Fig. 1

$$x^2 + 1 = 0$$

Let's rearrange this equation:

$$x^2 = -1$$

$$x = \pm \sqrt{(-1)} = \pm j \text{ (where } j = \sqrt{-1}\text{)}$$

The numbers $+j$ and $-j$ are examples of complex numbers.

We let j (or i) $= \sqrt{-1}$. Engineers use j but mathematicians use i.

Why do engineers use j?

Complex numbers are used in alternating current (a.c.) theory where i represents current.
We will use j to represent $\sqrt{-1}$.

Example 2

Solve
$$z^2 - 2z + 5 = 0$$

Solution

We need to use the quadratic formula

$$z = \frac{-b \pm \sqrt{b^2 - 4ac}}{2a}$$

What are the values of a, b and c in this example?

$a = 1$, $b = -2$ and $c = 5$. Therefore

$$z = \frac{-(-2) \pm \sqrt{(-2)^2 - (4 \times 1 \times 5)}}{2 \times 1}$$

$$= \frac{2 \pm \sqrt{4-20}}{2}$$

$$z = \frac{2 \pm \sqrt{-16}}{2}$$

Example 2 *continued*

Previously we would have thought of this problem as unsolvable, because we end up needing to take the square root of a negative number (-16). This is where we can use complex numbers. Hence

$$\sqrt{-16} = \sqrt{16 \times (-1)} = \sqrt{16} \times \sqrt{-1} = 4\sqrt{-1} = 4j$$

Therefore

$$z = \frac{2 \pm \sqrt{-16}}{2}$$

$$= \frac{2 \pm j4}{2} = \frac{2}{2} \pm j\frac{4}{2} = 1 \pm j2$$

The roots are $z = 1 + j2$ and $z = 1 - j2$.

A2 Complex numbers

A complex number is of the form $a + jb$ where a is called the **real part** and b is called the **imaginary part**.

For example, the complex number $3 + j9$, real part $= 3$, imaginary part $= 9$. In other books or in your notes $3 + j9$ might be written as $3 + 9j$. It doesn't make any difference!

Let z be a complex number then the real part of z is denoted by Re(z) and the imaginary part of z is denoted by Im(z). For example

$$\text{Re}(3 + j9) = 3 \qquad\qquad\qquad \text{Im}(3 + j9) = 9$$

A3 Addition and subtraction

For addition of complex numbers, add the real parts and add the imaginary parts:

10.5 $(a + jb) + (c + jd) = \underbrace{a + c}_{\substack{\text{add the}\\\text{real parts}}} + j\underbrace{(b + d)}_{\substack{\text{add the}\\\text{imaginary parts}}}$

Similarly for subtraction, subtract the real parts and subtract the imaginary parts.

10.6 $(a + jb) - (c + jd) = \underbrace{a - c}_{\substack{\text{subtract the}\\\text{real parts}}} + j\underbrace{(b - d)}_{\substack{\text{subtract the}\\\text{imaginary parts}}}$

Example 3

Evaluate

a $(2 + j7) + (3 + j12)$ **b** $(2 + j7) - (3 + j12)$ **c** $(3 + j4) - (3 - j4)$

Solution

a $(2 + j7) + (3 + j12) = (2 + 3) + j(7 + 12)$

$$= 5 + j19$$

b $(2 + j7) - (3 + j12) = (2 - 3) + j(7 - 12)$

$$= -1 + j(-5) = -1 - j5$$

c $(3 + j4) - (3 - j4) = (3 - 3) + j(4 - (-4))$

$$= 0 + j(4 + 4) = j8$$

A4 **Powers of j**

We can evaluate the first four powers of j by using the rules of indices:

$$j = \sqrt{-1}$$

$$j^2 = \left(\sqrt{-1}\right)^2 = (-1^{1/2})^2 = (-1)^1 = -1 \quad [\text{By } 10.2\,]$$

$$j^3 \underset{\text{by } 10.3}{=} j^2 \cdot j^1 = -1 \cdot j = -j$$

$$j^4 = (j^2)^2 = (-1)^2 = 1$$

These first four powers of j are important because we can write **all** the powers of j in terms of these first four powers:

10.7 $j = \sqrt{-1}$

10.8 $j^2 = -1$

10.9 $j^3 = -j$

10.10 $j^4 = 1$

10.2 $(a^m)^n = a^{m \times n}$ 10.3 $a^m a^n = a^{m+n}$

Example 4

Simplify **a** j^{12} **b** j^7 **c** j^{203}

Solution

We use the rules of indices and powers of j established above:

a $j^{12} = (j^4)^3 = (1)^3 = 1$
by 10.2   by 10.10

b $j^7 = j^4 \cdot j^3 = 1 \, j^3 = -j$
by 10.3   by 10.10   10.9

c $j^{203} = j^{200} \cdot j^3 = (j^4)^{50} \, j^3$
by 10.3   10.2

$$= (1)^{50}(-j) = -j$$

? To simplify higher powers of j you only need to use the first four powers of j. **Which power of j is the most useful in simplifying powers of j?**

From **Example 4** you should notice that it is better to write most powers in terms of j^4 because $j^4 = 1$, and 1 to the power of any real number is also 1, that is $1^x = 1$. These, 10.7 to 10.10, four powers of j are very important and are well worth remembering, in particular $j^2 = -1$.

A5 Multiplication

The easiest way to multiply two complex numbers is to use FOIL (First Outside Inside Last), as we used to expand brackets in Chapter 1:

10.11 $(a + jb)(c + jd) = \underset{F}{ac} + \underset{O}{jad} + \underset{I}{jbc} + \underset{L}{j^2bd}$

Collecting the j terms and using $j^2 = -1$ simplifies 10.11 to

$$(a + jb)(c + jd) = ac + j(ad + bc) - bd$$

$$= (ac - bd) + j(ad + bc)$$

10.2 $(a^m)^n = a^{m \times n}$ 10.3 $a^m a^n = a^{m+n}$ 10.9 $j^3 = -j$ 10.10 $j^4 = 1$

Example 5

Evaluate the following:

a $(2 + j3)(6 + j7)$ **b** $(2 + j3)(2 - j3)$

Solution

a $(2 + j3)(6 + j7) = \underbrace{(2 \times 6)}_{F} + \underbrace{j(2 \times 7)}_{O} + \underbrace{j(3 \times 6)}_{I} + \underbrace{j^2(3 \times 7)}_{L}$

$= 12 + \underbrace{j14 + j18}_{= \, j32} + j^2 21$

$= 12 + j32 - 21$ [Because $j^2 = -1$]

$= -9 + j32$

b $(2 + j3)(2 - j3) = 4 - \underbrace{j6 + j6}_{= \, 0} - j^2 9$

$= 4 - j^2 9$

$= 4 + 9 = 13$ [Because $j^2 = -1$]

A6 Complex conjugate

Example 6

Evaluate $(3 + j4)(3 - j4)$

Solution

$(3 + j4)(3 - j4) = 9 \underbrace{- j12 + j12}_{= \, 0} - j^2 16$

$= 9 - j^2 16$

$= 9 + 16$ [Remember that $j^2 = -1$]

$= 25$

The answer to **Example 6** contains **no** j, that is multiplication of two complex numbers sometimes gives a real number. Here are some more examples of complex number multiplication in which the answer contains no j.

i $(5 - j8)(5 + j8) = 89$

ii $(1 - j)(1 + j) = 2$

iii $(7 + j5)(7 - j5) = 74$

? **What do you notice about these complex numbers?**

They are identical apart from the signs of the imaginary part.

A pair of complex numbers like these are called **conjugate complex numbers** and the product of these is always **real** (no j).

Notice: $(3 + j4)(3 - j4) = 3^2 + 4^2 = 25$

Similarly $(4 + j5)(4 - j5) = 4^2 + 5^2 = 41$

$$(2 - j6)(2 + j6) = 2^2 + 6^2 = 40$$

Complex conjugate of $a + jb$ is

10.12 $a - jb$

Also $(a + jb)(a - jb) = a^2 + b^2$

If z is a complex number then we denote the complex conjugate of z by $\overline{z}$.

A7 **Division**

Division of a complex number by a real number is straightforward. For example

$$\frac{3 - j4}{2} = \frac{3}{2} - j\frac{4}{2} = 1.5 - j2$$

? **But what is** $\dfrac{(5 - j3)}{3 + j4}$ **equal to?**

Example 7

Evaluate $z = \dfrac{(5 - j3)}{3 + j4}$

Solution

We have just seen that dividing a complex number by a real number presents no problems, so if we could convert the denominator, $3 + j4$, into a real number then the problem can be solved.

? **How can we achieve this?**

Multiply by its complex conjugate $3 - j4$. We multiply numerator and denominator by $3 - j4$:

$$z = \frac{(5 - j3)(3 - j4)}{(3 + j4)(3 - j4)} = \frac{15 - j20 - j9 + j^2 12}{3^2 + 4^2} \quad \text{[Expanding]}$$

$$= \frac{15 - j29 + j^2 12}{25}$$

Example 7 *continued*

$$= \frac{15 - j29 - 12}{25} \quad [\text{Using } j^2 = -1]$$

$$= \frac{3 - j29}{25} = \frac{3}{25} - j\frac{29}{25} = 0.12 - j1.16$$

So to divide one complex number by another, we multiply numerator and denominator by the conjugate of the denominator. Hence

10.13 $\qquad \dfrac{a + jb}{c + jd} = \dfrac{(a + jb)(c - jd)}{c^2 + d^2}$

Remember that if we multiply a complex number by its conjugate, the result is a sum of squares, that is

$$(c + jd)(c - jd) = c^2 + d^2$$

That is why the denominator in 10.13 is $c^2 + d^2$.

A8 **Applications of complex numbers**

Example 8 *electrical principles*

The admittance, Y, in a circuit is given by

$$Y = \frac{1}{Z}$$

where Z is the impedance of the circuit. If $Z = 100 - j25$, find Y.

Solution

Substituting $Z = 100 - j25$ into $Y = \dfrac{1}{Z}$ gives

$$Y = \frac{1}{100 - j25}$$

To cut down the arithmetic we can take out a factor of 25 from the denominator because 25 is a common factor of 100 and -25:

$$Y = \frac{1}{25}\left[\frac{1}{4 - j}\right]$$

Example 8 *continued*

We need to use the rules of division.

What is the complex conjugate of 4 − *j*?

$$4 + j$$

Multiplying numerator and denominator by the complex conjugate, $4 + j$, gives

$$\frac{1}{25}\left[\frac{1}{4 - j}\right] = \frac{1}{25}\left[\frac{4 + j}{4^2 + 1^2}\right]$$

$$= \frac{1}{25}\left[\frac{4 + j}{17}\right]$$

$$= \frac{4 + j}{25 \times 17}$$

$$= \frac{4 + j}{425} = \frac{4}{425} + j\frac{1}{425} = 0.0094 + j0.0024$$

Thus

$$Y = 0.0094 + j0.0024$$

Normally Z is in ohm and Y is in siemen, that is $Z = (100 - j25)$ ohm (Ω) and $Y = (0.0094 + j0.0024)$ siemen (S).

Example 9 *electrical principles*

The total impedance, Z, of a circuit containing an inductor with inductance L and a resistor with resistance R in series is given by

$$Z = R + j\,2\pi f L \quad \text{where } f = \text{frequency}$$

If $Z = (50 + j200)\ \Omega$ and $f = 50$ Hz, find R and L.

Solution

We have

$$R + j2\pi f L = 50 + j200$$

Equating real and imaginary parts gives

$$R = 50\ \Omega$$

$$2\pi f L = 200$$

$$L = \frac{200}{2\pi \times 50} = \frac{2}{\pi}\text{H (using } f = 50)$$

SUMMARY

A number $a + jb$ is called a complex number, where $j = \sqrt{(-1)}$.
To add and subtract complex numbers:

10.5 $(a + jb) + (c + jd) = \underbrace{a + c}_{\substack{\text{add the}\\\text{real parts}}} + j\underbrace{(b + d)}_{\substack{\text{add the}\\\text{imaginary parts}}}$

10.6 $(a + jb) - (c + jd) = \underbrace{a - c}_{\substack{\text{subtract the}\\\text{real parts}}} + j\underbrace{(b - d)}_{\substack{\text{subtract the}\\\text{imaginary parts}}}$

Powers of j are given by

$$j^2 = -1, \quad j^3 = -j, \quad j^4 = 1$$

Multiplication of complex numbers is given by

10.11 $(a + jb)(c + jd) = \underbrace{ac}_{\text{F}} + \underbrace{jad}_{\text{O}} + \underbrace{jbc}_{\text{I}} + \underbrace{j^2 bd}_{\text{L}}$

The complex conjugate of $a + jb$ is

10.12 $a - jb$

Division of complex numbers is given by

10.13 $\dfrac{a + jb}{c + jd} = \dfrac{(a + jb)(c - jd)}{c^2 + d^2}$

Exercise 10(a)

Solutions at end of book. Complete solutions available at www.palgrave.com/science/engineering/singh

1 Let $u = 2 + j3$ and $v = 5 + j8$. Express the following in the form $a + jb$ where a and b are real:

a $u + v$ **b** $u - v$
c uv **d** vu
e u/v **f** v/u
g $2u + v$ **h** u^2
i v^2 **j** $u^2 + v^2$

2 State the complex conjugate of the following:

$3 - j15$, $3 + j15$, $j3$, j, 0, π and e.

3 If $z = 3 - j$ express $z^2 + 7z + 13$ in the form $a + jb$ where a and b are real.

4 Simplify j^{20}, j^9, j^{23}, j^{100} and j^{1007}.

5 **i** If $z = 2 + j$ find z^2, z^3, z^4 and
$z^4 - 6z^2 + 25$.
ii Find one of the roots of
$z^4 - 6z^2 + 25 = 0$.

6 [electrical principles] Determine the admittance, $Y = \dfrac{1}{Z}$, in a circuit, given that $Z = (75 - j40)\ \Omega$.

(Ω = ohm is the SI unit of impedance).

7 Assuming a and b are real, find them if

a $a + jb = 2 + j3$

b $a + jb = (2 + j3) - (2 - j3)$

c $a + jb = (1 + j)^2$

d $a + jb = \dfrac{1}{3 - j4}$

e $2a + jb = (1 + 2j)^2$

f $1.8a + j3.4b = \dfrac{7 + j5}{2 - j}$

Exercise **10(a) continued**

Solutions at end of book. Complete solutions available at www.palgrave.com/science/engineering/singh

8 Express the following in the form $a + jb$ where a and b are real:

a $-j5(3 + j4)$

b $\dfrac{3 + j4}{j5}$

c $(2 - j3)(1 + j)$

d $\dfrac{3 + j4}{j5} - \dfrac{2 - j3}{1 - j}$

9 [electrical principles] If $Z_1 = (1 + j)$ ohm and $Z_2 = (3 + j4)$ ohm are impedances connected in parallel in a circuit, find the total impedance, Z_t, in the circuit given that

$$\frac{1}{Z_t} = \frac{1}{Z_1} + \frac{1}{Z_2}$$

10 [electrical principles] A current, I, in a magnetically coupled circuit satisfies

$$20I + j100I = 200$$

Find the current I.

11 Find the exact values of z which satisfy the following equations:

a $z^2 + 2z + 26 = 0$

b $z^2 - 2z + 3 = 0$

c $3z^2 - 7z + 13 = 0$

By using a computer algebra system or a graphical calculator, plot on different axes the graphs $z^2 + 2z + 26$, $z^2 - 2z + 3$ and $3z^2 - 7z + 13$.

By looking at these graphs can you see why there are no real roots of these equations?

The remaining questions in this exercise are in the field of [*electrical principles*].

12 An a.c. voltage V across a circuit is given by $V = (3 + j6)$ volts. Given that the impedance Z of the circuit is $(8 + j)\,\Omega$ find the current I where

$$I = \frac{V}{Z}$$

13 Let I_p and I_s represent the primary and secondary currents through a mutually magnetically coupled circuit which satisfies the following equations

$$10 = 4I_p + j6I_p - j4I_s$$
$$0 = 50I_s + j12I_s + 8I_s - j4I_p$$

Determine I_p and I_s. (If you have difficulty with this question try it after Example 17.)

14 A resistance, R, and a capacitance, C, are connected in parallel. The impedance, Z, of the circuit is given by

$$\frac{1}{Z} = \frac{1}{R} + \frac{1}{X_c}$$

where $X_c = \dfrac{1}{j\omega C}$

(ω = angular frequency)

i Show that $Z = \dfrac{R}{1 + j\omega CR}$

ii Find the real and imaginary parts of Z. (Re(Z) and Im(Z) respectively.)

SECTION B **Representation of complex numbers**

By the end of this section you will be able to:
- ▶ show how a complex number can be represented on an Argand diagram
- ▶ determine the polar form of a complex number
- ▶ convert from polar to rectangular form, and vice versa
- ▶ use a calculator for conversion

B1 **Complex plane**

We know from our school days that a real number can be represented on a real line. For example, the number 5 can be represented as shown in Fig. 2.

Fig. 2

? **How can we represent a complex number such as $5 + j2$?**

Consider the x–y plane, where the y axis represents the imaginary numbers and the x axis represents the real numbers. Then $5 + j2$ is 5 across and 2 up (Fig. 3).

Fig. 3

? **Where are the following points on the plane?**

a $5 - j2$

b $-5 + j2$

c $-5 - j2$ (see Fig. 4)

Fig. 4

This plane is called the **complex plane** or the **Argand diagram**.

The Argand diagram is named after the amateur French mathematician **Jean Robert Argand** (1768–1822). He also developed the concept of the modulus of a complex number.

A complex number $x + jy$ can be represented by the point (x, y).

Figure 5 shows the x–y plane where the x **axis** is the **real axis**, and the y **axis** is the **imaginary axis**.

Fig. 5

A complex number, written $x + jy$, is said to be in rectangular form because x and y represent the rectangular co-ordinates of the complex number $x + jy$ (Fig. 6).

Fig. 6

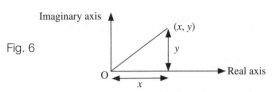

B2 Polar form of complex numbers

A complex number $z = x + jy$ can also be represented by its distance r from the origin and the angle θ it makes with the positive real axis. This is known as the **polar form** representation of z. You will see in **Section** C why this kind of representation is useful in that it allows us to multiply and divide complex numbers in a more straightforward manner than discussed earlier. First though, we need to familiarize ourselves with polar forms.

Fig. 7

By examining Fig. 7 and using trigonometry we have

$$x = r\cos(\theta) \text{ and } y = r\sin(\theta)$$

$$z = x + jy = r\cos(\theta) + j\,r\sin(\theta)$$

$$= r[\cos(\theta) + j\sin(\theta)] \quad \text{[Taking out } r\text{]}$$

So the polar form of $z = x + jy$ can be written as

10.14 $z = r[\cos(\theta) + j\sin(\theta)] \text{ or } r\angle\theta$

where r is the length OP and θ is the angle shown in Fig. 7. $r\angle\theta$ is a shorthand notation for $r[\cos(\theta) + j\sin(\theta)]$.

By using Pythagoras:

10.15 $r = \sqrt{x^2 + y^2}$

By trigonometry we have

10.16 $\theta = \tan^{-1}\left(\dfrac{y}{x}\right)$

We need to be **careful** in determining the angle θ. We plot the complex number on an Argand diagram to check that the angle θ lies in the correct quadrant.

The length r is called the **modulus** of z, denoted by |z| and

$$|z| = \sqrt{x^2 + y^2}$$

θ is called the **argument** of z, denoted by arg(z) and

$$\arg(z) = \tan^{-1}\left(\dfrac{y}{x}\right)$$

If θ satisfies $-180° < \theta \le 180°$, then θ is called the **principal argument**. (This convention gives slightly easier arithmetic and ensures that we only have **one** value for θ.)

Example 10

Express $2 + j3$ in polar form.

Solution
To write a complex number in polar form, it is good practice to sketch the number on an Argand diagram (Fig. 8) and then evaluate the modulus and argument. You will see in **Example 11** why this is so important.

Fig. 8

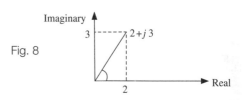

Using $|x + jy| = \sqrt{x^2 + y^2}$ with

$x = 2$ and $y = 3$, the modulus is

$$|2 + j3| = \sqrt{2^2 + 3^2} = \sqrt{13}$$

Using $\arg(x + jy) = \tan^{-1}\left(\dfrac{y}{x}\right)$ with $x = 2$ and $y = 3$ the argument is

$$\arg(2 + j3) = \tan^{-1}\left(\frac{3}{2}\right) = 56.31° \quad \text{[By calculator]}$$

In polar form, $2 + j3$ is (Fig. 9):

Fig. 9

$$\sqrt{13}\left[\cos(56.31°) + j\sin(56.31°)\right] \text{ or } \sqrt{13}\angle 56.31°$$

Example 11

Find the modulus and the principal argument of $-1 + j2$.

Solution

We first sketch $-1 + j2$ on an Argand diagram (Fig. 10).

Let r be the modulus and $\arg(-1 + j2)$ be the argument of $-1 + j2$. We have

Fig. 10

$$r = |-1 + j2| \underset{\text{by } 10.15}{=} \sqrt{(-1)^2 + 2^2} = \sqrt{5}$$

$$\text{angle} \underset{\text{by } 10.16}{=} \tan^{-1}\left(\frac{2}{-1}\right) = -63.43° \quad \text{[By calculator]}$$

This angle, $-63.43°$, is illustrated in Fig. 11.

This does not correspond to the complex number $-1 + j2$ in Fig. 10.

Fig. 11

10.15 $|x + jy| = \sqrt{x^2 + y^2}$ 10.16 $\theta = \tan^{-1}(y/x)$

Example 11 *continued*

? **Why is this?**

Well, the calculator gives the angle $-63.43°$ but we know from Fig. 10 that $\arg(-1 + j2)$ lies in a different quadrant. We can calculate the argument by subtracting $63.43°$ from $180°$ (Fig. 12).

Fig. 12 $180°$ ─── $63.43°$ ─── arg($-1+j2$) ─── $0°$
$-63.43°$

Hence

$$\arg(-1 + j2) = 180° - 63.43° = 116.57°$$

$-1 + j2$ in polar form is

$$\sqrt{5}[\cos(116.57°) + j\sin(116.57°)] \text{ or } \sqrt{5}\angle 116.57°$$

Note that the calculator did not in this example give the correct angle. Below we illustrate how to use a calculator to find the modulus and the correct principal argument.

? **How do we write 2 in polar form?**

Plotting 2 on an Argand diagram: ├────────┼──► Real
                                          2

Thus the modulus of 2 is 2 and the argument is $0°$, therefore $2 = 2\angle 0°$.

? **How do we write $j2$ in polar form?**

On an Argand diagram: Imaginary ▲
                              2 ┤
                                │ 90°
                                └────────► Real

The modulus of $j2$ is 2 and the argument is $90°$:

$$j2 = 2\angle 90° \quad \text{[Polar form]}$$

? **How do we write -2 in polar form?**

-2 can be represented by   ───┼───┬── 180° ──► Real
                                 -2

The modulus of -2 is 2 and the argument is $180°$:

$$-2 = 2\angle 180° \quad \text{[Polar form]}$$

? **How do we write $-j2$ in polar form?**

On an Argand diagram: ┌──────────► Real
                         │ $-90°$
                         │
                       -2 ┤

The modulus of $-j2$ is 2 and the argument is $-90°$:

$$- j2 = 2\underline{/(-90°)} \qquad \text{[Polar form]}$$

Let's do an example from polar to rectangular. This way round is easy!

Example 12

Express $2\underline{/53°}$ in rectangular form, $x + jy$ (x and y are real).

Solution

Putting $r = 2$ and $\theta = 53°$ into $r[\cos(\theta) + j\sin(\theta)]$ gives

$$2[\cos(53°) + j\sin(53°)] = 2(0.6 + j\,0.8)$$

$$= (2 \times 0.6) + j(2 \times 0.8)$$

$$= 1.2 + j1.6$$

When transforming from polar to rectangular form you can avoid making a sketch because you do not have to find an angle.

Most basic scientific calculators will transform a complex number from rectangular to polar or polar to rectangular.

For example, convert $3.1 - j6.5$ into polar form. On a calculator, PRESS

| POL(| | 3.1 | | , | | (–) | | 6.5 | |) | | = | which should show 7.201388755.

This is the modulus of $3.1 - j6.5$. Then PRESS
| ALPHA | | tan | | = | which should show -64.50244851.

This is the argument of $3.1 - j6.5$. Hence $3.1 - j6.5 = 7.2\underline{/(-64.5°)}$. Make sure the calculator is in degrees mode.

(If these instructions do not work, see the handbook of your calculator.)

Note that the calculator gives the **principal argument**. We will often use the shorthand notation $7.2\underline{/(-64.5°)}$ rather than $7.2[\cos(-64.5°) + j\sin(-64.5°)]$.

Let's consider polar to rectangular.

For example, convert $7.2\underline{/(-64.5°)}$ into rectangular form.

On a calculator, PRESS

| SHIFT | | Pol(| | 7.2 | | , | | (–) | | 64.5 | |) | | = | which should show 3.099679897.

Then PRESS

| ALPHA | | tan | | = | which should show -6.498614047.

This means $7.2\underline{/(-64.5°)} = 3.1 - j6.5$ (to 1 d.p.) in rectangular form.

The $\boxed{\text{SHIFT}}$ key gives access to a secondary function. Some calculators do not have a $\boxed{\text{SHIFT}}$ button but a $\boxed{\text{2nd Fun}}$ or $\boxed{\text{INV}}$ button instead.

Use your calculator to convert from polar to rectangular form, or vice versa.

Next we examine a mechanics example. A resultant force is the force which is equivalent to two or more forces. The following example uses the conversion from polar to rectangular form, and vice versa, to find the resultant of three forces. It is important that you realize forces are not complex numbers but we can use the concept of conversion (polar ↔ rectangular) to find the resultant force. The example can also be tackled by using vectors (Chapter 12).

Example 13 *mechanics*

Figure 13 shows (coplanar) forces F_1, F_2 and F_3 acting at a common point O. Find the magnitude and direction of the resultant force, $F = F_1 + F_2 + F_3$.

Fig. 13

Solution

The forces F_1, F_2 and F_3 can be written in polar form as

$$F_1 = 90\underline{/25°}, \; F_2 = 100\underline{/(32° + 25°)} \text{ and } F_3 = 120\underline{/(100° + 32° + 25°)}$$

To add complex numbers we need to write them in rectangular form. Use your calculator to place each number into rectangular form:

$$F_1 = 90\underline{/25°} = 81.57 + j38.04$$

$$F_2 = 100\underline{/(32° + 25°)} = 100\underline{/57°} = 54.46 + j83.87$$

$$F_3 = 120\underline{/(100° + 57°)} = 120\underline{/157°} = -110.46 + j46.89$$

$$\text{Resultant force} = \underbrace{81.57 + j38.04}_{F_1} + \underbrace{54.46 + j83.87}_{F_2} + \underbrace{-110.46 + j46.89}_{F_3}$$

$$= (81.57 + 54.46 - 110.46) + j(38.04 + 83.87 + 46.89)$$

adding the real and imaginary parts

$$= 25.57 + j168.8$$

$$= 170.73\underline{/81.39°} \quad \text{[By calculator]}$$

The resultant force, F, has a modulus of 170.73 N at an angle of 81.39°, as shown in Fig. 14.

Fig. 14

SUMMARY

If $z = x + jy$ then

10.15 $|z| = \sqrt{x^2 + y^2}$ [modulus]

10.16 $\arg(z) = \tan^{-1}\left(\dfrac{y}{x}\right)$ [argument]

We can rewrite z as

$$r[\cos(\theta) + j\sin(\theta)] \text{ or } r\angle\theta$$

where $r = |z|$ and $\theta = \arg(z)$.

Exercise **10(b)**

Solutions at end of book. Complete solutions available at www.palgrave.com/science/engineering/singh

1 Label the following complex numbers on an Argand diagram:

  a $2 + j3$ **b** $-1 - j$ **c** $-1 + j$
  d $-j2$ **e** -3

2 Without using a calculator, express the following in polar form, giving the principal argument:

  a $1 + j$ **b** $-1 + j$ **c** $\sqrt{3} - j$
  d $-1 - j\sqrt{3}$ **e** j **f** jx
  g -3 **h** 5

3 Express the following in rectangular form:

  a $2\angle30°$ **b** $5\angle23°$ **c** $2\angle(-45°)$

  d $1.4\angle(-130°)$

4 🚗 [mechanics]

Fig. 15

Figure 15 shows coplanar forces F_1, F_2, F_3 and F_4 acting at a common point O. Find the magnitude and direction of the resultant force F where

$$F = F_1 + F_2 + F_3 + F_4$$

5 🚗 [mechanics]

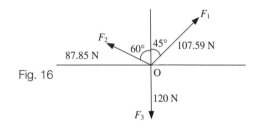

Fig. 16

Figure 16 shows three coplanar forces, F_1, F_2 and F_3, acting at a common point O. Find the resultant force $F = F_1 + F_2 + F_3$. From your solution, what can you conclude about F_1, F_2 and F_3?

SECTION C **Multiplication and division in polar form**

By the end of this section you will be able to:
► use the multiplication and division rules for complex numbers in polar form
► apply the polar form of complex numbers to engineering problems

C1 **Multiplication and division in polar form**

Example 14

Let $z = r\left[\cos(A) + j\sin(A)\right]$ and $w = q\left[\cos(B) + j\sin(B)\right]$ be any two complex numbers (Fig. 17). Show that

$$zw = rq\left[\cos(A + B) + j\sin(A + B)\right]$$

Fig. 17

Solution

Multiplying the two complex numbers gives

$$zw = rq\left\{\left[\cos(A) + j\sin(A)\right] \times \left[\cos(B) + j\sin(B)\right]\right\}$$

$$= rq\left\{\cos(A)\cos(B) + j\sin(B)\cos(A) + j\sin(A)\cos(B) + \underset{=-1}{j^2}\sin(A)\sin(B)\right\}$$
$$\underset{\text{using FOIL}}{}$$

$$= rq\left\{\underbrace{\left[\cos(A)\cos(B) - \sin(A)\sin(B)\right]}_{\text{real part}} + j\underbrace{\left[\sin(B)\cos(A) + \sin(A)\cos(B)\right]}_{\text{imaginary part}}\right\}$$

$$= rq\left\{\underbrace{\cos(A + B)}_{\text{by }4.39} + j\underbrace{\sin(A + B)}_{\text{by }4.37}\right\}$$

Hence if $z = r[\cos(A) + j\sin(A)]$ and $w = q[\cos(B) + j\sin(B)]$ then

Multiplication

10.17 $zw = rq\left[\cos(A + B) + j\sin(A + B)\right]$ or $r\angle A \times q\angle B = rq\angle(A + B)$

Division

10.18 $\dfrac{z}{w} = \dfrac{r}{q}\left[\cos(A - B) + j\sin(A - B)\right]$ or $\dfrac{r\angle A}{q\angle B} = \dfrac{r}{q}\angle(A - B)$

4.37 $\sin(A + B) = \sin(A)\cos(B) + \cos(A)\sin(B)$
4.39 $\cos(A + B) = \cos(A)\cos(B) - \sin(A)\sin(B)$

You are asked to show result 10.18 in **Exercise 10(c)**.

? **Can you see the advantages of putting a complex number in polar form?**

i 10.17 shows that multiplication of complex numbers is just a matter of **multiplying** their **moduli** and **adding** their **arguments**.

ii Similarly 10.18 shows that division of complex numbers is simply a matter of **dividing** their **moduli** and **subtracting** their **arguments**.

Generally it is better to use complex numbers in polar form if you want to multiply or divide two or more complex numbers.

Example 15

Express $(-1.2 + j)(-5 + j1.3)$ in polar form.

Solution

First use your calculator to transform each number into polar form:

$$-1.2 + j = 1.56\underline{/140.19°}$$

$$-5 + j1.3 = 5.17\underline{/165.43°}$$

Hence

$$(-1.2 + j)(-5 + j1.3) = (1.56\underline{/140.19°}) \times (5.17\underline{/165.43°})$$
$$= 1.56 \times 5.17\underline{/(140.19° + 165.43°)}$$
$$\text{by } 10.17$$
$$= 8.07\underline{/305.62°}$$

Since the question does not ask for the principal argument we can leave the argument as 305.62°.

C2 **Engineering applications**

Example 16 *electrical principles*

An a.c. voltage, $v = (7.6 - j6.8)$ volts, is applied across a circuit of impedance $z = (88 + j5)\,\Omega$. Find the current i through the circuit, given that $i = \dfrac{v}{z}$ (you may leave your answer in polar form).

Solution

Putting v and z into polar form via a calculator yields

$$7.6 - j6.8 = 10.20\underline{/(-41.82°)}$$
$$88 + j5 = 88.14\underline{/\,3.25°}$$

10.17 $r\underline{/A} \times q\underline{/B} = rq\underline{/(A + B)}$

Example 16 *continued*

Hence

$$i = \frac{7.6 - j6.8}{88 + j5} = \frac{10.20\underline{/(-41.82°)}}{88.14\underline{/3.25°}}$$

$$\underset{\text{by } 10.18}{=} \frac{10.20}{88.14}\underline{/(-41.82° - 3.25°)}$$

$$i = 0.12\underline{/(-45.07°)} \text{ A}   (\text{A = amp})$$

The next example is particularly challenging. We need to solve simultaneous equations where the coefficients of the variables are complex numbers. Addition or subtraction of complex numbers is carried out in rectangular form but multiplication and division is normally easier in polar form. Hence you will find that we interchange between polar ↔ rectangular as the arithmetic requires.

Example 17 *electrical principles*

Let I_1 and I_2 be the primary and secondary currents respectively, of a magnetically coupled circuit. By Kirchhoff we obtain the following mesh equations:

$$20I_1 + j100I_1 - j20I_2 = 200$$

$$20I_2 + j100I_2 + 100I_2 - j20I_1 = 0$$

Determine, in polar form, the primary and secondary currents, I_1 and I_2 respectively, through the circuit.

Solution

Since 20 is a common factor of both equations we divide through by 20:

$$I_1 + j5I_1 - jI_2 = 10$$

$$I_2 + j5I_2 + 5I_2 - jI_1 = 0$$

Adding jI_1 to both sides of the second equation gives

$$I_2 + j5I_2 + 5I_2 = jI_1$$

$$(6 + j5)I_2 = jI_1   \text{[Factorizing } I_2\text{]}$$

10.18 $$\frac{r\underline{/A}}{r\underline{/B}} = \frac{r}{q}\underline{/(A - B)}$$

Example 17 *continued*

Dividing by $6 + j5$ gives

$$\boxed{} \quad I_2 = \frac{jI_1}{6 + j5}$$

Substituting this for I_2 into the first equation, $10 = (1 + j5)I_1 - jI_2$, gives

$$10 = (1 + j5)I_1 - j\frac{jI_1}{6 + j5}$$

$$= \left[1 + j5 - \frac{j^2}{6 + j5} \right] I_1 \quad \text{[Taking out I_1]}$$

$$= \left[1 + j5 + \frac{6 - j5}{6^2 + 5^2} \right] I_1 \quad \text{[Because $j^2 = -1$]}$$

$$= \left[1 + j5 + \frac{6 - j5}{61} \right] I_1$$

$$10 = \left[\frac{67 + j300}{61} \right] I_1 \quad \text{[Adding with common denominator]}$$

$$I_1 = \frac{61 \times 10}{[67 + j300]} = \frac{610\angle 0°}{307.39\angle 77.41°} \underset{\text{by } \boxed{10.18}}{=} \frac{610}{307.39}\angle(0° - 77.41°)$$

$$I_1 = 1.98\angle(-77.41°)\ \text{A}$$

Substituting for I_1 into $\boxed{}$:

$$I_2 = \frac{j}{6 + j5}1.98\angle(-77.41°)$$

$$= \frac{1\angle 90° \times 1.98\angle(-77.41°)}{7.81\angle 39.81°} \quad \text{[Polar form]}$$

$$= \frac{1 \times 1.98}{7.81}\angle(90° + (-77.41°) - 39.81°) \quad \text{[By } \boxed{10.17} \text{ and } \boxed{10.18} \text{]}$$

$$I_2 = 0.25\angle(-27.22°)\ \text{A}$$

In a similar manner, try question **13** of **Exercise 10(a)**.

$\boxed{10.17}$ $r\angle A \times q\angle B = rq\angle(A + B)$ $\boxed{10.18}$ $\dfrac{r\angle A}{r\angle B} = \dfrac{r}{q}\angle(A - B)$

C3 Important angles

Some angles are more important than others. The sin, cos and tan of these angles, in degrees and radians, is given below. Some of these results were shown in Chapter 4.

TABLE 1

angle θ	0	$\dfrac{\pi}{6}$	$\dfrac{\pi}{4}$	$\dfrac{\pi}{3}$	$\dfrac{\pi}{2}$	$\dfrac{2\pi}{3}$	$\dfrac{3\pi}{4}$	$\dfrac{5\pi}{6}$	π
	$0°$	$30°$	$45°$	$60°$	$90°$	$120°$	$135°$	$150°$	$180°$
$\sin(\theta)$	0	$\dfrac{1}{2}$	$\dfrac{1}{\sqrt{2}}$	$\dfrac{\sqrt{3}}{2}$	1	$\dfrac{\sqrt{3}}{2}$	$\dfrac{1}{\sqrt{2}}$	$\dfrac{1}{2}$	0
$\cos(\theta)$	1	$\dfrac{\sqrt{3}}{2}$	$\dfrac{1}{\sqrt{2}}$	$\dfrac{1}{2}$	0	$-\dfrac{1}{2}$	$-\dfrac{1}{\sqrt{2}}$	$-\dfrac{\sqrt{3}}{2}$	-1
$\tan(\theta)$	0	$\dfrac{1}{\sqrt{3}}$	1	$\sqrt{3}$	un-def	$-\sqrt{3}$	-1	$-\dfrac{1}{\sqrt{3}}$	0

(In Table 1 un-def means undefined.) Remember that the table shows the exact trigonometric values.

The argument of a complex number might be in radians rather than degrees.

Example 18

Find the exact rectangular form of

$$\frac{1}{\sqrt{2}}\angle\left(\frac{\pi}{4}\right)$$

Solution
Using

$$r\angle\theta = r[\cos(\theta) + j\sin(\theta)]$$

with $r = \dfrac{1}{\sqrt{2}}$ and $\theta = \dfrac{\pi}{4}$ we have

$$\frac{1}{\sqrt{2}}\angle\left(\frac{\pi}{4}\right) = \frac{1}{\sqrt{2}}\left[\cos\left(\frac{\pi}{4}\right) + j\sin\left(\frac{\pi}{4}\right)\right] = \frac{1}{\sqrt{2}}\left[\frac{1}{\sqrt{2}} + j\left(\frac{1}{\sqrt{2}}\right)\right]$$

by Table 1

$$= \frac{1}{\sqrt{2}}\frac{1}{\sqrt{2}}[1 + j] \quad \left[\text{Taking out } \frac{1}{\sqrt{2}}\right]$$

$$= \frac{1}{2}[1 + j] \quad \left[\text{Because } \frac{1}{\sqrt{2}}\frac{1}{\sqrt{2}} = \frac{1}{2}\right]$$

SUMMARY

If $z = r\underline{/A}$ and $w = q\underline{/B}$ then

| 10.17 | $zw = rq\underline{/(A + B)}$ | [Multiplication] |

| 10.18 | $\dfrac{z}{w} = \dfrac{r}{q}\underline{/(A - B)}$ | [Division] |

Exercise 10(c)

Solutions at end of book. Complete solutions available at www.palgrave.com/science/engineering/singh

1 🔲 [*electrical principles*]   An a.c. voltage, $v = (2.9 + j6.2)$ volts, is applied across an impedance $Z = (8 + j1.9)$ ohms. Find the current, i, in polar form, through the circuit given that

$$i = \frac{v}{Z}$$

2 Find the modulus and argument of

$$z = \frac{2\underline{/45°} \times 2\underline{/(-60°)}}{4\underline{/(-22°)} \times 5\underline{/33°}}$$

3 Without using a calculator, express the following in rectangular form:

a $2\underline{/30°}$ **b** $2\underline{/\left(\dfrac{\pi}{6}\right)}$ **c** $240\underline{/0°}$

4 If $z = r\underline{/\theta}$ show that $\bar{z} = r\underline{/(-\theta)}$ ($\bar{z}$ is the complex conjugate of z.)

5 Find $\bar{z}$ (complex conjugate of z) in polar form, if

a $z = 6\underline{/45°}$ **b** $z = 2\underline{/131°}$
c $z = 15\underline{/(-26°)}$

6 i Show that, if $z = a + jb$ such that $|z| = 1$ then

$$\frac{1}{z} = \bar{z}$$

ii If $z = \dfrac{1}{3}\left(\sqrt{8} + j\right)$ then determine $\dfrac{1}{z}$.

7 Let $z = r[\cos(\alpha) + j\sin(\alpha)]$ and $w = q[\cos(\beta) + j\sin(\beta)]$. Show that

$$\frac{z}{w} = \frac{r}{q}\left[\cos(\alpha - \beta) + j\sin(\alpha - \beta)\right]$$

8 🔲 [*electronics*]   An amplifier has a transfer function, T, given by

$$T = \frac{1000}{1 + j\omega(0.5 \times 10^{-3})}$$

where ω is the angular frequency. The gain and phase of the amplifier is given by the modulus and argument of T respectively. Find the gain and phase at an angular frequency of $\omega = 2 \times 10^3$ rad/s.

9 🔲 [*control engineering*]   A system has a transfer function, T, given by

$$T = \frac{50}{j\omega(1 + j0.01\omega)(1 + j0.2\omega)}$$

Evaluate the gain (modulus of T) and phase (argument of T) of the system for $\omega = 45$ rad/s.

10 🔲 [*electrical principles*]   By applying Kirchhoff's laws on a circuit we obtain the following mesh equations:

$$(6 + j2)I_1 + (2 + j2)I_2 = 10$$
$$(2 + j2)I_1 + 7I_2 = 10$$

Determine, in polar form, the currents I_1 and I_2.

SECTION D **Powers and roots of complex numbers**

By the end of this section you will be able to:
► apply de Moivre's theorem
► find roots of complex numbers
► plot these roots on an Argand diagram

D1 **De Moivre's theorem**

? Let $z = r[\cos(\theta) + j\sin(\theta)]$ be a complex number. **What are z^2 and z^3 equal to?**

$$z^2 = z \cdot z \underset{\text{by } 10.17}{=} r \cdot r[\cos(\theta + \theta) + j\sin(\theta + \theta)]$$

$$z^2 = r^2[\cos(2\theta) + j\sin(2\theta)] \text{ or } r^2\underline{/2\theta}$$

Similarly we find

$$z^3 = r^3[\cos(3\theta) + j\sin(3\theta)] \text{ or } r^3\underline{/3\theta}$$

? **What do you think z^n will be?**

10.19 $z^n = r^n[\cos(n\theta) + j\sin(n\theta)]$ or $(r\underline{/\theta})^n = r^n\underline{/n\theta}$

This is known as de Moivre's theorem. To use de Moivre's theorem we first place the complex number in polar form.

> **Abraham de Moivre** (1667–1754) was born in France but spent most of his life in England. This rule was known to de Moivre for the case n is a whole number but the result also holds for a fraction.

We also have

$$\frac{1}{z^n} = z^{-n} = r^{-n}[\cos(-n\theta) + j\sin(-n\theta)]$$

$$= r^{-n}\big[\underbrace{\cos(n\theta)}_{\text{by } 4.51} + j(\underbrace{-\sin(n\theta)}_{\text{by } 4.50})\big] = r^{-n}[\cos(n\theta) - j\sin(n\theta)]$$

Thus we have

10.20 $z^{-n} = r^{-n}[\cos(n\theta) - j\sin(n\theta)]$

We will need to use the following definition of indices in this section:

10.21 $\sqrt[n]{z} = z^{1/n} \quad \left(\sqrt{z} = z^{1/2}\right)$

4.50 $\sin(-x) = -\sin(x)$ **4.51** $\cos(-x) = \cos(x)$

10.17 $zw = rq[\cos(A + B) + j\sin(A + B)]$

Example 19

Evaluate $(1 - j)^5$ and plot the result on an Argand diagram.

Solution

First we place $(1 - j)$ in polar form:

$$1 - j = \sqrt{2}\underline{/(-45°)}$$
$$= 2^{1/2}\underline{/(-45°)}$$

Taking to the power of 5:

$$(1 - j)^5 = \left(2^{1/2}\underline{/(-45°)}\right)^5$$
$$\underset{\text{by } \boxed{10.19}}{=} (2^{1/2})^5 \underline{/\left(5\times(-45°)\right)} \underset{\text{by } \boxed{10.2}}{=} 2^{5/2} \underline{/(-225°)}$$

Putting this, $2^{5/2}\underline{/(-225°)}$, into rectangular form by using a calculator gives

$$(1 - j)^5 = 2^{5/2}\underline{/(-225°)} = -4 + j4$$

The angle is measured in a clockwise direction from the positive real axis (Fig. 18).

Fig. 18

The next example is a leap in sophistication compared to the above example. It uses algebraic substitution, rules of complex division and powers of complex numbers. Also we don't end up with simple whole numbers after substitution. The only other hiccup might be writing $j6.28 \times 10^{-4}$ in polar form. Remember that jx (x is real) on an Argand diagram (Fig. 19) is

Fig. 19

$$jx = x\underline{/90°} \qquad \text{[Polar form]}$$

Similarly

$$j(6.28 \times 10^{-4}) = (6.28 \times 10^{-4})\underline{/90°} \quad \text{[Polar form]}$$

$\boxed{10.2}$ $\quad (a^m)^n = a^{m \times n}$ $\qquad$ $\boxed{10.19}$ $\quad (r\underline{/\theta})^n = r^n\underline{/n\theta}$

Example 20 *electrical principles*

The characteristic impedance, Z_0, of a transmission line is given by

$$Z_0 = \sqrt{\frac{R + j2\pi fL}{G + j2\pi fC}}$$

where $R = 5\ \Omega$, $L = 0.03 \times 10^{-3}\,\text{H}$, $C = 0.01 \times 10^{-6}\,\text{F}$, $G = 0$ siemen and $f = 10^4\,\text{Hz}$. Determine Z_0 where $\text{Re}(Z_0) \geq 0$.

Solution

In this example we use de Moivre's theorem with $n = 1/2$.

Substituting the given values into the numerator we have

$$R + j2\pi fL = 5 + j(2\pi \times 10^4 \times 0.03 \times 10^{-3})$$

$$= 5 + j1.88$$

$$= 5.34\angle 20.61° \qquad\qquad \text{[Polar form]}$$

Substituting the given values into the denominator:

$$G + j2\pi fC = 0 + j(2\pi \times 10^4 \times 0.01 \times 10^{-6})$$

$$= j(6.28 \times 10^{-4})$$

$$= (6.28 \times 10^{-4})\angle 90° \qquad\qquad \text{[Polar form]}$$

Substituting these gives

$$Z_0 = \left(\frac{R + j2\pi fL}{G + j2\pi fC}\right)^{\frac{1}{2}}$$

$$= \left(\frac{5.34\angle 20.61°}{6.28 \times 10^{-4}\angle 90°}\right)^{\frac{1}{2}}$$

$$\underset{\text{by } 10.18}{=} \left(\frac{5.34}{6.28 \times 10^{-4}}\angle(20.61° - 90°)\right)^{\frac{1}{2}}$$

$$= \left(8503.18\angle(-69.39°)\right)^{\frac{1}{2}}$$

$$\underset{\text{by } 10.19}{=} 8503.18^{1/2}\angle\left(\frac{1}{2} \times (-69.39°)\right)$$

$$= 92.21\angle(-34.70°)$$

$$Z_0 = 92.21\angle(-34.70°)\ \Omega = (75.81 - j52.49)\ \Omega \qquad \text{[Rectangular form]}$$

10.18 $\dfrac{r\angle A}{q\angle B} = \dfrac{r}{q}\angle(A - B)$ **10.19** $z^n = r^n\angle n\theta$

D2 **Finding roots**

De Moivre's theorem can be used to find roots of complex numbers. The **Fundamental Theorem of Algebra** says that the complex number

$$(x + jy)^{1/n} \text{ has } \textbf{exactly } n \text{ roots}$$

? **How many roots do $(1 + j)^{1/2}$, $(1 + j)^{1/5}$ and $(1 + j)^{1/12}$ have?**

2, 5 and 12 roots respectively.

? **Why have we only one root for Example 20?**

Since in the question we have $\text{Re}(Z_0) \geq 0$ which means the real part of Z_0 is positive or zero, so we do not look for the other root. We will show later that the other root is $(-75.81 + j52.49) \, \Omega$.

We can represent any complex number in polar form in many different ways. For example consider the real number 2 (Fig. 20).

Fig. 20

$2 = 2\angle 0°$

1 complete revolution
$2 = 2\angle 360°$

2 complete revolutions
$2 = 2\angle 720°$

Hence we can make complete revolutions and still get back to the same point. **How else can we
? represent the complex number $4\angle 30°$ in polar form?**

Fig. 21

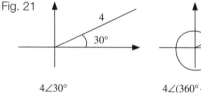

$4\angle 30°$

$4\angle(360° + 30°) = 4\angle 390°$

See Fig. 21. Thus $4\angle 30° = 4\angle 390°$.

We can use this concept to find **all** the roots of complex numbers, as the following example shows.

Example 21

Solve

$$z^2 - (2\sqrt{3} + j2) = 0$$

Solution

Adding $2\sqrt{3} + j2$ to both sides of the equation so that the complex number is on the Right-Hand Side:

$$z^2 = \left(2\sqrt{3} + j2\right) \text{ gives } z = \sqrt{(2\sqrt{3} + j2)}$$
$$\underline{= \left(2\sqrt{3} + j2\right)^{1/2}}$$
by 10.21

10.21 $\sqrt{z} = z^{1/2}$

Example 21 *continued*

? **How many roots does this equation have?**

Two roots.

1st root. To use de Moivre's theorem we need to put the complex number, $2\sqrt{3} + j2$, into polar form (via a calculator or otherwise):

$$2\sqrt{3} + j2 = 4\angle 30°$$

Applying de Moivre's theorem:

$$\left(2\sqrt{3} + j2\right)^{1/2} = (4\angle 30°)^{1/2}$$

Fig. 22

$$\underset{\text{by } 10.19}{=} 4^{1/2}\angle\left(\frac{1}{2} \times 30°\right)$$

$$= 2\angle 15°$$

? Hence one of the roots of $\left(2\sqrt{3} + j2\right)^{1/2}$ is $2\angle 15°$. **What about the other root?**

2nd root. If we make one complete revolution (360°) from $4\angle 30°$ we get back to the same point. Therefore

$$2\sqrt{3} + j2 \text{ is } 4\angle 30° \text{ or } 4\angle(30° + 360°) = 4\angle 390°$$

So for the second root we have

$$\left(2\sqrt{3} + j2\right)^{1/2} = \left(4\angle 390°\right)^{1/2}$$

$$\underset{\text{by } 10.19}{=} 4^{1/2}\angle\left(\frac{1}{2} \times 390°\right) = 2\angle 195°$$

So the **two** roots of $\left(2\sqrt{3} + j2\right)^{1/2}$ are $2\angle 15°$ and $2\angle 195°$ which can be represented on the Argand diagram as shown in Fig. 23.

In the above example, why not make 2, 3, 4, ... etc. revolutions from $4\angle 30°$?

If we do make 2, 3, 4, ... etc. revolutions we just reproduce one of the above two roots, $2\angle 15°$ or $2\angle 195°$.

Fig. 23

We can label the roots $z_1 = 2\angle 15°$, $z_2 = 2\angle 195°$ and they satisfy the equation given in the question:

$$z^2 - \left(2\sqrt{3} + j2\right) = 0$$

10.19 $(r\angle\theta)^n = r^n\angle n\theta$

Example 22

Find all the roots of

$$z^5 - 1 = 0$$

and plot them on an Argand diagram.

Solution

Adding 1 to both sides of $z^5 - 1 = 0$:

$$z^5 = 1 \text{ which gives } z = 1^{1/5}$$

? **How many roots does this equation have?**

? Five roots. **Can you find any of the roots of this equation?**

1 is an obvious root ($1^5 - 1 = 0$) which we need to place in polar form in order to use de Moivre's theorem. Since 1 is a real number (no j), therefore:

Fig. 24

$1 = 1\angle 0°$ [Polar form]

$$1^{1/5} = (1\angle 0°)^{1/5} = 1\angle 0°$$

? So $1\angle 0°$ is a root of $z^5 = 1$. **Where are the other four roots?**

If we make one, two, three and four complete revolutions from $1\angle 0°$ we will end up at the same point.

$1\angle 0°$, $1\angle 360°$ (one revolution), $1\angle(360° + 360°) = 1\angle 720°$ (two revolutions), $1\angle(720° + 360°) = 1\angle 1080°$ (three revolutions) and $1\angle(1080° + 360°) = 1\angle 1440°$ (four revolutions).

By applying de Moivre's theorem

10.19 $(r\angle\theta)^n = r^n\angle n\theta$

to each of these numbers we obtain

$$1^{1/5} = 1\angle 0°, \ 1^{1/5}\angle\left(\frac{1}{5} \times 360°\right), \ 1^{1/5}\angle\left(\frac{1}{5} \times 720°\right), \ 1^{1/5}\angle\left(\frac{1}{5} \times 1080°\right) \text{ and } 1^{1/5}\angle\left(\frac{1}{5} \times 1440°\right)$$

$$= 1\angle 0°, \ 1\angle 72°, \ 1\angle 144°, \ 1\angle 216° \text{ and } 1\angle 288°$$

We label the roots z_1, z_2, z_3, z_4 and z_5 and show them on an Argand diagram (Fig. 25).

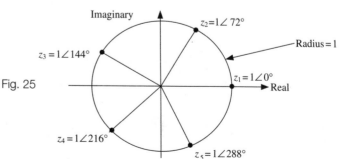

Fig. 25

Example 22 *continued*

Hence the five roots lie on a circle of radius 1 and are

$$z_1 = 1\angle 0°, \; z_2 = 1\angle 72°, \; z_3 = 1\angle 144°, \; z_4 = 1\angle 216° \text{ and } z_5 = 1\angle 288°$$

What do you notice about the size of each angle between the roots?

They are all equal to $72°$, that is

$$144° - 72° = 216° - 144° = \cdots = 72°$$

Hence a full revolution, $360°$, is divided into five equal size angles of

$$\left(\frac{360}{5}\right)° = 72° \left(\text{or in radians } \frac{2\pi}{5}\right)$$

These roots are normally called the five roots of unity.

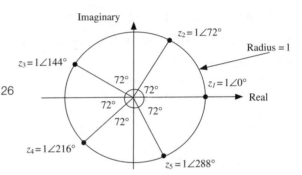

Fig. 26

Also in **Example 21** we evaluated $\left(2\sqrt{3} + j2\right)^{1/2}$ and found that the **two** roots were equally spaced at angles of size

$$\left(\frac{360}{2}\right)° = 180° \left(\text{or } \frac{2\pi}{2} = \pi \text{ radians}\right)$$

For **Example 20** we found one root

$$Z_0 = 92.21\angle(-34.70°)$$

What is the value of the other root?

Since Z_0 was the square root of a complex number we add $\left(\frac{360}{2}\right)° = 180°$:

$$92.21\angle(-34.70° + 180°) = 92.21\angle 145.3°$$

In rectangular form this is

$$-75.81 + j52.49$$

In general, the n complex roots of $z = r\left[\cos(\theta) + j\sin(\theta)\right] = r\angle\theta$ are equally spaced at angles of size $\left(\frac{360}{n}\right)° \left(\text{or } \frac{2\pi}{n} \text{ radians}\right)$. Hence the n roots of z are given by

10.22
$$z^{1/n} = r^{1/n}\angle\left(\frac{\theta + 360k°}{n}\right) \text{ or } z^{1/n} = r^{1/n}\angle\left(\frac{\theta + 2k\pi}{n}\right)$$

where $k = 0, 1, 2, \ldots, n - 1$.

Example 23

Solve

$$z^6 - 1 = j\sqrt{3}$$

and represent the roots on an Argand diagram.

Solution

Adding 1 to both sides:

$$z^6 = 1 + j\sqrt{3} = 2\angle 60° \qquad \text{[Polar form]}$$

Putting $n = 6$ into 10.22 gives

$$\boxed{\quad *\quad} \qquad z = 2^{1/6}\angle\left(\frac{60° + 360k°}{6}\right)$$

where $k = 0, 1, 2, 3, 4$ and 5. The roots are found by substituting these values of k into $\boxed{\;*\;}$:

$(k = 0) \qquad z_1 = 2^{1/6}\angle\left(\dfrac{60°}{6}\right) = 2^{1/6}\angle 10°$

$(k = 1) \qquad z_2 = 2^{1/6}\angle\left(\dfrac{60° + 360°}{6}\right) = 2^{1/6}\angle 70°$

$(k = 2) \qquad z_3 = 2^{1/6}\angle\left(\dfrac{60° + (360 \times 2)°}{6}\right) = 2^{1/6}\angle 130°$

$(k = 3) \qquad z_4 = 2^{1/6}\angle 190°$

$(k = 4) \qquad z_5 = 2^{1/6}\angle 250°$

$(k = 5) \qquad z_6 = 2^{1/6}\angle 310°$ Fig. 27

These roots are equally spaced at angles of

$$\left(\frac{360}{6}\right)° = 60° \text{ on a circle}$$

of radius $2^{1/6}$ (Fig. 27).

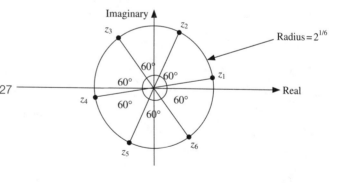

Normally it is easier to find the first root, $r^{1/n}\angle\theta$, and then add $\left(\dfrac{360}{n}\right)°\left(\text{or } \dfrac{2\pi}{n} \text{ radians}\right)$ to the angle θ rather than using formula 10.22.

In **Example 23** we had $z_1 = 2^{1/16}\angle 10°$ and so for the other roots just add 60°.

Hence

$$2^{1/6}\angle(10° + 60°), \; 2^{1/6}\angle(70° + 60°), \; 2^{1/6}\angle(130° + 60°), \dots, 2^{1/6}\angle(250° + 60°)$$

? Also note that we do **not** need k values greater than 5. **Why?**

Each of the k values greater than 5 will reproduce one of the above six roots; $k = 6$ gives the same root as $k = 0$ and $k = 7$ gives the same root as $k = 1$, and so on.

SUMMARY

If $z = r\angle\theta$ then

10.19 $\quad z^n = r^n\angle n\theta$

10.20 $\quad z^{-n} = r^{-n}[\cos(n\theta) - j\sin(n\theta)]$

The n roots of z are given by

10.22 $\quad z^{1/n} = r^{1/n}\angle\left(\dfrac{\theta + 360k°}{n}\right)$ or $z^{1/n} = r^{1/n}\angle\left(\dfrac{\theta + 2k\pi}{n}\right)$

where $k = 0, 1, 2, \ldots, n - 1$. Generally it is easier to find the first root, $r^{1/n}\angle\theta$,

and then add $\left(\dfrac{360}{n}\right)°\left(\text{or } \dfrac{2\pi}{n} \text{ radians}\right)$ to θ.

Exercise 10(d)

Solutions at end of book. Complete solutions available at www.palgrave.com/science/engineering/singh

1 By using your calculator or otherwise, evaluate the following, giving your answer in rectangular form:
   a $(1 + j)^2$ **b** $(1 - j)^3$ **c** $(1 + j)^4$
   d $(6 - j8)^7$ **e** $(-3 + j4)^7$ **f** $(-3 - j)^5$

2 Find all the roots of $z^4 - 1 = 0$.

3 Show that $(1 + j)^n = 2^{n/2}\angle\left(\dfrac{n\pi}{4}\right)$ and evaluate $(1 + j)^{12}$, giving your answer in rectangular form.

4 [electrical principles]   A cable has the following constants: $R = 10\,\Omega$, $L = 0.1 \times 10^{-3}$ H, $G = 1 \times 10^{-6}$ siemen and $C = 1 \times 10^{-9}$ F. For $\omega = 10{,}000$ rad/s, determine:
   i The characteristic impedance, Z_0, where

$$Z_0 = \sqrt{\dfrac{R + j\omega L}{G + j\omega C}}$$
$$(\text{Re}(Z_0) \geq 0)$$

   ii The propagation coefficient, γ, given by

$$\gamma = \sqrt{(R + j\omega L)(G + j\omega C)}$$

   $\text{Re}(\gamma) \geq 0$ and the units of γ are m^{-1}.

   iii The wavelength, $\lambda = \dfrac{2\pi}{\beta}$, where

   $\beta = \text{Im}(\gamma)$, (imaginary part of γ).

   iv The velocity of propagation, $v = \dfrac{\omega}{\beta}$, where $\beta = \text{Im}(\gamma)$.

5 [control engineering]   A robot arm has a transfer function, G, given by

$$G = \dfrac{40}{(\omega + j\omega)^4}$$

   where $\omega = $ angular frequency. Show that

$$G = -\dfrac{10}{\omega^4}$$

6 [control engineering]   A system has the following closed-loop transfer function:

$$G = \dfrac{K(z^4 + 16)}{z^3 + 64}$$

Exercise **10(d) continued**

Solutions at end of book. Complete solutions available at www.palgrave.com/science/engineering/singh

where K is a non-zero constant.
The **poles** of the system occur where the denominator of the transfer function is **zero**, that is $z^3 + 64 = 0$.
The **zeros** of the system occur where the numerator of the transfer function is **zero**, that is $z^4 + 16 = 0$.

  i Determine the **poles** of the system and label (∗) them on an Argand diagram.

  ii Determine the **zeros** of the system and label (o) them on the same Argand diagram.

7 Let

$$z = \cos\left(\frac{\pi}{3}\right) + j\sin\left(\frac{\pi}{3}\right)$$

  i Show that z is a root of $z^6 - 1 = 0$.

  ii Determine $z^4 + z$.

  iii Determine $z^6 + z^4 + z$.

8 Let a be a real number and satisfy

$$s^2 - (a + j\sqrt{3}a) = 0$$

Show that

$$s = \sqrt{\frac{a}{2}}\left(\sqrt{3} + j\right)$$

9 Simplify

$$\frac{(3 + j3)^3(1 - j)^4}{(1 + j\sqrt{3})^9}$$

10 i Solve $t^2 - 2t + 4 = 0$.

  ii Find all the roots of $z^6 - 2z^3 + 4 = 0$.

SECTION E **Exponential form of complex numbers**

By the end of this section you will be able to:
▸ convert a complex number from polar to exponential form
▸ convert from exponential to rectangular form
▸ apply the exponential form to engineering examples

E1 **Exponential form**

To write a complex number in exponential form, we need the argument of the complex number to be in **radians**. To convert from degrees → radians we use the following result:

10.23 $$\theta° = \frac{\theta \times \pi}{180} \text{ radians}$$

The exponential form of a complex number is

10.24 $$r\,[\cos(\theta) + j\sin(\theta)] = re^{j\theta} \text{ or } r\angle\theta = re^{j\theta}$$

where r = modulus and θ = argument in radians. You are asked to show this result in **Exercise 10(e)**. You need to use the Maclaurin series expansion (Chapter 7) of $\sin(x)$, $\cos(x)$ and e^x to show 10.24. Remember that the Maclaurin series is only valid if x is in radians and that is why we need the argument of the complex number to be in radians.

Example 24

Write $2[\cos(60°) + j \sin(60°)]$ in exponential form.

Solution

What is wrong with $2[\cos(60°) + j \sin(60°)] = 2e^{j60°}$?

Remember that we need to put the argument 60° into radians. We have

$$\theta° = \frac{\theta \times \pi}{180}$$

$$60° = \frac{6\cancel{0} \times \pi}{18\cancel{0}} = \frac{\pi}{3}$$

Applying

$\boxed{10.24}$ $r[\cos(\theta) + j \sin(\theta)] = re^{j\theta}$

with $r = 2$ and $\theta = \pi/3$ gives

$$2[\cos(60°) + j \sin(60°)] = 2\left[\cos\left(\frac{\pi}{3}\right) + j \sin\left(\frac{\pi}{3}\right)\right] = 2e^{j\frac{\pi}{3}}$$

Properties of the complex exponential function are similar to those of the real exponential function. For example if z_1 and z_2 are complex numbers then

$$e^{z_1 + z_2} = e^{z_1} e^{z_2}$$

and for a complex number z

$$(e^z)^n = e^{z \times n}$$

One advantage of using exponential form is that we do not need to remember special rules of the modulus and argument when multiplying, dividing and raising powers of complex numbers. We just use the laws of indices established earlier.

Example 25

Put $e^{2 + j\frac{\pi}{4}}$ into rectangular form, $a + jb$.

Solution

Using the rules of indices we have

$$e^{2 + j\frac{\pi}{4}} \underset{\text{by } \boxed{10.3}}{=} e^2 e^{j\frac{\pi}{4}}$$

$\boxed{10.3}$ $a^{m+n} = a^m a^n$

Example 25 *continued*

$$\underset{\text{by } 10.24}{=} e^2\left[\cos\left(\frac{\pi}{4}\right) + j\sin\left(\frac{\pi}{4}\right)\right]$$

$$\underset{\text{by Table 1}}{=} e^2\left[\frac{1}{\sqrt{2}} + j\frac{1}{\sqrt{2}}\right] = \frac{e^2}{\sqrt{2}}\,[1 + j] \quad \text{[Factorizing]}$$

There are many identities involving $re^{j\theta}$. For example $re^{j\theta} = r[\cos(\theta) + j\sin(\theta)]$.
If $r = 1$ then

10.25 $\qquad e^{j\theta} = \cos(\theta) + j\sin(\theta)$

This is known as Euler's formula. This formula is one of the most famous results in mathematics. One of the most beautiful formula in mathematics is

$$e^{j\pi} + 1 = 0$$

because it shows the relationship between the numbers e, j, π, -1 and 0. Try proving the result by substituting $\theta = \pi$ into Euler's formula.

Example 26

Show that $e^{-j\theta} = \cos(\theta) - j\sin(\theta)$

Solution

Remember $e^{-j\theta} = e^{j(-\theta)}$ so we have

$$e^{-j\theta} = \cos(-\theta) + j\sin(-\theta) \quad \text{[By Euler's formula]}$$

$$= \underbrace{\cos(\theta)}_{\text{by } 4.51} + j\underbrace{[-\sin(\theta)]}_{\text{by } 4.50}$$

$$= \cos(\theta) - j\sin(\theta)$$

This is another identity:

10.26 $\qquad e^{-j\theta} = \cos(\theta) - j\sin(\theta)$

4.50 $\quad \sin(-x) = -\sin(x)$ $\qquad$ **4.51** $\quad \cos(-x) = \cos(x)$ $\qquad$ **10.24** $\quad re^{j\theta} = r[\cos(\theta) + j\sin(\theta)]$

Example 27 *electrical principles*

Express the following voltage

$$v = 20\sin(1000t - \phi) \text{ volts where } \phi = 30°$$

into complex exponential form.

Solution

For exponential form, we need to put 30° into radians:

$$30° \underset{\text{using } 10.23}{=} \frac{3\cancel{0} \times \pi}{18\cancel{0}} = \frac{\pi}{6}$$

Substituting $r = 20$ and $\theta = 1000t - \dfrac{\pi}{6}$ into $re^{j\theta} = r[\cos(\theta) + j\sin(\theta)]$ gives

$$20e^{j\left(1000t - \frac{\pi}{6}\right)} = 20\left[\cos\left(1000t - \frac{\pi}{6}\right) + j\sin\left(1000t - \frac{\pi}{6}\right)\right]$$

$$= 20\cos\left(1000t - \frac{\pi}{6}\right) + j\underbrace{20\sin\left(1000t - \frac{\pi}{6}\right)}_{=v}$$

We only want the j part because v is just the imaginary part of $20e^{j\left(1000t - \frac{\pi}{6}\right)}$ and so we can write this as

$$v = 20\sin\left(1000t - \frac{\pi}{6}\right) = \text{Imaginary part of } 20e^{j\left(1000t - \frac{\pi}{6}\right)}$$

Normally this is denoted as

$$v = \text{Im}\left[20e^{j\left(1000t - \frac{\pi}{6}\right)}\right] \text{V} \quad (\text{V = volt})$$

Example 28 *control engineering*

A system transfer function, T, is given by

$$T = \frac{20 \times 6.32e^{j0.66}}{\left(3.87e^{j\frac{\pi}{2}}\right)(4.89e^{j0.91})(4e^{j1.32})}$$

Simplify T.

Example 28 *continued*

Solution

We can simplify T as an ordinary algebraic expression by using the rules of indices. Writing

$$T = \frac{20 \times 6.32e^{j0.66}}{3.87 \times 4.89 \times 4e^{\underbrace{j\left(\frac{\pi}{2}+0.91+1.32\right)}_{\text{by } 10.3}}}$$

$$= \frac{20 \times 6.32}{3.87 \times 4.89 \times 4} \underbrace{e^{j\left[0.66-\left(\frac{\pi}{2}+0.91+1.32\right)\right]}}_{\text{by } 10.4}$$

$$= 1.67e^{-j3.14}$$

E2 De Moivre's theorem in exponential form

Let $z = r[\cos(\theta) + j\sin(\theta)] = re^{j\theta}$ then

$$z^n = (re^{j\theta})^n = r^n e^{jn\theta}$$

$$= r^n[\cos(n\theta) + j\sin(n\theta)]$$

which is **de Moivre's theorem**, 10.19.

Note: $z^{-n} = (re^{j\theta})^{-n} = r^{-n}[\cos(n\theta) - j\sin(n\theta)]$

SUMMARY

The exponential form of a complex number is

10.24 $r[\cos(\theta) + j\sin(\theta)] = re^{j\theta}$

where r is the modulus and θ is the argument in radians. We can apply the rules of indices to complex exponential form.

Exercise 10(e)

Solutions at end of book. Complete solutions available at www.palgrave.com/science/engineering/singh

(Throughout this exercise x and y are real)

1 Express the following in complex exponential form:
   a $10\angle\pi$ **b** $10[\cos(180°) + j\sin(180°)]$
   c $j2$ **d** $1 + j$ **e** $\sqrt{3} - j$

2 Express the following in rectangular form, $x + jy$:
   a $e^{1+j\frac{\pi}{2}}$ **b** $e^{\pi(1+j)}$ **c** $2e^{2+j\frac{\pi}{6}}$
   d $e^{1000-j\pi/4}$

10.3 $a^m a^n = a^{m+n}$ 10.4 $\dfrac{a^m}{a^n} = a^{m-n}$

Exercise **10(e) continued**

Solutions at end of book. Complete solutions available at www.palgrave.com/science/engineering/singh

3 Express the following in rectangular form, $x + jy$:

a $e^{j2\pi}$ **b** $e^{j4\pi}$ **c** $e^{j6\pi}$

d $e^{-j2\pi}$ **e** $e^{-j8\pi}$ **f** $e^{-j20\pi}$

Do you notice anything about your results?

4 Express the following in rectangular form, $x + jy$:

a $e^{j\pi}$ **b** $e^{j3\pi}$ **c** $e^{j5\pi}$

d $e^{-j\pi}$ **e** $e^{-j11\pi}$ **f** $e^{-j21\pi}$

Do you notice anything about your results?

5 [electrical principles]   If $v = Ve^{j(\omega t + \alpha)}$ and $i = Ie^{j(\omega t + \beta)}$ then show that the impedance $z \left(= \dfrac{v}{i} \right)$ is given by

$$z = \frac{V}{I}\left[\cos(\alpha - \beta) + j\sin(\alpha - \beta) \right]$$

6 [control engineering]   A system transfer function, G, is given by

$$G = \frac{10}{5\, e^{j\frac{\pi}{2}}\, e^{j\frac{\pi}{3}}}$$

Simplify G.

7 [control engineering]   A system has an open-loop transfer function

$$G = \frac{100}{10 e^{j\frac{\pi}{2}} 1.02 e^{j0.2} 1.8 e^{j0.99}}$$

Simplify G.

8 Simplify:

a $(1 + j)e^{j\frac{\pi}{4}} + (1 - j)e^{-j\frac{\pi}{4}}$

b $(3 + j4)e^{(1+j)t} + (3 - j4)e^{(1-j)t}$ (t is real)

9 [electrical principles]

a A voltage, V, across an inductor with inductance L is given by

$$V = \text{Im}[\omega L\, Ie^{j(\omega t + \pi/2)}]$$

where Im is the imaginary part, ω = angular frequency, t = time and I = current. Show that

$$V = \omega LI \cos(\omega t)$$

b A voltage, V, across a capacitor with capacitance C is given by

$$V = \text{Im}\left[\frac{I}{\omega C} e^{j(\omega t - \pi/2)} \right]$$

Show that

$$V = -\frac{I}{\omega C}\cos(\omega t)$$

(ω, I and t are as in part **a**).

10 Let $z = \ln(3) + j\pi/2$ and a, b be real such that

$$a + jb = e^{z}$$

Determine a and b.

11 Show the identity

$$re^{j\theta} = r[\cos(\theta) + j\sin(\theta)]$$

[*Hint*: Consider the Maclaurin series expansion of e^x, $\cos(x)$ and $\sin(x)$]

Miscellaneous exercise **10**

Solutions at end of book. Complete solutions available at www.palgrave.com/science/engineering/singh

1 Mark on an Argand diagram the points z and $\bar{z}$ (complex conjugate of z) if

  a $z = 2 + j3$ **b** $z = 1 - j$ **c** $z = j$

  What do you notice?

2 Express the following in $x + jy$ form:

  a $(1 + j)^{10}$ **b** $\left(-1 - j\sqrt{3}\right)^7$

  c $\dfrac{\cos(5\theta) + j\sin(5\theta)}{\cos(\theta) + j\sin(\theta)}$

3 Find x and y if

  a $x + jy = \dfrac{1}{e^{j\theta}}$ **b** $x + jy = \dfrac{1}{1 + e^{j\theta}}$

4 Evaluate the following:

  a $\sqrt{2}e^{j\pi/4}$ **b** $-e^{-j\pi/6}$

5 Let z be any complex number. Show that
$$\bar{z}z = |z|^2$$

6 🚗 [mechanics]

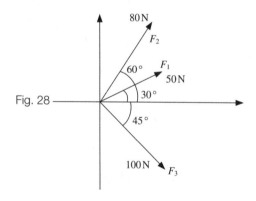

Fig. 28

Figure 28 shows three coplanar forces F_1, F_2 and F_3. Find the resultant magnitude and direction of F where
$$F = (F_1 + F_2 + F_3) - 155.8\underline{/8.7°}$$

🔲 **Questions 7 to 9, inclusive, are in the field of [electrical principles].**

7 In a three-phase supply, $V_R = 240$ V and $z = (10 + j10)\,\Omega$. Find the current I_R

through $z\left(I_R = \dfrac{V_R}{z}\right)$.

Find $I_Y\left(I_Y = \dfrac{V_Y}{z}\right)$ and $I_B\left(I_B = \dfrac{V_B}{z}\right)$.

for the following data:

$$V_Y = 240\underline{/120°}\ \text{V with } z = (5 + j8)\,\Omega$$
$$V_B = 240\underline{/240°}\ \text{V with } z = (2 + j9)\,\Omega$$

The neutral line current, I_N, is given by

$$I_N = I_R + I_Y + I_B$$

Determine I_N.

8 For a series circuit with resistance R, inductance L and capacitance C, the total impedance z is given by

$$z = R + j\omega L + \dfrac{1}{j\omega C}\quad \text{where}$$

$$\omega = \text{angular frequency}$$

Resonance occurs when $\text{Im}(z) = 0$. Show that the resonant angular frequency, ω_r, is given by $\omega_r = \dfrac{1}{\sqrt{LC}}$.

9 A cable of length L has the following impedances:

$$z_{oc} = 600e^{j\frac{\pi}{3}}$$
$$\text{(Open circuit impedance)}$$

$$z_{sc} = 400e^{-j\frac{\pi}{3}}$$
$$\text{(Short circuit impedance)}$$

Given that

$$\tanh(\gamma L) = \sqrt{\left(\dfrac{z_{sc}}{z_{oc}}\right)}$$

where γ is complex (propagation coefficient), show that

$$\tanh(\gamma L) = \dfrac{1}{\sqrt{6}}(1 - j\sqrt{3})$$

⚙ Questions 10 to 14, inclusive, belong to the field of [*control engineering*].

10 Let $T(s) = \dfrac{K}{s^2 + 4s + K}$ (K is real) be a transfer function of a system. Let $\sigma \pm j\omega$ be the roots of $s^2 + 4s + K = 0$ (poles of the system). The natural angular frequency ω_n satisfies

$$\omega_n^2 = \sigma^2 + \omega^2$$

Find the value of K required for the natural angular frequency $\omega_n = 5$ rad/s.

11 A control system has a transfer function, $\dfrac{\theta_0}{\theta_1}$, given by

$$\frac{\theta_0}{\theta_1} = \frac{1000}{200 + 2j\omega + (j\omega)^2}$$

where ω is the angular velocity.

Determine the gain $\left(\left| \dfrac{\theta_0}{\theta_1} \right| \right)$ and phase $\left(\arg\left(\dfrac{\theta_0}{\theta_1} \right) \right)$ of the function for $\omega = 10$ rad/s.

12 A system is said to be **stable** if all the poles of the transfer function lie within the unit circle, $|z| < 1$. The system is said to be **critically stable** if it has a pole on the unit circle, $|z| = 1$ (Fig. 29).

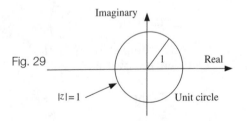

Fig. 29

$|z|=1$ Unit circle

Determine whether the systems with the following transfer functions are stable or critically stable:

a $G(z) = \dfrac{10}{3z^2 + 2z + 1}$

b $G(z) = \dfrac{z - 1}{z^3 + 1}$

(Poles occur where the denominator of the transfer function is zero.)

13 A system with feedback has a transfer function, $G(s)$, given by

$$G(s) = \frac{10}{(s + 1 + j)(s + 1 - j) + 10}$$

(s is complex)

i Show that the roots of the characteristic equation

$$(s + 1 + j)(s + 1 - j) + 10 = 0$$

are $s = -1 \pm j\sqrt{11}$.
Let $\sigma = \text{Re}(s)$ and $\omega = \text{Im}(s)$. Find σ and ω. For the remaining question use these values of σ and ω.

ii The undamped natural frequency, ω_n, of the system is given by

$$\omega_n = \sqrt{\sigma^2 + \omega^2}$$

Determine ω_n.

iii The damping ratio, ζ, of the system is given by

$$\zeta = \frac{|\sigma|}{\omega_n}$$

Determine ζ.

iv The response time, t_r, is given by

$$t_r = \frac{\pi}{2\omega_n}$$

Determine t_r.

14 Let $T = \dfrac{10}{10 + j\omega}$ be a transfer function of a system.

  i Evaluate the gain ($|T|$) and phase, $\arg(T)$, for $\omega = 1$, 10 and 100 rad/s.

  ii Sketch, on different axes, the graphs of $|T|$ and $\arg(T)$ as a function of ω.

📺 **Questions 15 to 17, inclusive, belong to the field of [electronics].**

15 The transfer function of a simple amplifier is given by

$$A = \frac{A_0}{1 + j\omega T_1}$$

where A_0 is the low frequency gain and T_1 is the amplifier time constant.

  i Given that $A_0 = 1000$,

  $T_1 = 0.5 \times 10^{-3}$ s, find the gain ($= |A|$) and phase ($= \arg(A)$) of A at an angular frequency $\omega = 2 \times 10^3$ rad/s.

  ii When negative feedback, β, is applied, the transfer function becomes

$$T = \frac{A_0/(1 + j\omega T_1)}{1 + \dfrac{A_0}{1 + j\omega T_1}\beta(1 + j\omega T_2)}$$

  (T = new transfer function, T_2 = feedback time constant)

  Show that

$$T = \frac{A_0}{1 + A_0\beta + j\omega(T_1 + A_0\beta T_2)}$$

  iii Find the gain and phase of T with $\beta = 0.1$ and $T_2 = 1 \times 10^{-4}$ s at $\omega = 2 \times 10^3$ rad/s.

16 The gain of a phase shifter circuit consisting of an op-amp, with resistance R and capacitance C, is given by $|A|$ where

$$A = \frac{1 - R/X_c}{1 + R/X_c} \text{ where } X_c = \frac{1}{j\omega C}$$

Show that $|A| = 1$.

17 The transfer function, G, of a Butterworth low pass filter is derived from the relationship

$$G = \frac{1}{1 - s^6}$$

where s is complex.
Find the poles of the filter, that is the values of s where $1 - s^6 = 0$, and label them on an Argand diagram.

18 🔊 [acoustics] The specific acoustic impedance, z, is given by

$$z = \frac{jp\omega}{\left(\dfrac{1}{r} + jk\right)}$$

where r is the radial distance, p is the acoustic pressure and k is the wavenumber. Show that:
Specific acoustic resistance,

$$\text{Re}(z) = \frac{p\omega k r^2}{1 + k^2 r^2}$$

Specific acoustic reactance,

$$\text{Im}(z) = \frac{p\omega r}{1 + k^2 r^2}$$

19 🔊 [acoustics] The specific impedance, z_s, of a fluid column is given by

$$z_s = z_0\left(\frac{e^{j\theta} + ke^{-j\theta}}{e^{j\theta} - ke^{-j\theta}}\right) \quad \text{where } \theta = \frac{2\pi}{\lambda}x$$

(z_0 is the characteristic impedance, k is a real constant and λ is the wavelength). Determine z_s at

  a $x = \lambda/2$ **b** $x = \lambda/4$

20 [*fluid mechanics*]   The complex potential function, $w = \phi + j\psi$, for a particular flow is given by

$$w = z^a + \frac{c^2}{z^a}$$

where a and c are non-zero real numbers. Show that the potential function ϕ is

$$\phi = \left(r^a + \frac{c^2}{r^a}\right)\cos(a\theta)$$

and the stream function ψ is

$$\psi = \left(r^a - \frac{c^2}{r^a}\right)\sin(a\theta)$$

where $r = |z|$ and $\theta = \arg(z)$.

21 [*fluid mechanics*]   The complex potential function $w = \phi + j\psi$ for a certain flow is given by

$$w = 3ze^{j\frac{\pi}{3}}$$

where $z = x + jy$. Determine the stream, ψ, and potential, ϕ, functions in terms of x and y.

22 We know from 10.19 that

$$\cos(3\theta) + j\sin(3\theta) = \left[\cos(\theta) + j\sin(\theta)\right]^3$$

By expanding the Right-Hand Side, show that

$$\cos(3\theta) = 4\cos^3(\theta) - 3\cos(\theta)$$

and $\sin(3\theta) = 3\sin(\theta) - 4\sin^3(\theta)$

23 Let $z = \cos(\theta) + j\sin(\theta)$. Show that

i $z - \dfrac{1}{z} = 2j\sin(\theta)$

ii $z^3 - \dfrac{1}{z^3} = 2j\sin(3\theta)$

iii $z^5 - \dfrac{1}{z^5} = 2j\sin(5\theta)$

iv By using the binomial theorem, or otherwise, show that

$$\left(z - \frac{1}{z}\right)^5 = \left(z^5 - \frac{1}{z^5}\right) - 5\left(z^3 - \frac{1}{z^3}\right) + 10\left(z - \frac{1}{z}\right)$$

v Show that

$$\sin^5(\theta) = \frac{1}{16}\left[\sin(5\theta) - 5\sin(3\theta) + 10\sin(\theta)\right]$$

24 By 10.19 we have

$$\cos(4\theta) + j\sin(4\theta) = \left[\cos(\theta) + j\sin(\theta)\right]^4$$

By expanding the Right-Hand Side and by equating the real and imaginary parts, show that

$$\cos(4\theta) = \cos^4(\theta) - 6\cos^2(\theta)\sin^2(\theta) + \sin^4(\theta)$$

and

$$\sin(4\theta) = 4\cos^3(\theta)\sin(\theta) - 4\cos(\theta)\sin^3(\theta)$$

CHAPTER 11
Matrices

SECTION A **Manipulation of matrices**

By the end of this section you will be able to:
▶ understand the arithmetic of matrices
▶ evaluate the determinant of a 2 × 2 matrix
▶ obtain the inverse matrix

A1 **Introduction**

Matrices are used in control theory, electrical principles, vibrations, structural analysis, etc. Also many engineering problems can be written in terms of simultaneous equations. If there are a large number of equations then in most cases it is easier to use matrices rather than the process of elimination discussed in Chapter 2.

Arthur Cayley (1821–95), an English lawyer who became a mathematician, was the first person to develop matrices as we know them today. He graduated in 1842 from Trinity College, Cambridge. In 1858 he published '*Memoir on the theory of matrices*' which contained the definition of a matrix, matrix addition, subtraction, etc.

Another application of matrices is in image processing. Many displays form images by lighting up tiny dots, called pixels, on their screens. Figure 1 shows a symbol represented by lighting up the appropriate pixels.

Fig. 1

If we let 1 = 'pixel on' and 0 = 'pixel off' then we can represent this in matrix form as

$$\begin{pmatrix} 1 & 1 & 1 & 1 & 1 \\ 0 & 0 & 1 & 0 & 0 \\ 0 & 0 & 1 & 0 & 0 \\ 1 & 1 & 1 & 1 & 1 \end{pmatrix}$$

This is an example of a 4 × 5 matrix.

A matrix is an array of numbers enclosed in a bracket. It is used to store information. For example $\begin{pmatrix} 12 & 6 \\ 3 & 7 \end{pmatrix}$ is a 2 × 2 matrix (2 rows and 2 columns) and is called a square matrix.

$\begin{pmatrix} 3 & 2 \\ 7 & 6 \\ 5 & 9 \\ 2 & 1 \end{pmatrix}$ is a 4 × 2 matrix (4 rows and 2 columns) and is **not** a square matrix.

An example of a 2 × 4 matrix is

$$\begin{pmatrix} 3 & 7 & 5 & 2 \\ 2 & 6 & 9 & 1 \end{pmatrix}$$

Note that when we say 2 × 4 matrix we state the number of rows first and then the number of columns. This is generally the case.

Matrices is the plural of matrix. A common notation of matrices is:

$$
\begin{array}{cc}
& \text{Column 1} \quad \text{Column 2} \\
\begin{array}{c} \text{Row 1} \\ \text{Row 2} \end{array} &
\begin{pmatrix} a_{11} & a_{12} \\ a_{21} & a_{22} \end{pmatrix}
\end{array}
$$

where a_{12} is the element in the first row and second column. Generally in this chapter, bold capital letters will denote matrices.

A2 **Arithmetic of matrices**

For example

$$
\begin{pmatrix} 2 & 5 \\ 4 & 1 \end{pmatrix} + \begin{pmatrix} 3 & 9 \\ 5 & 7 \end{pmatrix} = \begin{pmatrix} 2+3 & 5+9 \\ 4+5 & 1+7 \end{pmatrix} = \begin{pmatrix} 5 & 14 \\ 9 & 8 \end{pmatrix}
$$

Also

$$
\begin{pmatrix} 2 & 5 \\ 4 & 1 \end{pmatrix} - \begin{pmatrix} 3 & 9 \\ 5 & 7 \end{pmatrix} = \begin{pmatrix} 2-3 & 5-9 \\ 4-5 & 1-7 \end{pmatrix} = \begin{pmatrix} -1 & -4 \\ -1 & -6 \end{pmatrix}
$$

To add matrices, we add the corresponding locations in each matrix. Similarly for subtraction, we subtract the corresponding locations in each matrix. Matrices must be of the **same** size in order to add or subtract.

A scalar is a number which scales a matrix. The following are examples of scalar multiplication:

$$
10 \begin{pmatrix} 3 & 1 \\ 7 & -9 \end{pmatrix} = \begin{pmatrix} 3 \times 10 & 1 \times 10 \\ 7 \times 10 & -9 \times 10 \end{pmatrix} = \begin{pmatrix} 30 & 10 \\ 70 & -90 \end{pmatrix}
$$

and

$$
\frac{1}{5} \begin{pmatrix} 20 & -10 \\ 5 & -35 \end{pmatrix} = \begin{pmatrix} 20/5 & -10/5 \\ 5/5 & -35/5 \end{pmatrix} = \begin{pmatrix} 4 & -2 \\ 1 & -7 \end{pmatrix}
$$

Multiply each number in the matrix by the scalar. Also

$$
7 \begin{pmatrix} -1 \\ 5 \\ 3 \\ 9 \end{pmatrix} = \begin{pmatrix} -7 \\ 35 \\ 21 \\ 63 \end{pmatrix}
$$

Next we examine the multiplication of matrices. This is **not** how you multiply matrices:

$$
\begin{pmatrix} 2 & 1 \\ 3 & 7 \end{pmatrix} \times \begin{pmatrix} 4 & 9 \\ 5 & 8 \end{pmatrix} \neq \begin{pmatrix} 2 \times 4 & 1 \times 9 \\ 3 \times 5 & 7 \times 8 \end{pmatrix} \qquad \text{[Not equal]} \qquad \times
$$

Matrix multiplication is row × column:

$$
\begin{pmatrix} a & b \\ c & d \end{pmatrix} \begin{pmatrix} e & f \\ g & h \end{pmatrix} = \begin{pmatrix} ae+bg & af+bh \\ ce+dg & cf+dh \end{pmatrix}
$$

Applying this to the above matrices gives

$$\begin{pmatrix} 2 & 1 \\ 3 & 7 \end{pmatrix} \times \begin{pmatrix} 4 & 9 \\ 5 & 8 \end{pmatrix} = \begin{pmatrix} (2 \times 4) + (1 \times 5) & (2 \times 9) + (1 \times 8) \\ (3 \times 4) + (7 \times 5) & (3 \times 9) + (7 \times 8) \end{pmatrix}$$

$$= \begin{pmatrix} 13 & 26 \\ 47 & 83 \end{pmatrix}$$

We can **only** multiply matrices if the number of **columns** of the **first** matrix **equals** the number of **rows** of the **second** matrix.

$$\begin{array}{c} \text{Col. 1} \quad \text{Col. 2} \end{array}$$

$$\begin{array}{c} \text{Row 1} \\ \text{Row 2} \end{array} \begin{pmatrix} 2 & 1 \\ 3 & 7 \end{pmatrix} \times \begin{pmatrix} 4 & 9 \\ 5 & 8 \end{pmatrix} = \begin{pmatrix} \text{Row 1} \times \text{Col. 1} & \text{Row 1} \times \text{Col. 2} \\ \text{Row 2} \times \text{Col. 1} & \text{Row 2} \times \text{Col. 2} \end{pmatrix}$$

Example 1

Evaluate the following:

a $\begin{pmatrix} 2 & 3 \\ -1 & 5 \end{pmatrix}\begin{pmatrix} 2 \\ 1 \end{pmatrix}$ **b** $(3 \quad 4)\begin{pmatrix} 2 & 3 \\ -1 & 5 \end{pmatrix}$ **c** $(1 \quad 0)\begin{pmatrix} 3 \\ 4 \end{pmatrix}$

d $\begin{pmatrix} 2 & 3 & 6 \\ 1 & 5 & 7 \end{pmatrix}\begin{pmatrix} 3 & 7 \\ 4 & 2 \\ 1 & 3 \end{pmatrix}$ **e** $(1 \quad 2 \quad 3)\begin{pmatrix} 4 \\ 5 \\ 6 \end{pmatrix}$ **f** $\begin{pmatrix} 2 & 3 & 6 \\ 1 & 5 & 7 \end{pmatrix}\begin{pmatrix} 3 & 7 \\ 4 & 2 \end{pmatrix}$

Solution

Matrix multiplication is row × column.

a $\begin{pmatrix} 2 & 3 \\ -1 & 5 \end{pmatrix}\begin{pmatrix} 2 \\ 1 \end{pmatrix} = \begin{pmatrix} (2 \times 2) + (3 \times 1) \\ (-1 \times 2) + (5 \times 1) \end{pmatrix} = \begin{pmatrix} 7 \\ 3 \end{pmatrix}$

b $(3 \quad 4)\begin{pmatrix} 2 & 3 \\ -1 & 5 \end{pmatrix} = \begin{pmatrix} (3 \times 2) + (4 \times (-1)) & (3 \times 3) + (4 \times 5) \end{pmatrix}$

$$= (2 \quad 29)$$

c $(1 \quad 0)\begin{pmatrix} 3 \\ 4 \end{pmatrix} = \begin{pmatrix} (1 \times 3) + (0 \times 4) \end{pmatrix} = (3) = 3$

d $\begin{pmatrix} 2 & 3 & 6 \\ 1 & 5 & 7 \end{pmatrix}\begin{pmatrix} 3 & 7 \\ 4 & 2 \\ 1 & 3 \end{pmatrix} = \begin{pmatrix} (2 \times 3) + (3 \times 4) + (6 \times 1) & (2 \times 7) + (3 \times 2) + (6 \times 3) \\ (1 \times 3) + (5 \times 4) + (7 \times 1) & (1 \times 7) + (5 \times 2) + (7 \times 3) \end{pmatrix}$

$$= \begin{pmatrix} 24 & 38 \\ 30 & 38 \end{pmatrix}$$

e $(1 \quad 2 \quad 3)\begin{pmatrix} 4 \\ 5 \\ 6 \end{pmatrix} = \begin{pmatrix} (1 \times 4) + (2 \times 5) + (3 \times 6) \end{pmatrix} = (32) = 32$

Example 1 *continued*

f $\begin{pmatrix} 2 & 3 & 6 \\ 1 & 5 & 7 \end{pmatrix}\begin{pmatrix} 3 & 7 \\ 4 & 2 \end{pmatrix}$. Since the number of columns of the first matrix is 3 and number of rows of the second matrix is 2, we **cannot** multiply these matrices.

Example 2

Let

$$A = \begin{pmatrix} 2 & 1 \\ 3 & 7 \end{pmatrix} \text{ and } B = \begin{pmatrix} 4 & 9 \\ 5 & 8 \end{pmatrix}$$

Determine **AB** and **BA**. **What do you notice about your results?**

Solution

Multiplying matrices as described above gives

$$AB = \begin{pmatrix} 2 & 1 \\ 3 & 7 \end{pmatrix}\begin{pmatrix} 4 & 9 \\ 5 & 8 \end{pmatrix} = \begin{pmatrix} 13 & 26 \\ 47 & 83 \end{pmatrix}$$

$$BA = \begin{pmatrix} 4 & 9 \\ 5 & 8 \end{pmatrix}\begin{pmatrix} 2 & 1 \\ 3 & 7 \end{pmatrix} = \begin{pmatrix} 35 & 67 \\ 34 & 61 \end{pmatrix}$$

Note that **AB** ≠ **BA**, that is the order of multiplication of matrices **does** matter.

If **A** and **B** are matrices then in general **AB** ≠ **BA**.

A3 **Determinant of a matrix**

The determinant of a 2×2 matrix, $\begin{pmatrix} a & b \\ c & d \end{pmatrix}$, is given by

11.1 $\det\begin{pmatrix} a & b \\ c & d \end{pmatrix} = ad - cb$

$\det\begin{pmatrix} a & b \\ c & d \end{pmatrix}$ is used to represent the determinant. Another notation is $\begin{vmatrix} a & b \\ c & d \end{vmatrix}$. Note that the determinant is a number, **not** a matrix. For example

$$\det\begin{pmatrix} 5 & 4 \\ -2 & 1 \end{pmatrix} = (5 \times 1) - (-2 \times 4) = 13$$

The determinant is used to find the inverse of a matrix.

A4 **Identity matrix**

Evaluate

$$\begin{pmatrix} 5 & 3 \\ 1 & 2 \end{pmatrix}\begin{pmatrix} 1 & 0 \\ 0 & 1 \end{pmatrix} \text{ and } \begin{pmatrix} 6 & 7 \\ 2 & 3 \end{pmatrix}\begin{pmatrix} 1 & 0 \\ 0 & 1 \end{pmatrix}$$

? **What do you notice about your results?**

$$\begin{pmatrix} 5 & 3 \\ 1 & 2 \end{pmatrix}\begin{pmatrix} 1 & 0 \\ 0 & 1 \end{pmatrix} = \begin{pmatrix} 5 & 3 \\ 1 & 2 \end{pmatrix} \text{ and } \begin{pmatrix} 6 & 7 \\ 2 & 3 \end{pmatrix}\begin{pmatrix} 1 & 0 \\ 0 & 1 \end{pmatrix} = \begin{pmatrix} 6 & 7 \\ 2 & 3 \end{pmatrix}$$

We obtain the same matrix as the one used to multiply $\begin{pmatrix} 1 & 0 \\ 0 & 1 \end{pmatrix}$. In general for a square matrix

$$\begin{pmatrix} a & b \\ c & d \end{pmatrix}\begin{pmatrix} 1 & 0 \\ 0 & 1 \end{pmatrix} = \begin{pmatrix} a & b \\ c & d \end{pmatrix}$$

where $\begin{pmatrix} 1 & 0 \\ 0 & 1 \end{pmatrix}$ is called the identity matrix and is denoted by **I**, $\mathbf{I} = \begin{pmatrix} 1 & 0 \\ 0 & 1 \end{pmatrix}$.

The identity matrix, **I**, for matrices is analogous to the number 1 for numbers.

A5 **Inverse of a matrix**

Let **A** be the general 2×2 (square) matrix:

$$\mathbf{A} = \begin{pmatrix} a & b \\ c & d \end{pmatrix}$$

The inverse matrix of **A** is denoted by $\mathbf{A}^{-1}$ and has the property

| 11.2 | $\mathbf{A}^{-1} \times \mathbf{A} = \mathbf{I}$

The notation $\mathbf{A}^{-1}$ is the inverse matrix and not **A** to the index -1.

If the determinant of **A** is equal to 1 then $\mathbf{A}^{-1}$ is given by

| 11.3 | $\mathbf{A}^{-1} = \begin{pmatrix} d & -b \\ -c & a \end{pmatrix}$ [provided that det**A** = 1]

Remember this only works if det**A** = 1.

Example 3

For $\mathbf{A} = \begin{pmatrix} 3 & 7 \\ 2 & 5 \end{pmatrix}$, find $\mathbf{A}^{-1}$.

Example 3 *continued*

Solution

First we obtain the determinant by using

11.1 $\det\begin{pmatrix} a & b \\ c & d \end{pmatrix} = ad - cb$

$\det\begin{pmatrix} 3 & 7 \\ 2 & 5 \end{pmatrix} = (3 \times 5) - (2 \times 7) = 1$

Since the determinant of **A** is 1 we can find $\mathbf{A}^{-1}$ by using **11.3**, that is interchanging numbers 3 and 5 and putting a negative sign in front of the other numbers:

$$\mathbf{A}^{-1} = \begin{pmatrix} 5 & -7 \\ -2 & 3 \end{pmatrix}$$

We can confirm the result by checking the matrix multiplication $\mathbf{A}^{-1}\mathbf{A} = \mathbf{I}$:

$$\mathbf{A}^{-1}\mathbf{A} = \begin{pmatrix} 5 & -7 \\ -2 & 3 \end{pmatrix}\begin{pmatrix} 3 & 7 \\ 2 & 5 \end{pmatrix} = \begin{pmatrix} 1 & 0 \\ 0 & 1 \end{pmatrix} = \mathbf{I}$$

? What are the limitations of this method?

The determinant of **A** has to equal 1. Of course there are going to be many matrices where the determinant does not equal 1. Consider

$$\mathbf{A} = \begin{pmatrix} 11 & 4 \\ 7 & 3 \end{pmatrix}$$

The determinant is

$$\det\mathbf{A} = (11 \times 3) - (7 \times 4) = 5$$

Therefore we **cannot** use $\mathbf{A}^{-1} = \begin{pmatrix} d & -b \\ -c & a \end{pmatrix}$ to find $\mathbf{A}^{-1}$.

? However if we try using this, what matrix do we obtain?

Call this matrix **B**. Interchanging the numbers 3 and 11 and placing a negative sign in front of the other numbers:

$$\mathbf{B} = \begin{pmatrix} 3 & -4 \\ -7 & 11 \end{pmatrix}$$

Multiplying the two matrices gives

$$\mathbf{BA} = \begin{pmatrix} 3 & -4 \\ -7 & 11 \end{pmatrix}\begin{pmatrix} 11 & 4 \\ 7 & 3 \end{pmatrix} = \begin{pmatrix} 5 & 0 \\ 0 & 5 \end{pmatrix}$$

Of course $\begin{pmatrix} 5 & 0 \\ 0 & 5 \end{pmatrix} \neq \begin{pmatrix} 1 & 0 \\ 0 & 1 \end{pmatrix} = \mathbf{I}$ but it can be made into the identity matrix.

? **How?**

By multiplying with a scalar of 1/5:

$$\frac{1}{5}\begin{pmatrix} 5 & 0 \\ 0 & 5 \end{pmatrix} = \begin{pmatrix} 1 & 0 \\ 0 & 1 \end{pmatrix} = \mathbf{I}$$

? **Where did the figure 5 appear previously in this discussion?**

It is the determinant of the matrix **A**. Since $\frac{1}{5}\mathbf{B}$ multiplied by **A** gives the identity matrix, so $\frac{1}{5}\mathbf{B}$ must be the inverse of **A**. Thus

$$\mathbf{A}^{-1} = \frac{1}{5}\mathbf{B} = \frac{1}{5}\begin{pmatrix} 3 & -4 \\ -7 & 11 \end{pmatrix}$$

In general, if $\mathbf{A} = \begin{pmatrix} a & b \\ c & d \end{pmatrix}$ then the inverse matrix, $\mathbf{A}^{-1}$, is defined as

11.4 $\mathbf{A}^{-1} = \dfrac{1}{\det A}\begin{pmatrix} d & -b \\ -c & a \end{pmatrix} = \dfrac{1}{ad - cb}\begin{pmatrix} d & -b \\ -c & a \end{pmatrix}$ provided that det $\mathbf{A} \neq 0$

If the determinant of **A** is zero then matrix **A** does not have an inverse.

Example 4

Let

$$\mathbf{A} = \begin{pmatrix} 7 & 9 \\ 5 & 7 \end{pmatrix}$$

Find $\mathbf{A}^{-1}$.

Solution

First we have to find $\det\begin{pmatrix} 7 & 9 \\ 5 & 7 \end{pmatrix}$. By

11.1 $\det\begin{pmatrix} a & b \\ c & d \end{pmatrix} = ad - cb$

$\det\begin{pmatrix} 7 & 9 \\ 5 & 7 \end{pmatrix} = (7 \times 7) - (5 \times 9) = 4$

With det$\mathbf{A} = 4$:

$$\mathbf{A}^{-1} = \frac{1}{\det A}\begin{pmatrix} d & -b \\ -c & a \end{pmatrix}$$

$$= \frac{1}{4}\begin{pmatrix} 7 & -9 \\ -5 & 7 \end{pmatrix}$$

$$= \begin{pmatrix} 7/4 & -9/4 \\ -5/4 & 7/4 \end{pmatrix}$$

Example 4 *continued*

To check that $\mathbf{A}^{-1}$ is indeed the inverse matrix of $\mathbf{A}$, we carry out the following matrix multiplication:

$$\mathbf{A}^{-1} \times \mathbf{A} = \frac{1}{4}\begin{pmatrix} 7 & -9 \\ -5 & 7 \end{pmatrix}\begin{pmatrix} 7 & 9 \\ 5 & 7 \end{pmatrix}$$

$$= \frac{1}{4}\begin{pmatrix} 4 & 0 \\ 0 & 4 \end{pmatrix}$$

$$= \begin{pmatrix} 1 & 0 \\ 0 & 1 \end{pmatrix} = \mathbf{I}$$

Thus $\mathbf{A}^{-1}$ is the inverse matrix of $\mathbf{A}$.

It's good practice to check your inverse matrix and also you become familiar with matrix multiplication.

SUMMARY

When multiplying two matrices it is row × column.
If $\mathbf{A}$ and $\mathbf{B}$ represent matrices then in general

$$\mathbf{AB} \neq \mathbf{BA}$$

The determinant of a matrix $\begin{pmatrix} a & b \\ c & d \end{pmatrix}$ is defined as

11.1 $\det\begin{pmatrix} a & b \\ c & d \end{pmatrix} = ad - cb$

The inverse matrix, $\mathbf{A}^{-1}$, has the property

11.2 $\mathbf{A}^{-1}\mathbf{A} = \mathbf{I}$

where $\mathbf{I} = \begin{pmatrix} 1 & 0 \\ 0 & 1 \end{pmatrix}$ is the identity matrix.

The inverse matrix, $\mathbf{A}^{-1}$, of $\mathbf{A} = \begin{pmatrix} a & b \\ c & d \end{pmatrix}$ is defined as

11.4 $\mathbf{A}^{-1} = \dfrac{1}{ad - cb}\begin{pmatrix} d & -b \\ -c & a \end{pmatrix}$ $(\det\mathbf{A} \neq 0)$

Exercise **11(a)**

Solutions at end of book. Complete solutions available at www.palgrave.com/science/engineering/singh

1 For $\mathbf{A} = \begin{pmatrix} 1 & 2 \\ 3 & -1 \end{pmatrix}$, $\mathbf{B} = \begin{pmatrix} 6 & -1 \\ 5 & 3 \end{pmatrix}$, and $\mathbf{C} = \begin{pmatrix} -1 \\ 1 \end{pmatrix}$, evaluate the following:

a $\mathbf{A} + \mathbf{B}$ **b** $\mathbf{B} + \mathbf{A}$ **c** $\mathbf{A} + \mathbf{A} + \mathbf{A}$

d $3\mathbf{A} + 2\mathbf{B}$ **e** $\mathbf{A} + \mathbf{C}$ **f** $\mathbf{B} + \mathbf{C}$

g $\mathbf{AC}$ **h** $\mathbf{BC}$ **i** $5\mathbf{A} - 7\mathbf{BC}$

j $3\mathbf{AC} - 2\mathbf{BC}$

Exercise **11(a) continued**

Solutions at end of book. Complete solutions available at www.palgrave.com/science/engineering/singh

2 Compute det A and det B for the following:

a $A = \begin{pmatrix} 1 & 3 \\ 5 & 7 \end{pmatrix}$, $B = \begin{pmatrix} 1 & 5 \\ 3 & 7 \end{pmatrix}$

b $A = \begin{pmatrix} -1 & 2 \\ 5 & 3 \end{pmatrix}$, $B = \begin{pmatrix} -1 & 5 \\ 2 & 3 \end{pmatrix}$

c $A = \begin{pmatrix} \cos(-\pi) & \cos(\pi) \\ \sin(-\pi) & \sin(\pi) \end{pmatrix}$,

$B = \begin{pmatrix} \cos(-\pi) & \sin(-\pi) \\ \cos(\pi) & \sin(\pi) \end{pmatrix}$

What do you notice about matrices A and B and your results for detA and detB?

3 For $A = \begin{pmatrix} a & b \\ c & d \end{pmatrix}$ and $B = \begin{pmatrix} a & c \\ b & d \end{pmatrix}$, show that detA = detB.

4 Evaluate **i** detA **ii** detB
iii detA × detB **iv** det(AB) for the following:

a $A = \begin{pmatrix} 1 & 3 \\ 5 & 6 \end{pmatrix}$, $B = \begin{pmatrix} 3 & 7 \\ -1 & 5 \end{pmatrix}$

b $A = \begin{pmatrix} -1 & 10 \\ 170 & 1.5 \end{pmatrix}$, $B = \begin{pmatrix} -30 & -9.3 \\ 61 & -1.9 \end{pmatrix}$

c $A = \begin{pmatrix} 5 & -3 \\ 2.2 & -5.6 \end{pmatrix}$, $B = \begin{pmatrix} -7.1 & -2.1 \\ -3.5 & -12.2 \end{pmatrix}$

What do you notice about your results to parts **iii** and **iv**.

5 Find the inverse of the following matrices:

a $A = \begin{pmatrix} 9 & 2 \\ 13 & 3 \end{pmatrix}$ **b** $A = \begin{pmatrix} 17 & 7 \\ 12 & 5 \end{pmatrix}$

c $A = \begin{pmatrix} 1 & 0 \\ 0 & 1 \end{pmatrix}$

6 Find the inverse matrix of the following:

a $A = \begin{pmatrix} 5 & 4 \\ 3 & 1 \end{pmatrix}$ **b** $A = \begin{pmatrix} 3 & 6 \\ 7 & 8 \end{pmatrix}$

c $A = \begin{pmatrix} 7 & 1 \\ 14 & 2 \end{pmatrix}$

SECTION B **Applications**

By the end of this section you will be able to:
▶ apply matrices to solve equations
▶ use matrices in control engineering

B1 **Solving equations**

We can use matrices to solve simultaneous equations, as the following example shows.

Example 5 *electrical principles*

By applying Kirchhoff's law to a circuit we obtain the following equations:

$$7i_1 + 9i_2 = 3$$
$$5i_1 + 7i_2 = 1$$

where i_1 and i_2 represent current. Find the values of i_1 and i_2.

Example 5 *continued*

Solution

The product of matrices is row × column so the above equations can be written in matrix form as

$$\begin{pmatrix} 7 & 9 \\ 5 & 7 \end{pmatrix}\begin{pmatrix} i_1 \\ i_2 \end{pmatrix} = \begin{pmatrix} 3 \\ 1 \end{pmatrix}$$

Writing equations in this format is why the multiplication rule of matrices is useful.

Let $\mathbf{A} = \begin{pmatrix} 7 & 9 \\ 5 & 7 \end{pmatrix}$. Therefore we have

$$\mathbf{A}\begin{pmatrix} i_1 \\ i_2 \end{pmatrix} = \begin{pmatrix} 3 \\ 1 \end{pmatrix}$$

Since we want to find i_1 and i_2 we need to obtain

$$\begin{pmatrix} i_1 \\ i_2 \end{pmatrix} = \begin{pmatrix} ? \\ ? \end{pmatrix}$$

We need to remove the A from the Left-Hand Side. How?

Multiply both sides of ▮*▮ by the inverse matrix, $\mathbf{A}^{-1}$, which gives

$$\underbrace{\mathbf{A}^{-1}\mathbf{A}}_{=\mathbf{I}}\begin{pmatrix} i_1 \\ i_2 \end{pmatrix} = \mathbf{A}^{-1}\begin{pmatrix} 3 \\ 1 \end{pmatrix}$$

$$\mathbf{I}\begin{pmatrix} i_1 \\ i_2 \end{pmatrix} = \mathbf{A}^{-1}\begin{pmatrix} 3 \\ 1 \end{pmatrix}$$

Remember that **I** is the identity matrix, so on the Left-Hand Side we only have

$\begin{pmatrix} i_1 \\ i_2 \end{pmatrix}$. Thus $\begin{pmatrix} i_1 \\ i_2 \end{pmatrix} = \mathbf{A}^{-1}\begin{pmatrix} 3 \\ 1 \end{pmatrix}$. We need to find the inverse matrix, $\mathbf{A}^{-1}$, using

11.4
$$\begin{pmatrix} a & b \\ c & d \end{pmatrix}^{-1} = \frac{1}{ad - cb}\begin{pmatrix} d & -b \\ -c & a \end{pmatrix}$$

we have

$$\mathbf{A}^{-1} = \begin{pmatrix} 7 & 9 \\ 5 & 7 \end{pmatrix}^{-1} = \frac{1}{(7 \times 7) - (5 \times 9)}\begin{pmatrix} 7 & -9 \\ -5 & 7 \end{pmatrix} = \frac{1}{4}\begin{pmatrix} 7 & -9 \\ -5 & 7 \end{pmatrix}$$

$$\begin{pmatrix} i_1 \\ i_2 \end{pmatrix} = \frac{1}{4}\begin{pmatrix} 7 & -9 \\ -5 & 7 \end{pmatrix}\begin{pmatrix} 3 \\ 1 \end{pmatrix} = \frac{1}{4}\begin{pmatrix} (7 \times 3) + (-9 \times 1) \\ (-5 \times 3) + (7 \times 1) \end{pmatrix} = \frac{1}{4}\begin{pmatrix} 12 \\ -8 \end{pmatrix}$$

Multiplying each element of the last matrix by 1/4:

$$\begin{pmatrix} i_1 \\ i_2 \end{pmatrix} = \begin{pmatrix} 12/4 \\ -8/4 \end{pmatrix} = \begin{pmatrix} 3 \\ -2 \end{pmatrix}$$

Hence $i_1 = 3$ A (amp) and $i_2 = -2$ A. To check this solution is correct we substitute these into

$$7i_1 + 9i_2 = 3 \qquad\qquad 7(3) + 9(-2) = 3$$
$$5i_1 + 7i_2 = 1 \qquad\qquad 5(3) + 7(-2) = 1$$

Normally an $n \times 1$ matrix is called a column vector and is denoted by a bold small case letter (n is any positive whole number).

The above equations

$$7i_1 + 9i_2 = 3$$
$$5i_1 + 7i_2 = 1$$

can be written in matrix form as

$$\mathbf{Au} = \mathbf{b}$$

where $\mathbf{A} = \begin{pmatrix} 7 & 9 \\ 5 & 7 \end{pmatrix}$, $\mathbf{u} = \begin{pmatrix} i_1 \\ i_2 \end{pmatrix}$ and $\mathbf{b} = \begin{pmatrix} 3 \\ 1 \end{pmatrix}$. (In this case $\mathbf{u}$ and $\mathbf{b}$ are column vectors.)

For the general case

$$\mathbf{Au} = \mathbf{b}$$

where $\mathbf{A}$ is a 2×2 matrix with inverse $\mathbf{A}^{-1}$, $\mathbf{u} = \begin{pmatrix} x \\ y \end{pmatrix}$ (x and y are the variables whose values we are trying to find) and $\mathbf{b}$ is the value of the equations. To solve these equations we need to find the values of x and y which satisfy these equations, hence we need to find $\mathbf{u} = \begin{pmatrix} x \\ y \end{pmatrix}$.

This can be determined by

11.5 $\qquad \mathbf{u} = \mathbf{A}^{-1}\mathbf{b}$

To solve simultaneous equations we can also use the following algorithm:
1 Write the given equations in matrix form: $\mathbf{Au} = \mathbf{b}$.
2 Find $\mathbf{A}^{-1}$.
3 The solution $\mathbf{u}$ is given by $\mathbf{u} = \mathbf{A}^{-1}\mathbf{b}$.

Example 6

Solve

$$2x + 3y = 4$$
$$10x + 4y = 9$$

Solution

Following the above algorithm.

1 We write the given simultaneous equations in matrix form:

$$\begin{pmatrix} 2 & 3 \\ 10 & 4 \end{pmatrix} \begin{pmatrix} x \\ y \end{pmatrix} = \begin{pmatrix} 4 \\ 9 \end{pmatrix}$$

2 Let $\mathbf{A} = \begin{pmatrix} 2 & 3 \\ 10 & 4 \end{pmatrix}$ so we need to find $\mathbf{A}^{-1}$ by using

11.4 $\qquad \begin{pmatrix} a & b \\ c & d \end{pmatrix}^{-1} = \dfrac{1}{ad - cb}\begin{pmatrix} d & -b \\ -c & a \end{pmatrix}$

$$\begin{pmatrix} 2 & 3 \\ 10 & 4 \end{pmatrix}^{-1} = \frac{1}{8 - 30}\begin{pmatrix} 4 & -3 \\ -10 & 2 \end{pmatrix} = -\frac{1}{22}\begin{pmatrix} 4 & -3 \\ -10 & 2 \end{pmatrix}$$

Example 6 *continued*

3 Hence

$$\begin{pmatrix} x \\ y \end{pmatrix} = A^{-1}\begin{pmatrix} 4 \\ 9 \end{pmatrix}$$

$$= -\frac{1}{22}\begin{pmatrix} 4 & -3 \\ -10 & 2 \end{pmatrix}\begin{pmatrix} 4 \\ 9 \end{pmatrix} = -\frac{1}{22}\begin{pmatrix} -11 \\ -22 \end{pmatrix} = \begin{pmatrix} 1/2 \\ 1 \end{pmatrix}$$

Therefore $x = 1/2$, $y = 1$. Check for yourself that $x = 1/2$, $y = 1$ do indeed satisfy the given simultaneous equations.

B2 Applications in control engineering

Next we examine applications of matrices in control engineering. We use the following definitions of control:

Let **A**, **B** and **C** be matrices, where **A** is a 2×2 matrix, **B** is a 2×1 matrix and **C** is a 1×2 matrix. That is **A** is a square matrix, **B** is a column vector and **C** is a row vector.

The controllability matrix, C_T, is defined as

11.6 $C_T = (B \quad AB)$

where **B** forms the first column and the product **AB** forms the second column of the matrix C_T.

The observability matrix, O_T, is defined as

11.7 $O_T = \begin{pmatrix} C \\ CA \end{pmatrix}$

where **C** forms the first row and the product **CA** forms the second row of O_T.

Example 7 *control engineering*

The state–space equations of a system has the following matrices:

$$A = \begin{pmatrix} -1 & 1 \\ 2 & -3 \end{pmatrix}, B = \begin{pmatrix} 1 \\ 1 \end{pmatrix} \text{ and } C = (2 \quad 3)$$

Find C_T and O_T.

Solution

First we obtain the product **AB**:

$$AB = \begin{pmatrix} -1 & 1 \\ 2 & -3 \end{pmatrix}\begin{pmatrix} 1 \\ 1 \end{pmatrix} = \begin{pmatrix} (-1 \times 1) + (1 \times 1) \\ (2 \times 1) + (-3 \times 1) \end{pmatrix} = \begin{pmatrix} 0 \\ -1 \end{pmatrix}$$

Example 7 *continued*

Substituting $\mathbf{B} = \begin{pmatrix} 1 \\ 1 \end{pmatrix}$ and $\mathbf{AB} = \begin{pmatrix} 0 \\ -1 \end{pmatrix}$ into $\mathbf{C_T} = (\mathbf{B} \quad \mathbf{AB})$ gives

$$\mathbf{C_T} = \begin{pmatrix} 1 & 0 \\ 1 & -1 \end{pmatrix}$$

To find the observability matrix, $\mathbf{O_T}$, we first obtain the product $\mathbf{CA}$:

$$\mathbf{CA} = (2 \quad 3)\begin{pmatrix} -1 & 1 \\ 2 & -3 \end{pmatrix}\Big([\,2 \times (-1)] + (3 \times 2) \quad (2 \times 1) + [3 \times (-3)]\Big) = (4 \quad -7)$$

Using $\mathbf{O_T} = \begin{pmatrix} \mathbf{C} \\ \mathbf{CA} \end{pmatrix}$ with $\mathbf{C} = (2 \quad 3)$ and $\mathbf{CA} = (4 \quad -7)$ gives $\mathbf{O_T} = \begin{pmatrix} 2 & 3 \\ 4 & -7 \end{pmatrix}$

Example 8 *control engineering*

The system poles are defined to be those values of s which satisfy

$$\det(s\mathbf{I} - \mathbf{A}) = 0$$

These s are also called the eigenvalues of $\mathbf{A}$. We will discuss eigenvalues in **Section F**. Obtain the system poles for

$$\mathbf{A} = \begin{pmatrix} 3 & -4 \\ 2 & -1 \end{pmatrix}$$

Solution

Remember that $\mathbf{I}$ is the identity matrix. Therefore we have

$$s\mathbf{I} - \mathbf{A} = s\begin{pmatrix} 1 & 0 \\ 0 & 1 \end{pmatrix} - \begin{pmatrix} 3 & -4 \\ 2 & -1 \end{pmatrix} \qquad \left[\text{Because } \mathbf{I} = \begin{pmatrix} 1 & 0 \\ 0 & 1 \end{pmatrix} \right]$$

$$= \begin{pmatrix} s & 0 \\ 0 & s \end{pmatrix} - \begin{pmatrix} 3 & -4 \\ 2 & -1 \end{pmatrix}$$

$$= \begin{pmatrix} s - 3 & 0 - (-4) \\ 0 - 2 & s - (-1) \end{pmatrix} = \begin{pmatrix} s - 3 & 4 \\ -2 & s + 1 \end{pmatrix}$$

Hence

$$\det(s\mathbf{I} - \mathbf{A}) = \det\begin{pmatrix} s - 3 & 4 \\ -2 & s + 1 \end{pmatrix}$$

$$= (s - 3)(s + 1) - (-2 \times 4) \qquad [\text{By } 11.1\,]$$

$$= s^2 - 2s - 3 + 8$$

$$= s^2 - 2s + 5$$

We need to solve the quadratic equation, $s^2 - 2s + 5 = 0$, to find the system poles. Using the quadratic formula we have

$$s = \frac{2 \pm \sqrt{(-2)^2 - (4 \times 1 \times 5)}}{2} = \frac{2 \pm \sqrt{-16}}{2}$$

11.1 $\det\begin{pmatrix} a & b \\ c & d \end{pmatrix} = ad - cb$

⚡ Example 8 *continued*

? **How do we find $\sqrt{-16}$?**

We need to use complex numbers: $\sqrt{-16} = j4$

$$s = \frac{2 \pm j4}{2} = \frac{2}{2} \pm j\frac{4}{2} = 1 \pm j2$$

The system poles are at $s = 1 + j2$ or $s = 1 - j2$

SUMMARY

Solving simultaneous equations of the form

$$\mathbf{Au = b}$$

is given by

11.5 $\mathbf{u = A^{-1}b}$

In control engineering there are a number of formulae which can be used to find matrices such as controllability and observability matrices. We can also use matrices to find the system poles.

Exercise **11(b)**

Solutions at end of book. Complete solutions available at www.palgrave.com/science/engineering/singh

▦ **The first two questions are in the field of [*electrical principles*].**

1 We obtain the following equations from an electrical circuit:

$$2i_1 + 3i_2 = 5$$
$$6i_1 + 7i_2 = 10$$

where i_1 and i_2 represent current. Write these equations in matrix form and solve for i_1 and i_2.

2 By applying Kirchhoff's law to a circuit we obtain the following equations:

$$30i_1 - 10i_2 = 12$$
$$35i_2 - 10i_1 = 5$$

where i_1 and i_2 represent current. Find the exact values of i_1 and i_2.

⚡ **Questions 3 to 9, inclusive, are in the field of [*control engineering*].**

3 By using 11.6 and 11.7 determine the controllability matrix, $\mathbf{C_T}$, and the observability matrix, $\mathbf{O_T}$, for

$$\mathbf{A} = \begin{pmatrix} -5 & 6 \\ 2 & 3 \end{pmatrix}, \mathbf{B} = \begin{pmatrix} 1 \\ -4 \end{pmatrix} \text{ and } \mathbf{C} = (2 \quad 9)$$

4 By using 11.6 and 11.7 find the controllability matrix, $\mathbf{C_T}$, and the observability matrix, $\mathbf{O_T}$, for the system with

$$\mathbf{A} = \begin{pmatrix} 2 & 5 \\ -1 & 3 \end{pmatrix}, \mathbf{B} = \begin{pmatrix} 1 \\ 1 \end{pmatrix} \text{ and } \mathbf{C} = (-1 \quad 4)$$

A system is controllable if $\det\mathbf{C_T} \neq 0$ and is observable if $\det\mathbf{O_T} \neq 0$. Determine whether the above system is
i controllable   **ii** observable

5 By using definitions of controllable and observable of question **4**, check whether the following systems are controllable and/or observable:

a $\mathbf{A} = \begin{pmatrix} -1 & 2 \\ 9 & 3 \end{pmatrix}, \mathbf{B} = \begin{pmatrix} -1 \\ 5 \end{pmatrix}$ and
$$\mathbf{C} = (3 \quad 0)$$

b $\mathbf{A} = \begin{pmatrix} 6 & 2 \\ 1 & 5 \end{pmatrix}, \mathbf{B} = \begin{pmatrix} 1 \\ -1 \end{pmatrix}$ and
$$\mathbf{C} = (3 \quad 2)$$

Exercise 11(b) continued

Solutions at end of book. Complete solutions available at www.palgrave.com/science/engineering/singh

For questions 6 to 9 the system poles are defined in Example 8 (those values of s which satisfy $\det(sI - A) = 0$).

6 Determine the system poles for

a $A = \begin{pmatrix} -1 & 3 \\ 2 & 1 \end{pmatrix}$ **b** $A = \begin{pmatrix} 2 & 10 \\ 5 & -3 \end{pmatrix}$

c $A = \begin{pmatrix} -5.1 & 2.2 \\ 3.7 & 6.1 \end{pmatrix}$

7 The state matrix, **A**, of an *RLC* series circuit is given by

$$A = \begin{pmatrix} -\dfrac{R}{L} & -\dfrac{1}{L} \\ \dfrac{1}{C} & 0 \end{pmatrix}$$

Determine the system poles of **A**.

8 For a mechanical system the state matrix, **A**, is given by

$$A = \begin{pmatrix} 0 & 1 \\ -\dfrac{k}{m} & -\dfrac{c}{m} \end{pmatrix}$$

Find the system poles for **A**.

9 If the real part of the system poles, s, is negative, that is

$$\mathrm{Re}(s) < 0$$

then the system is said to be stable. Which of the following systems are stable?

a $A = \begin{pmatrix} -1 & -3 \\ 5 & 2 \end{pmatrix}$

b $A = \begin{pmatrix} -6 & -1 \\ 9 & -3 \end{pmatrix}$

c $A = \begin{pmatrix} 3.1 & 6.5 \\ 1.7 & 4.8 \end{pmatrix}$

10 Solve the following equations:

a $x + 3y = 14$
        $2x - 3y = -8$

b $-2x - 5y = -3$
        $-5x - 2y = 3$

c $4x + 6y = -1$
        $7x - \dfrac{3}{4}y = 2$

SECTION C **3 × 3 matrices**

By the end of this section you will be able to:
- ▶ evaluate the determinant of a 3 × 3 matrix
- ▶ find the inverse of a 3 × 3 matrix
- ▶ solve equations using the inverse matrix method

C1 **Determinants**

If

$$A = \begin{pmatrix} a & b & c \\ d & e & f \\ g & h & i \end{pmatrix}$$

is a general 3 × 3 matrix then

11.8 $\det A = a\left[\det\begin{pmatrix} e & f \\ h & i \end{pmatrix}\right] - b\left[\det\begin{pmatrix} d & f \\ g & i \end{pmatrix}\right] + c\left[\det\begin{pmatrix} d & e \\ g & h \end{pmatrix}\right]$

11.8 gives the determinant of a 3 × 3 matrix. The simplest way of evaluating the above determinant is **not** to remember all the details of **11.8** but to use the following procedure:

1 Delete the row and column containing the entry a:

$$\begin{pmatrix} (a) & & \\ & e & f \\ & h & i \end{pmatrix}$$

then evaluate the determinant of the remaining matrix $\begin{pmatrix} e & f \\ h & i \end{pmatrix}$.

2 Delete the row and column containing the entry b:

$$\begin{pmatrix} & (b) & \\ d & & f \\ g & & i \end{pmatrix}$$

then evaluate the determinant of the remaining matrix $\begin{pmatrix} d & f \\ g & i \end{pmatrix}$.

3 Delete the row and column containing the entry c:

$$\begin{pmatrix} & & (c) \\ d & e & \\ g & h & \end{pmatrix}$$

then evaluate the determinant of the remaining matrix $\begin{pmatrix} d & e \\ g & h \end{pmatrix}$.

4 Combine these steps as in **11.8** to obtain the determinant of the above matrix.

A general aid in finding the determinant is to place your thumb over the number and ignore the horizontal row and vertical column containing that number.

Example 9

Evaluate the determinant of

$$A = \begin{pmatrix} 1 & 7 & 5 \\ 3 & 2 & -7 \\ 6 & -8 & 9 \end{pmatrix}$$

Solution

By using steps 1–4 and **11.8** we have

$$\det A = 1\det\begin{pmatrix} 2 & -7 \\ -8 & 9 \end{pmatrix} - 7\det\begin{pmatrix} 3 & -7 \\ 6 & 9 \end{pmatrix} + 5\det\begin{pmatrix} 3 & 2 \\ 6 & -8 \end{pmatrix}$$

$$\underset{\text{by } \boxed{11.1}}{=} \left[(2 \times 9) - (-8 \times (-7))\right] - 7\left[(3 \times 9) - (6 \times (-7))\right] + 5\left[(3 \times (-8)) - (6 \times 2)\right]$$

$$= -701$$

11.1 $\det\begin{pmatrix} a & b \\ c & d \end{pmatrix} = ad - cb$

C2 **Cofactors**

A general 3×3 matrix is $\begin{pmatrix} a & b & c \\ d & e & f \\ g & h & i \end{pmatrix}$. The determinant of the remaining matrix after deleting the rows and columns of an entry is called the **minor** of that entry. For example, in the above case

$$\det\begin{pmatrix} e & f \\ h & i \end{pmatrix} \text{ is the minor of } a$$

$$\det\begin{pmatrix} d & f \\ g & i \end{pmatrix} \text{ is the minor of } b$$

? **What is the minor of c?**

$$\det\begin{pmatrix} d & e \\ g & h \end{pmatrix}$$

? **What is the minor of e?**

Deleting the rows and columns containing the entry e we have

$$\begin{pmatrix} a & & c \\ & (e) & \\ g & & i \end{pmatrix}$$

Hence $\det\begin{pmatrix} a & c \\ g & i \end{pmatrix}$ is the minor of e.

Example 10

Evaluate the minor of -1 in

$$\begin{pmatrix} 3 & 5 & 7 \\ -1 & 2 & 3 \\ -4 & 4 & -9 \end{pmatrix}$$

Solution

After deleting the rows and columns containing -1

$$\begin{pmatrix} & 5 & 7 \\ (-1) & & \\ & 4 & -9 \end{pmatrix}$$

we obtain the matrix $\begin{pmatrix} 5 & 7 \\ 4 & -9 \end{pmatrix}$. The minor of -1 is

$$\det\begin{pmatrix} 5 & 7 \\ 4 & -9 \end{pmatrix} \underset{\text{by } \boxed{11.1}}{=} [5 \times (-9)] - [4 \times 7] = -73$$

$\boxed{11.1}$ $\det\begin{pmatrix} a & b \\ c & d \end{pmatrix} = ad - cb$

The **cofactor** of −1 in **Example 10** is −(−73) = 73. The cofactor of an entry is the minor with a **place sign**. The place sign is found from the rule:

$$
\begin{array}{ccc}
+ & - & + \\
- & + & - \\
+ & - & +
\end{array}
$$

The first entry of a matrix has a positive place sign and then the place signs alternate.

Example 11

Find the cofactor of 5 of the matrix in **Example 10**.

Solution

The minor of 5 is

$$
\det\begin{pmatrix} -1 & 3 \\ -4 & -9 \end{pmatrix} \underset{\text{by } 11.1}{=} [-1 \times (-9)] - [(-4) \times 3] = 21
$$

According to the rule, the place sign is negative, so the cofactor of 5 is −21.

Note that the **minor** of 5 is 21 but the **cofactor** is −21 because the position of 5 in the matrix gives it a negative place sign.

Hence the cofactor is a minor with a place sign. We can rewrite the determinant of a 3×3 matrix in terms of its cofactors, that is if **A** is a general 3×3 matrix:

$$
\mathbf{A} = \begin{pmatrix} a & b & c \\ d & e & f \\ g & h & i \end{pmatrix}
$$

then

11.9 $\det \mathbf{A} = a(\text{cofactor of } a) + b(\text{cofactor of } b) + c(\text{cofactor of } c)$

We can find the determinant of a matrix by expanding along any of the rows or any of the columns. For example the formula for expanding along the second row is

$$\det \mathbf{A} = d(\text{cofactor of } d) + e(\text{cofactor of } e) + f(\text{cofactor of } f)$$

? **What is the formula for expanding along the third row?**

$$\det \mathbf{A} = g(\text{cofactor of } g) + h(\text{cofactor of } h) + i(\text{cofactor of } i)$$

We can also expand along any column. Expansion along the first column is

$$\det \mathbf{A} = a(\text{cofactor of } a) + d(\text{cofactor of } d) + g(\text{cofactor of } g)$$

and so on.

11.1 $\det\begin{pmatrix} a & b \\ c & d \end{pmatrix} = ad - cb$

Example 12

Find the determinant of

$$A = \begin{pmatrix} -1 & -6 & 3 \\ 5 & 6 & -7 \\ -2 & 0 & 1 \end{pmatrix}$$

Solution

Since there is a zero in the last row it is easier to expand along this row:

$$\det\begin{pmatrix} -1 & -6 & 3 \\ 5 & 6 & -7 \\ -2 & 0 & 1 \end{pmatrix} = -2\det\begin{pmatrix} -6 & 3 \\ 6 & -7 \end{pmatrix} - 0\det\begin{pmatrix} -1 & 3 \\ 5 & -7 \end{pmatrix} + 1\det\begin{pmatrix} -1 & -6 \\ 5 & 6 \end{pmatrix}$$

$$\underset{\text{by } \boxed{11.1}}{=} -2(42 - 18) - 0 + (-6 + 30)$$

$$= -24$$

Note that the middle term on the Right-Hand Side is zero and so we just need to evaluate the other two terms. We could also have found the determinant of **A** by expanding along the second column because this also contains (the same) zero.

If a row or column contains zero(s) then it is easier to expand along that row or column.

We can also obtain the determinant of a 4 × 4, 5 × 5, 6 × 6 etc. matrix but it becomes very laborious to do this by just using pen and paper. In these cases it is more convenient to use a graphical calculator or a computer algebra system.

C3 **Cofactor matrix**

Let **C** be the new matrix consisting of the cofactors of the general matrix **A**. If

$$A = \begin{pmatrix} a & b & c \\ d & e & f \\ g & h & i \end{pmatrix} \quad \text{then} \quad C = \begin{pmatrix} A & B & C \\ D & E & F \\ G & H & I \end{pmatrix}$$

where A is the cofactor of a, B is the cofactor of b, etc. **C** is called the cofactor matrix and it is used in finding the inverse matrix.

Example 13

Find the cofactor matrix **C** of

$$A = \begin{pmatrix} 1 & -1 & 5 \\ 3 & 9 & 7 \\ -2 & 1 & 0 \end{pmatrix}$$

$\boxed{11.1}$ $\det\begin{pmatrix} a & b \\ c & d \end{pmatrix} = ad - cb$

Example 13 *continued*

Solution

Cofactor of the first entry, 1, is

$$\det\begin{pmatrix} 9 & 7 \\ 1 & 0 \end{pmatrix} \underset{\text{by } 11.1}{=} [(9 \times 0) - (1 \times 7)] = -7$$

Cofactor of -1 is

$$\underset{\substack{\text{minus} \\ \text{place sign}}}{-}\det\begin{pmatrix} 3 & 7 \\ -2 & 0 \end{pmatrix} \underset{\text{by } 11.1}{=} -[(3 \times 0) - (-2 \times 7)] = -14$$

Cofactor of 5 is

$$\det\begin{pmatrix} 3 & 9 \\ -2 & 1 \end{pmatrix} \underset{\text{by } 11.1}{=} [(3 \times 1) - (-2 \times 9)] = 21$$

Cofactor of 3 is

$$\underset{\substack{\text{minus} \\ \text{place sign}}}{-}\det\begin{pmatrix} -1 & 5 \\ 1 & 0 \end{pmatrix} \underset{\text{by } 11.1}{=} -[(-1 \times 0) - (1 \times 5)] = 5$$

Cofactor of 9 is

$$\det\begin{pmatrix} 1 & 5 \\ -2 & 0 \end{pmatrix} \underset{\text{by } 11.1}{=} [(1 \times 0) - (-2 \times 5)] = 10$$

Cofactor of 7 is

$$\underset{\substack{\text{minus} \\ \text{place sign}}}{-}\det\begin{pmatrix} 1 & -1 \\ -2 & 1 \end{pmatrix} \underset{\text{by } 11.1}{=} -[(1 \times 1)-(-2 \times (-1))] = 1$$

Cofactor of -2 is

$$\det\begin{pmatrix} -1 & 5 \\ 9 & 7 \end{pmatrix} \underset{\text{by } 11.1}{=} [(-1 \times 7) - (9 \times 5)] = -52$$

Cofactor of 1 (the 1 on the bottom row of the given matrix) is

$$\underset{\substack{\text{minus} \\ \text{place sign}}}{-}\det\begin{pmatrix} 1 & 5 \\ 3 & 7 \end{pmatrix} \underset{\text{by } 11.1}{=} -[(1 \times 7)-(3 \times 5)] = 8$$

Cofactor of 0 is

$$\det\begin{pmatrix} 1 & -1 \\ 3 & 9 \end{pmatrix} \underset{\text{by } 11.1}{=} [(1 \times 9) - (3 \times (-1))] = 12$$

Hence by collecting these together and placing them in the corresponding position gives the cofactor matrix:

$$\mathbf{C} = \begin{pmatrix} -7 & -14 & 21 \\ 5 & 10 & 1 \\ -52 & 8 & 12 \end{pmatrix}$$

11.1 $\det\begin{pmatrix} a & b \\ c & d \end{pmatrix} = ad - cb$

C4 Inverse of a matrix

The **transpose** of a matrix is evaluated by interchanging the rows and columns. Let A be a matrix then the transpose is denoted by A^T. The first row of A becomes the first column of A^T, second row of A becomes the second column of A^T and so on. If

$$A = \begin{pmatrix} a & b & c \\ d & e & f \\ g & h & i \end{pmatrix} \text{ then } A^T = \begin{pmatrix} a & d & g \\ b & e & h \\ c & f & i \end{pmatrix}$$

Example 14

Find A^T and C^T of **Example 13**.

Solution

We have

$$A = \begin{pmatrix} 1 & -1 & 5 \\ 3 & 9 & 7 \\ -2 & 1 & 0 \end{pmatrix} \qquad C = \begin{pmatrix} -7 & -14 & 21 \\ 5 & 10 & 1 \\ -52 & 8 & 12 \end{pmatrix}$$

Interchanging rows and columns gives

$$A^T = \begin{pmatrix} 1 & 3 & -2 \\ -1 & 9 & 1 \\ 5 & 7 & 0 \end{pmatrix} \qquad C^T = \begin{pmatrix} -7 & 5 & -52 \\ -14 & 10 & 8 \\ 21 & 1 & 12 \end{pmatrix}$$

We can find the transpose of **any** matrix, it doesn't need to be a square matrix. For example if

$$A = \begin{pmatrix} 1 & 2 & 3 \\ 4 & 5 & 6 \end{pmatrix} \text{ then } A^T = \begin{pmatrix} 1 & 4 \\ 2 & 5 \\ 3 & 6 \end{pmatrix}$$

The **adjoint** of a square matrix A, denoted adj$A$, is defined as

11.10 $\text{adj}A = C^T$

where C is the matrix of cofactors of A. For **Example 14**:

$$\text{adj}A = C^T = \begin{pmatrix} -7 & 5 & -52 \\ -14 & 10 & 8 \\ 21 & 1 & 12 \end{pmatrix}$$

The inverse of any square matrix A, denoted by A^{-1}, is given by

11.11 $A^{-1} = \dfrac{1}{\det A}(\text{adj}A)$ [Provided that $\det A \neq 0$]

and it is defined as the matrix which satisfies

$$A^{-1} \times A = I$$

where I is the identity matrix.

Example 15

Determine A^{-1} of

$$A = \begin{pmatrix} 1 & -1 & 5 \\ 3 & 9 & 7 \\ -2 & 1 & 0 \end{pmatrix}$$

Solution

Since $A^{-1} = \dfrac{1}{\det A}(\text{adj}A)$ we need to find adjA and detA.

?   **What is adjA equal to?**

Because the matrix A is the same matrix A as in **Examples 13** and **14**, we have already evaluated adjA ($= C^T$, that is the cofactor matrix transposed). Hence from above we have

$$\text{adj}A = \begin{pmatrix} -7 & 5 & -52 \\ -14 & 10 & 8 \\ 21 & 1 & 12 \end{pmatrix} \qquad [= C^T]$$

We only need to find detA. Expanding along the **bottom row** of the given A because it contains a 0:

$$\det A = -2\det\begin{pmatrix} -1 & 5 \\ 9 & 7 \end{pmatrix} - 1\det\begin{pmatrix} 1 & 5 \\ 3 & 7 \end{pmatrix} + 0\det\begin{pmatrix} 1 & -1 \\ 3 & 9 \end{pmatrix}$$

$$\underset{\text{by } \boxed{11.1}}{=} -2[-7-45]-1[7-15]+0$$

$$= 112$$

Substituting these into $\boxed{11.11}$ gives

$$A^{-1} = \frac{1}{\det A}(\text{adj}A) = \frac{1}{112}\begin{pmatrix} -7 & 5 & -52 \\ -14 & 10 & 8 \\ 21 & 1 & 12 \end{pmatrix}$$

Example 16

Check that A^{-1} is indeed the inverse matrix of A in **Example 15**.

Solution

We need to check that

$$A^{-1} \times A = I = \begin{pmatrix} 1 & 0 & 0 \\ 0 & 1 & 0 \\ 0 & 0 & 1 \end{pmatrix}$$

$\boxed{11.1}$ $\det\begin{pmatrix} a & b \\ c & d \end{pmatrix} = ad - cb$

Example 16 *continued*

I is the identity matrix of a 3×3 matrix, with 1's along the main diagonal and zeros elsewhere. Also remember that multiplication of matrices is row $\times$ column.

$$\mathbf{A}^{-1} \times \mathbf{A} = \frac{1}{112} \begin{pmatrix} -7 & 5 & -52 \\ -14 & 10 & 8 \\ 21 & 1 & 12 \end{pmatrix} \begin{pmatrix} 1 & -1 & 5 \\ 3 & 9 & 7 \\ -2 & 1 & 0 \end{pmatrix}$$

$$= \frac{1}{112} \begin{pmatrix} 112 & 0 & 0 \\ 0 & 112 & 0 \\ 0 & 0 & 112 \end{pmatrix} = \begin{pmatrix} 1 & 0 & 0 \\ 0 & 1 & 0 \\ 0 & 0 & 1 \end{pmatrix} = \mathbf{I}$$

The identity matrix is a square matrix with 1's along the diagonal and zeros elsewhere.

Example 17 *electrical principles*

By applying Kirchhoff's law to a circuit we obtain the following equations:

$$i_1 + 2i_2 + i_3 = 4$$
$$3i_1 - 4i_2 - 2i_3 = 2$$
$$5i_1 + 3i_2 + 5i_3 = -1$$

where i_1, i_2 and i_3 represent current. By using the inverse matrix method, find the values of i_1, i_2 and i_3.

Solution

Writing the above equations in matrix form we obtain

$$\begin{pmatrix} 1 & 2 & 1 \\ 3 & -4 & -2 \\ 5 & 3 & 5 \end{pmatrix} \begin{pmatrix} i_1 \\ i_2 \\ i_3 \end{pmatrix} = \begin{pmatrix} 4 \\ 2 \\ -1 \end{pmatrix}$$

Let $\mathbf{A} = \begin{pmatrix} 1 & 2 & 1 \\ 3 & -4 & -2 \\ 5 & 3 & 5 \end{pmatrix}$, then we have

$$\mathbf{A} \begin{pmatrix} i_1 \\ i_2 \\ i_3 \end{pmatrix} = \begin{pmatrix} 4 \\ 2 \\ -1 \end{pmatrix}$$

Multiplying both sides by the inverse matrix, $\mathbf{A}^{-1}$, gives

$$\begin{pmatrix} i_1 \\ i_2 \\ i_3 \end{pmatrix} = \mathbf{A}^{-1} \begin{pmatrix} 4 \\ 2 \\ -1 \end{pmatrix}$$

Example 17 *continued*

We need to find $\mathbf{A}^{-1}$. **How do we obtain $\mathbf{A}^{-1}$?**

We need $\det\mathbf{A}$ because $\mathbf{A}^{-1} = \dfrac{1}{\det\mathbf{A}}(\text{adj}\mathbf{A})$. Thus

$$\det\begin{pmatrix} 1 & 2 & 1 \\ 3 & -4 & -2 \\ 5 & 3 & 5 \end{pmatrix} = \det\begin{pmatrix} -4 & -2 \\ 3 & 5 \end{pmatrix} - 2\det\begin{pmatrix} 3 & -2 \\ 5 & 5 \end{pmatrix} + \det\begin{pmatrix} 3 & -4 \\ 5 & 3 \end{pmatrix}$$

$$\underset{\text{by } \boxed{11.1}}{=} -14 - 50 + 29$$

$$\det\mathbf{A} = -35$$

What else do we need to find?

adj$\mathbf{A}$

adj$\mathbf{A}$ is the cofactor matrix $\mathbf{C}$ transposed. Evaluating the cofactors of each entry horizontally we obtain the following:

Cofactor of 1 (first entry) is

$$\det\begin{pmatrix} -4 & -2 \\ 3 & 5 \end{pmatrix} \underset{\text{by } \boxed{11.1}}{=} (-4 \times 5) - (3 \times (-2)) = -14$$

Cofactor of 2 is

$$-\det\begin{pmatrix} 3 & -2 \\ 5 & 5 \end{pmatrix} \underset{\text{by } \boxed{11.1}}{=} -[(3 \times 5) - (5 \times (-2))] = -25$$

Cofactor of 1 is

$$\det\begin{pmatrix} 3 & -4 \\ 5 & 3 \end{pmatrix} = 29$$

Cofactor of 3 (in the second row) is

$$-\det\begin{pmatrix} 2 & 1 \\ 3 & 5 \end{pmatrix} = -7$$

Cofactor of −4 is

$$\det\begin{pmatrix} 1 & 1 \\ 5 & 5 \end{pmatrix} = 0$$

Cofactor of −2 is

$$-\det\begin{pmatrix} 1 & 2 \\ 5 & 3 \end{pmatrix} = 7$$

Cofactor of 5 (the left 5 in the bottom row) is

$$\det\begin{pmatrix} 2 & 1 \\ -4 & -2 \end{pmatrix} = 0$$

$\boxed{11.1}$ $\det\begin{pmatrix} a & b \\ c & d \end{pmatrix} = ad - cb$

Example 17 *continued*

Cofactor of 3 (in the last row) is

$$-\det\begin{pmatrix} 1 & 1 \\ 3 & -2 \end{pmatrix} = 5$$

Cofactor of 5 (the right 5 in the bottom row) is

$$\det\begin{pmatrix} 1 & 2 \\ 3 & -4 \end{pmatrix} = -10$$

Thus the cofactor matrix $\mathbf{C} = \begin{pmatrix} -14 & -25 & 29 \\ -7 & 0 & 7 \\ 0 & 5 & -10 \end{pmatrix}$

Transposing (rows ↔ columns) this matrix gives the adjoint:

$$\text{adjA} = \mathbf{C}^{\mathrm{T}} = \begin{pmatrix} -14 & -7 & 0 \\ -25 & 0 & 5 \\ 29 & 7 & -10 \end{pmatrix}$$

Substituting this and detA $= -35$ into 11.11 yields

$$\mathbf{A}^{-1} = \frac{1}{\text{detA}} (\text{adjA}) = \frac{1}{-35} \begin{pmatrix} -14 & -7 & 0 \\ -25 & 0 & 5 \\ 29 & 7 & -10 \end{pmatrix}$$

Substituting this into ▨ * yields

$$\begin{pmatrix} i_1 \\ i_2 \\ i_3 \end{pmatrix} = \mathbf{A}^{-1} \begin{pmatrix} 4 \\ 2 \\ -1 \end{pmatrix}$$

$$= \frac{1}{-35} \begin{pmatrix} -14 & -7 & 0 \\ -25 & 0 & 5 \\ 29 & 7 & -10 \end{pmatrix} \begin{pmatrix} 4 \\ 2 \\ -1 \end{pmatrix}$$

$$= -\frac{1}{35} \begin{pmatrix} [-14 \times 4] + [-7 \times 2] + [0 \times (-1)] \\ [-25 \times 4] + [0 \times 2] + [5 \times (-1)] \\ [29 \times 4] + [7 \times 2] + [-10 \times (-1)] \end{pmatrix} = -\frac{1}{35} \begin{pmatrix} -70 \\ -105 \\ 140 \end{pmatrix}$$

$$\begin{pmatrix} i_1 \\ i_2 \\ i_3 \end{pmatrix} = -\frac{1}{35} \begin{pmatrix} -70 \\ -105 \\ 140 \end{pmatrix}$$

$$= \begin{pmatrix} -70/(-35) \\ -105/(-35) \\ 140/(-35) \end{pmatrix} = \begin{pmatrix} 2 \\ 3 \\ -4 \end{pmatrix}$$

Hence $i_1 = 2$ A, $i_2 = 3$ A and $i_3 = -4$ A.

You can apply the inverse matrix method as described in **Example 17** to solve 3, 4, 5, . . . simultaneous equations. It might seem long and tedious but generally solving three or more equations is laborious unless you use a graphical calculator or a computer algebra system.

SUMMARY

The determinant of a 3×3 matrix is

11.8 $\det\begin{pmatrix} a & b & c \\ d & e & f \\ g & h & i \end{pmatrix} = a\left[\det\begin{pmatrix} e & f \\ h & i \end{pmatrix}\right] - b\left[\det\begin{pmatrix} d & f \\ g & i \end{pmatrix}\right] + c\left[\det\begin{pmatrix} d & e \\ g & h \end{pmatrix}\right]$

The cofactor of a in **11.8** is the determinant of the matrix after deleting the row and column containing a with a place sign.

The cofactor matrix is the matrix of cofactors and the adjoint is defined as the cofactor matrix transposed.

The inverse of a square matrix **A** is

11.11 $\quad \mathbf{A}^{-1} = \dfrac{1}{\det\mathbf{A}}(\text{adj}\mathbf{A}) \quad$ [Provided that $\det\mathbf{A} \neq 0$]

Exercise 11(c)

Solutions at end of book. Complete solutions available at www.palgrave.com/science/engineering/singh

1 Calculate the determinants of the following matrices:

a $A = \begin{pmatrix} 1 & 3 & -1 \\ 2 & 0 & 5 \\ -6 & 3 & 1 \end{pmatrix}$

b $B = \begin{pmatrix} 2 & -10 & 11 \\ 5 & 3 & -4 \\ 7 & 9 & 12 \end{pmatrix}$

c $C = \begin{pmatrix} -12 & 9 & -5 \\ 3 & 9 & 1 \\ -7 & 2 & -2 \end{pmatrix}$

2 Find A^T (A transposed) for the following:

a $A = \begin{pmatrix} -2 & 3 \\ 1 & 5 \end{pmatrix}$ **b** $A = \begin{pmatrix} 1/3 & -2/5 \\ -3/7 & \pi \end{pmatrix}$

c $A = \begin{pmatrix} 7 & 3 & 4 \\ 2 & 6 & 1 \\ -3 & -3 & 1 \end{pmatrix}$

d $A = \begin{pmatrix} 1.17 & 1.36 \\ 9.39 & -1.45 \\ 2.11 & 5.20 \end{pmatrix}$

e $A = \begin{pmatrix} -7 & 2 & 1 & 5 \\ 3 & 6 & -4 & 7 \\ 8 & 3 & -3 & 5 \\ -4 & 6 & 7 & 0 \end{pmatrix}$

f $A = \begin{pmatrix} 1 & -3 & 7 \\ -9 & 4 & 6 \\ -4 & 2 & 8 \\ 9 & 19 & 11 \end{pmatrix}$

3 Show that

$$\det\begin{pmatrix} \mathbf{i} & \mathbf{j} & \mathbf{k} \\ 7 & 3 & -2 \\ 4 & 2 & 7 \end{pmatrix} = 25\mathbf{i} - 57\mathbf{j} + 2\mathbf{k}$$

4 Find the values of t so that

$$\det\begin{pmatrix} 1 & 0 & 3 \\ 5 & t & -7 \\ 3 & 9 & t-1 \end{pmatrix} = 0$$

5 Find the cofactor matrix C and C^T of

$$A = \begin{pmatrix} 1 & 0 & 5 \\ -2 & 3 & 7 \\ 6 & -1 & 0 \end{pmatrix}$$

6 [electrical principles] By applying Kirchhoff's law and Ohm's law to a circuit, we obtain the following equations:

$$3i_1 - 5i_2 + 3i_3 = 7.5$$
$$2i_1 + i_2 - 7i_3 = -17.5$$
$$-10i_1 + 4i_2 + 5i_3 = 16$$

where i_1, i_2 and i_3 represent current. By using the inverse matrix method, find the values of i_1, i_2 and i_3.

Solutions at end of book. Complete solutions available at www.palgrave.com/science/engineering/singh

Exercise **11(c) continued**

7 Find the determinant of the following:

a $A = \begin{pmatrix} 2 & 3 & 5 \\ 0 & 0 & 6 \\ 1 & 5 & 3 \end{pmatrix}$

b $B = \begin{pmatrix} 6 & 7 & 1 \\ 1 & 3 & 2 \\ 0 & 1 & 5 \end{pmatrix}$

c $C = \begin{pmatrix} 1 & 5 & 1 \\ 0 & 3 & 7 \\ 0 & 2 & 9 \end{pmatrix}$

d $D = \begin{pmatrix} 9 & 5 & 1 \\ 13 & 0 & 2 \\ 11 & 0 & 3 \end{pmatrix}$

For the following use a computer algebra system or a graphical calculator.

8 Evaluate the determinant of

a $A = \begin{pmatrix} 1 & 2 & 3 & 4 \\ 5 & 6 & 7 & 8 \\ 9 & 10 & 11 & 12 \\ 13 & 14 & 15 & 16 \end{pmatrix}$

b $B = \begin{pmatrix} -1.10 & 4.23 & 2.67 & 7.45 & 9.62 \\ 19.61 & 6.40 & 3.12 & 11.89 & 2.36 \\ -17.5 & -9.73 & 5.23 & 8.54 & 2.51 \\ 6.19 & 2.91 & 17.64 & 8.93 & 8.98 \\ 3.98 & 11.84 & 4.78 & 9.85 & 3.22 \end{pmatrix}$

SECTION D **Gaussian elimination**

By the end of this section you will be able to:
► understand the process of Gaussian elimination
► apply the Gaussian elimination procedure to a system

Solving three (or more) simultaneous equations by the inverse matrix method can be a strenuous task, as we observed in the previous section. The Gaussian elimination procedure is a more straightforward method for solving simultaneous linear equations.

D1 **Gaussian elimination**

The Gaussian elimination procedure is often used in computer software to solve systems of linear equations.

The Gaussian elimination procedure is named after the German mathematician **K.F. Gauss**, 1777–1855, who was one of the greatest mathematicians of all time.

Example 18

Solve the following equations using Gaussian elimination:

$$x + 5y - 3z = -9$$
$$2x - 3y - z = 19$$
$$3x + 4y + z = -13$$

Example 18 *continued*

Solution

This can be written in matrix form as

$$\begin{pmatrix} 1 & 5 & -3 \\ 2 & -3 & -1 \\ 3 & 4 & 1 \end{pmatrix} \begin{pmatrix} x \\ y \\ z \end{pmatrix} = \begin{pmatrix} -9 \\ 19 \\ -13 \end{pmatrix}$$

What are we trying to find?

The values of x, y and z which satisfy the given equations.

If we can somehow transform the above matrix to

11.12
$$\begin{array}{c} \text{Row 1} \\ \text{Row 2} \\ \text{Row 3} \end{array} \begin{pmatrix} * & * & * \\ 0 & * & * \\ 0 & 0 & A \end{pmatrix} \begin{pmatrix} x \\ y \\ z \end{pmatrix} = \begin{pmatrix} * \\ * \\ B \end{pmatrix}$$

then we can find z. **How?**

Look at the last row of the 3×3 matrix in 11.12 and multiply it out with the adjacent

column vector, $\begin{pmatrix} x \\ y \\ z \end{pmatrix}$. (An $n \times 1$ matrix is generally called a column vector.) We obtain

$$(0 \times x) + (0 \times y) + (A \times z) = B$$

$$Az = B \text{ gives } z = \frac{B}{A}$$

Therefore from the last row we can find z. Now using back substitution we can find y. Look at row 2. On the Left-Hand Side we get an equation in y and z. From above we know the value of $z = B/A$, so we can substitute $z = B/A$ and obtain y. Similarly from the first row we can find x by substituting the values of y and z.

The matrix in 11.12 shows that we need to achieve 0's in the position shown in order to create equations with fewer unknowns so that by substitution we can find the answers.

Let's try this method for the matrix in question:

$$\begin{pmatrix} 1 & 5 & -3 \\ 2 & -3 & -1 \\ 3 & 4 & 1 \end{pmatrix} \begin{pmatrix} x \\ y \\ z \end{pmatrix} = \begin{pmatrix} -9 \\ 19 \\ -13 \end{pmatrix}$$

We need to enter the Right-Hand Side within the matrix:

$$\begin{array}{c} \text{Row 1 } R_1 \\ \text{Row 2 } R_2 \\ \text{Row 3 } R_3 \end{array} \left(\begin{array}{ccc|c} 1 & 5 & -3 & -9 \\ 2 & -3 & -1 & 19 \\ 3 & 4 & 1 & -13 \end{array} \right)$$

This is called an **augmented** matrix. R_1, R_2 and R_3 just refer to row 1, row 2 and row 3 respectively. We need to obtain the matrix in 11.12 from the above matrix and therefore we require zeros in the bottom left-hand corner of this matrix; that is, 2 from the second row goes to zero and 3, 4 from the bottom row go to zero. Hence $2 \rightarrow 0$, $3 \rightarrow 0$ and $4 \rightarrow 0$.

Example 18 *continued*

Remember we can multiply an equation, and take one equation away from another. In terms of matrices this means that we can multiply a row and take one row away from another.

To get 0 in place of 2 we can multiply row 1, R_1, by 2 and take it away from row 2, R_2, that is $R_2 - 2R_1$. To get 0 in place of 3 we can multiply row 1, R_1, by 3 and take it away from row 3, R_3, that is $R_3 - 3R_1$. We have

$$
\begin{array}{c}
R_1 \\
R_2^\dagger = R_2 - 2R_1 \\
R_3^\dagger = R_3 - 3R_1
\end{array}
\left(
\begin{array}{ccc|c}
1 & 5 & -3 & -9 \\
2 - (2 \times 1) & -3 - (2 \times 5) & -1 - (2 \times (-3)) & 19 - (2 \times (-9)) \\
3 - (3 \times 1) & 4 - (3 \times 5) & 1 - (3 \times (-3)) & -13 - (3 \times (-9))
\end{array}
\right)
$$

which simplifies to

$$
\begin{array}{c}
R_1 \\
R_2^\dagger \\
R_3^\dagger
\end{array}
\left(
\begin{array}{ccc|c}
1 & 5 & -3 & -9 \\
0 & -13 & 5 & 37 \\
0 & -11 & 10 & 14
\end{array}
\right)
$$

The new row 2 and new row 3 are named $R_2^\dagger$ and $R_3^\dagger$ respectively.

We have nearly obtained the matrix of 11.12 . We need to get 0 in place of -11 in the bottom row, $R_3^\dagger$. **We can only use $R_2^\dagger$ and $R_3^\dagger$. Why?**

Because if we use R_1 and $R_2^\dagger$, we will get a non-zero number in place of the zero already established in $R_2^\dagger$. Similarly this holds for $R_3^\dagger$.

How can we obtain 0 in place of -11?

$$
R_3^\dagger - \frac{11}{13}R_2^\dagger \text{ so we have } -11 - \left(\frac{11}{13} \times (-13) \right) = 0 \text{ (gives 0 in place of } -11)
$$

Therefore

$$
\begin{array}{c}
R_1 \\
R_2^\dagger \\
R_3^{\dagger\dagger} = R_3^\dagger - \frac{11}{13}R_2^\dagger
\end{array}
\left(
\begin{array}{ccc|c}
1 & 5 & -3 & -9 \\
0 & -13 & 5 & 37 \\
0 & -11 - \left(\frac{11}{13} \times (-13)\right) & 10 - \left(\frac{11}{13} \times 5\right) & 14 - \left(\frac{11}{13} \times 37\right)
\end{array}
\right)
$$

which simplifies to

$$
\begin{array}{c}
R_1 \\
R_2^\dagger \\
R_3^{\dagger\dagger}
\end{array}
\left(
\begin{array}{ccc|c}
1 & 5 & -3 & -9 \\
0 & -13 & 5 & 37 \\
0 & 0 & 75/13 & -225/13
\end{array}
\right)
$$

From the last row, $R_3^{\dagger\dagger}$, we have

$$
(0 \times x) + (0 \times y) + \frac{75}{13}z = -\frac{225}{13}
$$

$$
\frac{75}{13}z = -\frac{225}{13} \text{ which gives } z = -3
$$

Example 18 *continued*

From the second row, $R_2{}^\dagger$, we have

$$-13y + 5z = 37$$

We know from above that $z = -3$, therefore substituting $z = -3$ gives

$$-13y + [5 \times (-3)] = 37$$
$$y = -4$$

From the first row, R_1, we have

$$x + 5y - 3z = -9$$

Substituting $y = -4$ and $z = -3$ yields

$$x + [5 \times (-4)] - [3 \times (-3)] = -9$$
$$x = 2$$

Hence $x = 2$, $y = -4$ and $z = -3$. Check that this solution is correct by substituting these values into the given equations:

$$x + 5y - 3z = -9 \qquad\qquad 2 + 5(-4) - 3(-3) = -9$$
$$2x - 3y - z = 19 \qquad\qquad 2(2) - 3(-4) - (-3) = 19$$
$$3x + 4y + z = -13 \qquad\qquad 3(2) + 4(-4) + (-3) = -13$$

The above process is called **Gaussian elimination**. We can:

1 interchange any two rows or columns (apart from the augmented column)
2 multiply any row by a non-zero scalar
3 add or subtract rows.

The aim of the Gaussian elimination process is to produce the 'triangular' matrix shown in 11.12, that is zeros in the lower triangle.

When interchanging two columns we need to be careful. Consider **Example 18:**

$$\begin{array}{ccc} x & y & z \\ \begin{pmatrix} 1 & 5 & -3 \\ 2 & -3 & -1 \\ 3 & 4 & 1 \end{pmatrix} \end{array}$$

Remember that column 1 represents the coefficients of x, column 2 represents the coefficients of y and column 3 represents the coefficients of z. So when interchanging any of these columns we need to remember the variable associated with that column. Interchanging column 1 and column 3 in the above gives

$$\begin{array}{ccc} z & y & x \\ \begin{pmatrix} -3 & 5 & 1 \\ -1 & -3 & 2 \\ 1 & 4 & 3 \end{pmatrix} \end{array}$$

Example 19 *mechanics*

A pulley system gives the following equations

$$\ddot{x}_1 + \ddot{x}_2 = 0$$
$$2\ddot{x}_1 = 20 - T$$
$$5\ddot{x}_2 = 50 - T$$

where $\ddot{x}_1$, $\ddot{x}_2$ represent acceleration and T represents tension in the rope. Determine $\ddot{x}_1$, $\ddot{x}_2$ and T by using Gaussian elimination.

Solution

Rearranging the above equations gives

$$\begin{aligned}\ddot{x}_1 + \ddot{x}_2 &= 0\\ 2\ddot{x}_1 \phantom{+ \ddot{x}_2} + T &= 20\\ 5\ddot{x}_2 + T &= 50\end{aligned}$$

Putting this in matrix form:

$$\begin{pmatrix} 1 & 1 & 0 \\ 2 & 0 & 1 \\ 0 & 5 & 1 \end{pmatrix}\begin{pmatrix} \ddot{x}_1 \\ \ddot{x}_2 \\ T \end{pmatrix} = \begin{pmatrix} 0 \\ 20 \\ 50 \end{pmatrix}$$

In augmented matrix form we have

$$\begin{array}{c}R_1\\R_2\\R_3\end{array}\left(\begin{array}{ccc|c} 1 & 1 & 0 & 0 \\ 2 & 0 & 1 & 20 \\ 0 & 5 & 1 & 50 \end{array}\right)$$

To get a 0 in place of 2 we execute $R_2 - 2R_1$:

$$\begin{array}{c}R_1\\R_2^* = R_2 - 2R_1\\R_3\end{array}\left(\begin{array}{ccc|c} 1 & 1 & 0 & 0 \\ 0 & -2 & 1 & 20 \\ 0 & 5 & 1 & 50 \end{array}\right)$$

We need to get 0 in place of 5 in the bottom row, R_3. How?

$$R_3 + \frac{5}{2}R_2^*$$

$$\begin{array}{c}R_1\\R_2^*\\R_3^* = R_3 + \dfrac{5}{2}R_2^*\end{array}\left(\begin{array}{ccc|c} 1 & 1 & 0 & 0 \\ 0 & -2 & 1 & 20 \\ 0 & 0 & 7/2 & 100 \end{array}\right)$$

From the last row, R_3^*, we have

$$\frac{7}{2}T = 100 \text{ which gives } T = \frac{200}{7}\text{ N}$$

Substituting $T = \dfrac{200}{7}$ into the second row, R_2^*, we have

$$-2\ddot{x}_2 + \frac{200}{7} = 20$$

Example 19 *continued*

Solving

$$\ddot{x}_2 = \frac{30}{7} \text{ m/s}^2$$

Similarly from the first row, R_1, we obtain $\ddot{x}_1 = -\frac{30}{7} \text{ m/s}^2$.

Check for yourself that these values do indeed satisfy the original equations.

Example 20 *electrical principles*

The mesh equations of a circuit are given by

$$i_1 + i_2 + 2i_3 = 9$$
$$4i_1 + 4i_2 - 3i_3 = 3$$
$$5i_1 + i_2 + 2i_3 = 13$$

Determine i_1, i_2 and i_3.

Solution

Putting these equations into matrix form:

$$\begin{pmatrix} 1 & 1 & 2 \\ 4 & 4 & -3 \\ 5 & 1 & 2 \end{pmatrix} \begin{pmatrix} i_1 \\ i_2 \\ i_3 \end{pmatrix} = \begin{pmatrix} 9 \\ 3 \\ 13 \end{pmatrix}$$

The augmented matrix is given by

$$\begin{array}{c} R_1 \\ R_2 \\ R_3 \end{array} \left(\begin{array}{ccc|c} 1 & 1 & 2 & 9 \\ 4 & 4 & -3 & 3 \\ 5 & 1 & 2 & 13 \end{array} \right)$$

To obtain 0's in place of 4 and 5 in the first column we execute $R_2 - 4R_1$ and $R_3 - 5R_1$:

$$\begin{array}{c} R_1 \\ R_2' = R_2 - 4R_1 \\ R_3' = R_3 - 5R_1 \end{array} \left(\begin{array}{ccc|c} 1 & 1 & 2 & 9 \\ 0 & 0 & -11 & -33 \\ 0 & -4 & -8 & -32 \end{array} \right)$$

Interchanging rows R_2' and R_3' yields

$$\begin{array}{c} R_1 \\ R_3' \\ R_2' \end{array} \left(\begin{array}{ccc|c} 1 & 1 & 2 & 9 \\ 0 & -4 & -8 & -32 \\ 0 & 0 & -11 & -33 \end{array} \right)$$

From the bottom row, R_2', we have

$$-11i_3 = -33 \text{ which gives } i_3 = 3$$

Substituting $i_3 = 3$ into the middle row, R_3':

$$-4i_2 - 8(3) = -32 \text{ which gives } i_2 = 2$$

Example 20 *continued*

Substituting $i_2 = 2$ and $i_3 = 3$ into the top row, R_1, we have

$$i_1 + 2 + 2(3) = 9 \text{ which gives } i_1 = 1$$

Hence $i_1 = 1$ A, $i_2 = 2$ A and $i_3 = 3$ A satisfy the given equations.

SUMMARY

The Gaussian elimination procedure obtains the following triangular matrix

11.12
$$\begin{pmatrix} * & * & * \\ 0 & * & * \\ 0 & 0 & * \end{pmatrix}$$

and then uses back substitution to solve the given linear equations.

Exercise 11(d)

Solutions at end of book. Complete solutions available at www.palgrave.com/science/engineering/singh

1 By using Gaussian elimination, solve

a $x + 2y - 3z = 3$
$2x - y - z = 11$
$3x + 2y + z = -5$

b $2x + y + 2z = 10$
$x + 4y - 3z = 0$
$3x + 6y - z = 12$

c $x + y + 2z = 1$
$2x + 3y + 2z = 2$
$5x + 2y + 8z = 4$

2 [*electrical principles*]   Mesh equations of a circuit are given by

$$(9 \times 10^3)i_1 - (3 \times 10^3)i_2 = 10$$
$$-(3 \times 10^3)i_1 + (13 \times 10^3)i_2 = 0$$

By using Gaussian elimination, solve these equations for the currents i_1 and i_2.

Hint: Work in kilo (lose the 10^3), then the answer comes out automatically in mA for i_1 and i_2.

3 By using Gaussian elimination solve

$$10x + y - 5z = 18$$
$$-20x + 3y + 20z = 14$$
$$5x + 3y + 5z = 9$$

4 [*mechanics*]   A simple pulley system gives the equations

$$\ddot{x}_1 = T - g$$
$$2\ddot{x}_2 = T - 2g$$
$$\ddot{x}_1 + \ddot{x}_2 = 0$$

where $\ddot{x}_1, \ddot{x}_2$ represent acceleration, T is the tension in the rope and g is the acceleration due to gravity. Determine $\ddot{x}_1, \ddot{x}_2$ and T.

5 [*mechanics*]   A simple pulley system gives the equations

$$\ddot{x}_1 + \ddot{x}_2 = 0$$
$$-3\ddot{x}_1 + T = 3g$$
$$-2\ddot{x}_2 + T = 2g$$

where $\ddot{x}_1, \ddot{x}_2$ represent acceleration, T is the tension in the rope and g is the acceleration due to gravity. Determine $\ddot{x}_1, \ddot{x}_2$ and T.

6 [*electrical principles*]   The mesh equations of a circuit are given by

$$(4 \times 10^3)I_1 - (3 \times 10^3)I_2 = 10$$
$$-(3 \times 10^3)I_1 + (18 \times 10^3)I_2$$
$$- (10 \times 10^3)I_3 = 0$$
$$-(10 \times 10^3)I_2 + (23 \times 10^3)I_3 = -15$$

By using Gaussian elimination, solve these equations for I_1, I_2 and I_3.

[See *Hint* for question **2**.]

SECTION E **Linear equations**

By the end of this section you will be able to:
- ▶ investigate equations which can be written in the form $\mathbf{Au} = \mathbf{b}$
- ▶ understand that a solution to $\mathbf{Au} = \mathbf{b}$ might still exist even if $\mathbf{A}^{-1}$ does not exist
- ▶ understand the concept of an infinite number of solutions

E1 **Properties of linear equations**

Linear equations are equations whose variables have an index of 1 or 0. So far we have looked at linear equations of the form

$$2x + 3y + 5z = 9$$
$$x + 2y + 7z = 13$$
$$4x + 11y + 3z = -1$$

These are called a system of linear equations.

? **How can we write these in matrix form?**

$$\begin{pmatrix} 2 & 3 & 5 \\ 1 & 2 & 7 \\ 4 & 11 & 3 \end{pmatrix} \begin{pmatrix} x \\ y \\ z \end{pmatrix} = \begin{pmatrix} 9 \\ 13 \\ -1 \end{pmatrix}$$

Also this can be written as

$$\mathbf{Au} = \mathbf{b}$$

where $\mathbf{A} = \begin{pmatrix} 2 & 3 & 5 \\ 1 & 2 & 7 \\ 4 & 11 & 3 \end{pmatrix}$, $\mathbf{u} = \begin{pmatrix} x \\ y \\ z \end{pmatrix}$ and $\mathbf{b} = \begin{pmatrix} 9 \\ 13 \\ -1 \end{pmatrix}$.

? **How do we find the values of x, y and z?**

We find the inverse matrix $\mathbf{A}^{-1}$ and multiply both sides by $\mathbf{A}^{-1}$ to obtain

$$\underbrace{\mathbf{A}^{-1} \mathbf{A}}_{=\mathbf{I}} \mathbf{u} = \mathbf{A}^{-1}\mathbf{b}$$
$$\mathbf{Iu} = \mathbf{A}^{-1}\mathbf{b} \qquad \text{[I is the identity matrix]}$$
$$\mathbf{u} = \mathbf{A}^{-1}\mathbf{b}$$

This solution, $\mathbf{u}$, is unique. We need to find $\mathbf{A}^{-1}$ for this method to work. We can only find $\mathbf{A}^{-1}$ if $\det \mathbf{A} \neq 0$ because $\mathbf{A}^{-1} = \dfrac{1}{\det \mathbf{A}}(\text{adj}\mathbf{A})$ and of course we **cannot** divide by 0.

In general if $\mathbf{A}$ is an $n \times n$ matrix, $\mathbf{u}$ and $\mathbf{b}$ are $n \times 1$ matrices ($n \times 1$ matrices are called vectors) such that

11.13 $\qquad \mathbf{Au} = \mathbf{b}$ with $\det \mathbf{A} \neq 0$

then we have a unique solution, that is we have a unique $\mathbf{u}$ which satisfies $\mathbf{Au} = \mathbf{b}$.

? **However if $\det \mathbf{A} = 0$ then what is the inverse matrix, $\mathbf{A}^{-1}$, of A?**

We **cannot** find A^{-1} because A^{-1} is defined with detA in the denominator.

Hence the inverse matrix, A^{-1}, does **not** exist.

? **How can we solve linear equations when A^{-1} does not exist? Is it possible?**

In the next example we examine these questions.

Example 21

Solve for x and y:

a $2x + 3y = 7$ **b** $2x + 10y = 10$
    $2x + 3y = 12$ $6x + 30y = 30$

Solution

a Writing equations **a** in matrix form:

$$\begin{pmatrix} 2 & 3 \\ 2 & 3 \end{pmatrix}\begin{pmatrix} x \\ y \end{pmatrix} = \begin{pmatrix} 7 \\ 12 \end{pmatrix}$$

Let $A = \begin{pmatrix} 2 & 3 \\ 2 & 3 \end{pmatrix}$, then

$$\det A = (2 \times 3) - (2 \times 3) = 0 \qquad [\text{By } \boxed{11.1}]$$

? **What is A^{-1}?**

Hence we do not have an inverse matrix, A^{-1}.

? **So what are the values of x and y?**

Look at these equations:

$\boxed{*}$ $2x + 3y = 7$
$\boxed{**}$ $2x + 3y = 12$

? **What do you notice?**

The coefficients of x and y are the same but the Right-Hand Side is different, 7 and 12.
Clearly they are inconsistent. Moreover if we subtract $\boxed{**}$ from $\boxed{*}$ then we obtain
$0 = -5$ which is false. Thus there is **no** solution to
these simultaneous equations.
The graphs of these equations are
shown in Fig. 2.

Fig. 2

We have parallel lines therefore
the lines don't cross, which
explains why there is no solution to these simultaneous equations.

$\boxed{11.1}$ $\det\begin{pmatrix} a & b \\ c & d \end{pmatrix} = ad - cb$

Example 21 *continued*

b Writing the equations **b** in matrix form:

$$\begin{pmatrix} 2 & 10 \\ 6 & 30 \end{pmatrix}\begin{pmatrix} x \\ y \end{pmatrix} = \begin{pmatrix} 10 \\ 30 \end{pmatrix}$$

Let $A = \begin{pmatrix} 2 & 10 \\ 6 & 30 \end{pmatrix}$ then

$$\det A = (2 \times 30) - (6 \times 10) = 0 \qquad [\text{By } \boxed{11.1}]$$

? **What is A^{-1}?**

? Again A^{-1} does not exist. **Are there any solutions to these equations?**

? Yes, try $x = 0$, $y = 1$. **Do these equations have any other solutions?**

Many:

$$x = 2, \quad y = 3/5$$
$$x = 3, \quad y = 2/5$$
$$x = -1, y = 6/5$$
$$\vdots \qquad \vdots$$

These equations have an infinite number of solutions. They can be found as follows:

The equations are

$\boxed{†}$ $2x + 10y = 10$

$\boxed{††}$ $6x + 30y = 30$

From $\boxed{†}$ we have

$$10y = 10 - 2x$$
$$y = \frac{10 - 2x}{10} = \frac{10}{10} - \frac{2}{10}x = 1 - \frac{x}{5} \qquad [\text{Cancelling}]$$

Let $x = a$, where a is any real number, then

$$y = 1 - \frac{a}{5}$$

Putting $a = 1$ we obtain $x = 1$, $y = 4/5$.

Putting $a = 2$ we obtain $x = 2$, $y = 3/5$ and so on.

? **What do you notice about equations b?**

$$2x + 10y = 10$$
$$6x + 30y = 30$$

The second equation is three times the first equation. The graphs of these are shown in Fig. 3.

$\boxed{11.1}$ $\det\begin{pmatrix} a & b \\ c & d \end{pmatrix} = ad - cb$

Example 21 *continued*

Since the lines lie on top of each other, we have an infinite number of solutions.

Fig. 3

$2x + 10y = 10$

$6x + 30y = 30$ Infinite number of solutions

If we consider the general n simultaneous equations written in matrix form as

11.14 $\mathbf{Au = b}$ (A is a square matrix, $\mathbf{b} \neq \mathbf{0}$)

with $\det A = 0$, then the inverse matrix, A^{-1}, does not exist. Furthermore 11.14 satisfies one of the following:

i there is no solution (there is no $\mathbf{u}$ satisfying $\mathbf{Au = b}$), or

ii there are an infinite number of solutions (there are an infinite number of $\mathbf{u}$'s which satisfy $\mathbf{Au = b}$)

In **Example 21a**: $A = \begin{pmatrix} 2 & 3 \\ 2 & 3 \end{pmatrix}$, $\mathbf{u} = \begin{pmatrix} x \\ y \end{pmatrix}$ and $\mathbf{b} = \begin{pmatrix} 7 \\ 12 \end{pmatrix}$ and there were **no** solutions.

In **Example 21b**: $A = \begin{pmatrix} 2 & 10 \\ 6 & 30 \end{pmatrix}$, $\mathbf{u} = \begin{pmatrix} x \\ y \end{pmatrix}$ and $\mathbf{b} = \begin{pmatrix} 10 \\ 30 \end{pmatrix}$ and we had an **infinite** number

of solutions of the form: $x = a$, $y = 1 - \dfrac{a}{5}$ where a is any real number.

In both cases we had $\det A = 0$.

Now we look at equations of the form

$$\mathbf{Au = 0} \text{ with } \det A = 0$$

Note that the Right-Hand Side is zero. Consider the equations

* $2x + 10y = 0$
** $6x + 30y = 0$

From * we have $x = -5y$, that is x is -5 times the y value. Hence $y = 1$, $x = -5$ is a solution to both equations * and ** . The following are also solutions of the simultaneous equations:

$$y = 2, \ x = -10$$
$$y = 3, \ x = -15$$
$$y = 4, \ x = -20$$
$$\vdots \qquad \vdots$$

There are an infinite number of solutions. The equations can be written in matrix form as

$$\begin{pmatrix} 2 & 10 \\ 6 & 30 \end{pmatrix} \begin{pmatrix} x \\ y \end{pmatrix} = \begin{pmatrix} 0 \\ 0 \end{pmatrix}$$

or $\mathbf{Au = 0}$

where $A = \begin{pmatrix} 2 & 10 \\ 6 & 30 \end{pmatrix}$, $\mathbf{u} = \begin{pmatrix} x \\ y \end{pmatrix}$, $\mathbf{0} = \begin{pmatrix} 0 \\ 0 \end{pmatrix}$ and $\det A = 0$.

In general if

11.15   **Au** = **0** with det**A** = 0

then 11.15 has an infinite number of solutions. That is an infinite number of **u**'s satisfy **Au** = **0**.

Now consider equations of the form

$$\mathbf{Au} = \mathbf{0} \text{ with } \det\mathbf{A} \neq 0$$

? Does **A**$^{-1}$ exist?

Yes. Multiply both sides by **A**$^{-1}$

$$\mathbf{A}^{-1}\mathbf{Au} = \mathbf{A}^{-1}\mathbf{0} = \mathbf{0}$$
$$\underset{\substack{\text{because}\\ \mathbf{A}^{-1}\mathbf{A}=\mathbf{I}}}{\mathbf{I}}\ \mathbf{u} = \mathbf{0}$$
$$\mathbf{u} = \mathbf{0} \qquad [\text{Because } \mathbf{Iu} = \mathbf{u}]$$

The solution **u** = **0** means $x = 0$, $y = 0$, $z = 0$, . . . This is known as the **trivial** solution.

In general if

11.16   **Au** = **0** with det**A** ≠ 0

then there is only the trivial solution $x = 0$, $y = 0$, $z = 0$, ...

Summarizing the above, we have four cases of

*   **Au** = **b**

11.13   If det**A** ≠ 0 then * has a unique solution

11.14   If **b** ≠ **0**, det**A** = 0 then either * has no solution or there are an infinite number of solutions

11.15   If **b** = **0**, det**A** = 0 then * has an infinite number of solutions

11.16   If **b** = **0**, det**A** ≠ 0 then * only has the trivial solution **u** = **0**; $x = 0$, $y = 0$, $z = 0$, . . .

Note that for the case with **b** = **0** we have non-trivial solutions only if det**A** = 0.

The next three examples look into the various types of solution outlined in 11.13 to 11.16 .

Example 22

Find all the solutions of

$$\begin{aligned} x + 2y + z &= 4 \\ 3x - 4y - 2z &= 2 \\ 5x + 3y + 5z &= -1 \end{aligned}$$

> **Example 22** *continued*

Solution

These equations can be written in matrix form as

$$\mathbf{Au} = \mathbf{b}$$

where $\mathbf{A} = \begin{pmatrix} 1 & 2 & 1 \\ 3 & -4 & -2 \\ 5 & 3 & 5 \end{pmatrix}$, $\mathbf{u} = \begin{pmatrix} x \\ y \\ z \end{pmatrix}$ and $\mathbf{b} = \begin{pmatrix} 4 \\ 2 \\ -1 \end{pmatrix}$. The equations were solved in

Example 17 with the variables i_1, i_2 and i_3 instead of x, y and z, hence

$$\boxed{} \qquad x = 2, y = 3 \text{ and } z = -4$$

Since $\det\mathbf{A} = -35 \neq 0$ (evaluated in **Example 17**), we can conclude from 11.13 that these solutions are unique, that is $x = 2$, $y = 3$ and $z = -4$ are the only values which satisfy the given equations.

> **Example 23**

Find all the values of x, y and z that satisfy

$$2x + 3y + 4z = 0$$
$$5x + 7y + 3z = 0$$
$$7x + 3y + 11z = 0$$

Solution

Putting these in matrix form gives

$$\begin{pmatrix} 2 & 3 & 4 \\ 5 & 7 & 3 \\ 7 & 3 & 11 \end{pmatrix} \begin{pmatrix} x \\ y \\ z \end{pmatrix} = \begin{pmatrix} 0 \\ 0 \\ 0 \end{pmatrix}$$

This is of the form $\mathbf{Au} = \mathbf{b}$ where $\mathbf{A} = \begin{pmatrix} 2 & 3 & 4 \\ 5 & 7 & 3 \\ 7 & 3 & 11 \end{pmatrix}$, $\mathbf{u} = \begin{pmatrix} x \\ y \\ z \end{pmatrix}$ and $\mathbf{b} = \begin{pmatrix} 0 \\ 0 \\ 0 \end{pmatrix}$.

We first obtain $\det\mathbf{A}$:

$$\det \begin{pmatrix} 2 & 3 & 4 \\ 5 & 7 & 3 \\ 7 & 3 & 11 \end{pmatrix} = 2\det \begin{pmatrix} 7 & 3 \\ 3 & 11 \end{pmatrix} - 3\det \begin{pmatrix} 5 & 3 \\ 7 & 11 \end{pmatrix} + 4\det \begin{pmatrix} 5 & 7 \\ 7 & 3 \end{pmatrix}$$

$$= 2[(7 \times 11) - (3 \times 3)] - 3[(5 \times 11) - (7 \times 3)] + 4[(5 \times 3) - (7 \times 7)]$$

$$= -102$$

Since $\det\mathbf{A} = -102 \neq 0$ we can conclude by 11.16 that we only have the trivial solution:

$$x = 0, y = 0 \text{ and } z = 0$$

Example 24

Find the values of λ that give a non-trivial solution of

$$(\mathbf{A} - \lambda\mathbf{I})\mathbf{u} = \mathbf{0}$$

where $\mathbf{A} = \begin{pmatrix} 2 & 3 \\ 7 & 1 \end{pmatrix}$, $\mathbf{u} = \begin{pmatrix} x \\ y \end{pmatrix}$ and $\mathbf{I} = \begin{pmatrix} 1 & 0 \\ 0 & 1 \end{pmatrix}$ is the identity matrix.

Solution

This is quite a difficult example to follow. It requires basic algebra as well as matrix manipulation.

By 11.15 we have a non-trivial solution if

* $\det(\mathbf{A} - \lambda\mathbf{I}) = 0$

Substituting the given information:

$$\det(\mathbf{A} - \lambda\mathbf{I}) = \det\left[\begin{pmatrix} 2 & 3 \\ 7 & 1 \end{pmatrix} - \lambda\begin{pmatrix} 1 & 0 \\ 0 & 1 \end{pmatrix}\right]$$

$$= \det\left[\begin{pmatrix} 2 & 3 \\ 7 & 1 \end{pmatrix} - \begin{pmatrix} \lambda & 0 \\ 0 & \lambda \end{pmatrix}\right]$$

$$= \det\begin{pmatrix} 2-\lambda & 3 \\ 7 & 1-\lambda \end{pmatrix}$$

$$\underset{\text{by } 11.1}{=} (2 - \lambda)(1 - \lambda) - (7 \times 3)$$

$$= 2 - 3\lambda + \lambda^2 - 21 = \lambda^2 - 3\lambda - 19$$

By * we equate this to zero:

$$\lambda^2 - 3\lambda - 19 = 0 \qquad \text{[Quadratic equation]}$$

? **How do we find λ?**

By using the quadratic formula

1.16 $\lambda = \dfrac{-b \pm \sqrt{b^2 - 4ac}}{2a}$

with $a = 1$, $b = -3$ and $c = -19$:

$$\lambda = \frac{3 \pm \sqrt{9 - (4 \times 1 \times (-19))}}{2}$$

$$= \frac{3 \pm \sqrt{85}}{2}$$

$$\lambda_1 = \frac{3 + \sqrt{85}}{2}, \lambda_2 = \frac{3 - \sqrt{85}}{2}$$

Different λ's are normally denoted by λ_1 and λ_2.

11.1 $\det\begin{pmatrix} a & b \\ c & d \end{pmatrix} = ad - cb$

These λ are called the **eigenvalues** of the matrix A. Eigenvalues are the subject of the next section.

SUMMARY

A system of linear equations can be written as

| ⋆ | $\mathbf{Au=b}$ |

It has the following properties:

11.13	If detA $\neq 0$ then	⋆	has a unique solution
11.14	If $\mathbf{b}\neq 0$, detA $=0$ then either	⋆	has no solution or there are an infinite number of solutions
11.15	If $\mathbf{b}=0$, detA $=0$ then	⋆	has an infinite number of solutions
11.16	If $\mathbf{b}=0$, detA $\neq 0$ then	⋆	only has the trivial solution $x = 0, y = 0, z = 0, \ldots$

Exercise 11(e)

Solutions at end of book. Complete solutions available at www.palgrave.com/science/engineering/singh

1 Solve:
$$x + 3y + 5z = 0$$
$$7x - y + z = 0$$
$$2x + 3y - 8z = 0$$

2 Solve and find whether the solution is unique:
$$2x + 3y = 5$$
$$4x + 5y = 9$$

3 Find x and y which satisfy:
$$2x + y = 0$$
$$4x + 2y = 0$$
Is the solution unique?

4 Test whether the solutions of questions **1**, **4** and **5** of **Exercise 11(d)** are unique.

5 Find the values of λ that give a non-trivial solution of

$$(\mathbf{A} - \lambda\mathbf{I})\mathbf{u} = \mathbf{0}$$

for

a $A = \begin{pmatrix} -1 & 1 \\ 9 & -1 \end{pmatrix}$ **b** $A = \begin{pmatrix} 1 & 3 \\ -6 & 5 \end{pmatrix}$

$\left(\mathbf{u} = \begin{pmatrix} x \\ y \end{pmatrix} \text{ and I is the identity matrix} \right).$

SECTION F Eigenvalues and eigenvectors

By the end of this section you will be able to:
▶ understand the terms eigenvalues and eigenvectors
▶ find the eigenvalues and eigenvectors of 2 × 2 matrices
▶ find the eigenvalues and eigenvectors of 3 × 3 matrices
▶ examine vibrations of buildings

This is a challenging section because it assumes a thorough understanding of matrices and it also requires a proficiency in algebra on topics such as expansion of brackets, solving equations, substitution, etc.

In some universities, eigenvalues and eigenvectors are part of the second year engineering mathematics course.

F1 Introduction

Example 25

If $A = \begin{pmatrix} 4 & -2 \\ 1 & 1 \end{pmatrix}$ and $\mathbf{u} = \begin{pmatrix} 2 \\ 1 \end{pmatrix}$ then evaluate $A\mathbf{u}$.

Solution

Multiplying the matrix and column vector:

$$A\mathbf{u} = \begin{pmatrix} 4 & -2 \\ 1 & 1 \end{pmatrix}\begin{pmatrix} 2 \\ 1 \end{pmatrix} = \begin{pmatrix} 6 \\ 3 \end{pmatrix} = 3\begin{pmatrix} 2 \\ 1 \end{pmatrix} = 3\mathbf{u}$$

Note that we have $A\mathbf{u} = 3\mathbf{u}$.

In general terms, this can be stated as

$$\boxed{*} \qquad A\mathbf{u} = \lambda\mathbf{u}$$

where A is a square matrix, $\mathbf{u}$ is a column vector and λ is a scalar (a number). **Can you think of a vector, u, which satisfies $\boxed{*}$?**

$$\mathbf{u} = 0$$

This is called the trivial solution: $x = 0$, $y = 0$, $z = 0$ . . .

$A\mathbf{u} = \lambda\mathbf{u}$ is only interesting if $\mathbf{u} \neq 0$, otherwise we always have the trivial solution. For $\mathbf{u} \neq 0$, λ is called an **eigenvalue** of the matrix A and $\mathbf{u}$ is called an **eigenvector** belonging to λ.

In this section the problem is that we need to find the eigenvalue, λ, and the corresponding eigenvector, $\mathbf{u}$, of a square matrix A. **How can we find λ?**

We have

$$A\mathbf{u} = \lambda\mathbf{u}$$
$$\boxed{\dagger} \qquad A\mathbf{u} - \lambda\mathbf{u} = 0$$

Remember the properties of the identity matrix, I:

$$\mathbf{u} = I\mathbf{u} \text{ and so } \lambda\mathbf{u} = \lambda I\mathbf{u}$$

$\boxed{\dagger}$ therefore becomes

$$A\mathbf{u} - \lambda I\mathbf{u} = 0$$

Factorizing we obtain

$$\boxed{11.17} \qquad (A - \lambda I)\mathbf{u} = 0$$

? **Are there any solutions of 11.17?**

Yes, the trivial solution $\mathbf{u} = \mathbf{0}$. For eigenvalues and eigenvectors, $\mathbf{u} \neq \mathbf{0}$.

? **What does matrix A need to satisfy in order to have a non-trivial solution to $(\mathbf{A} - \lambda\mathbf{I})\mathbf{u} = \mathbf{0}$?**

As discussed in **Example 24** of the previous section, for non-trivial solution

11.18 $\det(\mathbf{A} - \lambda\mathbf{I}) = 0$

Hence λ is an eigenvalue of $\mathbf{A}$ if and only if 11.18 holds (see **Example 24**).

F2 Finding eigenvalues and eigenvectors

Example 26

Find the eigenvalues of $\mathbf{A} = \begin{pmatrix} 2 & 0 \\ 1 & 3 \end{pmatrix}$.

Solution

We need to find λ in $\det(\mathbf{A} - \lambda\mathbf{I}) = 0$. First we obtain $\mathbf{A} - \lambda\mathbf{I}$:

$$\mathbf{A} - \lambda\mathbf{I} = \begin{pmatrix} 2 & 0 \\ 1 & 3 \end{pmatrix} - \lambda\begin{pmatrix} 1 & 0 \\ 0 & 1 \end{pmatrix} \qquad \left[\mathbf{I} = \text{identity matrix} = \begin{pmatrix} 1 & 0 \\ 0 & 1 \end{pmatrix} \right]$$

$$= \begin{pmatrix} 2 & 0 \\ 1 & 3 \end{pmatrix} - \begin{pmatrix} \lambda & 0 \\ 0 & \lambda \end{pmatrix}$$

$$= \begin{pmatrix} 2 - \lambda & 0 \\ 1 & 3 - \lambda \end{pmatrix}$$

Substituting this into $\det(\mathbf{A} - \lambda\mathbf{I})$ gives

$$\det(\mathbf{A} - \lambda\mathbf{I}) = \det\begin{pmatrix} 2 - \lambda & 0 \\ 1 & 3 - \lambda \end{pmatrix} \underset{\text{by } 11.1}{=} (2 - \lambda)(3 - \lambda) - 0$$

For eigenvalues we equate the determinant to zero:

$$(2 - \lambda)(3 - \lambda) = 0$$

$$\lambda = 2, \lambda = 3 \qquad\qquad \text{[Solving]}$$

11.1 $\det\begin{pmatrix} a & b \\ c & d \end{pmatrix} = ad - cb$

Example 27

Find the eigenvector corresponding to $\lambda = 2$ and $\lambda = 3$ for **A** in **Example 26**.

Solution

Let **u** be the eigenvector corresponding to $\lambda = 2$. Substituting $\mathbf{A} = \begin{pmatrix} 2 & 0 \\ 1 & 3 \end{pmatrix}$ and $\lambda = 2$ into

11.17 $[\mathbf{A} - \lambda \mathbf{I}]\mathbf{u} = \mathbf{0}$

gives

$$\left[\begin{pmatrix} 2 & 0 \\ 1 & 3 \end{pmatrix} - \begin{pmatrix} 2 & 0 \\ 0 & 2 \end{pmatrix} \right]\mathbf{u} = \mathbf{0}$$

$$\begin{pmatrix} 0 & 0 \\ 1 & 1 \end{pmatrix}\mathbf{u} = \mathbf{0} \qquad \text{[Simplifying]}$$

Remember the zero column vector, $\mathbf{0} = \begin{pmatrix} 0 \\ 0 \end{pmatrix}$. Let $\mathbf{u} = \begin{pmatrix} x \\ y \end{pmatrix}$ and so we have

$$\begin{pmatrix} 0 & 0 \\ 1 & 1 \end{pmatrix}\begin{pmatrix} x \\ y \end{pmatrix} = \begin{pmatrix} 0 \\ 0 \end{pmatrix}$$

Multiplying out gives

$$0 + 0 = 0$$
$$x + y = 0$$

Remember that the eigenvector cannot be zero; $\mathbf{u} \neq \mathbf{0}$ therefore at least one of the values, x or y, must be non-zero. From the second equation we have

$$x + y = 0$$
$$x = -y$$

Solutions are

$$x = 1, \; y = -1$$
$$x = 2, \; y = -2$$
$$x = \pi, \; y = -\pi$$

We have an infinite number of solutions. For the time being we need to write the general eigenvector **u**. **How?**

Let $x = a$ then $y = -a$ where $a \neq 0$ and is a number.

The eigenvectors corresponding to $\lambda = 2$ are

$$\mathbf{u} = \begin{pmatrix} a \\ -a \end{pmatrix} = a\begin{pmatrix} 1 \\ -1 \end{pmatrix} \text{ where } a \neq 0$$

Similarly we find the general eigenvector **v** corresponding to $\lambda = 3$. Putting $\lambda = 3$ into $[\mathbf{A} - \lambda \mathbf{I}]\mathbf{v} = \mathbf{0}$ gives

$$\left[\begin{pmatrix} 2 & 0 \\ 1 & 3 \end{pmatrix} - \begin{pmatrix} 3 & 0 \\ 0 & 3 \end{pmatrix} \right]\mathbf{v} = \mathbf{0}$$

$$\begin{pmatrix} -1 & 0 \\ 1 & 0 \end{pmatrix}\mathbf{v} = \mathbf{0}$$

Example 27 *continued*

By writing $\mathbf{v} = \begin{pmatrix} x \\ y \end{pmatrix}$ [different x and y from those above] and $\mathbf{0} = \begin{pmatrix} 0 \\ 0 \end{pmatrix}$ we obtain

$$\begin{pmatrix} -1 & 0 \\ 1 & 0 \end{pmatrix}\begin{pmatrix} x \\ y \end{pmatrix} = \begin{pmatrix} 0 \\ 0 \end{pmatrix}$$

Multiplying out:

$$-x + 0 = 0$$
$$x + 0 = 0$$

From these equations we must have $x = 0$. **What is y equal to?**

We can choose y to be any real number apart from zero. Thus

$$y = a \text{ where } a \neq 0$$

The general eigenvector of $\lambda = 3$ is found by substituting $x = 0$ and $y = a$:

$$\mathbf{v} = \begin{pmatrix} 0 \\ a \end{pmatrix} = a\begin{pmatrix} 0 \\ 1 \end{pmatrix}$$

Note that for one eigenvalue λ there are infinitely many eigenvectors. We can choose any one of these eigenvectors for the corresponding eigenvalue. By putting $a = 1$, an eigenvector can be selected. In **Example 27**:

$$\mathbf{u} = \begin{pmatrix} 1 \\ -1 \end{pmatrix} \text{ is an eigenvector corresponding to } \lambda = 2$$

$$\mathbf{v} = \begin{pmatrix} 0 \\ 1 \end{pmatrix} \text{ is an eigenvector corresponding to } \lambda = 3$$

Generally the values of the elements of the eigenvectors are scaled to the smallest possible size of whole numbers.

The resulting equation from $\det(\mathbf{A} - \lambda\mathbf{I}) = \mathbf{0}$ is called the characteristic equation.

F3 Applications to vibrations

Example 28

Consider a two-storey building subject to earthquake oscillations, as shown in Fig. 4.

The equations of motion are expressed as

$$\ddot{\mathbf{x}} = \mathbf{Ax} \text{ where } \mathbf{x} = \begin{pmatrix} x_1 \\ x_2 \end{pmatrix}.$$

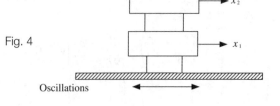

Fig. 4

Oscillations

Example 28 *continued*

The period, T, of natural vibrations is given by

$$T = \frac{2\pi}{\sqrt{-\lambda}} \text{ where } \lambda \text{ is the eigenvalue of the matrix A. Find the period(s) if } A = \begin{pmatrix} -20 & 10 \\ 10 & -10 \end{pmatrix}.$$

Solution

Substituting the given matrix A into $A - \lambda I$ yields

$$A - \lambda I = \begin{pmatrix} -20 & 10 \\ 10 & -10 \end{pmatrix} - \lambda \begin{pmatrix} 1 & 0 \\ 0 & 1 \end{pmatrix} = \begin{pmatrix} -20 & 10 \\ 10 & -10 \end{pmatrix} - \begin{pmatrix} \lambda & 0 \\ 0 & \lambda \end{pmatrix}$$

$$= \begin{pmatrix} -20 - \lambda & 10 \\ 10 & -10 - \lambda \end{pmatrix}$$

Therefore $\det(A - \lambda I)$ is given by

$$\det\begin{pmatrix} -20 - \lambda & 10 \\ 10 & -10 - \lambda \end{pmatrix} = (-20 - \lambda)(-10 - \lambda) - (10 \times 10) \quad \text{[By 11.1]}$$

$$= (20 + \lambda)(10 + \lambda) - 100 \qquad \text{[Taking out minus signs]}$$

$$= 200 + 30\lambda + \lambda^2 - 100 \qquad \text{[Expanding]}$$

$$= \lambda^2 + 30\lambda + 100$$

Equating this to zero gives the quadratic equation

$$\lambda^2 + 30\lambda + 100 = 0$$

? **How do we solve this quadratic equation?**

Apply the formula

1.16 $$\lambda = \frac{-b \pm \sqrt{b^2 - 4ac}}{2a}$$

with $a = 1$, $b = 30$ and $c = 100$:

$$\lambda = \frac{-30 \pm \sqrt{(-30)^2 - (4 \times 1 \times 100)}}{2} = \frac{-30 \pm \sqrt{500}}{2}$$

$$= -26.18, -3.82$$

The two eigenvalues are $\lambda_1 = -26.18$ and $\lambda_2 = -3.82$.

Substituting these into the given $T = \dfrac{2\pi}{\sqrt{-\lambda}}$:

$$T_1 = \frac{2\pi}{\sqrt{-(-26.18)}} = 1.23 \text{ s}, \quad T_2 = \frac{2\pi}{\sqrt{-(-3.82)}} = 3.21 \text{ s}$$

Natural periods of vibration are 1.23 s (2 d.p.) and 3.21 s (2 d.p.).

11.1 $\det\begin{pmatrix} a & b \\ c & d \end{pmatrix} = ad - cb$

F4 **3 × 3 Matrices**

Example 29

Determine the eigenvalues of

$$A = \begin{pmatrix} 1 & 0 & 4 \\ 0 & 4 & 0 \\ 3 & 5 & -3 \end{pmatrix}$$

Solution

We have

$$A - \lambda I = \begin{pmatrix} 1 & 0 & 4 \\ 0 & 4 & 0 \\ 3 & 5 & -3 \end{pmatrix} - \begin{pmatrix} \lambda & 0 & 0 \\ 0 & \lambda & 0 \\ 0 & 0 & \lambda \end{pmatrix} = \begin{pmatrix} 1-\lambda & 0 & 4 \\ 0 & 4-\lambda & 0 \\ 3 & 5 & -3-\lambda \end{pmatrix}$$

It's easier to remember that $A - \lambda I$ is actually A with $-\lambda$ along the leading diagonal (from top left to bottom right). We need to evaluate $\det(A - \lambda I)$. **What is the easiest way to evaluate this?**

From the properties of determinants we know that it will be easier to evaluate the determinant along the second row, containing the elements 0, $4 - \lambda$ and 0. **Why?**

Because it has two zeros so that we do not have to evaluate the 2×2 determinants associated with these zeros.

$$\det(A - \lambda I) = \det \begin{pmatrix} 1-\lambda & 0 & 4 \\ 0 & 4-\lambda & 0 \\ 3 & 5 & -3-\lambda \end{pmatrix}$$

$$= (4-\lambda)\left[\det \begin{pmatrix} 1-\lambda & 4 \\ 3 & -3-\lambda \end{pmatrix} \right] \qquad \text{[Expanding second row]}$$

$$= (4-\lambda)\left[(1-\lambda)(-3-\lambda) - (3 \times 4) \right] \qquad \text{[By } \boxed{11.1} \text{]}$$

$$= (4-\lambda)\left[(\lambda-1)(3+\lambda) - 12 \right] \qquad \text{[Taking out minus signs]}$$

$$= (4-\lambda)[3\lambda + \lambda^2 - 3 - \lambda - 12]$$

$$= (4-\lambda)[\lambda^2 + 2\lambda - 15] \qquad \text{[Simplifying]}$$

$$= (4-\lambda)(\lambda + 5)(\lambda - 3) \qquad \text{[Factorizing]}$$

This is equal to zero:

$$(4-\lambda)(\lambda + 5)(\lambda - 3) = 0$$

$$\lambda = 4, \quad \lambda = -5, \quad \lambda = 3$$

$\boxed{11.1}$ $\det \begin{pmatrix} a & b \\ c & d \end{pmatrix} = ad - cb$

Example 30

Find the eigenvectors associated with $\lambda = 3$ for the matrix A in **Example 29**.

Solution

Substituting $\lambda = 3$ and $\mathbf{A} = \begin{pmatrix} 1 & 0 & 4 \\ 0 & 4 & 0 \\ 3 & 5 & -3 \end{pmatrix}$ into $(\mathbf{A} - \lambda \mathbf{I})\mathbf{u} = \mathbf{0}$ gives (subtract 3 from the leading diagonal):

$$\begin{pmatrix} 1-3 & 0 & 4 \\ 0 & 4-3 & 0 \\ 3 & 5 & -3-3 \end{pmatrix} \mathbf{u} = \mathbf{0}$$

where $\mathbf{u}$ is the eigenvector corresponding to $\lambda = 3$. Remember that $\mathbf{0} = \begin{pmatrix} 0 \\ 0 \\ 0 \end{pmatrix}$ and let $\mathbf{u} = \begin{pmatrix} x \\ y \\ z \end{pmatrix}$, so we have

$$\begin{pmatrix} -2 & 0 & 4 \\ 0 & 1 & 0 \\ 3 & 5 & -6 \end{pmatrix} \begin{pmatrix} x \\ y \\ z \end{pmatrix} = \begin{pmatrix} 0 \\ 0 \\ 0 \end{pmatrix}$$

Expanding this gives

†	$-2x + 0 + 4z = 0$
††	$0 + y + 0 = 0$
†††	$3x + 5y - 6z = 0$

From †† we know that $y = 0$. From †

$$2x = 4z \text{ which gives } x = 2z$$

If $z = 1$ then $x = 2$; or more generally if $z = a$ then $x = 2a$ where $a \neq 0$.

The general eigenvector $\mathbf{u} = \begin{pmatrix} 2a \\ 0 \\ a \end{pmatrix} = a \begin{pmatrix} 2 \\ 0 \\ 1 \end{pmatrix}$ where $a \neq 0$ corresponds to $\lambda = 3$.

The eigenvectors corresponding to $\lambda = 4$, $\lambda = -5$ are part of **Exercise 11(f)**, question **4**.

SUMMARY

A scalar λ is called an eigenvalue of a square matrix $\mathbf{A}$ if it satisfies

11.17 $(\mathbf{A} - \lambda \mathbf{I})\mathbf{u} = 0$

and the corresponding $\mathbf{u}$ ($\neq \mathbf{0}$) is called an eigenvector.

We find λ by solving the resulting equation from

11.18 $\det(\mathbf{A} - \lambda \mathbf{I}) = 0$

Exercise **11(f)**

Solutions at end of book. Complete solutions available at www.palgrave.com/science/engineering/singh

1 Find the eigenvalues of the following matrices:

a $A = \begin{pmatrix} 7 & 3 \\ 0 & -4 \end{pmatrix}$ **b** $A = \begin{pmatrix} 5 & -2 \\ 4 & -1 \end{pmatrix}$

c $A = \begin{pmatrix} -1 & 4 \\ 2 & 1 \end{pmatrix}$

2 Find the particular eigenvectors of the matrices of question **1**.

3 🌾 [control engineering]   The system poles, λ, of a system are the eigenvalues of a given matrix A. Determine the system poles for

a $A = \begin{pmatrix} -1 & 1 \\ -2 & 1 \end{pmatrix}$

b $A = \begin{pmatrix} 1 & 2 \\ -4 & -3 \end{pmatrix}$

c $A = \begin{pmatrix} 5 & -5 \\ 5 & -2 \end{pmatrix}$

4 Obtain the general eigenvectors corresponding to $\lambda = 4$ and $\lambda = -5$ for the matrix in **Example 29**. (The eigenvector corresponding to $\lambda = 3$ was found in **Example 30**.)

5 Determine the eigenvalues and particular eigenvectors of

$$A = \begin{pmatrix} 5 & 0 & 0 \\ -9 & 4 & -1 \\ -6 & 2 & 1 \end{pmatrix}$$

6 〔Ψ〕 [vibrations]   The natural period, T, of vibrations of a building is given by

$$T = \frac{2\pi}{\sqrt{-\lambda}}$$

where λ is the eigenvalue of a given matrix A. Determine T for

$$A = \begin{pmatrix} -15 & 5 \\ 10 & -10 \end{pmatrix}$$

For the remaining questions use a computer algebra system or a graphical calculator.

7 🌾 [control engineering]   The state matrix, A, of a system is given by

$$A = \begin{pmatrix} 3.16 & 2.87 & 1.61 \\ 4.93 & 12.03 & 2.64 \\ 5.12 & 3.84 & 4.20 \end{pmatrix}$$

Determine the eigenvalues. The eigenvalues of the system are called system poles.

8 🌾 [control engineering]   Find the system poles for

$$A = \begin{pmatrix} -0.2 & 5.6 & 7.1 & 3.3 \\ 1.5 & -6.1 & -6.3 & -7.0 \\ -2.2 & 3.6 & 4.8 & 8.3 \\ -9.3 & 8.9 & 6.4 & -6.1 \end{pmatrix}$$

SECTION G **Applications in heat transfer**

By the end of this section you will be able to:
▶ form nodal equations
▶ solve the resulting equations by using a graphical calculator and a computer algebra system

The techniques described in this section are used in many engineering disciplines such as electric field, hydrodynamics and finite element analysis. However in this section we concentrate on the application to heat transfer.

G1 **Applications of matrices in heat transfer**

Figure 5 shows a body divided up into squares of size Δx.

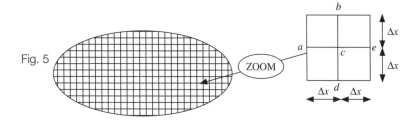

Fig. 5

The following formula gives the average temperature, T_c, close to a node (point) c in the body:

11.19 $T_a + T_b + T_d + T_e = 4T_c$

where T_a, T_b, T_d and T_e represent temperature at nodes a, b, d and e respectively.

The point c is an interior point of the body surrounded by the points a, b, d and e. Each of these points is Δx away from c.

11.19 is an equation used in heat transfer. We will not derive this equation but just apply it in the appropriate case.

Example 31

Figure 6 shows a square cross-section with the stated surface temperatures. Find the temperatures at the interior nodes 1, 2, 3, 4, 5 and 6.

Solution

Let T_1, T_2, T_3, T_4, T_5 and T_6 represent the temperatures at nodes 1, 2, 3, 4, 5 and 6 respectively.

Consider node 1. We have the situation shown in Fig. 7.

Applying 11.19 to a node means we sum the four surrounding temperatures, that is

$$450 + 450 + T_2 + T_3 = 4T_1$$

Similarly for node 2 (Fig. 8).

Examining Figure 8 and applying 11.19 we have

$$T_1 + 450 + T_1 + T_4 = 4T_2$$

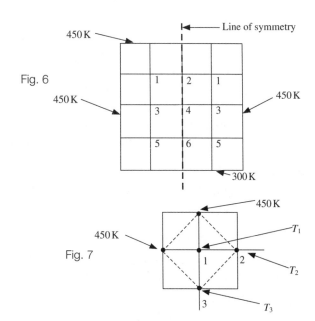

Fig. 6

Fig. 7

11.19 $T_a + T_b + T_d + T_e = 4T_c$

Example 31 *continued*

Following this procedure for the remaining
nodes we obtain:

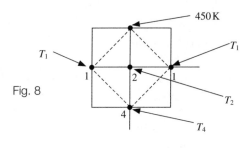

450 K

Fig. 8

Node 1: $450 + 450 + T_2 + T_3 = 4T_1$
Node 2: $T_1 + 450 + T_1 + T_4 = 4T_2$
Node 3: $450 + T_1 + T_4 + T_5 = 4T_3$
Node 4: $T_3 + T_2 + T_3 + T_6 = 4T_4$
Node 5: $450 + T_3 + T_6 + 300 = 4T_5$
Node 6: $T_5 + T_4 + T_5 + 300 = 4T_6$

Simplifying and rearranging these equations yields

$$
\begin{array}{rrrrrrl}
4T_1 & -T_2 & -T_3 & & & & = 900 \\
-2T_1 & +4T_2 & & -T_4 & & & = 450 \\
-T_1 & & +4T_3 & -T_4 & -T_5 & & = 450 \\
& -T_2 & -2T_3 & +4T_4 & & -T_6 & = 0 \\
& & -T_3 & & +4T_5 & -T_6 & = 750 \\
& & & -T_4 & -2T_5 & +4T_6 & = 300
\end{array}
$$

Writing this in matrix form gives

$$
\begin{pmatrix}
4 & -1 & -1 & 0 & 0 & 0 \\
-2 & 4 & 0 & -1 & 0 & 0 \\
-1 & 0 & 4 & -1 & -1 & 0 \\
0 & -1 & -2 & 4 & 0 & -1 \\
0 & 0 & -1 & 0 & 4 & -1 \\
0 & 0 & 0 & -1 & -2 & 4
\end{pmatrix}
\begin{pmatrix}
T_1 \\ T_2 \\ T_3 \\ T_4 \\ T_5 \\ T_6
\end{pmatrix}
=
\begin{pmatrix}
900 \\ 450 \\ 450 \\ 0 \\ 750 \\ 300
\end{pmatrix}
$$

Let **A** be the 6×6 matrix on the left-hand side, **t** be the temperature column vector and
b be the column vector on the right-hand side. We have

$$\mathbf{At = b}$$

Multiplying both sides by the inverse matrix, $\mathbf{A}^{-1}$, gives

$$\mathbf{t = A^{-1}b}$$

Using a graphical calculator:

Store the matrix **A** and the column vector **b** into a calculator and then find $\mathbf{A}^{-1}\mathbf{b}$.
The calculator should indicate the following results:

$T_1 = 439.29$, $T_2 = 435.27$, $T_3 = 421.88$, $T_4 = 412.5$, $T_5 = 385.71$ and $T_6 = 370.98$

(all temperatures are in Kelvin (K)).

For the procedure on a graphical calculator see the handbook of your calculator.

Normally to find the temperatures at these nodes
we have to use a graphical calculator or a computer
algebra system such as MAPLE, as shown above.

Fig. 9

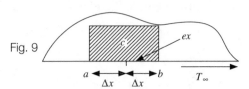

The average temperature, T_{ex}, close to an
exterior node *ex* (Fig. 9) is given by the formula

$$\boxed{11.20} \qquad T_a + 2T_c + T_b + \frac{2h\Delta x}{k}(T_\infty) = 2\left(\frac{h\Delta x}{k} + 2\right)T_{ex}$$

where h is the heat transfer coefficient, k is the thermal conductivity and T_∞ is the exterior temperature.

For the interior node we are looking at conduction heat transfer only while for the exterior node we also consider the effects of convective heat transfer from the surrounding fluid.

Example 32

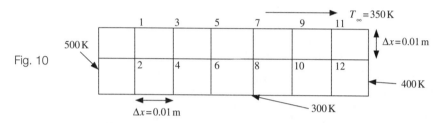

Fig. 10

Consider the rectangular bar of Fig. 10 with a thermal conductivity, $k = 1$ W/(m K) and heat transfer coefficient $h = 8$ W/(m² K). Find the temperatures at the nodes 1, 2, 3, 4, . . . , 11 and 12.

Solution

Let $T_1, T_2, T_3, \ldots$, and T_{12} represent the temperature at the nodes 1, 2, 3, . . . , and 12 respectively.

The even nodes 2, 4, 6, . . . , 12 are interior nodes so we use $\boxed{11.19}$ as in **Example 31.**

Node 2: $500 + T_1 + T_4 + 300 = 4T_2$
Node 4: $T_2 + T_3 + T_6 + 300 = 4T_4$
Node 6: $T_4 + T_5 + T_8 + 300 = 4T_6$
Node 8: $T_6 + T_7 + T_{10} + 300 = 4T_8$
Node 10: $T_8 + T_9 + T_{12} + 300 = 4T_{10}$
Node 12: $T_{10} + T_{11} + 400 + 300 = 4T_{12}$

The odd nodes 1, 3, 5, . . . , 11 are exterior nodes so we use $\boxed{11.20}$ by substituting $h = 8$, $\Delta x = 0.01$, $k = 1$, $T_\infty = 350$ and the given temperatures. Consider node 1 (Fig. 11).

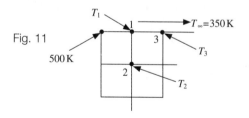

Fig. 11

Applying $\boxed{11.20}$ with T_1 as exterior node gives

$$500 + 2T_2 + T_3 + \left(\frac{2 \times 8 \times 0.01}{1} \times 350\right) = 2\left[\left(\frac{8 \times 0.01}{1}\right) + 2\right]T_1$$

$\boxed{11.19}$ $\quad T_a + T_b + T_d + T_e = 4T_c$ $\qquad$ $\boxed{11.20}$ $\quad T_a + 2T_c + T_b + \dfrac{2h\Delta x}{k}(T_\infty) = 2\left(\dfrac{h\Delta x}{k} + 2\right)T_{ex}$

Example 32 *continued*

This simplifies the equation at node 1 to

$$500 + 2T_2 + T_3 + 56 = 4.16T_1$$

Consider node 3 (Fig. 12).

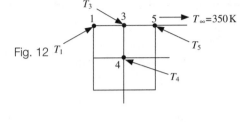

Fig. 12 T_1

Applying 11.20 to node 3 gives

$$T_1 + 2T_4 + T_5 + 56 = 4.16T_3$$

Similarly we have

Node 5: $T_3 + 2T_6 + T_7 + 56 = 4.16T_5$
Node 7: $T_5 + 2T_8 + T_9 + 56 = 4.16T_7$
Node 9: $T_7 + 2T_{10} + T_{11} + 56 = 4.16T_9$
Node 11: $T_9 + 2T_{12} + 400 + 56 = 4.16T_{11}$

Simplifying and rearranging these equations:

$$4.16T_1 - 2T_2 - T_3 = 556 \qquad \text{[Node 1]}$$
$$-T_1 + 4T_2 - T_4 = 800 \qquad \text{[Node 2]}$$
$$-T_1 + 4.16T_3 - 2T_4 - T_5 = 56 \qquad \text{[Node 3]}$$
$$-T_2 - T_3 + 4T_4 - T_6 = 300 \qquad \text{[Node 4]}$$
$$-T_3 + 4.16T_5 - 2T_6 - T_7 = 56 \qquad \text{[Node 5]}$$
$$-T_4 - T_5 + 4T_6 - T_8 = 300 \qquad \text{[Node 6]}$$
$$-T_5 + 4.16T_7 - 2T_8 - T_9 = 56 \qquad \text{[Node 7]}$$
$$-T_6 - T_7 + 4T_8 - T_{10} = 300 \qquad \text{[Node 8]}$$
$$-T_7 + 4.16T_9 - 2T_{10} - T_{11} = 56 \qquad \text{[Node 9]}$$
$$-T_8 - T_9 + 4T_{10} - T_{12} = 300 \qquad \text{[Node 10]}$$
$$-T_9 + 4.16T_{11} - 2T_{12} = 456 \qquad \text{[Node 11]}$$
$$-T_{10} - T_{11} + 4T_{12} = 700 \qquad \text{[Node 12]}$$

In matrix form we have

$$
\begin{bmatrix}
4.16 & -2 & -1 & 0 & 0 & 0 & 0 & 0 & 0 & 0 & 0 & 0 \\
-1 & 4 & 0 & -1 & 0 & 0 & 0 & 0 & 0 & 0 & 0 & 0 \\
-1 & 0 & 4.16 & -2 & -1 & 0 & 0 & 0 & 0 & 0 & 0 & 0 \\
0 & -1 & -1 & 4 & 0 & -1 & 0 & 0 & 0 & 0 & 0 & 0 \\
0 & 0 & -1 & 0 & 4.16 & -2 & -1 & 0 & 0 & 0 & 0 & 0 \\
0 & 0 & 0 & -1 & -1 & 4 & 0 & -1 & 0 & 0 & 0 & 0 \\
0 & 0 & 0 & 0 & -1 & 0 & 4.16 & -2 & -1 & 0 & 0 & 0 \\
0 & 0 & 0 & 0 & 0 & -1 & -1 & 4 & 0 & -1 & 0 & 0 \\
0 & 0 & 0 & 0 & 0 & 0 & -1 & 0 & 4.16 & -2 & -1 & 0 \\
0 & 0 & 0 & 0 & 0 & 0 & 0 & -1 & -1 & 4 & 0 & -1 \\
0 & 0 & 0 & 0 & 0 & 0 & 0 & 0 & -1 & 0 & 4.16 & -2 \\
0 & 0 & 0 & 0 & 0 & 0 & 0 & 0 & 0 & -1 & -1 & 4
\end{bmatrix}
\begin{bmatrix}
T_1 \\ T_2 \\ T_3 \\ T_4 \\ T_5 \\ T_6 \\ T_7 \\ T_8 \\ T_9 \\ T_{10} \\ T_{11} \\ T_{12}
\end{bmatrix}
=
\begin{bmatrix}
556 \\ 800 \\ 56 \\ 300 \\ 56 \\ 300 \\ 56 \\ 300 \\ 56 \\ 300 \\ 456 \\ 700
\end{bmatrix}
$$

11.20 $\quad T_a + 2T_c + T_b + \dfrac{2h\Delta x}{k}(T_\infty) = 2\left(\dfrac{h\Delta x}{k} + 2\right)T_{ex}$

Example 32 *continued*

In the 12×12 matrix the first column represents the T_1 coefficients, the second column represents the T_2 coefficients, the third column represents the T_3 coefficients, and so on. **Why are there so many zeros in this matrix?**

Consider the first equation

$$4.16T_1 - 2T_2 - T_3 = 556$$

Notice that there are no $T_4, T_5, T_6, \ldots, T_{12}$. These are represented by the zeros in the first row, that is zeros in the $T_4, T_5, T_6, \ldots, T_{12}$ columns. Similarly we have zeros in all the other rows.

We need to use a computer algebra system (MAPLE) or a graphical calculator to find the temperatures at nodes $1, 2, 3, \ldots, 11$ and 12. **It is best to use MAPLE rather than a graphical calculator. Why?**

Because with a matrix of this size, 12×12, it is very easy to make an incorrect entry and with MAPLE you can clearly see the whole matrix with a reasonable size monitor. Also if you do make a mistake you can alter the incorrect entry. The MAPLE procedure is shown below, and it gives the following results:

$T_1 = 404.68$, $T_2 = 368.42$, $T_3 = 354.61$, $T_4 = 341.00$, $T_5 = 332.50$, $T_6 = 322.98$, $T_7 = 326.60$, $T_8 = 318.44$, $T_9 = 333.29$, $T_{10} = 324.17$, $T_{11} = 355.57$ and $T_{12} = 344.93$ (all temperatures are in Kelvin (K)).

```
> A:=matrix([[4.16,-2,-1, 0,0,0,0,0,0,0,0,0], [-1, 4, 0,
-1,0,0,0,0,0,0,0,0], [-1, 0, 4.16,-2,-1,0,0,0,0,0,0,0], [0,-1,
-1, 4,0,-1,0,0,0,0,0,0], [0, 0,-1,0,4.16,-2,-1,0,0,0,0,0],
[0,0,0,-1,-1,4,0,-1,0,0,0,0],[0,0,0,0,-1,0,4.16,-2,-1,0,0,0],
[0,0,0,0,0,-1,-1,4,0,-1,0,0],[0,0,0,0,0,0,-1,0,4.16,-2,-1,0],
[0,0,0,0,0,0,0,1,-1,4,0,-1],[0,0,0,0,0,0,0,0,-1,0,4.16,-2],
[0,0,0,0,0,0,0,0,0,-1,-1,4]]);
```

$$A := \begin{bmatrix}
4.16 & -2 & -1 & 0 & 0 & 0 & 0 & 0 & 0 & 0 & 0 & 0 \\
-1 & 4 & 0 & -1 & 0 & 0 & 0 & 0 & 0 & 0 & 0 & 0 \\
-1 & 0 & 4.16 & -2 & -1 & 0 & 0 & 0 & 0 & 0 & 0 & 0 \\
0 & -1 & -1 & 4 & 0 & -1 & 0 & 0 & 0 & 0 & 0 & 0 \\
0 & 0 & -1 & 0 & 4.16 & -2 & -1 & 0 & 0 & 0 & 0 & 0 \\
0 & 0 & 0 & -1 & -1 & 4 & 0 & -1 & 0 & 0 & 0 & 0 \\
0 & 0 & 0 & 0 & -1 & 0 & 4.16 & -2 & -1 & 0 & 0 & 0 \\
0 & 0 & 0 & 0 & 0 & -1 & -1 & 4 & 0 & -1 & 0 & 0 \\
0 & 0 & 0 & 0 & 0 & 0 & -1 & 0 & 4.16 & -2 & -1 & 0 \\
0 & 0 & 0 & 0 & 0 & 0 & 0 & -1 & -1 & 4 & 0 & -1 \\
0 & 0 & 0 & 0 & 0 & 0 & 0 & 0 & -1 & 0 & 4.16 & -2 \\
0 & 0 & 0 & 0 & 0 & 0 & 0 & 0 & 0 & -1 & -1 & 4
\end{bmatrix}$$

```
> b:=vector([556,800,56,300,56,300,56,300,56,300,456,700]);
```
$b: = [556, 800, 56, 300, 56, 300, 56, 300, 56, 300, 456, 700]$

```
> linsolve(A,b);
```

404.6753628, 386.4197222, 354.6100646, 341.0035257, 332.4954545, 322.9843161,
   326.6023941, 318.4382843, 333.2939364, 324.1664271, 355.5675273, 344.9334886]

```
> evalf(%,5);
```

[404.68, 386.42, 354.61, 341.00, 332.50, 322.98, 326.60, 318.44, 333.29, 324.17,
   355.57, 344.93]

SUMMARY

The average interior temperature, T_c, close to a node c is given by

11.19 $T_a + T_b + T_d + T_e = 4T_c$

The average exterior temperature, T_{ex}, close to a node ex is given by

11.20 $T_a + 2T_c + T_b + \dfrac{2h\Delta x}{k}(T_\infty) = 2\left(\dfrac{h\Delta x}{k} + 2\right)T_{ex}$

Exercise **11(g)**

Solutions at end of book. Complete solutions available at www.palgrave.com/science/engineering/singh

For all the questions in this exercise use a computer algebra system or a graphical calculator.

1

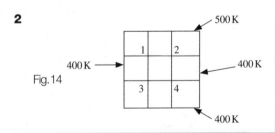

Fig. 13

Figure 13 shows a square cross-section with the surface temperatures as indicated. Determine the temperatures at nodes 1, 2, 3 and 4. Do you think your results are correct?

2

Fig. 14

Figure 14 shows a square cross-section with the surface temperatures as indicated.

  i Evaluate the temperatures at nodes 1, 2, 3 and 4.

  ii What do you notice about your results?

  iii Could you have found the temperatures of part **i** by using just two simultaneous equations?

3

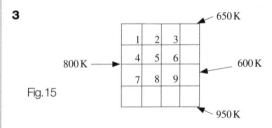

Fig. 15

Figure 15 shows a square cross-section with the surface temperatures as indicated. Determine the temperatures at nodes 1, 2, . . . , 8 and 9.

Exercise **11(g) continued**

4

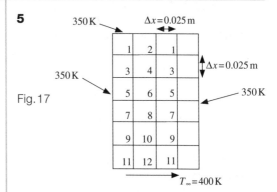

$\Delta x = 0.1\,\text{m}$

Fig. 16 $\Delta x = 0.1\,\text{m}$

600 K

600 K

$T_\infty = 400\,\text{K}$

600 K

600 K

Figure 16 shows a rectangular cross-section with the surface temperatures as indicated. For $h = 10$ W/(m² K) and $k = 1$ W/(m K), find the temperatures at nodes 1, 2, 3, . . . , 9 and 10. Are there any other solutions?

5

350 K $\Delta x = 0.025\,\text{m}$

350 K

Fig. 17

$\Delta x = 0.025\,\text{m}$

350 K

$T_\infty = 400\,\text{K}$

Figure 17 shows a rectangular cross-section of a bar. Determine the temperatures at nodes 1, 2, 3, . . . , 12 for $h = 50$ W/(m² K) and $k = 1$ W/(m K).

Miscellaneous exercise **11**

1 Evaluate the following:

i $\begin{pmatrix} 5 & 0 \\ 0 & 5 \end{pmatrix}\begin{pmatrix} 3 & 0 \\ 0 & 3 \end{pmatrix}$

ii $\begin{pmatrix} 6 & 0 \\ 0 & 6 \end{pmatrix}\begin{pmatrix} -7 & 0 \\ 0 & -7 \end{pmatrix}$

iii $\begin{pmatrix} a & 0 \\ 0 & b \end{pmatrix}\begin{pmatrix} c & 0 \\ 0 & d \end{pmatrix}$

2 Find a, b, c and d which satisfies

$$\begin{pmatrix} 1 & 2 \\ 3 & 4 \end{pmatrix}\begin{pmatrix} a & b \\ c & d \end{pmatrix} = \begin{pmatrix} 1 & 0 \\ 0 & 1 \end{pmatrix}$$

3 If $A = \begin{pmatrix} 2 & 0 \\ 0 & 2 \end{pmatrix}$, find A^2, A^3 and A^{10}.

4   **i** Find x and y which satisfy

$$\begin{pmatrix} 5 & 6 \\ 4 & 5 \end{pmatrix}\begin{pmatrix} x \\ y \end{pmatrix} = \begin{pmatrix} 1 \\ 0 \end{pmatrix}$$

ii Find a and b which satisfy

$$\begin{pmatrix} 5 & 6 \\ 4 & 5 \end{pmatrix}\begin{pmatrix} a \\ b \end{pmatrix} = \begin{pmatrix} 0 \\ 1 \end{pmatrix}$$

iii Obtain the inverse of $\begin{pmatrix} 5 & 6 \\ 4 & 5 \end{pmatrix}$.

5 a For $A = \begin{pmatrix} 2 & 1 \\ 0 & 3 \end{pmatrix}$ show that
$$A^2 - 5A + 6I = 0.$$

b For $A = \begin{pmatrix} 2 & 3 \\ 4 & 1 \end{pmatrix}$ show that
$$A^2 - 3A - 10I = 0.$$

6 Evaluate

$$\begin{pmatrix} 5 & -1 & -2 \\ 10 & -2 & -4 \\ 15 & -3 & -6 \end{pmatrix}\begin{pmatrix} 1 & 1 & 3 \\ 1 & -1 & -1 \\ 2 & 3 & 8 \end{pmatrix}$$

What do you notice? Could the same thing happen when multiplying two non-zero numbers?

7 Evaluate the following:

$$\det \begin{bmatrix} 1 & 15 & -6 & 7 & 77 \\ 0 & 2 & -973 & 533 & 207 \\ 0 & 0 & 3 & -66 & 1057 \\ 0 & 0 & 0 & 4 & 855 \\ 0 & 0 & 0 & 0 & 0 \end{bmatrix}$$

8 [electrical principles]   By using Kirchhoff's laws on a circuit we obtain the equations

$$100I_1 - 30I_2 - 40I_3 = -8$$
$$-30I_1 + 140I_2 - 60I_3 = -2$$
$$-40I_1 - 60I_2 + 170I_3 = 10$$

Write these equations in matrix form and solve for I_1, I_2 and I_3 by using Gaussian elimination.

9 Figure 18 shows the traffic flow in a network of one-way streets.

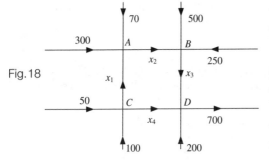

Fig. 18

The figure indicates the number of vehicles per hour during the morning rush hour and the arrows indicate the direction of flow. Assuming the number of vehicles entering a node (A, B, C and D) is equal to the number of vehicles leaving a node, show that the system of equations can be written in matrix form as

$$\begin{pmatrix} 1 & -1 & 0 & 0 \\ 0 & 1 & -1 & 0 \\ 1 & 0 & 0 & 1 \\ 0 & 0 & 1 & 1 \end{pmatrix} \begin{pmatrix} x_1 \\ x_2 \\ x_3 \\ x_4 \end{pmatrix} = \begin{pmatrix} -370 \\ -750 \\ 150 \\ 500 \end{pmatrix}$$

10 [electrical principles]   The equations of a circuit are given by

$$\begin{aligned} 9i_1 & -3i_2 & -5i_3 & = 12 \times 10^{-3} \\ -3i_1 & +16.5i_2 & -0.5i_3 & = 0 \\ -5i_1 & -0.5i_2 & +20.5i_3 & = 15 \times 10^{-3} \end{aligned}$$

Solve for i_1, i_2 and i_3 by using matrices. Is the solution unique?

11 [vibrations]   Consider the two-storey building shown in Fig. 19.

Fig. 19

Oscillations

The equations of motion are expressed as

$$\begin{pmatrix} 3k/2m & k/2m \\ k/m & k/m \end{pmatrix} \begin{pmatrix} x_1 \\ x_2 \end{pmatrix} = \begin{pmatrix} 0 \\ 0 \end{pmatrix}$$

where k is a negative constant and m is mass. Find the period, T, given by

$$T = \frac{2\pi}{\sqrt{-\lambda}} \text{ where } \lambda \text{ is an eigenvalue}$$

The remaining questions are in the field of [**control engineering**].

12 A system is described by the equations

$$\dot{x}_1 = 2x_1 + 2x_2 + u_1$$
$$\dot{x}_2 = -x_1 + 5x_2 + u_2$$
$$y = x_1 + 4x_2$$

Write these equations in the form

$$\dot{x} = Ax + Iu$$
$$y = Cx$$

where $x = \begin{pmatrix} x_1 \\ x_2 \end{pmatrix}$, $\dot{x} = \begin{pmatrix} \dot{x}_1 \\ \dot{x}_2 \end{pmatrix}$ is the time derivative, $u = \begin{pmatrix} u_1 \\ u_2 \end{pmatrix}$ and I is the identity matrix.

  i Determine the eigenvalues (system poles) of A.

  ii Determine the corresponding general eigenvectors.

13 By using $\boxed{11.6}$ and $\boxed{11.7}$, find the controllability matrix, C_T, and the observability matrix, O_T, for the system with

$$A = \begin{pmatrix} 2 & 3 \\ -1 & 5 \end{pmatrix}, B = \begin{pmatrix} 2 \\ 1 \end{pmatrix} \text{ and } C = (3 \quad 4)$$

The system is said to be controllable if $\det C_T \neq 0$ and is observable if $\det O_T \neq 0$. Determine whether the above system is controllable and/or observable.

14 By using

$$C_T = (B \quad AB \quad A^2B) \text{ and } O_T = \begin{pmatrix} C \\ CA \\ CA^2 \end{pmatrix}$$

find the controllability matrix, C_T, and the observability matrix, O_T, for the system with

$$A = \begin{pmatrix} -1 & 2 & 5 \\ 7 & 3 & 1 \\ 6 & 8 & -11 \end{pmatrix}, B = \begin{pmatrix} 1 \\ 5 \\ 7 \end{pmatrix} \text{ and }$$
$$C = (2 \quad 5 \quad 9)$$

15 A system has the following matrices:

$$A = \begin{pmatrix} -3 & 4 \\ 1 & -5 \end{pmatrix} \text{ and } B = \begin{pmatrix} 1 \\ -1 \end{pmatrix}$$

i Determine the controllability matrix, $C_T = (B \quad AB)$.

ii Evaluate C_T^{-1}.

iii We define $p(A)$ as

$$p(A) = A^2 + k_1A + k_0I \text{ where } I$$
is the identity matrix

Determine $p(A)$ for $k_0 = 2$ and $k_1 = 2$.

iv The feedback matrix, F, is determined by Ackermann's formula which is

$$F = (0 \quad 1)C_T^{-1} p(A)$$

Evaluate F for the above system.

16 The feedback matrix, F, of a system can be evaluated by using Ackermann's formula

$$F = (0 \quad 1)C_T^{-1} p(A)$$

where C_T and $p(A)$ are as defined in question **15**. Determine F for

$$A = \begin{pmatrix} -1 & 3 \\ 2 & 5 \end{pmatrix}, B = \begin{pmatrix} 1 \\ 2 \end{pmatrix},$$
$$k_0 = 3 \text{ and } k_1 = 4$$

17 The transfer function, TF, of a system is given by

$$TF = C(sI - A)^{-1}B$$

where A is the state matrix, B is the input matrix, C is the output matrix and I is the identity matrix.

For

$$A = \begin{pmatrix} 0 & -1 \\ 3 & -2 \end{pmatrix}, B = \begin{pmatrix} 1 \\ 0.6 \end{pmatrix} \text{ and }$$
$$C = (1 \quad 0)$$

obtain TF.

For the remaining questions use a computer algebra system or a graphical calculator.

18 For a system with

$$A = \begin{pmatrix} -5 & 2 & 1 \\ -6 & 5 & 9 \\ 7 & -1 & -4 \end{pmatrix}, B = \begin{pmatrix} -1 \\ 3 \\ 8 \end{pmatrix},$$
$$k_0 = 5, k_1 = -3 \text{ and } k_2 = 9$$

find the feedback matrix, F, by using Ackermann's formula:

$$F = (0 \quad 0 \quad 1)C_T^{-1} p(A)$$

where $C_T = (B \quad AB \quad A^2B)$ and $p(A) = A^3 + k_2A^2 + k_1A + k_0I$ (I is the 3×3 identity matrix).

Miscellaneous exercise **11 continued**   Solutions at end of book. Complete solutions available at www.palgrave.com/science/engineering/singh

19 By using $p(\mathbf{A}) = \mathbf{A}^3 + k_2\mathbf{A}^2 + k_1\mathbf{A} + k_0\mathbf{I}$, find $p(\mathbf{A})$ for

a $\mathbf{A} = \begin{pmatrix} 6 & 3 & -11 \\ -2 & 15 & 3 \\ 7 & 9 & 12 \end{pmatrix}$,

$k_2 = 5$, $k_1 = -7$ and $k_0 = 6$

b $\mathbf{A} = \begin{pmatrix} 9.31 & -6.11 & 2.30 \\ -2.75 & 3.93 & -6.51 \\ -8.99 & 12.23 & -1.75 \end{pmatrix}$,

$k_2 = -6.73$, $k_1 = 5.29$ and $k_0 = 6$

20 The transfer function, TF, of a system is given by
$$\text{TF} = \mathbf{C}(s\mathbf{I} - \mathbf{A})^{-1}\mathbf{B}$$
where $\mathbf{A}$ is the state matrix, $\mathbf{B}$ is the input matrix and $\mathbf{C}$ is the output matrix. For

$$\mathbf{A} = \begin{pmatrix} 2.2 & 3.1 & 6.7 & -2.7 \\ 1.1 & -2.8 & 5.6 & -5.5 \\ 3.4 & 2.7 & -1.9 & -4.6 \\ -7.6 & 8.2 & 1.5 & 4.2 \end{pmatrix},$$

$$\mathbf{B} = \begin{pmatrix} -1.1 \\ 2.3 \\ 1.7 \\ -4.1 \end{pmatrix} \text{ and}$$

$$\mathbf{C} = (9.1 \quad 6.2 \quad 4.7 \quad 5.5) \text{ find TF.}$$

CHAPTER 12
Vectors

SECTION A **Vector representation**

By the end of this section you will be able to:
▸ understand the terms vector and scalar
▸ apply vector addition and subtraction
▸ apply vectors to problems in mechanics

A1 **Vectors and scalars**

A vector is a quantity that has **size** (magnitude) and **direction**. Examples of vectors are velocity, acceleration, force, displacement and moment. A force of 10 N upwards is a vector.
? So what are scalars?

A scalar is a quantity that has size but no direction. Examples of scalars are mass, length, time, volume, speed and temperature.

? How do we write down vectors and scalars and how can we distinguish between them?

A vector **from** O **to** A is denoted by $\overrightarrow{OA}$ or written in bold typeface **a** (Fig. 1). You might use $\underline{a}$ or $\overrightarrow{OA}$ because handwriting bold face is not feasible. A scalar is denoted by a, not in bold, so that we can distinguish between vectors and scalars.

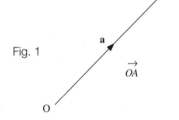

Fig. 1

Write the following vectors of Fig. 2 in the notation outlined.

a is $\overrightarrow{CD}$ **b** is $\overrightarrow{DC}$ **c** is $\overrightarrow{EF}$

d is $\overrightarrow{HG}$ **e** is $\overrightarrow{IJ}$

Fig. 2

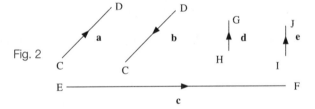

Two vectors are **equivalent** if they have the same direction and magnitude. In Fig. 2, $\overrightarrow{HG} = \overrightarrow{IJ}$ because $\overrightarrow{HG}$ and $\overrightarrow{IJ}$ have the same direction and magnitude but only differ in position.

Also note that the direction of the arrow gives the direction of the vector, that is $\overrightarrow{CD}$ is **different** from $\overrightarrow{DC}$.

The magnitude (or length) of the vector $\overrightarrow{AB}$ is denoted by $|\overrightarrow{AB}|$.

Fig. 3

There are many examples of vectors in engineering, particularly in dynamics:

a A displacement of 20 m to the horizontal right of an object from O to A (Fig. 3).

b A force on an object acting vertically downwards (Fig. 4).

Fig. 4

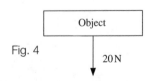

c The velocity and acceleration of a particle thrown vertically upwards, as illustrated in Fig. 5.

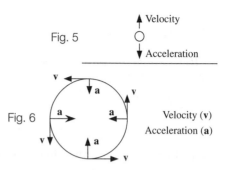

Fig. 5

d Figure 6 shows the various directions of the velocity and acceleration of a particle rotating in a circle. In each case the velocity and acceleration have magnitude and direction so they can be represented by vectors.

Fig. 6

Velocity (**v**)
Acceleration (**a**)

A2 **Vector addition**

Figure 7 shows how to add two vectors $\overrightarrow{AB}$ and $\overrightarrow{CD}$.

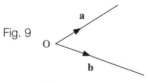

Fig. 7

To add vectors $\overrightarrow{AB}$ and $\overrightarrow{CD}$ we connect the ends B and C which results in the vector $\overrightarrow{AD}$. In other words, we can go directly from A to D or via B and C, $\overrightarrow{AB}$ and along $\overrightarrow{CD}$. The vectors $\overrightarrow{AB}$ and $\overrightarrow{CD}$ are examples of **free** vectors, which means that the vectors have direction and magnitude but are not at a fixed position.

Consider a robot which moves an object from A to B and then moves it to C. Thus the object can be moved from A to C directly or via B as shown in Fig. 8.

Fig. 8

Consider the addition of vectors **a** and **b** in Fig. 9.

We shift vector **b** to the end of vector **a** and the result is going from O to the end of vector **b**. Generally **a** + **b** is called the resultant vector of **a** and **b** and is represented by the line with two arrows. Let's do an example.

Fig. 9

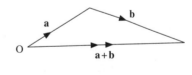

Example 1 *mechanics*

Two forces $\mathbf{F}_1$ and $\mathbf{F}_2$ are applied to an object O. Determine the magnitude and direction of the resultant force, $\mathbf{R} = \mathbf{F}_1 + \mathbf{F}_2$.

Fig. 10

Solution

Adding vectors $\mathbf{F}_1$ and $\mathbf{F}_2$ gives **R** (Fig. 11).

Example 1 *continued*

**How can we obtain the magnitude (i.e. the length)
of the vector R?**

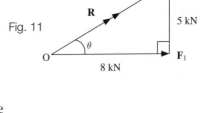

Fig. 11

We can use Pythagoras:

$$|\mathbf{R}| = \sqrt{5^2 + 8^2} = 9.43 \text{ kN}$$

We also need to find the direction of **R** which is given
by the angle θ in Fig. 11. By using trigonometry we have

$$\tan(\theta) = \frac{\text{opposite}}{\text{adjacent}} = \frac{5}{8}$$

Therefore taking inverse tan gives

$$\theta = \tan^{-1}\left(\frac{5}{8}\right) = 32.01° \quad \text{[by calculator]}$$

The resultant force, **R**, has a magnitude of 9.43 kN and an angle of 32.01° as shown.

A3 Subtraction of vectors

We define $\overrightarrow{OA} - \overrightarrow{OB}$ as $\overrightarrow{OA} + \left(-\overrightarrow{OB}\right)$

Example 2

By considering vectors $\overrightarrow{OA}$ and $\overrightarrow{OB}$ as in Fig. 12, show on a different diagram the vectors
$\overrightarrow{OA} - \overrightarrow{OB}$ and $\overrightarrow{OB} - \overrightarrow{OA}$. Find $\overrightarrow{OA} - \overrightarrow{OA}$.

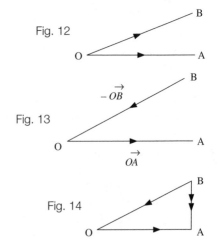

Fig. 12

Fig. 13

Fig. 14

Solution

$-\overrightarrow{OB}$ is the vector with the same length as $\overrightarrow{OB}$ but is
in the opposite direction. In fact $-\overrightarrow{OB}$ is the vector
$\overrightarrow{BO}$ (Fig. 13).

Therefore $\overrightarrow{OA} - \overrightarrow{OB}$ is given by

$$\overrightarrow{OA} - \overrightarrow{OB} = \overrightarrow{BA} \quad \text{(Fig. 14)}$$

We can justify this by using algebra:

$$\overrightarrow{OA} - \overrightarrow{OB} = \overrightarrow{OA} + (-\overrightarrow{OB})$$

$$= -\overrightarrow{OB} + \overrightarrow{OA}$$

$$= \overrightarrow{BO} + \overrightarrow{OA} = \overrightarrow{BA}$$

Example 2 *continued*

Similarly $-\vec{OA}$ is the vector $\vec{AO}$. Hence

$$\vec{OB} - \vec{OA} = \vec{AB} \quad \text{(Fig. 15)}$$

Clearly $\vec{OA} - \vec{OA} = \mathbf{0}$ [Zero vector]

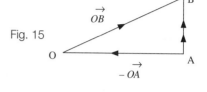

Fig. 15

The zero vector is denoted by $\mathbf{0}$, and has zero magnitude and direction.

Note that, if $\vec{AB}$ is not the zero vector then

$$\vec{AB} \neq \vec{BA} \quad \text{[Not equal vectors]}$$

$\vec{AB}$ and $\vec{BA}$ have the same magnitude but opposite directions, in fact

$$\vec{AB} = -\vec{BA}$$

Example 3

Write $\vec{AB}$, $\vec{BA}$ and $\vec{BC}$ in terms of $\mathbf{a}$, $\mathbf{b}$ and $\mathbf{c}$ (see Fig. 16).

Solution

$$\vec{AB} = \vec{AO} + \vec{OB}$$
$$= -\vec{OA} + \vec{OB} = -\mathbf{a} + \mathbf{b} = \mathbf{b} - \mathbf{a}$$

Fig. 16

The vector $\vec{BA}$ is given by

$$\vec{BA} = \vec{BO} + \vec{OA}$$
$$= -\vec{OB} + \vec{OA} = -\mathbf{b} + \mathbf{a} = \mathbf{a} - \mathbf{b}$$

The vector $\vec{BC}$ is

$$\vec{BC} = \vec{BO} + \vec{OC}$$
$$= -\vec{OB} + \vec{OC} = -\mathbf{b} + \mathbf{c} = \mathbf{c} - \mathbf{b}$$

A4 Scalar multiplication

Figure 17 shows the vector $\mathbf{a}$ and its scalar multiples $2\mathbf{a}$, $-2\mathbf{a}$ and $\frac{1}{2}\mathbf{a}$. Notice that $2\mathbf{a}$ has the same direction as vector $\mathbf{a}$ but is twice as long,

Fig. 17

and $\frac{1}{2}$**a** is half the length of **a**. Also −2**a** has the same magnitude as 2**a** but is in opposite direction.

Example 4

Figure 18 shows a triangle OAB with M as the midpoint of the line AB. Determine $\overrightarrow{OM}$ in terms of **a** and **b**.

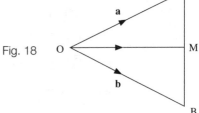

Fig. 18

Solution

Since M is the midpoint of BA we have

† $\overrightarrow{OM} = \overrightarrow{OB} + \frac{1}{2}\overrightarrow{BA}$

We have $\overrightarrow{OB} =$ **b** but we need to express $\overrightarrow{BA}$ in terms of **a** and **b**.

$$\overrightarrow{BA} = \overrightarrow{BO} + \overrightarrow{OA}$$

$$= -\overrightarrow{OB} + \overrightarrow{OA}$$

$$= \overrightarrow{OA} - \overrightarrow{OB} = \textbf{a} - \textbf{b}$$

Substituting $\overrightarrow{OB} =$ **b** and $\overrightarrow{BA} =$ **a** − **b** into † gives

$$\overrightarrow{OM} = \textbf{b} + \frac{1}{2}(\textbf{a} - \textbf{b})$$

$$= \textbf{b} + \frac{1}{2}\textbf{a} - \frac{1}{2}\textbf{b}$$

$$= \frac{1}{2}\textbf{b} + \frac{1}{2}\textbf{a} = \frac{1}{2}(\textbf{a} + \textbf{b}) \quad \left[\text{Taking out } \frac{1}{2}\right]$$

A5 Applications of vectors in mechanics

Example 5 *mechanics*

Find the magnitude, $|\textbf{R}|$, of the resultant force **R** shown in Fig. 19.

Solution

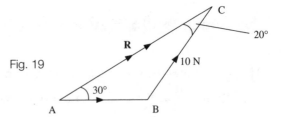

Fig. 19

What is angle *B* equal to?

angle $B = 180° − (20° + 30°) = 130°$

Example 5 *continued*

To find $|\mathbf{R}|$ we use the sine rule. Let a and b represent the sides opposite angles A and B respectively. Thus $a = 10$ and $b = |\mathbf{R}|$. We need to find b by using

$$\frac{b}{\sin(B)} = \frac{a}{\sin(A)}$$

Substituting $A = 30°$, $a = 10$ and $B = 130°$:

$$\frac{b}{\sin(130°)} = \frac{10}{\sin(30°)} = 20$$

$$b = 20 \times \sin(130°) = 15.32$$

Thus $|\mathbf{R}| = 15.32$ N.

Example 6 *mechanics*

The point C of Fig. 20 shows a particle rotating in a circle with the spin velocity represented by $\vec{CA}$ and the forward velocity represented by $\vec{CB}$. The resultant velocity, $\mathbf{v}$, is vertically downwards. Find $\mathbf{v}$ if $|\vec{CB}| = 5$ m/s.

Fig. 20

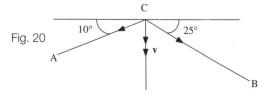

Solution

We shift $\vec{CA}$ to the end of $\vec{CB}$ to make the triangle shown in Fig. 21.

To obtain $\vec{CX}$ of Fig. 21 we use the sine rule:

Fig. 21

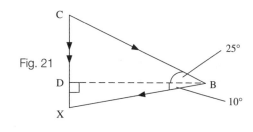

$$\frac{x}{\sin(X)} = \frac{b}{\sin(B)}$$

where x and b are the sides opposite angles X and B respectively.

Considering the triangle BDX, what is angle X equal to?

Recall that the sum of the angles inside a triangle add up to 180°, Thus

$$\text{angle } X = 180° - (90° + 10°) = 80°$$

In our triangle CBX:

$$\text{angle } X = 80°, \text{ angle } B = 10° + 25° = 35° \text{ and } x = 5 \text{ because } |\vec{CB}| = 5$$

Example 6 *continued*

By substituting these values into $\dfrac{x}{\sin(X)} = \dfrac{b}{\sin(B)}$ we have

$$\frac{5}{\sin(80°)} = \frac{b}{\sin(35°)}$$

$$b = \frac{5}{\sin(80°)} \times \sin(35°) = 2.912$$

Therefore the velocity $\mathbf{v} = 2.91$ m/s vertically downwards.

SUMMARY

A vector has direction and magnitude, and a scalar has magnitude only but no direction. Vectors are basic tools used in engineering.

Exercise **12(a)**

Solutions at end of book. Complete solutions available at www.palgrave.com/science/engineering/singh

1 [*mechanics*]

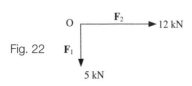

Fig. 22

Figure 22 shows two forces $\mathbf{F}_1$ and $\mathbf{F}_2$ acting on an object O. Determine the resultant force, $\mathbf{R} = \mathbf{F}_1 + \mathbf{F}_2$.

2 [*mechanics*]

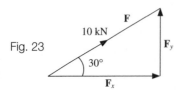

Fig. 23

Determine the horizontal, $\mathbf{F}_x$, and vertical, $\mathbf{F}_y$, components of the force $\mathbf{F}$ shown in Fig. 23.

3 [*mechanics*]   A force of 12 000 N is applied to a body at an angle of 45° to the horizontal. Determine the vertical and horizontal components of the force.

4 [*mechanics*]

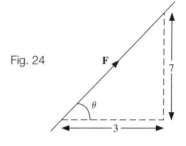

Fig. 24

Figure 24 shows a force, $\mathbf{F}$, of 10 kN applied at an angle θ. Obtain the horizontal and vertical components of the force.

Exercise **12(a) continued**

Solutions at end of book. Complete solutions available at www.palgrave.com/science/engineering/singh

5

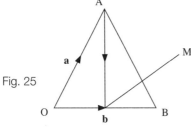

Fig. 25

In the triangle OAB shown in Fig. 25, the point M is the midpoint of OB. Determine AM in terms of **a** and **b**.

6

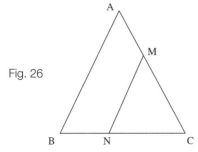

Fig. 26

Figure 26 shows a triangle ABC where M is the midpoint of AC and N is the midpoint of BC. Show that

$$\overrightarrow{MN} = \frac{1}{2}\overrightarrow{AB}.$$

🚗 **The remaining questions are in the field of [*mechanics*].**

7 Figure 27 shows forces F_1 and F_2 exerted on a particle O. The resultant force, $F = F_1 + F_2$, is horizontal. Determine **F** if F_2 has a magnitude of 10 N.

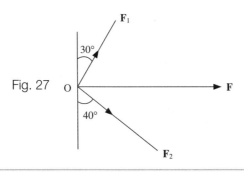

Fig. 27

8 Figure 28 shows forces F_1 and F_2 applied to an object O. The resultant force **R** is vertical.

Fig. 28

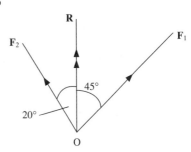

Determine the magnitude of **R** if F_1 has a magnitude of 12 kN.

9 Figure 29 shows forces F_1 and F_2 of magnitude 7 kN and 20 kN respectively. Determine the resultant force.

Fig. 29

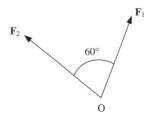

10 Determine the resultant force, $F_1 + F_2$, of the forces F_1 and F_2 shown in Fig. 30.

Fig. 30

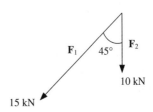

SECTION B **Vectors in Cartesian co-ordinates**

By the end of this section you will be able to:
▶ write vectors of two dimensions in **i** and **j** notation
▶ use a calculator to find the resultant vector
▶ examine engineering applications

B1 **Vectors in two dimensions**

We define vectors **i** and **j** in the *x*–*y* plane as the unit vectors in the *x* and *y* directions respectively. The term 'unit vector' means that the magnitude (size) of the vector is equal to 1. So **i** is one unit in the direction of the *x* axis and **j** is one unit in the direction of the *y* axis. The vectors **i** and **j** are shown in Fig. 31.

Fig. 31

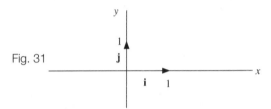

Do not confuse the unit vector **j** with the complex number $\sqrt{-1}$ of Chapter 10. The unit vector **j** is denoted by bold typeface.

Consider the vector $\overrightarrow{OA}$ shown in Fig. 32.

To get from O to A we can go 3 units along the *x* axis and 5 units up the *y* axis. We can write this as

Fig. 32

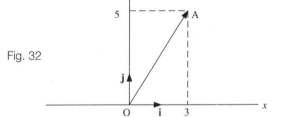

$$3\mathbf{i} + 5\mathbf{j} \text{ or } \overrightarrow{OA} = 3\mathbf{i} + 5\mathbf{j}$$

Write the vectors $\overrightarrow{OA}, \overrightarrow{OB}$ and $\overrightarrow{OC}$, shown in Fig. 33, in terms of the unit vectors **i** and **j**.

The vector $\overrightarrow{OA}$ is 7 units across and 10 units up, so

$$\overrightarrow{OA} = 7\mathbf{i} + 10\mathbf{j}$$

Similarly

Fig. 33

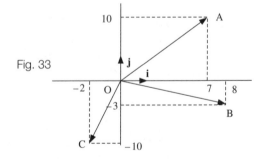

$$\overrightarrow{OB} = 8\mathbf{i} - 3\mathbf{j}$$

Also the vector $\overrightarrow{OC}$ is 2 units to the left and 10 units down, therefore

$$\overrightarrow{OC} = -2\mathbf{i} - 10\mathbf{j}$$

Generally a vector from the origin, O, to a point, P, is called the position vector of P and is denoted by $\overrightarrow{OP}$. Moreover the vector $\overrightarrow{OP}$ from the origin to the point P = (*x*, *y*) can be written as

12.1 $$\overrightarrow{OP} = x\mathbf{i} + y\mathbf{j}$$

We can solve problems in mechanics by using this new notation. The advantage of this approach is that we can find the resultant vector of two or more vectors without using the sine and cosine rules. Consider the following examples.

The displacement, **s**, of an object by 5 m is shown in Fig. 34.

This displacement **s** can be represented in
i and **j** notation as

$$s = 5i + 0j = 5i$$

Fig. 34

O   5 m   A

 Example 7 *mechanics*

A projectile is fired with an initial velocity of 20 m/s and makes an angle of 60° with the horizontal (Fig. 35).

Write the velocity in **i** and **j** notation.

Fig. 35

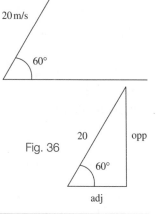

20 m/s

60°

Solution

We can construct a right-angled triangle (Fig. 36).
By using trigonometry we have

$$adj = 20 \cos(60°) = 10$$

$$opp = 20 \sin(60°) = 17.32$$

Fig. 36

20   opp

60°

adj

Thus the velocity can be written as

$$10i + 17.32j$$

 Example 8 *mechanics*

Figure 37 shows an object O subject to forces F_1 and F_2.
Find the resultant force **R**.

Solution

Writing F_1 and F_2 in **i** and **j** components:

$$F_1 = 10i \quad \text{[Because it is only in the } \mathbf{i} \text{ direction]}$$

The 12 kN force at an angle of 45° is made up of 12cos(45°) in the **i** direction (the adjacent) and 12sin(45°) in the **j** direction (the opposite), as shown in Fig. 38.

We have

$$F_2 = \left[12\cos(45°)\right]i + \left[12\sin(45°)\right]j$$

Adding the two vectors gives

$$R = F_1 + F_2$$
$$= 10i + 12\cos(45°)i + 12\sin(45°)j$$
$$= \left[10 + 12\cos(45°)\right]i + 12\sin(45°)j \quad \text{[Collecting like terms]}$$
$$= 18.485i + 8.485j \quad \text{[Evaluating]}$$

Fig. 37

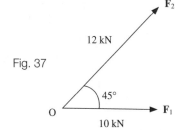

F_2

12 kN

45°

O F_1

10 kN

Fig. 38

F_2

12 kN

12 sin(45°)

45°

12 cos(45°)

Example 8 *continued*

Hence the vector **R** is 18.485 units in the horizontal direction and 8.485 units up (Fig. 39).

We can find the magnitude by using Pythagoras:

$$|\mathbf{R}| = \sqrt{18.485^2 + 8.485^2} = 20.34 \text{ kN}$$

Fig. 39

Now we need to find the direction of **R**. **What is the value of angle θ?**

By trigonometry

$$\tan(\theta) = \frac{\text{opposite}}{\text{adjacent}} = \frac{8.485}{18.485}$$

Taking inverse tan gives

$$\theta = \tan^{-1}\left(\frac{8.485}{18.485}\right) = 24.66°$$

The resultant force **R** has a magnitude of 20.34 kN at an angle of 24.66° inclined to the horizontal.

Example 9 *mechanics*

Figure 40 shows three forces **F**₁, **F**₂ and **F**₃ applied to an object O. Determine the resultant force **R**.

Solution

We can measure each angle from the positive horizontal axis. Therefore the 8 kN force is at an angle of 135° (=180° − 45°). Putting each force into its **i** and **j** components by using trigonometry gives

Fig. 40

$$\mathbf{F}_1 = [10\cos(60°)]\mathbf{i} + [10\sin(60°)]\mathbf{j}$$

$$\mathbf{F}_2 = [8\cos(135°)]\mathbf{i} + [8\sin(135°)]\mathbf{j}$$

$$\mathbf{F}_3 = -20\mathbf{j}$$

Adding the forces gives the resultant force, **R**:

$$\mathbf{R} = \mathbf{F}_1 + \mathbf{F}_2 + \mathbf{F}_3$$

$$= [10\cos(60°)]\mathbf{i} + [10\sin(60°)]\mathbf{j} + [8\cos(135°)]\mathbf{i} + [8\sin(135°)]\mathbf{j} - 20\mathbf{j}$$

$$= [10\cos(60°) + 8\cos(135°)]\mathbf{i} + [10\sin(60°) + 8\sin(135°) - 20]\mathbf{j}$$

$$= -0.657\mathbf{i} - 5.683\mathbf{j}$$

The magnitude of the resultant force, $|\mathbf{R}|$, is found by using Pythagoras:

$$|\mathbf{R}| = \sqrt{(-0.657)^2 + (-5.683)^2} = 5.72 \text{ kN}$$

Example 9 *continued*

The angle can be determined by plotting $-0.657\mathbf{i} - 5.683\mathbf{j}$ and then using trigonometry and a calculator.

So **R** has magnitude 5.72 kN and is at an angle of $\theta = 83.41°$, as shown in Fig. 41.

Fig. 41

$\theta = \tan^{-1}\left(\dfrac{5.683}{0.657}\right) = 83.41°$

B2 Use of calculator

The calculations in examples can be simplified by using a calculator.

Consider **Example 9.** We can also write forces $\mathbf{F}_1$, $\mathbf{F}_2$ and $\mathbf{F}_3$ in polar form, that is

$$\mathbf{F}_1 = 10\angle 60°$$

Measuring the angle anticlockwise from the positive horizontal axis:

$$\mathbf{F}_2 = 8\angle 135°$$

Measuring the angle clockwise from the positive horizontal axis:

$$\mathbf{F}_3 = 20\angle(-90°)$$

$10\angle 60°$ means the vector has a magnitude of 10 units and is at an angle of 60°. We used this polar form notation for complex numbers in Chapter 10.

On some calculators you can add these forces directly in polar form, in others you need to place each force into $\mathbf{i}$ and $\mathbf{j}$ notation and then carry out the addition.

To put $\mathbf{F}_1 = 10\angle 60°$ in $\mathbf{i}$ and $\mathbf{j}$ notation on a calculator we PRESS

$$\boxed{\text{SHIFT}}\ \boxed{\text{Pol(}}\ \boxed{10}\ ,\ \boxed{60}\ \boxed{)}\ \boxed{=}$$

The display should show 5. To find the $\mathbf{j}$ part we PRESS

$$\boxed{\text{ALPHA}}\ \boxed{\text{tan}}\ \boxed{=}$$

and the display should show 8.660254038. If your calculator does not have a $\boxed{\text{SHIFT}}$ button, then try the $\boxed{\text{INV}}$ or $\boxed{\text{2nd FUN}}$ button instead. Thus

$$10\angle 60° = 5\mathbf{i} + 8.66\mathbf{j}$$

Similarly

$$8\angle 135° = -5.657\mathbf{i} + 5.657\mathbf{j}$$
$$20\angle(-90°) = -20\mathbf{j}$$

Adding these together as before gives

$$(5\mathbf{i} + 8.66\mathbf{j}) + (-5.657\mathbf{i} + 5.657\mathbf{j}) + (-20\mathbf{j}) = (5 - 5.657)\mathbf{i} + (8.66 + 5.657 - 20)\mathbf{j}$$
$$= -0.657\mathbf{i} - 5.683\mathbf{j}$$

Putting this, $-0.657\mathbf{i} - 5.683\mathbf{j}$, back into polar form on a calculator we PRESS

$$\boxed{\text{Pol(}}\ \boxed{(-)}\ \boxed{0.657}\ ,\ \boxed{(-)}\ \boxed{5.683}\ \boxed{)}\ \boxed{=}$$

which should show 5.72085116. Then to find the angle we PRESS

$$\boxed{\text{ALPHA}}\ \boxed{\text{tan}}\ \boxed{=}$$

which should show −96.59457254. (This is the angle of the resultant vector.) The minus sign in front of 96.59° means it is measured in a clockwise direction. Note that the calculator gives the angle measured from the positive horizontal axis.

Fig. 42

The angle 96.59° in Fig. 42 is the same as the angle 83.41° in Fig. 41.

The resultant vector, **R**, has a magnitude of 5.72 kN and an angle of 96.59° as shown in Fig. 42.

It is important to note that this use of the calculator only works for vectors in two dimensions and we cannot use this technique for three-dimensional vectors which are discussed in the next section.

SUMMARY

Unit vectors in the direction of the x and y axes are denoted by **i** and **j** respectively.
We apply trigonometry to find the resultant vector.
We can also use a calculator to find the resultant vector.

Exercise 12(b)

Solutions at end of book. Complete solutions available at www.palgrave.com/science/engineering/singh

🚗 **All questions are in the field of [*mechanics*].**

1

Fig. 43

a   b

Figure 43**a** and **b** shows forces F_1 and F_2 applied to an object O.

Determine the resultant force, $R = F_1 + F_2$, in each case.

2 The following forces have units of kN:

$$F_1 = 4i − 6j$$
$$F_2 = 7i − 12j$$

Determine the magnitude of the force $F = 31F_1 − 3F_2$.

3

Fig. 44

Figure 44 shows forces F_1, F_2 and F_3 acting on a particle P. Determine the resultant force **R**.

4 Determine the resultant force of the forces shown in Fig. 45.

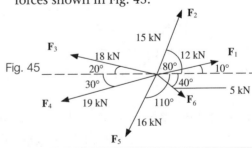

Fig. 45

SECTION C **Three-dimensional vectors**

By the end of this section you will be able to:
▶ write vectors of three dimensions in **i**, **j** and **k** notation
▶ find the magnitude of a vector in three dimensions
▶ evaluate the unit vector

C1 **Vectors in three dimensions**

The x–y plane can be extended to cover three dimensions by including a third axis called the z axis. This axis is at right angles to the other two, x and y, axes. The position of a point in three dimensions is given by three co-ordinates (x, y, z). Real life is three dimensional, hence many engineering problems require the application of three dimensions.

Let's examine vectors in three dimensions. We define **i**, **j** and **k** as the unit vectors in the x, y and z directions respectively (Fig. 46).

Consider the vector $\overrightarrow{OA}$ in Fig. 47.

Fig. 46

To get from O to A we can go 3 units along the x axis, 5 units along the y axis and then 7 units up the z axis. Therefore

$$\overrightarrow{OA} = 3\mathbf{i} + 5\mathbf{j} + 7\mathbf{k}$$

Thus the position vector of the point with co-ordinates (3, 5, 7) is

Fig. 47

$$3\mathbf{i} + 5\mathbf{j} + 7\mathbf{k}$$

? How would you write the vectors from the origin to the points:

$$\mathbf{B} = (-3, 5, 7), \quad \mathbf{C} = (1, 1, -1) \text{ and } \mathbf{D} = (-5, -11, 23)?$$

We have

$$\overrightarrow{OB} = -3\mathbf{i} + 5\mathbf{j} + 7\mathbf{k}, \quad \overrightarrow{OC} = \mathbf{i} + \mathbf{j} - \mathbf{k} \text{ and } \overrightarrow{OD} = -5\mathbf{i} - 11\mathbf{j} + 23\mathbf{k}$$

In general the vectors from the origin to the point A = (x, y, z) are given by

$$\overrightarrow{OA} = x\mathbf{i} + y\mathbf{j} + z\mathbf{k}$$

Let **p** and **q** be vectors from the origin in three-dimensional space such that

$$\mathbf{p} = a\mathbf{i} + b\mathbf{j} + c\mathbf{k} \text{ and } \mathbf{q} = d\mathbf{i} + e\mathbf{j} + f\mathbf{k}$$

Then the addition of **p** and **q** is defined by

12.2 $$\mathbf{p} + \mathbf{q} = (a + d)\mathbf{i} + (b + e)\mathbf{j} + (c + f)\mathbf{k}$$

All you need to do is add the **i**, **j** and **k** components separately.

Consider the vector $\mathbf{r} = a\mathbf{i} + b\mathbf{j} + c\mathbf{k}$.

? **What is the magnitude (length) of the vector r?**

Figure 48 shows the vector **r**. By Pythagoras we can find the length OA²:

$$OA^2 = a^2 + b^2$$

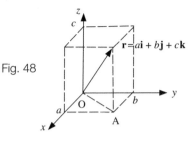

Fig. 48

The length of the vector $|\mathbf{r}|$ is found by applying Pythagoras again:

$$|\mathbf{r}|^2 = OA^2 + c^2$$
$$= a^2 + b^2 + c^2$$

? **How can we find $|\mathbf{r}|$?**

Take the square root of both sides:

$$|\mathbf{r}| = \sqrt{a^2 + b^2 + c^2}$$

Hence the magnitude, $|\mathbf{r}|$, of a vector

$$\mathbf{r} = a\mathbf{i} + b\mathbf{j} + c\mathbf{k}$$

is given by

12.3 $$|\mathbf{r}| = \sqrt{a^2 + b^2 + c^2}$$

In two-dimensional vectors we always found the direction of the vector by evaluating the angle that the vector made with the horizontal axis. In three-dimensional vectors there is **no one** angle between a three-dimensional vector and the horizontal axis or any other axes which specifies the direction.

🚗 Example 10 *mechanics*

Three forces F_1, F_2 and F_3 (in newton) act on a particle. Obtain the magnitude of the resultant force, $\mathbf{R} = \mathbf{F}_1 + \mathbf{F}_2 + \mathbf{F}_3$, in each of the following:

a $F_1 = 5\mathbf{i} - \mathbf{j} - \mathbf{k}$, $F_2 = 3\mathbf{i} + \mathbf{j}$, $F_3 = 2\mathbf{i} - 7\mathbf{j} + 3\mathbf{k}$
b $F_1 = 2\mathbf{i} - 3\mathbf{j} + 2\mathbf{k}$, $F_2 = \mathbf{i}$, $F_3 = \mathbf{i} + 3\mathbf{j} - \mathbf{k}$
c $F_1 = 7\mathbf{i} - \mathbf{j}$, $F_2 = 2\mathbf{i} + \mathbf{j} - \mathbf{k}$, $F_3 = 7\mathbf{i} - \mathbf{j} - 21\mathbf{k}$

Solution

a Adding the forces, $F_1 + F_2 + F_3$, gives

$$\mathbf{R} = (5\mathbf{i}-\mathbf{j}-\mathbf{k}) + (3\mathbf{i} + \mathbf{j}) + (2\mathbf{i}-7\mathbf{j} + 3\mathbf{k})$$
$$= (5 + 3 + 2)\mathbf{i} + (-1 + 1 - 7)\mathbf{j} + (-1 + 0 + 3)\mathbf{k} \quad \text{[Collecting like terms]}$$
$$= 10\mathbf{i}-7\mathbf{j} + 2\mathbf{k}$$

To find the magnitude we use $|\mathbf{r}| = \sqrt{a^2 + b^2 + c^2}$ with $a = 10$, $b = -7$ and $c = 2$:

$$|\mathbf{R}| = \sqrt{10^2 + (-7)^2 + 2^2} = 12.37 \text{ N (2 d.p.)}$$

⊞ Example 10 *continued*

b We have

$$R = (2\mathbf{i} - 3\mathbf{j} + 2\mathbf{k}) + \mathbf{i} + (\mathbf{i} + 3\mathbf{j} - \mathbf{k})$$
$$= (2 + 1 + 1)\mathbf{i} + (-3 + 0 + 3)\mathbf{j} + (2 + 0 - 1)\mathbf{k}$$
$$= 4\mathbf{i} + \mathbf{k} \qquad \text{[No } \mathbf{j} \text{ term]}$$

The magnitude is

$$|R| = \sqrt{4^2 + 0^2 + 1^2} = 4.12 \text{ N} \qquad \text{(2 d.p.)}$$

c Similarly we have

$$R = (7\mathbf{i} - \mathbf{j}) + (2\mathbf{i} + \mathbf{j} - \mathbf{k}) + (7\mathbf{i} - \mathbf{j} - 21\mathbf{k})$$
$$= (7 + 2 + 7)\mathbf{i} + (-1 + 1 - 1)\mathbf{j} + (0 - 1 - 21)\mathbf{k}$$
$$= 16\mathbf{i} - \mathbf{j} - 22\mathbf{k}$$

and the magnitude is

$$|R| = \sqrt{16^2 + (-1)^2 + (-22)^2} = 27.22 \text{ N} \qquad \text{(2 d.p.)}$$

C2 Unit vectors

Unit vectors are vectors which have a magnitude of 1. If a vector, $\mathbf{r}$, has a magnitude of 5 then the unit vector, denoted by $\mathbf{u}$, in the direction of $\mathbf{r}$ is given by

$$\mathbf{u} = \frac{1}{5}\mathbf{r}$$

For a general vector, $\mathbf{r}$, we have the situation shown in Fig. 49.

The unit vector, $\mathbf{u}$, along the direction of a vector $\mathbf{r}$ is given by

12.4 $$\mathbf{u} = \frac{\mathbf{r}}{|\mathbf{r}|}$$

provided that $\mathbf{r}$ is not the zero vector.

A unit vector is a vector which has a magnitude of 1.

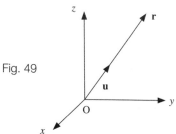

Fig. 49

❓ **What is the unit vector, u, along the direction of the vector r = $a\mathbf{i}$ + $b\mathbf{j}$ + $c\mathbf{k}$?**

The magnitude, $|\mathbf{r}|$, of the vector $\mathbf{r} = a\mathbf{i} + b\mathbf{j} + c\mathbf{k}$ is given by

$$|\mathbf{r}| = \sqrt{a^2 + b^2 + c^2}$$

Thus

12.5 $$\mathbf{u} = \frac{1}{\sqrt{a^2 + b^2 + c^2}}(a\mathbf{i} + b\mathbf{j} + c\mathbf{k})$$

Find the unit vector, $\mathbf{u}$, in the direction of the vector $\mathbf{r} = 2\mathbf{i} - 2\mathbf{j} + \mathbf{k}$.

The magnitude is $|\mathbf{r}| = \sqrt{2^2 + (-2)^2 + 1^2} = 3$. Therefore the unit vector is given by

$$\mathbf{u} = \frac{1}{3}(2\mathbf{i} - 2\mathbf{j} + \mathbf{k})$$

? **What is the magnitude (size) of the vector $\mathbf{u} = \frac{1}{3}(2\mathbf{i} - 2\mathbf{j} + \mathbf{k})$?**

It's 1 because it is a unit vector. [You can check this by using 12.3 .]

Example 11

Determine the unit vector in the direction of $\overrightarrow{AB}$ where A = (1, 3, 5) and B = (7, 1, 2).

Solution

In Fig. 50:

$$\overrightarrow{AB} = \overrightarrow{AO} + \overrightarrow{OB}$$

* $\qquad = -\overrightarrow{OA} + \overrightarrow{OB} = \overrightarrow{OB} - \overrightarrow{OA}$

Fig. 50

The vector, $\overrightarrow{OA}$, from the origin to A = (1, 3, 5) is written as

$$\overrightarrow{OA} = \mathbf{i} + 3\mathbf{j} + 5\mathbf{k}$$

Similarly for B = (7, 1, 2) we have

$$\overrightarrow{OB} = 7\mathbf{i} + \mathbf{j} + 2\mathbf{k}$$

Substituting these into * gives

$$\begin{aligned}
\overrightarrow{AB} &= \overrightarrow{OB} - \overrightarrow{OA} \\
&= (7\mathbf{i} + \mathbf{j} + 2\mathbf{k}) - (\mathbf{i} + 3\mathbf{j} + 5\mathbf{k}) \\
&= (7 - 1)\mathbf{i} + (1 - 3)\mathbf{j} + (2 - 5)\mathbf{k} \\
&= 6\mathbf{i} - 2\mathbf{j} - 3\mathbf{k}
\end{aligned}$$

The unit vector in the direction of $\overrightarrow{AB} = 6\mathbf{i} - 2\mathbf{j} - 3\mathbf{k}$ is determined by

12.5 $$\mathbf{u} = \frac{1}{\sqrt{a^2 + b^2 + c^2}}(a\mathbf{i} + b\mathbf{j} + c\mathbf{k})$$

$$\begin{aligned}
\text{unit vector} &= \frac{1}{\sqrt{6^2 + (-2)^2 + (-3)^2}}(6\mathbf{i} - 2\mathbf{j} - 3\mathbf{k}) \\
&= \frac{1}{7}(6\mathbf{i} - 2\mathbf{j} - 3\mathbf{k})
\end{aligned}$$

12.3 $\quad |\mathbf{r}| = \sqrt{a^2 + b^2 + c^2}$

Example 12 *mechanics*

Determine the force, **F**, which has a magnitude of 28 kN in the direction of the vector $\overrightarrow{AB}$ where

$$A = (1, 3, 5) \text{ and } B = (7, 1, 2)$$

Solution

We first obtain the unit vector in the direction of $\overrightarrow{AB}$ and then multiply the result by 28 kN because the unit vector gives the direction of $\overrightarrow{AB}$ and 28 kN gives the magnitude. **How do we find the unit vector?**

$A = (1, 3, 5)$ and $B = (7, 1, 2)$ are the same co-ordinates as those of **Example 11**, so the unit vector $= \dfrac{1}{7} (6\mathbf{i} - 2\mathbf{j} - 3\mathbf{k})$. Thus

$$\mathbf{F} = 28 \times \frac{1}{7} (6\mathbf{i} - 2\mathbf{j} - 3\mathbf{k})$$

$$= 4 (6\mathbf{i} - 2\mathbf{j} - 3\mathbf{k})$$

$$= (24\mathbf{i} - 8\mathbf{j} - 12\mathbf{k}) \text{ kN}$$

SUMMARY

Unit vectors in the direction of the x, y and z axes are denoted by **i**, **j** and **k** respectively. The magnitude, $|\mathbf{r}|$, of a vector $\mathbf{r} = a\mathbf{i} + b\mathbf{j} + c\mathbf{k}$ is

12.3 $\qquad |\mathbf{r}| = \sqrt{a^2 + b^2 + c^2}$

The unit vector, **u**, in the direction of the non-zero vector $\mathbf{r} = a\mathbf{i} + b\mathbf{j} + c\mathbf{k}$ is given by

12.5 $\qquad \mathbf{u} = \dfrac{1}{\sqrt{a^2 + b^2 + c^2}} (a\mathbf{i} + b\mathbf{j} + c\mathbf{k})$

Exercise **12(c)**

Solutions at end of book. Complete solutions available at www.palgrave.com/science/engineering/singh

🚗 All questions belong to the field of [*mechanics*].

1 Forces F_1, F_2, F_3 and F_4 act on a particle so that it is in equilibrium, that is $F_1 + F_2 + F_3 + F_4 = 0$. Determine F_4 for each of the following:

a $F_1 = 2\mathbf{i} - \mathbf{j} - 3\mathbf{k}$, $F_2 = 7\mathbf{i} - 7\mathbf{j} - 7\mathbf{k}$, $F_3 = 9\mathbf{i} - 6\mathbf{j} - \mathbf{k}$

b $F_1 = 61\mathbf{i} - 64\mathbf{j} - 89\mathbf{k}$, $F_2 = 93\mathbf{i} - 98\mathbf{j}$, $F_3 = 22\mathbf{i} - 41\mathbf{j} + 43\mathbf{k}$

2 Forces F_1, F_2, F_3 and F_4 all act through the same point in a body such that the body is in equilibrium. Obtain the force F_1 given that $F_2 = 3\mathbf{i} + \mathbf{j} - \mathbf{k}$, $F_3 = -2\mathbf{i} - 3\mathbf{j} - 10\mathbf{k}$ and $F_4 = -\mathbf{i} - 5\mathbf{j} + 3\mathbf{k}$.

3 Forces

$F_1 = 2\mathbf{i} - \mathbf{j} + 10\mathbf{k}$, $\quad F_2 = 12\mathbf{i} - 3\mathbf{j} + 22\mathbf{k}$, $F_3 = 7\mathbf{i} + \mathbf{j} - 7\mathbf{k}$, $\quad F_4 = -9\mathbf{i} - 4\mathbf{j} + 5\mathbf{k}$

are applied on a particle. Determine $F = F_1 + F_2 + F_3 + F_4$ in terms of **i**, **j** and **k**.

Solutions at end of book. Complete solutions available at www.palgrave.com/science/engineering/singh

Exercise **12(c) continued**

4 Determine the resultant, $F_1 + F_2 + F_3 + F_4$, of the following forces:

$F_1 = i - j + 3k$

F_2 has a magnitude of 26 N in the direction of $12i + 5k$

F_3 has a magnitude of 18 N in the direction of $2i - 2j + k$

F_4 has a magnitude of 12 N in the direction of $4i - 4j - 2k$

5 A force **F** of magnitude 36 kN is applied in the direction of the vector $\overrightarrow{AB}$ where $A = (7, 5, 14)$ and $B = (-1, 1, 15)$. Find the force **F**.

6 Find the force **F** which has a magnitude of 11.3 kN in the direction of the vector $\overrightarrow{AB}$ where $A = (1.1, -2.6, -4.3)$ and $B = (2.2, 5.6, 7.1)$.

7 Let **A**, **B** and **C** have co-ordinates $(-16, 30, 19)$, $(-12, -17, -23)$ and $(-1, 27, -31)$ respectively. Let F_{AB} be the force applied in the direction of $\overrightarrow{AB}$ and F_{AC} be the force applied in the direction of $\overrightarrow{AC}$. Given that the magnitude of F_{AB} is 850 kN and the magnitude of F_{AC} is 700 kN, determine

$$F = F_{AB} + F_{AC}$$

Also find the magnitude of **F**.

SECTION D **Scalar products**

By the end of this section you will be able to:
▶ find scalar products
▶ examine engineering applications of a scalar product

D1 **Scalar product**

Figure 51 shows a constant force **F** applied to an object and as a result the object moves from A to B. The work done by the force is the product of the magnitude of the force in the line of action, $|F|\cos(\theta)$, and the distance the object has moved, $|\overrightarrow{AB}|$. Hence

Fig. 51

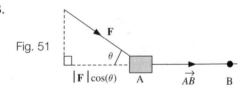

$$\text{Work done} = |F|\cos(\theta)|\overrightarrow{AB}|$$

This $|F|\cos(\theta)|\overrightarrow{AB}|$ is the scalar product of the vectors **F** and $\overrightarrow{AB}$.

In general if a constant force, **F**, moves a particle from A to B through a displacement vector **r**, then the work done by the force **F** is given by

$$\text{Work done} = F \cdot r = |F||r|\cos(\theta)$$

F·r is called the scalar product of **F** and **r**. We will return to this application of scalar product later in this section. Let's consider the scalar product of any two vectors, **p** and **q**.

Fig. 52

Figure 52 shows two vectors **p** and **q** with an angle θ between them, measured from **p** to **q**. The scalar product, denoted by **p·q**, is defined by

| 12.6 | $\mathbf{p} \cdot \mathbf{q} = |\mathbf{p}||\mathbf{q}|\cos(\theta)$ |

where $|\mathbf{p}|$ is the magnitude of **p** and $|\mathbf{q}|$ is the magnitude of **q**.

The dot in **p·q** represents the scalar product of **p** and **q**. Note that **p·q** is a scalar quantity, that is a vector times a vector gives a scalar.

The scalar product has the following properties:

| 12.7 | $\mathbf{p} \cdot \mathbf{q} = \mathbf{q} \cdot \mathbf{p}$ |

| 12.8 | $(\mathbf{p} + \mathbf{q}) \cdot \mathbf{r} = \mathbf{p} \cdot \mathbf{r} + \mathbf{q} \cdot \mathbf{r}$ |

| 12.9 | $\mathbf{p} \cdot (\mathbf{q} + \mathbf{r}) = \mathbf{p} \cdot \mathbf{q} + \mathbf{p} \cdot \mathbf{r}$ |

| 12.10 | $k(\mathbf{p} \cdot \mathbf{q}) = k\mathbf{p} \cdot \mathbf{q} = \mathbf{p} \cdot k\mathbf{q}$ [k is a constant] |

We will apply these properties in evaluating a scalar product of two vectors in component form.

Example 13

Determine the following scalar products:

$$\mathbf{i} \cdot \mathbf{i}, \ \ \mathbf{i} \cdot \mathbf{j}, \ \ \mathbf{j} \cdot \mathbf{j}, \ \ \mathbf{j} \cdot \mathbf{k}, \ \ \mathbf{k} \cdot \mathbf{k} \text{ and } \mathbf{k} \cdot \mathbf{i}$$

where **i**, **j** and **k** are unit vectors in the x, y and z directions respectively.

Solution

Since **i**, **j** and **k** are unit vectors, they all have a magnitude of 1. Furthermore **i**, **j** and **k** are at an angle of 90° to each other (Fig. 53).

We use $\mathbf{p} \cdot \mathbf{q} = |\mathbf{p}||\mathbf{q}| \cos(\theta)$.

The angle between **i** and **i** is 0° therefore

$$\mathbf{i} \cdot \mathbf{i} = |\mathbf{i}||\mathbf{i}| \underbrace{\cos(0°)}_{=1} = 1 \times 1 \times 1 = 1$$

Similarly $\mathbf{j} \cdot \mathbf{j} = |\mathbf{j}||\mathbf{j}|\cos(0°) = 1$ and $\mathbf{k} \cdot \mathbf{k} = |\mathbf{k}||\mathbf{k}|\cos(0°) = 1$.

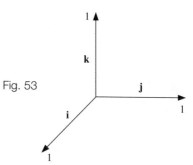

Fig. 53

Also

$$\mathbf{i} \cdot \mathbf{j} = |\mathbf{i}||\mathbf{j}| \underbrace{\cos(90°)}_{=0} = 1 \times 1 \times 0 = 0$$

$$\mathbf{j} \cdot \mathbf{k} = |\mathbf{j}||\mathbf{k}|\cos(90°) = 0$$

$$\mathbf{k} \cdot \mathbf{i} = |\mathbf{k}||\mathbf{i}|\cos(90°) = 0$$

We have

12.11 $\mathbf{i} \cdot \mathbf{i} = 1, \mathbf{j} \cdot \mathbf{j} = 1$ and $\mathbf{k} \cdot \mathbf{k} = 1$

12.12 $\mathbf{i} \cdot \mathbf{j} = 0, \mathbf{j} \cdot \mathbf{k} = 0$ and $\mathbf{k} \cdot \mathbf{i} = 0$

Notice that the scalar product of two different unit vectors which are at right angles to each other is 0, and if the unit vectors are the same then the scalar product is 1.

Example 14

Let $\mathbf{p}$ and $\mathbf{q}$ be vectors such that

$$\mathbf{p} = a\mathbf{i} + b\mathbf{j} + c\mathbf{k}$$
$$\mathbf{q} = d\mathbf{i} + e\mathbf{j} + f\mathbf{k}$$

Show that the scalar product of $\mathbf{p}$ and $\mathbf{q}$ is given by

$$\mathbf{p} \cdot \mathbf{q} = ad + be + cf$$

Solution

Expanding as algebraic brackets gives

$$\mathbf{p} \cdot \mathbf{q} = (a\mathbf{i} + b\mathbf{j} + c\mathbf{k}) \cdot (d\mathbf{i} + e\mathbf{j} + f\mathbf{k})$$
$$= (ad)\mathbf{i} \cdot \mathbf{i} + (ae)\mathbf{i} \cdot \mathbf{j} + (af)\mathbf{i} \cdot \mathbf{k} + (bd)\mathbf{j} \cdot \mathbf{i} + (be)\mathbf{j} \cdot \mathbf{j} + (bf)\mathbf{j} \cdot \mathbf{k} + (cd)\mathbf{k} \cdot \mathbf{i}$$
$$+ (ce)\mathbf{k} \cdot \mathbf{j} + (cf)\mathbf{k} \cdot \mathbf{k}$$

Since the scalar product of two different unit vectors which are at right angles to each other is 0, the only terms that do not vanish in the above are the ones containing $\mathbf{i} \cdot \mathbf{i}, \ \mathbf{j} \cdot \mathbf{j}$ and $\mathbf{k} \cdot \mathbf{k}$, and all these are equal to 1. Thus

$$\mathbf{p} \cdot \mathbf{q} = ad + be + cf$$

Example 14 shows that the scalar product of two vectors is defined as

12.13 $(a\mathbf{i} + b\mathbf{j} + c\mathbf{k}) \cdot (d\mathbf{i} + e\mathbf{j} + f\mathbf{k}) = ad + be + cf$

To find the scalar product of two vectors you multiply the corresponding coefficients in the $\mathbf{i}, \mathbf{j}$ and $\mathbf{k}$ directions and then add them together.

Example 15

Determine $(2\mathbf{i} - 3\mathbf{j} + \mathbf{k}) \cdot (-5\mathbf{i} + \mathbf{j} - 4\mathbf{k})$

Solution

Applying

$$(a\mathbf{i} + b\mathbf{j} + c\mathbf{k}) \cdot (d\mathbf{i} + e\mathbf{j} + f\mathbf{k}) = ad + be + cf$$

gives

$$(2\mathbf{i} - 3\mathbf{j} + \mathbf{k}) \cdot (-5\mathbf{i} + \mathbf{j} - 4\mathbf{k}) = [2 \times (-5)] + [(-3) \times 1] + [1 \times (-4)] = -17$$

Notice that the scalar product of two vectors gives a scalar, -17.

The scalar product can also be implemented on a calculator. (See the handbook of your calculator.)

We can use the scalar product formula

12.6 $\mathbf{p} \cdot \mathbf{q} = |\mathbf{p}||\mathbf{q}|\cos(\theta)$

to find the angle θ between vectors $\mathbf{p}$ and $\mathbf{q}$. Rearranging 12.6 gives

12.14 $\cos(\theta) = \dfrac{\mathbf{p} \cdot \mathbf{q}}{|\mathbf{p}||\mathbf{q}|}$

where $\mathbf{p}$ and $\mathbf{q}$ are not the zero vectors.

D2 Applications of scalar product

 Example 16 *mechanics*

The velocity, $\mathbf{v}$, and acceleration, $\mathbf{a}$, of a particle are given by

$$\mathbf{v} = 18\mathbf{i} - 6\mathbf{j} + \mathbf{k}$$
$$\mathbf{a} = 8\mathbf{i} + 4\mathbf{j} + \mathbf{k}$$

Determine the angle between the velocity, $\mathbf{v}$, and acceleration, $\mathbf{a}$.

Solution

We have

† $\cos(\theta) = \dfrac{\mathbf{v} \cdot \mathbf{a}}{|\mathbf{v}||\mathbf{a}|}$

We use

12.3 $|a\mathbf{i} + b\mathbf{j} + c\mathbf{k}| = \sqrt{a^2 + b^2 + c^2}$

to find the magnitude of the velocity and acceleration:

$$|\mathbf{v}| = |18\mathbf{i} - 6\mathbf{j} + \mathbf{k}| = \sqrt{18^2 + (-6)^2 + 1^2} = 19$$

$$|\mathbf{a}| = |8\mathbf{i} + 4\mathbf{j} + \mathbf{k}| = \sqrt{8^2 + 4^2 + 1^2} = 9$$

$$\mathbf{v} \cdot \mathbf{a} = (18\mathbf{i} - 6\mathbf{j} + \mathbf{k}) \cdot (8\mathbf{i} + 4\mathbf{j} + \mathbf{k}) \quad \text{[Scalar product]}$$

$$= \underbrace{(18 \times 8) + [(-6) \times 4] + [1 \times 1]}_{\text{by } 12.13} = 121$$

Substituting $\mathbf{v} \cdot \mathbf{a} = 121$, $|\mathbf{v}| = 19$ and $|\mathbf{a}| = 9$ into † gives

$$\cos(\theta) = \frac{121}{(19 \times 9)}$$

Taking inverse cos gives the angle

$$\theta = \cos^{-1}\left(\frac{121}{(19 \times 9)}\right) = 44.96°$$

The angle between velocity and acceleration is 44.96°.

12.13 $(a\mathbf{i} + b\mathbf{j} + c\mathbf{k}) \cdot (d\mathbf{i} + e\mathbf{j} + f\mathbf{k}) = ad + be + cf$

As seen earlier in this section, the scalar product also gives the work done by a force **F**. If a constant force **F** moves a particle through a displacement vector **r**, then the work done by the force **F** is given by

12.15   Work done = **F**·**r**

where **F**·**r** is the scalar product of **F** and **r**.

Example 17 *mechanics*

Determine the work done by a force, **F**, in moving an object through a displacement **r** in each of the following:

a F = 3i − j − k, r = i − j − k
b F = 6i + 2j − 6k, r = 9i + 6j + k
c F = 37i + 35j − k, r = 61i + 64j + 74k

Solution

We need to use

12.15   Work done = **F**·**r**

to find the work done for each of **a**, **b** and **c**.

a (3i − j − k)·(i − j − k) = (3 × 1) + (−1 × (−1)) + (−1 × (−1))
$$= 5\,\text{J (J = joule is the SI unit of work)}$$

b (6i + 2j − 6k)·(9i + 6j + k) = (6 × 9) + (2 × 6) + (−6 × 1)
$$= 60\,\text{J}$$

c (37i + 35j − k)·(61i + 64j + 74k) = (37 × 61) + (35 × 64) + (−1 × 74)
$$= 4423\,\text{J}$$

D3 Differentiation of vectors

Remember $\dot{x}$ means the time derivative, $\dfrac{dx}{dt}$, that is differentiate x with respect to t.

Suppose a particle moves in two dimensions, then its position can be defined by the x, y co-ordinates which in turn are functions of time, t. The position vector, **r**, of a particle can be specified by

$$\mathbf{r} = x\mathbf{i} + y\mathbf{j}$$

where x and y are functions of time, t. The velocity, **v**, and acceleration, **a**, of a particle can be found by differentiating **r** with respect to t, that is

$$\mathbf{v} = \dot{\mathbf{r}} = \dot{x}\mathbf{i} + \dot{y}\mathbf{j} \quad \text{[First derivative]}$$

$$\mathbf{a} = \ddot{\mathbf{r}} = \ddot{x}\mathbf{i} + \ddot{y}\mathbf{j} \quad \text{[Second derivative]}$$

Remember $\dot{\mathbf{r}} = \dfrac{d\mathbf{r}}{dt}$, $\dot{x} = \dfrac{dx}{dt}$, $\dot{y} = \dfrac{dy}{dt}$, $\ddot{x} = \dfrac{d^2x}{dt^2}$ and $\ddot{y} = \dfrac{d^2y}{dt^2}$

Example 18 *mechanics*

An object is dropped and its position vector, **r**, is given by

$$\mathbf{r} = (t^4 - 3t^3)\mathbf{i} + (6t^2 - 2t)\mathbf{j}$$

Find the velocity, **v**, and acceleration, **a**, and the angle between **v** and **a** at $t = 1$.

Solution

Differentiating **r** with respect to t:

$$\mathbf{r} = (t^4 - 3t^3)\mathbf{i} + (6t^2 - 2t)\mathbf{j}$$

$$\mathbf{v} = (4t^3 - 9t^2)\mathbf{i} + (12t - 2)\mathbf{j} \quad \text{[Differentiating]}$$

Differentiating again:

$$\mathbf{a} = (12t^2 - 18t)\mathbf{i} + 12\mathbf{j}$$

Substituting $t = 1$ into **v** and **a** yields

$$\mathbf{v} = (4 - 9)\mathbf{i} + (12 - 2)\mathbf{j}$$

$$= -5\mathbf{i} + 10\mathbf{j}$$

$$\mathbf{a} = (12 - 18)\mathbf{i} + 12\mathbf{j}$$

$$= -6\mathbf{i} + 12\mathbf{j}$$

To find the angle between the velocity and acceleration we need to use

 $$\cos(\theta) = \frac{\mathbf{a} \cdot \mathbf{v}}{|\mathbf{a}||\mathbf{v}|}$$

We have

$$|\mathbf{v}| = |-5\mathbf{i} + 10\mathbf{j}| = \sqrt{(-5)^2 + 10^2} = \sqrt{125}$$

$$|\mathbf{a}| = |-6\mathbf{i} + 12\mathbf{j}| = \sqrt{(-6)^2 + 12^2} = \sqrt{180}$$

$$\mathbf{a} \cdot \mathbf{v} = (-6\mathbf{i} + 12\mathbf{j}) \cdot (-5\mathbf{i} + 10\mathbf{j})$$

$$= [-6 \times (-5)] + (12 \times 10)$$

$$= 150$$

Substituting these into * gives

$$\cos(\theta) = \frac{150}{\sqrt{180}\,\sqrt{125}} = 1$$

Taking inverse cos yields

$$\theta = \cos^{-1}(1) = 0°$$

The angle between velocity and acceleration is 0°.

SUMMARY

Let $\mathbf{p}$ and $\mathbf{q}$ be vectors with angle θ between them. Then the scalar product $\mathbf{p} \cdot \mathbf{q}$ is given by

12.6 $\mathbf{p} \cdot \mathbf{q} = |\mathbf{p}||\mathbf{q}| \cos(\theta)$

Also

12.13 $(a\mathbf{i} + b\mathbf{j} + c\mathbf{k}) \cdot (d\mathbf{i} + e\mathbf{j} + f\mathbf{k}) = ad + be + cf$

The scalar product is applied in mechanics to find the work done:

12.15   Work done $= \mathbf{F} \cdot \mathbf{r}$

Exercise **12(d)**

Solutions at end of book. Complete solutions available at www.palgrave.com/science/engineering/singh

1 Determine the scalar products of the following:

  a $(3\mathbf{i} + \mathbf{j} + \mathbf{k}) \cdot (5\mathbf{i} + \mathbf{j} + 7\mathbf{k})$

  b $(-\mathbf{i} + \mathbf{j} - \mathbf{k}) \cdot (3\mathbf{i} + 2\mathbf{j} - \mathbf{k})$

  c $(10\mathbf{i} - 15\mathbf{j} + \mathbf{k}) \cdot (-21\mathbf{i} - \mathbf{k})$

2 🚗 [*mechanics*]   A force $\mathbf{F} = 100\mathbf{i} + 220\mathbf{j} - 250\mathbf{k}$ moves an object through a displacement $\mathbf{s} = 150\mathbf{i} + 200\mathbf{j}$. Find the work done by the force $\mathbf{F}$.

3 🚗 [*mechanics*]   A force $\mathbf{F} = 5\mathbf{i} + 3\mathbf{j} - 2\mathbf{k}$ is applied to an object which moves from $A = (1, 1, 1)$ to $B = (5, -1, 2)$. Determine the work done by the force $\mathbf{F}$.

4 🚗 [*mechanics*]   A force of magnitude 20 N is applied to an object in the direction $3\mathbf{i} + \mathbf{j} + \mathbf{k}$. Determine the work done by the force in moving the object from $A = (3, -1, 5)$ to $B = (-1, 7, 10)$.

5 Let $\mathbf{r} = x\mathbf{i} + y\mathbf{j} + z\mathbf{k}$ be a vector. Determine

  i $\mathbf{r} \cdot \mathbf{r}$ **ii** $|\mathbf{r}|^2$

where $|\mathbf{r}|$ is the magnitude of $\mathbf{r}$.

What do you notice about your results?

6 🚗 [*mechanics*]   The velocity, $\mathbf{v}$, and acceleration, $\mathbf{a}$, of a particle are given by

$$\mathbf{v} = 2\mathbf{i} + 3\mathbf{j}, \quad \mathbf{a} = 2\mathbf{i} + 2\mathbf{j}$$

Find the angle between these vectors.

7 🚗 [*mechanics*]   The position vector, $\mathbf{r}$, of a particle moving in curvilinear motion is given by

$$\mathbf{r} = (t^3 - t^2)\mathbf{i} + t^4\mathbf{j}$$

  i Find expressions for $\mathbf{v} = \dot{\mathbf{r}}$ and $\mathbf{a} = \ddot{\mathbf{r}}$.
  ii Determine the angle between $\mathbf{v}$ and $\mathbf{a}$ for $t = 1$.
  ($\mathbf{v}$ is velocity and $\mathbf{a}$ is acceleration).

SECTION E **Vector products**

By the end of this section you will be able to:
► find vector products
► examine engineering applications of vector products

E1 **Vector product**

The vector product of vectors **p** and **q**, denoted by **p** × **q**, is defined as the vector of size

$$|\mathbf{p}||\mathbf{q}|\sin(\theta)$$

in the direction 90° to the plane containing the vectors **p** and **q**. We have

| 12.16 | $\mathbf{p} \times \mathbf{q} = |\mathbf{p}||\mathbf{q}|\sin(\theta)\mathbf{u}$ |

where θ is the angle between the vectors **p** and **q**, and **u** is the unit vector in the direction of **p** × **q** shown in Fig. 54. The vector product **p** × **q** is a

Fig. 54

vector quantity, that is a vector times a vector gives a vector. The direction of the vector **p** × **q** is determined by the direction of a right-handed screw turned from **p** to **q** as shown in Fig. 55.

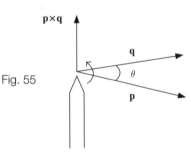

Fig. 55

Note that the vector **q** × **p** is in the direction of a right-handed screw turned from **q** to **p** (Fig. 56).

? **What do you notice about p × q and q × p?**

$$\mathbf{p} \times \mathbf{q} \neq \mathbf{q} \times \mathbf{p} \quad \text{[Not equal]}$$

In fact

| 12.17 | $\mathbf{p} \times \mathbf{q} = -(\mathbf{q} \times \mathbf{p})$ |

The vector product also has the following properties:

| 12.18 | $\mathbf{p} \times (\mathbf{q} + \mathbf{r}) = (\mathbf{p} \times \mathbf{q}) + (\mathbf{p} \times \mathbf{r})$ |

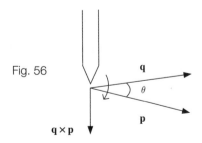

Fig. 56

| 12.19 | $k(\mathbf{p} \times \mathbf{q}) = (k\mathbf{p}) \times \mathbf{q} = \mathbf{p} \times (k\mathbf{q})$ |
| | [k is a constant] |

| 12.20 | If $\mathbf{p} = 0$ then $\mathbf{p} \times \mathbf{q} = 0$ |

Example 19

Determine the following vector products:

$$\mathbf{i} \times \mathbf{i}, \ \mathbf{i} \times \mathbf{j}, \ \mathbf{k} \times \mathbf{i}, \ \mathbf{j} \times \mathbf{j}, \ \mathbf{j} \times \mathbf{k} \text{ and } \mathbf{k} \times \mathbf{k}$$

where **i**, **j** and **k** are unit vectors in the directions of the *x*, *y* and *z* axes respectively.

Solution

Plot the unit vectors, as shown in Fig. 57.

Clearly i × i = 0. Why?

Because the angle between **i** and **i** is 0° and $\sin(0°) = 0$.
Thus

Fig. 57

$$\mathbf{i} \times \mathbf{i} = |\mathbf{i}||\mathbf{i}|\sin(0°)\mathbf{j} = \mathbf{0}$$

Similarly $\mathbf{j} \times \mathbf{j} = \mathbf{0}$ and $\mathbf{k} \times \mathbf{k} = \mathbf{0}$.

The unit vectors **i, j** and **k** are all 90° to each other.
We have

$$\mathbf{i} \times \mathbf{j} = \underbrace{|\mathbf{i}||\mathbf{j}|}_{=1\ =1} \underbrace{[\sin(90°)]}_{=1} \mathbf{k}$$

because **k** is 90° to the plane
containing the vectors i and j

$$= (1)(1)(1)\mathbf{k} = \mathbf{k}$$

Likewise

$$\mathbf{k} \times \mathbf{i} = \underbrace{|\mathbf{k}||\mathbf{i}|}_{=1\ =1} \underbrace{[\sin(90°)]}_{=1}\mathbf{j} = \mathbf{j}$$

By the same token:

$$\mathbf{j} \times \mathbf{k} = \underbrace{|\mathbf{j}||\mathbf{k}|}_{=1\ =1} \underbrace{[\sin(90°)]}_{=1}\mathbf{i} = \mathbf{i}$$

What are j × i, i × k and k × j equal to?

By

12.17 $\mathbf{p} \times \mathbf{q} = -(\mathbf{q} \times \mathbf{p})$

and **Example 19** we have $\mathbf{j} \times \mathbf{i} = -\mathbf{k}$, $\mathbf{i} \times \mathbf{k} = -\mathbf{j}$ and $\mathbf{k} \times \mathbf{j} = -\mathbf{i}$.

We have the following vector products of the unit vectors:

12.21 $\mathbf{i} \times \mathbf{i} = 0, \ \mathbf{j} \times \mathbf{j} = 0$ and $\mathbf{k} \times \mathbf{k} = 0$

12.22 $\mathbf{i} \times \mathbf{j} = \mathbf{k}, \ \mathbf{j} \times \mathbf{k} = \mathbf{i}$ and $\mathbf{k} \times \mathbf{i} = \mathbf{j}$

12.23 $\mathbf{j} \times \mathbf{i} = -\mathbf{k}, \ \mathbf{k} \times \mathbf{j} = -\mathbf{i}$ and $\mathbf{i} \times \mathbf{k} = -\mathbf{j}$

Compare these results with those given for scalar products in 12.11 and 12.12 .

Let **p** and **q** be vectors such that

$$\mathbf{p} = a\mathbf{i} + b\mathbf{j} + c\mathbf{k}$$
$$\mathbf{q} = d\mathbf{i} + e\mathbf{j} + f\mathbf{k}$$

Then the vector product of **p** and **q** can be shown to be

$$\boxed{\star} \qquad (a\mathbf{i} + b\mathbf{j} + c\mathbf{k}) \times (d\mathbf{i} + e\mathbf{j} + f\mathbf{k}) = (bf - ce)\mathbf{i} + (cd - af)\mathbf{j} + (ae - bd)\mathbf{k}$$

Try verifying this result by expanding brackets and using the preceding results for **i**, **j** and **k**.

Rather than applying this formula to find the vector product we can use a determinant as a mnemonic device for evaluating the vector product:

12.24 $\qquad \mathbf{p} \times \mathbf{q} = (a\mathbf{i} + b\mathbf{j} + c\mathbf{k}) \times (d\mathbf{i} + e\mathbf{j} + f\mathbf{k}) = \det\begin{pmatrix} \mathbf{i} & \mathbf{j} & \mathbf{k} \\ a & b & c \\ d & e & f \end{pmatrix}$

This can be shown by expanding the determinant:

$$\det\begin{pmatrix} \mathbf{i} & \mathbf{j} & \mathbf{k} \\ a & b & c \\ d & e & f \end{pmatrix} = \mathbf{i}\left[\det\begin{pmatrix} b & c \\ e & f \end{pmatrix}\right] - \mathbf{j}\left[\det\begin{pmatrix} a & c \\ d & f \end{pmatrix}\right] + \mathbf{k}\left[\det\begin{pmatrix} a & b \\ d & e \end{pmatrix}\right]$$

$$= \mathbf{i}(bf - ec) - \mathbf{j}(af - dc) + \mathbf{k}(ae - db) \quad [\text{By } \boxed{11.1}]$$

$$= (bf - ec)\mathbf{i} + (dc - af)\mathbf{j} + (ae - db)\mathbf{k}$$

$$= (a\mathbf{i} + b\mathbf{j} + c\mathbf{k}) \times (d\mathbf{i} + e\mathbf{j} + f\mathbf{k}) \quad [\text{By } \boxed{\star}]$$

Hence we evaluate this determinant to find the vector product.

Example 20

Obtain the vector product of

$$\mathbf{r} = 5\mathbf{i} + 7\mathbf{j} + 6\mathbf{k}$$
$$\mathbf{s} = 3\mathbf{i} - \mathbf{j} + \mathbf{k}$$

Solution

Using the determinant in 12.24 to find $\mathbf{r} \times \mathbf{s}$ gives

$$(5\mathbf{i} + 7\mathbf{j} + 6\mathbf{k}) \times (3\mathbf{i} - \mathbf{j} + \mathbf{k}) = \det\begin{pmatrix} \mathbf{i} & \mathbf{j} & \mathbf{k} \\ 5 & 7 & 6 \\ 3 & -1 & 1 \end{pmatrix}$$

$$= \mathbf{i}\left[\det\begin{pmatrix} 7 & 6 \\ -1 & 1 \end{pmatrix}\right] - \mathbf{j}\left[\det\begin{pmatrix} 5 & 6 \\ 3 & 1 \end{pmatrix}\right] + \mathbf{k}\left[\det\begin{pmatrix} 5 & 7 \\ 3 & -1 \end{pmatrix}\right]$$

$$= \mathbf{i}(7 + 6) - \mathbf{j}[(5 \times 1) - (3 \times 6)] + \mathbf{k}[(5 \times (-1)) - (3 \times 7)]$$
by 11.1

$$= 13\mathbf{i} + 13\mathbf{j} - 26\mathbf{k}$$

11.1 $\quad \det\begin{pmatrix} a & b \\ c & d \end{pmatrix} = ad - cb$

We can also calculate the vector product of **r** and **s** of **Example 20** on a calculator.
Of course $\mathbf{s} \times \mathbf{r} = -13\mathbf{i} - 13\mathbf{j} + 26\mathbf{k}$ (that is $-(\mathbf{r} \times \mathbf{s})$).

E2 Applications of vector product

Next we look at an application of vector products in mechanics.

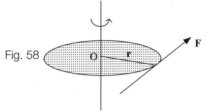

A force, **F**, is applied to an object, making it rotate around a fixed axis as shown in Fig. 58. The moment vector, **M**, of a force, **F**, about the origin O is defined as

Fig. 58

| 12.25 | $\mathbf{M} = \mathbf{r} \times \mathbf{F}$

where **r** is the position vector from O of any point on the line of action of **F**.

Note that the moment, **M**, is a vector. (The moment vector is called the torque in many books on mechanics.)

? **What is the direction of the vector M?**

It is 90° to the plane
containing the vectors **r**
and **F** (Fig. 59).

Fig. 59

Example 21 *mechanics*

Determine the moment vector, **M**, about the origin, O, of a force $\mathbf{F} = 10\mathbf{i} - \mathbf{j} + 2\mathbf{k}$, passing through the point with position vector $\mathbf{r} = 2\mathbf{i} - 3\mathbf{j} + 5\mathbf{k}$.

Solution

By 12.25 we have

$$\mathbf{M} = \mathbf{r} \times \mathbf{F}$$
$$= (2\mathbf{i} - 3\mathbf{j} + 5\mathbf{k}) \times (10\mathbf{i} - \mathbf{j} + 2\mathbf{k})$$
$$\underset{\text{by } 12.24}{=} \det \begin{pmatrix} \mathbf{i} & \mathbf{j} & \mathbf{k} \\ 2 & -3 & 5 \\ 10 & -1 & 2 \end{pmatrix}$$

12.24 $(a\mathbf{i} + b\mathbf{j} + c\mathbf{k}) \times (d\mathbf{i} + e\mathbf{j} + f\mathbf{k}) = \det \begin{pmatrix} \mathbf{i} & \mathbf{j} & \mathbf{k} \\ a & b & c \\ d & e & f \end{pmatrix}$

Example 21 *continued*

$$= \mathbf{i}\left[\det\begin{pmatrix} -3 & 5 \\ -1 & 2 \end{pmatrix}\right] - \mathbf{j}\left[\det\begin{pmatrix} 2 & 5 \\ 10 & 2 \end{pmatrix}\right] + \mathbf{k}\left[\det\begin{pmatrix} 2 & -3 \\ 10 & -1 \end{pmatrix}\right]$$

$$= -\mathbf{i} + 46\mathbf{j} + 28\mathbf{k} \quad \text{[Evaluating determinant]}$$

The next example is more complicated than the previous examples. The complication is to determine vector moments about other points rather than just the origin.

Example 22 *mechanics*

Let A and B have position vectors
$$\mathbf{a} = 2\mathbf{i} - 3\mathbf{j} + \mathbf{k}$$
$$\mathbf{b} = 7\mathbf{i} - 4\mathbf{j} - \mathbf{k}$$
respectively. The force $\mathbf{F} = 3\mathbf{i} - \mathbf{j} + 5\mathbf{k}$ passes through the point A. Determine the vector moment of $\mathbf{F}$ about the point B.

Solution

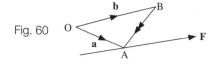

Fig. 60

Draw the vectors as in Fig. 60.
Find the resultant vector first, $\vec{BA}$:

$$\vec{BA} = \vec{OA} - \vec{OB}$$
$$= (2\mathbf{i} - 3\mathbf{j} + \mathbf{k}) - (7\mathbf{i} - 4\mathbf{j} - \mathbf{k}) = -5\mathbf{i} + \mathbf{j} + 2\mathbf{k}$$

Let $\mathbf{M}$ be the moment of $\mathbf{F}$, then by using ⌈12.25⌉ and substituting $\vec{BA} = -5\mathbf{i} + \mathbf{j} + 2\mathbf{k}$, $\mathbf{F} = 3\mathbf{i} - \mathbf{j} + 5\mathbf{k}$ we have

$$\mathbf{M} = \vec{BA} \times \mathbf{F}$$

$$= (-5\mathbf{i} + \mathbf{j} + 2\mathbf{k}) \times (3\mathbf{i} - \mathbf{j} + 5\mathbf{k})$$

$$\underset{\text{by }\boxed{12.24}}{= \det}\begin{pmatrix} \mathbf{i} & \mathbf{j} & \mathbf{k} \\ -5 & 1 & 2 \\ 3 & -1 & 5 \end{pmatrix}$$

$$= \mathbf{i}\left[\det\begin{pmatrix} 1 & 2 \\ -1 & 5 \end{pmatrix}\right] - \mathbf{j}\left[\det\begin{pmatrix} -5 & 2 \\ 3 & 5 \end{pmatrix}\right] + \mathbf{k}\left[\det\begin{pmatrix} -5 & 1 \\ 3 & -1 \end{pmatrix}\right]$$

$$= \mathbf{i}(5 + 2) - \mathbf{j}(-25 - 6) + \mathbf{k}(5 - 3)$$

$$= 7\mathbf{i} + 31\mathbf{j} + 2\mathbf{k}$$

⌈12.24⌉ $(a\mathbf{i} + b\mathbf{j} + c\mathbf{k}) \times (d\mathbf{i} + e\mathbf{j} + f\mathbf{k}) = \det\begin{pmatrix} \mathbf{i} & \mathbf{j} & \mathbf{k} \\ a & b & c \\ d & e & f \end{pmatrix}$

SUMMARY

The vector product of **p** and **q** is defined as

12.16 $\mathbf{p} \times \mathbf{q} = |\mathbf{p}||\mathbf{q}| \sin(\theta)\, \mathbf{u}$

where **u** is the unit vector in the direction of $\mathbf{p} \times \mathbf{q}$. Also

12.24 $(a\mathbf{i} + b\mathbf{j} + c\mathbf{k}) \times (d\mathbf{i} + e\mathbf{j} + f\mathbf{k}) = \det \begin{pmatrix} \mathbf{i} & \mathbf{j} & \mathbf{k} \\ a & b & c \\ d & e & f \end{pmatrix}$

The moment vector, **M**, about the origin is defined as

12.25 $\mathbf{M} = \mathbf{r} \times \mathbf{F}$

where **r** is the position vector from the origin of any point on the line of action of **F**.

Exercise 12(e)

Solutions at end of book. Complete solutions available at www.palgrave.com/science/engineering/singh

1 If $\mathbf{r} = 5\mathbf{i} + \mathbf{j} + 3\mathbf{k}$ and $\mathbf{s} = 2\mathbf{i} - 3\mathbf{j} - 4\mathbf{k}$, determine the vector products

  i $\mathbf{r} \times \mathbf{s}$ **ii** $\mathbf{s} \times \mathbf{r}$

2 🚗 [*mechanics*] Forces $\mathbf{F}_1$, $\mathbf{F}_2$ and $\mathbf{F}_3$ act at a point on a body with the corresponding position vectors:

$\mathbf{F}_1 = 2\mathbf{i} + \mathbf{j} - 3\mathbf{k},\ \mathbf{r}_1 = \mathbf{i} + 3\mathbf{j} + \mathbf{k}$
$\mathbf{F}_2 = 3\mathbf{i} - 4\mathbf{j} + 6\mathbf{k},\ \mathbf{r}_2 = \mathbf{i} - \mathbf{j} + 2\mathbf{k}$
$\mathbf{F}_3 = \mathbf{i} + 10\mathbf{j} + 7\mathbf{k},\ \mathbf{r}_3 = \mathbf{i} + 4\mathbf{j} + 2\mathbf{k}$

Show that the body is in equilibrium, that is

$(\mathbf{r}_1 \times \mathbf{F}_1) + (\mathbf{r}_2 \times \mathbf{F}_2) + (\mathbf{r}_3 \times \mathbf{F}_3) = 0$

3 🚗 [*mechanics*] Forces $\mathbf{F}_1$, $\mathbf{F}_2$, $\mathbf{F}_3$ and $\mathbf{F}_4$ act at a point on a body with the corresponding position vectors:

$\mathbf{F}_1 = 2\mathbf{i} - \mathbf{j} + \mathbf{k},\quad \mathbf{r}_1 = 3\mathbf{i} + \mathbf{j} + 5\mathbf{k}$
$\mathbf{F}_2 = 8\mathbf{i} - 6\mathbf{j} + 6\mathbf{k},\quad \mathbf{r}_2 = 5\mathbf{i} - 2\mathbf{j} - \mathbf{k}$
$\mathbf{F}_3 = 4\mathbf{i} + 3\mathbf{k},\qquad \mathbf{r}_3 = \mathbf{i} - 6\mathbf{j} + 7\mathbf{k}$
$\mathbf{F}_4 = \mathbf{i} - 5\mathbf{j},\qquad\ \ \mathbf{r}_4 = \mathbf{i} + 6\mathbf{k}$

Show that the body is in equilibrium, that is

$(\mathbf{r}_1 \times \mathbf{F}_1) + (\mathbf{r}_2 \times \mathbf{F}_2)$
$\qquad + (\mathbf{r}_3 \times \mathbf{F}_3) + (\mathbf{r}_4 \times \mathbf{F}_4) = 0$

4 🚗 [*mechanics*] Obtain the moment, M, about the point B of a force, F, passing through the point A for the following:

  a $\mathbf{F} = 2\mathbf{i} - \mathbf{j} + \mathbf{k}$, A is (1, 2, 3), B is (5, 7, 9)

  b $\mathbf{F} = 4\mathbf{i} - 3\mathbf{j} - 6\mathbf{k}$, A is (1, −1, 0), B is (3, −1, −1)

5 🚗 [*mechanics*] The points R and S have position vectors

$\mathbf{r} = \mathbf{i} - \mathbf{j} - \mathbf{k},\ \mathbf{s} = 7\mathbf{i} + \mathbf{j} - 9\mathbf{k}$

respectively. A force $\mathbf{F} = 14\mathbf{i} + 2\mathbf{j} - 18\mathbf{k}$ passes through R. Calculate the vector moment of F about the point S.

6 Let $\mathbf{r} = a\mathbf{i} + b\mathbf{j}$ and $\mathbf{s} = c\mathbf{i} + d\mathbf{j}$ be any vectors. Show that

$\mathbf{r} \times \mathbf{s} = (ad - cb)\mathbf{k}$

7 Let $\mathbf{r} = a\mathbf{i} + b\mathbf{j} + c\mathbf{k}$ and $\mathbf{s} = d\mathbf{i} + e\mathbf{j} + f\mathbf{k}$ be any vectors. Show that

$\mathbf{r} \times \mathbf{s} = -(\mathbf{s} \times \mathbf{r})$

Miscellaneous exercise **12**

Solutions at end of book. Complete solutions available at www.palgrave.com/science/engineering/singh

1 Find the exact unit vector in the direction of the vector $\mathbf{F} = 5\mathbf{i} + 7\mathbf{j} + 12\mathbf{k}$.

2 Let $\mathbf{a} = 2\mathbf{i} + 3\mathbf{j} - \mathbf{k}$, $\mathbf{b} = 7\mathbf{i} + 12\mathbf{k}$, $\mathbf{c} = -\mathbf{i} - \mathbf{j} - 6\mathbf{k}$ and $\mathbf{d} = 12\mathbf{i} - \mathbf{j} - 11\mathbf{k}$. Determine:

  i $\mathbf{a} + \mathbf{c}$ **ii** $\mathbf{a} + \mathbf{b} + \mathbf{c} + \mathbf{d}$ **iii** $\mathbf{a} \cdot \mathbf{b}$
  iv $\mathbf{b} \cdot \mathbf{a}$ **v** $\mathbf{a} \cdot (\mathbf{b} + \mathbf{c})$
  vi $(\mathbf{c} \cdot \mathbf{d})\hat{\mathbf{a}}$ where $\hat{\mathbf{a}}$ is the unit vector in the direction of $\mathbf{a}$
  vii the magnitude of $\mathbf{a} + \mathbf{b} + \mathbf{c} + \mathbf{d}$

🚗 **Questions 3 to 14 are in the field of [*mechanics*].**

3 Let the velocity, $\mathbf{v}$, and acceleration, $\mathbf{a}$, be given by

$$\mathbf{v} = -4\mathbf{i} - 8\mathbf{j} + 4\mathbf{k} \text{ and}$$

$$\mathbf{a} = 2\mathbf{i} + \mathbf{j} + 4\mathbf{k}$$

Find the angle between the velocity and acceleration.

4 The velocity, $\mathbf{v}$, and acceleration, $\mathbf{a}$, of a particle are given by

$$\mathbf{v} = 6\mathbf{i} + 2\mathbf{j}, \quad \mathbf{a} = 3\mathbf{i} + 7\mathbf{j}$$

Find the angle between $\mathbf{v}$ and $\mathbf{a}$.

5 Determine the resultant force, $\mathbf{R}$, of the forces shown in Fig. 61.

Fig. 61

6 Determine the resultant force, $\mathbf{R}$, of the forces shown in Fig. 62.

Fig. 62

7 Find the horizontal, $\mathbf{F}_x$, and vertical, $\mathbf{F}_y$, components of the force $\mathbf{F}$ shown in Fig. 63.

Fig. 63

8 Find the resultant force, $\mathbf{R} = \mathbf{F}_1 + \mathbf{F}_2 + \mathbf{F}_3 + \mathbf{F}_4$, in each of the following cases:

  a $\mathbf{F}_1 = 2\mathbf{i} - \mathbf{j} + 7\mathbf{k}$,
    $\mathbf{F}_2 = -3\mathbf{i} - 2\mathbf{j} + 11\mathbf{k}$,
    $\mathbf{F}_3 = 61\mathbf{i} + 64\mathbf{j} + 19\mathbf{k}$ and
    $\mathbf{F}_4 = 3\mathbf{i} + 8\mathbf{j} + 11\mathbf{k}$

  b $\mathbf{F}_1 = \mathbf{i} + \mathbf{j} + \mathbf{k}$,
    $\mathbf{F}_2 = 16\mathbf{i} + 99\mathbf{j} + 19\mathbf{k}$,
    $\mathbf{F}_3 = 47\mathbf{i} + 84\mathbf{j} + 88\mathbf{k}$ and
    $\mathbf{F}_4 = 22\mathbf{i} + 41\mathbf{j} + 3\mathbf{k}$

9 A force $\mathbf{F} = \mathbf{i} + \mathbf{j} + 3\mathbf{k}$ moves an object through a displacement $\mathbf{s} = 15\mathbf{i} + 20\mathbf{j}$. Find the work done by the force $\mathbf{F}$.

10 A force $\mathbf{F} = 2\mathbf{i} - 7\mathbf{j} + 12\mathbf{k}$ acts on a particle P. Calculate the work done by $\mathbf{F}$ as the particle P moves from the point $A = (2, 3, 7)$ to the point $B = (7, -11, 37)$.

11 Determine the work done by a force $\mathbf{F}$ on a particle P moving from point A to point B where

  a $\mathbf{F} = 3\mathbf{i} - \mathbf{j} - 2\mathbf{k}$,
    $A = (1, 7, 1)$, $B = (3, 4, -1)$

  b $\mathbf{F} = 2\mathbf{i} + \mathbf{j} - 7\mathbf{k}$,
    $A = (3, 4, 3)$, $B = (6, 7, -17)$

Miscellaneous exercise **12 continued**

Solutions at end of book. Complete solutions available at www.palgrave.com/science/engineering/singh

12 A force $\mathbf{F} = 5\mathbf{i} - \mathbf{j} - \mathbf{k}$ passes through the point A whose position vector is $\mathbf{a} = 3\mathbf{i} + \mathbf{j} + 8\mathbf{k}$. Determine the vector moment, $\mathbf{M}$, of $\mathbf{F}$ about the point B with position vector $\mathbf{b} = 2\mathbf{i} - 3\mathbf{j} + 7\mathbf{k}$. Also find the magnitude of $\mathbf{M}$.

13 Determine the force F which has a magnitude of 39 kN in the direction of the vector $\overrightarrow{AB}$ where A = (−1, 2, 3) and B = (5, 1, −2).

14 Forces $\mathbf{F}_1$, $\mathbf{F}_2$ and $\mathbf{F}_3$ act at a point in a body with the corresponding position vectors:

$$\mathbf{F}_1 = \mathbf{i} + 3\mathbf{j} - 6\mathbf{k}, \quad \mathbf{r}_1 = 2\mathbf{i} + 2\mathbf{j} + 2\mathbf{k}$$
$$\mathbf{F}_2 = -7\mathbf{i} + 2\mathbf{j} + 5\mathbf{k}, \; \mathbf{r}_2 = 3\mathbf{i} - 2\mathbf{j} - \mathbf{k}$$
$$\mathbf{F}_3 = 2\mathbf{j} + 3\mathbf{k}, \qquad \mathbf{r}_3 = 2\mathbf{i} + 2\mathbf{j} - 10\mathbf{k}$$

Show that the body is in equilibrium, that is

$$(\mathbf{r}_1 \times \mathbf{F}_1) + (\mathbf{r}_2 \times \mathbf{F}_2) + (\mathbf{r}_3 \times \mathbf{F}_3) = \mathbf{0}$$

15 Let $\mathbf{r} = a\mathbf{i} + b\mathbf{k}$ and $\mathbf{s} = c\mathbf{i} + d\mathbf{k}$ be any vectors. Determine the vector product $\mathbf{r} \times \mathbf{s}$.

16 Let $\mathbf{a} = 2\mathbf{i} - 3\mathbf{j} + x\mathbf{k}$ and $\mathbf{b} = -3\mathbf{i} + \mathbf{j} + x\mathbf{k}$ be vectors. Find the value(s) of x for which the vectors $\mathbf{a}$ and $\mathbf{b}$ are at right angles.

17 Consider the vectors

$$\mathbf{a} = 2\mathbf{i} + 2\mathbf{j} + \mathbf{k} \text{ and } \mathbf{b} = x\mathbf{i} + \mathbf{j} - \mathbf{k}$$

Find the value of x for which the angle between $\mathbf{a}$ and $\mathbf{b}$ is 45°.

First Order Differential Equations

> ### SECTION A **Solving differential equations**
>
> By the end of this section you will be able to:
> - ▶ find the order of a differential equation
> - ▶ solve first order differential equations by using direct integration
> - ▶ solve first order differential equations by using separation of variables
> - ▶ solve engineering applications with initial conditions

A1 **Introduction**

? **What is a differential equation?**

An equation which contains derivatives.

For example

$$3\frac{dy}{dx} + x = \cos(y) \text{ is a first order differential equation}$$

$$3\frac{d^2y}{dx^2} + \frac{dy}{dx} + x = \cos(y) \text{ is a second order differential equation}$$

$$3\frac{d^3y}{dx^3} + \frac{dy}{dx} + x = \cos(y) \text{ is a third order differential equation}$$

A differential equation can be classified according to its order. The **order** is the order of the highest derivative appearing in the differential equation.

Example 1

If
$$y = e^{5x}$$
Show that
$$\frac{dy}{dx} - 5y = 0$$

Solution

By using

| 6.11 | $\frac{d}{dx}(e^{kx}) = ke^{kx}$ |

on $y = e^{5x}$ we have

$$\frac{dy}{dx} = \frac{d}{dx}(e^{5x}) = 5e^{5x}$$

Substituting $y = e^{5x}$, so $5y = 5e^{5x}$ and we have

$$5e^{5x} - 5y = 5e^{5x} - 5e^{5x} = 0$$

Hence $\frac{dy}{dx} - 5y = 0$ holds.

Up until now we have dealt with the situation of **Example 1**, where we are given a solution, $y = e^{5x}$, and asked to verify that it satisfies the differential equation, $\dfrac{dy}{dx} - 5y = 0$. In this chapter we are given the differential equation, $\dfrac{dy}{dx} - 5y = 0$, and we are asked to find a solution, $y = e^{5x}$. **If we are asked to solve $\dfrac{dy}{dx} - 5y = 0$ then what are we trying to find?**

A function of x, $y = y(x)$, which satisfies $\dfrac{dy}{dx} - 5y = 0$.

Note that x is the independent variable and y is the dependent variable.

A2 Direct integration

If we are given the general first order differential equation

13.1 $$\frac{dy}{dx} = f(x) \quad \text{[Function of } x\text{]}$$

then we can use direct integration to find y.

Example 2

Solve

$$\frac{dy}{dx} = 4x$$

Solution

We have

$$dy = 4x\,dx$$

because $dy = \dfrac{dy}{dx}\,dx$. Integrating gives

$$\int dy = \int 4x\,dx$$

$$y = \frac{4x^2}{2} + C$$

$$y = 2x^2 + C$$

$y = 2x^2 + C$ is called the general solution of $\dfrac{dy}{dx} = 4x$.

Why is it called the general solution?

Because the solution has a constant, C, which does not have a particular value. The solution, $y = 2x^2 + C$, satisfies the differential equation, $\dfrac{dy}{dx} = 4x$, for any value of C.

We can sketch these solutions, $y = 2x^2 + C$, for different values of C.

Figure 1 shows the graphs of $y = 2x^2 + C$.

Fig. 1

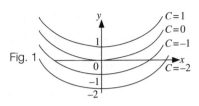

The differential equation, $\dfrac{dy}{dx} = 4x$, has an infinite number of solutions which are called the **family of solutions.**

Example 3 *materials*

The bending moment, M, of a beam of length l is given by

$$\frac{dM}{dx} = w(x + l)$$

where w is the constant load. Find an expression for M in terms of x.

Solution

For the given differential equation, M is the dependent variable and x is the independent variable. We have

$$dM = w(x + l)\,dx$$
$$= (wx + wl)\,dx \qquad \text{[Expanding brackets]}$$

Integrating both sides:

$$\int dM = \int (wx + wl)\,dx$$
$$M = \frac{wx^2}{2} + wlx + C$$

This M is the general solution of the differential equation $\dfrac{dM}{dx} = w(x + l)$.

Example 4 *mechanics*

The velocity, $v = u + at$, of an object is defined as

   $v = \dfrac{ds}{dt}$

where s is the displacement, u is the initial velocity and a is the acceleration. Given that when $t = 0$, $s = 0$ show that

$$s = ut + \frac{1}{2}at^2$$

Solution

Substituting the given equation $v = u + at$ into ⬛ ⋆ yields

$$\frac{ds}{dt} = u + at$$
$$ds = (u + at)\,dt$$

Example 4 *continued*

Integrating:

$$\int ds = \int (u + at)\, dt$$

$$s = ut + \frac{at^2}{2} + C$$

Substituting the given conditions $t = 0$, $s = 0$:

$$0 = 0 + 0 + C \text{ gives } C = 0$$

Hence we have

$$s = ut + \frac{1}{2}at^2$$

Notice that $s = ut + \dfrac{1}{2}at^2$ is a **particular solution** of $\dfrac{ds}{dt} = u + at$ because we have a particular value for $C\ (= 0)$. A particular solution is **one** member of the family of general solutions. Generally if we can find a value for C by using some given conditions then the solution is called a particular solution and the conditions are called the **initial conditions** or **boundary conditions** of the differential equation.

A3 **Separating variables**

A general first order differential equation of the form

$$\boxed{13.2} \qquad \frac{dy}{dx} = f(x)g(y)$$

is called a separable equation because we can separate the variables by dividing through by $g(y)$. In $\boxed{13.2}$, $f(x)$ is a function of x only and $g(y)$ is a function of y only.

Example 5

Solve

$$\frac{dy}{dx} = (1 + x)(1 + y) \quad (y > -1)$$

Solution

Since

$$dy = \frac{dy}{dx}\, dx$$

Example 5 *continued*

substituting $\dfrac{dy}{dx} = (1 + x)(1 + y)$ gives

$$dy = (1 + x)(1 + y)\,dx$$

Dividing through by $1 + y$ so that the y's and dy's are on one side (separating variables) gives

$$\dfrac{dy}{1 + y} = (1 + x)\,dx$$

? **What are we trying to find?**

We need to find y as a function of x which satisfies the differential equation.

? Therefore we eliminate dy and dx. **How?**

We integrate both sides:

$$\int \dfrac{dy}{1 + y} = \int (1 + x)\,dx$$

$$\underbrace{\ln(1+y)}_{\text{by } 8.42} = x + \dfrac{x^2}{2} + C$$

? We need to extract y from the left-hand side. **How?**

Take exponentials of both sides:

$$e^{\ln(1+y)} = e^{x + (x^2/2) + C}$$

By using the rules of indices and exponentials we have

$$\underbrace{(1 + y)}_{\text{by } 5.17} = e^x e^{x^2/2} e^C$$

$$= Ae^x\, e^{x^2/2} \text{ where } A = e^C \text{ is a constant}$$

Subtracting 1 from both sides gives

$$y = Ae^x\, e^{x^2/2} - 1$$

Again this $y = Ae^x e^{x^2/2} - 1$ is the general solution of $\dfrac{dy}{dx} = (1 + x)(1 + y)$ because we do not have a particular value for A.

Sometimes it can be difficult to extract y from the resulting integration.

5.17 $e^{\ln(x)} = x$ 8.42 $\displaystyle\int \dfrac{f'(x)}{f(x)}\,dx = \ln|f(x)|$

Example 6 *fluid mechanics*

The streamlines of a fluid flow are given by the first order differential equation

$$\frac{dy}{dx} = -\frac{x}{y}$$

Solve the differential equation.

Solution

Separating the variables of the given differential equation:

$$y\,dy = -x\,dx$$

$$\int y\,dy = \int -x\,dx$$

$$\frac{y^2}{2} = -\frac{x^2}{2} + C \qquad\qquad \text{[Integrating]}$$

Multiplying both sides by 2:

$$y^2 = -x^2 + 2C$$
$$y^2 + x^2 = r^2 \text{ where } r^2 = 2C$$

The equation

$$x^2 + y^2 = r^2$$

is an equation of a circle with centre origin and radius r. By using this equation we can sketch the streamlines of the above differential equation.

[A streamline is a curve in the x–y plane given by $y = f(x)$.]

The streamlines are circles with centre at the origin.

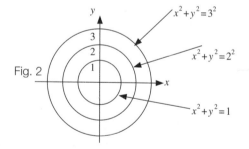

Fig. 2

Figure 2 shows the streamlines for $2C = 1$, $2C = 4$ and $2C = 9$ where $2C = r^2$. There are an infinite number of these streamlines.

Example 7 *fluid mechanics*

Consider a tank full of water which is being drained out through an outlet. The height H (m) of water in the tank at time t (s) is given by

$$\frac{dH}{dt} = -(2.8 \times 10^{-3})\sqrt{H}$$

Given that when $t = 0$, $H = 4$, find an expression for H in terms of t.

Example 7 *continued*

Solution

Separating variables:

$$\frac{dH}{\sqrt{H}} = -(2.8 \times 10^{-3})\, dt$$

$$H^{-1/2}\, dH = -(2.8 \times 10^{-3})\, dt$$

Integrating:

$$\int H^{-1/2}\, dH = -\int (2.8 \times 10^{-3})\, dt$$

$$\underbrace{\frac{H^{1/2}}{1/2}}_{\text{by } 8.1} = -(2.8 \times 10^{-3})t + C$$

$$2H^{1/2} = -(2.8 \times 10^{-3})t + C$$

Substituting the given initial condition $t = 0$, $H = 4$ yields

$$2 \times 4^{1/2} = 0 + C \text{ which gives } C = 4$$

Substituting $C = 4$ and dividing by 2 gives

$$H^{1/2} = \frac{4 - (2.8 \times 10^{-3})t}{2}$$

$$= 2 - (1.4 \times 10^{-3})t$$

Squaring both sides:

$$H = \left[2 - (1.4 \times 10^{-3})t\right]^2$$

SUMMARY

Differential equations are equations which contain a derivative. Solving a differential equation containing $\frac{dy}{dx}$ involves finding an expression for the dependent variable, y, in terms of the independent variable, x. Two methods were discussed in this section – direct integration and separating variables.

8.1 $$\int x^n\, dx = \frac{x^{n+1}}{n+1}$$

Exercise **13(a)**

Solutions at end of book. Complete solutions available at www.palgrave.com/science/engineering/singh

≈ **Questions 1 to 4, inclusive, are in the field of [*fluid mechanics*].**

1 The pressure, p, on an object under a fluid of density ρ is given by

$$\frac{dp}{dz} = -\rho g$$

where z represents depth and g is the acceleration due to gravity. Find an expression for p.

2 Consider a fluid rotating about a vertical axis. The pressure p at radius r from the axis is given by

$$\frac{dp}{dr} = -\rho \omega^2 r$$

where ρ is the density and ω is the angular velocity of the fluid. Find an expression for p.

3 The streamlines of a fluid flow are given by

$$\frac{dy}{dx} = C$$

where C is a constant. Plot the streamline that passes through the origin and $C = 5$.

4 The streamlines of a fluid flow are given by

$$\frac{dy}{dx} = e^x$$

Solve the differential equation and sketch the streamlines.

🚗 **Questions 5 to 7, inclusive, are in the field of [*mechanics*].**

5 The acceleration, a, of an object is defined as

$$a = \frac{dv}{dt}$$

where v is the velocity of the object at time t. Given that when $t = 0$, $v = u$, and the acceleration, a, is constant, show that $v = u + at$.

6 The acceleration, a, of a particle is given by

$$a = 5 - 3t$$

Given that the initial displacement at $t = 0$ is -2.1 m and the initial velocity is 8 m/s, find expressions for the velocity and displacement.

(*Hint*: We have, $a = \dfrac{dv}{dt}$, $v = \dfrac{ds}{dt}$

where s is the displacement.)

7 An object rotates with angular acceleration, α, defined as

$$\alpha = \frac{d\omega}{dt}$$

where ω is the angular velocity. Given that when $t = 0$, $\omega = \omega_0$, show that

$$\omega = \omega_0 + \alpha t$$

Also the angular velocity ω is defined as

$$\omega = \frac{d\theta}{dt}$$

where θ is the angular displacement. Given that when $t = 0$, $\theta = 0$ show that

$$\theta = \omega_0 t + \frac{1}{2}\alpha t^2$$

Solutions at end of book. Complete solutions available at www.palgrave.com/science/engineering/singh

≈ **Questions 8 to 10, inclusive, belong to [*fluid mechanics*].**

8 The streamlines of a fluid flow are given by

$$\frac{dy}{dx} = -\frac{x}{y}$$

Show that $x^2 + y^2 = A$ where A is a constant. Sketch the streamlines for $A = 1$, 25 and 100.

9 The streamlines of a fluid flow are given by

$$\frac{dy}{dx} = -\frac{y}{x} \quad (y > 0 \text{ and } x > 0)$$

Show that $y = \frac{A}{x}$ and sketch on the

same axes the streamlines for $A=1$, 5 and 8 (A is a constant).

10 The streamlines of a fluid flow are given by the first order differential equation

$$\frac{dy}{dx} = \frac{y+1}{x+1} \quad (y > -1 \text{ and } x > -1)$$

Show that $y = Ax + A - 1$ where A is a constant.

11 ⛴ [*heat transfer*]   The temperature gradient, $\frac{d\theta}{dx}$, of a slab of thickness

t and with thermal conductivity k is given by

$$\frac{d\theta}{dx} = C$$

where C is a constant and θ is a function of x.

By using the conditions $\theta(0)=\theta_1$, $\theta(t)=\theta_2$ and Fourier's law

$$Q = -kA\frac{d\theta}{dx}$$

show that

$$Q = -kA\left(\frac{\theta_2 - \theta_1}{t}\right)$$

[$\theta(0)=\theta_1$ means that when x is 0 then $\theta = \theta_1$, and $\theta(t)=\theta_2$ means that when x is t then $\theta = \theta_2$.]

12 ✗ [*aerodynamics*]   The pressure, $p > 0$, of a gas in isothermal condition is given by

$$\frac{dp}{dz} = -\frac{pmg}{RT}$$

where z is the altitude, T (constant) is the temperature, m is the molar mass and R is a gas constant. Show that

$$p = Ae^{-\frac{mg}{RT}z}$$

(where A is a constant).

13 ✗ [*aerodynamics*]   The pressure, p, of the atmosphere at an altitude z is given by

$$\frac{dp}{dz} = -kp^{\frac{1}{\gamma}} \quad (k \neq 0)$$

where γ is the specific heat constant ($\gamma > 1$) and k is a constant. Show that

$$z = \frac{\gamma p^{\frac{\gamma-1}{\gamma}}}{k(1 - \gamma)} + A$$

(where A is a constant).

SECTION B **Using the integrating factor**

By the end of this section you will be able to:
▶ understand the integrating factor method for solving first order differential equations
▶ apply the integrating factor method to examples

B1 **Integrating factor**

Example 8 is on differentiation and should ideally be placed in Chapter 6 but we establish a result in this example which is used to explain the integrating factor.

> **Example 8**
>
> Find $\dfrac{d}{dx}(ye^{Px})$ where y is a function of x and P is a constant.
>
> Solution
>
> We can differentiate ye^{Px} by using the product rule:
>
> **6.31** $\dfrac{d}{dx}(uv) = u'v + uv'$
>
> with
>
> $$u = y \qquad v = e^{Px}$$
> $$u' = \frac{dy}{dx} \qquad v' = Pe^{Px} \qquad\qquad \text{[Differentiating]}$$
>
> Substituting these into **6.31** gives
>
> $$\frac{d}{dx}(ye^{Px}) = \frac{dy}{dx}e^{Px} + Pye^{Px}$$

Suppose we have a general first order differential equation of the form

13.3 $\dfrac{dy}{dx} + Py = Q(x)$

which we need to solve, that is find y which satisfies the differential equation. Multiply both sides of **13.3** by e^{Px}:

$$\frac{dy}{dx}e^{Px} + Pye^{Px} = e^{Px}\,Q(x)$$

[?] **What do you notice about the Left-Hand Side?**

By the result of **Example 8**, the Left-Hand side $= \dfrac{d}{dx}(ye^{Px})$. Hence we have

$$\frac{d}{dx}(ye^{Px}) = e^{Px}\,Q(x)$$

? **How can we obtain y from this?**

Integrate both sides with respect to x.

Integration removes the derivative, $\dfrac{d}{dx}$, on the Left-Hand Side:

$$ye^{Px} = \int e^{Px} Q(x)\, dx$$

Since P is constant, $Px = \int P\, dx$. Thus replacing Px with $\int P\, dx$ we have

$$ye^{\int P dx} = \int e^{\int P dx} Q(x)\, dx$$

$e^{\int P dx}$ is called the integrating factor (IF). Summarizing the above;
The solution of the differential equation

13.3 $\dfrac{dy}{dx} + Py = Q(x)$

is given by

13.4 $y(\text{IF}) = \int \big[(\text{IF})Q(x)\big]\, dx$ where $\text{IF} = e^{\int P dx}$

In the above we illustrated the integrating factor method for a constant, P, but it may be generalized to cover P as function of x, that is $P = P(x)$. We still obtain formula 13.4 . We will not go over the same treatment for a non-constant P but just apply 13.4 in the appropriate case.

Example 9

Solve

$$\frac{dy}{dx} + 3y = e^x$$

with the initial condition that when $x = 0$, $y = 0$.

Solution
The given equation is of the same format as 13.3 , so we can use 13.4 with $P = 3$:

$$\text{IF} = e^{\int P dx} = e^{\int 3 dx} = e^{3x}$$

Applying $y(\text{IF}) = \int \big[(\text{IF})Q(x)\big]\, dx$ with $\text{IF} = e^{3x}$ and $Q(x) = e^x$ we have

$$ye^{3x} = \int [e^{3x} e^x]\, dx$$

$$= \int [e^{3x+x}]\, dx$$

$$= \int e^{4x}\, dx$$

$$= \frac{e^{4x}}{4} + C \qquad \left[\text{Integrating by } \int e^{kx} dx = \frac{e^{kx}}{k}\right]$$

Example 9 *continued*

Substituting the initial condition $x = 0$, $y = 0$ into $ye^{3x} = \dfrac{e^{4x}}{4} + C$ gives

$$0 = \frac{1}{4} + C \text{ and so } C = -\frac{1}{4} \qquad \text{[Remember that } e^0 = 1\text{]}$$

Putting $C = -1/4$:

$$ye^{3x} = \frac{e^{4x}}{4} - \frac{1}{4}$$

Remember we need to find y. **How?**

Divide through by e^{3x} and then use the rules of indices to simplify:

$$
\begin{aligned}
y &= \frac{e^{4x}}{4e^{3x}} - \frac{1}{4e^{3x}} \\
&= \frac{e^{4x-3x}}{4} - \frac{e^{-3x}}{4} \\
&= \frac{e^x}{4} - \frac{e^{-3x}}{4} \\
&= \frac{1}{4}(e^x - e^{-3x}) \qquad \left[\text{Taking out } \frac{1}{4}\right]
\end{aligned}
$$

B2 **Engineering applications**

In this section we examine electrical engineering applications. In electrical principles R, L and C are constants representing resistance, inductance and capacitance respectively. E represents the e.m.f. and $v = v(t)$, $i = i(t)$ represent voltage and current respectively at time t.

Example 10 *electrical principles*

The first order differential equation

$$RC\frac{dv}{dt} + v = Ee^{-\frac{t}{RC}}$$

describes a series RC circuit. By using the initial condition $v(0) = 1$, show that

$$v = e^{-\frac{t}{RC}}\left(\frac{Et}{RC} + 1\right)$$

Solution

Divide the given differential equation by RC:

$$\frac{dv}{dt} + \frac{v}{RC} = \frac{E}{RC}e^{-\frac{t}{RC}}$$

Example 10 *continued*

For this differential equation, can we use the integrating factor method?

Yes, because we can write the differential equation as

$$\frac{dv}{dt} + Pv = Q(t)$$

where $P = \dfrac{1}{RC}$ and $Q(t) = \dfrac{E}{RC}\,e^{-\frac{t}{RC}}$

With $P = \dfrac{1}{RC}$ and integrating with respect to t:

$$\text{IF} = e^{\int P dt} = e^{\int \frac{1}{RC} dt} = e^{\frac{t}{RC}}$$

Applying

13.4 $$v(\text{IF}) = \int \big[(\text{IF})Q(t)\big]\,dt$$

with $\text{IF} = e^{t/RC}$ and $Q(t) = \dfrac{E}{RC}e^{-t/RC}$ we have

$$ve^{\frac{t}{RC}} = \int \left[\frac{E}{RC}e^{\frac{t}{RC}}e^{-\frac{t}{RC}}\right] dt$$

$$= \frac{E}{RC}\int \left(e^{\frac{t}{RC}-\frac{t}{RC}}\right) dt \qquad \left[\text{Taking out } \frac{E}{RC}\right]$$

$$= \frac{E}{RC}\int \underbrace{(e^{0})}_{=1}\, dt$$

$$= \frac{E}{RC}\int dt$$

$$ve^{\frac{t}{RC}} = \frac{Et}{RC} + D \qquad\qquad \text{[Integrating]}$$

where D is the constant of integration. The initial condition $v(0) = 1$ means that when $t = 0$, $v = 1$. Substituting this condition, $t = 0$, $v = 1$, gives

$$1 = 0 + D \qquad\qquad \text{[Remember that } e^0 = 1]$$

Hence $D = 1$ and therefore we have

$$ve^{\frac{t}{RC}} = \frac{Et}{RC} + 1$$

Dividing through by $e^{t/RC}$ to find v on its own gives

$$v = \frac{1}{e^{\frac{t}{RC}}}\left(\frac{Et}{RC} + 1\right)$$

$$= e^{-\frac{t}{RC}}\left(\frac{Et}{RC} + 1\right)$$

So we have the required result.

SUMMARY

The solution of the general first order differential equation of the form

13.3 $\dfrac{dy}{dx} + Py = Q(x)$

where $P = P(x)$ is given by

13.4 $y(\text{IF}) = \displaystyle\int \big[(\text{IF})Q(x) \big]\, dx \quad$ where $\text{IF} = e^{\int P dx}$

Exercise 13(b)

Solutions at end of book. Complete solutions available at www.palgrave.com/science/engineering/singh

1 Solve the differential equation

$$\frac{dy}{dx} + 2y = e^{-x}$$

with the initial condition $x = 0$, $y = 0$.

2 Solve the differential equation

$$\frac{dy}{dx} - y = e^{2x}$$

3 Solve the differential equation

$$(1 - x^2)\frac{dy}{dx} - 2xy = 1 \quad (x < 1)$$

with the initial condition $x = 0$, $y = 0$.

🚗 **Questions 4 and 5 belong to the field of [mechanics].**

4 The velocity, v, of an object of mass m in free fall can be described by the first order differential equation

$$m\frac{dv}{dt} = mg - kv$$

where k is a positive constant and g is the acceleration due to gravity. For the initial condition, when $t = 0$, $v = 0$, show by using the integrating factor method that

$$v = \frac{mg}{k}\left(1 - e^{-\frac{k}{m}t}\right)$$

5 The velocity, v, of an object is given by the differential equation

$$\frac{dv}{dt} = \frac{1}{t}(1 - v) \qquad (t > 0)$$

By using the integrating factor method, determine an expression for v with the initial condition that when $t = 1$, $v = 0$.

What is the terminal velocity (the velocity as $t \to \infty$)?

Sketch the graph of v versus t for $t > 0$.

📟 **The remaining questions are in the field of [electrical principles].**

6 By applying Kirchhoff's voltage law to a series RL circuit we obtain the first order differential equation

$$L\frac{di}{dt} + Ri = E(t)$$

where $E(t)$ is the applied voltage and $i = i(t)$ is the current through the circuit.

i Show that $i = \dfrac{e^{-\frac{R}{L}t}}{L}\displaystyle\int \left(E(t)e^{\frac{R}{L}t} \right) dt.$

ii If $E(t)$ is constant, $E(t) = E$, and $i(0) = i_0$ then show that

$$i = \frac{E}{R} + \left(i_0 - \frac{E}{R} \right)e^{-\frac{R}{L}t}$$

iii Sketch the graph of i versus t (assuming $i_0 < E/R$). What value does i tend to as t becomes large?

7 A circuit consists of a resistor, R, and a capacitor, C, connected in series, and is

Exercise **13(b) continued**

Solutions at end of book. Complete solutions available at www.palgrave.com/science/engineering/singh

described by the first order differential equation

$$RC\frac{dv}{dt} + v = E$$

where E is the constant e.m.f. and v is the voltage across the capacitor. Given that $v(0) = 0$, show by using the integrating factor method that

$$v = E(1 - e^{-t/(RC)})$$

8 By applying Kirchhoff's voltage law to a series RL circuit we obtain the differential equation

$$(1 \times 10^{-3})\frac{di}{dt} + (3 \times 10^{3})i = 10e^{t}$$

where $i = i(t)$ is the current through the circuit. Given the initial condition $i(0) = 0$, determine an expression for the current i.

9 The charge, q, on a capacitor, C, of a series RC circuit is given by

$$R\frac{dq}{dt} + \frac{q}{C} = Ve^{-t}$$

Show that when $t = 0$, $q = 0$ gives

$$q = \frac{VC}{1 - RC}\left(e^{-t} - e^{-\frac{t}{RC}}\right)$$

SECTION C **Applications to electrical principles**

By the end of this section you will be able to:
► model electrical circuits which give rise to first order differential equations
► solve the resulting first order differential equation
► use a computer algebra system to solve a first order differential equation

C1 **Electrical laws**

Below are some of the laws regarding basic electrical components such as a resistor, capacitor and inductor. Kirchhoff's voltage law says:

In a closed path

13.5   sum of the voltage rises = sum of the voltage drops

With reference to Fig. 3, Kirchhoff's law says

$$v = v_1 + v_2 + v_3$$

The voltage, v, across a resistor, R (Fig. 4), is given by

13.6 $v = iR$ (Ohm's law)

The voltage, v, across an inductor, L (Fig. 5), is given by

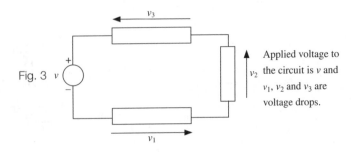

Fig. 3

Applied voltage to the circuit is v and v_1, v_2 and v_3 are voltage drops.

$$\boxed{13.7} \qquad v = L\frac{di}{dt}$$

Fig. 4   [Resistor]

The voltage, v, across a capacitor, C (Fig. 6), is related to the current by

$$\boxed{13.8} \qquad i = C\frac{dv}{dt}$$

Fig. 5   [Inductor]

Remember that R, L and C are positive constants.

Fig. 6   [Capacitor]

Next we use these rules to form differential equations of electrical circuits.

The following example is complicated because it involves a number of concepts such as forming and solving a differential equation, using algebraic techniques, graphing functions, etc.

C2 **Electrical circuits**

Example 11 *electrical principles*

Figure 7 shows a resistor, R, connected in series with an inductor, L, and an applied constant voltage E.

  i Obtain a first order differential equation for the current i at time t.

  ii Solve this differential equation for the initial condition, when $t = 0$, $i = 0$.

  iii What is the value of i as $t \to \infty$?

  iv Sketch the graph of i versus t for $t \geq 0$.

Fig. 7

Solution

  i By applying Kirchhoff's law to the circuit of Fig. 7 we have:

$$E = \text{(voltage drop across resistor, } R\text{)} + \text{(voltage drop across inductor, } L\text{)}$$

$$\boxed{*} \qquad = \underbrace{iR}_{\text{by } \boxed{13.6}} + \underbrace{L\frac{di}{dt}}_{\text{by } \boxed{13.7}}$$

    This, $\boxed{*}$, is a first order differential equation.

 $v = iR$ $\boxed{13.7}$ $v = Ldi/dt$

Example 11 *continued*

ii This differential equation can be solved by separating variables as discussed in **Section 13A**. By subtracting iR from both sides of ⬛ ✱ ⬛ we have

$$L\frac{di}{dt} = E - iR$$

Separating variables:

$$L\frac{di}{E - iR} = dt$$

Integrating both sides:

$$L\int \frac{di}{E - iR} = \int dt$$

⬛ † ⬛
$$\underbrace{-\frac{L}{R}\ln(E - iR) = t + C}_{\text{by } 8.42}$$

Substituting the given initial condition $t = 0$, $i = 0$ into ⬛ † ⬛:

$$-\frac{L}{R}\ln(E) = C$$

Placing $C = -\dfrac{L}{R}\ln(E)$ into ⬛ † ⬛ gives

$$-\frac{L}{R}\ln(E - iR) = t - \frac{L}{R}\ln(E)$$

$$-t = \frac{L}{R}\left[\ln(E - iR) - \ln(E)\right]$$

Using the properties of logs inside the square brackets and multiplying both sides by $\dfrac{R}{L}$:

$$-\frac{Rt}{L} = \underbrace{\ln\left(\frac{E - iR}{E}\right)}_{\text{by } 5.13}$$

Taking exponentials of both sides:

$$e^{-\frac{Rt}{L}} = \underbrace{e^{\ln\left(\frac{E - iR}{E}\right)}}_{\text{by } 5.17} = \frac{E - iR}{E}$$

$$Ee^{-\frac{Rt}{L}} = E - iR \quad \text{[Multiplying by } E\text{]}$$

5.13 $\ln(A) - \ln(B) = \ln\left(\dfrac{A}{B}\right)$ 5.17 $e^{\ln(x)} = x$ 8.42 $\displaystyle\int \frac{f'(x)}{f(x)}\,dx = \ln|f(x)|$

Example 11 *continued*

Rearranging gives

$$iR = E - Ee^{-\frac{Rt}{L}}$$

$$= E\left(1 - e^{-\frac{Rt}{L}}\right) \qquad \text{[Taking out } E]$$

Dividing through by R yields i:

$$i = \frac{E}{R}\left(1 - e^{-\frac{Rt}{L}}\right)$$

iii Since R and L are positive, we have $\dfrac{R}{L} > 0$, $e^{-\frac{R}{L}t} \to 0$ as $t \to \infty$ and so $i \to \dfrac{E}{R}$.

iv At $t = 0$, $i = 0$ therefore the graph goes through the origin and is asymptotic to the line $i = \dfrac{E}{R}$ (Fig. 8).

Fig. 8

The next example uses the application of a computer algebra system such as MAPLE.

Example 12 *electrical principles*

Figure 9 shows a series *RL* circuit with an applied voltage, $E = 240(1 - e^{-t/3})$.

Fig. 9

  i Obtain a differential equation giving the current i at time t.
 ii Solve the differential equation for the initial condition, when $t = 0$, $i = 0$.
iii Plot the graph of i versus t. [Use MAPLE for parts **ii** and **iii**.]

Solution

i By applying Kirchhoff's law we have

$$E = \text{(voltage drop across resistor)} + \text{(voltage drop across inductor)}$$

$$= \underbrace{iR}_{\text{by } 13.6} + \underbrace{L\frac{di}{dt}}_{\text{by } 13.7}$$

Substituting the given values [remember that k = kilo = 10^3 and m = milli = 10^{-3}], $R = 3 \times 10^3$, $L = 5 \times 10^{-3}$, $E = 240(1 - e^{-t/3})$ into the differential equation:

$$240(1 - e^{-t/3}) = (3 \times 10^3)i + (5 \times 10^{-3})\frac{di}{dt}$$

13.6 $v = IR$ **13.7** $v = L\, di/dt$

Example 12 *continued*

ii We solve this first order differential equation, by using MAPLE:

```
> de_1:=(3*10^3)*i(t)+(5*10^(-3))*diff(i(t),t)=240
  *(1-exp(-t/3));
```

$$de\_1: = 3000\,i(t) + \frac{1}{200}\left(\frac{\partial}{\partial t}i(t)\right) = 240 - 240\,\mathbf{e}^{(-1/3t)}$$

```
> soln:=dsolve({de_1,i(0)=0},i(t));
```

$$soln: = i(t) = \frac{2}{25} - \frac{144000}{1799999}\,\mathbf{e}^{(-1/3t)} + \frac{2}{44999975}\,\mathbf{e}^{(-600000t)}$$

iii

```
plot(rhs(soln),t=0..100);
```

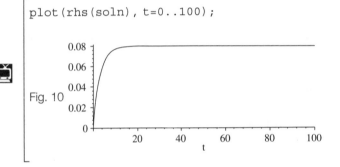

Fig. 10

In ⬛* of Example 12, the extreme right term is called the transient term. Transient is a brief disturbance in a circuit. Because it happens to have a large negative exponent, it dies away very rapidly, and cannot be shown on the scale of Fig. 10. The first two terms show the time variation of the applied voltage. The current is eventually E/R, or $2/25 = 0.08$ in this case, which is shown in Fig. 10.

SUMMARY

13.6 The voltage, v, across a resistor, R, is given by $v = iR$ (Ohm's law)

13.7 The voltage, v, across an inductor, L, is given by $v = L\dfrac{di}{dt}$

13.8 The voltage, v, across a capacitor, C, is related to the current by $i = C\dfrac{dv}{dt}$

Exercise **13(c)**

Solutions at end of book. Complete solutions available at www.palgrave.com/science/engineering/singh

⚏ **All questions belong to the field of [*electrical principles*].**

1

Fig. 11

Figure 11 shows a series *RL* circuit.
i Show that

$$\frac{di}{dt} = -\frac{iR}{L}$$

where *i* represents the current through the circuit.
ii Show that for the initial condition, $t = 0$ s, $i = 1$ A, then

$$i = e^{-\frac{R}{L}t}$$

iii Sketch the graph of *i* against *t*.
iv What is the value of *i* at $t = \tau$ where

$$\tau = \frac{L}{R} \; (\tau \text{ is called the time constant for this circuit)?}$$

2

Fig. 12

Figure 12 shows a series *RC* circuit.
i Show that the differential equation relating the voltage, *v*, across the capacitor *C* at time *t* is given by

$$\frac{dv}{dt} = -\frac{v}{RC}$$

ii Show that for the initial condition when $t = 0$, $v = E \; (> 0)$ that

$$v = Ee^{-t/RC}$$

iii Sketch the graph of *v* against *t* for $t \geq 0$.

iv What is the value of *v* at $t = \tau$ where $\tau = RC$ (τ is called the time constant)?

3

Fig. 13

Figure 13 shows a series *RC* circuit with a constant applied e.m.f. = *E*. Show that the differential equation of this circuit is given by

$$\frac{dv}{dt} = \frac{1}{RC}(E - v)$$

where *v* is the voltage across the capacitor. Solve this differential equation for the initial condition, $t = 0$, $v = 0$.

4

Fig. 14 15 V ⎓   10 kΩ   10 μF | *v*

i For the circuit of Fig. 14 show that

$$0.1\frac{dv}{dt} = 15 - v$$

where *v* is the voltage across the capacitor at time *t*.

ii Solve this differential equation for the initial condition, when $t = 0$, $v = 0$.

iii Sketch the graph of *v* against *t* for $t \geq 0$.

iv What percentage of the final voltage is gained at each of the following stages: $t = 0.1, 0.2$ and 0.3?
(0.1 is the time constant for this circuit and 0.2, 0.3 are 2 and 3 time constants respectively.)

Exercise **13(c) continued**

Solutions at end of book. Complete solutions available at www.palgrave.com/science/engineering/singh

Use a computer algebra system for the remaining questions.

5

Fig. 15 $15t^2$ 15 kΩ $3\,\mu F$ v

  i Show that the differential equation of the circuit in Fig. 15 is given by

$$0.045\frac{dv}{dt} = 15t^2 - v$$

  ii Solve this differential equation for the initial condition, $t = 0$, $v = 1$.
  iii Plot the graph of v against t for $0 \le t \le 0.5$.

6

Fig. 16 $9\sin(t)$ 3 kΩ i 0.1 mH

Figure 16 shows a series resistor–inductor circuit.

  i Show that
$$\frac{di}{dt} = (90 \times 10^3)\sin(t) - (30 \times 10^6)i$$

  where i is the current through the circuit.

  ii Solve for the initial condition, $t = 0$, $i = 1 \times 10^{-3}$.

  iii Plot, on different axes, the graphs of i against t for

    a $0 \le t \le 20$

    b $0 \le t \le 2 \times 10^{-7}$.

  Comment upon your results.

7

Fig. 17 $5\cos(100\pi t)$ i 9 kΩ   0.1 mH

  i For the circuit of Fig. 17, show that

$$\frac{di}{dt} = (50 \times 10^3)\cos(100\pi t) - (90 \times 10^6)i$$

  ii Solve the differential equation for the initial condition, $t = 0$, $i = 0$.

  iii Plot the graph of i against t. Choose a range that includes 5 complete cycles. Also plot the range $0 \le t \le 10^{-7}$.

  Comment upon your results.

8

Fig. 18 $240\sin(\omega t)$ i 9 kΩ   3 mH

  i Show that the differential equation of Fig. 18 is given by

$$\frac{di}{dt} = (80 \times 10^3)\sin(\omega t) - (3 \times 10^6)i$$

  ii Solve this differential equation for the initial condition, $t = 0$, $i = 0$.

  iii Plot the graph of i against ωt for $\omega t \ge 0$. Initially choose $\omega = 10\,000$ and $\omega = 10^6$. Plot about 5 to 10 complete cycles. From the results you get, choose other values of ω which give interesting results.

SECTION D **Further engineering applications**

By the end of this section you will be able to:

▶ apply first order differential equations to other fields of engineering such as fluid mechanics and heat transfer

▶ understand Torricelli's law and derive and solve the relevant first order differential equation

▶ apply Newton's law of cooling

D1 **Torricelli's law (fluid mechanics)**

Figure 19 shows a water tank with an outlet of cross-sectional area A_o (m^2) and water filled in the tank to a height of h (m). Torricelli's law states that the velocity, v (m/s), of water exiting through the outlet is given by

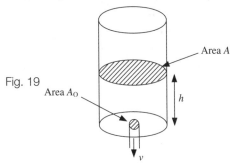

Fig. 19

$$v = \sqrt{2gh}$$

where g is the acceleration due to gravity

(This is obtained by equating energies, $\dfrac{1}{2}mv^2 = mgh$ and solving for v.)

Let the area of the outlet be A_o (m^2) then the volume of water leaving the tank per second is $A_o\sqrt{2gh}$. The volume of water leaving the tank in time Δt is

$$\left(A_o\sqrt{2gh}\right)\Delta t$$

Hence the change of volume, ΔV, of water in the tank in time Δt is given by

$$\Delta V = -\left(A_o\sqrt{2gh}\right)\Delta t$$

where the negative sign indicates that the volume V is decreasing. We have

$$\frac{\Delta V}{\Delta t} = -\left(A_o\sqrt{2gh}\right)$$

Taking limits gives

13.9 $$\frac{dV}{dt} = -A_o\sqrt{2gh}$$

If the volume of water in the tank can be written as Ah where A is the constant cross-sectional area of the tank then

$$\frac{dV}{dt} = \frac{d}{dt}(Ah) = A\frac{dh}{dt} \qquad [A \text{ is constant}]$$

Substituting $\dfrac{dV}{dt} = A\dfrac{dh}{dt}$ into 13.9 gives

$$A\frac{dh}{dt} = -A_o\sqrt{2gh}$$

Dividing both sides by A:

13.10 $$\frac{dh}{dt} = -\frac{A_o}{A}\sqrt{2gh}$$

We take g to be 9.81 m/s².

Example 13 *fluid mechanics*

A cubical tank of side 3 m is filled with water to a height of h (m). The water is drained through a circular hole of diameter 0.1 m.

  i Obtain the differential equation relating the height h of water at time t.
 ii Solve this differential equation for the initial condition, $t = 0$, $h = 2$.
iii How long (in minutes) does it take to empty the tank which is 2 m full?
 iv By using a computer algebra system or a graphical calculator, plot the graph of h against t for $t \geq 0$. Comment upon your graph.

Solution

i The outlet area, A_o, is a circle of diameter 0.1 m and so the radius $r = 0.1/2$:

$$A_o = \pi \times r^2$$
$$= \pi \times \left(\frac{0.1}{2}\right)^2$$
$$= \pi \times 0.05^2 = (2.5 \times 10^{-3})\pi$$

Since we have a cubical tank of side 3 m, therefore
$$\text{cross-sectional area of tank } A = 3^2 = 9$$
Substituting $g = 9.81$, $A_o = (2.5 \times 10^{-3})\pi$ and $A = 9$ into

13.10 $$\frac{dh}{dt} = -\frac{A_o}{A}\sqrt{2gh}$$
$$\frac{dh}{dt} = -\frac{(2.5 \times 10^{-3})\pi}{9}\sqrt{2 \times 9.81 \times h}$$
$$= -\frac{(2.5 \times 10^{-3})\pi\sqrt{2 \times 9.81}}{9}\sqrt{h}$$
$$= -(3.87 \times 10^{-3})h^{1/2} \qquad\qquad \text{[Evaluating]}$$

ii The remaining evaluation is very similar to **Example 7**. Separating variables of

$$\frac{dh}{dt} = -(3.87 \times 10^{-3})h^{1/2}$$

gives

$$\frac{dh}{h^{1/2}} = -(3.87 \times 10^{-3})\, dt$$

13.9 $\dfrac{dV}{dt} = -A_o\sqrt{2gh}$

Example 13 *continued*

Integrating:

$$\int h^{-1/2}dh = \int -(3.87 \times 10^{-3})\, dt$$

$$\frac{h^{1/2}}{1/2} = -(3.87 \times 10^{-3})t + C \qquad\qquad \text{[Integrating]}$$

$$2h^{1/2} = -(3.87 \times 10^{-3})t + C$$

Substituting the given initial condition, $t = 0$, $h = 2$:

$$2(2)^{1/2} = 0 + C$$

Putting $C = 2(2)^{1/2}$:

$$2h^{1/2} = -(3.87 \times 10^{-3})t + 2(2)^{1/2}$$
$$h^{1/2} = -(1.935 \times 10^{-3})t + 2^{1/2} \qquad \text{[Dividing by 2]}$$

How can we find h?

Square both sides:

$$† \qquad h = \left[-(1.935 \times 10^{-3})t + 2^{1/2}\right]^2$$

iii What are we trying to find?

The value of t when $h = 0$ because the tank is empty at $h = 0$.

Substituting $h = 0$ and then taking the square root of † gives

$$0 = -(1.935 \times 10^{-3})t + 2^{1/2}$$

$$t = \frac{2^{1/2}}{(1.935 \times 10^{-3})} = 731\,\text{s} \quad (\text{s} = \text{seconds})$$

What is t in minutes (mins)?

$$t = \frac{731}{60} = 12.18 \text{ mins}$$

It takes approximately just over 12 mins to drain the water in the tank.

iv We plot $h = \left[-(1.935 \times 10^{-3})t + 2^{1/2}\right]^2$ against t using MAPLE:

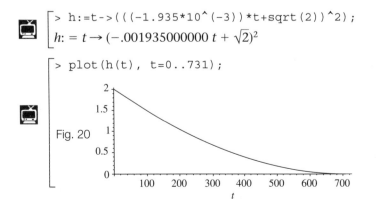

```
> h:=t->(((-1.935*10^(-3))*t+sqrt(2))^2);
  h: = t → (-.001935000000 t + √2)²

> plot(h(t), t=0..731);
```

Fig. 20

≈ Example 13 *continued*

The graph of Fig. 20 is only valid for $0 \leq t \leq 731$. When $t > 731$ the tank is empty and hence the height of the water in the tank is **zero**. Moreover the expression plotted is only valid for the tank which is 2 m full.

From the graph it is clear that the height of the water is decreasing with time t.

We can use MAPLE to solve the differential equation in **Example 13** and find how long it takes to empty the tank. It is in forming the differential equation [part **i**] where we cannot use MAPLE.

D2 Newton's law of cooling (heat transfer)

Newton's law of cooling states that the rate at which a body cools is proportional to the difference in temperature between the body and its surrounding. If $\theta = \theta(t)$ is the temperature of the body at time t and T is the constant surrounding temperature then

$$\frac{d\theta}{dt} \propto \theta - T \quad (\propto \text{ means directly proportional to})$$

Note that θ is a function of time t. We have

13.11
$$\frac{d\theta}{dt} = k(\theta - T)$$

where k is called the constant of proportionality.

⇕ Example 14 *heat transfer*

A body is exposed to a constant temperature of 280 K. After 1 minute the temperature of the body is 350 K and after 5 minutes it is 310 K. Find an expression for the temperature θ at time t. Sketch the graph of θ against t for $t \geq 0$. Comment upon your graph.

Solution

Using

13.11
$$\frac{d\theta}{dt} = k(\theta - T)$$

with $T = 280$ gives

$$\frac{d\theta}{dt} = k(\theta - 280)$$

Separating variables:

$$\frac{d\theta}{\theta - 280} = k \, dt$$

Example 14 *continued*

Integrating both sides:

$$\int \frac{d\theta}{\theta - 280} = \int k\, dt$$

$$\underbrace{\ln(\theta - 280)}_{\text{by } 8.42} = kt + C$$

Since t is measured in seconds we have $t = 1 \times 60 = 60$ and $\theta = 350$.

Substituting these values into $\ln(\theta - 280) = kt + C$ gives

† $\ln(350 - 280) = 60k + C$

We are also given $t = 5 \times 60 = 300$ and $\theta = 310$. Similarly we have

†† $\ln(310 - 280) = 300k + C$

Solving the simultaneous equations **†** and **††** gives $k = -3.53 \times 10^{-3}$ and $C = 4.46$.

Hence substituting these values into $\ln(\theta - 280) = kt + C$ yields

$$\ln(\theta - 280) = -(3.53 \times 10^{-3})t + 4.46$$

Taking exponentials of both sides to find θ we have

$$\theta - 280 = e^{-(3.53 \times 10^{-3})t + 4.46}$$ [Because $e^{\ln(\theta - 280)} = \theta - 280$]

$$\theta = 280 + e^{4.46}e^{-(3.53 \times 10^{-3})t}$$

$$\theta = 280 + \underbrace{86.49}_{= e^{4.46}}e^{-(3.53 \times 10^{-3})t}$$

For sketching the graph we need to find the value of θ at $t = 0$:

$$\theta = 280 + 86.49e^{0}$$
$$= 280 + 86.49 = 366.49$$

Also as $t \to \infty$, $e^{-(3.53 \times 10^{-3})t} \to 0$, thus $\theta \to 280$.

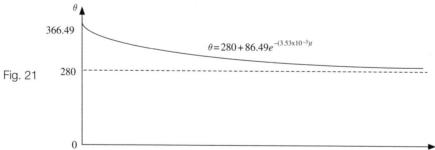

Fig. 21

The graph in Fig. 21 shows that the temperature of the body will eventually reach very close to the temperature of the surroundings at 280 K. This is because $86.49e^{-(3.53 \times 10^{-3})t}$ is the **transient term** and decays to zero as t gets large.

8.42 $\int \dfrac{f'(\theta)}{f(\theta)}\, d\theta = \ln|f(\theta)|$

SUMMARY

The differential equation describing the tank filled to a height h (m) at time t is given by

$$\boxed{13.10} \quad \frac{dh}{dt} = -\frac{A_o}{A} \sqrt{2gh}$$

where A_o and A are the cross-sectional areas of the outlet and the tank respectively. Newton's law of cooling gives a first order differential equation which describes the temperature θ at time t by

$$\boxed{13.11} \quad \frac{d\theta}{dt} = k(\theta - T)$$

where k is a constant and T is the constant surrounding temperature.

Fig. 22   Area A_o Area A

Plot the graphs in this exercise by using a computer algebra system or a graphical calculator.

≈ Questions 1 to 4, inclusive, are in the field of [*fluid mechanics*].

1 A cubical tank of side 2 m is filled with water to a height of h (m). The water is drained through a circular hole of diameter 0.2 m.

  i Obtain a differential equation giving the height h at any time t.

  ii Given that when $t = 0$, $h = 1.3$, solve this differential equation to give an expression for h in terms of t.

  iii How long (in min) does it take to empty the tank if it is 1.3 m full?

  iv Plot the graph of h against t for $t \geq 0$. Comment upon your graph.

2 A cylindrical tank of diameter 1.5 m contains water of height h. The water is drained through a square hole of length 0.05 m.

  i Obtain a differential equation representing the height h at any time t.

  ii Given that when $t = 0$, $h = 1.6$, solve this differential equation to give an expression for h in terms of t.

  iii How long (in min) does it take to empty the tank if it is 1.6 m full?

  iv Plot the graph of h against t for $t \geq 0$.

3

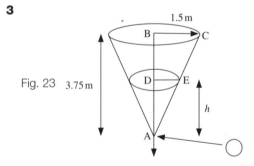

Fig. 23   3.75 m   1.5 m

Figure 23 shows a conical tank containing water of height h. The outlet is a circular hole of diameter 0.05 m.

  i Obtain the differential equation relating the height h of water at time t.

Exercise **13(d) continued**

Solutions at end of book. Complete solutions available at www.palgrave.com/science/engineering/singh

(*Hint*: Consider similar triangles, ABC and ADE. Use the property

$$\frac{AB}{BC} = \frac{AD}{DE}$$

This is a similarity property of triangles.)

ii Solve this differential equation for the initial condition, when $t = 0$, $h = 3$.

iii How long (in min) does it take for the tank to empty if it is initially 3 m full?

iv Plot the graph of h against t for $t \geq 0$. Comment upon your graph.

v How long does it take to empty the tank if it is 3.8 m full?

4 The general differential equation giving the height, h, of water in a tank at time t is given by

$$\frac{dh}{dt} = -k\sqrt{2gh}$$

where k is a constant. Solve this differential equation for the initial condition, $h(0) = 1$.

Plot the graphs of the solutions of height, h, against time, t, for

$$k = 0.1, 0.01 \text{ and } 0.001$$

Take g to be 9.81 m/s².

What effect does k have on the rate at which water is drained from the tank?

⚡ The remaining questions are in the field of [*heat transfer*].

5 A body is at a temperature of 373 K. After 5 minutes the temperature of the body is 330 K. Find an expression for $\theta = \theta(t)$ given that the constant surrounding temperature is 300 K. Sketch the graph of θ against t for $t \geq 0$.

What does θ tend to as $t \to \infty$?

6 An object is initially at 400 K, the constant surrounding temperature is 300 K. Determine an expression that gives the temperature $\theta = \theta(t)$ at time t.

7 Newton's law of cooling gives

$$\frac{d\theta}{dt} = k(\theta - T)$$

where θ is the temperature at time t, T is the constant surrounding temperature and k is a constant. Given that $\theta(0) = T_0$, show that

$$\theta = (T_0 - T)e^{kt} + T$$

8 By applying Newton's law of cooling to an object we obtain

$$\frac{d\theta}{dt} = k(\theta - 320)$$

where θ is the temperature at time t and k is a constant. Given that when $t = 0$, $\theta = 348$ K, find an expression for θ. Plot, on the same axes, the graphs of θ against t for $t \geq 0$ and $k = -0.001$, -0.01 and -0.1. How does the temperature, θ, change with k?

9 Stefan's law of radiation gives the rate of change of temperature, $\dfrac{d\theta}{dt}$, of a body by

$$\frac{d\theta}{dt} = k(\theta^4 - T^4)$$

where k is a constant, θ ($= \theta(t)$) is the temperature of the body at time t and T is the constant ambient temperature. Show, by solving the differential equation, that

$$\frac{1}{4T^3}\left[\ln\left(\frac{\theta - T}{\theta + T}\right) - 2\tan^{-1}\left(\frac{\theta}{T}\right)\right] = kt + C$$

where C is the constant of integration.

SECTION E **Euler's numerical method**

By the end of this section you will be able to:
▶ apply Euler's numerical formula
▶ use a computer algebra system to find the analytical and numerical solutions of differential equations
▶ use a spreadsheet to find numerical solutions

Up until now we have solved differential equations by employing analytical methods such as separating variables and integrating factor. However many differential equations **cannot** be solved by using these analytical techniques. Sometimes it is difficult, or impossible, to find a solution to a given differential equation using these methods. With the advancement in technology it is becoming easier to use numerical methods. These numerical methods involve using a repetitive process, hence the use of computers is beneficial in evaluating the laborious repetition. This repetitive mathematical process is called an **iteration**.

You can use a spreadsheet such as EXCEL or a computer algebra system like MAPLE, DERIVE, MATHEMATICA, etc.

In this chapter we cover three main numerical methods: Euler's, improved Euler's and 4th order Runge–Kutta. There are others but we will confine ourselves to these three. In this section we examine Euler's method.

Leonard Euler was born in 1707 to a Swiss clergyman and become a leading mathematician of the 18th century. Euler attended Basle University and took private lessons in mathematics. He contributed to mathematical notation, such as π, e, $\sum$ and $\sin(x)$, and also to other fields of mathematics.

He had 13 children and started to lose his eyesight at the age of 60. However he carried on working in mathematics until his death at the age of 76 in 1783.

E1 **Euler's method**

Euler's method approximates the solution of the differential equation by using tangent lines. Remember that the solution of a differential equation is a function of x which we can plot on a graph.

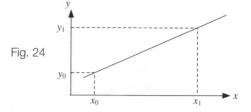

Fig. 24

The gradient, m, of the line in the graph of Fig. 24 is given by

$$m = \frac{y_1 - y_0}{x_1 - x_0}$$

$$y_1 - y_0 = m(x_1 - x_0)$$

Transposing:

$$y_1 = y_0 + m(x_1 - x_0)$$

We use this result, ┼ , in deriving Euler's numerical formula.

Suppose we want to solve the first order differential equation

$$\frac{dy}{dx} = f(x, y)$$

which satisfies the initial condition, at $x = x_0$ we have $y = y_0$. The notation $f(x, y)$ means a function of x and y. **What do we mean by 'solve'?**

Ideally we need to find a solution of the form, $y =$ function of x.

Suppose we know the solution, $y = g(x)$, and it has the graph shown in Fig. 25.

We know that $g(x_0) = y_0$ because this is the given initial condition.

Can we find the equation of the tangent line at (x_0, y_0) shown in Fig. 25?

Fig. 25   Approximating by a tangent

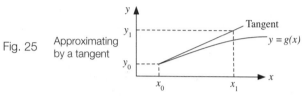

It is a straight line so we can use ┼ . **How do we find the gradient, m?**

Remember the derivative gives the gradient. Hence the gradient at $x = x_0$ is $\frac{dy}{dx}$. **What is $\frac{dy}{dx}$ equal to at $x = x_0$?**

From the given differential equation we have

$$\frac{dy}{dx} = f(x, y)$$

and at $x = x_0$

$$\frac{dy}{dx} = f(x_0, y_0) = m \qquad \text{[Gradient]}$$

That means by the equation of a tangent

┼ $\qquad y_1 = y_0 + m(x_1 - x_0)$

we have

$$y_1 = y_0 + f(x_0, y_0)[x_1 - x_0] \qquad \text{[Substituting for } m\text{]}$$

Thus y_1 is an approximate value of the actual solution, $g(x_1)$.

We can repeat this procedure to find y_2 and y_3, the y values at x_2 and x_3 respectively. Moreover we can similarly show that

$$y_2 = y_1 + f(x_1, y_1)[x_2 - x_1]$$

and

$$y_3 = y_2 + f(x_2, y_2)[x_3 - x_2]$$

What would be the formula for y_{n+1}?

＊ $\qquad y_{n+1} = y_n + f(x_n, y_n)[x_{n+1} - x_n]$

If we assume a uniform step size of h (Fig. 26)

then $x_{n+1} - x_n = h$ and replacing this in formula we have

| 13.12 | $y_{n+1} = y_n + h[f(x_n, y_n)]$ |

Fig. 26

or

$$y_{n+1} = y_n + hK_1 \qquad \text{where } K_1 = f(x_n, y_n)$$

13.12 is known as Euler's numerical formula. This formula approximates the actual solution by tangent lines. By using this formula we can find an approximate solution to a differential equation. Note that

$$x_{n+1} = x_n + h \qquad\qquad y_{n+1} = y_n + h[f(x_n, y_n)]$$

which means

$$\text{(new } x) = \text{(old } x) + h \qquad \text{(new } y) = \text{(old } y) + h[f(\text{old } x, \text{old } y)]$$

We use the old x and y values to find the new x and y values. By repeating this procedure we obtain the collection of lines shown in Fig. 27.

Fig. 27

? **How can we get better accuracy for our solution?**

Well, the smaller the step size, h, the better the approximation to the actual solution. Let's do an example.

Example 15

Consider the differential equation

$$\frac{dy}{dx} = x^2 + y^2 \text{ with } y(0) = 1$$

Use Euler's method to find the approximate values of y at $x = 0.1, \ 0.2$ and 0.3 with step size $h = 0.1$.

Solution

Putting $h = 0.1$ into 13.12 gives

$$y_{n+1} = y_n + 0.1[f(x_n, y_n)]$$

? **What is $f(x_n, y_n)$?**

Remember that $f(x_n, y_n)$ is the given function of x and y:

$$f(x_n, y_n) = x_n^2 + y_n^2$$

Substituting this into y_{n+1}:

| ★ | $y_{n+1} = y_n + 0.1(x_n^2 + y_n^2)$ |

Example 15 *continued*

Starting with $n = 0$ gives

$$y_1 = y_0 + 0.1(x_0^2 + y_0^2)$$

What are x_0 and y_0?

These values are given by the initial condition, $y(0) = 1$, which means that $x_0 = 0$ and $y_0 = 1$. Substituting these into y_1 gives

$$y_1 = y_0 + 0.1(x_0^2 + y_0^2)$$
$$= 1 + 0.1(0^2 + 1^2) = 1.1$$

The y value at $x = 0.1$ is 1.1.

We need to evaluate y_2 and y_3, the y values at $x = 0.2$ and 0.3 respectively (Fig. 28).

Fig. 28

In this example $h = 0.1$.

How do we calculate y_2?

By substituting $n = 1$ into $y_{n+1} = y_n + 0.1(x_n^2 + y_n^2)$

$$y_2 = y_1 + 0.1(x_1^2 + y_1^2)$$

What are x_1 and y_1 equal to?

$$x_1 = 0.1 \text{ and } y_1 = 1.1 \qquad \text{[Evaluated above]}$$

Substituting these into y_2 gives

$$y_2 = y_1 + 0.1(x_1^2 + y_1^2)$$
$$= 1.1 + 0.1(0.1^2 + 1.1^2) = 1.222$$

Similarly, substituting $n = 2$ into ⬛ * ⬛ gives y_3:

$$y_3 = y_2 + 0.1(x_2^2 + y_2^2)$$

What are x_2 and y_2 equal to?

$$x_2 = 0.2 \text{ and } y_2 = 1.222$$

Putting these into y_3 gives

$$y_3 = y_2 + 0.1(x_2^2 + y_2^2)$$
$$= 1.222 + 0.1(0.2^2 + 1.222^2) = 1.375\ 3284$$

We have $y_1 = 1.1$, $y_2 = 1.222$ and $y_3 = 1.375\ 3284$, and these are the approximate values of y at $x = 0.1$, 0.2 and 0.3 respectively.

Note that the **old** values of x and y are used to find the **new** values. For example to find y_3 we need to use x_2 and y_2.

We can apply MAPLE or any other computer algebra system to find the exact solution of **Example 15** and determine the error in using Euler's method at y_1, y_2 and y_3. The commands for using MAPLE are given in the next example. If you find difficulty in following these commands, have a look at the help menu. You can copy and paste examples from the help menu and make the appropriate changes for your example.

Example 16

By using a computer algebra system, solve the differential equation

$$\frac{dy}{dx} = x^2 + y^2 \quad \text{with } y(0) = 1$$

Determine y at $x = 0.1$, $x = 0.2$ and $x = 0.3$, and plot your solution. Compare your answers with y_1, y_2 and y_3 respectively, obtained in **Example 15**.

Solution

By using MAPLE we have:

```
> de_1:=diff(y(x),x)=(x^2)+(y(x)^2);
```

$$de\_1: = \frac{\partial}{\partial x} y(x) = x^2 + y(x)^2$$

```
> soln:=dsolve({de_1,y(0)=1},y(x)):
> evalf(subs(x=0.1,soln));
```

$y(.1) = 1.111463377$

```
> evalf(subs(x=0.2,soln));
```

$y(.2) = 1.253016934$

```
> evalf(subs(x=0.3,soln));
```

$y(.3) = 1.439670778$

```
> e_soln:=dsolve({de_1,y(0)=1},y(x),type=numeric,
method=classical,output=array([0.1,0.2,0.3]),stepsize=0.1);
```

$$e\_soln:= \begin{bmatrix} & [x,y(x)] \\ \begin{bmatrix} .1 & 1.10000000000000008 \\ .2 & 1.22200000000000020 \\ .3 & 1.37532840000000034 \end{bmatrix} \end{bmatrix}$$

```
> with(plots):
> p_1:=plots[odeplot](e_soln,[x,y(x)],linestyle=6):
> p_2:=plot(rhs(soln),x=0.1..0.3):
> display({p_1,p_2});
```

Fig. 29

Figure 29 shows the Euler solution and the actual solution; notice the discrepancy. We explore this discrepancy by evaluating the percentage error, % error, which is defined as

13.13 $\% \text{ error} = \dfrac{(\text{actual value}) - (\text{approximate value})}{\text{actual value}} \times 100$

The percentage error for y_3 for the above example is

$$\left(\frac{1.439670778 - 1.3753284}{1.439670778} \right) \times 100 = 4.47\%$$

Note that the actual value in this example is only correct to 9 decimal places.

So Euler's method gives an error of more than 4%.

? **How can we reduce this error?**

By considering a smaller value of h.

? **Are there any penalties if we consider a smaller value of h?**

We will need more iterations to evaluate y_1, y_2 and y_3. We need 30 iterations to evaluate y at $x = 0.3$ if $h = 0.01$ and $x_0 = 0$.

? **How many iterations are required for y at $x = 0.3$ if $h = 0.001$ and $x_0 = 0$?**

300. Thus to do justice to Euler's method, or any other numerical method for differential equations, we need to use mathematical software or write a procedure in a programmable calculator. Nobody would want to perform 300 iterations by hand.

By using MAPLE with different values of h we obtain Table 1 for **Example 16**.

TABLE 1	x	$h = 0.1$	$h = 0.01$	$h = 0.001$	Solution (correct to 9 d.p.)
	0.1	1.1	1.110129764	1.111327762	1.111463377
	0.2	1.222	1.249332251	1.252641232	1.253016934
	0.3	1.3753284	1.431804205	1.438865618	1.439670778

The percentage error for $h = 0.01$ in evaluating y at $x = 0.3$ is

$$\left(\frac{1.439670778 - 1.431804205}{1.439670778} \right) \times 100 = 0.55\%$$

The percentage error for $h = 0.001$ in evaluating y at $x = 0.3$ is

$$\left(\frac{1.439670778 - 1.438865618}{1.439670778} \right) \times 100 = 0.056\%$$

Note how the error decreases as h gets smaller.

Most students are more familiar with spreadsheets than a computer algebra system. Finding numerical solutions of differential equations is a straightforward task for spreadsheets such as EXCEL. These numerical solutions are evaluated in a worksheet document. A worksheet is an array of cells labelled A1, A2, ..., B666, etc.

We will not give detailed instructions on how to use a spreadsheet but just assume that the reader is familiar with the process.

Table 2 shows an EXCEL spreadsheet for the above example:

$$\frac{dy}{dx} = x^2 + y^2 \quad \text{with } y(0) = 1$$

and $h = 0.1$ for $x = 0.1$ to $x = 1$.

TABLE 2	xn	yn	K1	yn + 1
	0	1	1	1.1
	0.1	1.1	1.22	1.222
	0.2	1.222	1.533284	1.3753284
	0.3	1.3753284	1.98152821	1.57348122
	0.4	1.57348122	2.63584315	1.83706554
	0.5	1.83706554	3.62480978	2.19954651
	0.6	2.19954651	5.19800487	2.719347
	0.7	2.719347	7.88484811	3.50783181
	0.8	3.50783181	12.944884	4.80232022
	0.9	4.80232022	23.8722794	7.18954816
	1	7.18954816	52.6896027	12.4585084

The yn + 1 column shows the y values by using Euler's method with step size of 0.1.

SUMMARY

Euler's numerical method gives an approximate solution to the first order differential equation

$$\frac{dy}{dx} = f(x, y) \quad \text{with } y(x_0) = y_0$$

This is obtained by using the formula

13.12 $y_{n+1} = y_n + h\big[f(x_n, y_n) \big]$

The smaller the value of h the better the accuracy, but then we need more iterations.

Exercise **13(e)**

Solutions at end of book. Complete solutions available at www.palgrave.com/science/engineering/singh

1 Consider the differential equation

$$\frac{dy}{dx} = x + y^2 \quad \text{with } y(0) = 1$$

Use Euler's numerical formula with $h = 0.1$ to determine approximate solutions of y at $x = 0.1$, 0.2 and 0.3.

2 Consider the initial value problem

$$\frac{dy}{dx} = x^2 + y^2 \quad \text{with } y(0) = 1$$

Use Euler's numerical formula with $h = 0.1$ to determine the approximate solution of y at $x = 0.5$ (work to 7 d.p.).

3 For the differential equation

$$\frac{dy}{dt} = t \quad \text{with } y(0) = 0$$

complete Table 3.

TABLE 3	t	Euler y with $h = 1$	Exact $y = 0.5t^2$
	0	0	0
	1		
	2		
	3		
	4		
	5		

4 🚗 [*mechanics*]   The velocity, v, of an object falling under gravity is given by

$$\frac{dv}{dt} = 10 - 0.1v^2 \quad \text{with } v(0) = 0$$

Determine $v(0.3)$ by using Euler's numerical formula with $h = 0.05$ (work to 5 d.p.).

Use a computer algebra system or a spreadsheet for the remaining questions.

5 🚗 [*mechanics*]   The velocity, v, of a sphere falling under gravity is given by

$$\frac{dv}{dt} = 10 - v^2 \quad \text{with } v(0) = 0$$

i Use Euler's numerical formula with $h = 0.1$ and $h = 0.01$ to obtain approximate values of v at $t = 0.5$, 1, 1.5 and 2.

ii Determine the solution of the given differential equation.

iii Plot, on the same axes, the graphs for $h = 0.1$, $h = 0.01$ and the solution of part **ii**.

6 Consider the initial value problem

$$\frac{dy}{dx} = \sqrt{x + y} \quad \text{with } y(0) = 0$$

Complete Table 4.

TABLE 4	x	0	0.1	0.2	0.3	0.4	0.5
	Euler y with $h = 0.1$						
	Euler y with $h = 0.01$						
	Euler y with $h = 0.005$						

7 Consider the differential equation

$$\frac{dx}{dt} = t^2 + xe^{-t} \text{ with } x(0.1) = -1$$

Exercise **13(e) continued**

Solutions at end of book. Complete solutions available at www.palgrave.com/science/engineering/singh

i Determine the solution of the given differential equation.

ii Use Euler's formula with $h = 0.25$ and $h = 0.005$ to obtain approximate values of x at $t = 0.5, 1, 1.5, 2$ and 2.5.

iii Plot, on the same axes, the graphs for $h = 0.25$, $h = 0.005$ and the solution in part **i**.

iv Determine the percentage error in x for $h = 0.25$ and $h = 0.005$ at $t = 2$.

SECTION F **Improved Euler's method**

By the end of this section you will be able to:
▶ use improved Euler's formula
▶ compare errors between improved Euler's and Euler's methods
▶ use a computer algebra system and a spreadsheet to evaluate and compare solutions

F1 **Improved Euler's method (Heun's method)**

Say we want to solve the general first order differential equation

$$\frac{dy}{dx} = f(x, y) \text{ with } y(x_n) = y_n$$

using a numerical method. With Euler's method we have the situation shown in Fig. 30.

Euler's numerical formula from **Section E** is

Fig. 30

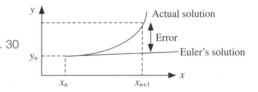

13.12 $$y_{n+1} = y_n + h\big[f(x_n, y_n) \big]$$

and at $x = x_n$ we have $f(x_n, y_n) = \dfrac{dy}{dx} = y'_n$. This is the slope at $x = x_n$. Thus substituting $f(x_n, y_n) = y'_n$ into 13.12 yields

* $$y_{n+1} = y_n + hy'_n$$

Euler's method involves finding the slope, y'_n, at the beginning of the interval, $[x_n, x_{n+1}]$. Generally this leads to a large error unless h is very small.

? **Where can we consider the slope in order to reduce this error?**

We consider two slopes, one at the beginning and the other at the end of the interval. Then taking the average of these two slopes gives an improved estimate of the slope for that interval.

Let y'_n be the slope at x_n, and y'_{n+1} be the slope at x_{n+1}. We can replace the slope, y'_n, by the average slope

$$\frac{y'_n + y'_{n+1}}{2}$$

We thus have the situation shown in Fig. 31.

Fig. 31

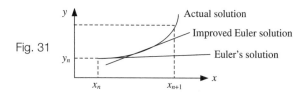

Replacing y'_n with $\dfrac{y'_n + y'_{n+1}}{2}$ in

Euler's numerical formula, [*], gives

[†] $y_{n+1} = y_n + h\left(\underbrace{\dfrac{y'_n + y'_{n+1}}{2}}_{\text{average slope}}\right)$

We need to replace the y'_{n+1} term on the Right-Hand Side because we can only find y'_{n+1} by using the old x and y values (x_n and y_n). We have

$$y'_{n+1} = f(x_{n+1}, y_{n+1}) = f\left(x_{n+1}, \underbrace{y_n + hy'_n}_{\text{by } *}\right)$$

Remember that $x_{n+1} = x_n + h$. Substituting this, $x_{n+1} = x_n + h$, into y'_{n+1}:

$$y'_{n+1} = f(x_n + h, \ y_n + hy'_n)$$

We have now established (new) y'_{n+1} in terms of the nth (old) iteration. Replacing this term in [†] and taking out a factor of $1/2$ gives

13.14 $y_{n+1} = y_n + \dfrac{h}{2}\Big[y'_n + f(x_n + h, \ y_n + hy'_n)\Big]$

13.14 is called the improved Euler's formula or Heun's formula.

This formula is complicated compared to Euler's formula of **Section E**.

Example 17

For the differential equation

$$\frac{dy}{dx} = x^2 + y^2 \quad \text{with } y(0) = 1$$

find the approximate values of y at $x = 0.1$ and $x = 0.2$ by using improved Euler's method with $h = 0.1$ (work to 6 d.p.).

Example 17 *continued*

Solution

Applying 13.14 with $h = 0.1$ gives

$$y_{n+1} = y_n + \frac{0.1}{2}\left[y_n' + f(x_n + 0.1,\ y_n + 0.1y_n')\right]$$

† $\qquad y_{n+1} = y_n + 0.05\left[y_n' + f(x_n + 0.1,\ y_n + 0.1y_n')\right]$

? What are we trying to find?

First we need to obtain y_1, the y value at $x = 0.1$ (Fig. 32).

Fig. 32

Step size is 0.1

Substituting $n = 0$ into ⌷† gives

$$y_1 = y_0 + 0.05[y_0' + f(x_0 + 0.1,\ y_0 + 0.1y_0')]$$

? What are the values of x_0 and y_0?

These are the given initial condition, $y(0) = 1$, which means $x_0 = 0$ and $y_0 = 1$.

Substituting these into y_1:

★ $\qquad y_1 = 1 + 0.05[y_0' + f(0.1,\ 1 + 0.1y_0')]$

? What is y_0' equal to?

From the question we have

$$y' = \frac{dy}{dx} = x^2 + y^2 = f(x, y)$$
$$y_0' = x_0^2 + y_0^2$$
$$= 0^2 + 1^2 = 1$$

Substituting $y_0' = 1$ into ★ gives

$$
\begin{aligned}
y_1 &= 1 + 0.05\left[y_0' + f(0.1, 1 + 0.1y_0')\right]\\
&= 1 + 0.05[1 + f(0.1, 1 + (0.1 \times 1))]\\
&= 1 + 0.05[1 + f(0.1, 1.1)]
\end{aligned}
$$

$$= 1 + 0.05\left[1 + \underbrace{(0.1^2 + 1.1^2)}_{\text{by using } f(x,y) = x^2 + y^2}\right] = 1.111$$

Next we determine y_2, the y value at $x = 0.2$. Substituting $n = 1$ into ⌷† gives

$$y_2 = y_1 + 0.05[y_1' + f(x_1 + 0.1,\ y_1 + 0.1y_1')]$$

? What are the values of x_1 and y_1?

$$x_1 = x_0 + 0.1 = 0 + 0.1 = 0.1$$
$$y_1 = 1.111 \qquad \text{[Evaluated above]}$$

Example 17 *continued*

Putting these into $y_2 = y_1 + 0.05[y_1' + f(x_1 + 0.1, y_1 + 0.1y_1')]$ gives

** $\qquad\qquad y_2 = 1.111 + 0.05[y_1' + f(0.2, 1.111 + 0.1y_1')]$

What is the value of y_1'?

$$y_1' = x_1^2 + y_1^2 \quad \text{[Given function of } x, y]$$
$$= 0.1^2 + 1.111^2 = 1.244321$$

Substituting $y_1' = 1.244321$ into ****** gives

$$y_2 = 1.111 + 0.05[1.244321 + f(0.2, \ 1.111 + (0.1 \times 1.244321))]$$
$$= 1.111 + 0.05[1.244321 + f(0.2, \ 1.235432)]$$
$$= 1.111 + 0.05\big[1.244321 + (0.2^2 + 1.235432^2)\big] \quad \text{[Using } f(x, y) = x^2 + y^2]$$
$$y_2 = 1.251531$$

We have calculated the following values: at $x = 0.1$, $y = 1.111$ and at $x = 0.2$, $y = 1.251531$.

Normally we would not work to so many significant figures but because of the effect of cumulative rounding errors, we must keep many more significant figures as guard digits.

The solutions to the differential equation of **Example 17** at $x = 0.1$ and $x = 0.2$ are

$$y(0.1) = 1.111463 \text{ and } y(0.2) = 1.253017$$

These were found by using MAPLE as shown in **Example 16**. The solutions are given correct to 6 d.p. Hence the percentage error at $x = 0.2$ is

$$\left(\frac{\text{actual value} - \text{approximate value}}{\text{actual value}} \right) \times 100 = \left(\frac{1.253017 - 1.251531}{1.253017} \right) \times 100 = 0.12\%$$

The error produced by the improved method is approximately 0.12%.

What is the size of the error developed by Euler's technique?

Euler's technique gave $y_2 = 1.222$ as evaluated in **Example 15**. The percentage error with this process at $x = 0.2$ is

$$\left(\frac{1.253017 - 1.222}{1.253017} \right) \times 100 = 2.48\%$$

Hence the improved method gives an error of 0.12% while Euler's numerical formula produces an error of 2.48%. For this example, Euler's formula gives an error of more than 20 times the error produced by the improved Euler's method. Generally there is a substantial reduction in error by using the improved approach.

Next we use MAPLE to compare exact, Euler's and improved Euler's solutions. You could use any other computer algebra system such as MATHEMATICA, DERIVE, etc. Moreover you could employ a spreadsheet such as EXCEL to find the numerical solution.

> ## Example 18

Consider the initial value problem

$$\frac{dy}{dx} = x^2 + y^2 \quad \text{with } y(0) = 1$$

i Solve this differential equation.

ii By employing Euler's numerical formula with $h = 0.1$, determine y at $x = 0.1$, 0.2, 0.3, 0.4 and 0.5.

iii By using improved Euler's formula with $h = 0.1$, determine y at $x = 0.1, 0.2, 0.3, 0.4$ and 0.5.

iv Plot, on the same axes, the graphs for Euler's, improved Euler's and the solution in part **i**.

Some commands might be difficult to follow, have a look at the help menu in MAPLE.

Solution

Using MAPLE we have

i

```
> de_18:=diff(y(x),x)=(y(x))^2+x^(2);
```

$$de\_18: = \frac{\partial}{\partial x} y(x) = y(x)^2 + x^2$$

```
> soln:= dsolve({de_18,y(0)=1},y(x)):
```

ii

```
> soln_Euler:=evalf(dsolve({de_18,y(0)=1},y(x),type=numeric,
  method=classical,output=array([0.1,0.2,0.3,0.4,0.5]),
  stepsize=0.1));
```

$$soln\_Euler := \begin{bmatrix} [x, \, y(x)] \\ \begin{bmatrix} .1 & 1.100000000 \\ .2 & 1.222000000 \\ .3 & 1.375328400 \\ .4 & 1.573481221 \\ .5 & 1.837065536 \end{bmatrix} \end{bmatrix}$$

iii

```
> soln_IMPROVEul:=evalf(dsolve({de_18, y(0)=1}, y(x),
  type=numeric,method=classical[heunform],output=array([0.1,0.2,
  0.3,0.4,0.5]),stepsize=0.1));
```

$$Soln\_IMPROVEul := \begin{bmatrix} [x, \, y(x)] \\ \begin{bmatrix} .1 & 1.111000000 \\ .2 & 1.251530674 \\ .3 & 1.436057423 \\ .4 & 1.688007332 \\ .5 & 2.048770723 \end{bmatrix} \end{bmatrix}$$

Example 18 *continued*

iv
```
> with(plots):
> p_1:=plots[odeplot](soln_Euler,[x,y(x)],linestyle=6,
  color=blue):
> p_2:=plots[odeplot](soln_IMPROVEul,[x,y(x)],linestyle=4,
  color=black):
> p_3:=plot(rhs(soln),x=0..0.5):
> display({p_1,p_2,p_3});
```

Fig. 33

Legend
---- Improved Euler   ---- Euler   —— Exact

The numerical results in the graph of Fig. 33 are plotted as points which are joined by straight lines.

Improved Euler's formula is sometimes written as

$$y_{n+1} = y_n + \frac{h}{2}\left[K_1 + K_2\right]$$

where $K_1 = f(x_n, y_n)$ and $K_2 = f(x_n + h, y_n + hK_1)$.

We can use a spreadsheet such as EXCEL to obtain the numerical solution of the above example: $\frac{dy}{dx} = x^2 + y^2$, $y(0) = 1$.

TABLE 5

xn	yn	xn + h	K1	yn + h*K1	K2	yn + 1
0	1	0.1	1	1.1	1.22	1.111
0.1	1.111	0.2	1.244321	1.2354321	1.56629247	1.25153067
0.2	1.25153067	0.3	1.60632903	1.41216358	2.08420597	1.43605742
0.3	1.43605742	0.4	2.15226092	1.65128352	2.88673725	1.68800733
0.4	1.68800733	0.5	3.00936875	1.98894421	4.20589906	2.04877072
0.5	2.04877072	0.6	4.44746147	2.49351687	6.57762638	2.60002512
0.6	2.60002512	0.7	7.1201306	3.31203818	11.4595969	3.52901149
0.7	3.52901149	0.8	12.9439221	4.8234037	23.9052232	5.37146876
0.8	5.37146876	0.9	29.4926766	8.32073641	70.0446545	10.3483353
0.9	10.3483353	1	107.898044	21.1381397	447.820949	38.1342849
1	38.1342849	1.1	1455.22369	183.656654	33730.9765	1797.44429

The yn + 1 column in Table 5 shows the approximate y values for x = 0.1, 0.2, . . . , 1.1 using improved Euler's method.

Look at the last figure in the yn + 1 column, 1797.44429. This figure is too large (it also has the wrong sign). The reason is that there is a zero in the denominator of the analytical solution at about $x = 0.96$. If we plot the graph (in MAPLE) of the analytical solution between 0.9 and 1 we can see that there is a pole at $x = 0.96$.

Euler and other numerical methods step over the pole, so its effect could be missed. The moral is, don't believe everything your computer says. Euler's methods are intrinsically unstable. If you go far enough the error in the Euler methods will always diverge. Runge–Kutta methods, discussed in the next section, are stable; errors are finite.

SUMMARY

Improved Euler's method gives an approximate solution to the first order differential equation

$$\frac{dy}{dx} = f(x, y) \text{ with } y(x_n) = y_n$$

This is attained by using the formula

13.14 $\quad y_{n+1} = y_n + \dfrac{h}{2}[y_n' + f(x_n + h, \ y_n + hy_n')]$

This formula produces a substantial reduction in error compared with Euler's numerical formula, 13.12 .

Exercise 13(f)

Solutions at end of book. Complete solutions available at www.palgrave.com/science/engineering/singh

1 Consider the differential equation

$$\frac{dy}{dx} = x + y \text{ with } y(0) = 2$$

Use improved Euler's formula with $h = 0.1$ to obtain approximate solutions of y at $x = 0.1$ and $x = 0.2$.

2 For the initial value problem

$$\frac{dy}{dx} = x^2 + y^2, \quad y(0) = 1$$

approximate $y(0.3)$ by using improved Euler's method with $h = 0.1$ (work to 6 d.p.).

3 For the differential equation

$$\frac{dy}{dx} = \ln|x + y|, \quad y(1.2) = 1$$

approximate $y(1.6)$ by employing improved Euler's formula with $h = 0.2$ (work to 4 d.p.).

Use a computer algebra system or a spreadsheet for the remaining questions.

4 Consider the initial value problem

$$\frac{dx}{dt} = t^2 + xe^{-t} \text{ with } x(0.1) = -1$$

i Solve this differential equation.

ii Determine the approximate values of x at $t = 0.5, 1, 1.5, 2$ and 2.5 by using Euler's numerical formula with $h = 0.25$.

iii Determine the approximate values of x at $t = 0.5, 1, 1.5, 2$ and 2.5 by using improved Euler's formula with $h = 0.25$.

iv Plot, on the same axes, the graphs of Euler's, improved Euler's and the exact solutions.

Solutions at end of book. Complete solutions available at www.palgrave.com/science/engineering/singh

5 Consider the initial value problem

$$\frac{dy}{dx} = x^3 + y^2, \; y(0) = -1$$

  i Solve this differential equation.

  ii Determine the approximate values of y at $x = 0.5$, 1, 1.5 and 2 by using Euler's numerical formula with $h = 0.01$.

  iii Determine the approximate values of y at $x = 0.5$, 1, 1.5 and 2 by using improved Euler's formula with $h = 0.01$.

  iv Plot, on the same axes, the graphs of Euler's, improved Euler's and the solution of part **i**.

6 Consider the initial value problem

$$\frac{dx}{dt} = 10 - x^2 \text{ with } x(0) = 0$$

  i Solve this differential equation.

  ii Determine the approximate values of x at $t = 0.1, 0.2, 0.3, 0.4$ and 0.5 by using Euler's numerical formula with $h = 0.09$.

  iii Determine the approximate values of x at $t = 0.1, 0.2, 0.3, 0.4$ and 0.5 by using improved Euler's formula with $h = 0.09$.

  iv Plot, on the same axes, the graphs of Euler's, improved Euler's and the exact solutions.

  v Determine the percentage error at $t = 0.5$ for Euler's and improved Euler's methods. What do you notice?

SECTION G **Fourth order Runge–Kutta**

By the end of this section you will be able to:
▶ apply the 4th order Runge–Kutta method
▶ use a computer algebra system to find the Runge–Kutta solutions
▶ compare errors between Euler's, improved Euler's and the Runge–Kutta methods

We have discussed Euler's and improved Euler's methods in the previous sections. In this part we examine another numerical method called 4th order Runge–Kutta. 1st, 2nd and 3rd order Runge–Kutta methods also exist but the 4th order seems to be one of the most important and accurate numerical methods.

Carle Runge was born in Germany in 1856. He studied mathematics at the University of Munich. He obtained a chair at Hannover and 18 years later he was offered a chair of Applied Mathematics at the University of Gottingen.

Martin Kutta was born in Poland in 1867 and also studied at the University of Munich. He become professor at Stuttgart in 1911.

The Runge–Kutta method was developed in 1901.

G1 Fourth order Runge–Kutta

The 4th order Runge–Kutta method is used to determine approximate solutions to a first order differential equation. Consider the **initial value problem**:

$$\frac{dy}{dx} = f(x, y) \text{ with } y(x_n) = y_n$$

Sometimes a differential equation with initial conditions is called an initial value problem.

The 4th order Runge–Kutta method is defined as

13.15 $$y_{n+1} = y_n + \frac{h}{6}[K_1 + 2(K_2 + K_3) + K_4]$$

where

$$K_1 = f(x_n, y_n)$$

$$K_2 = f\left(x_n + \frac{1}{2}h,\ y_n + \frac{1}{2}hK_1\right)$$

$$K_3 = f\left(x_n + \frac{1}{2}h,\ y_n + \frac{1}{2}hK_2\right)$$

$$K_4 = f(x_n + h,\ y_n + hK_3)$$

Example 19

Consider the differential equation

$$\frac{dy}{dx} = x^2 + y^2 \text{ with } y(0) = 1$$

Use the 4th order Runge–Kutta method to determine the approximate values of y at $x = 0.1$ and $x = 0.2$ with $h = 0.1$ (work to 6 d.p.).

Solution

We need to ascertain y_1 and y_2, the y values at $x = 0.1$ and $x = 0.2$ respectively. Substituting $n = 0$, $h = 0.1$ into **13.15** produces

* $$y_1 = y_0 + \frac{0.1}{6}\left[K_1 + 2(K_2 + K_3) + K_4\right]$$

For $n = 0$ we have $K_1 = f(x_0, y_0)$.

? **What are the values of x_0 and y_0?**

These are given by the initial condition, $y(0) = 1$, which means $x_0 = 0$ and $y_0 = 1$. Putting these into $K_1 = f(x_0, y_0)$ gives

† $$K_1 = f(0, 1)$$

? **What does f represent?**

This is the function of x and y on the Right-Hand Side of the given differential equation:

$$f(x,y) = x^2 + y^2$$

$$f(0,1) = 0^2 + 1^2 = 1$$

Example 19 *continued*

It follows from ⬛ † that $K_1 = 1$. We need to find K_2. Putting $h = 0.1$, $n = 0$ and $K_1 = 1$ into
$K_2 = f\left(x_n + \dfrac{1}{2}h,\ y_n + \dfrac{1}{2}hK_1\right)$ gives

$$K_2 = f\left(x_0 + \frac{1}{2}(0.1),\ y_0 + \frac{1}{2}[\,0.1 \times 1]\right)$$
$$= f(0 + 0.05,\ 1 + 0.05) \qquad \text{[Substituting } x_0 = 0 \text{ and } y_0 = 1]$$
$$= f(0.05,\ 1.05)$$
$$K_2 = 0.05^2 + 1.05^2 = 1.105$$

? **What else do we need to find to use ⬛ * ?**

K_3 and K_4. Evaluating K_3 is similar to finding K_2. Substituting $h = 0.1$, $n = 0$ and $K_2 = 1.105$
into $K_3 = f\left(x_n + \dfrac{1}{2}h,\ y_n + \dfrac{1}{2}hK_2\right)$ gives

$$K_3 = f\left(x_0 + \frac{1}{2}(0.1),\ y_0 + \frac{1}{2}[\,0.1 \times 1.105]\right)$$
$$= f(0 + 0.05,\ 1 + 0.05525) \qquad \text{[Substituting } x_0 = 0 \text{ and } y_0 = 1]$$
$$= f(0.05,\ 1.05525)$$
$$K_3 = 0.05^2 + 1.05525^2 = 1.116053$$

? **How do we calculate K_4?**

By placing $n = 0$, $h = 0.1$ and $K_3 = 1.116053$ into $K_4 = f(x_n + h,\ y_n + hK_3)$:

$$K_4 = f(x_0 + 0.1,\ y_0 + [\,0.1 \times 1.116053])$$
$$= f(0 + 0.1,\ 1 + 0.111605)$$
$$= f(0.1,\ 1.111605)$$
$$K_4 = 0.1^2 + 1.111605^2 = 1.245666$$

Substituting the calculated values, $y_0 = 1$, $K_1 = 1$, $K_2 = 1.105$, $K_3 = 1.116053$ and
$K_4 = 1.245666$, into ⬛ * gives

$$y_1 = y_0 + \frac{0.1}{6}[\,K_1 + 2(K_2 + K_3) + K_4]$$
$$= 1 + \frac{0.1}{6}[\,1 + 2(1.105 + 1.116053) + 1.245666]$$
$$= 1.111463$$

Similarly we evaluate y_2. Putting $n = 1$ and $h = 0.1$ into ⬛13.15 results in

** $\qquad y_2 = y_1 + \dfrac{0.1}{6}[\,K_1 + 2(K_2 + K_3) + K_4]$

? **What is K_1 equal to in this case?**

$$K_1 = f(x_1,\ y_1)$$

Example 19 *continued*

What are the values of x_1 and y_1?

$$x_1 = 0.1 \text{ and } y_1 = 1.111463 \qquad \text{[Evaluated above]}$$

Substituting these into $K_1 = f(x_1, y_1)$ yields

$$K_1 = f(0.1, 1.111463)$$
$$= 0.1^2 + 1.111463^2 = 1.245350$$

We need to determine K_2 next. Putting $n = 1$, $h = 0.1$ and $K_1 = 1.245350$ into

$$K_2 = f\left(x_n + \frac{1}{2}h, \ y_n + \frac{1}{2}hK_1\right) \text{ gives}$$

$$K_2 = f\left(x_1 + \frac{1}{2}(0.1), \ y_1 + \frac{1}{2}[\,0.1 \times 1.245350]\right)$$

$$= f(0.1 + 0.05, \ 1.111463 + 0.062268) \quad \text{[Substituting } x_1 = 0.1, \ y_1 = 1.111463\text{]}$$

$$= f(0.15, \ 1.173731)$$

$$K_2 = 0.15^2 + 1.173731^2 = 1.400144$$

Similarly by the definition of K_3 we have

$$K_3 = f\left(x_1 + \frac{1}{2}(0.1), \ y_1 + \frac{1}{2}[\,0.1 \times 1.400144]\right)$$

$$= f(0.1 + 0.05, \ 1.111463 + 0.070007)$$

$$= f(0.15, \ 1.181470)$$

$$K_3 = 0.15^2 + 1.181470^2 = 1.418371$$

Next we determine K_4:

$$K_4 = f(x_1 + 0.1, \ y_1 + [\,0.1 \times 1.418371])$$

$$= f(0.1 + 0.1, \ 1.111463 + 0.141837)$$

$$= f(0.2, \ 1.2533)$$

$$K_4 = 0.2^2 + 1.2533^2 = 1.610761$$

Substituting $y_1 = 1.111463$, $K_1 = 1.245350$, $K_2 = 1.400144$, $K_3 = 1.418371$ and $K_4 = 1.610761$ into ▓▓ ** ▓▓ gives

$$y_2 = 1.111463 + \frac{0.1}{6}[\,1.245350 + 2(1.400144 + 1.418371) + 1.610761]$$

$$= 1.253015$$

The calculated values are $x = 0.1$, $y = 1.111463$ and at $x = 0.2$, $y = 1.253015$.

Next we compare the errors for the differential equation of **Example 19** between Euler's, improved Euler's and the 4th order Runge–Kutta solutions.

The y values at $x = 0.2$ for Euler's, improved Euler's and the 4th order Runge–Kutta solutions are 1.222, 1.251531 and 1.253015 respectively. The actual value (correct to 6 d.p.)

is 1.253017 as evaluated in **Example 16**. The Runge–Kutta value might well be the same as, or closer to, the actual value if we had performed the calculations to more decimal places. However, notice how close the Runge–Kutta approximation is to the actual value compared to Euler's and improved Euler's.

We can use MAPLE to plot solutions of Euler's, improved Euler's and the 4th order Runge–Kutta methods, as the next example shows.

Example 20

Consider the differential equation

$$\frac{dy}{dx} = x^2 + y^2 \text{ with } y(0) = 1$$

i Solve this differential equation.
ii Find y at $x = 0.2, 0.4, 0.6$ and 0.8 by using Euler's method with $h = 0.1$.
iii Repeat **ii** for improved Euler's method with $h = 0.1$.
iv Repeat **ii** for the 4th order Runge–Kutta method with $h = 0.1$.
v Plot, on the same axes, the graphs of Euler's, improved Euler's and the 4th order Runge–Kutta solutions, and the solution of part **i**.
  What do you notice about your graphs?

Solution

Using Maple we have

i
```
> de_20:=diff(y(x),x)=(x^2)+(y(x)^2);
```

$$de\_20: = \frac{\partial}{\partial x}y(x) = x^2 + y(x)^2$$

```
> soln:=dsolve({de_20,y(0)=1},y(x)):
```

ii
```
> e_soln:=dsolve({de_20,y(0)=1}, y(x), type=numeric, method=
  classical, output=array ([0.2, 0.4, 0.6, 0.8]), stepsize=0.1);
```

$$e\_soln: = \begin{bmatrix} [x, y(x)] \\ \begin{bmatrix} .2 & 1.22200000000000020 \\ .4 & 1.57348122078465646 \\ .6 & 2.19954651435706426 \\ .8 & 3.50783181255390319 \end{bmatrix} \end{bmatrix}$$

iii
```
> improve_soln:=dsolve({de_20, y(0)=1},y(x),type=numeric,
  method= classical [heunform], output=array([.2,.4,.6,.8]),
  stepsize=0.1);
```

$$improve\_soln: = \begin{bmatrix} [x, y(x)] \\ \begin{bmatrix} .2 & 1.25153067368552052 \\ .4 & 1.68800733199093366 \\ .6 & 2.60002511535832781 \\ .8 & 5.37146875547270320 \end{bmatrix} \end{bmatrix}$$

Example 20 *continued*

iv
```
> rk_soln:=dsolve({de_20,y(0)=1},y(x),type=numeric,
  method=classical[rk4], output=array([.2,.4,.6,.8]),
  stepsize=0.1);
```

$$rk\_soln := \begin{bmatrix} [x, y(x)] \\ \begin{bmatrix} .2 & 1.25301517460353452 \\ .4 & 1.69609790372281698 \\ .6 & 2.64386019707933162 \\ .8 & 5.84201333527994748 \end{bmatrix} \end{bmatrix}$$

v
```
> with(plots):p_1:=plots[odeplot](e_soln,[x,y(x)],
  linestyle=4,color=black):
```
```
> p_2:=plots[odeplot](improve_soln,[x,y(x)],linestyle=2,
  color=blue):
```
```
> p_3:=plots[odeplot](rk_soln,[x,y(x)],linestyle=6, color=coral):
```
```
> p_4:=plot(rhs(soln),x=0..0.8):
```
```
> display({p_1,p_2,p_3,p_4});
```

Fig. 34

Legend
```
-·-·-·-·-  Euler solution
─────────  Improved Euler solution
- - - - -  Runge–Kutta solution
─────────  Exact solution
```

The Runge–Kutta solution seems to be the closest to the actual solution. Improved Euler's is a close second and Euler's solution increases in error as you move away from 0. Remember that the numerical results are plotted as points. So when comparing exact and numerical solutions, we should ignore the portions between the kinks in the lines.

We can also find the Runge–Kutta solution of differential equations by using a spreadsheet. Try the above example on a spreadsheet available to you.

SUMMARY

The fourth order Runge–Kutta solution of the differential equation

$$\frac{dy}{dx} = f(x, y) \text{ with } y(x_n) = y_n$$

is given by

13.5 $y_{n+1} = y_n + \dfrac{h}{6}[K_1 + 2(K_2 + K_3) + K_4]$

where

$$K_1 = f(x_n, y_n)$$

$$K_2 = f\left(x_n + \frac{1}{2}h, \ y_n + \frac{1}{2}hK_1\right)$$

$$K_3 = f\left(x_n + \frac{1}{2}h, \ y_n + \frac{1}{2}hK_2\right)$$

$$K_4 = f(x_n + h, \ y_n + hK_3)$$

Exercise 13(g)

Solutions at end of book. Complete solutions available at www.palgrave.com/science/engineering/singh

1 Consider the differential equation

$$\frac{dy}{dx} = x + y \text{ with } y(0) = 2$$

Use the 4th order Runge–Kutta method with $h = 0.1$ to obtain approximate solutions of y at $x = 0.1$ and $x = 0.2$ (work to 6 d.p.).

2 For the initial value problem

$$\frac{dy}{dx} = \sqrt{x + y}, \quad y(0) = 0$$

find $y(0.4)$ using the 4th order Runge–Kutta formula with $h = 0.2$ (work to 4 d.p.).

3 Consider the differential equation

$$\frac{dy}{dx} = \ln(x + y) \text{ with } y(1.2) = 1$$

Approximate $y(1.6)$ by employing the 4th order Runge–Kutta formula with $h = 0.2$ (work to 4 d.p.).

4 [mechanics]   The velocity, v, of an object dropped in air is given by

$$\frac{dv}{dt} = 9.8 - \frac{v^2}{10} \text{ with } v(0) = 0$$

Apply the 4th order Runge–Kutta method with $h = 1$ to find an approximate v at $t = 2$ s (work to 4 d.p.).

Use a computer alegbra system for the remaining questions.

5 Consider the differential equation

$$\frac{dy}{dx} = x^4 + y^2 \text{ with } y(0) = 1$$

  i Solve this differential equation.
  ii Determine approximate values of y at $x = 0.2, \ 0.4, \ 0.6$ and 0.8 by using Euler's method with $h = 0.05$.
  iii Determine approximate values of y at $x = 0.2, \ 0.4, \ 0.6$ and 0.8 by using the 4th order Runge–Kutta method with $h = 0.05$.
  iv Plot, on the same axes, the graphs of Euler's, Runge–Kutta and the exact solutions.

6 Consider the differential equation

$$\frac{dy}{dx} = e^y \sin(x), \ y(0) = -1$$

Solutions at end of book. Complete solutions available at www.palgrave.com/science/engineering/singh

Exercise **13(g) continued**

i Solve this differential equation.
ii Determine approximate values of y at $x = 1, 2, 3$ and 4 by using improved Euler's method with $h = 0.2$.
iii Determine the approximate values of y at $x = 1, 2, 3$ and 4 by using the

4th order Runge–Kutta method with $h = 0.2$.
iv Plot, on the same axes, the graphs of improved Euler's, Runge–Kutta and the exact solutions.

Solutions at end of book. Complete solutions available at www.palgrave.com/science/engineering/singh

Miscellaneous exercise **13**

1 [thermodynamics]   In thermodynamics the relationship between mass, m, volume, V, and pressure, P, is given by

$$mCd(pV) = pVdm$$

where C is a constant. Show that

$$m = (pV)^C$$

(Assume the constant of integration to be zero.)

2 [mechanics]   The acceleration, a, of a vehicle is given by

$$a = v^2$$

where v is the velocity. Find an expression for v given that $a = \dfrac{dv}{dt}$ and when $t = 0$, $v = 10$.

3 [mechanics]   The acceleration, $a = \dfrac{dv}{dt}$, of a vehicle is given by

$$a = -kv^2$$

where k is a positive constant and v is the velocity. Given that when $t = 0$, $v = 0.1$, determine v as a function of time t. By using a graphical calculator or a computer algebra system, plot graphs of v against t for $k = 0.1$, 1 and 10 ($0 \le t \le 10$). What effect does k have?

4 [mechanics]   The velocity, v, of a vehicle during the application of brakes is given by

$$\frac{dv}{dt} = -kv$$

where t is time and k is a constant. If the velocity at time $t = 0$ is V_0, show that $v = V_0\,e^{-kt}$.

5 [thermodynamics]   The PV relationship of an ideal gas is given by

$$\frac{dP}{dV} = -\frac{kP}{V}$$

where P is pressure, V is volume and k is a constant. Show that

$$PV^k = \text{constant}$$

6 [thermodynamics]   The PV relationship of an ideal gas is given by

$$\frac{dP}{P} + k\frac{dV}{V} = 0$$

where P is pressure, V is volume and k is a constant. Show that

$$PV^k = \text{constant}$$

7 [electrical principles]   The voltage, v, across an inductor, L, is given by

$$v = L\frac{di}{dt}$$

where i is the current through the inductor. For $v = 240\sin(100t)$, $L = 1 \times 10^{-3}$ H and the initial condition $t = 0$, $i = 0$, show that

$$i = 2400\big[1 - \cos(100t)\big]$$

Solutions at end of book. Complete solutions available at www.palgrave.com/science/engineering/singh

Miscellaneous exercise **13 continued**

8 [electrical principles]   A circuit consists of a resistor, R, and a capacitor, C, in series, and is described by the differential equation

$$RC\frac{dV}{dt} + V = E$$

where E is the constant applied e.m.f. and V is the voltage across the capacitor. Show that, if $V(0) = 0$ then

$$V = E(1 - e^{-t/RC})$$

Questions 9 to 11, inclusive, are in the field of [mechanics].

9 The velocity, v, of an object initially at rest ($t = 0$, $v = 0$) is given by

$$100\frac{dv}{dt} = 200 - 10v$$

i Find an expression for v in terms of t.
ii Find v for large t ($t \to \infty$). (This v is called the terminal velocity.)
iii Sketch the graph of v against t for $t \geq 0$.

10 An object of mass, m, falls from rest ($t = 0$, $v = 0$) and its velocity v is given by

$$m\frac{dv}{dt} = mg - kv \qquad \left(g > \frac{k}{m}v\right)$$

where k is a positive constant. Show that

$$v = \frac{mg}{k}(1 - e^{-kt/m})$$

Determine the (terminal) velocity as $t \to \infty$.

11 By applying Newton's law to a parachutist falling from rest, we obtain the first order differential equation

$$m\frac{dv}{dt} = mg - kv^2 \qquad \left(v > \sqrt{\frac{mg}{k}}\right)$$

where m is mass, v is velocity, t is time, g is acceleration due to gravity and k is a constant. Show that

$$v = \sqrt{\frac{mg}{k}\left(\frac{1 + Ae^{-2\beta t}}{1 - Ae^{-2\beta t}}\right)}$$

where $\beta = \sqrt{\dfrac{gk}{m}}$ and A is a constant.

What is the value of the terminal velocity?

$$\left(\text{Hint: Use} \int \frac{du}{u^2 - a^2} = \frac{1}{2a}\ln\left|\frac{u - a}{u + a}\right|.\right)$$

Questions 12 to 14, inclusive, are in the field of [electrical principles].

12 The current, i, through a series RL circuit is given by

$$L\frac{di}{dt} + Ri = t$$

Show that, if $t = 0$, $i = 0$, then

$$i = \frac{t}{R} + \frac{L}{R^2}\left(e^{-\frac{R}{L}t} - 1\right)$$

13

Fig. 35

Figure 35 shows a voltage, $5e^{-t}\cos(100\pi t)$, applied to a circuit with a capacitance, (0.9×10^{-6}) F, in series with a resistance, (13×10^3) Ω. Show that

$$\frac{dv}{dt} = 427.35e^{-t}\cos(100\pi t) - 85.47v$$

where v is the voltage across the capacitor.

14 By applying Kirchhoff's law to a circuit we obtain the differential equation

$$L\frac{di}{dt} + Ri = E\cos(\omega t)$$

where L is inductance, R is resistance, E is e.m.f., i is current, t is time and ω is angular frequency.

Miscellaneous exercise **13 continued**   Solutions at end of book. Complete solutions available at www.palgrave.com/science/engineering/singh

For the initial condition, when $t = 0$, $i = 0$, show that

$$i = \frac{E}{R^2 + \omega^2 L^2} \Big[R\cos(\omega t)$$

$$+ \omega L\sin(\omega t) - Re^{-Rt/L} \Big]$$

Use a computer algebra system for the remaining questions.

15 Consider the differential equation

$$\frac{dy}{dx} = x^2 + y^2 \text{ with } y(0.5) = 0$$

i Solve this differential equation.
ii Determine approximate values of y at $x = 1$, 1.2, 1.4, 1.6 and 1.8 by using Euler's method with $h = 0.1$.
iii Determine approximate values of y at $x = 1$, 1.2, 1.4, 1.6 and 1.8 by using the 4th order Runge–Kutta method with $h = 0.1$.
iv Plot, on the same axes, the graphs of Euler's, Runge–Kutta and the exact solutions.
What do you notice about your results?

16 Consider the differential equation

$$\frac{dy}{dx} = x^2 + ye^x, \quad y(0) = -1$$

i Solve this differential equation.
ii Determine approximate values of y at $x = 0.2$, 0.4, 0.6 and 1.2 by using improved Euler's method with $h = 0.2$.
iii Determine approximate values of y at $x = 0.2$, 0.4, 0.6 and 1.2 by using the 4th order Runge–Kutta method with $h = 0.2$.
iv Plot, on the same axes, the graphs of improved Euler's, Runge–Kutta and the exact solutions.

17 [*electrical principles*] Applying Kirchhoff's law to a circuit we obtain the differential equation

$$(2 \times 10^{-3})\frac{di}{dt} + (1 \times 10^3)i = 10$$

where i is the current through the circuit. Given the initial condition $i(0) = 0$:
i determine the current i
ii plot the graph of i against t between $t = 0$ and $t = 10$ μs.

18 [*electrical principles*]

Fig. 36   9 V

Figure 36 shows a variable capacitor, C, connected in series with a resistor of resistance 10 kΩ and a 9 V supply. The differential equation for this circuit is

$$(10 \times 10^3)C\frac{dv}{dt} + v = 9$$

where v is the voltage across the capacitor, C.
i Solve this differential equation for the initial condition, $t = 0$, $v = 0$.
ii Substitute the following values of C into your solution:

$$C = 10 \times 10^{-6}, 20 \times 10^{-6}, 30 \times 10^{-6},$$
$$40 \times 10^{-6} \text{ and } 50 \times 10^{-6}$$

iii Plot, on the same axes, v against t for the above values of C and $0 \le t \le 1.5$.
iv What value does the voltage, v, tend to as $t \to \infty$? (This is called the 'final value'.)
v What effect does the change in capacitance, C, have on the voltage, v, across it?

19 [*electrical principles*]

Fig. 37   9 V

Miscellaneous exercise **13 continued**

Solutions at end of book. Complete solutions available at www.palgrave.com/science/engineering/singh

Figure 37 shows a variable inductor, L, connected in series with a resistor of value 10 kΩ and an applied voltage of 9 V. The differential equation for this circuit is

$$(10 \times 10^3)i + L\frac{di}{dt} = 9$$

i Solve this differential equation for the initial condition, $t = 0, i = 0$.

ii Substitute the following values of L into your solution:

$$L = \frac{1}{2}, \frac{1}{4}, \frac{1}{8}, \frac{1}{16} \text{ and } \frac{1}{32}$$

iii Plot, on the same axes, i against t for the values of L in part **ii** and $0 \le t \le 0.0001$.

iv What value does the current, i, tend to as $t \to \infty$? (This is called the 'final value'.)

v What effect does increasing the inductance, L, have on the current, i?

CHAPTER 14
Second Order Linear Differential Equations

> ## SECTION A **Homogeneous differential equations**
>
> By the end of this section you will be able to:
> ▶ understand the term *homogeneous second order differential equation*
> ▶ state the relevant solution to the second order differential equation by examining solutions of the corresponding characteristic equation

A1 **Homogeneous equations**

Generally

14.1 $\qquad a\dfrac{d^2y}{dx^2} + b\dfrac{dy}{dx} + cy = f(x) \quad (a \neq 0)$

?
is called a second order differential equation (2nd order DE), where a, b and c are constants and $f(x)$ is a function of x. Also y is a function of the independent variable, x. **What does the term 'second order' mean?**

The highest order of the derivative in the differential equation is 2. If $f(x) = 0$ then **14.1** becomes

14.2 $\qquad a\dfrac{d^2y}{dx^2} + b\dfrac{dy}{dx} + cy = 0$

This, **14.2**, is called a homogeneous second order differential equation because the Right-Hand Side is zero.

? **If we are asked to solve such an equation then what are we trying to find?**

We need to find y, a **function** of x, which satisfies **14.2**. Note that the solution is a function of x, that is $y = y(x)$.

? **Try the function $y = e^{mx}$. Why?**

Because if you differentiate e^{mx} you end up with a e^{mx} term again. Thus

$$\frac{dy}{dx} = me^{mx}$$

Differentiating again: $\dfrac{d^2y}{dx^2} = m^2 e^{mx}$

Substituting these into

14.2 $\qquad a\dfrac{d^2y}{dx^2} + b\dfrac{dy}{dx} + cy = 0$

gives

$$am^2 e^{mx} + bme^{mx} + ce^{mx} = 0$$

$$e^{mx}(am^2 + bm + c) = 0 \quad \text{[Factorizing out } e^{mx}\text{]}$$

> **So we have $e^{mx} = 0$ or $am^2 + bm + c = 0$. What about $e^{mx} = 0$?**

For any real values of mx, $e^{mx} \neq 0$. (One of the properties of the exponential function is that it cannot equal zero.) Therefore we must have

$$am^2 + bm + c = 0$$

which is a quadratic equation. Since every quadratic equation is solvable, $y = e^{mx}$ is a solution of 14.2 provided that m satisfies

14.3 $am^2 + bm + c = 0$

This quadratic equation is called the **characteristic equation** of the differential equation

$$a\frac{d^2y}{dx^2} + b\frac{dy}{dx} + cy = 0$$

> **What is the characteristic equation of $2\dfrac{d^2y}{dx^2} + 3\dfrac{dy}{dx} + 7y = 0$?**

$$2m^2 + 3m + 7 = 0$$

 14.3 is also known as the **auxiliary equation**.

The solution of 14.3 is given by the quadratic equation formula:

$$m = \frac{-b \pm \sqrt{b^2 - 4ac}}{2a}$$

or by factorizing the quadratic where appropriate.

> **How many different types of roots are there of a quadratic equation?**

Three

   i   Real and different, m_1 and m_2
   ii Real and equal, $m_1 = m_2 = m$
   iii Complex, $m = \alpha \pm j\beta$

The theory of differential equations says that, if $y_1(x)$ and $y_2(x)$ are two independent solutions of the differential equation

$$a\frac{d^2y}{dx^2} + b\frac{dy}{dx} + cy = 0$$

then $Ay_1(x) + By_2(x)$ is also a solution (A and B are any real constants). This is only true if the differential equation is linear, that is y has index 1 in the differential equation.

For the case of real and different roots we have $e^{m_1x}\ [=y_1(x)]$ and $e^{m_2x}\ [=y_2(x)]$ as solutions of the differential equation. Hence

$$y = Ae^{m_1x} + Be^{m_2x}$$

is also a solution. This is called the **general solution** because we do not have specific values for the constants A and B.

Hence the solution, y, of the differential equation

$$a\frac{d^2y}{dx^2} + b\frac{dy}{dx} + cy = 0$$

depends on the type of root of the characteristic equation

$$am^2 + bm + c = 0$$

If the roots are real and different, m_1 and m_2, then

14.4 $y = Ae^{m_1x} + Be^{m_2x}$

If the roots are real and equal, m_1 (or m_2), then

14.5 $y = (A + Bx)e^{m_1x}$

If the roots are complex, $m = \alpha \pm j\beta$, then

14.6 $y = e^{\alpha x}\left[A\cos(\beta x) + B\sin(\beta x)\right]$

(A and B are constants in 14.4 to 14.6.)

We will not show results 14.5 and 14.6 but just use them whenever appropriate.

Every homogeneous differential equation has a **trivial solution**, $y = 0$, because differentiating $y = 0$ once and twice still gives 0, therefore it satisfies the homogeneous differential equation. So if the question says solve a differential equation it means find the **non-trivial** solution.

Example 1

Solve

$$\frac{d^2y}{dx^2} + 2\frac{dy}{dx} - 3y = 0$$

Solution

We need to find y which satisfies the above equation. **What is the characteristic equation in this case?**

$$m^2 + 2m - 3 = 0$$
$$(m + 3)(m - 1) = 0 \quad \text{[Factorizing]}$$
$$m_1 = -3, \, m_2 = 1 \quad \text{[Solving]}$$

Hence the solution to the characteristic equation is two real and different roots, $m_1 = -3$ and $m_2 = 1$. Substituting these into

14.4 $y = Ae^{m_1x} + Be^{m_2x}$

gives

$$y = Ae^{-3x} + Be^x$$

This is the general solution of the given differential equation.

Example 2

Solve

$$\frac{d^2y}{dx^2} + 2\frac{dy}{dx} + y = 0$$

Example 2 *continued*

Solution

The characteristic equation is

$$m^2 + 2m + 1 = 0$$
$$(m + 1)^2 = 0 \qquad \text{[Factorizing]}$$
$$m_1 = m_2 = -1 \qquad \text{[Solving]}$$

The solution to the characteristic equation is two equal roots, $m_1 = m_2 = -1$. Substituting $m_1 = -1$ into

14.5 $\qquad y = (A + Bx)e^{m_1 x}$

gives the general solution

$$y = (A + Bx)e^{-x}$$

For **Example 3** we need to use the following result of complex numbers:

14.7 $\qquad \sqrt{-k^2} = \pm jk$

where k is any real number.

Example 3

Find the general solution of

$$\frac{d^2 y}{dx^2} + \frac{dy}{dx} + y = 0$$

Solution

The characteristic equation is

$$m^2 + m + 1 = 0$$

To solve this quadratic equation we use the formula:

1.16 $\qquad m = \dfrac{-b \pm \sqrt{b^2 - 4ac}}{2a}$

Putting $a = 1$, $b = 1$ and $c = 1$ into 1.16 gives

$$m = \frac{-1 \pm \sqrt{1^2 - 4}}{2}$$
$$= \frac{-1 \pm \sqrt{-3}}{2} = -\frac{1}{2} \pm \frac{\sqrt{-3}}{2}$$
$$m = -\frac{1}{2} \pm j\,\frac{\sqrt{3}}{2} \qquad \text{[Complex roots]}$$
$$\text{by } 14.7$$

Equating this with $m = \alpha \pm j\beta$ gives $\alpha = -\dfrac{1}{2}$ and $\beta = \dfrac{\sqrt{3}}{2}$.

Example 3 *continued*

Substituting this, $\alpha = -\dfrac{1}{2}$ and $\beta = \dfrac{\sqrt{3}}{2}$, into

14.6 $y = e^{\alpha x}\left[A \cos(\beta x) + B \sin(\beta x)\right]$

gives the general solution

$$y = e^{-x/2}\left[A \cos\left(\frac{\sqrt{3}}{2}x\right) + B \sin\left(\frac{\sqrt{3}}{2}x\right)\right]$$

We started this section with the differential equation

$$a\frac{d^2 y}{dx^2} + b\frac{dy}{dx} + cy = f(x) \quad (a \neq 0)$$

A differential equation written in this format is said to be in **standard form**. Sometimes we have to rearrange a given differential equation to place it into standard form, for example

$$\frac{d^2 y}{dx^2} + 3\frac{dy}{dx} + \cos(x) = 0$$

is not in standard form. **How do we write this in standard form?**

Subtract $\cos(x)$ from both sides:

$$\frac{d^2 y}{dx^2} + 3\frac{dy}{dx} = -\cos(x)$$

The function of x, $f(x)$, is on the Right-Hand side of the equation.

SUMMARY

The solution of the differential equation

$$a\frac{d^2 y}{dx^2} + b\frac{dy}{dx} + cy = 0$$

depends on the type of roots of the characteristic equation

$$am^2 + bm + c = 0$$

If the roots are real and different, m_1 and m_2, then

14.4 $y = Ae^{m_1 x} + Be^{m_2 x}$

If the roots are real and equal, m_1, then

14.5 $y = (A + Bx)e^{m_1 x}$

If the roots are complex, $m = \alpha \pm j\beta$, then

14.6 $y = e^{\alpha x}[A \cos(\beta x) + B \sin(\beta x)]$

Solutions at end of book. Complete solutions available at www.palgrave.com/science/engineering/singh

Exercise **14(a)**

1 Find the general solutions of the following equations:

a $\dfrac{d^2y}{dx^2} + 5\dfrac{dy}{dx} + 6y = 0$

b $\dfrac{d^2y}{dx^2} + 4\dfrac{dy}{dx} + 4y = 0$

c $\dfrac{d^2y}{dx^2} - 2\dfrac{dy}{dx} + 4y = 0$

2 Solve:

a $\dfrac{d^2y}{dx^2} + 3\dfrac{dy}{dx} + 2y = 0$

b $\dfrac{d^2y}{dx^2} - 7\dfrac{dy}{dx} + 6y = 0$

c $\dfrac{d^2y}{dx^2} - 6\dfrac{dy}{dx} + 9y = 0$

d $\dfrac{d^2y}{dx^2} - 4\dfrac{dy}{dx} + 5y = 0$

e $\dfrac{d^2y}{dx^2} - 8\dfrac{dy}{dx} + 16y = 0$

3 Solve:

a $\dfrac{d^2y}{dx^2} + 6\dfrac{dy}{dx} + 9y = 0$

b $10\dfrac{d^2y}{dx^2} + 50\dfrac{dy}{dx} + 250y = 0$

c $-\dfrac{d^2y}{dx^2} - 3\dfrac{dy}{dx} + 8y = 0$

SECTION B **Engineering applications**

By the end of this section you will be able to:
▶ solve engineering applications of second order differential equations with initial conditions
▶ solve boundary value problems

This is a particularly challenging section because there are lots of examples from engineering which involve complicated symbols and notation. However the concepts are similar in nature to **Section A**, the major difference being that we are not using simple numbers as coefficients of the derivatives.

B1 **Particular homogeneous solutions**

In engineering applications we are normally given some initial conditions of the problem. For example, in the motion of a projectile we might be given that at time $t = 0$, the displacement and velocity are zero. These values are the initial conditions. The differential equation with initial conditions is called an **initial value problem**.

Given initial conditions we can evaluate unknowns A and B in the general solution of the equation, and this is then known as the **particular solution**.

Example 4 *electrical principles*

By applying Kirchhoff's voltage law to a circuit we obtain the second order differential equation

$$\frac{d^2 i}{dt^2} + 7\frac{di}{dt} + 10i = 0$$

where i is the current and t is the time. Find the current i in terms of t for the initial conditions, when $t = 0$, both $i = 0$ and $\frac{di}{dt} = 3$.

Solution

The characteristic equation is

$$m^2 + 7m + 10 = 0$$
$$(m + 5)(m + 2) = 0 \quad \text{[Factorizing]}$$
$$m_1 = -5, \ m_2 = -2 \quad \text{[Solving]}$$

We have real and different roots, so putting these into

14.4 $\quad i = Ae^{m_1 t} + Be^{m_2 t}$

gives

† $\quad i = Ae^{-5t} + Be^{-2t}$

Substituting the given initial condition $t = 0$, $i = 0$ into **†**

$$Ae^0 + Be^0 = 0$$
$$A + B = 0 \quad \text{[Because } e^0 = 1\text{]}$$

Using the other given initial condition $t = 0$, $\frac{di}{dt} = 3$: to find $\frac{di}{dt}$ we need to differentiate $i = Ae^{-5t} + Be^{-2t}$:

$$\frac{di}{dt} = -5Ae^{-5t} - 2Be^{-2t} \quad \left[\text{Because } \frac{d}{dt}(e^{kt}) = ke^{kt}\right]$$
$$3 = -5Ae^0 - 2Be^0 = -5A - 2B$$

Thus we have two simultaneous equations:

$$A + B = 0$$
$$-5A - 2B = 3$$

From the first equation we have

$$A = -B$$

Substituting this into the second equation gives

$$-5(-B) - 2B = 5B - 2B = 3B = 3$$

Therefore $B = 1$. Since $A = -B$ so $A = -1$.

Substituting $A = -1$ and $B = 1$ into $i = Ae^{-5t} + Be^{-2t}$ gives

$$i = -e^{-5t} + e^{-2t} = e^{-2t} - e^{-5t}$$

This is a particular solution since we have found particular values of A and B.

B2 Differential equation of the form $\dfrac{d^2y}{dx^2} + k^2y = 0$

Consider $\dfrac{d^2y}{dx^2} + k^2y = 0$ where k is a constant. Differential equations of this type arise in vibrational problems. One of the simplest cases is known as simple harmonic motion (SHM). Consider a particle, P, moving to and fro about its equilibrium position O (Fig. 1).

Fig. 1

The equation of motion is given by

$$\frac{d^2x}{dt^2} + k^2x = 0$$

where x is displacement, k is a constant and t is time.

In mechanics the differential equation might be written as

$$\ddot{x} + k^2x = 0$$

where $\ddot{x}$ represents the second derivative, $\dfrac{d^2x}{dt^2}$. Also $\dot{x} = \dfrac{dx}{dt}$.

The characteristic equation of the general second order differential equation $\dfrac{d^2y}{dx^2} + k^2y = 0$ is

$$m^2 + k^2 = 0$$
$$m^2 = -k^2$$
$$m = \sqrt{-k^2} \underset{\text{by }14.7}{=} \pm jk \text{ hence } m = 0 \pm jk$$

Equating $m = 0 \pm jk$ with $m = \alpha \pm j\beta$ gives $\alpha = 0$ and $\beta = k$.

Putting $\alpha = 0$ and $\beta = k$ into

14.6 $y = e^{\alpha x}\left[A\cos(\beta x) + B\sin(\beta x)\right]$

gives

$$y = \underset{=1}{e^0}\left[A\cos(kx) + B\sin(kx)\right]$$
$$= A\cos(kx) + B\sin(kx)$$

Hence the general solution of the differential equation $\dfrac{d^2y}{dx^2} + k^2y = 0$ is

14.8 $y = A\cos(kx) + B\sin(kx)$

14.7 $\sqrt{-k^2} = \pm jk$

Example 5 *vibrations*

The motion of a spring–mass system is described by

$$\frac{d^2x}{dt^2} + 25x = 0$$

where x is displacement and t is time. Determine the particular solution for this differential equation with initial conditions, when $t = 0$, both $x = 1$ and $\frac{dx}{dt} = 10$.

Solution

Since $25 = 5^2$, the given differential equation can be written as

$$\frac{d^2x}{dt^2} + 5^2x = 0$$

Putting $k = 5$ into

14.8 $x = A\cos(kt) + B\sin(kt)$

gives the general solution

$$x = A\cos(5t) + B\sin(5t)$$

Substituting the initial condition, $t = 0$, $x = 1$, yields

$$1 = A\underbrace{\cos(0)}_{=1} + B\underbrace{\sin(0)}_{=0}$$
$$1 = A$$

To use the other initial condition we need to differentiate first:

$$x = A\cos(5t) + B\sin(5t)$$
$$\frac{dx}{dt} = \underbrace{-5A\sin(5t)}_{\text{by } 6.13} + \underbrace{5B\cos(5t)}_{\text{by } 6.12}$$

Substituting $t = 0$, $\frac{dx}{dt} = 10$ gives

$$10 = -5A\sin(0) + 5B\cos(0)$$
$$10 = 0 + 5B \text{ it follows that } B = 2$$

Putting $A = 1$ and $B = 2$ into $x = A\cos(5t) + B\sin(5t)$ yields the particular solution

$$x = \cos(5t) + 2\sin(5t)$$

B3 Boundary value problems

If the given conditions are stated at different values of the (independent) variable, x, then we have **boundary conditions**. In **Example 6** which follows, the y values are given for two different values of x, $x = 0$ and $x = L$, so the given conditions are boundary conditions. Whilst in **Example 4** we were given the following conditions, $i = 0$ and $\frac{di}{dt} = 3$,

---

6.12 $[\sin(kt)]' = k\cos(kt)$ **6.13** $[\cos(kt)]' = -k\sin(kt)$

at the same value of the (independent) variable $t = 0$. Hence these are called initial conditions.

The differential equation associated with boundary conditions is called the **boundary value problem**.

The following example is more sophisticated than previous examples because it involves engineering symbols. However the concepts are the same as above.

The example is on structures and for all the problems on structures, EI represents flexural rigidity and $EI \neq 0$.

Example 6 *structures*

A column of length L is subject to a compressive load P (Fig. 2).

The displacement, y, at a distance x is given by the solution of

Fig. 2

$$EI\frac{d^2y}{dx^2} + Py = 0$$

i Show that $y = A\cos(kx) + B\sin(kx)$ where $k = \sqrt{\dfrac{P}{EI}}$.

ii The critical load is the value of the load, P, that will cause buckling.
The critical load, P_{cr}, occurs for the boundary conditions $x = 0$, $y = 0$ and $x = L$, $y = 0$.
Show that the critical load is given by

$$P_{cr} = \frac{n^2\pi^2\,EI}{L^2}$$

where n is any whole number.

(*Hint*: Use the result $\sin(n\pi) = 0$ where n is any whole number.)

Solution

i Dividing the given differential equation by EI:

$$\frac{d^2y}{dx^2} + \frac{P}{EI}y = 0$$

Compare this with

$$\frac{d^2y}{dx^2} + k^2y = 0 \text{ where } k^2 = \frac{P}{EI}$$

Taking square roots gives $k = \sqrt{\dfrac{P}{EI}}$. Hence by 14.8

$$*\qquad y = A\cos(kx) + B\sin(kx)$$

ii Substituting the given boundary condition, $x = 0, y = 0$, into ⁎ :

$$0 = A\underset{=1}{\underbrace{\cos(0)}} + B\underset{=0}{\underbrace{\sin(0)}} = A$$

We have $A = 0$. Using the other boundary condition, $x = L, y = 0$, gives

$$0 = 0 + B\sin(kL) \quad \text{so } B\sin(kL) = 0$$

Example 6 *continued*

$B \neq 0$, why?

If $B = 0$ then $y = 0$ because putting $A = 0$ and $B = 0$ into ▮ ✱ ▮ gives the trivial solution, $y = 0$. Therefore we must have

$$\sin(kL) = 0$$

$$kL \underset{\text{by hint}}{=} n\pi \text{ gives } k = \frac{n\pi}{L} \quad \text{[Dividing by } L]$$

Squaring both sides:

$$k^2 = \left(\frac{n\pi}{L}\right)^2 = \frac{n^2\pi^2}{L^2}$$

Remember that $k^2 = \dfrac{P}{EI}$ and for the given boundary conditions, $P = P_{cr}$:

$$k^2 = \frac{P_{cr}}{EI} = \frac{n^2\pi^2}{L^2}$$

$$P_{cr} = \frac{n^2\pi^2 \, EI}{L^2} \quad \text{[Multiplying by } EI]$$

This non-trivial solution is **only** valid at the critical loads. For loads below P_{cr} the only solution which satisfies the differential equation with the given boundary conditions is the trivial solution $y = 0$. That is, for loads less than $\pi^2 EI/L^2$ the beam does not buckle, it is only compressed. Buckling will not occur until the load equals P_{cr}.

B4 Differential equation of the form $\dfrac{d^2y}{dx^2} - k^2y = 0$

Now consider the differential equation of the form

$$\frac{d^2y}{dx^2} - k^2y = 0$$

The characteristic equation is

$$m^2 - k^2 = 0$$
$$m^2 = k^2$$
$$m = \pm k \qquad \text{[Taking square root]}$$
$$m_1 = k, \ m_2 = -k \quad \text{[Real and different]}$$

Putting these into

14.4 $\qquad y = Ae^{m_1 x} + Be^{m_2 x}$

gives

$$y = Ae^{kx} + Be^{-kx}$$

The general solution of the differential equation $\dfrac{d^2y}{dx^2} - k^2y = 0$ is

14.9 $\qquad y = Ae^{kx} + Be^{-kx}$

SUMMARY

If the characteristic equation is $m^2 + k^2 = 0$ then

14.8 $y = A\cos(kx) + B\sin(kx)$

If the characteristic equation is $m^2 - k^2 = 0$ then

14.9 $y = Ae^{kx} + Be^{-kx}$

If the initial and boundary conditions are given then A and B can be evaluated and we obtain a particular solution.

Exercise 14(b)

Solutions at end of book. Complete solutions available at www.palgrave.com/science/engineering/singh

1 [electrical principles]   Consider the second order differential equation

$$\frac{d^2i}{dt^2} + 7\frac{di}{dt} + 10i = 0$$

where i represents current. Determine the current, i, in terms of t for the initial conditions, when $t = 0$, both $i = 0$ and $\dfrac{di}{dt} = 6$.

2 [vibrations]   For the following simple harmonic motions, state the general solution:

a $\dfrac{d^2x}{dt^2} + 9x = 0$ **b** $\dfrac{d^2x}{dt^2} + 16x = 0$

c $\dfrac{d^2x}{dt^2} + 2x = 0$ **d** $\dfrac{d^2x}{dt^2} + 5x = 0$

3 [vibrations]

Fig. 3

Figure 3 shows a body of mass m attached to a spring of stiffness k. The displacement, x, is related by

$$\frac{d^2x}{dt^2} + \frac{k}{m}x = 0$$

Show that $x = A\cos(\omega t) + B\sin(\omega t)$ where A, B are constants and $\omega = \sqrt{\dfrac{k}{m}}$.

4 [electrical principles]   The voltage, v, of an RLC circuit is given by the second order differential equation

$$\frac{d^2v}{dt^2} + (4 \times 10^{-6})v = 0$$

with the initial conditions that when $t = 0$, both $v = 1$ and $\dfrac{dv}{dt} = 2 \times 10^{-3}$.

Show that

$$v = \sqrt{2}\cos\left[(2 \times 10^{-3})t - \frac{\pi}{4}\right]$$

[Hint: Use 4.76 to convert into a cosine term only.]

5 [structures]   A load, P, is applied at both ends of a column of length L. The second order differential equation

$$EI\frac{d^2y}{dx^2} + Py = 0$$

gives the relationship between the deflection y and the distance x. Show that if $y = e$ (eccentricity) at both $x = L/2$ and $x = -L/2$ then

$$y = e\sec\left(\frac{kL}{2}\right)\cos(kx) \text{ where } k = \sqrt{P/EI}$$

$$\left[\text{Assume that } \sin\left(\frac{kL}{2}\right) \neq 0\right]$$

[Hint: Use the solution ★ to **Example 6**.]

Exercise **14(b) continued**

Solutions at end of book. Complete solutions available at www.palgrave.com/science/engineering/singh

("Y") **Questions 6 to 8, inclusive, are in the field of [*vibrations*].**

6

Fig. 4

Figure 4 shows a body of mass m attached to a spring of stiffness k and a viscous damper with a damping coefficient c. The displacement, x, is given by

$$\frac{d^2x}{dt^2} + 2\zeta\omega\frac{dx}{dt} + \omega^2 x = 0$$

where $\omega = \sqrt{\dfrac{k}{m}}$ and the viscous

damping factor $\zeta = \dfrac{c}{2m\omega}$.

Show that the solution of this differential equation for $\zeta > 1$ is given by

$$x = Ae^{r_1 t} + Be^{r_2 t}$$
$$(A \text{ and } B \text{ are constants})$$

where $r_1 = \omega\left(-\zeta + \sqrt{\zeta^2 - 1}\right)$ and
$r_2 = -\omega\left(\zeta + \sqrt{\zeta^2 - 1}\right)$.

7 The equation of motion for a vertical spring–mass system is given by

$$\ddot{x} + 2\zeta\omega\dot{x} + \omega^2 x = 0$$

where ζ is the damping factor, ω is the angular frequency and x is the displacement of mass from its equilibrium position. When $t = 0$, both $x = 5$ and $\dot{x} = 0$. Find an expression for x in terms of t with $\zeta = 1$.

(*Hint*: Note that we have the same characteristic equation as in question **6**.) Plot the graph of x versus ωt for $\omega t > 0$. (Use a graphical calculator or a computer algebra system for this part.)

8 The equation of motion for a spring–mass system is given by

$$\ddot{x} + 2\zeta\omega\dot{x} + \omega^2 x = 0$$

where x is the displacement of mass from its equilibrium level, ζ is the damping factor and ω is the angular frequency. For $\zeta < 1$ and the initial conditions, when $t = 0$, both $x = 0$ and $\dot{x} = \beta\omega$ where $\beta = \sqrt{1 - \zeta^2}$, show that

$$x = e^{-\zeta\omega t}\sin(\beta\omega t)$$

Questions 9 to 11, inclusive, are in the field of [*electrical principles*].
(R is resistance, L is inductance and C is capacitance)

9 The voltage, v, in a parallel *RLC* circuit satisfies

$$C\frac{d^2v}{dt^2} + \frac{1}{R}\frac{dv}{dt} + \frac{v}{L} = 0$$

Find an expression for the voltage, v, in terms of t, R, L and C, for the following cases:

a $L = 4CR^2$ (Critically damped)
b $L > 4CR^2$ (Over damped)
c $L < 4CR^2$ (Under damped)

10 A capacitance $C = (1 \times 10^{-9})$ F, a resistance $R = 10$ kΩ and a variable inductance L are connected in parallel. The voltage, v, is governed by the second order differential equation

$$C\frac{d^2 v}{dt^2} + \frac{1}{R}\frac{dv}{dt} + \frac{v}{L} = 0$$

The inductance, L, is varied so that the roots of the characteristic equation become $-(50 \times 10^3) \pm j(30 \times 10^3)$. Determine this value of L given that $L < 0.4$ H.

11 A circuit consisting of a resistor R, inductor L and capacitor C has the

Exercise **14(b) continued**

Solutions at end of book. Complete solutions available at www.palgrave.com/science/engineering/singh

following second order differential equation:

$$C\frac{d^2v}{dt^2} + \frac{1}{R}\frac{dv}{dt} + \frac{v}{L} = 0$$

By considering the characteristic equation $m^2 + 2\zeta\omega m + \omega^2 = 0$, find expressions for the natural frequency, ω, and damping ratio, ζ.

SECTION C **Non-homogeneous (inhomogeneous) differential equations**

By the end of this section you will be able to:
▶ understand the term *non-homogeneous*
▶ understand why we need a particular integral $Y(x)$
▶ use a trial function to find a particular integral
▶ evaluate general solutions of non-homogeneous second order differential equations

C1 **Non-homogeneous equations**

What does the term 'non-homogeneous second order differential equation' mean?

Differential equations of the type

$$a\frac{d^2y}{dx^2} + b\frac{dy}{dx} + cy = f(x) \text{ where } f(x) \neq 0$$

Remember that homogeneous is where $f(x) = 0$, therefore non-homogeneous is where $f(x) \neq 0$, does **not** equal zero. Another name for these equations is **inhomogeneous**.

Example 7

Solve

$$★ \qquad \frac{d^2y}{dx^2} - \frac{dy}{dx} - 2y = 5e^{4x}$$

Solution

What are we trying to find?

We need to find y (a function of x) which satisfies ★ . If the Right-Hand Side is 0 then we have

$$\frac{d^2y}{dx^2} - \frac{dy}{dx} - 2y = 0$$

Example 7 *continued*

and the characteristic equation is

$$m^2 - m - 2 = 0$$
$$(m - 2)(m + 1) = 0 \quad \text{[Factorizing]}$$
$$m_1 = 2, \; m_2 = -1 \quad \text{[Solving]}$$

Since we have two real and different roots we substitute these, $m_1 = 2$ and $m_2 = -1$, into

14.4 $\qquad y = Ae^{m_1 x} + Be^{m_2 x}$

which gives

$$y_c = Ae^{2x} + Be^{-x}$$

(y_c denotes complementary function and will be explained later.)

If we substitute y_c into the Left-Hand Side of ▮ * ▮ and then simplify, what will be the final answer?

ZERO

Why do you think we obtain zero?

Because we have just solved a homogeneous differential equation

$$\frac{d^2 y}{dx^2} - \frac{dy}{dx} - 2y = 0$$

and not $\quad \dfrac{d^2 y}{dx^2} - \dfrac{dy}{dx} - 2y = 5e^{4x} \quad$ [the given equation]

So there must be an extra term, function $Y = Y(x)$ (a function of x) to the existing homogeneous solution, $y_c = Ae^{2x} + Be^{-x}$.

What is this function Y going to be?

Y must satisfy the given differential equation

$$\frac{d^2 Y}{dx^2} - \frac{dY}{dx} - 2Y = 5e^{4x}$$

Try the function $Y = Ce^{4x}$, where C is a constant. Why?

Because when we differentiate Ce^{4x} we still have a term in e^{4x} and on the Right-Hand Side we have $5e^{4x}$.

We need to find a value for the constant C. How can we find C?

Since

$$Y = Ce^{4x}$$

satisfies the given equation, we differentiate this:

$$\frac{dY}{dx} \underset{\text{by } \boxed{6.11}}{=} 4Ce^{4x}$$

$$\frac{d^2 Y}{dx^2} \underset{\text{by } \boxed{6.11}}{=} 16Ce^{4x}$$

6.11 $\quad (e^{kx})' = ke^{kx}$

> ### Example 7 continued
>
> and substitute into
>
> $$\frac{d^2Y}{dx^2} - \frac{dY}{dx} - 2Y = 5e^{4x}$$
>
> which gives
>
> $$16Ce^{4x} - 4Ce^{4x} - 2Ce^{4x} = 5e^{4x}$$
> $$(16 - 4 - 2)Ce^{4x} = 5e^{4x} \qquad \text{[Taking out } Ce^{4x}\text{]}$$
> $$10Ce^{4x} = 5e^{4x} \qquad \text{[Simplifying]}$$
> $$10C = 5$$
> $$C = \frac{5}{10} = \frac{1}{2}$$
>
> Hence putting $C = \dfrac{1}{2}$ into $Y = Ce^{4x}$ gives $Y = \dfrac{e^{4x}}{2}$
>
> The general solution of the given differential equation
>
> $$\frac{d^2y}{dx^2} - \frac{dy}{dx} - 2y = 5e^{4x}$$
>
> is found by adding together the homogeneous solution $y_c = Ae^{2x} + Be^{-x}$ and the
>
> particular integral $Y = \dfrac{e^{4x}}{2}$
>
> Therefore the general solution is $y = Ae^{2x} + Be^{-x} + \dfrac{e^{4x}}{2}$

The homogeneous solution, y_c, is called the **complementary function** and Y is called the **particular integral**. To find the particular integral we use a trial function which depends on $f(x)$.

Summarizing the above

The non-homogeneous differential equation

$$a\frac{d^2y}{dx^2} + b\frac{dy}{dx} + cy = f(x) \quad \text{where } f(x) \neq 0$$

is solved by:

1 finding the complementary function y_c (homogeneous solution)

$$a\frac{d^2y_c}{dx^2} + b\frac{dy_c}{dx} + cy_c = 0$$

2 determining the particular integral Y by using a trial function (depends on $f(x)$)
3 adding together y_c and Y. The solution is

14.10 $y = y_c + Y$

Property **3** only holds if the given differential equation is **linear,** that is the power of y is 1, we do **not** have $\sqrt{y}$, y^2, etc. in the differential equation. Thus the general solution of a linear non-homogeneous differential equation is

14.10a   (complementary function) + (particular integral)

C2 **Particular integrals**

> **Example 8**

Determine the particular integral, Y, of

$$\frac{d^2y}{dx^2} - \frac{dy}{dx} - 2y = 2x$$

Solution

Since on the Right-Hand Side we have a linear function, $f(x) = 2x$, we try the general linear function:

$$Y = ax + b$$

Do not confuse the coefficients a and b in Y with the coefficients of the differential equation, they are not the same.

This Y satisfies the given differential equation

$$\boxed{*} \qquad \frac{d^2Y}{dx^2} - \frac{dY}{dx} - 2Y = 2x$$

Differentiating Y:

$$Y = ax + b$$
$$\frac{dY}{dx} = a$$
$$\frac{d^2Y}{dx^2} = 0$$

Substituting these into $\boxed{*}$ gives

$$0 - a - 2(ax + b) = 2x$$
$$\boxed{**} \qquad -a - 2ax - 2b = 2x$$

We need to find the values of a and b. How?

We could equate coefficients of $\boxed{**}$. Thus equating x:

$$-2a = 2 \text{ gives } a = -1$$

constants (putting $x = 0$):

$$-a - 2b = 0$$
$$-(-1) - 2b = 0 \qquad [\text{Because } a = -1]$$
$$1 - 2b = 0 \text{ gives } b = \frac{1}{2}$$

Substituting $a = -1$ and $b = \frac{1}{2}$ into the trial function, $Y = ax + b$, gives the particular integral:

$$Y = -x + \frac{1}{2} = \frac{1}{2} - x$$

Example 9

Determine the particular integral, Y, of

$$\frac{d^2y}{dx^2} - \frac{dy}{dx} - 2y = 5x^2$$

Solution

We know the Right-Hand Side is a quadratic, $f(x) = 5x^2$, so what is Y?

A trial function, Y, is the general quadratic, that is

$$Y = ax^2 + bx + c$$

Y satisfies

$$★ \qquad \frac{d^2Y}{dx^2} - \frac{dY}{dx} - 2Y = 5x^2$$

So we need to differentiate Y to find a, b and c:

$$Y = ax^2 + bx + c$$
$$\frac{dY}{dx} = 2ax + b$$
$$\frac{d^2Y}{dx^2} = 2a$$

Substituting these into ★ gives

$$\frac{d^2Y}{dx^2} - \frac{dY}{dx} - 2Y = 2a - (2ax + b) - 2(ax^2 + bx + c)$$
$$= 2a - 2ax - b - 2ax^2 - 2bx - 2c$$
$$= -2ax^2 + (-2a - 2b)x + (2a - b - 2c) \quad \text{[Factorizing]}$$

How can we find a, b and c?

One way is by equating coefficients. The above is equal to $5x^2$, thus equating coefficients of the highest power of x first:

$$-2ax^2 + (-2a - 2b)x + (2a - b - 2c) = 5x^2$$

x^2: $\qquad -2a = 5$ gives $a = -\dfrac{5}{2}$

x: $\qquad -2a - 2b = 0$
$$-2b = 2a$$
$$b = -a = -\left(-\frac{5}{2}\right) = \frac{5}{2} \quad \left[\text{Substituting } a = -\frac{5}{2}\right]$$

constants: $\qquad 2a - b - 2c = 0$

$$2\left(-\frac{5}{2}\right) - \frac{5}{2} - 2c = 0, \text{ solving gives } c = -\frac{15}{4}$$

Example 9 *continued*

Substituting $a = -\dfrac{5}{2}$, $b = \dfrac{5}{2}$ and $c = -\dfrac{15}{4}$ into $Y = ax^2 + bx + c$ gives

$$Y = -\frac{5}{2}x^2 + \frac{5}{2}x - \frac{15}{4} = -\frac{10}{4}x^2 + \frac{10}{4}x - \frac{15}{4}$$

Taking out the common factor $-1/4$ gives the particular integral:

$$Y = -\frac{1}{4}(10x^2 - 10x + 15)$$

The particular integral, Y, of a general second order differential equation

$$L\frac{d^2y}{dx^2} + M\frac{dy}{dx} + Ny = f(x)$$

depends on $f(x)$. Table 1 shows trial functions that can be used to find the particular integral Y. (We are using the letters L, M and N for the coefficients of the derivatives so that you do not get confused with the letters a, b and c used for the trial functions in Table 1.)

TABLE 1

Reference number	$f(x)$	Trial function
14.11	A (constant)	C (constant)
14.12	$Ax + B$	$ax + b$
14.13	$Ax^2 + Bx + C$	$ax^2 + bx + c$
14.14	$A_n x^n + A_{n-1} x^{n-1} + \ldots + A_0$	$a_n x^n + a_{n-1} x^{n-1} + \ldots + a_0$
14.15	$A\cos(kx)$	$a\cos(kx) + b\sin(kx)$
14.16	$A\sin(kx)$	$a\cos(kx) + b\sin(kx)$
14.17	Ae^{kx}	Ce^{kx}

Note the trial function for the trigonometric functions $\cos(kx)$ and $\sin(kx)$. If $f(x) = A\cos(kx)$ then we consider the trial function $Y = a\cos(kx) + b\sin(kx)$ and **not** $Y = a\cos(kx)$ because the presence of first and second derivatives in the equation will **produce** sine and cosine terms from either sine **or** cosine. Also if $f(x)$ is an nth degree polynomial, 14.14 , then a trial function is the general nth degree polynomial. For example, if $f(x) = x^3 + 1$ then our trial function is

$$Y = ax^3 + bx^2 + cx + d$$

Example 10 *electrical principles*

By applying Kirchhoff's voltage law to a circuit we obtain the following second order differential equation:

$$\frac{d^2i}{dt^2} + (12 \times 10^6)\frac{di}{dt} + (36 \times 10^{12})\, i = 5 \times 10^9$$

where i is the current through the circuit. Determine the current, i, at any time t.

Solution

What do we evaluate first?

Complementary function, i_c (place the Right-Hand Side equal to 0):

$$\frac{d^2i}{dt^2} + (12 \times 10^6)\frac{di}{dt} + (36 \times 10^{12})i = 0$$

The characteristic equation is

$$m^2 + (12 \times 10^6)\, m + (36 \times 10^{12}) = 0$$

To solve this characteristic equation we use the quadratic equation formula:

1.16 $$m = \frac{-b \pm \sqrt{b^2 - 4ac}}{2a}$$

with $a = 1$, $b = 12 \times 10^6$ and $c = 36 \times 10^{12}$ which gives

$$m = \frac{-(12 \times 10^6) \pm \sqrt{(12 \times 10^6)^2 - (4 \times 1 \times 36 \times 10^{12})}}{2 \times 1}$$

$$= \frac{-(12 \times 10^6)}{2} \pm \frac{1}{2}\underbrace{\sqrt{(144 \times 10^{12}) - (144 \times 10^{12})}}_{=\,0}$$

$$m = -(6 \times 10^6) \quad \text{[Equal roots]}$$

Since we have equal roots, $m = -6 \times 10^6$, of the characteristic equation so we substitute this into

14.5 $$i = (A + Bt)e^{mt}$$

which gives

$$i_c = (A + Bt)e^{-(6 \times 10^6)t}$$

Is this the solution of the given differential equation

$$\frac{d^2i}{dt^2} + (12 \times 10^6)\frac{di}{dt} + (36 \times 10^{12})i = 5 \times 10^9?$$

No, because the Right-Hand Side is **not** zero (i_c is the homogeneous solution).

Hence we need to find the particular integral I. Since $f(t) = 5 \times 10^9$, a constant, so by 14.11 a trial function is

$$I = C \text{ where } C \text{ is a constant}$$

$$\frac{dI}{dt} = 0 \text{ and } \frac{d^2I}{dt^2} = 0 \quad \text{[Differentiating a constant]}$$

Example 10 *continued*

So the first and second derivatives are zero. Substituting $I = C$ into the given differential equation

$$\frac{d^2I}{dt^2} + (12 \times 10^6)\frac{dI}{dt} + (36 \times 10^{12})I = 5 \times 10^9$$

yields

$$0 + 0 + (36 \times 10^{12})C = 5 \times 10^9$$
$$C = \frac{5 \times 10^9}{36 \times 10^{12}} = 1.39 \times 10^{-4}$$

Hence $I = C = 1.39 \times 10^{-4}$. Since $i = i_c + I$ so we have the general solution

$$i = (A + Bt)e^{-(6 \times 10^6)t} + (1.39 \times 10^{-4})$$

The next example might seem like a colossal leap from previous examples. The difficulty is the complicated notation and engineering symbols used. Don't be put off because the concepts are the same as above. Remember that $\ddot{x}$ is a second order derivative.

Example 11 *vibrations*

An undamped spring–mass system is subject to a force, $F\sin(2t)$. The equation of motion is

$$m\ddot{x} = -kx + F\sin(2t) \qquad (k - 4m \neq 0)$$

where m is the mass, k is the stiffness of the spring, x is the displacement and t is the time. Show that

$$x = A\cos(\omega t) + B\sin(\omega t) + \frac{F}{k - 4m}\sin(2t) \text{ where } \omega = \sqrt{\frac{k}{m}}$$

Solution

By adding kx to both sides, the differential equation becomes

$$m\ddot{x} + kx = F\sin(2t)$$

The differential equation is now in standard form. (Standard form is covered at the end of **Section A**.)

Hence the complementary function x_c (Right-Hand Side $= 0$) is given by

$$m\ddot{x}_c + kx_c = 0$$
$$\ddot{x}_c + \frac{k}{m}x_c = 0 \quad \text{[Dividing by } m\text{]}$$

Thus the characteristic equation is

$$r^2 + \left(\sqrt{\frac{k}{m}}\right)^2 = 0 \quad \left[\text{Remember that } \frac{k}{m} = \left(\sqrt{\frac{k}{m}}\right)^2\right]$$

Example 11 *continued*

By 14.8 the complementary function is

$$x_c = A\cos(\omega t) + B\sin(\omega t) \text{ where } \omega = \sqrt{\frac{k}{m}}$$

Particular integral X:

Since the Right-Hand Side is equal to $F\sin(2t)$, by 14.16 we consider the trial function:

$$X = a\cos(2t) + b\sin(2t)$$

Differentiating gives

$$\dot{X} = \underbrace{-2a\sin(2t)}_{\text{by } 6.13} + \underbrace{2b\cos(2t)}_{\text{by } 6.12}$$

$$= 2[-a\sin(2t) + b\cos(2t)] \quad [\text{Taking out } 2]$$

$$\ddot{X} = -4a\cos(2t) - 4b\sin(2t) \quad [\text{Differentiating}]$$

$$= -4[a\cos(2t) + b\sin(2t)] \quad [\text{Taking out } -4]$$

Substituting into

$$m\ddot{X} + kX = F\sin(2t)$$

gives

$$-4m[a\cos(2t) + b\sin(2t)] + k[a\cos(2t) + b\sin(2t)] = F\sin(2t)$$

$$(k - 4m)a\cos(2t) + (k - 4m)b\sin(2t) = F\sin(2t)$$

Equating coefficients of

$\cos(2t)$: $(k - 4m)a = 0$, gives $a = 0$ because $k - 4m \neq 0$
$\sin(2t)$: $(k - 4m)b = F$

$$b = \frac{F}{k - 4m} \quad [\text{Dividing by } k - 4m]$$

Substituting $a = 0$ and $b = \dfrac{F}{k - 4m}$ into our trial function $X = a\cos(2t) + b\sin(2t)$ gives the particular integral

$$X = \frac{F}{k - 4m}\sin(2t)$$

The general solution, $x = x_c + X$, is

$$x = A\cos(\omega t) + B\sin(\omega t) + \frac{F}{k - 4m}\sin(2t) \quad \text{where } \omega = \sqrt{\frac{k}{m}}$$

6.12 $[\sin(kt)]' = k\cos(kt)$ 6.13 $[\cos(kt)]' = -k\sin(kt)$

Table 1 is a good guide for finding the particular integral, Y, but sometimes we need to adjust the trial function to find the particular integral, as the next example shows.

Example 12

Solve

$$* \qquad \frac{d^2y}{dx^2} - \frac{dy}{dx} - 2y = 3e^{2x}$$

Solution

The complementary function, y_c, is the same as **Example 7** because we have the **same** characteristic equation. Hence

$$y_c = Ae^{2x} + Be^{-x}$$

We need to find the particular integral Y. By 14.17 of Table 1 $Y = Ce^{2x}$, because $f(x) = 3e^{2x}$. Let's find C:

$$\frac{dY}{dx} = 2Ce^{2x} \quad \left[\text{Differentiating by } \frac{d}{dx}(e^{kx}) = ke^{kx} \right]$$

$$\frac{d^2Y}{dx^2} = 4Ce^{2x} \quad \left[\text{Differentiating by } \frac{d}{dx}(e^{kx}) = ke^{kx} \right]$$

Substituting these into $\quad \dfrac{d^2Y}{dx^2} - \dfrac{dY}{dx} - 2Y = 3e^{2x} \quad$ gives

$$4Ce^{2x} - 2Ce^{2x} - 2Ce^{2x} = 3e^{2x}$$

$$(4 - 2 - 2)Ce^{2x} = 3e^{2x} \quad \text{[Taking out } Ce^{2x}\text{]}$$

$$0 = 3e^{2x}$$

This is impossible because $3e^{2x}$ cannot be zero.

? **Where have we gone wrong in the above derivation?**

? **Are you surprised by the result?**

? You shouldn't be surprised because by taking the trial function, Y, to be Ce^{2x} was clearly going to give zero. **Why?**

Because Ae^{2x} is part of the complementary function, y_c, so Ce^{2x} is also part of the complementary function, that is part of the solution with $f(x) = 0$.

? In the case when the trial function in Table 1 is part of the complementary function then we need to change our trial function. We cannot make a dramatic change since we need an e^{2x} term because $f(x) = 3e^{2x}$. **What is the simplest function we can try with an e^{2x} term?**

$$xe^{2x}$$

We try the function

$$Y = Cxe^{2x}$$

(Call this function Y from now on, rather than $Y = Ce^{2x}$, as above.)

Example 12 *continued*

How do we differentiate $Y = Cxe^{2x}$?

Use the product rule with the constant C outside:

$$u = x \qquad v = e^{2x}$$

$$u' = 1 \qquad v' = 2e^{2x}$$

$$\frac{dY}{dx} = \frac{d}{dx}(Cxe^{2x}) = [u'v + uv']C \qquad \text{[Taking out constant } C\text{]}$$

$$= [(1)e^{2x} + x(2e^{2x})]C$$

$$= [e^{2x} + 2xe^{2x}]C$$

Using the product rule again:

$$\frac{d^2Y}{dx^2} = [2e^{2x} + 2(e^{2x} + 2xe^{2x})]C$$

$$= \left[\underbrace{2e^{2x} + 2e^{2x}}_{= 4e^{2x}} + 4xe^{2x} \right]C$$

$$= [4e^{2x} + 4xe^{2x}]C$$

Substituting these, $\dfrac{dY}{dx} = [e^{2x} + 2xe^{2x}]C$ and $\dfrac{d^2Y}{dx^2} = [4e^{2x} + 4xe^{2x}]C$, into the differential equation

$$\frac{d^2Y}{dx^2} - \frac{dY}{dx} - 2Y = 3e^{2x}$$

gives

$$[4e^{2x} + 4xe^{2x} - (e^{2x} + 2xe^{2x}) - 2(xe^{2x})]C = 3e^{2x}$$

$$[4e^{2x} + 4xe^{2x} - e^{2x} - 2xe^{2x} - 2xe^{2x}]C = 3e^{2x} \qquad \text{[Expanding]}$$

$$\underbrace{(4 + 4x - 1 - 2x - 2x)}_{= 3}Ce^{2x} = 3e^{2x} \qquad \text{[Taking out } e^{2x}\text{]}$$

$$3Ce^{2x} = 3e^{2x} \qquad \text{[Simplifying]}$$

$$3C = 3 \quad \text{gives } C = 1$$

Putting $C = 1$ into our trial function $Y = Cxe^{2x}$ gives $Y = xe^{2x}$.

Adding together $Y = xe^{2x}$ and $y_c = Ae^{2x} + Be^{-x}$ gives the general solution:

$$y = Ae^{2x} + Be^{-x} + xe^{2x}$$

The above example shows that if $f(x)$ is part of the complementary function then the particular integral Y is found by selecting the appropriate trial function from Table 1 and multiplying it by x.

This snag rarely occurs but, as shown above, when it does occur we can still find the solution of the differential equation.

SUMMARY

The non-homogeneous differential equation

$$a\frac{d^2y}{dx^2} + b\frac{dy}{dx} + cy = f(x) \text{ where } f(x) \neq 0$$

is solved by:

1 finding the complementary function y_c (homogeneous solution)

$$a\frac{d^2y_c}{dx^2} + b\frac{dy_c}{dx} + y_c = 0$$

2 determining the particular integral Y by using a trial function.
Then the complete solution is $y = y_c + Y$.
If $f(x)$ is already part of the complementary function, y_c, then choose the appropriate
trial function from Table 1 and multiply by x.

Exercise **14(c)**

Solutions at end of book. Complete solutions available at www.palgrave.com/science/engineering/singh

1 Determine the particular integrals of

$$\frac{d^2y}{dx^2} - \frac{dy}{dx} - 2y = f(x) \text{ with}$$

a $f(x) = 18$ **b** $f(x) = 2x + 3$
c $f(x) = 130\sin(3x)$ **d** $f(x) = \cos(x)$
e $f(x) = e^{3x}$

(Note that the Left-Hand Side of the differential equation is the same as **Example 7**.)

2 Find the general solutions of the following differential equations:

a $\dfrac{d^2y}{dx^2} + 3\dfrac{dy}{dx} + 2y = 6$

b $3\dfrac{d^2y}{dx^2} - 2\dfrac{dy}{dx} - y = 2x - 3$

c $\dfrac{d^2y}{dx^2} + 2\dfrac{dy}{dx} + y = 36e^{5x}$

d $\dfrac{d^2y}{dx^2} + 3\dfrac{dy}{dx} - 4y = -34\sin(x)$

Ψ Questions 3 to 7, inclusive, belong to the field of [*vibrations*].

3 The equation of motion of mass, m, on one end of a cantilever beam is given by

$$m\ddot{x} = mg - kx$$

where x is the displacement, k is a constant and g is acceleration due to

gravity. Find an expression for x in terms of t. $\left(\text{Remember that } \ddot{x} = \dfrac{d^2x}{dt^2}.\right)$

4 A vehicle of mass $m = (1 \times 10^3)$ kg is brought to rest by a force F given by

$$F = (0.5 + 4x) \times 10^3$$

where x is distance. Given that $F = m\dfrac{d^2x}{dt^2}$, find an expression for x in terms of t.

5

Fig. 5

Figure 5 shows an undamped spring–mass system subject to a force $F\sin(\alpha t)$. The equation of motion is

$$m\ddot{x} = -kx + F\sin(\alpha t) \quad \left(\alpha^2 \neq \frac{k}{m}\right)$$

where m is the mass, k is the stiffness of the spring, x is the displacement and t is the time. Show that

Exercise **14(c) continued**

Solutions at end of book. Complete solutions available at www.palgrave.com/science/engineering/singh

$$x = A\cos(\omega t) + B\sin(\omega t) + \frac{F}{k - m\alpha^2}\sin(\alpha t)$$

where $\omega = \sqrt{\dfrac{k}{m}}$

6 The spring–mass system of Fig. 5 is subject to a force $F\cos(\alpha t)$. Show that

$$x = A\cos(\omega t) + B\sin(\omega t) + \frac{F}{k - m\alpha^2}\cos(\alpha t)$$

where $\omega = \sqrt{\dfrac{k}{m}}$

7

Fig. 6

Figure 6 shows a body of mass m suspended by a spring of stiffness k in a fluid. The equation of motion is given by

$$m\ddot{x} = mg - k(x - L) - c\dot{x}$$

where x is the length shown, L is the unstretched length of the spring and c is a constant. Show that, if $c^2 < 4km$ then

$$x = e^{-\frac{c}{2m}t}\left[A\cos(\beta t) + B\sin(\beta t)\right] + \frac{mg}{k} + L$$

where $\beta = \dfrac{1}{2m}\sqrt{4km - c^2}$ and A, B are constants.

8 🔲 [electrical principles]   By applying Kirchhoff's law to a parallel *RLC* circuit (all three are in parallel with each other)

we obtain the second order differential equation

$$CL\frac{d^2i}{dt^2} + \frac{L}{R}\frac{di}{dt} + i = I$$

where i is the current through the circuit and I is the d.c. current source applied to the circuit. The capacitance $C = 10$ nF, inductance $L = 50$ mH, resistance $R = 1$ kΩ and $I = 50$ mA. Find an expression for i in terms of t.

9 Find the general solution of

$$\frac{d^2y}{dx^2} - \frac{dy}{dx} - 2y = 5e^{-x}$$

10 Determine the form of the trial function, Y, for the following:

a $\dfrac{d^2y}{dx^2} + 4\dfrac{dy}{dx} + 3y = 6e^{-3x}$

b $\dfrac{d^2y}{dx^2} + 9y = 6\cos(3x)$

11 🔲 [structures]   A load, P, is applied to a column of length L. The differential equation relating the deflection y at a distance x along the column is given by

$$EI\frac{d^2y}{dx^2} + P\left[y + m\sin\left(\frac{\pi x}{L}\right)\right] = 0$$
$$\left(P \neq \frac{EI\pi^2}{L^2}\right)$$

where m is the initial maximum deflection. Show that

$$y = A\cos(kx) + B\sin(kx)$$
$$+ \frac{PL^2 m}{EI\pi^2 - PL^2}\sin\left(\frac{\pi x}{L}\right) \text{ where } k = \sqrt{\frac{P}{EI}}$$

SECTION D **Particular solutions**

By the end of this section you will be able to:
► find particular solutions of non-homogeneous equations

D1 **Particular solutions of non-homogeneous equations**

In the previous section we only found **general solutions** of non-homogeneous differential equations. In **Example 7**

$$\frac{d^2y}{dx^2} - \frac{dy}{dx} - 2y = 5e^{4x}$$

the general solution was

$$y = Ae^{2x} + Be^{-x} + \frac{e^{4x}}{2}$$

If we are given initial (or boundary) conditions then we can find values of the constants A and B. In this case y is called the **particular solution**. The names **particular solution** and **particular integral** can be a cause of confusion. These two terms are **not** the same and therefore cannot be interchanged. The particular integral is the term added to the complementary function to give a solution, while the particular solution is the solution of the differential equation which has particular values for the constants.

Example 13 *mechanics*

A vehicle is brought to rest by a buffer stop. By applying Newton's second law we obtain the second order differential equation

$$(2 \times 10^3)\ddot{x} + (18 \times 10^3)x + (5 \times 10^3) = 0$$

where x is the distance by which the buffer is compressed. Given the initial conditions, when $t = 0$, both $x = 0$ and $\dot{x} = 0$, find the expression for x in terms of t.

(Remember that $\dot{x}$ represents the first derivative and $\ddot{x}$ represents the second derivative.)

Solution

Dividing the given differential equation by 10^3 yields

$$2\ddot{x} + 18x + 5 = 0$$
$$2\ddot{x} + 18x = -5$$

The differential equation is now in standard form.

First we find the complementary function, x_c:

$$2\ddot{x}_c + 18x_c = 0 \qquad \text{[Right-Hand Side = 0]}$$

Example 13 *continued*

The characteristic equation is

$$2m^2 + 18 = 0$$

Dividing through by 2:

$$m^2 + 9 = 0$$

$$m^2 + 3^2 = 0$$

By 14.8 :

$$x_c = A\cos(3t) + B\sin(3t)$$

Now we need to find the particular integral, X:

Since $f(t) = -5$ is a constant, we use the trial function $X = C$ where C is a constant. Differentiating $X = C$ gives

$$\dot{X} = \ddot{X} = 0$$

Substituting into $2\ddot{X} + 18X = -5$ yields

$$18C = -5, \text{ giving } C = -\frac{5}{18}$$

which means

$$X = C = -\frac{5}{18}$$

The general solution is given by $x = x_c + X$, thus

** $$x = A\cos(3t) + B\sin(3t) - \frac{5}{18}$$

Substituting the initial condition, when $t = 0$, $x = 0$, into ** gives

$$0 = A\underbrace{\cos(0)}_{=1} + B\underbrace{\sin(0)}_{=0} - \frac{5}{18}$$

$$0 = A - \frac{5}{18} \text{ giving } A = \frac{5}{18}$$

Using the other initial condition, when $t = 0$, $\dot{x} = 0$, means we need to differentiate

$$x = A\cos(3t) + B\sin(3t) - \frac{5}{18}$$

$$\dot{x} = \underbrace{-3A\sin(3t)}_{\text{by } 6.13} + \underbrace{3B\cos(3t)}_{\text{by } 6.12}$$

Putting $\dot{x} = 0$ and $t = 0$ gives

$$0 = -3A\sin(0) + 3B\cos(0)$$

6.12 $\dfrac{d}{dt}\left[\sin(kt)\right] = k\cos(kt)$ 6.13 $\dfrac{d}{dt}\left[\cos(kt)\right] = -k\sin(kt)$

14.8   If $m^2 + k^2 = 0$ then $x = A\cos(kt) + B\sin(kt)$

Example 13 *continued*

$$0 = 0 + 3B \text{ which gives } B = 0$$

Substituting $A = \dfrac{5}{18}$ and $B = 0$ into ** yields

$$x = \frac{5}{18}\cos(3t) - \frac{5}{18}$$

$$x = \frac{5}{18}(\cos(3t) - 1) \qquad \left[\text{Taking out } \frac{5}{18}\right]$$

In the above example we have found the particular solution to a non-homogeneous differential equation.

What is the general solution of this differential equation?

It is ** , that is

** $\qquad x = A\cos(3t) + B\sin(3t) - \dfrac{5}{18}$

 Example 14 *electrical principles*

By applying Kirchhoff's voltage law to a series *RLC* circuit we obtain the differential equation

$$\frac{d^2i}{dt^2} + 3\frac{di}{dt} + 2i = 5e^{-3t}$$

Solve this differential equation for the initial conditions, when $t = 0$, both $i = 0$ and $\dfrac{di}{dt} = 0$.

Solution

For the complementary function we first consider the homogeneous equation by putting the Right-Hand Side equal to zero:

$$\frac{d^2i}{dt^2} + 3\frac{di}{dt} + 2i = 0$$

The characteristic equation is

$$m^2 + 3m + 2 = 0$$
$$(m + 2)(m + 1) = 0 \qquad \text{[Factorizing]}$$
$$m_1 = -2, m_2 = -1 \qquad \text{[Real and different]}$$

Substituting these into

14.4 $\qquad i = Ae^{m_1t} + Be^{m_2t}$

gives the complementary function

$$i_c = Ae^{-2t} + Be^{-t}$$

For particular integral *I*: since $f(t) = 5e^{-3t}$ so our trial function is

$$I = Ce^{-3t}$$

Example 14 *continued*

Differentiating:

$$\frac{dI}{dt} = -3Ce^{-3t} \qquad \left[\text{By } \frac{d}{dt}(e^{kt}) = ke^{kt} \right]$$

$$\frac{d^2I}{dt^2} = 9Ce^{-3t}$$

Substituting into $\dfrac{d^2I}{dt^2} + 3\dfrac{dI}{dt} + 2I = 5e^{-3t}$ gives

$$9Ce^{-3t} + 3(-3Ce^{-3t}) + 2Ce^{-3t} = 5e^{-3t}$$

$$\underbrace{9Ce^{-3t} - 9Ce^{-3t}}_{=0} + 2Ce^{-3t} = 5e^{-3t}$$

$$2Ce^{-3t} = 5e^{-3t}$$

$$2C = 5 \text{ thus } C = 2.5$$

Substituting $C = 2.5$ into $I = Ce^{-3t}$ yields

$$I = 2.5e^{-3t}$$

The general solution, $i = i_c + I$, is

† $i = Ae^{-2t} + Be^{-t} + 2.5e^{-3t}$

To find A and B we use the given initial conditions:

Substituting $t = 0$, $i = 0$ into †

$$0 = Ae^0 + Be^0 + 2.5e^0$$

$$= A + B + 2.5 \qquad [\text{Because } e^0 = 1]$$

$$A + B = -2.5$$

Differentiating † yields

$$\frac{di}{dt} = -2Ae^{-2t} - Be^{-t} - 7.5e^{-3t}$$

Putting $t = 0$ and $\dfrac{di}{dt} = 0$ gives

$$0 = -2A - B - 7.5$$

$$-2A - B = 7.5$$

We have the simultaneous equations

$$A + B = -2.5$$
$$-2A - B = 7.5$$

Solving these gives

$$A = -5 \text{ and } B = 2.5$$

Substituting $A = -5$ and $B = 2.5$ into $i = Ae^{-2t} + Be^{-t} + 2.5e^{-3t}$ yields

$$i = -5e^{-2t} + 2.5e^{-t} + 2.5e^{-3t}$$

$$= 2.5(e^{-3t} + e^{-t}) - 5e^{-2t}$$

The next example is more sophisticated than the above examples. It uses unfamiliar symbols as coefficients rather than simple numbers. Again the technique of finding a particular solution is the same.

Example 15 *structures*

A load, P, is applied to each end of a column of length L and is also subject to a force, F, at $x = L/2$. The differential equation is given by

$$EI\frac{d^2y}{dx^2} + Py + \frac{F}{2}x = 0 \qquad 0 \le x \le \frac{L}{2}$$

where y is the deflection at a distance x. If at $x = 0$, $y = 0$ and at $x = \frac{L}{2}$, $\frac{dy}{dx} = 0$ then show that

$$y = \frac{F}{2kP}\left[\sin(kx)\sec\left(\frac{kL}{2}\right) - kx\right] \quad \text{where } k = \sqrt{\frac{P}{EI}}$$

Solution

Rearranging the given differential equation to standard form:

$$EI\frac{d^2y}{dx^2} + Py = -\frac{Fx}{2}$$

Dividing through by EI:

$$\frac{d^2y}{dx^2} + \frac{P}{EI}y = -\frac{Fx}{2EI}$$

We found the complementary function y_c in **Example 6**:

$$y_c = A\cos(kx) + B\sin(kx) \quad \text{where } k = \sqrt{P/EI}$$

The particular integral Y: since the Right-Hand Side is a linear expression, our trial function is

$$Y = ax + b$$

$$\frac{dY}{dx} = a, \quad \frac{d^2Y}{dx^2} = 0 \qquad \text{[Differentiating]}$$

Substituting into $\dfrac{d^2Y}{dx^2} + \dfrac{P}{EI}Y = -\dfrac{F}{2EI}x$ gives

$$0 + \frac{P}{EI}(ax + b) = -\frac{F}{2EI}x$$

Equating coefficients of x: $\qquad \dfrac{P}{EI}a = -\dfrac{F}{2EI}$ which gives $a = -\dfrac{F}{2P}$

Equating constants: $\qquad \dfrac{Pb}{EI} = 0$, hence $b = 0$

Example 15 *continued*

Placing $a = -\dfrac{F}{2P}$ and $b = 0$ into $Y = ax + b$ yields $Y = -\dfrac{Fx}{2P}$.

Since $y = y_c + Y$, we have

$$y = A\cos(kx) + B\sin(kx) - \frac{Fx}{2P}$$

We are given boundary conditions so we can find the constants A and B. Substituting the given boundary conditions, $x = 0$, $y = 0$:

$$0 = A + 0 + 0, \text{ hence } A = 0$$

Substituting $A = 0$ into $y = A\cos(kx) + B\sin(kx) - \dfrac{Fx}{2P}$ gives

$$y = B\sin(kx) - \frac{Fx}{2P}$$

Differentiating:

$$\frac{dy}{dx} = \underbrace{kB\cos(kx)}_{\text{by } 6.12} - \frac{F}{2P}$$

Substituting the other boundary condition, $x = \dfrac{L}{2}$, $\dfrac{dy}{dx} = 0$:

$$0 = kB\cos\left(\frac{kL}{2}\right) - \frac{F}{2P}$$

$$kB\cos\left(\frac{kL}{2}\right) = \frac{F}{2P}$$

$$B = \frac{F}{2kP\cos\left(\dfrac{kL}{2}\right)} \qquad \left[\text{Dividing by } k\cos\left(\frac{kL}{2}\right)\right]$$

$$= \frac{F}{2kP}\sec\left(\frac{kL}{2}\right) \qquad \left[\text{Remember that } \frac{1}{\cos(x)} = \sec(x)\right]$$

Substituting $B = \dfrac{F}{2kP}\sec\left(\dfrac{kL}{2}\right)$ into $y = B\sin(kx) - \dfrac{Fx}{2P}$ gives

$$y = \frac{F}{2kP}\sec\left(\frac{kL}{2}\right)\sin(kx) - \frac{Fx}{2P}$$

$$= \frac{F}{2kP}\sec\left(\frac{kL}{2}\right)\sin(kx) - \frac{Fkx}{2kP} \qquad \left[\text{Writing } \frac{Fx}{2P} = \frac{Fkx}{2kP}\right]$$

$$= \frac{F}{2kP}\left[\sec\left(\frac{kL}{2}\right)\sin(kx) - kx\right] \qquad \left[\text{Taking out } \frac{F}{2kP}\right]$$

6.12 $\dfrac{d}{dx}[\sin(kx)] = k\cos(kx)$

SUMMARY

If we are given initial or boundary conditions then we can find the particular solution of the given differential equation.

1 [electrical principles]   The current, i, in a circuit is given by the second order differential equation

$$\frac{d^2 i}{dt^2} + 8\frac{di}{dt} + 15i = 150$$

When $t = 0$, both $i = 0$ and $\frac{di}{dt} = 0$. Find the current, i, in terms of t.

2 [electrical principles]   By applying Kirchhoff's voltage law to a series RLC circuit we obtain the differential equation

$$\frac{d^2 i}{dt^2} + 3\frac{di}{dt} + 2i = 5e^{-3t}$$

Solve this differential equation for the initial conditions, when $t = 0$, both $i = 0$ and $\frac{di}{dt} = 5$.

3 [mechanics]   By applying Newton's second law to a body of mass m at a distance x, we obtain

$$\ddot{x} + 9x = -18$$

Solve this differential equation for the initial conditions, when $t = 0$, both $x = 0$ and $\dot{x} = 3$.

4 [vibrations]

Fig. 7

The undamped mass–spring system in Fig. 7 is subject to an external force $F\sin(\alpha t)$. The equation of motion is

$$m\ddot{x} + kx = F\sin(\alpha t) \quad \left(\alpha \neq \sqrt{\frac{k}{m}}\right)$$

where x is the displacement of mass m from its equilibrium level and k is the spring stiffness. By assuming the trial function $X = C\sin(\alpha t)$ and the initial conditions, when $t = 0$, both $x = 0$ and $\dot{x} = 0$, show that

$$x = \frac{F}{\omega(k - m\alpha^2)}[\omega\sin(\alpha t) - \alpha\sin(\omega t)]$$

where $\omega = \sqrt{\frac{k}{m}}$

The remaining questions are in the field of [**structures**].

5 A strut of length L is subject to a load P. The deflection y at a distance x is related by the second order differential equation

$$EI\frac{d^2 y}{dx^2} + Py = Fx$$

where F is a force. By using the boundary conditions, at $x = 0$, $y = 0$ and at $x = L$, $\frac{dy}{dx} = 0$, show that

$$y = \frac{F}{kP}\Big[kx - \sec(kL)\sin(kx)\Big]$$

where $k = \sqrt{\frac{P}{EI}}$

6 A strut of length L is fixed at one end and is subject to a load P at the other

Exercise **14(d) continued**

Solutions at end of book. Complete solutions available at www.palgrave.com/science/engineering/singh

end (free end). The differential equation relating deflection y at a distance x from the free end is given by

$$EI\frac{d^2y}{dx^2} + P(y - e - d) = 0$$

where e is the eccentricity and d is the deflection at the free end.

Using the initial conditions at $x = 0$, both $y = 0$, and $\frac{dy}{dx} = 0$, show that

$$y = (e + d)[1 - \cos(kx)] \text{ where } k = \sqrt{\frac{P}{EI}}$$

7 A load, P, is applied to each end of a strut of length L. The deflection y at a distance x is related by

$$EI\frac{d^2y}{dx^2} + Py = M$$

where M is the moment at each end of the strut.

By using the initial conditions, at $x = 0$, both $y = 0$ and $\frac{dy}{dx} = 0$, show that

$$y = \frac{M}{P}[1 - \cos(kx)] \text{ where } k = \sqrt{\frac{P}{EI}}$$

Miscellaneous exercise **14**

Solutions at end of book. Complete solutions available at www.palgrave.com/science/engineering/singh

1 Find the general solutions of

a $\dfrac{d^2y}{dx^2} + 8\dfrac{dy}{dx} + 16y = 0$

b $\dfrac{d^2y}{dx^2} - 2\dfrac{dy}{dx} - 3y = 0$

c $\dfrac{d^2y}{dx^2} - 6\dfrac{dy}{dx} + 7y = 0$

2 Find the general solutions of the following:

a $\dfrac{d^2y}{dx^2} + 8\dfrac{dy}{dx} + 16y = 8$

b $\dfrac{d^2y}{dx^2} + 8\dfrac{dy}{dx} + 16y = 8x$

c $\dfrac{d^2y}{dx^2} + 8\dfrac{dy}{dx} + 16y = x^2 + x + 1$

3 【Y】 [vibrations]   A shaft of length l with modulus of rigidity G has angular displacement θ, a function of time t, given by

$$\frac{d^2\theta}{dt^2} + \frac{GJ}{Il}\theta = 0$$

where I is moment of inertia and J is second moment of area. Show that

$$\theta = A\cos(\omega t) + B\sin(\omega t) \text{ where } \omega = \sqrt{\frac{GJ}{Il}}$$

4 【Ü】 [heat transfer]   The temperature, $T (=T(x))$, of a fin of perimeter P with cross-sectional area A is given by

$$\frac{d^2T}{dx^2} - \frac{hP}{kA}T = 0$$

where h is the convection heat transfer coefficient, k is the thermal conductivity and x is the distance along the fin. Find an expression for T in terms of x.

5 【Y】 [vibrations]   An undamped spring–mass system is subject to a force $F\cos(\omega t)$ where $\omega = \sqrt{\dfrac{k}{m}}$. The equation of motion is given by

$$m\ddot{x} = -kx + F\cos(\omega t) \quad (\omega \neq 0)$$

where m is mass, k is stiffness of spring, x is displacement and t is time. Show that

$$x = A\cos(\omega t) + B\sin(\omega t) + \frac{Ft}{2m\omega}\sin(\omega t)$$

6 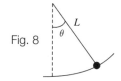 [*vibrations*]

Fig. 8

The equation of motion of a simple pendulum of length L is given by

$$L\ddot{\theta} + g\theta = 0$$

where θ is the angle shown, $\ddot{\theta} = \dfrac{d^2\theta}{dt^2}$, $\dot{\theta} = \dfrac{d\theta}{dt}$ and $\omega = \sqrt{\dfrac{g}{L}}$ (angular frequency). For the initial conditions $\theta = 1$ and $\dot{\theta} = \sqrt{3}\omega$ when $t = 0$, show that

$$\theta = 2\cos\left(\omega t - \frac{\pi}{3}\right)$$

and sketch the graph of θ versus ωt for $\omega t \geq 0$.

Questions 7 to 11, inclusive, are in the field of [*electrical principles*].

7 A parallel LC circuit has the second order differential equation

$$C\frac{d^2v}{dt^2} + \frac{v}{L} = 0$$

If $\omega^2 = \dfrac{1}{LC}$, show that the voltage, v, can be written as

$$v = r\cos(\omega t - \alpha)$$

8 A second order differential equation for a parallel RLC circuit is given by

$$C\frac{d^2v}{dt^2} + \frac{1}{R}\frac{dv}{dt} + \frac{v}{L} = \frac{di}{dt}$$

i Determine the characteristic equation of ⁣ * ⁣.

ii The characteristic equation can be written as

$$m^2 + 2\zeta\omega m + \omega^2 = 0$$

where ζ is the damping ratio and ω is the angular frequency. Critical damping occurs when $\zeta = 1$. We define the critical resistance, R_{cr}, as the resistance value which gives critical damping. Determine R_{cr} for the characteristic equation of part **i**.

9 A second order differential equation for an RLC series circuit is given by

$$\boxed{}\qquad L\frac{d^2 i}{dt^2} + R\frac{di}{dt} + \frac{i}{C} = \frac{dv}{dt}$$

i Determine the characteristic equation of ⁣ * ⁣.

ii Determine the critical resistance, R_{cr}, where R_{cr} is defined as in question **8**.

10

Fig. 9

Figure 9 shows a series RLC circuit. By applying Kirchhoff's voltage law, we obtain the second order differential equation

$$L\frac{d^2i}{dt^2} + R\frac{di}{dt} + \frac{i}{C} = \frac{dv}{dt}$$

i By using the solution to question **9**, determine the value of the critical resistance, R_{cr}.

ii For $R = R_{cr}$, obtain the homogeneous solution of the given second order differential equation.

11 The differential equation governing the voltage, v, of a parallel RLC circuit is given by

$$C\frac{d^2v}{dt^2} + \frac{1}{R}\frac{dv}{dt} + \frac{v}{L} = 0$$

The resistance $R = 500\ \Omega$, inductance $L = 100$ mH and capacitance $C = 0.5\ \mu$F. Determine an expression for the voltage, v, in terms of t with the initial conditions, when $t = 0$, both $v = 9$ and $\dfrac{dv}{dt} = 0$.

12 🌡 [*heat transfer*]   Consider the differential equation of question **4**:

$$\frac{d^2T}{dx^2} - \frac{hP}{kA}T = 0 \qquad \left(\frac{hP}{kA} \neq 0\right)$$

Given the boundary conditions, at $x = 0$, $T = T_B$ (base temperature) and at $x = L$, $\dfrac{dT}{dx} = 0$, show that

$$T = T_B\frac{\cosh[m(L - x)]}{\cosh(mL)} \text{ where } m = \sqrt{\frac{hP}{kA}}$$

13 🌡 [*heat transfer*]   The second order differential equation of question **4**:

$$\frac{d^2T}{dx^2} - \frac{hP}{kA}T = 0 \qquad \left(\frac{hP}{kA} \neq 0\right)$$

has boundary conditions, at $x = 0$, $T = T_B$ and at $x = L$, $T = 0$. Show that

$$T = T_B\frac{\sinh[m(L - x)]}{\sinh(mL)} \text{ where } m^2 = \frac{hP}{kA}$$

🏔 **Questions 14 to 17, inclusive, belong to the field of [*structures*].**

14 Figure 10 shows an elastic strut with a load P applied at both ends with an eccentricity e. The relationship between the deflection y at a distance x is given by

$$EI\frac{d^2y}{dx^2} + Py = 0$$

Fig. 10

If $y = e$ at $x = 0$ and $x = L$, show that

$$y = e\left[\cos(kx) + \tan\left(\frac{kL}{2}\right)\sin(kx)\right]$$

where $k = \sqrt{\dfrac{P}{EI}}$

$$\left[Hint: \frac{1 - \cos(x)}{\sin(x)} = \tan\left(\frac{x}{2}\right)\right]$$

15 A column of length L has a uniform load w and an external load P. The deflection, y, at a distance, x, is related by the second order differential equation

$$EI\frac{d^2y}{dx^2} + Py + \frac{w}{2}(Lx - x^2) = 0$$

i Show that

$$y = A\cos(kx) + B\sin(kx)$$
$$+ \frac{w}{2P}\left[x^2 - Lx - \frac{2}{k^2}\right]$$

where $k^2 = \dfrac{P}{EI}$.

ii By using the boundary conditions, at $x = 0$, $y = 0$ and at $x = L$, $y = 0$, and the identity

$$\tan\left(\frac{x}{2}\right) = \frac{1 - \cos(x)}{\sin(x)}$$

show that

$$y = \frac{w}{Pk^2}\left[\tan\left(\frac{kL}{2}\right)\sin(kx) + \cos(kx)\right]$$
$$+ \frac{w}{2P}\left[x^2 - Lx - \frac{2}{k^2}\right]$$

16 A load, P, is applied to a strut of length L which is fixed at one end and pinned at the other end. The deflection y is given by

$$EI\frac{d^2y}{dx^2} + Py + Fx + M = 0$$

where F is the applied force, M is the moment and x is distance. By using the boundary conditions, at $x = 0$, $y = 0$ and at $x = L$, $y = 0$, show that

$Py = M\cos(kx)$

$+ \left(FL\text{cosec}(kL) + M\tan\left(\dfrac{kL}{2}\right)\right)\sin(kx)$

$- Fx - M$

where $k = \sqrt{\dfrac{P}{EI}}$. (You may use the hint of question **14**.)

17 A load, P, is applied to a column of length L fixed at both ends and with a uniform load q. The differential equation

$$EI\dfrac{d^2y}{dx^2} + Py = \dfrac{q}{2}(x^2 - xL) - M$$

gives the relationship between the deflection y and the distance x (M is the fixed end moment). At $x = 0$, both $y = 0$ and $\dfrac{dy}{dx} = 0$. Show that

$$y = \dfrac{1}{2P}\left[\dfrac{qL}{k}\sin(kx) + 2\left(\dfrac{q}{k^2} + M\right)\cos(kx)\right.$$
$$\left. + qx^2 - qLx - 2M - \dfrac{2q}{k^2}\right]$$

where $k^2 = \dfrac{P}{EI}$.

You may use a computer algebra system for question 18.

18 [vibrations] The equation of motion for a mass–spring system is given by

$$\ddot{x} + 6\dot{x} + 500x = 1000t + 400$$

Solve this differential equation for the initial conditions, when $t = 0$, both $x = 0$ and $\dfrac{dx}{dt} = 0$. Plot the graph of x versus t for $0 \le t \le 1$.

Partial Differentiation

SECTION A **Partial derivatives**

By the end of this section you will be able to:
▶ distinguish between partial and ordinary differentiation
▶ understand the notation used for partial differentiation
▶ find the first and second partial derivatives

So far we have only differentiated functions of one variable $f(x)$, which means f is a function of x only. In this chapter we look at differentiating functions of two or more variables, $f(x, y)$, a function of x and y, or $f(x, y, z)$, a function of x, y and z. Most engineering problems involve two or more variables. For example:

a Thermodynamics

$$P = \frac{nRT}{V}$$

The pressure, P, of a gas in a cylinder will change with temperature, T, volume, V, and the number of moles of gas present, n (R is a constant).

b Electronics

$$\alpha = K\frac{BL}{\sqrt{V}}$$

The angular deflection, α, of a beam of electrons depends on magnetic field, B, length, L, and accelerating voltage, V (K is a constant).

c Materials

$$\sigma = K\left(\frac{r}{L}\right)^2$$

The axial stress, σ, varies with radius of gyration, r, and length, L.

A1 **Notation**

The volume, V, of a right circular cylinder is given by

$$V = \pi r^2 h$$

where r is the radius and h is the height.

If we change the radius r and keep the height h fixed, the rate of change in volume V is denoted by

$$\left[\frac{\mathrm{d}V}{\mathrm{d}r}\right]_{h\text{ constant}}$$

This means the rate of change in volume V with respect to radius r, whilst keeping the height h constant. This is normally written as

$$\frac{\partial V}{\partial r}$$

Hence $\left[\dfrac{dV}{dr}\right]_{h\text{ constant}} = \dfrac{\partial V}{\partial r}$

'Curly dees' are used to distinguish between partial and ordinary differentiation.

A function of two variables is written as $f(x, y)$, that is f is a function of x and y. For example

$$f(x, y) = x^2 + e^{2y} + 2\cos(xy^2)$$

A function of two variables, $f(x, y)$, is graphed similar to that of one variable, $f(x)$. However in this case we have a three-dimensional co-ordinate system where a point is identified by three co-ordinates (x, y, z).

The dependent variable, z, is plotted on the vertical axis and the two independent variables on the horizontal axes, as shown in Fig. 1. Notice that a function in two variables can be visualized as a surface, and that z is a function of x and y.

Fig. 1

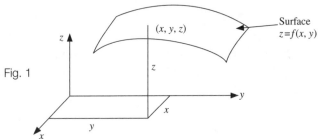

The partial derivative of $f(x, y)$ with respect to x is denoted by $\dfrac{\partial f}{\partial x}$, which means differentiate f with respect to x while keeping y fixed (or treating y as a constant) (Fig. 2). Thus

$$\frac{\partial f}{\partial x} = \lim_{h \to 0} \frac{f(x + h, y) - f(x, y)}{h}$$

$\dfrac{\partial f}{\partial x}$ is stated verbally as 'partial dee f by dee x'.

Similarly $\dfrac{\partial f}{\partial y}$ is the partial derivative of $f(x, y)$ with respect to y, which means differentiate f with respect to y keeping x fixed (Fig. 3). Hence

$$\frac{\partial f}{\partial y} = \lim_{h \to 0} \frac{f(x, y + h) - f(x, y)}{h}$$

Fig. 2

Fig. 3

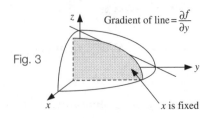

Because of these definitions, most of the rules of ordinary differentiation hold, such as

$$\frac{\partial}{\partial x}(u + v) = \frac{\partial u}{\partial x} + \frac{\partial v}{\partial x}$$

where u and v are functions of x and y.

Example 1

Consider the function
$$f(x, y) = x^2 + 2xy^2 + y^2$$
Determine $\dfrac{\partial f}{\partial x}$ and $\dfrac{\partial f}{\partial y}$.

Solution

$\dfrac{\partial f}{\partial x}$ means differentiate f with respect to x and think of y as a constant. As stated above, you can partially differentiate each part and then add your results:

$$f(x, y) = x^2 + 2xy^2 + y^2$$
$$\frac{\partial f}{\partial x} = 2x + 2y^2 + \underset{\substack{\text{because } y \\ \text{is treated as} \\ \text{a constant}}}{0} = 2x + 2y^2$$

Similarly

$$\frac{\partial f}{\partial y} = \underset{\substack{\text{because } x \\ \text{is treated as} \\ \text{a constant}}}{0} + 4xy + 2y = 4xy + 2y$$

Example 2 *materials*

The second moment of area, I, of a rectangle with dimensions b and d is given by
$$I = \frac{bd^3}{12}$$
Determine $\dfrac{\partial I}{\partial b}$ and $\dfrac{\partial I}{\partial d}$.

Solution

Just because the function is not in terms of x and y does not mean it is more difficult. We have
$$\frac{\partial I}{\partial b} = \frac{d^3}{12} \quad \text{and} \quad \frac{\partial I}{\partial d} = \frac{3bd^2}{12} = \frac{bd^2}{4}$$

Example 3 *thermodynamics*

The pressure, P, of a gas is given by
$$P = \frac{KT}{V}$$
where K is a constant, T is temperature and V is volume. Find
$$\frac{\partial P}{\partial T} \quad \text{and} \quad \frac{\partial P}{\partial V}$$

Example 3 *continued*

Solution

We have

$$\frac{\partial P}{\partial T} = \frac{\partial}{\partial T}\left(\frac{KT}{V}\right) = \frac{K}{V}$$

We rewrite P as $P = \dfrac{KT}{V} = KTV^{-1}$. Differentiating partially with respect to V gives

$$\frac{\partial P}{\partial V} = -KTV^{-2}$$

$$= -\frac{KT}{V^2}$$

Example 4 *thermodynamics*

From the ideal gas equation, $PV = RT$, show that

$$\left(\frac{\partial P}{\partial T}\right)\left(\frac{\partial T}{\partial V}\right)\left(\frac{\partial V}{\partial P}\right) = -1$$

Solution

By rearranging the given equation, $PV = RT$, and partially differentiating we get

$$P = \frac{RT}{V} \text{ and } \frac{\partial P}{\partial T} = \frac{R}{V}$$

$$T = \frac{PV}{R} \text{ and } \frac{\partial T}{\partial V} = \frac{P}{R}$$

$$V = \frac{RT}{P} = RTP^{-1} \text{ and } \frac{\partial V}{\partial P} = -RTP^{-2} = -\frac{RT}{P^2}$$

Substituting these partial derivatives, $\dfrac{\partial P}{\partial T} = \dfrac{R}{V}, \dfrac{\partial T}{\partial V} = \dfrac{P}{R}$ and $\dfrac{\partial V}{\partial P} = -\dfrac{RT}{P^2}$, gives

$$\left(\frac{\partial P}{\partial T}\right)\left(\frac{\partial T}{\partial V}\right)\left(\frac{\partial V}{\partial P}\right) = \frac{R}{V}\frac{P}{R}\left(-\frac{RT}{P^2}\right)$$

$$= -\frac{RT}{VP} \qquad \text{[Cancelling]}$$

$$= -\frac{RT}{V\left(\dfrac{RT}{V}\right)} \qquad \left[\text{Because } P = \frac{RT}{V}\right]$$

$$= -1 \qquad \text{[Cancelling]}$$

A function of three variables is denoted by $f(x, y, z)$.

Example 5 *electronics*

The resistance, R, of a wire of length L and diameter D is given by

$$R = \frac{k\rho L}{D^2}$$

where k is constant and ρ is resistivity. Find $\dfrac{\partial R}{\partial L}$, $\dfrac{\partial R}{\partial \rho}$ and $\dfrac{\partial R}{\partial D}$.

Solution

Note that R is a function of ρ, L and D, that is $R = f(\rho, L, D)$.

The first two, $\dfrac{\partial R}{\partial L}$ and $\dfrac{\partial R}{\partial \rho}$, are straightforward:

$$\frac{\partial R}{\partial L} = \frac{k\rho}{D^2} \text{ and } \frac{\partial R}{\partial \rho} = \frac{kL}{D^2}$$

For $\dfrac{\partial R}{\partial D}$ we rewrite $R = \dfrac{k\rho L}{D^2} = k\rho L D^{-2}$, hence considering $k\rho L$ as a constant we differentiate

$$\frac{\partial R}{\partial D} = -2k\rho L D^{-3} = -\frac{2k\rho L}{D^3}$$

A2 **Higher derivatives**

Let f be a function of x and y then $\dfrac{\partial^2 f}{\partial x^2}$ means partially differentiate $\dfrac{\partial f}{\partial x}$ with respect to x.

Example 6

Let

$$f = f(x, y) = x^3 + y^2\cos(x)$$

Determine $\dfrac{\partial^2 f}{\partial x^2}$ and $\dfrac{\partial^2 f}{\partial y^2}$.

Solution

How can we find $\dfrac{\partial^2 f}{\partial x^2}$?

First find $\dfrac{\partial f}{\partial x}$ by differentiating f with respect to x and treating y as a constant:

$$\frac{\partial f}{\partial x} = 3x^2 - y^2\sin(x)$$

Example 6 *continued*

Note that $\dfrac{\partial f}{\partial x}$ is a function of x and y. We can partially differentiate this,

$\dfrac{\partial f}{\partial x} = 3x^2 - y^2 \sin(x)$, with respect to x to give $\dfrac{\partial^2 f}{\partial x^2}$:

$$\frac{\partial^2 f}{\partial x^2} = 6x - y^2 \cos(x)$$

To find $\dfrac{\partial^2 f}{\partial y^2}$ we need to first obtain $\dfrac{\partial f}{\partial y}$:

$$f = x^3 + y^2 \cos(x)$$
$$\frac{\partial f}{\partial y} = 0 + 2y\cos(x) = 2y\cos(x)$$

Partially differentiating this, $\dfrac{\partial f}{\partial y} = 2y\cos(x)$, with respect to y gives $\dfrac{\partial^2 f}{\partial y^2}$:

$$\frac{\partial^2 f}{\partial y^2} = 2\cos(x)$$

Hence we define $\dfrac{\partial^2 f}{\partial x^2}$ and $\dfrac{\partial^2 f}{\partial y^2}$ for $f = f(x, y)$, a general function of x and y, as

15.1
$$\frac{\partial^2 f}{\partial x^2} = \frac{\partial}{\partial x}\left(\frac{\partial f}{\partial x}\right)$$

15.2
$$\frac{\partial^2 f}{\partial y^2} = \frac{\partial}{\partial y}\left(\frac{\partial f}{\partial y}\right)$$

15.1 says we first find $\dfrac{\partial f}{\partial x}$ and then partially differentiate this again with respect to x to give

$\dfrac{\partial^2 f}{\partial x^2}$. In each case we treat y as a constant. 15.2 is similar with the variables interchanged.

In this chapter we will use the term 'differentiate' to mean partially differentiate.

A3 Mixed partial derivatives

There are two other second derivatives, $\dfrac{\partial^2 f}{\partial x \partial y}$ and $\dfrac{\partial^2 f}{\partial y \partial x}$, which are defined as

15.3
$$\frac{\partial^2 f}{\partial x \partial y} = \frac{\partial}{\partial x}\left(\frac{\partial f}{\partial y}\right)$$

15.4
$$\frac{\partial^2 f}{\partial y \partial x} = \frac{\partial}{\partial y}\left(\frac{\partial f}{\partial x}\right)$$

15.3 says differentiate f with respect to y and then differentiate this result with respect to x. In each case we treat the other variables as constants. 15.4 is similar with variables interchanged.

Example 7

With $f = f(x, y)$ as in **Example 6**, find $\dfrac{\partial^2 f}{\partial x \partial y}$ and $\dfrac{\partial^2 f}{\partial y \partial x}$.

Solution

We have already established

$$\frac{\partial f}{\partial x} = 3x^2 - y^2 \sin(x)$$

Differentiating this with respect to y:

$$\frac{\partial^2 f}{\partial y \partial x} = \frac{\partial}{\partial y} [3x^2 - y^2 \sin(x)]$$
$$= 0 - 2y \sin(x)$$
$$= -2y \sin(x)$$

From **Example 6** we have $\dfrac{\partial f}{\partial y} = 2y \cos(x)$ and differentiate this with respect to x:

$$\frac{\partial^2 f}{\partial x \partial y} = \frac{\partial}{\partial x} [2y \cos(x)]$$
$$= -2y \sin(x)$$

? **What do you notice about your results to Example 7?**

15.5 $$\frac{\partial^2 f}{\partial x \partial y} = \frac{\partial^2 f}{\partial y \partial x}$$

The result, 15.5 , is usually but not always true. For all the functions we deal with we will assume it is true. Many of the rules of ordinary differentiation can be used in a similar manner for partial differentiation. For example the product and quotient rules are defined by

15.6 $$\frac{\partial}{\partial x}(uv) = v\frac{\partial u}{\partial x} + u\frac{\partial v}{\partial x} \qquad \text{[Product rule]}$$

15.7 $$\frac{\partial}{\partial x}\left(\frac{u}{v}\right) = \frac{v\dfrac{\partial u}{\partial x} - u\dfrac{\partial v}{\partial x}}{v^2} \qquad \text{[Quotient rule]}$$

where u and v are functions x, y, z etc. Similarly we can find

$$\frac{\partial}{\partial y}(uv), \ \frac{\partial}{\partial y}\left(\frac{u}{v}\right), \ \frac{\partial}{\partial z}(uv) \text{ etc.}$$

Moreover we can apply the differentiation table (Table 3 of Chapter 6) by replacing the ordinary dee with a curly dee. For example

6.18 $$\frac{d}{dx}[\ln(u)] = \frac{1}{u}\frac{du}{dx}$$

becomes $\dfrac{\partial}{\partial x} [\ln(u)] = \dfrac{1}{u} \dfrac{\partial u}{\partial x}$

where u is a function of x, y, z etc.

Example 8 *fluid mechanics*

For a fluid flow to be a potential flow it must satisfy Laplace's equation which is given by

$$\frac{\partial^2 \psi}{\partial x^2} + \frac{\partial^2 \psi}{\partial y^2} = 0$$

where $\psi = \psi(x, y)$ is the stream function. Show that if

$$\psi = \ln(x^2 + y^2)$$

then ψ is a potential flow.

Solution

Don't be put off by the Greek symbol, ψ (psi).

First we need to find $\dfrac{\partial \psi}{\partial x}$. How?

Applying

$$\frac{\partial}{\partial x} \left[\ln(u) \right] = \frac{1}{u} \frac{\partial u}{\partial x}$$

with $u = x^2 + y^2$ we have

$$\frac{\partial}{\partial x} \left[\ln(x^2 + y^2) \right] = \frac{1}{x^2 + y^2} \frac{\partial}{\partial x} (x^2 + y^2)$$

$$= \frac{1}{x^2 + y^2} 2x = \frac{2x}{x^2 + y^2}$$

Hence $\dfrac{\partial \psi}{\partial x} = \dfrac{2x}{x^2 + y^2}$. To find $\dfrac{\partial^2 \psi}{\partial x^2}$ we need to differentiate $\dfrac{\partial \psi}{\partial x}$. Applying the quotient rule, 15.7 , with

$$u = 2x \qquad v = x^2 + y^2$$
$$\frac{\partial u}{\partial x} = 2 \qquad \frac{\partial v}{\partial x} = 2x$$

we have

$$\frac{\partial^2 \psi}{\partial x^2} = \frac{\dfrac{\partial u}{\partial x} v - u \dfrac{\partial v}{\partial x}}{v^2}$$

$$= \frac{2(x^2 + y^2) - 2x(2x)}{(x^2 + y^2)^2} \qquad \text{[Substituting]}$$

$$= \frac{2x^2 + 2y^2 - 4x^2}{(x^2 + y^2)^2} \qquad \text{[Expanding numerator]}$$

$$= \frac{2y^2 - 2x^2}{(x^2 + y^2)^2} \qquad \text{[Simplifying numerator]}$$

Example 8 *continued*

We go through the same procedure for $\dfrac{\partial^2 \psi}{\partial y^2}$.

$$\psi = \ln(x^2 + y^2)$$

$$\frac{\partial \psi}{\partial y} = \frac{1}{x^2 + y^2} \frac{\partial}{\partial y}\left(x^2 + y^2\right) = \frac{1}{x^2 + y^2} 2y = \frac{2y}{x^2 + y^2}$$

Applying the quotient rule, 15.7 , with

$$u = 2y \qquad v = x^2 + y^2$$
$$\frac{\partial u}{\partial y} = 2 \qquad \frac{\partial v}{\partial y} = 2y$$

we have

$$\frac{\partial^2 \psi}{\partial y^2} = \frac{\dfrac{\partial u}{\partial y} v - u \dfrac{\partial v}{\partial y}}{v^2}$$

$$= \frac{2(x^2 + y^2) - 2y(2y)}{(x^2 + y^2)^2}$$

$$= \frac{2x^2 + 2y^2 - 4y^2}{(x^2 + y^2)^2} = \frac{2x^2 - 2y^2}{(x^2 + y^2)^2}$$

Adding both the second derivatives gives

$$\frac{\partial^2 \psi}{\partial x^2} + \frac{\partial^2 \psi}{\partial y^2} = \frac{2y^2 - 2x^2}{(x^2 + y^2)^2} + \frac{2x^2 - 2y^2}{(x^2 + y^2)^2}$$

$$= \frac{2y^2 - 2x^2 + 2x^2 - 2y^2}{(x^2 + y^2)^2}$$

$$= \frac{0}{(x^2 + y^2)^2} = 0$$

Hence $\psi = \ln(x^2 + y^2)$ is a potential flow.

Example 9

Let

$$f = f(x, y) = \cos(xy)$$

Determine $\dfrac{\partial^2 f}{\partial x^2}, \dfrac{\partial^2 f}{\partial y^2}$ and $\dfrac{\partial^2 f}{\partial x \partial y}$.

Solution

To find $\dfrac{\partial^2 f}{\partial x^2}$ we have first to obtain $\dfrac{\partial f}{\partial x}$:

Example 9 *continued*

$$f = \cos(xy)$$

$$\frac{\partial f}{\partial x} = -y\sin(xy) \quad \left[\text{Differentiating } \frac{\partial}{\partial x}\left[\cos(kx)\right] = -k\sin(kx)\right]$$

$$\frac{\partial^2 f}{\partial x^2} = -y^2\cos(xy) \quad \left[\text{Differentiating } \frac{\partial}{\partial x}\left[\sin(kx)\right] = k\cos(kx)\right]$$

Similarly we find $\dfrac{\partial^2 f}{\partial y^2}$:

$$f = \cos(xy)$$

$$\frac{\partial f}{\partial y} = -x\sin(xy) \quad \left[\text{Differentiating } \frac{\partial}{\partial y}\left[\cos(ky)\right] = -k\sin(ky)\right]$$

$$\frac{\partial^2 f}{\partial y^2} = -x^2\cos(xy) \quad \left[\text{Differentiating } \frac{\partial}{\partial y}\left[\sin(ky)\right] = k\cos(ky)\right]$$

To find the mixed derivative $\dfrac{\partial^2 f}{\partial x \partial y}$ means we need to differentiate $\dfrac{\partial f}{\partial y} = -x\sin(xy)$ with respect to x. Using the product rule we have

$$u = x \qquad v = \sin(xy)$$
$$u' = 1 \qquad v' = y\cos(xy)$$
$$\frac{\partial^2 f}{\partial x \partial y} = \frac{\partial}{\partial x}\left[-x\sin(xy)\right]$$
$$= -[u'v + uv']$$
$$= -[(1)\sin(xy) + (x)y\cos(xy)]$$
$$= -[\sin(xy) + xy\cos(xy)]$$

SUMMARY

In this section we considered differentiating functions of two or more variables.

For example let $f = f(x, y)$ then $\dfrac{\partial f}{\partial x}$ means differentiate f with respect to x while treating y as a constant. Higher derivatives are defined as

15.1 $\qquad \dfrac{\partial^2 f}{\partial x^2} = \dfrac{\partial}{\partial x}\left(\dfrac{\partial f}{\partial x}\right)$

15.2 $\qquad \dfrac{\partial^2 f}{\partial y^2} = \dfrac{\partial}{\partial y}\left(\dfrac{\partial f}{\partial y}\right)$

15.3 $\qquad \dfrac{\partial^2 f}{\partial x \partial y} = \dfrac{\partial}{\partial x}\left(\dfrac{\partial f}{\partial y}\right)$

15.4 $\qquad \dfrac{\partial^2 f}{\partial y \partial x} = \dfrac{\partial}{\partial y}\left(\dfrac{\partial f}{\partial x}\right)$

We can modify the rules of ordinary differentiation by replacing 'dee' by a 'curly dee'.

Exercise 15(a)

Solutions at end of book. Complete solutions available at www.palgrave.com/science/engineering/singh

1 Find $\dfrac{\partial f}{\partial x}$ and $\dfrac{\partial f}{\partial y}$ for each of the following functions:

a $f(x, y) = x^2 + y^2$

b $f(x, y) = \sin(x) + \cos(y)$

c $f(x, y) = x^3 + y^3 + 3xy$

d $f(x, y) = 2y - \dfrac{1000}{x}$

2 ▨ [*materials*]   A beam is subject to a uniform load of w per unit length and a concentrated load P. The bending moment M at a distance x from one end is given by

$$M = Px + \frac{wx^3}{3}$$

Determine $\dfrac{\partial M}{\partial P}$ and $\dfrac{\partial M}{\partial x}$.

3 ▨ [*materials*]   Castigliano's theorem says the displacement, Δ, under a force P is given by

$$\Delta = \frac{\partial U}{\partial P}$$

where U is the internal strain energy of the body. For the following, find Δ:

a $U = \dfrac{P^2 L}{2AE}$ b $U = \dfrac{9P^2 L^3}{96EI}$

4 ⛰ [*structures*]   Given the stress function, $\Omega(x, y)$, as (A and B are constants)

$$\Omega(x, y) = Ax^2 y^4 + Bx^4 y^2$$

find the stresses, σ_x and σ_y, where

$$\sigma_x = \frac{\partial^2 \Omega}{\partial y^2} \text{ and } \sigma_y = \frac{\partial^2 \Omega}{\partial x^2}$$

5 ≈ [*fluid mechanics*]   The continuity equation is defined as

$$\frac{\partial u}{\partial x} + \frac{\partial v}{\partial y} = 0$$

where u, v are velocities of the fluid in the x, y directions respectively. Show that each of the following satisfy the continuity equation:

a $u = y^2 - x^2$, $v = 2xy$

b $u = \tan^{-1}\left(\dfrac{y}{x}\right)$, $v = \dfrac{1}{2}\ln(x^2 + y^2)$

c $u = \dfrac{-2y}{(1 + x)^2 + y^2}$, $v = \dfrac{1 - x^2 - y^2}{(1 + x)^2 + y^2}$

6 ≈ [*fluid mechanics*]   The stream function, $\psi(x, y)$, is related to the velocity components u and v of the fluid flow by

$$u = \frac{\partial \psi}{\partial y} \text{ and } v = -\frac{\partial \psi}{\partial x}$$

If $\psi = \dfrac{1}{2}\ln(x^2 + y^2)$, find u and v.

The flow is irrotational if ψ satisfies Laplace's equation which is given by

$$\frac{\partial^2 \psi}{\partial x^2} + \frac{\partial^2 \psi}{\partial y^2} = 0$$

Show that the flow is irrotational for

$$\psi = \frac{1}{2}\ln(x^2 + y^2)$$

⛭ **Questions 7 to 9, inclusive, are on [*thermodynamics*].**

(P is pressure, V is volume, T is temperature and R is the gas constant)

7 The following terms are used in thermodynamics:

Isothermal compressibility $\kappa = -\dfrac{1}{V}\left(\dfrac{\partial V}{\partial P}\right)$

Thermal expansion coefficient

$$\beta = \frac{1}{V}\left(\frac{\partial V}{\partial T}\right)$$

Show that for the ideal gas equation, $PV = RT$, we have

$$\kappa = \frac{1}{P} \text{ and } \beta = \frac{1}{T}$$

8 An equation involving specific heats c_p, c_v is defined as

$$c_p - c_v = T\left(\frac{\partial V}{\partial T}\right)\left(\frac{\partial P}{\partial T}\right)$$

Exercise **15(a) continued**

Solutions at end of book. Complete solutions available at www.palgrave.com/science/engineering/singh

For the equation

$$\frac{RT}{P} = V + \frac{K}{RT}$$

(K is a constant)

find $c_p - c_v$.

9 Van der Waals equation is given by

$$P = \frac{RT}{V - b} - \frac{a}{V^2}$$

where a and b are constants. The critical isotherm occurs when

$$\frac{\partial P}{\partial V} = 0 \text{ and } \frac{\partial^2 P}{\partial V^2} = 0$$

A point which satisfies these two equations is called a 'critical point'. At the critical point the volume and temperature is V_c and T_c respectively. Find a and b in their simplest form for a critical point.

10 ⓨ [vibrations]   The displacement, $u(x, t)$, of a rod is a function of position x and time t:

$$u(x, t) = \left(A\sin\left(\frac{\omega}{\alpha}x\right) + B\cos\left(\frac{\omega}{\alpha}x\right) \right)$$

$$\times [C\sin(\omega t) + D\cos(\omega t)]$$

where $\alpha = \sqrt{\dfrac{E}{\rho}}$, (E is modulus of elasticity, ρ is density of rod and ω is natural frequency of vibration). Show that $u(x, t)$ satisfies the wave equation:

$$\alpha^2 \frac{\partial^2 u}{\partial x^2} = \frac{\partial^2 u}{\partial t^2}$$

11 Find $\left(\dfrac{\partial^2 f}{\partial x^2}\right)\left(\dfrac{\partial^2 f}{\partial y^2}\right) - \left(\dfrac{\partial^2 f}{\partial x \partial y}\right)^2$ for the following functions:

a $f(x, y) = x^3 + y^3 - 3xy$

b $f(x, y) = 2xy + \dfrac{2000}{x} + \dfrac{2000}{y}$

SECTION B **Applications**

By the end of this section you will be able to:
▶ understand total differential and small change
▶ apply small change to engineering applications
▶ apply total differential formula to thermodynamics

B1 **Small changes**

Engineering components are subject to tolerance limits. For example a resistor might be 10 kΩ ±5% where ±5% refers to the tolerance limit of the resistor. Also physical dimensions are prone to error measurements. In this section we examine the effects of such small changes.

Consider a cylinder with volume, V (Fig. 4).

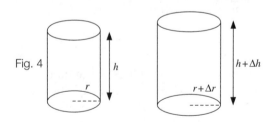

Fig. 4

Let Δr be a small change in the radius and Δh be a small change in the height, then the change in volume, ΔV, can be shown to be

$$\Delta V \approx \frac{\partial V}{\partial r}\Delta r + \frac{\partial V}{\partial h}\Delta h$$

Let's consider a general function of x and y, $f = f(x, y)$. Let Δx and Δy be small changes in x and y respectively. The change in f, Δf, is given by

$$\begin{aligned}
\Delta f &= f(x + \Delta x, y + \Delta y) - f(x, y) \\
&= f(x + \Delta x, y + \Delta y) - f(x, y + \Delta y) + f(x, y + \Delta y) - f(x, y) \\
&= \frac{f(x + \Delta x, y + \Delta y) - f(x, y + \Delta y)}{\Delta x}\Delta x + \frac{f(x, y + \Delta y) - f(x, y)}{\Delta y}\Delta y
\end{aligned}$$

$$\Delta f \approx \frac{\partial f}{\partial x}\Delta x + \frac{\partial f}{\partial y}\Delta y$$

Note the approximation sign. This is because

$$\frac{\partial f}{\partial x} = \lim_{\Delta x \to 0} \frac{f(x + \Delta x, y) - f(x, y)}{\Delta x}$$
$$\frac{\partial f}{\partial y} = \lim_{\Delta y \to 0} \frac{f(x, y + \Delta y) - f(x, y)}{\Delta y}$$

Hence the change in Δf is given by

15.8 $$\Delta f \approx \frac{\partial f}{\partial x}\Delta x + \frac{\partial f}{\partial y}\Delta y$$

B2 Engineering applications

Example 10 *electrical principles*

The power, p, consumed by a resistor, R, is given by

$$p = \frac{E^2}{R}$$

Given that $E = 20$ V and $R = 1$ kΩ, find the approximate change in power if E is decreased by 0.5 V and R is increased by 50 Ω.

Solution

What are we trying to find?

The change in power, Δp.

Since p is a function of E and R:

$$\Delta p \approx \frac{\partial p}{\partial E}\Delta E + \frac{\partial p}{\partial R}\Delta R$$

Example 10 *continued*

From $p = \dfrac{E^2}{R}$ we have

$$\frac{\partial p}{\partial E} = \frac{2E}{R} \qquad\qquad \text{[Partially differentiating]}$$

Rewrite $p = \dfrac{E^2}{R} = E^2 R^{-1}$:

$$\frac{\partial p}{\partial R} = -E^2 R^{-2} = -\frac{E^2}{R^2} \qquad \text{[Partially differentiating]}$$

Substituting these, $\dfrac{\partial p}{\partial E} = \dfrac{2E}{R}$ and $\dfrac{\partial p}{\partial R} = -\dfrac{E^2}{R^2}$, into

$$\Delta p \approx \frac{\partial p}{\partial E}\Delta E + \frac{\partial p}{\partial R}\Delta R$$

gives

$$\Delta p \approx \frac{2E}{R}\Delta E - \frac{E^2}{R^2}\Delta R$$

We are given $\Delta E = -0.5$, with minus sign because E is decreased by 0.5 V. Also $\Delta R = 50$. Thus we have

$$\Delta p \approx \frac{2E}{R}(-0.5) - \frac{E^2}{R^2}(50)$$

$$= \frac{2 \times 20}{1000}(-0.5) - \frac{20^2}{1000^2}(50) \qquad \left[\begin{array}{l}\text{Substituting}\\ E = 20 \text{ and } R = 1000\end{array}\right]$$

$$= -0.04 = -40 \times 10^{-3}$$

The power is reduced by approximately 40 mW (milliwatts).

For the above example we can find the exact change in power by subtracting the original power, p, from the new power (new p):

Since E is reduced by 0.5 V and R is increased by 50 Ω, we have

$$\text{new } E = 20 - 0.5 = 19.5$$
$$\text{new } R = (1 \times 10^3) + 50 = 1050$$

The new power is given by

$$\text{new } p = \frac{19.5^2}{1050} = 0.362$$

Before the change we had

$$p = \frac{20^2}{1000} = 0.4$$

The change in power is

$$(\text{new } p) - p = 0.362 - 0.4 = -0.038 \quad \text{(3 d.p.)}$$

or -38 mW. Hence the approximation of -40 mW is very close to the change of -38 mW.

? **Why do we find an approximate change when we can evaluate the exact change?**

Generally it is difficult or impossible to find the exact change because we are not given enough data. For the next example we are only given the percentage error in measuring the physical dimensions, therefore we cannot evaluate the exact error in h.

Example 11 *fluid mechanics*

The head loss, h, in a pipeline is given by

$$h = \frac{\lambda L V^2}{2gD}$$

where λ is a constant, L is the pipe length, V is the mean velocity of fluid, D is the diameter of pipe and g is the acceleration due to gravity.

If the error in measuring L, V and D is $+0.1\%$, $+1.5\%$ and $+1\%$ respectively, find the approximate percentage error in h.

Solution

Note that h is a function of L, V and D because g and λ are constants. We can apply 15.8 to functions of more than two variables. In this case we have

$$\Delta h \approx \frac{\partial h}{\partial L}\Delta L + \frac{\partial h}{\partial V}\Delta V + \frac{\partial h}{\partial D}\Delta D$$

We are given $h = \dfrac{\lambda L V^2}{2gD}$:

$$\frac{\partial h}{\partial L} = \frac{\lambda V^2}{2gD}, \quad \frac{\partial h}{\partial V} = \frac{2\lambda L V}{2gD} \qquad \text{[Partially differentiating]}$$

Rewriting h as

$$h = \frac{\lambda L V^2 D^{-1}}{2g}$$

$$\frac{\partial h}{\partial D} = -\frac{\lambda L V^2 D^{-2}}{2g} \qquad \text{[Partially differentiating]}$$

$$= -\frac{\lambda L V^2}{2gD^2}$$

Substituting these into Δh gives

$$\Delta h \approx \frac{\lambda V^2}{2gD}\Delta L + \frac{2\lambda L V}{2gD}\Delta V - \frac{\lambda L V^2}{2gD^2}\Delta D$$

 How do we find ΔL, ΔV and ΔD?

 Since the error in measuring L is $+0.1\%$, how do we write this in terms of a mathematical expression?

% means out of 100 therefore

$$0.1\% \text{ of } L = \frac{0.1}{100}L = 0.001L$$

Example 11 *continued*

Thus we have

$$\Delta L = 0.001L$$

Similarly we have

$$\Delta V = \frac{1.5}{100} V = 0.015V \text{ and } \Delta D = \frac{1}{100} D = 0.01D$$

Substituting these into

 *

$$\Delta h \approx \frac{\lambda V^2}{2gD} \Delta L + \frac{2\lambda LV}{2gD} \Delta V - \frac{\lambda LV^2}{2gD^2} \Delta D$$

gives

$$\Delta h \approx \frac{\lambda V^2}{2gD}(0.001L) + \frac{2\lambda LV}{2gD}(0.015V) - \frac{\lambda LV^2}{2gD^2}(0.01D)$$

$$= \frac{\lambda V^2 L}{2gD}(0.001) + \frac{\lambda LV^2}{2gD}(2 \times 0.015) - \frac{\lambda LV^2}{2gD}(0.01)$$

$$= \frac{\lambda V^2 L}{2gD}[0.001 + (2 \times 0.015) - 0.01] \quad \left[\text{Taking out } \frac{\lambda V^2 L}{2gD}\right]$$

$$= h(0.021) \quad \left[\text{Because } h = \frac{\lambda V^2 L}{2gD}\right]$$

Hence $\Delta h \approx 0.021h$. The approximate percentage error is 2.1%.

The error in measuring is more likely to be $\pm 0.1\%$, $\pm 1.5\%$ and $\pm 1\%$ rather than $+0.1\%$, $+1.5\%$ and $+1\%$. With $\pm$ change the method of **Example 11** is likely to be wrong because the error in D is subtracted. The method of **Example 12** shows one way in which this problem can be overcome.

Example 12 *materials*

The coefficient of rigidity, η, of a wire of length L and diameter D is given by

$$\eta = \frac{KL}{D^4}$$

where K is a constant. Find the largest percentage error in η if errors of $\pm 0.3\%$ and $\pm 0.4\%$ occur in measuring L and D respectively.

Solution

We are given $\eta = \frac{KL}{D^4}$ where K is a constant. Partially differentiating:

$$\frac{\partial \eta}{\partial L} = \frac{K}{D^4} \text{ and } \frac{\partial \eta}{\partial D} = -\frac{4KL}{D^5}$$

Example 12 *continued*

Putting these into

$$\Delta\eta \approx \frac{\partial\eta}{\partial L}\Delta L + \frac{\partial\eta}{\partial D}\Delta D$$

$$= \frac{K}{D^4}\Delta L - \frac{4KL}{D^5}\Delta D$$

$$= K\left(\frac{\Delta L}{D^4} - \frac{4L}{D^5}\Delta D\right) \qquad\qquad \text{[Taking out } K]$$

The percentage errors are given by

$$\Delta L = \pm 0.003L, \qquad \Delta D = \pm 0.004D$$

Substituting these into the above gives

$$\Delta\eta \approx K\left[\frac{\pm 0.003L}{D^4} - \frac{4L}{D^5}(\pm 0.004D)\right]$$

$$= K\left[\frac{\pm 0.003L}{D^4} - \frac{4L}{D^4}(\pm 0.004)\right]$$

$$= \frac{KL}{D^4}[\pm 0.003 - 4(\pm 0.004)] \qquad \left[\text{Taking out } \frac{L}{D^4}\right]$$

$$= \eta[\pm 0.003 - 4(\pm 0.004)] \qquad \left[\text{Because } \eta = \frac{KL}{D^4}\right]$$

The largest error occurs when the signs are different, that is $+0.003$ and -0.004 or vice versa:

$$\Delta\eta \approx \eta[0.003 - 4(-0.004)]$$
$$= \eta[0.019]$$

The largest error in η is 1.9%.

B3 Total differential

In the above we established

15.8 $\qquad \Delta f \approx \dfrac{\partial f}{\partial x}\Delta x + \dfrac{\partial f}{\partial y}\Delta y$

where f is a function of x and y, Δx and Δy are small but measurable changes in x and y respectively. By using the limiting process on 15.8 we have

15.9 $\qquad \mathrm{d}f = \dfrac{\partial f}{\partial x}\,\mathrm{d}x + \dfrac{\partial f}{\partial y}\,\mathrm{d}y$

where $\mathrm{d}f$ is called the **total differential**.

Example 13 *thermodynamics*

From the ideal gas equation

$$P = \frac{RT}{V}$$

show that

$$\frac{dP}{P} = \frac{dT}{T} - \frac{dV}{V}$$

where P, V, T and R represent pressure, volume, temperature and gas constant respectively.

Solution

From $P = \dfrac{RT}{V}$ we have P is a function of T and V (remember that R is a constant). So by 15.9

$$\star \qquad dP = \frac{\partial P}{\partial T} dT + \frac{\partial P}{\partial V} dV$$

Use

$$P = \frac{RT}{V} = RTV^{-1}$$

Partially differentiating:

$$\frac{\partial P}{\partial T} = \frac{R}{V} \text{ and } \frac{\partial P}{\partial V} = -RTV^{-2} = -\frac{RT}{V^2}$$

Substituting these, $\dfrac{\partial P}{\partial T} = \dfrac{R}{V}$ and $\dfrac{\partial P}{\partial V} = -\dfrac{RT}{V^2}$, into $\star$ yields

$$dP = \frac{R}{V} dT - \frac{RT}{V^2} dV$$

By transposing the given equation, $P = \dfrac{RT}{V}$, we have

$$\frac{P}{T} = \frac{R}{V} \text{ and } \frac{P}{V} = \frac{RT}{V^2}$$

Putting these into $dP = \dfrac{R}{V} dT - \dfrac{RT}{V^2} dV$ gives

$$dP = \frac{P}{T} dT - \frac{P}{V} dV$$

$$= P\left(\frac{dT}{T} - \frac{dV}{V}\right) \qquad \text{[Taking out } P]$$

Dividing through by P gives the required result:

$$\frac{dP}{P} = \frac{dT}{T} - \frac{dV}{V}$$

SUMMARY

Let $f(x, y)$ be a function of x and y, and Δx, Δy be small changes in x and y respectively, then the change in f is approximately

15.8 $\Delta f \approx \dfrac{\partial f}{\partial x}\Delta x + \dfrac{\partial f}{\partial y}\Delta y$

We define the total differential, $\mathrm{d}f$, as

15.9 $\mathrm{d}f = \dfrac{\partial f}{\partial x}\mathrm{d}x + \dfrac{\partial f}{\partial y}\mathrm{d}y$

Exercise 15(b)

Solutions at end of book. Complete solutions available at
www.palgrave.com/science/engineering/singh

1 [materials] The moment of inertia, I, of a body is given by

$$I = \frac{1}{12}\,bd^3$$

where b and d are dimensions of the body. If $b = 5$ m, $d = 0.8$ m and the change in b is 5 cm and in d is 0.8 cm, find the approximate change in I. (The SI units of I are m^4.)

2 [materials] The modulus of rigidity, G, of a shaft of length L and radius R is given by

$$G = \frac{R^4\theta}{L}$$

where θ is the angle of twist.
After heat treatment during a forging process, R is reduced by 0.5%, θ is increased by 2% and L is increased by 1.5%. Find the approximate percentage change in G caused by the treatment.

3 [fluid mechanics] The rate of discharge, Q, of a liquid through an orifice of area A is given by

$$Q = 0.7A\sqrt{2gh}$$

where h is the depth of the centre of the orifice and g is the acceleration due to gravity.

Given that $A = (2 \times 10^{-3})$ m^2, $h = 1.75$ m and $g = 9.81$ m/s^2, find the approximate change in Q if A is increased by (0.2×10^{-3}) m^2 and h is decreased by 0.25 m. (The SI units of Q are m^3/s.)

4 [electrical principles] The frequency, f, of an LC circuit is given by

$$f = \frac{1}{2\pi\sqrt{LC}}$$

where L represents an inductance and C a capacitance. If L is decreased by 1.5% and C is decreased by 0.5%, find the approximate percentage change in f.

5 [aerodynamics] The lift, L, of a body in a fluid of density ρ is given by

$$L = \frac{C\rho V^2 A}{2}$$

where C is the lift coefficient, V is the free stream velocity and A is the area.

If C can be measured to within 1%, ρ to within 0.5%, V to within 0.6% and A to within 0.1%, find the largest percentage error in the value of L.

6 [fluid mechanics] The rate of flow, Q, of a fluid through a pipe of diameter d and length L is given by

$$Q = \frac{p\pi d^4}{128\eta L}$$

where p is the pressure difference between the ends and η is the coefficient of viscosity.

If Q can be measured to within 0.5%, p to within 1.1%, L to within 0.3% and d to within 0.1%, calculate the maximum percentage error in the value of η.

7 [thermodynamics]   The ideal gas equation is given by

$$T = \frac{PV}{R}$$

where T is the temperature, V is the volume, P is the pressure and R is the gas constant. Show that

$$dT = \frac{VdP + PdV}{R}$$

SECTION C **Optimization**

By the end of this section you will be able to:
► find where stationary points occur
► distinguish between the three types of stationary points – maximum, minimum and saddle point
► identify the nature of stationary points
► find the solution to a practical optimization problem

C1 **Stationary points**

We have differentiated functions of two or more variables but what do the graphs of these functions look like?

We can use a computer algebra system or a graphical calculator to plot these functions. As discussed in **Section A** the graph of $z = f(x, y)$ is a surface. We can make analogies between the graphs of two independent variables, $z = f(x, y)$, and one independent variable, $y = f(x)$.

Compare the graphs of $y = x^2$ and the surface $z = x^2$ as shown in Fig. 5.

Fig. 5

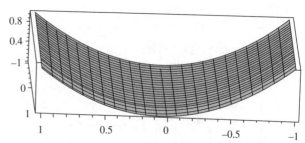

Notice the similarities.

In MAPLE we have four choices for axes – boxed, framed, normal or none. In the plot for $z = f(x, y)$ the normal axes are difficult to visualize, hence the use of boxed axes. Figure 6 shows other functions of two variables plotted with the aid of MAPLE.

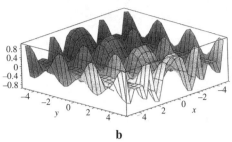

Fig. 6

a   b

Figure **6a** shows a portion of the surfaces of
$f(x, y) = y^4 + x^4 - xy$ and
Fig. **6b** shows $h(x, y) = \sin^2(y)\cos(x^2)$.

Let $f(x, y)$ be a function of two variables. The
stationary points of $f(x, y)$ occur (Fig. 7) when

Fig. 7

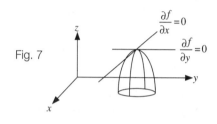

15.10 $\dfrac{\partial f}{\partial x} = 0$ and $\dfrac{\partial f}{\partial y} = 0$

A stationary point is a point on the surface where both slopes in the x and y directions are
zero. Remember that the stationary point in $y = f(x)$ is where

$$\frac{dy}{dx} = 0 \qquad [\text{Slope} = 0]$$

Example 14

Find where the stationary point(s) of

$$f(x, y) = 2x^2 + xy - 16x + y^2 - 11y$$

occur.

Solution

Differentiating $f = f(x, y)$ we have

$$f(x, y) = 2x^2 + xy - 16x + y^2 - 11y$$
$$\frac{\partial f}{\partial x} = 4x + y - 16 + 0 - 0, \qquad \frac{\partial f}{\partial y} = 0 + x - 0 + 2y - 11$$

Putting both these slopes to zero:

$$4x + y - 16 = 0$$
$$x + 2y - 11 = 0$$

Thus we have two simultaneous equations. Solving these

$$4x + y = 16$$
$$x + 2y = 11$$

gives $x = 3$ and $y = 4$. Hence the function $f(x, y)$ has a stationary point at (3, 4).

C2 **Maximum, minimum and saddle points**

With the one independent variable case, $y = f(x)$, the stationary point is either a maximum, a minimum or a point of inflexion. We have the same concepts for functions of more than one variable, that is a stationary point is either a maximum, a minimum or a saddle point.

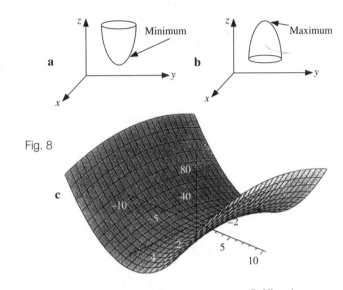

Fig. 8

Let $z = f(x, y)$. Figure 8 shows a maximum, a minimum and saddle point. Note that a saddle point is where a curve in one direction is a maximum and the curve in the other direction is a minimum (Fig. 9)

We can examine the graphs of these stationary points by using a computer algebra system, as the next example shows.

Fig. 9

Example 15

By using a computer algebra system (MAPLE) or a graphical calculator, plot the surfaces:

a $f(x, y) = y^2 - x^2$ for x between -5 and 5 and y between -10 and 10.
b $l(x, y) = e^{-x^2 - y^2}$ for x between -3 and 3 and y between -3 and 3.
c $F(x,y) = x^3 + y^3 - 3xy$ for x between 0 and 2 and y between 0 and 2.

Solution

a This is the same as the graph of Fig. **8c** above. The MAPLE commands are

```
> f:=(x,y)->y^2-x^2;
```
$f := (x, y) \rightarrow y^2 - x^2$

```
> plot3d(f,-5..5,-10..10);
```

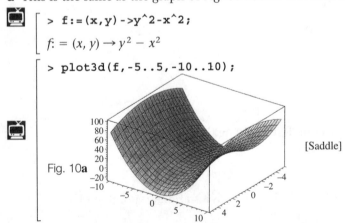

Fig. 10a

[Saddle]

Example 15 *continued*

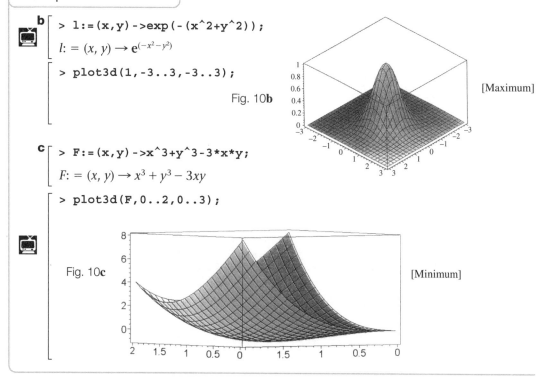

b
```
> l:=(x,y)->exp(-(x^2+y^2));
```
$l := (x, y) \rightarrow e^{(-x^2 - y^2)}$

```
> plot3d(l,-3..3,-3..3);
```

Fig. 10**b**

[Maximum]

c
```
> F:=(x,y)->x^3+y^3-3*x*y;
```
$F := (x, y) \rightarrow x^3 + y^3 - 3xy$

```
> plot3d(F,0..2,0..3);
```

Fig. 10**c**

[Minimum]

Figures 10**a, b** and **c** show functions plotted in MAPLE. Notice the maximum, the minimum and the saddle point.

What are the limitations of 15.10 ?

15.10 tell us where the stationary point(s) occur but does **not** tell us whether the points are a maximum, a minimum or a saddle point. We have to use the following test for $f = f(x, y)$. We will not show these results but just apply them in the appropriate cases.

15.11   If $\left(\dfrac{\partial^2 f}{\partial x^2}\right)\left(\dfrac{\partial^2 f}{\partial y^2}\right) - \left(\dfrac{\partial^2 f}{\partial x \partial y}\right)^2 > 0$ and $\dfrac{\partial^2 f}{\partial x^2} > 0$

then the stationary point is a **minimum**.

15.12   If $\left(\dfrac{\partial^2 f}{\partial x^2}\right)\left(\dfrac{\partial^2 f}{\partial y^2}\right) - \left(\dfrac{\partial^2 f}{\partial x \partial y}\right)^2 > 0$ and $\dfrac{\partial^2 f}{\partial x^2} < 0$

then the stationary point is a **maximum**.

15.13   If $\left(\dfrac{\partial^2 f}{\partial x^2}\right)\left(\dfrac{\partial^2 f}{\partial y^2}\right) - \left(\dfrac{\partial^2 f}{\partial x \partial y}\right)^2 < 0$

then the stationary point is a **saddle point**.

However if

15.14 $\left(\dfrac{\partial^2 f}{\partial x^2}\right)\left(\dfrac{\partial^2 f}{\partial y^2}\right) - \left(\dfrac{\partial^2 f}{\partial x \partial y}\right)^2 = 0$

at a stationary point, then no conclusion can be drawn.

In 15.11 and 15.12 we can replace $\dfrac{\partial^2 f}{\partial x^2} > 0$ with $\dfrac{\partial^2 f}{\partial y^2} > 0$ and $\dfrac{\partial^2 f}{\partial x^2} < 0$ with $\dfrac{\partial^2 f}{\partial y^2} < 0$

respectively.

Sometimes 15.11 to 15.14 are called the second derivative tests.

Example 16

Find where the stationary points of

$$f(x, y) = x^3 + y^3 - 3xy$$

occur and state their nature.

Solution

Let $f = f(x, y)$. First we find where the stationary points occur:

$$\frac{\partial f}{\partial x} = 3x^2 - 3y \text{ and } \frac{\partial f}{\partial y} = 3y^2 - 3x \qquad \text{[Partial differentiating]}$$

By equating these to zero we have the simultaneous equations

† $3x^2 - 3y = 0$

†† $3y^2 - 3x = 0$

From † we have

$$3x^2 = 3y$$

$$x^2 = y$$

Substituting $y = x^2$ into †† gives

$$3x^4 - 3x = 0$$

$$3x(x^3 - 1) = 0 \qquad \text{[Taking out } 3x\text{]}$$

$$x = 0, \quad x = 1 \qquad \text{[Solving]}$$

If $x = 0$ then $y = 0$ and if $x = 1$ then $y = 1$ because $y = x^2$. Stationary points of f are at $(0, 0)$ and $(1, 1)$.

At $(0, 0)$ and at $(1, 1)$, do we have a maximum, a minimum or a saddle point?

We need to use 15.11 to 15.14 . We have $f = x^3 + y^3 - 3xy$:

$$\frac{\partial f}{\partial x} = 3x^2 - 3y$$

$$\frac{\partial^2 f}{\partial x^2} = 6x - 0 = 6x \qquad \text{[2nd derivative]}$$

Example 16 *continued*

$$\frac{\partial f}{\partial y} = 3y^2 - 3x$$

$$\frac{\partial^2 f}{\partial y^2} = 6y - 0 = 6y \qquad \text{[2nd derivative]}$$

$$\frac{\partial^2 f}{\partial x \partial y} = \frac{\partial}{\partial x}\left(\frac{\partial f}{\partial y}\right) \qquad \text{[Mixed derivative]}$$

$$= \frac{\partial}{\partial x}\left(3y^2 - 3x\right) = 0 - 3 = -3$$

We have

$$\left(\frac{\partial^2 f}{\partial x^2}\right)\left(\frac{\partial^2 f}{\partial y^2}\right) - \left(\frac{\partial^2 f}{\partial x \partial y}\right)^2 = (6x)(6y) - (-3)^2 \qquad \text{[Substituting]}$$

$$= 36xy - 9$$

At $(0, 0)$, which means $x = 0$ and $y = 0$, substituting these gives

$$36xy - 9 = 0 - 9 = -9 < 0 \qquad \text{[Negative]}$$

By 15.13 , at $(0, 0)$ we have a saddle point.

At $(1, 1)$, $x = 1$ and $y = 1$ so we have

$$36xy - 9 = (36 \times 1 \times 1) - 9 = 36 - 9 > 0$$

$$\frac{\partial^2 f}{\partial x^2} = 6x = 6 \times 1 > 0 \qquad \text{[Positive]}$$

By 15.11 , at $(1, 1)$ we have a minimum.

C3 **Optimization problems**

We can tackle practical engineering problems involving maximum and minimum by finding values of variables which maximize or minimize a particular problem.

Example 17

Consider a closed rectangular box with dimensions x, y and z. If the fixed volume of the box is 1000 cm³, find the dimensions so that it has a minimum total surface area. What conclusions can you draw from your result?

Solution

Figure 11 shows the box.

The volume $xyz = 1000$. By transposing, we can write this in terms of x and y only. We write the expressions in terms of two independent variables because the established formulae only apply to functions of two variables. Dividing $xyz = 1000$ by xy gives

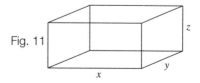

Fig. 11

Example 17 *continued*

$$\boxed{*} \qquad z = \frac{1000}{xy}$$

We first obtain the values of x and y which give minimum surface area and then we substitute these into $\boxed{*}$ to find z.

The surface area, A, is given by

$$A = 2xy + 2yz + 2xz$$

We need to replace the z term so we can obtain a function in terms of x and y only.

Substituting $z = \dfrac{1000}{xy}$ into A gives

$$
\begin{aligned}
A &= 2xy + 2yz + 2xz \\
&= 2xy + 2y\left(\frac{1000}{xy}\right) + 2x\left(\frac{1000}{xy}\right) \\
&= 2xy + \frac{2000}{x} + \frac{2000}{y} \qquad\qquad \text{[Cancelling]} \\
A &= 2xy + 2000x^{-1} + 2000y^{-1}
\end{aligned}
$$

At what values of x and y does A have a minimum surface area?

First we find where the stationary points occur by differentiating:

$$\frac{\partial A}{\partial x} = 2y - 2000x^{-2} + 0$$

$$\frac{\partial A}{\partial y} = 2x + 0 - 2000y^{-2}$$

By equating these to zero we have the simultaneous equations

$$2y - 2000x^{-2} = 0$$
$$2x - 2000y^{-2} = 0$$

From the first equation we have

$$2y = \frac{2000}{x^2}$$
$$x^2y = 1000$$

From the second equation we have

$$2x = \frac{2000}{y^2}$$
$$xy^2 = 1000$$

Combining both equations yields

$$x^2y = xy^2 \ (=1000)$$

Dividing through by xy gives

$$x = y$$

Example 17 *continued*

Substituting $x = y$ into $xy^2 = 1000$ we have

$$xx^2 = x^3 = 1000$$

Taking the cube root gives $x = 10$.

Since $x = y$ therefore $y = 10$.

?　**Do these values $x = 10$ and $y = 10$ give the minimum surface area?**

We need to check this by using tests 15.11 to 15.14 :

$$\frac{\partial A}{\partial x} = 2y - 2000x^{-2}$$

$$\frac{\partial^2 A}{\partial x^2} = 0 - 2(-2000)x^{-3} = 4000x^{-3} \qquad \text{[2nd derivative]}$$

Similarly we have

$$\frac{\partial A}{\partial y} = 2x - 2000y^{-2}$$

$$\frac{\partial^2 A}{\partial y^2} = 0 - 2(-2000)y^{-3} = 4000y^{-3} \qquad \text{[2nd derivative]}$$

$$\frac{\partial^2 A}{\partial x \partial y} = \frac{\partial}{\partial x}\left(\frac{\partial A}{\partial y}\right) \qquad \text{[Mixed derivative]}$$

$$= \frac{\partial}{\partial x}\left(2x - 2000y^{-2}\right) = 2 - 0 = 2$$

Hence we have

$$\left(\frac{\partial^2 A}{\partial x^2}\right)\left(\frac{\partial^2 A}{\partial y^2}\right) - \left(\frac{\partial^2 A}{\partial x \partial y}\right)^2 = (4000x^{-3})(4000y^{-3}) - 2^2$$

Substituting $x = 10$ and $y = 10$ gives

$$(4000 \times 10^{-3})(4000 \times 10^{-3}) - 4 = 12 > 0 \qquad \text{[Positive]}$$

Hence the point (10, 10) gives a maximum or a minimum.

We need to check for a minimum:

$$\frac{\partial^2 A}{\partial x^2} = 4000x^{-3} = 4000 \times 10^{-3} = 4 > 0 \qquad \text{[Positive]}$$

Hence by 15.11 , $x = 10$, $y = 10$ gives minimum surface area. To find z, use $z = \dfrac{1000}{xy}$:

$$z = \frac{1000}{xy} = \frac{1000}{10 \times 10} = 10$$

Conclusion – the minimum surface area of a rectangular box containing a fixed volume of 1000 cm^3 is a cube of 10 cm × 10 cm × 10 cm.

15.11　$\left(\dfrac{\partial^2 f}{\partial x^2}\right)\left(\dfrac{\partial^2 f}{\partial y^2}\right) - \left(\dfrac{\partial^2 f}{\partial x \partial y}\right)^2 > 0$ and $\dfrac{\partial^2 f}{\partial x^2} > 0$ then minimum

SUMMARY

Stationary points of a function $f(x, y)$ occur when

15.10 $\dfrac{\partial f}{\partial x} = 0$ and $\dfrac{\partial f}{\partial y} = 0$

There are three types of stationary points – maximum, minimum and saddle point. We can decide what type of a stationary point we have by using the second derivative tests:

15.11   If $\left(\dfrac{\partial^2 f}{\partial x^2}\right)\left(\dfrac{\partial^2 f}{\partial y^2}\right) - \left(\dfrac{\partial^2 f}{\partial x \partial y}\right)^2 > 0$ and $\dfrac{\partial^2 f}{\partial x^2} > 0$

then the stationary point is a **minimum**.

15.12   If $\left(\dfrac{\partial^2 f}{\partial x^2}\right)\left(\dfrac{\partial^2 f}{\partial y^2}\right) - \left(\dfrac{\partial^2 f}{\partial x \partial y}\right)^2 > 0$ and $\dfrac{\partial^2 f}{\partial x^2} < 0$

then the stationary point is a **maximum**.

15.12   If $\left(\dfrac{\partial^2 f}{\partial x^2}\right)\left(\dfrac{\partial^2 f}{\partial y^2}\right) - \left(\dfrac{\partial^2 f}{\partial x \partial y}\right)^2 < 0$

then the stationary point is a **saddle point**.

Exercise 15(c)

Solutions at end of book. Complete solutions available at www.palgrave.com/science/engineering/singh

1 Find where the stationary points of the following functions occur:

a $f(x, y) = xy + \dfrac{1}{2}y^2 - 7y - 4x$

b $f(x, y) = x^2 + y^2 + 4xy - 5x - 4y$

c $f(x, y) = x^3 + y^2 + xy + 22y$

2 Find where the stationary points of the following functions occur and determine whether they are maximum, minimum or saddle points:

a $f(x, y) = x^2 + y^2 + 6xy - 10x - 14y$

b $f(x, y) = x^3 + y^2 - x + y$

c $f(x, y) = x^2 + y^3 + 4xy - 11x - 18y$

3 Find the dimensions of a rectangular tank, without a top, with minimum surface area and a fixed volume of 0.9 m³. What can you conclude about your results?

4 An open rectangular tank has volume V. Find the dimensions of the tank for minimum surface area A.

5 An open top rectangular storage tank has a volume of 9 m³.
   i If the base of the tank has dimensions x and y then show that the total surface area A is given by

$$A = xy + \frac{18}{x} + \frac{18}{y}$$

   ii Find the dimensions of the tank for the surface area, A, to be a minimum.

6 i The temperature, T, in a circular plate $x^2 + y^2 < 1$ is given by

$$T = x^2 + y^2 - x - y + 100$$

   Find the minimum temperature.
   ii If T is the same expression as part **i** then find where the stationary points occur if

$$x^2 + y^2 = 1$$

Exercise **15(c) continued**

Solutions at end of book. Complete solutions available at www.palgrave.com/science/engineering/singh

7 A length of channel is made from a rectangular sheet of metal of length 9 m. The metal is bent as shown in Fig. 12. Find the values of x and θ for which the channel's capacity is a maximum.

Fig. 12

Miscellaneous exercise **15**

Solutions at end of book. Complete solutions available at www.palgrave.com/science/engineering/singh

1 [materials]   A beam of length L has a bending moment M given by

$$M = Kx + Px\left(1 - \frac{x}{L}\right)$$

where K is a constant, x is the distance along the beam from one end and P is the load. Determine $\dfrac{\partial M}{\partial P}$ and $\dfrac{\partial M}{\partial x}$.

2 [materials]   The bending moment M at a distance a of a beam of length L is given by

$$M = \frac{Kx}{L} + Pa\left(1 - \frac{x}{L}\right) - P(a - x)$$

(P is load and x is distance). Determine $\dfrac{\partial M}{\partial P}$ and $\dfrac{\partial M}{\partial x}$.

3 [materials]   Given the stress function, $\Omega(x, y)$, as

$$\Omega(x, y) = Ax^3 y^2 + Bxy^2 \quad (A \neq 0, B \neq 0)$$

find

$$\frac{\partial^4 \Omega}{\partial x^4} + 2\frac{\partial^4 \Omega}{\partial x^2 \partial y^2} + \frac{\partial^4 \Omega}{\partial y^4} \ (= \nabla^4 \Omega)$$

If $\nabla^4 \Omega = 0$ then $\Omega(x, y)$ is called the Airy stress function. Test whether the above $\Omega(x, y)$ is an Airy stress function for $x \neq 0$.

4 [thermodynamics]   The pressure, P, of a gas is given by

$$P = \frac{RT}{V}$$

where T is the temperature, V is the volume and R is the gas constant. Find the approximate percentage change in P, if T is decreased by 1.3% and V is increased by 0.2%.

5 [electrical principles]   The resistance, R, of a wire of length L and diameter d is given by

$$R = \frac{cL}{d^2} \quad (c \text{ is a constant})$$

If the error in measuring L and d are 3% and 2% respectively, find the largest possible error in R.

6 [materials]   The axial stress, σ, is given by

$$\sigma = \frac{kr^2}{l^2} \quad (k = \text{constant})$$

where r is the radius of gyration and l is the length. Find the approximate percentage change in σ if r is increased by 1.3% and l is decreased by 0.6%.

7 [fluid mechanics]   The fluid velocity, V, around a sphere of radius r is given by

$$V = V_0\left(1 + \frac{r^3}{x^3}\right)$$

The streamline acceleration, a, is defined by

$$a = V\frac{\partial V}{\partial x}$$

Show that

$$a = -3V_0^2\left(1 + \frac{r^3}{x^3}\right)\frac{r^3}{x^4}$$

At what value of x is $a = 0$?

8 [vibrations]   A mass, m, is placed on a soil column, which is excited to vibrate at a natural frequency ω. Given that

$$u(x, t) = \left[A\sin(\omega t) + B\cos(\omega t)\right]C\sin\left(\frac{\omega x}{k}\right)$$

where A, B, C and k are constants, t is time and x is displacement along the soil column and also that

$$\frac{F}{SE} = \frac{\partial u}{\partial x} \text{ and } F = -m\frac{\partial^2 u}{\partial t^2}$$

where F is the force acting on the soil column, S is the cross-sectional area and E is the Young modulus, show that

$$SE = mk\omega\tan\left(\frac{\omega x}{k}\right)$$

9 [structures]   A beam of length L (fixed at one end) and shear load S has load P given by

$$P = \frac{SA}{\alpha}\left[\frac{\sinh(\mu x)}{\mu\cosh(\mu L)} - x\right]$$

where α and μ are constants, A is the cross-sectional area and x is the length along the beam. The shear flow, q, is defined by

$$q = -\frac{1}{2}\frac{\partial P}{\partial x}$$

Show that

$$q = \frac{SA}{2\alpha}\left[1 - \frac{\cosh(\mu x)}{\cosh(\mu L)}\right]$$

10 [structures]   A beam of length L has load P given by

$$P = A\cosh(\mu x) + B\sinh(\mu x) + \alpha$$

where A, B, μ and α are constants and x is the distance along the beam. Given that

at $x = 0$, $P = 0$ and at $x = L$, $\dfrac{\partial P}{\partial x} = 0$, show that

$$P = \alpha\left[\tanh(\mu L)\sinh(\mu x) - \cosh(\mu x) + 1\right]$$

11 [materials]   Radial, σ_r, tangential σ_θ and shearing $\tau_{r\theta}$ stresses are defined by

$$\sigma_r = \frac{1}{r}\left(\frac{\partial\Omega}{\partial r}\right) + \frac{1}{r^2}\left(\frac{\partial^2\Omega}{\partial\theta^2}\right)$$

$$\sigma_\theta = \frac{\partial^2\Omega}{\partial r^2}$$

$$\tau_{r\theta} = \frac{1}{r^2}\left(\frac{\partial\Omega}{\partial\theta}\right) - \frac{1}{r}\left(\frac{\partial^2\Omega}{\partial r\partial\theta}\right)$$

For $\Omega = r^2\left[A\cos(2\theta) + B\sin(2\theta)\right]$ find σ_r, σ_θ and $\tau_{r\theta}$.

12 [heat transfer]   The one-dimensional heat equation is given by

$$\frac{\partial u}{\partial t} = k\frac{\partial^2 u}{\partial x^2}$$

where k is a constant, $u = u(x, t)$ is the temperature at time t and distance x of an insulated rod. Show that the following functions satisfy the one-dimensional heat equation:

a $u(x, t) = e^{-kt}\sin(x)$

b $u(x, t) = e^{-kt}\cos(x)$

c $u(x, t) = e^{-kt}\left[\cos(x) + \sin(x)\right]$

d $u(x, t) = e^{-\eta^2 kt}\left[\cos(\eta x) + \sin(\eta x)\right]$
where η is a constant.

13 Figure 13 shows an open top triangular container with a volume of 1 m³. Find the values of x, l and θ which give the minimum surface area A.

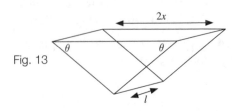

Fig. 13

(*Hint*: First find x, l and θ, which gives the stationary values of A, and then use MAPLE to confirm that these values do indeed give a minimum.)

You may use a computer algebra system for the remaining questions.

14 [*thermodynamics*]   The Dieterici equation is given by

$$P = \frac{RT}{V - b}\, e^{-a/RVT}$$

A critical point occurs at $\dfrac{\partial P}{\partial V} = 0$ and $\dfrac{\partial^2 P}{\partial V^2} = 0$.

Find an expression for a and b in terms of R, V and T at a critical point.

15 [*thermodynamics*]   The Redlich–Kwong equation is given by

$$P = \frac{RT}{V - b} - \frac{a}{V(V + b)T^{0.5}}$$

At a critical point find expressions for a and b in terms of R, V and T. (Critical point defined in question **14**.)

CHAPTER 16
Probability and Statistics

SECTION A **Data representation**

By the end of this section you will be able to:
▶ distinguish between continuous and discrete data
▶ construct frequency distributions
▶ draw a histogram
▶ plot a frequency polygon

A1 Discrete and continuous data

The following are examples of discrete data:
▶ Number of people in a room
▶ Number of rejects on an assembly line
▶ Shoe size of children.

? **What do you think is the definition of 'discrete data'?**

Data which can only take **certain** values. The number of people in a room can only be 1, 2, 3, . . . and not 1.23, 1.57, 10.11, ...

The following are examples of continuous data:
▶ Heights of people
▶ Output voltage of an analogue system
▶ Loads on a beam.

? **What do you think is the definition of 'continuous data'?**

Data which can take **any** values between the two end points.

The weights of people can be 60.28 kg, 70.1 kg, ...

A2 Frequency distribution

? **What does the term 'frequency' mean?**

It is the number of times a particular value occurs in some data. The combination of particular values and their frequency is called a frequency distribution.

One way of representing the distribution of data is by a frequency table. This is a table that summarizes the data into some order.

Example 1

The number of rejects, in the last 30 days, from an assembly line has yielded the following data:

Example 1 *continued*

36	37	49	30	36	35
40	42	44	37	33	42
37	41	44	42	30	36
37	37	36	30	44	31
30	42	41	44	39	42

Construct a frequency distribution table.

Solution

Remember that frequency of a value is the number of times it occurs in the data. For example there are 30 rejects on 4 days. We say the frequency of 30 is 4. We can summarize the above data as detailed in Table 1.

TABLE 1

Number of rejects	Frequency f
30	4
31	1
33	1
35	1
36	4
37	5
39	1
40	1
41	2
42	5
44	4
49	1

The representation of data is a lot clearer in the table.

? **We can check that all the data has been placed in the table. How?**

The sum of the frequencies should add up to 30 because there are 30 data values, that is $\sum f = 30$. This is **no** guarantee that our frequency distribution is correct but it is a good guide.

? **This table is the frequency distribution for what sort of data?**

Discrete data (number of rejects)

We can use a similar idea to form a frequency distribution for continuous data. One way of representing continuous data is to group it into particular 'classes' or 'intervals', as the next example shows. This is particularly useful for a large amount of data.

Example 2

The diameters, in mm, of 20 pipes is as follows:

40.6	40.7	40.9	41.0	41.1
41.4	41.5	41.7	41.2	41.2
41.9	41.3	41.4	41.6	41.8
41.6	41.2	40.5	40.5	41.9

Form a frequency distribution table, by grouping the data into five classes.

Solution

How do we form a frequency distribution for this data? We can group the data into classes, but classes of what size?

That depends on the data. The smallest value is 40.5 and the largest value is 41.9. If we use classes of size 0.3 then we will get five classes. Let's form the frequency distribution table for classes of size 0.3 (Table 2).

TABLE 2

Diameter d (mm)	Frequency
$40.45 \leq d < 40.75$	4
$40.75 \leq d < 41.05$	2
$41.05 \leq d < 41.35$	5
$41.35 \leq d < 41.65$	5
$41.65 \leq d < 41.95$	4

In this example 40.45, 40.75, 41.05, ... are called **class boundaries**.

These boundaries are normally chosen to lie between the observed values.

The **class width** is the difference between the upper class boundary and the lower class boundary.

Class width = (Upper class boundary) − (Lower class boundary)

For the above example:

$$\text{Class width} = 40.75 - 40.45$$
$$= 0.3$$

If we use classes of size 0.2 then we will need eight classes.

How many classes do we need if we use classes of size 0.1?

15 classes

This is an example of a frequency distribution with equal class width. A frequency distribution may have unequal class widths.

A3 **Histogram**

A histogram is a graphical representation of a frequency distribution.

Example 3

Considering the data of **Example 2**, draw a histogram for this data.

Solution

The frequency is plotted along the vertical axis and the grouped diameter of pipes along the horizontal axis.

Figure 1 shows a histogram of the data contained in Table 2.

The symbol, N, used in Fig. 1 means that there is no data before the specified value, in this case 40.45.

Fig. 1

This is one of the simplest histograms to draw because it has equal class widths and so the height represents the frequency.

In a histogram the **area** of the rectangle is proportional to the **frequency**.

Only in the case of equal class width do we have the height of each rectangle representing the frequency.

For histograms with unequal class widths we need to be careful, as the following example shows.

Example 4

The resistances of 100 resistors are given in Table 3.

	Resistance R (kΩ)	Frequency
TABLE 3	$1.1 \leq R < 1.2$	9
	$1.2 \leq R < 1.4$	20
	$1.4 \leq R < 1.5$	8
	$1.5 \leq R < 1.6$	12
	$1.6 \leq R < 1.8$	18
	$1.8 \leq R < 2.0$	12
	$2.0 \leq R < 2.3$	21

Draw a histogram for this data.

Example 4 *continued*

Solution

Remember that the frequency is proportional to the area. We need to choose a standard class width. By looking down the left-hand column of the table, we find the class widths are of the size 0.1, 0.2, 0.1, 0.1, 0.2, 0.2, and 0.3. **Which class width would you choose to be standard?**

It really doesn't make much difference, but the most suitable seems to be 0.1 because there are three intervals with this width and it keeps the arithmetic easy, that is $0.1 \times 2 = 0.2$, $0.1 \times 3 = 0.3$.

If we choose our standard width = 0.1 then the second interval is twice the standard width and so we halve the frequency height. Similarly for the last interval, we take 1/3 of the frequency height (Table 4). This new figure is known as the standard frequency.

	Resistance $R(k\Omega)$	Class width (Standard width, SW)	Frequency	Standard frequency
TABLE 4	$1.1 \le R < 1.2$	0.1 (SW)	9	9
	$1.2 \le R < 1.4$	0.2 (2 × SW)	20	20/2 = 10
	$1.4 \le R < 1.5$	0.1 (SW)	8	8
	$1.5 \le R < 1.6$	0.1 (SW)	12	12
	$1.6 \le R < 1.8$	0.2 (2 × SW)	18	18/2 = 9
	$1.8 \le R < 2.0$	0.2 (2 × SW)	12	12/2 = 6
	$2.0 \le R < 2.3$	0.3 (3 × SW)	21	21/3 = 7

Plotting standard frequency against resistance values of Table 4 gives the histogram of Fig. 2.

Again note that the data does not start at zero.

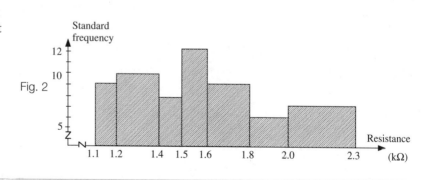

Fig. 2

For unequal class widths the vertical axis gives the standard frequency. The frequency is evaluated by the area. Consider **Example 4**: to evaluate the frequency from the histogram of the last interval, we multiply the height by 3. **Why?**

Because it is three times the standard width. Hence $3 \times 7 = 21$.

A4 **Frequency polygons**

Another graphical representation of a frequency distribution is a frequency polygon. There are two ways of constructing a frequency polygon:

1 Draw a histogram and join the midpoints of the tops of the rectangles.

2 Plot the standard frequency on the vertical axis against the midpoint of the interval.

Example 5

Plot the frequency polygon for the data of **Example 4**.

Solution

Which method should we use?

1

because we have already drawn a histogram for the data.

Fig. 3

Figure 3 shows the frequency polygon of resistance values.

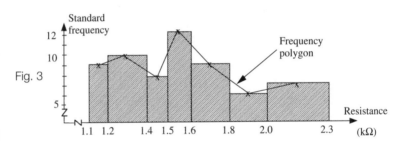

SUMMARY

Discrete data can only take certain values while continuous data can take any value between the two end points.

A frequency table is one way of representing the distribution of data.

A histogram is a graphical representation of a frequency distribution. The frequency is proportional to the area.

A frequency polygon is another graphical representation of a frequency distribution.

Exercise **16(a)**

Solutions at end of book. Complete solutions available at www.palgrave.com/science/engineering/singh

1 State whether the following are discrete or continuous data:
  a Weights of people.
  b Marks in an examination.
  c The number of motors in a batch.
  d Lifetime of light bulbs.
  e The resistance value of a variable resistor.

Exercise **16(a) continued**

Solutions at end of book. Complete solutions available at www.palgrave.com/science/engineering/singh

2 The temperatures, to the nearest degree Centigrade, for the last 30 days are as follows:

22	23	23	16	17	20
21	24	18	17	19	21
23	25	26	21	20	20
16	15	17	22	24	23
23	19	20	22	21	20

Construct a frequency distribution.

3 The heights, in m, of 40 students is shown below:

1.68 1.67 1.53 1.70 1.69 1.71 1.71 1.80
1.81 1.85 1.76 1.66 1.91 1.95 1.87 1.80
1.82 1.76 1.84 1.55 1.61 1.85 1.93 1.88
1.95 1.57 1.97 1.56 1.99 1.83 1.74 1.83
1.86 1.88 1.93 1.64 1.89 1.90 1.72 1.95

Construct a frequency distribution with an equal class width of 0.1.

4 Draw a histogram for question **3** with the same class width.

5 Draw a frequency polygon for question **3**.

6 The table below shows the time taken, in ms, for 105 op-amps to become fully operational:

Time taken t (ms)	Frequency
$10 \leq t < 20$	10
$20 \leq t < 30$	14
$30 \leq t < 40$	18
$40 \leq t < 45$	16
$45 \leq t < 50$	12
$50 \leq t < 55$	8
$55 \leq t < 70$	15
$70 \leq t < 100$	12

Draw a histogram to illustrate this data.

7 The following table shows the lifetime (in hours) of 88 light bulbs:

Lifetime L (hours)	Frequency
$150 \leq L < 200$	5
$200 \leq L < 250$	7
$250 \leq L < 300$	10
$300 \leq L < 350$	12
$350 \leq L < 400$	14
$400 \leq L < 450$	11
$450 \leq L < 500$	10
$500 \leq L < 550$	8
$550 \leq L < 600$	7
$600 \leq L < 650$	4

Draw a frequency polygon to illustrate this data.

8 The diameter of 100 piston rings are measured and recorded as follows:

Diameter d (mm)	Frequency
$86.00 \leq d < 86.10$	9
$86.10 \leq d < 86.15$	9
$86.15 \leq d < 86.25$	12
$86.25 \leq d < 86.40$	15
$86.40 \leq d < 86.45$	10
$86.45 \leq d < 86.50$	7
$86.50 \leq d < 86.70$	20
$86.70 \leq d < 87.00$	18

Draw a histogram and a frequency polygon for this data.

SECTION B Data summaries

By the end of this section you will be able to:
▶ evaluate the mean
▶ understand what *standard deviation* means
▶ evaluate the standard deviation
▶ derive properties of mean and standard deviation
▶ evaluate the mean and standard deviation of data in a frequency distribution

B1 Averages

The sample mean, or average, of n numbers, $x_1, x_2, x_3, \ldots, x_n$, denoted by $\bar{x}$, is given by

16.1
$$\bar{x} = \frac{x_1 + x_2 + x_3 + \ldots + x_n}{n} = \frac{\sum_{j=1}^{n} x_j}{n}$$
$$\left(= \frac{\text{sum of observations}}{\text{number of observations}} \right)$$

The notation $\sum_{j=1}^{n} x_j$ means sum x_j from 1 to n, that is $x_1 + x_2 + x_3 + \ldots + x_n$.

Example 6

The lifetime (to the nearest hour) of 20 light bulbs is given by: 599, 601, 586, 604, 623, 624, 584, 613, 581, 600, 605, 592, 613, 617, 591, 599, 584, 616, 580 and 603.

i Without evaluating, estimate the mean from the data.

ii Evaluate the mean lifetime of these light bulbs.

Solution

i Since all the data is quite close to 600, a good estimate would be 600.

ii Adding the data values and dividing by 20 gives

$$\text{Mean} = \frac{599 + 601 + 586 + \ldots + 603}{20} = 600.75$$

The easiest way to find the mean for data of **Example 6** is to use a calculator. We first place the calculator into SD mode and then enter our data values. PRESS

MODE 2 SHIFT MODE 1 = 599 M+ 601 M+ 586 M+ ⋯ 603 M+

Then press SHIFT 2 1 = to give the mean 600.75.

The above instructions might seem to be calculator specific. On some calculators there is a $\boxed{\text{STO}}$ key rather than $\boxed{\text{M+}}$ which means you press this, $\boxed{\text{STO}}$, after each data entry. You might need to see the handbook of your calculator to evaluate the mean because most basic calculators can find this average.

Example 7

Consider the data of **Example 6** again. If only the ninth piece of data, 581, changes to 518, what will be the new mean?

Solution

Remember that the mean is just an average.

With this change of data, is the new mean higher than 600.75 or lower?

Lower because we have entered a lower value, 518, rather than 581. The change can be evaluated by $\dfrac{581 - 518}{20} = 3.15$

The new mean $= 600.75 - 3.15 = 597.6$

Other averages to consider are the mode and the median. The mode is the most popular or most frequently occurring observation. The mode of

$$2, 2, 3, 4, 6, 6, 6, 7$$

is 6.

The median is the middle value of the data arranged in ascending (or descending) order. The median of

$$2, 5, 3, 6, 10, 4, 15$$

is evaluated by placing the data in order of magnitude

$$2, 3, 4, 5, 6, 10, 15$$

Hence the median is 5.

The median of the following eight pieces of data:

$$101, 137, 112, 67, 93, 12, 107, 152$$

is evaluated by first placing these data pieces into some order:

$$12, 67, 93, \underbrace{101, 107}, 112, 137, 152$$

$$\frac{101 + 107}{2} = 104$$

The mean of the two middle values, 101 and 107, is taken, thus

$$\text{median} = 104$$

In this chapter we will concentrate on the mean, as it is the most widely used average in statistics.

B2 **Standard deviation (*s* or SD)**

? The standard deviation is a 'measure of the spread' of data from the mean value. **How can we evaluate this?**

Consider the data 3, 4 and 5.

? **How can we evaluate the standard deviation for this data?**

Since we want to find out how far the data is away from the mean, we must first obtain the mean:

$$\text{mean } \bar{x} = \frac{3 + 4 + 5}{3} = 4$$

? **How can we evaluate how far each piece of data is from the mean?**

By calculating the difference (deviation from the mean):

$$3 - 4 = -1, 4 - 4 = 0 \text{ and } 5 - 4 = 1$$

To find a value for the spread we add these:

$$-1 + 0 + 1 = 0$$

By using this method we obtain a value for spread = 0.

? **What does this 0 signify?**

This signifies that **all** the data is centred on the mean and there is no deviation from the mean.

? As you can see this is **not** the case, since 3 and 5 are 1 away from the mean. **So how did we get an answer of zero?**

$$-1 \text{ and } 1 \text{ cancelled each other out in the addition}$$

? **How can we overcome this problem?**

We can square each difference from the mean, giving a positive or zero answer (squared deviation):

$$(-1)^2, \ 0^2 \text{ and } \ 1^2$$

So the spread can be measured by the sum of the squared deviations:

$$(-1)^2 + 0^2 + 1^2 = 2$$

Dividing this by the number of pieces of data gives us the 'measure of spread', called the **variance**:

$$\text{variance} = \frac{2}{3}$$

The **standard deviation** (*s*) is the square root of the variance. Hence

$$s = \sqrt{\frac{2}{3}}$$

Generally the variance of $x_1, x_2, \ldots, x_n$ is given by

16.2

$$\text{variance} = \frac{(x_1 - \bar{x})^2 + (x_2 - \bar{x})^2 + \ldots + (x_n - \bar{x})^2}{n}$$

$$\left[\text{variance} = \frac{\sum\limits_{j=1}^{n}(x_j - \bar{x})^2}{n} \right]$$

and the standard deviation, s, is given by

16.3

$$s = \sqrt{\frac{(x_1 - \bar{x})^2 + (x_2 - \bar{x})^2 + \ldots + (x_n - \bar{x})^2}{n}}$$

$$\left[s = \sqrt{\frac{\sum\limits_{j=1}^{n}(x_j - \bar{x})^2}{n}} \right]$$

The standard deviation, $s = \sqrt{\text{variance}}$. A slightly easier formula in evaluating the variance is $\text{variance} = \dfrac{\sum x_j^2}{n} - (\bar{x})^2$

By referring to formula 16.2 :

$x_j - \bar{x}$ is the deviation from the mean

$(x_j - \bar{x})^2$ is the squared deviation

The mean of the squared deviations gives us the **variance**. The square root of the variance is the **standard deviation**.

Remember that the standard deviation (SD) is a 'measure of spread' of data from the mean. **What does a small SD tell you about the data?**

That most of the data is close to the mean.

For a large SD, much of the data is far from the mean (Fig. 4).

Fig. 4

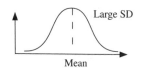

In other literature you might see the standard deviation and variance formulae with division by $n - 1$ rather than n, because the sample of data values has been taken from a population. In a nutshell, if we want to find the standard deviation of a given set of data then we divide by n, but if the data represents a sample from a population then we divide by $n - 1$. Throughout this book we will use the formula stated above: divide by n.

Example 8

The marks of an examination are given by the following data:

12, 27, 13, 21, 36, 56, 53, 55, 59, 83, 92, 75, 67, 80, 91, 99, 84 and 77.

Evaluate the standard deviation for this data.

> ### Example 8 *continued*
>
> Solution
>
> **?** **To use** 16.3 **, what do we need to evaluate first?**
>
> the mean ($\bar{x}$)
>
> By 16.1 : $\bar{x} = \dfrac{12 + 27 + \ldots + 77}{18} = 60$
>
> First find the variance, s^2, by 16.2 and then take the square root:
>
> $$s^2 = \frac{(12 - 60)^2 + (27 - 60)^2 + \ldots + (77 - 60)^2}{18} = 742.444$$
>
> Hence standard deviation, $s = \sqrt{742.444} = 27.25$ (2 d.p.). This is considered a large standard deviation, as you can observe that the exam marks are quite widespread from the mean of 60.
>
> To find the standard deviation by hand is very tedious, as the above example shows. The easiest way to determine the standard deviation is to use your calculator. We first place the calculator into SD mode. PRESS $\boxed{\text{MODE}}$ $\boxed{2}$ $\boxed{\text{SHIFT}}$ $\boxed{\text{MODE}}$ $\boxed{1}$ $\boxed{=}$ $\boxed{12}$ $\boxed{\text{M+}}$ $\boxed{27}$ $\boxed{\text{M+}}$ ... $\boxed{77}$ $\boxed{\text{M+}}$ then press $\boxed{\text{SHIFT}}$ $\boxed{2}$ $\boxed{2}$ $\boxed{=}$, which gives the standard deviation of 27.247 833 76.
>
> This gives the standard deviation of the data of **Example 8**. To evaluate the mean, press $\boxed{\text{SHIFT}}$ $\boxed{2}$ $\boxed{1}$ $\boxed{=}$ which shows 60. Generally on the calculator $\bar{x}$ refers to mean and σ_n refers to the standard deviation.

> ### Example 9
>
> Find the standard deviation of the following:
>
> **a** 1, 2, 3, 4, 5 and 6 **b** 1001, 1002, 1003, 1004, 1005 and 1006
> **c** 0.1, 0.2, 0.3, 0.4, 0.5 and 0.6 **d** 1.1, 1.2, 1.3, 1.4, 1.5 and 1.6
>
> Solution
>
> By calculator:
>
> **a** $s = 1.71$ (2 d.p.) **b** $s = 1.71$ (2 d.p.) **c** $s = 0.17$ (2 d.p.) **d** $s = 0.17$ (2 d.p.)

? **What do you notice about the results of Example 9?**

The standard deviation of **a** = standard deviation of **b**.

? **Are you surprised?**

This is not surprising since clearly, by looking at the data, we can see that the spread of data is the same. Data of **b** has only been shifted up by 1000.

? **What about c?**

$$\text{SD of } \mathbf{c} = \frac{\text{SD of } \mathbf{a}}{10}$$

In **Example 10** we will show that if the standard deviation of $x_1, x_2, \ldots, x_n$ is s then standard deviation of $\dfrac{x_1}{k}, \dfrac{x_2}{k}, \dfrac{x_3}{k}, \ldots, \dfrac{x_n}{k}$ is $\dfrac{s}{|k|}$, where $k \neq 0$ and $|k| = \sqrt{k^2}$.

Example 10

Let $y_i = \dfrac{x_i}{k}$ where $k \neq 0$ and $i = 1, 2, \ldots, n$. Show that:

i $\bar{y} = \dfrac{\bar{x}}{k}$ where $\bar{y}$ and $\bar{x}$ represent the mean values of y_i and x_i respectively.

ii $s_y = \dfrac{s_x}{|k|}$ where s_y and s_x represent the SD of y_i and x_i respectively.

(Note: $|a| = \sqrt{a^2}$.)

Solution

i By 16.1 with $y_1 = \dfrac{x_1}{k}$, $y_2 = \dfrac{x_2}{k}, \ldots$ we have

$$\bar{y} = \frac{y_1 + y_2 + \ldots + y_n}{n}$$

$$= \frac{\dfrac{x_1}{k} + \dfrac{x_2}{k} + \ldots + \dfrac{x_n}{k}}{n} \qquad \left[\text{Substituting } y_j = \frac{x_j}{k}\right]$$

$$= \frac{1}{k}\left(\frac{x_1 + x_2 + \ldots + x_n}{n}\right) = \frac{1}{k}\;\underset{\text{by } 16.1}{\underbrace{\bar{x}}} = \frac{\bar{x}}{k}$$

ii By 16.2:

$$s_y^2 = \frac{\displaystyle\sum_{j=1}^{n}(y_j - \bar{y})^2}{n} = \frac{\displaystyle\sum_{j=1}^{n}\left(\frac{x_j}{k} - \frac{\bar{x}}{k}\right)^2}{n} \qquad \left[\text{Substituting } \bar{y} = \frac{\bar{x}}{k}\right]$$

$$= \frac{1}{k^2}\frac{\displaystyle\sum_{j=1}^{n}(x_j - \bar{x})^2}{n} \qquad \left[\text{Taking out } \left(\frac{1}{k}\right)^2 = \frac{1}{k^2}\right]$$

$$s_y^2 = \frac{1}{k^2}\;\underset{\text{by } 16.2}{\underbrace{s_x^2}}$$

$$s_y = \sqrt{\frac{1}{k^2}}\,s_x = \frac{s_x}{\sqrt{k^2}} = \frac{s_x}{|k|}$$

Consider the data of **Example 9**. Let

$$x_1 = 1, \ x_2 = 2, \ x_3 = 3, \ x_4 = 4, \ x_5 = 5 \text{ and } x_6 = 6$$

Dividing these by 10 gives

$$y_1 = 0.1, \ y_2 = 0.2, \ y_3 = 0.3, \ y_4 = 0.4, \ y_5 = 0.5 \text{ and } y_6 = 0.6$$

The x mean is

$$\bar{x} = \frac{1 + 2 + 3 + 4 + 5 + 6}{6} = 3.5$$

By property **i** of **Example 10** we have the y mean, $\bar{y} = 3.5/10 = 0.35$.

The standard deviation of the x data is 1.71.

? **What is the standard deviation of the y data?**

Using property **ii** we have standard deviation of y data is $1.71/10 = 0.171$.

B3 Data in a frequency distribution

In this subsection we look at finding the standard deviation of data where each piece of data occurs one or more times and is therefore represented in a table. Consider the general frequency distribution (Table 5).

TABLE 5	Data	Frequency
	x_1	f_1
	x_2	f_2
	x_3	f_3
	. . .	. . .
	. . .	. . .
	. . .	. . .
	x_n	f_n

The data in Table 5 says that x_1 occurs f_1 times, x_2 occurs f_2 times etc.

The mean, $\bar{x}$, of this data is given by

16.4
$$\bar{x} = \frac{f_1 x_1 + f_2 x_2 + f_3 x_3 + \ldots + f_n x_n}{f_1 + f_2 + f_3 + \ldots + f_n} \ \text{ or } \ \bar{x} = \left[\frac{\sum\limits_{j=1}^{n} f_j x_j}{\sum\limits_{j=1}^{n} f_j} \right]$$

and the variance, s^2, is given by

16.5
$$s^2 = \frac{f_1(x_1 - \bar{x})^2 + f_2(x_2 - \bar{x})^2 + f_3(x_3 - \bar{x})^2 + \ldots + f_n(x_n - \bar{x})^2}{f_1 + f_2 + f_3 + \ldots + f_n}$$

$$\text{or} \left[s^2 = \frac{\displaystyle\sum_{j=1}^{n} f_j(x_j - \bar{x})^2}{\displaystyle\sum_{j=1}^{n} f_j} \right]$$

To find the standard deviation, s, we take the square root of the variance.

Example 11

The number of faulty telephones in various departments of an institution is as shown in Table 6.

TABLE 6	Number of faulty telephones (x)	Frequency (f)
	1	7
	2	3
	4	5
	5	8
	6	2
	8	3

Evaluate the standard deviation.

Solution

To find the standard deviation what do we need to evaluate first?

the mean $(\bar{x})$

By 16.4

$$\bar{x} = \frac{(1 \times 7) + (2 \times 3) + (4 \times 5) + \ldots + (8 \times 3)}{7 + 3 + 5 + \ldots + 3} = 3.893 \quad \text{(3 d.p.)}$$

To find the standard deviation, we construct Table 7.

TABLE 7	Column A x	Column B $\bar{x}$	Column C $(x - \bar{x})^2$	Column D f	Col C × Col D $f(x - \bar{x})^2$
	1	3.893	$(1 - 3.893)^2 = 8.369$	7	58.583
	2	3.893	$(2 - 3.893)^2 = 3.583$	3	10.749
	4	3.893	$(4 - 3.893)^2 = 0.011$	5	0.055
	5	3.893	$(5 - 3.893)^2 = 1.225$	8	9.800
	6	3.893	$(6 - 3.893)^2 = 4.440$	2	8.880
	8	3.893	$(8 - 3.893)^2 = 16.869$	3	50.607
				$\sum f = 28$	$\sum f(x - \bar{x})^2 = 138.674$

Example 11 *continued*

By 16.5

$$s^2 = \frac{138.674}{28} = 4.953$$

Hence the standard deviation is found by taking the square root of both sides:
$s = \sqrt{4.953} = 2.23$ (2 d.p.).

As **Example 11** shows, the standard deviation calculation of data in a frequency distribution is very long and tedious. It's easier to use a calculator. To find the standard deviation of the data of **Example 11**, proceed as follows:

Place the calculator into SD mode and clear memory, then PRESS 1 SHIFT , 7 M+ ... 8 SHIFT , 3 M+
Then PRESS SHIFT 2 2 = , which should show 2.225490086, the standard deviation of the data. Hence $s = 2.23$ (2 d.p.).

SHIFT 2 1 = shows 3.892857143, which is the mean of the data.

Example 12

Table 8 shows the maximum loads on various cables.

Max load L (kN)	Number of cables (frequency)
$49.5 \leq L < 54.5$	2
$54.5 \leq L < 59.5$	3
$59.5 \leq L < 64.5$	6
$64.5 \leq L < 69.5$	10
$69.5 \leq L < 74.5$	8
$74.5 \leq L < 79.5$	11
$79.5 \leq L < 84.5$	5
$84.5 \leq L < 89.5$	1
$89.5 \leq L < 94.5$	7
$94.5 \leq L < 99.5$	3
$99.5 \leq L < 104.5$	4

TABLE 8

Determine the mean and standard deviation of this data.

Example 12 *continued*

Solution

? We have grouped continuous data. **To use formula** 16.5 , **what value of L do we consider?**

? **For $49.5 \leq L < 54.5$ is $L = 49.5$ or 54.5 or...?** We take L to be the midpoint value, $L = 52$. Hence we have the solution shown in Table 9.

TABLE 9

Max load L (kN)	Midpoint L	Frequency f
$49.5 \leq L < 54.5$	52	2
$54.5 \leq L < 59.5$	57	3
$59.5 \leq L < 64.5$	62	6
$64.5 \leq L < 69.5$	67	10
$69.5 \leq L < 74.5$	72	8
$74.5 \leq L < 79.5$	77	11
$79.5 \leq L < 84.5$	82	5
$84.5 \leq L < 89.5$	87	1
$89.5 \leq L < 94.5$	92	7
$94.5 \leq L < 99.5$	97	3
$99.5 \leq L < 104.5$	102	4

By using a calculator we get mean $\bar{L} = 76.33$ kN (2 d.p.) and standard deviation $s = 13.21$ kN (2 d.p.).

SUMMARY

The average mean, $\bar{x}$, is defined by

16.1 $$\text{mean, } \bar{x} = \frac{x_1 + x_2 + x_3 + \ldots + x_n}{n}$$

Standard deviation is a 'measure of spread' of data.

16.2 $$\text{variance} = \frac{(x_1 - \bar{x})^2 + (x_2 - \bar{x})^2 + \ldots + (x_n - \bar{x})^2}{n}$$

The standard deviation, $s = \sqrt{\text{variance}}$. For a frequency distribution, mean

16.4 $$\bar{x} = \frac{f_1 x_1 + f_2 x_2 + f_3 x_3 + \ldots + f_n x_n}{f_1 + f_2 + f_3 + \ldots + f_n}$$

Variance, s^2, is given by

16.5 $$s^2 = \frac{f_1(x_1 - \bar{x})^2 + f_2(x_2 - \bar{x})^2 + f_3(x_3 - \bar{x})^2 + \ldots + f_n(x_n - \bar{x})^2}{f_1 + f_2 + f_3 + \ldots + f_n}$$

Standard deviation is s – take the square root of s^2.

Solutions at end of book. Complete solutions available at www.palgrave.com/science/engineering/singh

Exercise 16(b)

1 The number of rejects on an assembly line are as follows:

12, 27, 34, 50, 5, 19, 39, 15, 8, 36, 29

Determine the median, mean and standard deviation for this data.

2 Determine the mean and standard deviation for the following data:
a 5, 7, 12, 15, 37
b 105, 107, 112, 115, 137
c 0.5, 0.7, 1.2, 1.5, 3.7
d 53, 73, 123, 153, 373
What do you notice about your results?
[For questions **3** and **4**: $\bar{x}$ and $\bar{y}$ represent x mean and y mean respectively. s_x and s_y represent the standard deviations for x and y respectively. Also $j = 1, 2, \ldots, n$.]

3 If $y_j = kx_j$ then show that $\bar{y} = k\bar{x}$ and $s_y = |k| s_x$.

4 If $y_j = kx_j + a$ then show that $\bar{y} = k\bar{x} + a$ and $s_y = |k| s_x$.

5 Find the mean and standard deviation for the following:
i 27, 53, 61, 64, 89, 93, 98
ii 270, 530, 610, 640, 890, 930, 980
iii 0.27, 0.53, 0.61, 0.64, 0.89, 0.93, 0.98
iv 32, 58, 66, 69, 94, 98, 103
v 2800, 5400, 6200, 6500, 9000, 9400, 9900

6 The following data shows the mean critical load on various struts:

Critical load (kN)	150.0	157.0	159.0	162.0
Frequency	7	8	12	15
Critical load (kN)	163.5	166.0	168.3	
Frequency	9	4	2	

Evaluate the mean and standard deviation for this data.

7 A Goods-In department measures a sample of 50 resistors from a batch which claims that they are all 10 kΩ.

The data is displayed in a frequency distribution as follows:

Resistance R (kΩ)	Frequency
$9.6 \le R < 9.7$	1
$9.7 \le R < 9.8$	2
$9.8 \le R < 9.9$	5
$9.9 \le R < 10.0$	17
$10.0 \le R < 10.1$	18
$10.1 \le R < 10.2$	5
$10.2 \le R < 10.3$	1
$10.3 \le R < 10.4$	1

i Estimate the mean, without evaluating it.
ii Do you think this data will have a large SD or a small SD?
iii Determine the actual mean and standard deviation.

8 Evaluate the mean and standard deviation for the following.
a The tensile strength (MN/m²) of aluminium alloy is given by:

Tensile strength T (MN/m²)	Frequency
$320 \le T < 350$	4
$350 \le T < 380$	12
$380 \le T < 410$	18
$410 \le T < 440$	16
$440 \le T < 470$	15
$470 \le T < 500$	10
$500 \le T < 530$	3
$530 \le T < 560$	7

b The Young modulus (GN/m²) for stainless steel is given by:

Young modulus (GN/m²)	193	194	195	196
Frequency	7	7	10	19
Young modulus (GN/m²)	197	198	199	200
Frequency	13	9	8	3

SECTION C **Probability rules**

By the end of this section you will be able to:
▶ understand the term *probability*
▶ understand some of the probability rules
▶ use these rules

C1 **Introduction**

? **What does the term probability mean?**

Probability is a numerical value which measures the likelihood of an event A occurring. It is normally denoted by $P(A)$.

Let X be the event that a test rig conforms to specification. If it is known that 97% of the rigs comply with specification then we say that

$$P(X) = \frac{97}{100} = 0.97$$

This means the probability that X conforms to specification is 0.97.

The probability of any event lies between 0 and 1, that is $0 \le P(A) \le 1$. $P(A) = 0$ means that the event A **cannot** happen.

? **Can you think of any examples where $P(A) = 0$?**

A man being pregnant, the sun rising from the west, etc.

$P(A) = 1$ means that the event A **will** certainly happen.

? **Can you think of any examples where $P(A) = 1$?**

You will pass away at some time (unless science manages to conquer death), the sun will rise each day, etc.

If all possibilities are divided into n equally likely outcomes and the event A occurs for r out of the n outcomes then the probability of A occurring, $P(A)$, is given by

$$P(A) = \frac{r}{n}$$

where $n \ge 2$ and $0 \le r \le n$.

? **What is the probability of obtaining a head on tossing a coin?**

There are two possible outcomes – heads or tails, hence $n = 2$. Since the event of getting a head can only occur once, therefore $r = 1$ and

$$P(\text{head}) = \frac{1}{2}$$

C2 **Notation**

Probability theory can be represented by using Venn diagrams. The sample space, S, is the representation of all the possible outcomes of an event. For example, if the event is throwing a six-sided die (Fig. 5) then

$$S = \{1, 2, 3, 4, 5, 6\}$$

Fig. 5

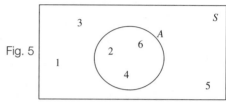

If A is the event of getting an even number by throwing a die then $A = \{2, 4, 6\}$. The event, A, is called a subset of S.

What is the probability of getting an even number on throwing a die?

There are three even numbers, 2, 4 and 6, and there are six possibilities, therefore

$$P(\text{even number}) = \frac{3}{6} = \frac{1}{2}$$

Fig. 6

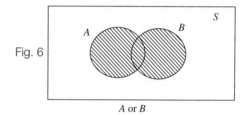

A or B

Let B be the event of getting 5 or 6 on throwing a die, that is $B = \{5, 6\}$.

We first consider the general case with A and B being two different events. $A \cup B$ (read as 'A union B') are all the possible outcomes of A **or** B occurring (Fig. 6).

Fig. 7

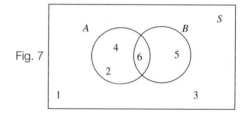

For our example with $A = \{2, 4, 6\}$ and $B = \{5, 6\}$ as defined above we have the situation shown in Fig. 7.

Thus A or B has four possible outcomes, $\{2, 4, 5, 6\}$, and so

$$P(A \text{ or } B) = \frac{4}{6} = \frac{2}{3}$$

Fig. 8

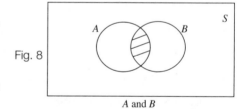

A and B

$A \cap B$ (read as 'A intersection B') are all the possible outcomes of both A **and** B occurring (Fig. 8).

From our example with $A = \{2, 4, 6\}$ and $B = \{5, 6\}$ we consider Fig. 7. Since 6 is the **only** element in both A and B, the intersection, we have

$$P(A \text{ and } B) = \frac{1}{6}$$

Fig. 9

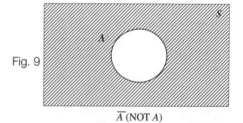

$\overline{A}$ (NOT A)

$\overline{A}$ (read as 'not A') are all the possible outcomes of **not** A occurring (Fig. 9).

For our example with $A = \{2, 4, 6\}$ we have the situation shown in Fig. 10.

Fig. 10

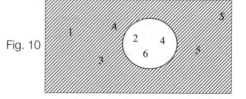

Hence the event not A has three outcomes, $\{1, 3, 5\}$, therefore

$$P(\text{not } A) = \frac{3}{6} = \frac{1}{2}$$

C3 **The *and* rule**

If the occurrence of event A does not effect the occurrence of event B then A and B are said to be **independent** events.

If we toss a coin twice then the probability of getting a head on the second toss is 0.5, and is independent of what happened on the first toss. Thus the tossing of coins are independent events.

If A and B are independent events then

16.6 $P(A \text{ and } B) = P(A) \times P(B)$

The probability of getting two heads when tossing a coin twice is

$$P(\text{Head and Head}) = P(\text{Head}) \times P(\text{Head})$$
$$= \frac{1}{2} \times \frac{1}{2} = \frac{1}{4}$$

If events A and B are independent then so are **i** $\overline{A}$ and B, **ii** A and $\overline{B}$ and **iii** $\overline{A}$ and $\overline{B}$. Remember the notation: $\overline{A}$ means not A.

We can extend formula 16.6 for any sequence of independent events $E_1, E_2, \ldots, E_k$:

16.7 $P(E_1 \text{ and } E_2 \text{ and } E_3 \ldots \text{ and } E_k) = P(E_1) \times P(E_2) \times \ldots \times P(E_k)$

C4 **The *not* rule**

Either event A occurs or it does not occur, therefore

$$P(A) + P(\text{not } A) = 1$$

That is the event A or not A covers **all** possibilities. Rearranging gives

16.8 $P(\text{not } A) = 1 - P(A)$

Example 13

A component is subject to two independent tests A and B. The probability that a component passes test A is 0.7 and that it passes test B is 0.85. Determine the probability that a randomly chosen component:

  i does not pass test A
 ii does not pass test B
iii does not pass test A and does not pass test B
 iv passes both tests, A and B.

Solution

Let A and B denote the events of passing test A and test B respectively, then $P(A) = 0.7$ and $P(B) = 0.85$.

Example 13 *continued*

i $P(\text{a component does not pass test } A) = P(\text{not } A)$

$$\underset{\text{by } 16.8}{=} 1 - P(A)$$

$$= 1 - 0.7 = 0.3$$

ii Similarly $P(\text{not } B) = 1 - P(B) = 1 - 0.85 = 0.15$

iii $P(\text{a component [does not pass test } A] \text{ and [does not pass test } B])$

$$= P([\text{not } A] \text{ and } [\text{not } B])$$

$$\underset{\text{by } 16.6}{=} P(\text{not } A) \times P(\text{not } B) = 0.3 \times 0.15 = 0.045$$

iv $P(A \text{ and } B) = P(A) \times P(B) = 0.7 \times 0.85 = 0.595$

Example 14

Let A, B and C denote three independent sections of a machine. The probabilities of each one working is given by

$$P(A) = 0.6, \quad P(B) = 0.55 \quad \text{and } P(C) = 0.8$$

Determine the probabilities that:

a all three sections are working
b none are working.

Solution

Let A, B and C denote the events of A, B and C working respectively.

a $P(\text{All three sections working}) = P(A \text{ and } B \text{ and } C)$

$$= P(A) \times P(B) \times P(C) \quad [\text{by } 16.7]$$

$$= 0.6 \times 0.55 \times 0.8$$

$$= 0.264$$

b $P(\text{None are working}) = P([\text{not } A] \text{ and } [\text{not } B] \text{ and } [\text{not } C])$

$$\underset{\text{by } 16.7}{=} P(\text{not } A) \times P(\text{not } B) \times P(\text{not } C)$$

$$\underset{\text{by } 16.8}{=} (1 - 0.6)(1 - 0.55)(1 - 0.8) = 0.036$$

16.6 $P(A \text{ and } B) = P(A) \times P(B)$

16.7 $P(E_1 \text{ and } E_2 \text{ and } E_3 \ldots \text{ and } E_k) = P(E_1) \times P(E_2) \times P(E_3) \ldots P(E_k)$

16.8 $P(\text{not } A) = 1 - P(A)$

Example 15

In a batch of 100 bearings, 10 are defective. If 5 are chosen at random (without replacement), what is the probability that all 5 are good?

Solution

The probability of the first bearing being good is $\dfrac{90}{100}$.

What is the probability of the second bearing being good?

Since we are not replacing the bearing there are now a total of 99 bearings and we have already selected a good bearing in the first case and therefore there are only 89 good bearings remaining, hence

$$P(\text{Second bearing is good}) = \frac{89}{99}$$

$$P(\text{2 good bearings}) = P\Big([\text{1st good bearing}]\text{ and }[\text{2nd good bearing}]\Big)$$

$$= P(\text{1st good bearing}) \times P(\text{2nd good bearing})$$

$$= \frac{90}{100} \times \frac{89}{99}$$

We want to choose 5 good bearings so repeating this process we have

$$P(\text{All good}) = \frac{90}{100} \times \frac{89}{99} \times \frac{88}{98} \times \frac{87}{97} \times \frac{86}{96} = 0.584 \text{ (3 d.p.)}$$

Example 16

An amplifier consists of two independent stages, A and B, connected in series (Fig. 11). The output of A is the input to B. The amplifier works if both stages A and B work. If the probability of A working is 0.72 and the probability of the amplifier working is 0.61 then find the probability that stage B works, given that stage A works.

Fig. 11

Solution

Let A and B denote stages A and B working respectively. Then we have $P(A) = 0.72$ and $P(A \text{ and } B) = 0.61$ because the probability of the amplifier working is 0.61. Since A and B are independent

$$P(A \text{ and } B) = P(A) \times P(B)$$

Rearranging gives

$$P(B) = \frac{P(A \text{ and } B)}{P(A)} = \frac{0.61}{0.72} = 0.847 \quad \text{(3 d.p.)}$$

C5 The *or* rule

For any events, *A* and *B*

16.9 $P(A \text{ or } B) = P(A) + P(B) - P(A \text{ and } B)$

A or *B* means *A* occurs or *B* occurs or both *A* and *B* occur.

We can justify 16.9 by the Venn diagrams shown in Fig. 12.

Fig. 12

A or *B*   =   *A*   +   *B*   −   *A* and *B*

The intersection (*A* and *B*) is counted twice, and that is why we have to subtract the extra (*A* and *B*).

A and *B* are said to be mutually exclusive events if there is **no** element which is common between *A* and *B* (Fig. 13).

Fig. 13

Mutually exclusive events

If *A* and *B* are mutually exclusive events then $P(A \text{ and } B) = 0$ and

$$P(A \text{ or } B) = P(A) + P(B) - P(A \text{ and } B)$$
$$= P(A) + P(B) - 0$$
$$= P(A) + P(B)$$

16.10 $P(A \text{ or } B) = P(A) + P(B)$

? **What is the probability of getting a 3 or 4 with a single throw of a die?**

Since 3 or 4 cannot occur together they are mutually exclusive events. (We cannot get a 3 and a 4 with a single throw of a die.) Thus

$$P(\text{Getting a 3 or 4}) = P(\text{Getting a 3}) + P(\text{Getting a 4})$$
$$= \frac{1}{6} + \frac{1}{6} = \frac{2}{6} = \frac{1}{3}$$

We can extend the formula 16.10 for any sequence of mutually exclusive events $E_1, E_2, \ldots, E_k$:

16.11 $P(E_1 \text{ or } E_2 \text{ or } E_3 \ldots \text{ or } E_k) = P(E_1) + P(E_2) + \ldots + P(E_k)$

If events *A* and *B* are independent then

16.12 $P(A \text{ or } B) = P(A) + P(B) - [P(A) \times P(B)]$

This is shown by 16.9 and 16.6 :

$$P(A \text{ or } B) = P(A) + P(B) - P(A \text{ and } B)$$
$$= P(A) + P(B) - \underline{P(A) \times P(B)}$$
$$\text{by } 16.6$$

16.6 $P(A \text{ and } B) = P(A) \times P(B)$

Example 17

An electrical circuit has two switches A and B connected in parallel as shown in Fig. 14.

The probability of switch A working is 0.6 and of switch B working is 0.7. Find the probability that there is a connection between X and Y.

Solution

Let A denote the event that switch A works. Let B denote the event that switch B works. For a connection between X and Y we need A or B to work:

Fig. 14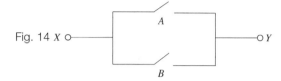

$$P(A \text{ or } B) = P(A) + P(B) - P(A \text{ and } B)$$

$$\underset{\substack{\text{because } A \text{ and } B \\ \text{are independent}}}{=} P(A) + P(B) - [P(A) \times P(B)]$$

$$= 0.6 + 0.7 - [\,0.6 \times 0.7\,] = 0.88$$

C6 De Morgan's laws

In evaluating probabilities it is useful to have the following laws from set theory. We assume these laws and apply them to find probabilities.

For any two sets A and B:

16.13
$$\overline{A \cup B} = \overline{A} \cap \overline{B}$$
$$\text{not}(A \text{ or } B) = (\text{not } A) \text{ and } (\text{not } B)$$

16.14
$$\overline{A \cap B} = \overline{A} \cup \overline{B}$$
$$\text{not}(A \text{ and } B) = (\text{not } A) \text{ or } (\text{not } B)$$

16.13 and 16.14 are known as de Morgan's laws.

In probability terms, where A and B are events we have

> English mathematician, **Augustus de Morgan** (1806–71), was Professor of Mathematics at University College, London.

16.15
$$P(\text{not}[A \text{ or } B]) = P([\text{not } A] \text{ and } [\text{not } B])$$

and

16.16
$$P(\text{not}[A \text{ and } B]) = P([\text{not } A] \text{ or } [\text{not } B])$$

We can see how these rules work in practice below.

Example 18

An engineering module consists of coursework and an examination. It is known that 70% passed the examination, 85% passed coursework and 60% passed both. For a randomly chosen student, determine the probabilities that the student has

i passed the examination or coursework
ii failed both the coursework and examination.

Solution

Let E and C denote passes in examination and coursework respectively. We can visualize this example by Venn diagrams (Figs 15 and 16).

i P(passed the examination or coursework)

$$= P(E \text{ or } C)$$

$$\underset{\text{by } 16.9}{=} P(E) + P(C) - P(E \text{ and } C)$$

$$= 0.7 + 0.85 - 0.6$$
$$= 0.95$$

Fig. 15 E or C

ii P(failed both) $= P([\text{not } E] \text{ and } [\text{not } C])$

$$\underset{\text{by } 16.15}{=} P(\text{not } [E \text{ or } C])$$

$$P(\text{failed both}) \underset{\text{by } 16.8}{=} 1 - P(E \text{ or } C)$$

$$= 1 - 0.95 = 0.05$$

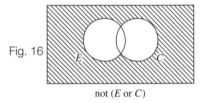

Fig. 16   not $(E$ or $C)$

which means a 5% chance of failing both, the coursework and the examination, which seems reasonable given the data.

C7 *At least* rule

If an event has two possible outcomes then we call one of the outcomes a **success** and the other a **failure**. For example the event of tossing a coin gives two possible outcomes – heads and tails. If we are asked to find the probability of obtaining at least n heads on tossing a coin then we consider heads as success and tails as failure.

The probability of at least n successful outcomes can be calculated by

16.17 P(at least n successful outcomes)
$$= 1 - P(\text{less than } n \text{ successful outcomes})$$

Example 19

An inspector fails 5% of the motors from a batch. If two motors are chosen at random from the batch, evaluate the probabilities that

a both motors are good
b at least one motor is faulty.

Example 19 *continued*

Solution

$$P(\text{good motor}) = \frac{95}{100} = 0.95$$

a $P(\text{Both motors good})\;=\;P([\text{good motor}]\text{ and }[\text{good motor}])$

$$= P(\text{good motor}) \times P(\text{good motor})$$

$$= 0.95 \times 0.95 = 0.9025$$

b Let the selection of a faulty motor be a successful outcome, then

$$P(\text{At least one motor is faulty}) \underset{\text{by }16.17}{=}\; 1 - P(\text{less than one motor is faulty})$$

$$= 1 - P(\text{both motors are good})$$

$$P(\text{At least one motor is faulty}) = 1 - 0.9025 = 0.0975$$

SUMMARY

If A and B are independent events then

16.6 $P(A \text{ and } B) = P(A) \times P(B)$

For any events A and B:

16.8 $P(\text{not } A) = 1 - P(A)$

16.9 $P(A \text{ or } B) = P(A) + P(B) - P(A \text{ and } B)$

If A and B are mutually exclusive events then

16.10 $P(A \text{ or } B) = P(A) + P(B)$

For independent events A and B:

16.12 $P(A \text{ or } B) = P(A) + P(B) - [P(A) \times P(B)]$

For any events A and B:

16.15 $P(\text{not } [A \text{ or } B]) = P([\text{not } A] \text{ and } [\text{not } B])$

16.16 $P(\text{not } [A \text{ and } B]) = P([\text{not } A] \text{ or } [\text{not } B])$

16.17 $P(\text{at least } n \text{ successful outcomes})$
$$= 1 - P(\text{less than } n \text{ successful outcomes})$$

Exercise **16(c)**

Solutions at end of book. Complete solutions available at www.palgrave.com/science/engineering/singh

1 In a large selection of electrical components, 10% of resistors are faulty, 5% of inductors are faulty and 3% of capacitors are faulty.
   a From a batch a resistor is chosen at random. Find the probability that it is not faulty.
   b By choosing randomly each of the three components, what is the probability that all three are faulty?

2 Three motors are chosen at random from a batch of which 2% are defective. Find the probability that all three are defective.

3 Two switches A and B are connected in parallel as shown in Fig. 17.

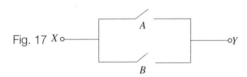

Fig. 17

Given that the probabilities of A and B functioning are

$$P(A) = 0.8 \text{ and } P(B) = 0.65$$

find the probability of a connection between X and Y.

4 The lifetimes, L, of 10,000 light bulbs produced by a certain manufacturer are as follows:

Lifetime L (hours)	Number of bulbs
$L \leq 200$	3000
$200 < L \leq 500$	3500
$500 < L \leq 1000$	2750
$L > 1000$	750

For a randomly selected bulb, find the probability that the bulb has
   i $L \leq 200$ hours
   ii $L \leq 500$ hours
   iii $L > 500$ hours.

5 A calculator manufacturer finds that 5% of the calculators do not have a memory button, 3% do not have a log button and 1% do not have the log and memory buttons. For a randomly selected calculator, find the probability that the calculator
   a does not have a memory button or a log button
   b does have a memory button and a log button.

6 A communication network has two systems, A and B, connected in parallel and it only fails if both systems fail. The probability of A and B functioning properly is given by

$$P(A) = 0.88, \ P(B) = 0.93$$

What is the probability that the communication network fails?

7 There are 50 components in a batch of which 4 are defective. Given that two components are randomly chosen, without replacement, and that the first component chosen is good, what is the probability that the second component is defective?

8 A toy is assembled from three different parts A, B and C. The probabilities of parts A, B and C being defective are given by:

$$P(A) = 0.15, P(B) = 0.25 \text{ and } P(C) = 0.02$$

Find the probability that a toy picked at random contains
   i no defective parts
   ii at least one defective part.

Exercise **16(c) continued**

Solutions at end of book. Complete solutions available at www.palgrave.com/science/engineering/singh

9 An electrical circuit consist of three switches A, B and C connected as shown in Fig. 18.

Fig. 18

The probability of A, B and C working are 0.7, 0.75 and 0.85 respectively. Find the probability that there is a connection between X and Y.

10 Three components A, B and C are connected in parallel as shown in Fig. 19.

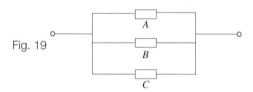

Fig. 19

A, B and C are independent. The probability of each component working is as follows:

$$P(A) = 0.6, P(B) = 0.75 \text{ and } P(C) = 0.55$$

Find the probability that at least one component is working.

11 There are 60 components in a batch of which 3 are defective. Given that two components are randomly selected, without replacement, and that the first component is good, what is the probability that the second component is defective?

12 A box contains 100 fuses of which 5 are defective. If two fuses are chosen at random, without replacement, find the probabilities that
   a both are defective
   b none are defective
   c only one is defective.

13 A system consists of two stages, A and B, in series (Fig. 20).

Fig. 20

The output of stage A acts as the input to stage B. Given that the probability of A working is 0.92 and the probability of the system working is 0.88, find the probability of B working given that A works.

14 5% of a batch of 120 components are faulty. If two components are chosen at random, without replacement, find the probability that the second component is faulty given that the first component is faulty.

15 A batch of 150 motors contains the following defects:
   6 of type I defect
   4 of type II defect
   7 of type III defect
   From these defects we also have:
   3 of type I and type II defects
   2 of type II and type III defects
   1 of type I, type II and type III defects.
   One motor is chosen at random. If this motor has a defect of type II, find the probabilities that it also has
   a a defect of type I
   b a defect of type III
   c defects of type I and type III.

16 A batch of 96 transformers contains 7 that are defective. If two transformers are selected at random, without replacement, find the probability that both are defective.

17 The reliability of an engine working to specification is 0.85. Two engines are tested at random. Find the probability that at least one of the engines is working to specification.

Exercise **16(c) continued**

Solutions at end of book. Complete solutions available at www.palgrave.com/science/engineering/singh

18 Switches A, B, C and D are connected as shown in Fig. 21.

Fig. 21

The probabilities of each switch working is given by

$$P(A) = 0.7, P(B) = 0.8,$$
$$P(C) = 0.77 \text{ and } P(D) = 0.95$$

What is the probability of a connection between X and Y?

SECTION D **Permutations and combinations**

By the end of this section you will be able to:
▶ evaluate permutations
▶ distinguish between permutations and combinations
▶ evaluate combinations

D1 **Permutations**

? **Do you remember what $n!$ (n factorial) means?**

16.18 $n! = 1 \times 2 \times 3 \times 4 \times \ldots \times (n-1) \times n$

For example $6! = 1 \times 2 \times 3 \times 4 \times 5 \times 6 = 720$. Also we define $0! = 1$. We use $n!$ to count the number of ways of arranging (arrangements) n different items, as the next example shows.

Example 20

If car number plates are to contain the three letters A, B and C, how many different number plates can we form without repetition?

Solution

The first letter can be chosen in three ways – A, B or C. The second letter can be chosen in two ways and the third letter can be chosen in only one way. Hence the number of different plates $= 3 \times 2 \times 1 = 3! = 6$. The six different plates are ABC, ACB, BAC, BCA, CAB and CBA.

Doing the same example with four letters A, B, C and D gives us $4! = 24$ different plates.

The number of ways of arranging n different items is

16.19 $n!$

On a calculator we can evaluate $n!$ as follows. For 6! PRESS

| 6 | | ! | or | 6 | | **SHIFT** | | **X⁻¹** | | **=** | which should show 720.

If some of the n items are the same then the number of arrangements reduces.
In **Example 20**, if we have the letters AAB, that is two letters out of 3 are the same, then the only arrangements would be AAB, ABA and BAA. This can be evaluated directly by

$$\frac{3!}{2!} = \frac{6}{2} = 3$$

In general, if there are n items of which p are the same then the number of arrangements is

16.20 $$\frac{n!}{p!}$$

A **permutation** is an arrangement where order matters. For example AB is a different permutation from BA. The number of permutations of r items chosen from n different items is denoted by ^nP_r and given as

16.21 $$^nP_r = \frac{n!}{(n-r)!}$$

Remember that order does matter for permutations.

Example 21

Consider A, B, C and D. If two letters are chosen then find the number of all the possible permutations without repetition.

Solution

Permutations are AB, BA, AC, CA, AD, DA, BC, CB, BD, DB, CD and DC: 12 altogether. Alternately if we use

16.21 $$^nP_r = \frac{n!}{(n-r)!}$$

with $r = 2$ and $n = 4$ we obtain

$$^4P_2 = \frac{4!}{2!} = 12$$

Example 22

The security code of an alarm consists of three single digit numbers followed by three letters, without any repetition. Evaluate how many different codes can by formed.

Solution

The number of permutations of choosing 3 numbers from 10 is given by

16.21 $$^nP_r = \frac{n!}{(n-r)!}$$

Example 22 *continued*

With $r = 3$ and $n = 10$:
$$^{10}P_3 = \frac{10!}{(10 - 3)!} = \frac{10!}{7!}$$

The number of permutations of choosing 3 letters from 26 is given by
$$^{26}P_3 = \frac{26!}{23!}$$

Evaluating each of these on a calculator and multiplying gives

Number of codes $= {}^{10}P_3 \times {}^{26}P_3 = 11{,}232{,}000$

Thus more than 11 million different codes can be formed.

Most calculators will evaluate permutations directly. To evaluate $^{26}P_3$ PRESS
[26] [SHIFT] [nCr] [3] [=] which should show 15600.

On some calculators you might **not** need to press the [SHIFT] button because ^nP_r is not a secondary function. In this case PRESS [26] [nPr] [3] [=] Also to access a secondary function on your calculator might not involve a [SHIFT] button but an [INV] or a [2nd Fun] button instead. You press this in place of the [SHIFT] key.

D2 **Combinations**

Example 23

Consider the letters A, B, C and D. Without considering order, find the number of selections if

a 2 letters are chosen at random
b 3 letters are chosen at random.

Solution

a Selections are AB, AC, AD, BC, BD and CD; hence 6 selections altogether. Since order does not matter, AB is the same as BA and we count this as **one** selection. The number of selections can be evaluated by using the permutation formula 16.21 (where order does matter) and then halving it:

$$\frac{1}{2}{}^4P_2 = \frac{1}{2}\left(\frac{4!}{2!}\right) = \frac{4!}{2!2!} = 6 \quad \text{[Remember that } 2! = 2\text{]}$$

b For three letters chosen at random the selections are ABC, ACD, BCD and ABD; hence 4 selections. Remember that order does **not** matter: ABC, ACB, BAC, BCA, CAB and CBA all count as one selection rather than 6 (or 3!). We can use the permutation formula 16.21 and then divide by 3! (= 6).

Number of selections $= \dfrac{1}{3!}{}^4P_3 = \dfrac{1}{3!}\left(\dfrac{4!}{(4 - 3)!}\right) = 4.$

A combination is a selection of items where order does **not** matter. In the above example AB is the same combination as BA but AC is a different combination.

The number of combinations of r items chosen from n different items is denoted by ^nC_r and given by

$$\frac{1}{r!} \, ^nP_r = \frac{1}{r!} \underbrace{\left(\frac{n!}{(n-r)!} \right)}_{\text{by } \boxed{16.21}}$$

$$= \frac{1 \times n!}{r! \times (n-r)!} = \frac{n!}{r!(n-r)!}$$

Therefore

$$\boxed{16.22} \qquad ^nC_r = \frac{n!}{r!(n-r)!}$$

(^nC_r can be read as 'choosing r out of n'.)

Remember in combinations that order does **not** matter.

We can evaluate ^nC_r directly on a calculator. To evaluate $^{21}C_{15}$ PRESS $\boxed{21}$ $\boxed{\text{nCr}}$ $\boxed{15}$ $\boxed{=}$ which should show 54264.

Example 24

In playing the lottery the object is to choose 6 numbers between 1 to 49. If these 6 numbers match the 6 numbers drawn from the machine (it does not matter about order) then you win the jackpot. What is the probability of winning this prize if you have bought 100 lottery tickets and have chosen different numbers on each ticket?

Solution

The number of ways of choosing 6 numbers from 49 is

$$^{49}C_6$$

because order does not matter. Since we have selected 100 different numbers, therefore

$$P(\text{Winning the jackpot}) = \frac{100}{^{49}C_6} = \frac{100}{13{,}983{,}816}$$

Approximately 100 in 14 million chance of winning the jackpot.

If you had brought only one lottery ticket then there is approximately one in 14 million chance of winning the jackpot.

$\boxed{16.21}$ $^nP_r = \dfrac{n!}{(n-r)!}$

Example 25

In a box there are 25 CD-ROM disks of which 4 are defective. If 3 disks are chosen at random, without replacement, from this box, what is the probability that

a all 3 disks are good?
b only 1 is defective?
c all 3 are defective?

Solution

The order of choosing disks is of no interest. So the number of ways of choosing 3 disks from 25 is given by $^{25}C_3$.

a For all 3 disks to be good, we choose 3 out of the 21 good disks, $^{21}C_3$. Hence

$$P(\text{All 3 good}) = \frac{^{21}C_3}{^{25}C_3} = 0.578 \quad (3 \text{ s.f.})$$

$$\left[\text{Alternatively } P(\text{All 3 good}) = \frac{21}{25} \times \frac{20}{24} \times \frac{19}{23} = 0.578 \right]$$

b Only 1 is defective and so 2 are good. We choose 1 out of the 4 defective disks, $^{4}C_1$, and 2 out of the 21 good disks, $^{21}C_2$. The probability of one defective is given by

$$P(\text{Only one defective}) = \frac{^{4}C_1 \times {}^{21}C_2}{^{25}C_3} = 0.365 \quad (3 \text{ s.f.})$$

We multiply $^{4}C_1$ and $^{21}C_2$ because we require 1 defective disk **and** 2 good disks.

c For all 3 defective, we choose 3 out of the 4 defective disks, $^{4}C_3$:

$$P(\text{All 3 defective}) = \frac{^{4}C_3}{^{25}C_3} = 1.74 \times 10^{-3} \quad (3 \text{ s.f.})$$

SUMMARY

The number of permutations of r items chosen from n is given by

16.21 $$^{n}P_r = \frac{n!}{(n-r)!}$$

In permutations, order **does** matter.

The number of combinations of r items chosen from n is given by

16.22 $$^{n}C_r = \frac{n!}{r!(n-r)!}$$

In combinations, order does **not** matter.

| Exercise **16(d)** | Solutions at end of book. Complete solutions available at www.palgrave.com/science/engineering/singh |

1 There are 60 light bulbs of which 10 are faulty. 7 bulbs are chosen at random, without replacement. Find the probability that 3 of these are faulty.

2 There are 60 fuses in a box of which 9 are defective. If 8 fuses are chosen at random, without replacement, then find the probability that
　i all 8 are good
　ii all 8 are defective
　iii at least 1 is defective.

3 In a game of lottery one has to select six different numbers between 1 and 40. If your six numbers match the six numbers drawn from the machine then you win the jackpot (order does not matter).

a Find the probability of winning the jackpot if you have bought 150 lottery tickets.

b If each lottery ticket cost £1 then how much would you have to spend to obtain a probability of more than 0.5 of winning the jackpot?

4 A licence plate consists of 1 letter followed by 3 single digits and then 3 letters. Assuming all 26 letters and 10 numbers can be used, find the number of plates if

a no number or letter is duplicated
b numbers and letters can be duplicated.

SECTION E **Binomial distribution**

By the end of this section you will be able to:
▸ understand the term *discrete random variable*
▸ use the binomial distribution to find probabilities
▸ apply the binomial distribution formula

E1 **Discrete random variables**

Let $E_1, E_2, \ldots, E_k$ be mutually exclusive events, that is they have no elements in common. Then a random variable, X, gives each of these $E_1, E_2, \ldots, E_k$ a numerical value. The random variable is usually denoted by a capital letter. For example, let X represent the sum of points on throwing two dice. We have

　　　$X = 2$ is the numerical value for throwing $(1, 1)$
　　　$X = 3$ is the numerical value for throwing $(1, 2)$ or $(2, 1)$
　　　　　　　　　$\vdots$
　　　$X = 12$ is the numerical value for throwing $(6, 6)$

The word **random** is used because it is associated with an experiment and the value is determined by chance. The particular value a random variable takes is usually denoted by a small letter, $x, y, z, \ldots$.

The function which gives the probabilities of the random variable, X, is called the **probability density function** and denoted by $P(X = x)$.

In this and the next section we examine discrete random variables. A discrete random variable is a random variable, X, which can only take certain values $x_1, x_2, \ldots, x_n$ and the associated probabilities total 1. If

$$P(X = x_1) = p_1, \; P(X = x_2) = p_2, \ldots, \; P(X = x_n) = p_n$$

where $p_1, p_2, \ldots, p_n$ are probabilities then

16.23 $\qquad p_1 + p_2 + \ldots + p_n = 1$

Example 26

The probability density function of a discrete random variable X is given by the formula

$$P(X = x) = kx^2 \quad \text{for } x = 0, 1, 2, 3$$

Evaluate the constant k.

Solution

Substituting the given values of x into $P(X = x) = kx^2$:

$$P(X = 0) = k(0^2) = 0$$
$$P(X = 1) = k(1^2) = k$$
$$P(X = 2) = k(2^2) = 4k$$
$$P(X = 3) = k(3^2) = 9k$$

Summing these results and equating to 1, because total probabilities $= 1$, gives

$$9k + 4k + k = 1$$
$$14k = 1 \text{ so } k = 1/14$$

E2 Binomial distribution

Example 27

On average, 7% of the motors produced by a machine are faulty. A random sample of 5 motors is selected from a large batch so that the probability remains constant. Determine the probabilities that

a 1 motor is faulty
b 2 motors are faulty
c 3 motors are faulty
d 4 motors are faulty
e all 5 motors are faulty
f no motor is faulty.

Do you notice any relationship between the number of faulty motors and their associated probabilities?

Example 27 *continued*

Solution

Let *F* and *G* denote faulty and good motor respectively.

$$P(F) = \frac{7}{100} = 0.07$$

Since 7% are faulty, therefore 93% must be good motors:

$$P(G) = \frac{93}{100} = 0.93$$

a $P(1 \text{ motor faulty out of } 5) = P(FGGGG)$

$$= P(F) \times P(G) \times P(G) \times P(G) \times P(G)$$
$$= 0.07 \times 0.93 \times 0.93 \times 0.93 \times 0.93$$
$$= 0.07 \times (0.93)^4$$
$$= 0.052$$

So the probability of getting 1 faulty motor among a sample of 5 is 0.052 – approximately 1 in 20 chance of getting a faulty motor among 5.

Does this answer seem correct to you?

No.

Is it too high or too low?

Too low.

You would expect a higher chance of selecting 1 faulty motor among 5. This probability of 0.052 is the probability of getting the first motor faulty and the remaining 4 good. But we do **not** care about the order of the faulty motor, it can be the first motor or the second motor or the third motor etc.

FGGGG or *GFGGG* or *GGFGG* or . . .

How many ways can we choose 1 faulty motor from a sample of 5?

$^5C_1$

Hence the probability of a faulty motor is

$$P(\text{one motor faulty}) = {}^5C_1 \times 0.052$$
$$= 5 \times 0.052 = 0.26$$

Alternatively we can evaluate this more accurately by

$$P(\text{one motor faulty}) = {}^5C_1 \times 0.07 \times (0.93)^4 = 0.262 \quad (3 \text{ s.f.})$$

Similarly working to an appropriate number of places we have:

b $P(2 \text{ motors faulty}) = {}^5C_2(0.07)^2(0.93)^3$
$$= 0.039$$
c $P(3 \text{ motors faulty}) = {}^5C_3(0.07)^3(0.93)^2$
$$= 0.00297$$
d $P(4 \text{ motors faulty}) = {}^5C_4(0.07)^4(0.93)$
$$= 1.116 \times 10^{-4}$$

Example 27 *continued*

e $P(5 \text{ motors faulty}) = {}^5C_5(0.07)^5(0.93)^0$
$$= 0.07^5 = 1.681 \times 10^{-6}$$
f $P(\text{no motors faulty}) = {}^5C_0(0.07)^0(0.93)^5$
$$= 0.93^5 = 0.696$$

Let X be the discrete random variable representing the number of faulty motors and x be the value X takes. Then

$$P(X = x) = {}^5C_x(0.07)^x(0.93)^{5-x}$$

We can represent the probabilities of **Example 27** as in Table 10.

TABLE 10

Number of faulty motors	Probability
0	0.696
1	0.262
2	0.039
3	0.00297
4	1.116×10^{-4}
5	1.681×10^{-6}

This table gives the **probability distribution** for the number of faulty motors. We can denote the above probabilities as follows:

$$P(X = 0) = 0.696, P(X = 1) = 0.262, \ldots$$

where X is the discrete random variable.

? **What is the sum of the probabilities equal to?**

1.

Remember that a discrete random variable is a discrete variable whose associated probabilities all add up to 1.

For **Example 27**, you might also notice that all the terms are from the binomial expansion of

$$(0.93 + 0.07)^5$$

By 7.23 :

$$(0.93 + 0.07)^5 = \underbrace{0.93^5}_{P(X=0)} + \underbrace{{}^5C_1(0.93)^4(0.07)}_{P(X=1)} + \underbrace{{}^5C_2(0.93)^3(0.07)^2}_{P(X=2)}$$

$$+ \underbrace{{}^5C_3(0.93)^2(0.07)^3}_{P(X=3)} + \underbrace{{}^5C_4(0.93)(0.07)^4}_{P(X=4)} + \underbrace{0.07^5}_{P(X=5)}$$

The probability distribution shown in Table 10 is an example of a binomial distribution.

7.23 $(a + b)^n = a^n + {}^nC_1 a^{n-1}b + \ldots + b^n$

? **Why do we call this binomial?**

Because there are only two possible outcomes – a faulty motor or a good motor. Generally we denote two possible outcomes by success and failure. The probabilities associated with these outcomes are normally given by the letters p and q where p is the probability of success and $q = 1 - p$ is the probability of failure. For **Example 27**, a successful outcome is the selection of a faulty motor:

$$p = 0.07 \text{ and } q = 0.93$$

We always have $p + q = 1$.

It might seem confusing to associate success with the selection of a faulty motor, but if we nominate p(success) as the probability required by the question then it keeps each application of the binomial formula consistent. Since in **Example 27** we are interested in finding the probability of a faulty motor, therefore the selection of a faulty motor is considered as a success.

In probability theory the term 'trial' means a single execution of an experiment. Let n denote the number of independent trials ($n = 5$ for **Example 27**). Generally the probability of obtaining x success in n independent trials is given by

16.24 $\qquad P(X = x) = {}^{n}C_{x}p^{x}q^{n-x} = \dfrac{n!}{x!(n - x)!}p^{x}q^{n-x}$

Remember that $q = 1 - p$ and that X is the random variable representing the number of successes in n trials. The probability distribution of 16.24 is called the **binomial distribution** and denoted by $B(n, p)$. Also note that

16.25 $\qquad (q + p)^{n} = P(X = 0) + P(X = 1) + P(X = 2) + \ldots + P(X = n)$
$$= 1$$

We can justify this by

$$(q + p)^{n} = \underbrace{\left(1 - p}_{= q} + p\right)^{n} = (1)^{n} = 1$$

E3 **Applications**

Example 28

From previous tests it is known that the probability of a resistor being out of tolerance is 0.15. From a large batch, a random sample of 20 resistors is chosen so that the probability remains constant. Evaluate the probabilities that

a exactly 7 will be out of tolerance
b at least 3 will be out of tolerance.

Example 28 *continued*

Solution

Let X be the random variable denoting the number of resistors out of tolerance. Consider 'resistor being out of tolerance' as success, so $p = 0.15$, $q = 1 - 0.15 = 0.85$ and $n = 20$. We need to use

16.24 $P(X = x) = {}^nC_x p^x q^{n-x}$

with these values:

⬛ ⋆ $P(X = x) = {}^{20}C_x(0.15)^x(0.85)^{20-x}$

a Substituting $x = 7$ into ⬛⋆ gives

$$P(X = 7) = {}^{20}C_7(0.15)^7(0.85)^{13} = 0.016$$

The probability of exactly 7 resistors being out of tolerance is 0.016 (3 d.p.).

b By 16.17 :

P(at least 3 will be out of tolerance)
$= 1 - P$(less than 3 will be out of tolerance)
$= 1 - P$(0 or 1 or 2 will be out of tolerance)
⬛† $= 1 - \left[P(X = 0) + P(X = 1) + P(X = 2) \right]$

Using ⬛⋆ with values of x equal to 0, 1 and 2 respectively:

$P(X = 0) = 0.85^{20} = 0.039$ $[{}^{20}C_0 = 1]$
$P(X = 1) = 20(0.15)(0.85)^{19} = 0.137$ $[{}^{20}C_1 = 20]$
$P(X = 2) = {}^{20}C_2(0.15)^2(0.85)^{18} = 0.229$

Substituting these into ⬛† gives

P(at least 3 will be out of tolerance) $= 1 - (0.039 + 0.137 + 0.229)$
$= 0.595$ (3 d.p.)

Note that ${}^nC_n = 1, {}^nC_0 = 1$ and ${}^nC_1 = n$. For **Example 28**:

$${}^{20}C_{20} = 1, {}^{20}C_0 = 1 \text{ and } {}^{20}C_1 = 20$$

Example 29

The probability of passing an examination is 0.7. Out of 15 students, evaluate the probabilities that
a all 15 will pass
b none will pass
c at least 12 will pass.

16.17 P(at least n successes) $= 1 - P$(less than n successes)

Example 29 *continued*

Solution

Using

16.24 $P(X = x) = {}^nC_x p^x q^{n-x}$

with $p = 0.7$, $q = 0.3$, $n = 15$ and X is the random variable which represents 'the number of students who pass':

★ $P(X = x) = {}^{15}C_x (0.7)^x (0.3)^{15-x}$

a With $x = 15$:

$$P(X = 15) = {}^{15}C_{15}(0.7)^{15}(0.3)^0$$
$$= 0.7^{15} = 4.748 \times 10^{-3} \quad \text{(4 s.f.)}$$

b Putting $x = 0$ into ★ :

$$P(\text{none will pass}) = P(X = 0) = 0.3^{15} = 1.435 \times 10^{-8} \quad \text{(4 s.f.)}$$

c $P(\text{at least 12 will pass}) = P(12 \text{ pass or } 13 \text{ pass or } 14 \text{ pass or } 15 \text{ pass})$

† $= P(X = 12) + P(X = 13) + P(X = 14) + P(X = 15)$

Evaluating each of these probabilities by ★ gives

$$P(X = 12) = {}^{15}C_{12}(0.7)^{12}(0.3)^3 = 0.170$$
$$P(X = 13) = {}^{15}C_{13}(0.7)^{13}(0.3)^2 = 0.092$$
$$P(X = 14) = {}^{15}C_{14}(0.7)^{14}(0.3) = 0.031$$
$$P(X = 15) = (0.7)^{15} = 4.748 \times 10^{-3}$$

Substituting these into † gives

$$P(\text{at least 12 will pass}) = 0.170 + 0.092 + 0.031 + (4.748 \times 10^{-3})$$
$$= 0.298 \quad \text{(3 s.f.)}$$

SUMMARY

The sum of the probabilities, $p_1, p_2, \ldots, p_n$, that a discrete random variable takes is 1. That is

16.23 $p_1 + p_2 + \ldots + p_n = 1$

Binomial distribution is a probability distribution of a discrete random variable, X, such that

16.24 $P(X = x) = \dfrac{n!}{x!(n-x)!} p^x q^{n-x} \quad (\text{or } {}^nC_x p^x q^{n-x})$

where p is the probability of success, $q = 1 - p$ is the probability of failure and n is the number of independent trials. Also

16.25 $(q + p)^n = P(X = 0) + P(X = 1) + P(X = 2) + \ldots + P(X = n)$

$$= 1$$

Solutions at end of book. Complete solutions available at www.palgrave.com/science/engineering/singh

Exercise 16(e)

1 10% of the resistors produced by a manufacturer are out of tolerance. If a sample of 5 resistors is selected at random, find the probability that
   a exactly 3 resistors are out of tolerance
   b 3 or fewer than 3 resistors are out of tolerance.

2 A component has a 1 in 25 chance of failing. Five components are chosen from a large batch so that the probability of failure remains constant. Determine the probability of
   a 1 component failing
   b more than 3 components failing.

3 The probability of an integrated circuit passing a quality test is 0.86. In a batch of 10 integrated circuits find the probabilities that
   i 2 will fail
   ii 8 will pass
   iii at least 2 will fail
   iv fewer than 2 will fail.

4 Determine the probability of having more boys than girls in a family of 7 children. Assume the probability of having a boy is 0.48.

5 The probability that a student passes a subject is 0.85. If the student takes 8 subjects, what is the probability that he passes more than 6 subjects?

6 a If X is the random variable which follows the binomial distribution with $n = 600$ and $p = 1 \times 10^{-3}$, evaluate (to 4 d.p.)

   i $P(X = 1)$
   ii $P(X = 2)$
   iii $P(X = 3)$

   The Poisson distribution is described by

   $$P(X = x) = \frac{e^{-np}(np)^x}{x!}$$

   b For this distribution, evaluate (consider the above values of n and p)
   i $P(X = 1)$ **ii** $P(X = 2)$ **iii** $P(X = 3)$
   c What do you notice about the results of **a** and **b**?

7 The probability of an England player scoring a penalty is 0.85. What is the probability that an England team will score more than 3 penalties out of 5?

8 The probability density function of a discrete random variable, X, is given by

   $$P(X = x) = kx^4 \quad x = 0, 1, 2$$

   Determine k.

9 The probability density function of a discrete random variable, X, is given by

   $$P(X = x) = kx^3 \quad x = 0, 1, 2, 3$$

   Evaluate the constant k.

10 Let X be the discrete random variable which follows the binomial distribution with parameters n and p. Show that

   $$\sum_{x=1}^{n} xP(X = x) = np$$

SECTION F **Properties of discrete random variables**

By the end of this section you will be able to:
▶ find the mean and variance of a discrete random variable
▶ state the mean and standard deviation of the binomial distribution
▶ apply the Poisson distribution

F1 **Expected value**

Example 30

An experiment is carried out by throwing a die 100 times, and the results are recorded in Table 11.

Score (x)	1	2	3	4	5	6
Frequency	15	12	21	19	17	16

TABLE 11

Determine the probability of each score and the mean score.

Solution

Let p_1, p_2, ..., p_6 denote the probabilities for scores 1, 2, ..., 6 respectively. Since a score of 1 appears 15 out of 100 times we have

$$p_1 = \frac{15}{100} = 0.15$$

Similarly

$$p_2 = \frac{12}{100} = 0.12, p_3 = \frac{21}{100} = 0.21, p_4 = \frac{19}{100} = 0.19, p_5 = \frac{17}{100} = 0.17, p_6 = \frac{16}{100} = 0.16$$

To find the mean we use

16.4 $$\bar{x} = \frac{f_1 x_1 + f_2 x_2 + f_3 x_3 + \ldots + f_n x_n}{f}$$

where $f = f_1 + f_2 + f_3 + \ldots + f_n$ = sum of frequencies. We can rewrite 16.4 as

$$\bar{x} = x_1 \left(\frac{f_1}{f} \right) + x_2 \left(\frac{f_2}{f} \right) + \ldots + x_n \left(\frac{f_n}{f} \right)$$

$$= x_1(p_1) + x_2(p_2) + \ldots + x_n(p_n) \quad \left[\text{Because } \frac{f_k}{f} = p_k \right]$$

Example 30 *continued*

Substituting $x_1 = 1$, $x_2 = 2, \ldots, x_6 = 6$ and $p_1 = 0.15$, $p_2 = 0.12, \ldots, p_6 = 0.16$ into $\bar{x} = x_1(p_1) + x_2(p_2) + \ldots + x_6(p_6)$ gives

$$\bar{x} = (1 \times 0.15) + (2 \times 0.12) + (3 \times 0.21) + (4 \times 0.19) + (5 \times 0.17) + (6 \times 0.16)$$

$$= 3.59$$

We have evaluated the mean score of a die by an experimental approach. In this subsection we look at the mean from a theoretical viewpoint.

The expected value, $E(X)$, is the mean value of a random variable X and is normally denoted by μ, pronounced 'mew'. The symbol $\bar{x}$ is used for the mean of a sample and μ is used for the mean of a random variable.

The expected value, $E(X)$, of a discrete random variable X with values $x_1, x_2, \ldots, x_n$ is defined as

16.26 $E(X) = x_1\, P(X = x_1) + x_2\, P(X = x_2) + \ldots + x_n\, P(X = x_n)$

Notice the similarity with the formula for the mean $\bar{x}$ above:

$$\bar{x} = x_1(p_1) + x_2(p_2) + \ldots + x_n(p_n)$$

Example 31

Evaluate $E(X)$ for $P(X = x) = \dfrac{1}{6}$ for $x = 1, 2, 3, 4, 5$ and 6 where X represents the score on a die.

Solution

Writing the probability for each x gives Table 12.

TABLE 12	x	1	2	3	4	5	6
	$P(X = x)$	$\dfrac{1}{6}$	$\dfrac{1}{6}$	$\dfrac{1}{6}$	$\dfrac{1}{6}$	$\dfrac{1}{6}$	$\dfrac{1}{6}$

Multiplying each x with its associated probability and summing yields:

$$E(X) = \left(1 \times \frac{1}{6}\right) + \left(2 \times \frac{1}{6}\right) + \left(3 \times \frac{1}{6}\right) + \left(4 \times \frac{1}{6}\right) + \left(5 \times \frac{1}{6}\right) + \left(6 \times \frac{1}{6}\right) = 3.5$$

Since the probability distribution of Table 12 is symmetric, the mean value, $E(X)$ ($=\mu$), is going to lie midway between 3 and 4, hence 3.5.

Notice the discrepancy between the mean, $\bar{x} = 3.59$, and the expected value, $E(X) = 3.5$, of the score on a die.

The value of $E(X)$ is determined by the underlying theoretical probabilities but $\bar{x}$ is calculated in an experiment by using observed frequencies which are only approximately correct.

Example 32 *signal processing*

The probability distribution of a sampled signal is as given in Table 13.

TABLE 13									
x (volts)	1	2	3	4	5	6	7	8
$P(X = x)$	0.10	0.12	0.15	0.20	0.17	0.13	0.07	0.06

Determine the mean voltage, μ.

Solution

As in the above example, we have

$$\mu = (1 \times 0.1) + (2 \times 0.12) + (3 \times 0.15) + (4 \times 0.2) + (5 \times 0.17)$$
$$+ (6 \times 0.13) + (7 \times 0.07) + (8 \times 0.06)$$

$$= 4.19 \text{ volts}$$

The mean voltage is $\mu = 4.19$ V (2 d.p.).

Generally we can write the formula **16.26** in shorthand notation as

16.27 $$E(X) = \sum_{j=1}^{n} x_j\, P(X = x_j)$$

The Greek symbol $\displaystyle\sum_{j=1}^{n}$ means sum from $j = 1$ to n.

The expected value of a function $g(X)$ of the discrete random variable X with particular values $x_1, x_2, \ldots, x_n$ is defined by

16.28 $$E[g(X)] = g(x_1)P(X = x_1) + g(x_2)P(X = x_2) + \ldots + g(x_n)P(X = x_n)$$

Notice the similarity with the formula for $E(X)$. This time we multiply the probabilities by the function $g(x_1), g(x_2), \ldots, g(x_n)$. Formula **16.28** might seem difficult to comprehend but the following example should make it more digestible.

Example 33

Consider the discrete random variable X with a probability density function

$$P(X = x_j) = \frac{x_j}{10} \text{ for } x_j = 1, 2, 3 \text{ and } 4$$

Determine $E(X^2)$.

Solution

Substituting the given values of x_j yields Table 14.

Example 33 *continued*

x_j	1	2	3	4
$P(X = x_j)$	$\dfrac{1}{10}$	$\dfrac{2}{10}$	$\dfrac{3}{10}$	$\dfrac{4}{10}$

TABLE 14

To find $E(X^2)$ we sum the product of x_j^2 and its associated probability:

$$E(X^2) = \left(1^2 \times \frac{1}{10}\right) + \left(2^2 \times \frac{2}{10}\right) + \left(3^2 \times \frac{3}{10}\right) + \left(4^2 \times \frac{4}{10}\right) = 10$$

The expected value function has the following properties for the random variables X and Y:

$$E(X + Y) = E(X) + E(Y)$$

The expected value of $X + Y$ is equal to the expected value of X plus the expected value of Y. Similarly

$$E(X - Y) = E(X) - E(Y)$$
$$E(kX) = kE(X) \text{ where } k \text{ is a constant}$$

The general formula is given by

16.29 $$E\big[kf(X) + m\big] = kE\big[f(X)\big] + m$$

where k and m are constants and f is any function of the random variable X. Formula 16.29 states that we can take out a constant, k, and the expected value of a constant, $E(m) = m$.

F2 **Variance**

The variance, σ^2, of a discrete random variable, X, is defined by

16.30 $$\sigma^2 = E(X - \mu)^2$$

where μ is the mean of X. (σ^2 is pronounced 'sigma squared'.)

Example 34

Show that

$$\sigma^2 = E(X^2) - \mu^2$$

where μ is the mean of X.

Solution

Applying 16.30 and expanding gives

$$E(X - \mu)^2 = E(X^2 - 2\mu X + \mu^2)$$
$$\underset{\text{by } 16.29}{=} E(X^2) - 2\mu E(X) + \mu^2$$
$$= E(X^2) - 2\mu\mu + \mu^2 \quad [\text{Remember that } \mu = E(X)]$$
$$= E(X^2) - \mu^2 \quad\quad\quad [\text{Simplifying}]$$

This is an important result:

16.31 $\qquad \sigma^2 = E(X^2) - \mu^2$

where

16.32 $\qquad E(X^2) = \sum_{\text{all } x} x^2 P(X = x)$

Example 35

A discrete random variable X has the probability distribution shown in Table 15.

TABLE 15	x_j	5	6	7	8
	$P(X = x_j)$	0.17	0.3	0.35	0.18

Find the mean, μ, and variance, σ^2.

Solution

Remember that $\mu = E(X)$ and so multiplying each x_j by the associated probability and adding together gives

$$\mu = (5 \times 0.17) + (6 \times 0.3) + (7 \times 0.35) + (8 \times 0.18)$$
$$= 6.54$$

Does this seem reasonable?

Since most of the probability is concentrated between 6 and 7, a mean value of 6.54 seems reasonable.

Variance: $\qquad \sigma^2 = E(X^2) - 6.54^2$

Using Table 15 we have

$$E(X^2) = (5^2 \times 0.17) + (6^2 \times 0.3) + (7^2 \times 0.35) + (8^2 \times 0.18)$$
$$= 43.72$$

Substituting $E(X^2) = 43.72$ into $\sigma^2 = E(X^2) - 6.54^2$ gives

$$\sigma^2 = 43.72 - 6.54^2 = 0.948 \qquad \text{(3 d.p.)}$$

We can also use a calculator to evaluate the mean and variance of the distribution in Table 15. We first place the calculator into SD mode and PRESS SHIFT MODE 1 = 5 SHIFT , 0.17 M+ ... 8 SHIFT , 0.18 M+ then PRESS SHIFT 2 1 = to give the mean, and SHIFT 2 2 = to give the SD. Square this value to find the variance.

F3 **Poisson distribution**

The Poisson distribution can be viewed as an approximation to the binomial distribution for large n and small p, that is a large sample size with small probability. It is also a useful distribution in its own right.

> The term Poisson comes from the French mathematician **Simeon-Denis Poisson** (1781–1840). He discovered the distribution towards the end of his life.

The probability $P(X = x)$ of the discrete random variable, X, is defined as

16.33 $\qquad P(X = x) = \dfrac{e^{-\mu}\mu^x}{x!}$ for $x = 0, 1, 2, 3, \ldots$

where μ is the mean of the distribution. In this case we say that the discrete random variable, X, follows the Poisson distribution.

Example 36

Let X be the number of failures of a system. If X follows a Poisson distribution with a mean of 2.5 failures per year, determine the probabilities of

a no failures in a year
b at most 2 failures in a year.

Solution
Using

16.33 $\qquad P(X = x) = \dfrac{e^{-\mu}\mu^x}{x!}$

with $\mu = 2.5$ gives

$$P(X = x) = \frac{e^{-2.5}2.5^x}{x!}$$

a $P(\text{no failures}) = P(X = 0) = \dfrac{e^{-2.5}(2.5)^0}{0!} = 0.082 \quad$ (3 d.p.)

b $P(\text{at most 2 failures}) = P(X = 0) + P(X = 1) + P(X = 2)$

$$= \frac{e^{-2.5}(2.5)^0}{0!} + \frac{e^{-2.5}(2.5)^1}{1!} + \frac{e^{-2.5}(2.5)^2}{2!}$$
$$= 0.544 \quad \text{(3 d.p.)}$$

Remember that $0! = 1$.

The mean value, μ, for the binomial distribution with parameters n and p is given by

16.34 $\qquad \mu = np$

This result is shown in question **10** of **Exercise 16(e)**:

$$\mu = \sum xP(X = x)$$

where $x = 0, 1, 2, \ldots, n$ and $P(X = x) = {}^nC_x\, p^x\, q^{n-x}$ (remember that $q = 1 - p$).

The variance, σ^2, is given by

16.35 $\sigma^2 = npq$

By taking the square root of both sides we obtain the standard deviation, σ:

16.36 $\sigma = \sqrt{npq}$

Example 37

A manufacturer produces a batch of 1000 calculators. The probability that a calculator is defective is 1×10^{-3}. Evaluate the probability that the batch contains 3 defective calculators by

i binomial distribution   **ii** Poisson distribution.

Comment upon your results.

Solution

Since n is large (1000) and p is small (1×10^{-3}) we can use the Poisson approximation to the binomial.

i Let X be the number of defective calculators. Applying the binomial distribution:

16.24 $P(X = x) = {}^nC_x p^x q^{n-x}$

with $n = 1000$, $p = 1 \times 10^{-3}$ and $q = 1 - (1 \times 10^{-3}) = 0.999$ gives

$$P(X = x) = {}^{1000}C_x (1 \times 10^{-3})^x (0.999)^{1000-x}$$

Since we want to find the probability of 3 defective calculators, therefore $x = 3$:

$$P(X = 3) = {}^{1000}C_3 (1 \times 10^{-3})^3 (0.999)^{997}$$
$$= 0.06128 \quad \text{(4 s.f.)}$$

ii Using the Poisson approximation, we will first need to find μ (mean value). **What is μ in this case?**
By 16.34:

$$\mu = np$$
$$= 1000 \times 1 \times 10^{-3} = 1$$

Using

16.33 $P(X = x) = \dfrac{e^{-\mu}\mu^x}{x!}$

with $\mu = 1$ and $x = 3$ gives

$$P(X = 3) = \dfrac{e^{-1}1^3}{3!} = 0.06131 \quad \text{(4 s.f.)}$$

Note that the answers to **i** and **ii** agree to 3 s.f.

Example 37 illustrates that for small p and large n it is easier to use the Poisson distribution as an approximation to the binomial distribution. The mean and variance of the Poisson distribution are both equal to μ.

SUMMARY

The expected value, $E(X)$, is the mean value of a random variable, X, and is given by

16.26 $E(X) = x_1 P(X = x_1) + x_2 P(X = x_2) + \ldots + x_n P(X = x_n)$

The variance, σ^2, of a random variable, X, is given by

16.31 $\sigma^2 = E(X^2) - \mu^2$

where $\mu = E(X)$.

We say a discrete random variable, X, follows a Poisson distribution if

16.33 $P(X = x) = \dfrac{e^{-\mu} \mu^x}{x!}$ $(x = 0, 1, 2, 3, \ldots)$

Exercise 16(f)

Solutions at end of book. Complete solutions available at www.palgrave.com/science/engineering/singh

1 The probability density function of a discrete random variable, X, is given by

$$P(X = x) = \frac{1}{5} \qquad x = 1, 2, 3, 4 \text{ and } 5$$

Find the expected value $E(X)$. What do you notice about your result?

2 The probability distribution of a discrete random variable, X, is given by:

x	1	2	3	4	5	6
$P(X = x)$	0.15	0.21	0.11	0.36	0.04	0.13

Determine $E(X^2)$.

3 ▦ [signal processing]   The probability distribution of a sampled signal is given by:

x (volts)	1	2	3	4	5
$P(X = x)$	0.1	0.2	0.4	0.2	0.1

Determine the mean, μ, and standard deviation, σ.

What do you notice about the mean value?

4 ▦ [signal processing]   The probability distribution of a sampled signal is given by:

x (volts)	1	2	3	4	5	6	7	8
$P(X = x)$	0.10	0.12	0.15	0.20	0.17	0.13	0.07	0.06

Determine the mean, μ, and standard deviation, σ.

5 ▦ [signal processing]   A random signal has the following probability distribution:

x (volts)	0	1	2	3	4
$P(X = x)$	0.35	0.33	0.19	0.12	0.01

Estimate a value for the mean.

Calculate the mean and standard deviation of the distribution.

6 The probability density function of a discrete random variable is given by

$$P(X = x) = kx^2 \qquad x = 1, 2, 3 \text{ and } 4$$

Determine

i k **ii** $E(X)$
iii standard deviation   **iv** variance

Solutions at end of book. Complete solutions available at www.palgrave.com/science/engineering/singh

Exercise **16(f) continued**

7 The probability distribution of a discrete random variable is given by:

x	1	2	3	4
$P(X = x)$	0.1	0.3	0.2	0.4

Evaluate

i $E(X)$ **ii** $E(X^2)$ **iii** $E(X^2 + X)$
iv $E(5X^2 + 7X + 3)$

8 i Show that $E(k) = k$ where k is a constant.
ii Show 16.29:

$$E[af(X) + b] = aE[f(X)] + b$$

9 Let X follow a Poisson distribution with a mean of 1.6. Find the following probabilities:

a $P(X = 0)$ **b** $P(X = 1)$ **c** $P(X \le 2)$

d $P(X = 10)$

10 Consider a communication network where the number of errors follow the Poisson distribution. If the probability of an error in a bit is 1×10^{-3}, find the probability of
a no errors in 4096 bits
b more than 2 errors in 4096 bits.

SECTION G **Applications of continuous random variables**

By the end of this section you will be able to:
▶ understand the term *continuous random variable*
▶ see the relationship between probability and its density function
▶ apply the above to reliability engineering
▶ find various functions of reliability engineering

This section is difficult compared to the previous sections. There are a number of functions related to reliability engineering which are used here. It is very easy to get confused between the various functions in this particular field.

G1 **Continuous random variable**

? **Can you think of any variables that are continuous?**

Output of an analogue system, mass, lifetime of a component etc.

A continuous random variable, T, is a variable that can have any values between two end points, unlike discrete random variables which can only take certain values.

The probability of a discrete random variable is given by a sum.

The probability of a continuous random variable is given by an integral.

The probability density function of T, with a particular value t, is normally denoted by $f(t)$. The probability of T having values between a and b is given by the area under the probability density function, $f(t)$, as shown shaded in Figure 22.

Fig. 22

The area can be evaluated by integrating:

16.37
$$P(a \leq T \leq b) = \int_a^b f(t)dt$$

which means that the probability of T having values between a and b is the integral $\int_a^b f(t)dt$. Consider the following example.

Let T represent the time that a component lasts, or the failure time of a component. Then $P(a \leq T \leq b)$ is the probability that the component will fail between (or at) the times a and b, and is evaluated by $\int_a^b f(t)dt$.

The other property is

16.38
$$P(-\infty < T < \infty) = \int_{-\infty}^{\infty} f(t)dt = 1$$

The probability of T having any value is 1. If T represents the failure time of a component then T can only have positive values and zero. Hence 16.38 becomes

16.39
$$P(0 \leq T < \infty) = \int_0^{\infty} f(t)dt = 1$$

This means that the component will eventually fail.

G2 Reliability engineering

Many of the engineering applications in this area lie in the field of reliability engineering.

Example 38 *reliability engineering*

The density function of failure time of a component is given by
$$f(t) = 0.3e^{-0.3t}$$
where t is in years. In reliability engineering, $f(t)$ is normally called the failure density function rather than the probability density function.

Find the probability that the component will fail within 5 years of operation.

Solution

Let T be the failure time of the component. Using

16.37
$$P(a \leq T \leq b) = \int_a^b f(t)dt$$

Example 38 *continued*

with $a = 0$, $b = 5$ and $f(t) = 0.3e^{-0.3t}$ gives

$$P(0 \leq T \leq 5) = 0.3 \int_0^5 e^{-0.3t}dt$$

$$= 0.3 \left[\frac{e^{-0.3t}}{-0.3} \right]_0^5 \quad \left[\text{Integrating by} \int e^{kt}dt = \frac{e^{kt}}{k} \right]$$

$$= -\left[e^{-0.3 \times 5} - e^0 \right] \quad \text{[Substituting]}$$

$$= -\left[e^{-1.5} - 1 \right] \quad \text{[Because } e^0 = 1]$$

Remember that taking a minus inside the brackets changes the signs:

$$P(0 \leq T \leq 5) = 1 - e^{-1.5} = 0.777 \quad \text{(3 d.p.)}$$

Hence there is more than a 77% chance that the component will fail within 5 years.

Example 39 *reliability engineering*

The failure density function, $f(t)$, for a particular set of components is given by

$$f(t) = \lambda e^{-\lambda t} \quad (t \text{ in years})$$

where $\lambda > 0$ is a constant. Show that the probability of a component failing within 10 years is given by

$$1 - e^{-10\lambda}$$

Solution

Using

16.37 $\quad P(a \leq T \leq b) = \int_a^b f(t)dt$

with $a = 0$, $b = 10$ and $f(t) = \lambda e^{-\lambda t}$ yields

$$P(0 \leq T \leq 10) = \lambda \int_0^{10} e^{-\lambda t}\, dt$$

$$= \lambda \left[\frac{e^{-\lambda t}}{-\lambda} \right]_0^{10} \quad \text{[Integrating]}$$

$$= \frac{-\lambda}{\lambda} \left[e^{-\lambda t} \right]_0^{10} = -(e^{-10\lambda} - e^0) = 1 - e^{-10\lambda}$$

The cumulative distribution function, $F(t)$, of a continuous random variable, T, is defined as

16.40 $\quad F(t) = P(T \leq t)$

For example let T be the continuous random variable of failure time of a component. Then the cumulative distribution function, $F(t)$, is given by

16.41 $\quad F(t) = P(0 \leq T \leq t)$

because T **cannot** have negative values. This distribution function in reliability engineering is called the **failure** distribution function. $F(t)$ gives the probability of a component failing within (or at) t years. We will use the word 'within' to mean $0 \le T \le t$ rather than $0 < T < t$.

For **Example 38**, we evaluated $F(5)$ – the probability of the component failing within 5 years.

Example 40 *reliability engineering*

The failure distribution function of a system is given by

$$F(t) = 1 - e^{-0.5t}$$

Determine the probability that the system will fail within 6 years.

Solution

Substituting $t = 6$ into our given function

$$F(t) = 1 - e^{-0.5t}$$

yields

$$F(6) = 1 - e^{-0.5 \times 6}$$
$$= 1 - e^{-3}$$
$$= 0.95 \quad \text{(2 d.p.)}$$

The probability that the system will fail within 6 years is 0.95 (2 d.p.).

? **What is the relationship between $f(t)$ and $F(t)$?**

Since $F(t)$ is the area under the graph of $f(x)$ between 0 and 1 we have

Fig. 23

16.42 $$F(t) = \int_0^t f(x)\,dx$$

Remember that $F(t)$ gives the probability of a component failing within t years.

Differentiating this we have

16.43 $$f(t) = \frac{dF}{dt}$$

G3 Reliability functions

In reliability engineering the continuous random variable, T, represents the failure time of a component. There are three particular functions which need to be defined:

1 Reliability function, $R(t)$, which gives the probability of a component lasting more than t years, that is does not fail within t years.

$$P(\text{does not fail within } t \text{ years}) = 1 - P(\text{fails within } t \text{ years})$$
$$= 1 - F(t)$$

where $F(t)$ is the failure distribution function. Hence

16.44 $R(t) = 1 - F(t)$

2 Hazard function, $h(t)$, is the failure rate of the component and is defined as

16.45 $h(t) = \dfrac{f(t)}{R(t)}$

Remember that $f(t)$ is the derivative of $F(t)$.

3 The Mean Time To Failure, $MTTF$, is defined as

16.46 $MTTF = \displaystyle\int_0^\infty R(t)\mathrm{d}t$

Generally these functions are **only** defined over a strict domain.

Example 41 *reliability engineering*

Let the failure distribution function, $F(t)$, for a system be given by

$$F(t) = 1 - e^{-t/2}$$

Determine the reliability function, $R(t)$, for t equal to 5 years. This gives the probability that the system will last more than 5 years.

Solution

Using

16.44 $R(t) = 1 - F(t)$

with $F(t) = 1 - e^{-t/2}$, is given by

$$R(t) = 1 - (1 - e^{-t/2})$$
$$= 1 - 1 + e^{-t/2}$$
$$= e^{-t/2}$$
$$R(5) = e^{-5/2} = 0.082 \quad (3 \text{ d.p.})$$

The probability that the component will last more than 5 years is 0.082.

Example 42 *reliability engineering*

The failure distribution function, $F(t)$, of a building system is given by

$$F(t) = 1 - e^{-t/100}$$

where t is in years. Determine

i the reliability function, $R(t)$
ii the hazard function, $h(t)$.

Example 42 *continued*

Solution

i By using $R(t) = 1 - F(t)$ we have

$$R(t) = 1 - (1 - e^{-t/100}) = 1 - 1 + e^{-t/100} = e^{-t/100}$$

ii $h(t) = \dfrac{f(t)}{R(t)}$. **What do we need to find first?**

$$f(t)$$

What is $f(t)$?

Using 16.43 , it means we need to differentiate $F(t)$:

$$f(t) = \frac{d}{dt}\left(1 - e^{-t/100}\right) = 0 - \underbrace{\left(-\frac{1}{100}\right)e^{-t/100}}_{\text{by } 6.11} = \frac{1}{100}e^{-t/100}$$

Substituting these into $h(t) = \dfrac{f(t)}{R(t)}$ gives

$$h(t) = \frac{\dfrac{1}{100}e^{-t/100}}{e^{-t/100}} = \frac{1}{100} \quad \text{[Cancelling } e^{-t/100}\text{]}$$

We can say the system has a constant failure rate because there are no variables in the answer.

Example 43 *reliability engineering*

Determine the *MTTF* (Mean Time To Failure) of the failure density function

$$f(t) = 0.2 - 0.02t \quad 0 \le t \le 10$$

where t is in years.

Solution

What do we need to evaluate first?

$R(t)$ and $F(t)$ where these are defined for $0 \le t \le 10$.

This time we have $f(t)$ and need $F(t)$, therefore we need to integrate $f(t)$ to find $F(t)$. This is the reverse of the previous example.

6.11 $(e^{kt})' = ke^{kt}$ 16.43 $f(t) = dF/dt$

Example 43 *continued*

By

16.42 $$F(t) = \int_0^t f(x)\,dx$$

with $f(x) = 0.2 - 0.02x$ (replacing t with x in the given function), we have

$$F(t) = \int_0^t (0.2 - 0.02x)\,dx$$

$$= \left[0.2x - \frac{0.02x^2}{2} \right]_0^t \quad \text{[Integrating]}$$

$$F(t) = 0.2t - 0.01t^2 \quad \text{[Substituting]}$$

Using $R(t) = 1 - F(t)$ with

$$F(t) = 0.2t - 0.01t^2$$

gives

$$R(t) = 1 - (0.2t - 0.01t^2)$$
$$= 1 - 0.2t + 0.01t^2$$

Next we use $MTTF = \int_0^\infty R(t)\,dt$ with $R(t) = 1 - 0.2t + 0.01t^2$ and the functions are only valid for $0 \le t \le 10$:

$$MTTF = \int_0^\infty (1 - 0.2t + 0.01t^2)\,dt$$

$$= \int_0^{10} (1 - 0.2t + 0.01t^2)\,dt \quad \text{[Changing limits]}$$

$$= \left[t - \frac{0.2t^2}{2} + \frac{0.01t^3}{3} \right]_0^{10} \quad \text{[Integrating]}$$

$$= 10 - \left(\frac{0.2 \times 10^2}{2} \right) + \left(\frac{0.01 \times 10^3}{3} \right) \quad \text{[Substituting]}$$

$$= \frac{10}{3} \text{ years}$$

The Mean Time to Failure, $MTTF$, is just over 3 years.

Example 44 *reliability engineering*

The failure density function for a set of components is given by

$$f(t) = 0.2(1 - 0.1t) \quad 0 \le t \le 10$$

where t is in years.

  i Determine the reliability function, $R(t)$, and the hazard function, $h(t)$.
 ii Find the probability that a component does not fail within 6 years.
iii In what time period will **all** the components fail?

Example 44 *continued*

Solution

i Note that by multiplying the brackets by 0.2 we have the same failure density function as **Example 43**:

$$f(t) = 0.2 - 0.02t \quad 0 \le t \le 10$$

Thus from above we have

$$R(t) = 1 - 0.2t + 0.01t^2$$

Substituting these into $h(t) = \dfrac{f(t)}{R(t)}$ gives the hazard function:

$$h(t) = \frac{0.2 - 0.02t}{1 - 0.2t + 0.01t^2} \quad 0 \le t \le 10$$

ii To find the probability that a component does not fail within 6 years we substitute $t = 6$ into $R(t) = 1 - 0.2t + 0.01t^2$:

$$P(\text{does not fail within 6 years}) = R(6)$$

$$= 1 - (0.2 \times 6) + (0.01 \times 6^2) = 0.16$$

iii The given density function is

$$f(t) = 0.2(1 - 0.1t) \quad 0 \le t \le 10$$

which says that a component will not last more than 10 years. So the fact that our limits are set at maximum 10 tells us that all components will fail within 10 years.

SUMMARY

The probability of a continuous random variable, T, having values between a and b is given by

16.37 $\qquad P(a \le T \le b) = \displaystyle\int_a^b f(t)dt$

where $f(t)$ is the density function.

For the cumulative distribution function, $F(t)$, we have

16.40 $\qquad F(t) = P(T \le t)$

In reliability engineering $f(t)$ and $F(t)$ are called the failure density function and failure distribution function respectively.

The reliability function, $R(t)$, is defined by

16.44 $\qquad R(t) = 1 - F(t)$

The hazard function is defined as

16.45 $\qquad h(t) = \dfrac{f(t)}{R(t)}$

Mean Time To Failure, $MTTF$, is defined as

16.46 $\qquad MTTF = \displaystyle\int_0^\infty R(t)dt$

Solutions at end of book. Complete solutions available at www.palgrave.com/science/engineering/singh

Exercise **16(g)**

All questions are in the field of [reliability engineering].

1 Find the probability that a component will fail within 2 years for the following density functions:

  a $f(t) = 0.5e^{-0.5t}$ **b** $f(t) = e^{-t}$

  c $f(t) = 0.1e^{-0.1t}$

2 For the following failure distribution functions of a system determine the probabilities that the system will fail within 5 years:

  a $F(t) = 1 - e^{-0.2t}$ **b** $F(t) = 1 - e^{-0.6t}$

  c $F(t) = 1 - e^{-t/25}$

3 For the following failure distribution functions of a system determine the probabilities that the system will last more than 6 years:

  a $F(t) = 1 - e^{-t/25}$ **b** $F(t) = 1 - e^{-t/2}$

  c $F(t) = 1 - e^{-0.1t}$

4 The reliability function, $R(t)$, of a component is given by

$$R(t) = 1 - 0.04t^2 \quad 0 \le t \le 5$$

where t is in years.

  i Evaluate the Mean Time To Failure, *MTTF*

  ii Find the hazard function, $h(t)$.

5 The hazard function

$$h(t) = 0.1t$$

Another definition of the reliability function is

$$R(t) = \exp\left[-\int_0^t h(x)\,dx \right]$$

Determine $R(t)$.

6

Fig. 24

Figure 24 shows a failure density function $f(t)$. Write $f(t)$ as an equation.

Also determine the failure distribution function, $F(t)$, reliability function, $R(t)$, hazard function, $h(t)$, and *MTTF*.

7 Determine *MTTF* for the following failure density functions:

  a $f(t) = 0.4\left(1 - \dfrac{t}{5} \right) \qquad 0 \le t \le 5$

  b $f(t) = \dfrac{1}{3}e^{-t/3} \qquad t \ge 0$

where t is in years. (Remember that $\lim\limits_{x \to \infty} (e^{-x}) = 0$.)

8 The median time to failure, t_{med}, is defined as

$$R(t_{\text{med}}) = 0.5$$

where $R(t)$ is the reliability function. Determine t_{med} for

$$f(t) = 0.1e^{-0.1t} \quad (t \text{ is in years})$$

9 For the reliability function, $R(t) = e^{-\lambda t}$, show that

$$MTTF = \frac{1}{\lambda} \text{ where } \lambda > 0$$

> ### SECTION H **Normal distribution**
>
> By the end of this section you will be able to:
> ▶ use the standard normal distribution to find probabilities
> ▶ examine examples which follow the normal distribution
> ▶ apply the standard normal distribution table to questions of probability

H1 Introduction

Suppose we measured the heights of 500 male adults.

What would the probability distribution curve look like?

It would be symmetric about the mean because it seems equally likely that an individual will be 1 and 2 inches lower or 1 and 2 inches higher than the mean.

What happens as you move further away from the mean?

It is more likely that an individual is 1 or 2 inches lower than the mean rather than 1 or 2 feet lower than the mean. The distribution might look like Fig. 25.

Fig. 25

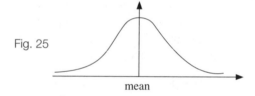

mean

As you move further away from the mean, there is less likelihood of finding an individual of that height. There could be many curves which fit this description but the one we examine in this section is called the normal distribution. The curve of Fig. 25 is a normal distribution which means it is **symmetric** about the mean and tails away at each extreme, and the probability density function, $f(x)$, is defined by

16.47
$$f(x) = \frac{1}{\sigma\sqrt{2\pi}} e^{-\frac{(x-\mu)^2}{2\sigma^2}}$$

where μ is the mean and σ is the standard deviation. Hence the graph of $f(x)$ is as shown in Fig. 26.

Fig. 26

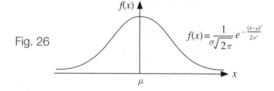

$$f(x) = \frac{1}{\sigma\sqrt{2\pi}} e^{-\frac{(x-\mu)^2}{2\sigma^2}}$$

Trying plotting $f(x)$ for different values of μ and σ on a graphical calculator. If you have MATLAB you can examine the normal distribution graph in one of the demonstrations. Try changing the values of the variables to see what happens to the graph.

H2 Standard normal distribution

Different μ and σ give different normal distribution curves.

Figure 27 shows normal distribution curves with the same mean, μ, but different standard deviations, $\sigma = 1$ and $\sigma = \sqrt{2}$.

Dealing with formula 16.47 is difficult. It is easier to use tables that are produced for the standard normal distribution which has mean = 0 and standard deviation = 1. We convert our normal distribution to the **standard** normal distribution by using the formula

16.48 $z = \dfrac{x - \mu}{\sigma}$

where z represents the number of standard deviations away from the mean and x is our continuous variable which follows a normal distribution.

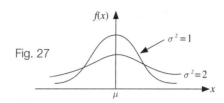

Fig. 27

We have carried out two operations in moving from **a** to **b** in Fig. 28:

1 From our given x we have subtracted μ so that the mean is shifted to 0.
2 Then we have divided our result of **1** by the standard deviation, σ, which gives $\sigma = 1$.

Fig. 28

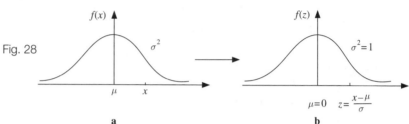

With these two operations we obtain the standard normal distribution.

The standard normal distribution table (see **Appendix**, page 821) gives the area under the curve as shown in Fig. 29.

Fig. 29

The area A represents the probability of $P(z < a)$ and is given by

16.49 $P(z < a) = \text{Area } A$

because the normal distribution is a probability distribution.

Suppose we want to find the area corresponding to $z < 2.5$, we:

1 look down the z (left-hand) column of the **Appendix** and find 2.5
2 look across at the 0 column and we find .99379.

Hence the area under the normal distribution curve for $z < 2.5$ is 0.99379.

If we want to find the area corresponding to $z < 2.54$ then it is the same procedure as above but this time we look at the 4 column across from 2.5 and the area is 0.99446.

The total area covered under the curve is 1. Moreover $f(z)$ is symmetric about $z = 0$. Hence 0.5 of the area lies to the left of 0 and 0.5 to the right.

Example 45

Suppose that z follows the standard normal distribution. Calculate the following probabilities by using the table in the **Appendix**:

a $P(z < 0.5)$ **b** $P(z > 0.5)$ **c** $P(z < 1)$

Example 45 *continued*

Solution

We can illustrate each of these as the standard normal distribution.

a Using the table of the **Appendix** with $z < 0.5$ gives 0.6915. Thus

$$P(z < 0.5) = 0.6915 \text{ (see Fig. 30)}$$

The value from the table gives the probability for $z < 0.5$.

b Since the total area under the curve is 1 we have

$$P(z < 0.5) + P(z > 0.5) = 1$$
$$P(z > 0.5) = 1 - P(z < 0.5) = 1 - 0.6915 = 0.3085.$$

This represents the area to the right of $z = 0.5$ in Fig. 30.

c Similarly we have Fig. 31.

$$P(z < 1) = 0.8413$$

Remember that the area found from the table corresponds to the probability.

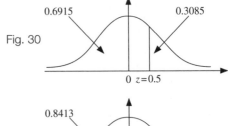

Fig. 30

Fig. 31

Example 46

Suppose z follows the standard normal distribution. Evaluate the following probabilities:

a $P(z > -1)$ **b** $P(z < -1)$ **c** $P(-0.5 < z < 0.1)$

Solution

The table in the Appendix *only* gives the area for positive values of z, so how are we going to evaluate $P(z > -1)$?

Since the normal distribution is symmetric we can examine $z < 1$.

a $z = -1$ where the minus sign means that the z value is below the mean (Fig. 32).

Fig. 32

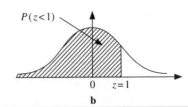

Example 46 *continued*

Because of symmetry, the area represented by $P(z > -1)$ is the same as the area represented by $P(z < 1)$. Thus by using the table we have

$$P(z > -1) = P(z < 1) = 0.8413$$

b $P(z < -1)$ is equal to the area on the left of $z = -1$ in Fig. 32**a**. By **a** and knowing that the total area under the curve is 1, we have

$$P(z < -1) = 1 - 0.8413 = 0.1587$$

c See Fig. 33.

Fig. 33

We can find the shaded area from Fig. 34.

Fig. 34

Thus

$$P(-0.5 < z < 0.1) = P(z < 0.1) - P(z < -0.5)$$

By using the table

$$P(z < 0.1) = 0.5398$$

We also need to determine $P(z < -0.5)$. Remember that the table does not give the probability for negative z values, therefore

$$P(z < -0.5) = P(z > 0.5)$$
$$= 1 - P(z < 0.5)$$
$$= 1 - 0.6915 = 0.3085$$

Substituting these into ▪ * gives

$$P(-0.5 < z < 0.1) = P(z < 0.1) - P(z < -0.5)$$
$$= 0.5398 - 0.3085 = 0.2313$$

Example 47

The examination marks of an engineering module follow a normal distribution with mean $\mu = 40$ and the standard deviation $\sigma = 5$. If 1000 students took the examination and the pass mark is 35, how many students passed the examination?

Solution

This time we do not have the standard normal distribution as in the above examples because the mean, $\mu = 40$ and standard deviation, $\sigma = 5$. We need to convert our data

Example 47 *continued*

into the standard normal distribution so that we can use the table. **How?**

By using

16.48 $$z = \frac{x - \mu}{\sigma}$$

Our x value is the pass mark 35. Thus we have

$$z = \frac{x - 40}{5}$$

$$= \frac{35 - 40}{5}$$

$$z = -1$$

Fig. 35

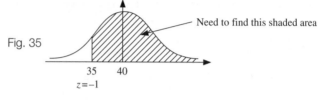

Need to find this shaded area

35   40
z=−1

We therefore have the situation shown in Fig. 35.

The area to the right of 35 in Fig. 35 represents the students who have passed the examination. Using the table of the **Appendix** with $z = 1$ gives 0.8413. We consider $z = 1$ rather than $z = -1$ because the normal distribution is symmetric and the minus sign only indicates that z is to the left of the mean. The area above $z = -1$ is 0.8413.

The probability of a student passing the examination is 0.8413, therefore the number of students who passed the examination is

$$= 0.8413 \times 1000$$

$$= 841.3$$

841 candidates passed the examination.

Example 48

The resistance values of resistors in a batch follow a normal distribution with a mean of (1×10^3) Ω and a standard deviation of 100 Ω. Evaluate the probability of a resistor having a resistance of

a more than 1.2 kΩ

b less than 0.8 kΩ

c between 0.9 kΩ and 1.1 kΩ.

Solution

a Using

16.48 $$z = \frac{x - \mu}{\sigma}$$

with $\mu = 1 \times 10^3$ and $\sigma = 100$ we have

Example 48 *continued*

$$z = \frac{x - (1 \times 10^3)}{100}$$

$$= \frac{(1.2 \times 10^3) - (1 \times 10^3)}{100} \quad \text{[Substituting } x = 1.2 \times 10^3\text{]}$$

$$z = 2$$

The table of the **Appendix** gives an area of 0.9772 for $z < 2$. Remember that the area corresponds to probability. Hence the probability of a resistance value of more than 1.2 kΩ is

$$P(\text{Resistance} > 1.2 \text{ k}\Omega) = 1 - P(\text{Resistance} < 1.2 \text{ k}\Omega)$$

$$= 1 - 0.9772 = 0.0228$$

See Fig. 36.

b 0.8 kΩ is below the mean by the same amount as 1.2 kΩ is above.
For $x = 0.8$ kΩ, $z = -2$: since the normal distribution is symmetric, the probability of a resistance value less than 0.8 kΩ is 0.0228. [Same as **a**]

Fig. 36
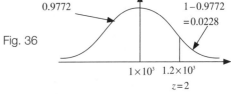

c The probability of resistance values between 0.9 kΩ and 1.1 kΩ is represented in Fig. 37.
By 16.48 with $\mu = 1 \times 10^3$, $x = 1.1 \times 10^3$ and $\sigma = 100$ we have

Fig. 37

$$z = \frac{(1.1 \times 10^3) - (1 \times 10^3)}{100} = 1$$

The table gives an area of 0.8413 for $z < 1$ (Fig. 38).

Fig. 38
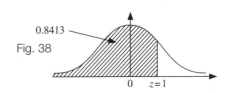

For $x = 0.9$:

$$z = \frac{(0.9 \times 10^3) - (1 \times 10^3)}{100} = -1$$

By symmetry for $z > -1$ we have Fig. 39.

Fig. 39
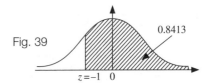

Since the normal distribution is symmetrical we can find the area between $z = -1$ and $z = 0$ in Fig. 39 by subtracting 0.5 from 0.8413. So the area between 0.9 and 1.1 is as shown in Fig. 40.

Fig. 40

The probability of a resistor having a value between 0.9 kΩ and 1.1 kΩ is

$$2 \times 0.3413 = 0.6826$$

Example 49

The lengths of certain components produced by a machine are normally distributed with a mean of 1.107 m and standard deviation of 0.2 m. Find the values that constitute the central 90% of the lengths.

Solution

Since the normal distribution is symmetric, each half contains an area of 0.45, that is half of 0.9. We need to find x_1 and x_2 of Fig. 41. The area less than the x_2 value is $0.5 + 0.45 = 0.95$.

Fig. 41

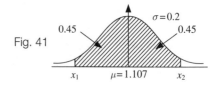

Next we determine the corresponding z value from the **Appendix** for 0.95. Within the table we try to find the number 0.95. The closest number seems to be 0.9495.

z	0	1	2	3	4	5
•	•	•	•	•	•	•
•	•	•	•	•	•	•
1.6	•	•	•	•	0.9495	•

We need to add 0.0005 to this, 0.9495, to get 0.95, hence we look at the ADD columns on the right-hand side:

z	4	1 2 3 4 5 6 7 ADD
1.6	0.9495	• • • • • 5

Therefore the z value for an area of 0.95 is 1.645.

Applying

16.48 $$z = \frac{x - \mu}{\sigma}$$

with $\mu = 1.107$, $z = 1.645$ and $\sigma = 0.2$ we have

$$\frac{x_2 - 1.107}{0.2} = 1.645$$
$$x_2 = 1.107 + (0.2 \times 1.645)$$
$$x_2 = 1.436$$

Similarly by symmetry:

$$\frac{x_1 - 1.107}{0.2} = -1.645$$
$$x_1 = 0.778$$

0.778 m to 1.436 m contains 90% of the central lengths.

SUMMARY

A normal distribution is a bell-shaped curve which is symmetric about the mean. We convert our normal distribution to a standard normal distribution so that we can use an established table to find the required probabilities. The conversion is carried out by

16.48 $$z = \frac{x - \mu}{\sigma}$$

where z is the number of standard deviations away from the mean.

Exercise 16(h)

Solutions at end of book. Complete solutions available at www.palgrave.com/science/engineering/singh

1 Suppose that z follows the standard normal distribution. Calculate the following probabilities:

  a $P(z < 1.5)$

  b $P(z < 1.645)$

  c $P(z < 1.96)$

  d $P(z > 1.645)$

  e $P(z < -1.645)$
  f $P(z < -2)$
  g $P(-1.645 < z < 1.645)$

  h $P(-0.1 < z < 0.2)$

  i $P(-0.25 < z < 0.17)$

2 The lifetime of light bulbs follows a normal distribution with a mean of 500 hours and a standard deviation of 22 hours. Find the probability of a bulb lasting less than 540 hours.

3 The wearout times of a machine are normally distributed with a mean of 200 hours and a standard deviation of 10 hours. What is the probability that a machine has a wearout time of more than 220 hours?

4 The wearout times of a product has a normal distribution with a mean of 600 hours and a standard deviation of 20 hours. Determine the probability that such a product will have a wearout time of

  a more than 660 hours
  b less than 550 hours.

5 Fuses follow a normal distribution with a mean current of 2 amp and a standard deviation of 0.15 amp. Evaluate the proportion of fuses blowing with currents between 1.75 amp and 2.13 amp.

6 The heights of 500 students follow a normal distribution with a mean of 1.70 m and a standard deviation of 0.15 m. Determine the number of students who have a height

  a greater than 1.80 m

  b less than 1.50 m

  c between 1.65 m and 1.85 m

7 The masses of students follow a normal distribution with a mean μ and a standard deviation of 1.2 kg. If 90% of the students have a mass of less than 70 kg, find the mean μ.

8 The lengths of rods conform to a normal distribution with a mean of 0.5 m. If 95% of the lengths are less than 0.9 m, find the standard deviation σ.

9 The diameters of cylinders are normally distributed with a mean of 0.604 m and a standard deviation of 0.01 m. Find the values of diameters that contain the central 99% of the cylinders.

Solutions at end of book. Complete solutions available at www.palgrave.com/science/engineering/singh

1 A manufacturer checks a sample of 50 capacitors from a batch which is labelled as 1 μF capacitors. The results are as follows:

1.091	1.020	0.901	0.905	0.943
1.010	1.047	0.984	1.084	1.093
1.061	0.953	0.961	0.999	1.089
1.064	0.951	0.947	1.042	1.021
1.019	1.003	0.910	1.067	0.932
1.089	1.086	0.988	1.031	1.022
0.954	1.002	1.014	0.998	1.035
1.065	1.057	0.996	1.072	1.080
1.008	1.013	1.007	0.998	1.069
1.012	1.006	0.997	1.098	1.003

Plot a histogram and a frequency polygon for this data.

2 The marks of students on an engineering module follow a normal distribution with mean of 45.5 and standard deviation of 5. Find the probability that a student selected at random doing this engineering module has

a fewer than 35 marks

b more than 50 marks

c between 50 and 60 marks.

3 The reliability of an engine working to specification is 0.9. If two engines are tested, what is the probability that at least one of them is working to specification?

4 An amplifier consists of two stages. Each stage has three distinct circuits, through which a signal can pass. Each stage is working if at least one of the circuits is activated by the signal. If the probability that a circuit is activated is 0.85, then determine the probability that

a the signal will pass the first stage

b the signal passes both stages.

5 Suppose there are n students in a room. What is the probability that no two of them have their birthday on the same day of the year for

i $n = 30$?

ii $n = 60$?

Are you surprised by your results?

(Assume there are 365 days in the year and a student is equally likely to have his/her birthday on any day of the year.)

6 In a multi-choice test there are 10 questions and 5 selections for each question. Only 1 selection is correct. If a student guesses all his/her answers, find the probability that the student will obtain

a all answers correct

b none correct

c exactly 4 correct.

7 The probability density function of a discrete random variable, X, is given by

$$P(X = x) = \frac{x^2}{k} \text{ for } x = 0, 1, 2, 3 \text{ and } 4$$

Determine

i k **ii** mean   **iii** standard deviation
iv variance

8 A mechanical device has a wearout time distributed normally with a mean of 1100 hours and a standard deviation of 50 hours. Find the values that constitute the central 95% of the wearout times.

🔧 **The remaining questions, apart from 12, belong to [*reliability engineering*].**

9 For the reliability function

$$R(t) = \frac{k^2}{(k + t)^2} \quad t \geq 0$$

where k is a positive constant, show that the hazard function

$$h(t) = \frac{2}{k + t} \quad t \geq 0$$

10 A particular set of components has a constant failure rate, $h(t)$, given by

$$h(t) = \lambda \quad (\lambda \text{ is a constant})$$

By assuming the constant of integration to be zero, show that

$$R(t) = e^{-\lambda t}$$

11 The cumulative hazard function, $H(t)$, in reliability engineering is defined as

$$H(t) = -\int_0^t \frac{\mathrm{d}}{\mathrm{d}x}\Big[\ln[\,R(x)]\Big]\mathrm{d}x$$

where $R(x)$ is the reliability function, $R(0) = 1$ and t is time. Show that

$$H(t) = -\ln[\,R(t)]$$

12 2% of the transformers in a batch are defective. Determine the largest sample size for the probability of no defectives to be > 0.8.

13

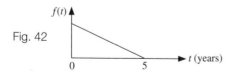

Fig. 42

Figure 42 shows a failure density function. What is the value of $f(t)$ at $t = 0$? Determine following functions of reliability engineering: failure distribution function, $F(t)$, reliability function, $R(t)$, and $MTTF$.

Sketch $F(t)$ and $R(t)$ on the same axes for $0 \leq t \leq 5$. What is the value of $F(t)$ and $R(t)$ for $t > 5$ and why?

14 The reliability function, $R(t)$, of a component is given by

$$R(t) = e^{-\lambda t}$$

where $\lambda > 0$ is a constant and t is in years.

Plot, on the same axes, $R(t)$ against t for

a $\lambda = 1$ **b** $\lambda = 0.1$ **c** $\lambda = 0.01$

(Use a computer algebra system or a graphical calculator.)

$R(t)$ represents the probability of the component surviving up to t years. What happens to $R(t)$ as λ gets smaller?

Note from your graphs that $R(0) = 1$ and $\lim_{t \to \infty} R(t) = 0$. Would you expect these results?

15 The failure density function, $f(t)$, for a particular set of components is given by

$$f(t) = \frac{4}{(1 + t)^5}$$

where t is in years. Determine the corresponding failure distribution function, $F(t)$, reliability function, $R(t)$, and the hazard function, $h(t)$. Determine the probability that the component lasts for more than 10 years.

By using a computer algebra system, or otherwise, plot $R(t)$ and $F(t)$ on the same axes for $t \geq 0$.

(You may use a computer algebra system or a graphical calculator for parts **a** and **b** of question **16**.)

16 The distribution function, $F(t)$, for a component is given by

$$F(t) = 4 - \frac{20}{5 + t/3} \quad 0 \leq t \leq 5$$

where t is in years.

a Determine the reliability function, $R(t)$, density function, $f(t)$, and the hazard function $h(t)$.

b Plot the graphs of $F(t)$, $R(t)$, $f(t)$ and $h(t)$ on different axes for $0 \leq t \leq 5$.

c Determine the probabilities of a component lasting more than

    i 1 year   **ii** 2 years   **iii** 3 years
    iv 4 years   **v** 5 years

d What does your answer to part **v** of **c** mean?

Solutions

For complete solutions see www.palgrave.com/science/engineering/singh

I(a) **1 a** $>$ **b** $<$ **c** $<$ **d** $\leqslant$ or $\geqslant$ **e** $<$

2 a $+$ **b** $-$ **c** $\div$ **d** $\times$ **e** $-$

3 a -5 **b** -2 **c** -2 **d** -22 **e** 0

4 a 3 **b** -12 **c** 12 **d** -3 **e** 3
f -3 **g** -16

5 a $-117\,649$ **b** $117\,649$ **c** -9 **d** -9
e 1279 **f** 700

I(b) **1 a** 64 **b** 512 **c** $16\,777\,216$ **d** 2025
e $69\,696$ **f** 1 **g** -243

2 a ± 7 **b** ± 41 **c** 428 **d** 13 **e** 9
f ± 2 **g** 7 **h** ± 2 **i** 3 **j** -3

3 a 8 **b** 13 **c** 36 **d** 6 **e** 10 **f** -10

4 a 20 **b** 108 **c** 2 **d** 3 **e** 2 **f** -2

I(c) **1 a** 30 **b** 42 **c** 54 **d** 42

2 a $2 \times 3^2 \times 5$ **b** $2^4 \times 3^2$ **c** 2×47
d $3^2 \times 5 \times 11$

3 a 216 **b** 1728 **c** $2\,178\,000$

I(d) **1** $3\frac{16}{113}, 3, 2\frac{232}{323}, 1\frac{169}{408}, 9\frac{1}{7}$

2 a $\frac{1}{3}$ **b** $\frac{2}{5}$ **c** $\frac{18}{25}$ **d** $\frac{56}{75}$ **e** 1

3 a $\frac{16}{64} = \frac{1}{4}$ **b** $\frac{26}{65} = \frac{2}{5} \neq \frac{1}{5}$; not equivalent

c $\frac{49}{89}$ and $\frac{4}{8}$ are not equivalent

4 a $\frac{10}{3}$ **b** $\frac{139}{69}$ **c** $\frac{987}{100}$

I(e) **1 a** $\frac{5}{6}$ **b** $\frac{13}{36}$ **c** $\frac{11}{60}$ **d** $\frac{12}{35}$ **e** $\frac{17}{12}$
2 a $\frac{1}{6}$ **b** $1\frac{23}{63}$ **c** $\frac{1}{145}$ **d** $\frac{61}{1271}$

3 a 1 **b** $\frac{47}{60}$ **c** $\frac{5}{84}$

4 a $\frac{1}{6}$ **b** $1\frac{59}{60}$ **c** 94

5 a $1\frac{1}{2}$ **b** $\frac{9}{20}$ **c** 94

6 a $1\frac{497}{1200}$ **b** $\frac{2}{97}$ **c** $\frac{1}{39780}$ **d** $\frac{1}{6360}$

e $2\frac{1}{288}$ **f** $2\frac{1}{144}$ **g** $\frac{2870}{2871}$ **h** $\frac{2639}{2640}$

I(f) **1 a i** 1.618 **ii** 1.62 **b i** 4.6692 **ii** 4.67
c i 2.503 **ii** 2.5 **d i** 0.374 **ii** 0.37

2 a i 1.62 **ii** 1.6 **b i** 2.9732 **ii** 3.0
c i 0.1100 **ii** 0.1 **d i** 9.9 **ii** 10

3 a 1700 **b** $100\,000$ **c** $110\,000\,000\,000$

4 a $3.14, 2.72, 1.41$ **b** $3.1, 2.7, 1.4$

I(g) **1 a** 1.86×10^5 **b** 1.392×10^6 **c** 1.36×10^5
d 3.439×10^{-8} **e** 9.51×10^{-8}
f 9.29×10^{-3} **g** 2.58×10^{-5}
h 1.496×10^7 **i** 2.7315×10^2 **j** 7.6×10^2

2 a $6\,400\,000$ **b** $0.000\,000\,003\,3$
c $0.000\,072\,92$ **d** $300\,000\,000$

3 a $12.75 \times 10^{-3}, 12.75, 12.75 \times 10^2, 12750$
b $3.14 \div 10^3, 3.14 \times 10^{-2}, 3.14 \times 10^3$

4 a 0.04 **b** 1.58 **c** 0.49

5 a $100\,\text{pF}$ **b** $30\,\text{k}\Omega$ **c** $0.3\,\text{mA}$

6 a $8.536\,\text{kN}$ **b** $75\,\text{MW}$ **c** $200\,\text{GPa}$

7 a $3\,\text{m}$ **b** $573 \times 10^3\,\text{N}$ **c** $25 \times 10^6\,\text{J}$
d $12 \times 10^{-12}\,\text{s}$ **e** $25 \times 10^{-3}\,\text{W}$

8 a 3 **b** 3 **c** $10\,000$ **d** 30

I(h) **1** $111.6\,\text{km/h}$

2 $9.81\,\text{m/s}^2$

3 765 miles per hour

For complete solutions see www.palgrave.com/science/engineering/singh

4 1.206×10^{-3} g/cm³

5 6.3 nm⁴

6 0.465 m²

7 a 13.4 m/s   **b** 22.22 m/s   **c** 2.99×10^8 m/s

8 2×10^6 N/cm²

I(i) **1 a** 135   **b** 144   **c** 441   **d** 175
   e 239   **f** 2311

2 a 2047   **b** −2   **c** 223/480

3 a 1   **b** 3   **c** 2 or 3   **d** −5 or 1
   e −5 or 5/14

4 a 461.81   **b** 83.78   **c** 523.60
   d 527.79   **e** 22.21

5 a 496   **b** 257

I(j) **1 a** 10%   **b** $8\frac{1}{3}$%   **c** 87.5%   **d** $23\frac{1}{13}$%

   e 16.7%   **f** 258.3%

2 a $\frac{1}{25}$ **b** $\frac{9}{100}$ **c** $\frac{7}{40}$ **d** $\frac{1}{40}$

3 a 810 Ω to 990 Ω   **b** 1140 Ω to 1260 Ω
   c 26 730 Ω to 27 270 Ω   **d** 18 430 Ω to
   19 570 Ω   **e** 4 992 500 Ω to 5 007 500 Ω

4 114.6 km/h to 125.4 km/h

5 3.572 m

6 0.88%

7 4.97×10^{-3} m³

8 8.6%

9 26.1 mA

I(k) **1 a** 10:1   **b** 3:1   **c** 1:5:9   **d** 1:7:9:12
   e 4:5

2 a 8:15   **b** 42:29

3 a 4:5   **b** 13:18   **c** 1:2

4 0.128 m, 0.192 m and 0.32 m

5 Copper mass = 48 kg and zinc mass = 18 kg.

6 Copper mass = 10.5 kg, zinc mass = 14 kg and
   nickel mass = 21 kg.

7 Manufacturing = 180, building services = 270,
   vehicle = 450 and control = 360.

MEI **1** 1×10^{-9} m³

2 a 125   **b** 10   **c** −2   **d** 2   **e** 13

3 a $\frac{19}{15}$ **b** $\frac{8}{15}$ **c** $\frac{10}{9}$

4 a i 3.142   **ii** 3.14   **b i** 3.16   **ii** 3.162
   c i 1.64   **ii** 1.645   **d i** 1 000 000
   ii 1 050 000
   e i 65 500   **ii** 66 000

5 a 1.79   **b** 3 or 4   **c** 8.19   **d** 118.08

6 a 1061.03   **b** 1×10^6 or 4×10^6

7 a 0.378 MV   **b** 10 μA   **c** 1.3 kΩ

8 a 95 Ω to 105 Ω   **b** 4.875 kΩ to 5.125 kΩ
   c 12.987 MΩ to 13.013 MΩ

9 6:43

10 $\frac{15}{4}$

11 15%

12 1.2 kW

13 a 1 333 333.33   **b** 407.45   **c** 335.12
   d 34 090.91   **e** 2.26

1(a) **1** $c = 25$

2 $I = 0.015$ A

3 $V_2 = 0.5 \times 10^{-4}$ m³

4 $m = 0.094$ kg

5 $a = 1.6$ m/s²

6 $\alpha = 6.8 \times 10^{-4}/°C$

7 $V = 11.4$ V

8 i $u = v - at$ **ii** $a = \dfrac{v - u}{t}$

9 $R = \dfrac{V}{I}$

For complete solutions see www.palgrave.com/science/engineering/singh

10 $T = \dfrac{PV}{mR}$

11 $\omega = \dfrac{v}{r}\left(1 - \dfrac{S}{100}\right)$

1(b) **1** $V = \sqrt{PR}$

2 $P = \dfrac{\rho c^2}{\gamma}$

3 $A = \dfrac{2D}{\rho C v^2}$

4 $R = 73\ \Omega$

5 $l = 62$ mm

6 $C = 1.4 \times 10^{-5}$ F $= 14\ \mu$F

7 $D = \left(\dfrac{P}{2\pi k\rho n^3}\right)^{\frac{1}{5}}$

9 $r = \left(\dfrac{2T}{\pi\tau}\right)^{\frac{1}{3}}$

10 a $r = \dfrac{ER}{V} - R$ **b** $u = \sqrt{v^2 - 2as}$

  c $K = \rho v^2 - \dfrac{4a}{3}$

  d $r = \left(\dfrac{8vL\eta}{\pi Pt}\right)^{\frac{1}{4}}$ **e** $l = \dfrac{T^2 g}{4\pi^2}$

  f $P = \dfrac{RT}{V - b} - \dfrac{a}{V^2}$

  g $V_1 = \dfrac{W(n - 1) + P_2 V_2}{P_1}$

  h $r = \left(\dfrac{C}{P}\right)^{\frac{1}{n}}$ **i** $u = \dfrac{C^2 W^2 h}{f^2 D^2}$

1(c) **1 a** x^7 **b** $x^{\frac{7}{10}}$ **c** 1 **d** $\dfrac{1}{x^2}$ **e** $x^{\frac{11}{15}}$

2 a $(1 + y)^3$ **b** $(1 + x^2)^2$

  c We *cannot* simplify this any further because under $\sqrt[3]{\ }$ we have different expressions.

  d $(x^2 + x + 1)^2$

3 0.734 m^3

1(d) **3 a** Incorrect **b** Correct **c** Correct
  d Incorrect **e** Incorrect

4 $ML^{-1}T^{-1}$

1(e) **1 a** $6x + 2$ **b** $-2x - 1$ **c** $-15y - 3$

  d $3x^2 + 5x$ **e** $y - 4$ **f** $4x^2 - x$

2 a $\dfrac{wLx^3}{2EI} - \dfrac{wx^4}{2EI}$ **b** $\dfrac{wLx^3}{4EI} - \dfrac{3wx^4}{8EI}$

  c $\dfrac{wL^2x^2}{16EI} - \dfrac{wx^4}{24EI}$

  d $-\dfrac{wLx^3}{12EI} + \dfrac{wx^4}{24EI} + \dfrac{wL^3 x}{24EI}$

3 a $x^2 + 3x + 2$ **b** $6x^2 + 19x + 15$
  c $4x^2 - 4x + 1$ **d** $4ab$ **e** $x^2y^2 + xy - x + 1$

5 a $R^2 - \omega^2 L^2$ **b** $\dfrac{1}{R^2} + \omega^2 C^2 - \dfrac{2C}{L} + \dfrac{1}{\omega^2 L^2}$

6 0 (zero)

1(f) **1 a** $4(x + y + z)$ **b** $8x(1 + y)$ **c** $2(x - 2y)$
  d $x(3 - 2x)$ **e** $x(x - y)$

2 a $t\left(u + \dfrac{1}{2}at\right)$ **b** $\dfrac{m}{t}(v_2 - v_1)$ **c** $\rho A v_1(v_2 - v_1)$

3 $\pi r\left[r + (r^2 + h^2)^{\frac{1}{2}}\right]$

4 a $(x + 5)(x + 2)$ **b** $(x + 4)(x + 1)$
  c $(x - 4)(x - 1)$ **d** $(x + 2)(x - 6)$
  e $(2x - 1)(x + 1)$ **f** $(x - 4)(x + 1)$
  g $(3x + 5)(7x - 2)$

5 a $(Z - R)(Z + R)$ **b** $\left(\omega L - \dfrac{1}{\omega C}\right)\left(\omega L + \dfrac{1}{\omega C}\right)$

7 a $\dfrac{wx^2}{6EI}(3L - x)$ **b** $\dfrac{wx^3}{8EI}(2L - 3x)$

  c $\dfrac{wx^2}{24EI}(x - L)^2$

1(g) **1 a** $x = \dfrac{1}{2}$ **b** $x = -2,\ x = -3$

  c $x = 7,\ x = 3$ **d** $x = -\dfrac{1}{3},\ x = \dfrac{5}{2}$

  e $x = \dfrac{1}{5},\ x = -3$

2 $u = 5.31$ m/s

3 $x = \dfrac{L}{2}$ or $x = L$

4 $t = 3.2$ s

5 $l = 12$ m, width $= 7$ m

6 $x = 5$

7 $x = 1.01$ m, $x = 0.25$ m

For complete solutions see www.palgrave.com/science/engineering/singh

8 $t = 11.44$ s

9 $t = \dfrac{u \pm \sqrt{u^2 + 2gh_0}}{g}$

1(h) **1** $x = 1, y = 1$

2 $a = 0.15, b = 35$ N

3 $a = 10$ m/s^2 and $u = 6.5$ m/s

4 $I_1 = 50$ mA and $I_2 = 73$ mA

5 $l_0 = 20$ m and $\alpha = 1 \times 10^{-4}/°$C

6 $R_1 = 1$ kΩ and $R_2 = 5$ kΩ

7 $F = K\rho A v^2$

8 $a = 1, b = 3$ and $c = 5$, hence $P = K\rho n^3 D^5$

ME1 **1** $C = 1 - \left(\dfrac{u}{v}\right)^2$

4 41, 43, 47, 53, 61 and 71. All are prime numbers

5 **a** $f_0 = 2.25$ kHz   **b** $C = 25.3$ μF

7 $R = 1$ kΩ

8 $v = \sqrt{\dfrac{2D}{\rho A(C_D + kC_L^2)}}$

9 $I = \dfrac{2E}{v_1^2 - v_2^2}$

10 $v_1 = \left(\dfrac{2(p_2 - p_1)}{\rho} + 2g(h_2 - h_1) + v_2^2\right)^{\frac{1}{2}}$

11 **a** $x = \dfrac{1}{5}$ **b** $x = 2$ **c** $x = 1, x = -2$

    d $x = \dfrac{1}{3}, x = -\dfrac{3}{2}$

12 $t = 1.43$ s

13 $u = 140$ m/s

14 $R = 3.42$ Ω

15 **a** $x = 5, x = 2$ **b** $x = 1, x = -1$

    c $x = \dfrac{1}{2}, x = 1$

    d $x = \dfrac{2}{5}, x = -\dfrac{1}{3}$ **e** $x = 3, x = 1$

17 $y = \dfrac{w}{36\,EI}(x^4 - 4Lx^3 + L^4)$

18 **i** $P = \dfrac{n^2\pi^2\,EI}{L^2}$ **ii** $P = 12 \times 10^6$ N

19 **a** $x = -0.38, x = -2.62$
    b $x = -0.59, x = -3.41$
    c $x = 0.29, \quad x = -0.69$
    d $x = 0.28, \quad x = -1.78$

21 $I_1 = 22$ mA and $I_2 = 70$ mA

24 **a** $x = 1, y = 1$ **b** $x = 58/5, y = -33/5$
    c $x = 1.77, y = 0.85$

26 $\zeta = 2\sqrt{mk}$

28 $\omega_1 = 20$ Hz, $\omega_2 = -20$ Hz, $\omega_3 = \sqrt{2}$ Hz
    and $\omega_4 = -\sqrt{2}$ Hz

31 $x = L, x = 0$

2(a) **1** **a** $m = 1, c = 1$ **b** $m = 1, c = -1$
    c $m = 3, c = 5$ **d** $m = \dfrac{1}{2}, c = -7$
    e $m = 0, c = \pi$ **f** $m = 1, c = -\pi$

2(b) **1** **a** Acceleration $= 0$
    b Total displacement $= 301.5$

2 $v = \begin{cases} \dfrac{1}{6}t & 0 \le t \le 60 \\ 10 & 60 < t \le 240 \\ -\dfrac{1}{12}t + 30 & 240 < t \le 360 \end{cases}$

3 $v = \begin{cases} 0.3t + 2 & 0 \le t \le 10 \\ 5 & 10 < t \le 20 \\ \dfrac{5}{3}(23 - t) & 20 < t \le 23 \end{cases}$

5 **ii** $t = 14$ s

6 **a** When $0 \le t \le 6$ and $14 \le t \le 20$
    b Acceleration $= -3.125$

2(c) **5** Negative acceleration = deceleration, that is the particle is slowing down.

The initial deceleration is zero, and deceleration is continuously increasing.

2(d) **1** **a** $(x - 2)^2 - 1$ **b** $25 - (x - 4)^2$

    c $\left(x + \dfrac{7}{2}\right)^2 - \dfrac{45}{4}$ **d** $2\left(x + \dfrac{7}{4}\right)^2 - \dfrac{41}{8}$

For complete solutions see www.palgrave.com/science/engineering/singh

2 a $x = 1, 3$ **b** $x = -1, 9$

c $x = \dfrac{-7 + \sqrt{45}}{2}, \dfrac{-7 - \sqrt{45}}{2}$

d $x = \dfrac{-7 \pm \sqrt{41}}{4}$

3 Minimum voltage is -0.25 V

5 The maximum height is 36

6 The maximum height is 2041 m

7 Maximum value $= \dfrac{WL^2}{96EI}$

2(e)

8 The larger the k, the larger the velocity, v

9 As k increases, R decreases and as k goes to infinity, R goes to zero

10 As A increases, R decreases

13 a $R = 10$ **b** $R = 10 \times 10^3$ **c** $R = 1500$

Maximum power occurs at the value of the number inside the brackets.

14 a $V = 30$ volts   **b** $V = 7.5$ volts

c $V = 5$ volts. Relationship is $V = E/2$.

2(f)

1 a $x^6 + 6x^5 + 15x^4 + 20x^3 + 15x^2 + 6x + 1$

b $a^6 + 6a^5b + 15a^4b^2 + 20a^3b^3 + 15a^2b^4$
$\qquad\qquad + 6ab^5 + b^6$

2 a $x^4 + 20x^3 + 150x^2 + 500x + 625$

b $32 + 240x + 720x^2 + 1080x^3 + 810x^4 + 243x^5$

c $4096 - 18432x + 34560x^2 - 34560x^3$
$\qquad + 19440x^4 - 5832x^5 + 729x^6$

3 Amplitudes are 1, 9, 36, 84 and 126

4 $\dfrac{w^7}{16384} - \dfrac{7w^6x}{12288} + \dfrac{7w^5x^2}{3072} - \dfrac{35w^4x^3}{6912}$

$\qquad + \dfrac{35w^3x^4}{5184} - \dfrac{7w^2x^5}{1296} + \dfrac{7wx^6}{2916} - \dfrac{x^7}{2187}$

ME2

6 $v = \begin{cases} t/2 & 0 \le t \le 20 \\ 2(t + 55)/15 & 20 < t \le 35 \end{cases}$

8 $s = 20t + 5$

10 As a increases y becomes less steep

11 $Q = \begin{cases} 0.4t & 0 < t \le 5 \\ 2 & t > 5 \end{cases}$

3(a)

1 1, 4, 9, 1, 4, 9. Many to one

2 32, 212, 75.2

3 $1, -1, 8, -8, 27, -27$

4 $-4.9, -48.36, -1$

5 Not a function

6 a 0, 9.8, 19.6, 49   **b** $9.8t^2, 9.8t + 9.8$

7 a 5, 6.125, 14.375   **b** $t^3 + 2t^2 + t + 5$

8 I^2R

9 $\dfrac{E}{2}, \dfrac{2E}{3}$

10 a $t = 0, t = 40.82$ **b** $380.4 + 180.4t - 4.9t^2$

3(b)

1 a $f^{-1}(x) = \dfrac{x + 2}{7}$ **b** $f^{-1}(x) = \dfrac{1}{x}$ $(x \ne 0)$

c $f^{-1}(x) = 1 - x$ **d** $f^{-1}(x) = 5 - \dfrac{3}{x}$ $(x \ne 0)$

2 a $g^{-1}(t) = \dfrac{2t}{1 - t}$ $(t \ne 1)$

b $h^{-1}(x) = \dfrac{2x + 1}{x + 1}$ $(x \ne -1)$

c $f^{-1}(x) = \dfrac{x + 1}{2 - x}$ $(x \ne 2)$

3 ii $g^{-1}(t) = \dfrac{t + 1}{3t + 1}$

4 ii $g^{-1}(t) = \dfrac{t^3 + 3}{2 - t^3}$ $(2 - t^3 \ne 0)$

3(c)

2 The graphs are reflections of each other in the horizontal axis

4 The f graph is a stretch of the g graph (or vice versa)

3(d)

1 i $2(x + 2)$ **ii** $2x^2 + 5x + 3$

iii $2x + 2$ **iv** $3x + 4$ **v** $-x - 2$

vi $\dfrac{x + 1}{2x + 3}$ $(2x + 3 \ne 0)$

For complete solutions see www.palgrave.com/science/engineering/singh

2 i $4x + 9$ **ii** $8x + 21$ **iii** -3

3 i c
   ii $ax^2 + (b + 2a)x + (a + b + c)$
   iii $2ax + a + b$

4 i $x + 2$ **ii** $x^2 + 2x + 1$ **iii** $x^2 + 1$

5 i $x^2 - 1$ **ii** $x^2 - 1$ **iv** x **v** $x^2 (x^2 - 2)$

6 i $\dfrac{3x - 6}{x}$ $(x \neq 0)$ **ii** x **iii** x

7 i x **ii** x

8 $f \circ f^{-1} = f^{-1} \circ f = x$

9 i $f \circ g$ or $g - 1$ **ii** $g \circ f$ **iii** $(g \circ f) + 7$
   iv f or $g(f^2)$

10 $F(t) = \dfrac{t}{5}$, $R(t) = 1 - \dfrac{t}{5}$ and $h(t) = \dfrac{1}{5 - t}$

11 a $\dfrac{k}{s^2 + s + 0.01 k}$ **b** $\dfrac{1}{s + (k_1 + k_2)}$

   c $\dfrac{1}{s + 2.3}$

12 $\dfrac{10s}{s^3 + 10s^2 + 21s + 10}$

3(e) **1 a** 1   **b** 0

2 10 m/s

3 a 6   **b** $-3/2$ **c** 2

4 2/3

5 0

6 $1/k$

7 0

8 $2x$

9 $3x^2$

3(f) **1** Same graph $|x - 1| = |1 - x|$

ME3 **1 i** 0   **ii** 5   **iii** $\dfrac{5x - 3}{2 - 3x}$ $(2 - 3x \neq 0)$

2 i $4.9t^2 + 9.8ht + 4.9h^2$ **ii** $9.8t + 4.9h$
   iii $9.8 + 4.9h$ **iv** 9.8 **v** $9.8t$

3 i $3t^2 + 3th + h^2$ **ii** $3t^2$

7 $x = 5$, $x = 1$

8 i -2.625 **ii** 5.375

9 $f^{-1}(x) = \dfrac{1}{x}$ $(x \neq 0)$

10 $f^{-1}(x) = \dfrac{2x - 3}{5x - 2}$ $(5x - 2 \neq 0)$

11 $x = -2, -1$ and 2

12 ii 1   **iii** $2.718\ 281\ 828^x$

15 a $s = -3, s = -1$ **b** $s = 1.14, s = -6.14$
   c no poles

16 $F(t) = \dfrac{t}{10}$, $R(t) = 1 - \dfrac{t}{10}$ and $h(t) = \dfrac{1}{10 - t}$

17 $R(t) = 1 - 0.3t + 0.015t^2$

18 $T(s) = 1000 \dfrac{s}{500s^3 + 4000s + 8501s + 5000}$

4(a) **1** 0.57, 0.68 and 0.38

2 i $\dfrac{1}{2}$ **ii** $\dfrac{4}{\sqrt{3}}$ **iii** $\dfrac{3\sqrt{3}}{2}$ **iv** 1   **v** $\dfrac{\sqrt{2}+\sqrt{3}}{2}$
   vi 1

3 AC $= 9.46$ m, span AB $= 17.16$ m

5 26.57°

6 4.62 m

7 42.1×10^3 mm² (3 s.f.)

8 $2\sqrt{3}$ m

9 i $\cos(\theta) = \sqrt{\dfrac{2}{3}}$, $\tan(\theta) = \dfrac{1}{\sqrt{2}}$

   ii $\sin(\theta) = \dfrac{\sqrt{5}}{3}$, $\tan(\theta) = \dfrac{\sqrt{5}}{2}$

   iii $\sin(\theta) = \dfrac{3}{5}$, $\cos(\theta) = \dfrac{4}{5}$

10 1.242, 1.791 and 1.428

4(b) **1**

x	0°	30°	60°	90°	120°	150°	180°
$y = \cos(x)$	1	0.866	0.5	0	-0.5	-0.866	-1

x	210°	240°	270°	300°	330°	360°
$y = \cos(x)$	-0.866	-0.5	0	0.5	0.866	1

2

x	0°	30°	60°	90°	120°	150°	180°
$y = \tan(x)$	0	0.577	1.732		-1.732	-0.577	0

x	210°	240°	270°	300°	330°	360°
$y = \tan(x)$	0.577	1.732		-1.732	-0.577	0

For complete solutions see www.palgrave.com/science/engineering/singh

4(c) **1 a** 45°, 135° **b** 60°, 300° **c** 60°, 240°
d 60°, 120°, 240°, 300° **e** 60°, 300°
f 0°, 180°, 360° **g** 25.38°, 154.62°

2 a $(180 \times n)° + 38.66°$
b $(360 \times n)° \pm 97.18°$
c $(180 \times n)° + \left[(-1)^n \times 12.71\right]°$
d $(60 \times n)° + \left[(-1)^n \times 15.35\right]°$ **e** No solution
f $(180 \times n)°$, $(180 \times n)° - 60°$

3 a 45°, 225°, 405°, 585° **b** 90°, 270°
c 150.5°
d 58.31°, 148.20°, 238.31°, 328.20°
e 49.81°, 130.19°, 409.81°, 490.19°
f No solution

4 a 45°, 90°, 135°, 225°, 270°, 315°
b 0°, 15°, 165°, 180°, 360°
c 30°, 60°, 120°, 150°, 210°, 240°, 300°, 330°

4(d) **1** $BC = 197$ mm, $AC = 135$ mm

2 41 mm

3 110°

4 2.47 m

5 $BC = 6.45$ m, $AC = 7.10$ m

6 101°

7 2.6 m

8 77 mm

9 451 m

10 $CD = 2.75$ m, $DE = 1.27$ m

11 a $B = 56.44°$, $C = 93.56°$, $AB = 11.98$ or
$B = 123.56°$, $C = 26.44°$, $AB = 5.34$

b $A = 47.52°$, $B = 97.48°$, $AC = 12.1$ or
$A = 132.48°$, $B = 12.52°$, $AC = 2.65$

4(e) **1** **i** 2.15 rad **ii** 0.23 rad **iii** 2.2981 rad
iv 5.817 rad

2 i $\dfrac{\pi}{2}$ rad **ii** $\dfrac{\pi}{6}$ rad **iii** $\dfrac{11\pi}{6}$ rad **iv** $\dfrac{\pi}{8}$ rad

v $\dfrac{\pi}{18}$ rad **vi** $\dfrac{3\pi}{20}$ rad **vii** $\dfrac{\pi}{60}$ rad

viii $\dfrac{4\pi}{5}$ rad

3 i 30° **ii** 54° **iii** 83.1° **iv** $\left(\dfrac{10}{7}\right)°$

v $(1.15 \times 10^4)°$ **vi** 20° **vii** $\left(\dfrac{140}{9}\right)°$
viii 19.5°

4 i 0.31 **ii** 0 **iii** −0.5

5 i $\dfrac{\pi}{2\sqrt{2}}$ **ii** 1 **iii** $\dfrac{1}{2}$

6 i 261.8 rad/s **ii** 52.4 rad/s **iii** 659.7 rad/s

4(f) In these solutions A = amplitude, T = period,
f = frequency.

1 a $A = 10$, $T = \pi$, $f = 1/\pi$
b $A = 0.1$, $T = 20\pi$, $f = 1/(20\pi)$
c $A = H$, $T = 2$, $f = 0.5$
d $A = 1/\pi$, $T = 2/\pi$, $f = \pi/2$
e $A = 10000$, $T = 2\pi^3$, $f = \dfrac{1}{2\pi^3}$
f $A = \dfrac{1}{\pi e}$, $T = 2\pi^2 e$, $f = \dfrac{1}{2\pi^2 e}$
g $A = 220$, $T = 0.002$, $f = 500$
h $A = 1$, $T = \dfrac{\pi^2}{3}$, $f = \dfrac{3}{\pi^2}$

2 a $A = 4$, $T = 2\pi$, phase = $\pi/3$ rad lead
b $A = 1$, $T = 0.02$, phase = 0.25 rad lag
c $A = 20$, $T = 0.02$, phase = $\pi/5$ rad lead
d $A = 315$, $T = 2\pi/7$, phase = 30° lead
e $A = 2$, $T = 1$, phase = $\pi/2$ rad lag

3 a 0 **b** $\pi/7$ lead **c** $\pi/4$ lead **d** 0.01 lag
e $\pi/2$ lag **f** 10π lag **g** π lag **h** 1 lead

4 a $12\pi/175$ **b** 0.487 **c** $\pi/12$

5 a $2\cos(2t)$ **b** $7\sin(8t)$ **c** $10\sin(t - \pi)$

8 a $A = 5$ mA, $T = 16.67$ ms, $f = 60$ Hz,
phase = 29.79° lead
b 2.48 mA for all values
c 2.79 ms **d** 1.38 ms

4(g) **1 i** $\dfrac{1}{2\sqrt{2}}\left(\sqrt{3} + 1\right)$ **ii** $\dfrac{\sqrt{3}}{2}$ **iii** $\dfrac{1}{2\sqrt{2}}\left(1 - \sqrt{3}\right)$

iv $-\dfrac{\sqrt{3}}{2}$ **v** $\dfrac{\sqrt{3} - 1}{\sqrt{3} + 1}$

4(i) **1 i** $\sqrt{200}\cos(\omega t - 45°)$

ii $\sqrt{6}\cos(\omega t - 54.74°)$

iii $\sqrt{29}\cos(\omega t + 21.80°)$

iv $\sqrt{2}\cos(\omega t + 45°)$

v $2\cos(\omega t - 240°)$

vi $\sqrt{50}\cos(\omega t - 135°)$

3 Amplitude $= 2\sqrt{3}$ or $\sqrt{12}$, period $= 2\pi/3$

4 Amplitude $= 1.12$, period $= 1.99$ s, phase $= 26.57°$ lagging

5 i $x = 2\cos\left(10t - \dfrac{\pi}{6}\right)$, amplitude $= 2$

6 $r = 4\sqrt{2}\cos\left(2t - \dfrac{\pi}{4}\right)$, amplitude $= 4\sqrt{2}$ cm, period $= \pi$ s

ME4 **7** 29.30, 26.48

8 58.41°, 48.19° and 73.4°

10 $x = 22.5°$

11 a $\pi/4, 3\pi/4, 5\pi/4, 7\pi/4$ **b** $\pi/6, 7\pi/6$

c $\pi/6, 5\pi/6, 7\pi/6, 11\pi/6$

d $\pi/8, 5\pi/8, 9\pi/8, 13\pi/8$

12 i Amplitude $= 20$, period $= 1/25$, frequency $= 25$

ii 3.82×10^{-3}

14 20 s

15 0.75

17 i 128.17°, 231.83° **ii** No solution

iii 128.17°, 231.83°

18 a $\sqrt{3}$ **b** $1/\sqrt{2}$ **c** $\sqrt{3}/2$

19 $x = n\pi/3$ (n is a whole number)

5(a) **1** $V_2 = 0.14$ m^3

2 $V_2 = 0.081$ m^3

3 $V_2 = 0.046$ m^3

5(b) **1** 1520.26 N

2

x	-2	-1	0
$y = e^{-x}$	$e^2 = 7.39$	$e^1 = 2.72$	$e^0 = 1.00$
x	1	2	
$y = e^{-x}$	$e^{-1} = 0.37$	$e^{-2} = 0.14$	

3

v	0	0.1	0.2
i	0	6.39×10^{-7}	5.36×10^{-6}
v	0.3	0.4	0.5
i	4.02×10^{-5}	2.93×10^{-4}	2.2×10^{-3}
v	0.6	0.7	
i	0.0163	0.12	

4 a 10705 Pa **b** 8337 Pa **c** 5730 Pa

5 400°C, 390°C, 382°C, 361°C

6 $(0.5 \times 10^{-6})e^{-(2\times10^3)t}$

7 a $0.4(e^{-t/2\times10^{-6}} - e^{-t/1\times10^{-6}})$

b $-0.05e^{-(10\times10^3)(t-1)}$

8 $5e^{-400t}(1 + 399e^{-600t} - 400e^{-1200t})$

10 $v = 20$ m/s

5(c) **1 a** 2.30 nep **b** 4.61 nep **c** -1.10 nep

2 a 0.80 nep **b** 1.50 nep

3 60 dB

4 0.316 W

7 0.01 A

9 a $2, \sqrt{2}$ **b** $10, \sqrt{10}$ **c** $100, 10$

10 0.105

11 4.05

12 a 2.926 **b** 0.342 **c** 0.946 **d** 1.057

13 a 1.93 **b** 3.14 **c** 4.75

18 $v = \dfrac{\eta T}{11600}\ln\left(\dfrac{i}{I_s} + 1\right)$, 0.62 V

5(d) **1** $k = 3.14$, $b = 0$, circle

2 $k = 10 \times 10^{-9}$

3 iii $k = 43$

4 $k = 0.075$, $n = -0.2$

5 $k = 1.46$, $n = 0.676$

5(e) **1** 11013.23, −10.02, 11013.23, 1, 1, 0.46

8 18.39 m/s

9 1.24 W

ME5 **1** 0.18 V

4 63.2%, 86.5%, 95%, 98.2% and 99.3%

5 i 1   **ii** 0.632, 0.865, 0.950, 0.982, 0.993

iii 63.2%, 86.5%, 95%, 98.2%, 99.3%

6 ii $v \to 10/k$ **iv** Yes   **v** 10, 5, 2 (m/s)

7 $n = -0.2$ and $k = 0.058$

8 27.9 kN

9 $\lim_{t \to \infty} \theta = 300$

12 ii $R(t) \to 0$ and $F(t) \to 1$

14 Cable becomes more taut

6(a) **1 a** 1   **b** 5   **c** $2x + 1$ **d** $-\dfrac{1}{x^2}$

4 $V = 0$

5 iii 455.625

6(b) **1 a** $3x^2$ **b** $3x^2(5x^2 + 1)$ **c** $\dfrac{1}{2}x^{-1/2} + \cos(x)$

d $-\dfrac{1}{3}x^{-4/3} + e^x - \sin(x)$

e $-\dfrac{3}{2}x^{-3/2} - \dfrac{5}{3}x^{-4/3}$

2 i $75 - 3t^2$ **ii** 5 s

3 $-\dfrac{1}{\rho^2}$

4 a $-3\sin(3x)$ **b** $\pi\cos(\pi x)$ **c** $11e^{11x}$

5 $-2\pi f K N \cos(2\pi ft)$

6 a $2x\cos(x^2)$ **b** $e^{\sin(x)}\cos(x)$ **c** $2/x$

d $(3 - 6x^2)\sin(2x^3 - 3x)$ **e** $20x(x^2 + 1)^9$

f $(35x^6 + 12x^3)\sec^2(5x^7 + 3x^4)$

6(c) **1 a** $6x(x^2 + 2)^2$ **b** $2x\cos(x^2)$

c $-(2x + 2)\sin(x^2 + 2x)$

d $\dfrac{3x^2 + 1}{x^3 + x}$ **e** $(15x^4 + 3x^2)\cosh(3x^5 + x^3)$

2 a $\cot(x)$ **b** $\tanh(x)$ **c** $-\sin(x)\sec^2[\cos(x)]$

d $2x(10)^{x^2}\ln(10)$ **e** $\dfrac{2\cos[\ln(x^2)]}{x}$

3 $(15 \times 10^{-3})e^{-(1 \times 10^3)t}$

4 $-\dfrac{E}{R}e^{-t/RC}$

5 $\dfrac{\eta T e^{-11600V/\eta T}}{11600 I_s}$

6 $\dfrac{\alpha}{\eta}\tau I_s e^{\alpha V/\eta}$

7 $\dfrac{1}{2\sqrt{t}\,e^{\sqrt{t}}}$

8 $\dfrac{20}{3(5 + t/3)^2}$

6(d) **1** $5(1 - 5t)e^{-5t}$

2 $i = -(1 \times 10^{-6})[999 + 1000t]e^{-1000t}$

3 a $\sin(t) + t\cos(t)$ **b** $\ln(x) + 1$ **c** $e^\theta(1 + \theta)$

d $e^u[\sin(u) + \cos(u)]$ **e** $\cos(2\alpha)$

f $\dfrac{1}{(1 + q)^2}$ **g** $-\dfrac{1}{1 + \sin(a)}$

h $\dfrac{e^z[\cos(z) + \sin(z)]}{\cos^2(z)}$ **i** $M^2(3\ln(M) + 1)$

j $\dfrac{\beta[2\sin(\beta) - \beta\cos(\beta)]}{\sin^2(\beta)}$ **k** $\dfrac{1}{(1 + h^2)^{3/2}}$

l $\ln[\sin(J)] + J\cot(J)$

m $e^\Sigma[\sin(2\Sigma) + 2\cos(2\Sigma)]$

n $-\sqrt{\dfrac{1}{(Z+1)(Z-1)^3}}$

o $e^{\Sigma}[(1+\Sigma)\sin(5\Sigma) + 5\Sigma\cos(5\Sigma)]$

p $\dfrac{-4\delta}{\delta^4 - 1}$

4 $-e^{-kt}[k\cos(\omega t) + \omega\sin(\omega t)]$

6(e) **1 a** $v = 3t^2 - 20t + 40,\ a = 6t - 20$

b $v = 21t^2 - 100t + 50,\ a = 42t - 100$

c $v = -10\sin(2t),\ a = -20\cos(2t)$

d $v = -30\pi\sin(10\pi t) + 50\pi\cos(10\pi t),$

$a = -100\pi^2[3\cos(10\pi t) + 5\sin(10\pi t)]$

2 i $v = 6t^2 - 30t - 36,\ a = 12t - 30$

ii $t = 6$ s

3 $a = -9.8\,\text{m/s}^2$. Constant acceleration

4 $v = 14.7$ m/s, $a = 9.8\,\text{m/s}^2$

5 a $v = 3 - 3\cos(3t),\ a = 9\sin(3t)$

b $v = \dfrac{e^{-t/5}}{25},\ a = -\dfrac{e^{-t/5}}{125}$

c $v = (1 - t)e^{-t},\ a = (t - 2)e^{-t}$

d $v = \dfrac{6t}{1 + 3t^2},\ a = \dfrac{6 - 18t^2}{(1 + 3t^2)^2}$

e $v = \sqrt{5}\tanh(\sqrt{20}t),\ a = 10\,\text{sech}^2(\sqrt{20}t)$

7 0

6(f) **1 a** $-\tan(t)$ **b** $\dfrac{20 - 3t}{6t - 2}$ **c** $\dfrac{3t^2}{2(t - 3)}$ **d** $\dfrac{1}{2e^{t/2}}$

2 b $\dfrac{\cos(t) - t\sin(t)}{\sin(t) + t\cos(t)}$

3 ii $v_x = 2(t - 2),\ v_y = 3t^2,$

$v = \sqrt{9t^4 + 4t^2 - 16t + 16}$

4 ii $v_x = (t^2 + 2)^{1/2}.t,\ v_y = 1$ **iii** $v = t^2 + 1$

5 i $v_x = 16 - 24t^2,\ v_y = 3t^2$

ii $v = \sqrt{256 - 768t^2 + 585t^4}$

6 $v_x = -r\sin(t),\ v_y = r\cos(t),$

$a_x = -r\cos(t),\ a_y = -r\sin(t)$

7 i $v_x = 3t^2 - 2t,\ v_y = 10t - 3t^2,$

$a_x = 6t - 2,\ a_y = 10 - 6t$

ii $v = 11.31$ m/s, $a = 10.20\,\text{m/s}^2$

6(g) **1 a** $-\dfrac{x}{y}$ **b** $\dfrac{2 - 3x^2}{3y^2}$ **c** $-\dfrac{4x}{y}$ **d** $\dfrac{2 - x}{y - 3}$

2 a $\dfrac{y^2 - 2x}{2y - 2xy}$

b $-\dfrac{2[y + x\sin(x^2 + y)]}{2x + \sin(x^2 + y)}$ **c** $y\dfrac{2 - x}{x(x - 1)}$

d $-\dfrac{y}{x(2y + 1)}$ **e** $-[e^{-(x + y)}\cos(x) + 1]$

(The denominator is not zero in any of the above cases.)

4 a $x^{\sin(x)}\left[\ln(x).\cos(x) + \dfrac{\sin(x)}{x}\right]$

b $(x^2 + 1)^{1/2}(3x + 1)^{1/3}\left[\dfrac{x}{x^2 + 1} + \dfrac{1}{3x + 1}\right]$

c $\dfrac{(x^2 + 1)^{1/2}(3x + 1)^{1/3}}{(x^4 - 2)^{1/5}}\left[\dfrac{x}{x^2 + 1}\right.$

$\left. + \dfrac{1}{3x + 1} - \dfrac{4x^3}{5(x^4 - 2)}\right]$

d $e^{(x^x)}x^x[\ln(x) + 1]$

e $x^{x^x + x - 1}\left[x[\ln(x)]^2 + x\ln(x) + 1\right]$

ME6 **1** 0

4 $Q = 210t^2 + 3500$

5 $Q = (3.49 \times 10^3)e^{-3t}$

6 $\theta = \sin^{-1}\left(\dfrac{mg}{kl}\right),\ \theta = \dfrac{\pi}{2}$

7 $x = \dfrac{mg}{k}$

8 $i = 0.036e^{-(2 \times 10^3)t} - 0.24e^{-(8 \times 10^3)t}$

10 $i = 0.0375\cos(75t)$

15 $v_x = 2t\sin(t) + t^2\cos(t),$

$a_x = 2\sin(t) + 4t\cos(t) - t^2\sin(t)$

For complete solutions see www.palgrave.com/science/engineering/singh

16 i $v_x = \cos(t) - t\sin(t)$,

$a_x = -2\sin(t) - t\cos(t)$

$v_y = \sin(t) + t\cos(t)$,

$a_y = 2\cos(t) - t\sin(t)$

iii $\dfrac{dy}{dx} = \dfrac{\sin(t) + t\cos(t)}{\cos(t) - t\sin(t)}$

17 $10^{(x^x)}x^x[\ln(x) + 1]\ln(10)$

7(a) 1 a $(-3/2, -5/4)$

    b $(1, -2), (-1, 2)$

    c $(1, -2/3), (-1, 2/3)$

    d $(0,0), (5,625), (-5,625)$

3 $(-5/3, 175/27)$max, $(1, -3)$min

4 $(2, 8)$ pt of inflexion

9 i $p \to 0$ **ii** $0, 2$ **iii** -0.288 **iv** 0.0086

10 i $i \to 0$ **ii** e^{-1}

7(b) 1 $x = y = 25$ m

2 $x = 60$ m, $y = 120$ m

3 $C_L = \sqrt{\dfrac{C_D}{k}}$

4 $r = h = 2/\pi^{1/3}$

6 $x = L/2$, max $M = WL^2/8$

7 $x = L$, max $y = WL^3/3EI$

8 $WL^4/24EI$

10 i $\pi/3$ **ii** 7.8

7(c) 1 $(0,0)$ min, $(2,16)$ and $(-2,16)$ max

2 $r = 25$

5 $2\pi fk$

7(d) 1 i 1 s, 3 s   **ii** $v = 2t - 4$, $a = 2$ m/s^2

2 $v = 25 - 9.8t$, $a = -9.8$ m/s^2,

    max height $= 31.9$ m

3 0s, 2s for $v = 0$ and 1s for $a = 0$

4 $v = 9t^2 - 20t + 1$, $a = 18t - 20$

5 min velocity $= 0$ m/s.

6 $v = 3(t^2 - 8t + 6)$

7 $\dfrac{ds}{dv} = -\dfrac{1}{v^2}$

8 $\omega = -248$ rad/s, $\alpha = 6$ rad/s^2

9 $\alpha = -\sin\left(2t + \dfrac{\pi}{6}\right)$, $t = \dfrac{5\pi}{12}$s

7(e) 1 a $y = 4x - 7$, $y = \dfrac{1}{4}(6 - x)$

    b $y = \dfrac{\pi}{2} - x$, $y = x - \dfrac{\pi}{2}$

    c $y = ex$, $y = e + e^{-1}(1 - x)$

    d $y = x - 1$, $y = 1 - x$

2 $i = 7.68v - 8.19$

3 $i = 541.34t + 1.19$

7(f) 1 $1 - \dfrac{x^2}{2!} + \dfrac{x^4}{4!} - \cdots$

2 $x - \dfrac{x^2}{2} + \dfrac{x^3}{3} - \dfrac{x^4}{4} + \cdots$

4 $\cos^2(x) = 1 - x^2 + \dfrac{x^4}{3} - \dfrac{2x^6}{45} + \cdots$

    $\sin^2(x) = x^2 - \dfrac{x^4}{3} + \dfrac{2x^6}{45} - \cdots$

5 a 1 **b** $1/2$

6 $\ln(x)$ is not defined at 0

7 $1 - \dfrac{1}{2}x^2 + \dfrac{x^4}{8} - \dfrac{x^6}{48} + \cdots$

8 $x + \dfrac{x^3}{3} + \dfrac{x^5}{5} + \cdots$

7(g) 1 a $1 + 20x + 190x^2 + 1140x^3 + \cdots$

    b $x^{15} + 15x^{14}y + 105x^{13}y^2 + 455x^{12}y^3 + \cdots$

    c $1024 - 5120Z + 11520Z^2 - 15360Z^3 + \cdots$

2 $1 + Z + Z^2 + Z^3 + Z^4 + \cdots$

3 $1 + \dfrac{W^2x^2}{2T^2} - \dfrac{W^4x^4}{8T^4} + \dfrac{W^6x^6}{16T^6}$

    where $-1 < \left(\dfrac{Wx}{T}\right)^2 < 1$

7(h) **1 a** -2.26 **b** -1.89 **c** -2.46

2 $-4.018, -8.967$ and -15.015

3 10.196 m/s

4 2.325 m

ME7 **1** $x = 10$ m, $y = 5$ m

2 $x = 1/3$ m, $y = 2/3$ m

4 $x = 25/36$ m

5 $x = 2.16$ m, $y = 4.05 \times 10^{-3}$ m

6 $\theta = \pi/4$

7 $\theta = \pi/4$

8 $v = (t - 1)e^{-t}$, $a = (2 - t)e^{-t}$

11 $n = 7.58$

12 1

13 $y = x + \dfrac{2 - \pi}{4}$, $y = \dfrac{2 + \pi}{4} - x$

14 e^{-1}

15 $v = -5e^{-500t}$

16 $\dfrac{r}{2}$

18 i $x + \dfrac{x^3}{3!} + \dfrac{x^5}{5!}$

19 iii $\dfrac{\pi}{4}$

20 ii 0.446

21 $1 - \dfrac{v^2}{2c^2} - \dfrac{v^4}{8c^4} - \dfrac{v^6}{16c^6} - \cdots$

8(a) **1 a** $\dfrac{x^2}{2} + C$ **b** $\dfrac{u^3}{3} + C$ **c** $\dfrac{z^4}{4} + C$

d $\dfrac{2t^{3/2}}{3} + C$ **e** $\dfrac{(\omega t)^2}{2} + C$ **f** $3t + C$

g $-t^{-1} + C$ **h** $2x^{1/2} + C$

2 a $\sin(\omega t) + C$ **b** $\dfrac{10^x}{\ln(10)} + C$

c $\cosh(t) + C$ **d** $\ln|\sec(x) + \tan(x)| + C$

3 a $\dfrac{PV^{2.35}}{2.35} + C$ **b** $\dfrac{PV^{2.61}}{2.61} + C$

4 a $\dfrac{x^3}{3} + x^2 + C$ **b** $2x^{1/2} + \ln|\sec(x)| + C$

c $\tan(x) - 5e^x + x + C$

d $\dfrac{4x^{3/2}}{3} - \cos(x) - x + C$

5 ii $\dfrac{3 - 4x}{1 + x^2} + C$

8(b) **1 a** $-\dfrac{\cos(7x + 1)}{7} + C$

b $\dfrac{\sin(7x + 1)}{7} + C$

2 a $\dfrac{\sin(\omega t)}{\omega} + C$ **b** $\dfrac{\sin(\omega t + \theta)}{\omega} + C$

3 a $-\dfrac{\cos(\omega t)}{\omega} + C$ **b** $-\cos(\omega t) + C$

4 a $\ln|x^2 - 1| + C$ **b** $\ln|x^3 - 3x^2 + 1| + C$

7 a $\dfrac{1}{7}\ln|7t - 1| + C$ **b** $\dfrac{1}{4}\ln|t^4 - 1| + C$

c $-\dfrac{1}{3}\ln|5 - t^3| + C$

8 a $\dfrac{e^{11x+5}}{11} + C$ **b** $-\dfrac{e^{-2x+1000}}{2} + C$

10 $v = u - gt$

12 i $v = 48 - 3t^2$ **ii** 4 seconds

13 $\dfrac{k^{1/\gamma}\gamma P^{\frac{\gamma-1}{\gamma}}}{\gamma - 1} + D$ (D = constant of integration)

8(c) **1 vii** 0.25

2 $-\dfrac{4}{3}$ m

3 12.5 J

4 45.1 J

5 22.71 m

6 90.85 m

7 $P = 4.96$ kN, $R = 15.2$ kN m

8 1.3 kJ

9 8.1 minutes

10 9.9 kJ

For complete solutions see www.palgrave.com/science/engineering/singh

15 $\dfrac{mr^2}{2}$

16 $\dfrac{49kr^{16/7}}{144}$

17 $\pi u R^2$

18 $\dfrac{V}{\omega T}\Big[\sin(\omega T + \theta) - \sin(\theta)\Big]$

19 $\dfrac{a}{\omega T}\Big[\sin(\omega T + \phi) - \sin(\theta)\Big]$

22 $\dfrac{\rho}{2\pi}\ln\Big(\dfrac{b}{a}\Big)$

23 $-(2 \times 10^3)\big[e^{-(5 \times 10^3)t} - 1\big]$

24 $\dfrac{Ir\sin(\beta)}{2(d^2 + r^2)}$

25 $\dfrac{I}{2\pi r}$

27 $(t^2 - 4t + 9)\exp\Big[-\Big(\dfrac{t^3}{3} - 2t^2 + 9t\Big)\Big]$

28 $(3 - t)\exp\Big[\dfrac{t^2}{2} - 3t\Big]$

29 $R(t) = 0.01t^2 - 0.2t + 1,$

$h(t) = \dfrac{0.02(10 - t)}{0.01t^2 - 0.2t + 1}$

30 $e^{-(4 \times 10^{-3})t^{1/2}}$

32 $\dfrac{nC}{n - 1}\Big[p_2^{\,1-\frac{1}{n}} - p_1^{\,1-\frac{1}{n}}\Big]$

8(d) **1 a** $2\Big[e^x(x - 1)\Big] + C$

  b $\sin(t) - t\cos(t) + C$

2 a $\dfrac{1}{9}\Big[3q\sin(3q) + \cos(3q)\Big] + C$

  b $\dfrac{s^3}{9}(3\ln(s) - 1) + C$

3 $\dfrac{e^{2t}}{4}(2t - 1) + C$

6 $500(1 - e^{-t} - te^{-t})$

7 $50[t\sin(t) + \cos(t) - 1]^2$

8 161 A

8(e) **1 a** $\dfrac{1}{x + 1} + \dfrac{2}{x + 2}$ **b** $\dfrac{1}{t - 1} + \dfrac{1}{t + 1}$

  c $\dfrac{3}{s - 1} - \dfrac{1}{s + 2}$ **d** $\dfrac{2}{2u + 1} - \dfrac{7}{u - 3}$

2 a $x - 1 + \dfrac{1}{x + 1}$

  b $x^3 + x - 2 + \dfrac{1}{2}\Big[\dfrac{3}{x + 1} - \dfrac{1}{x - 1}\Big]$

3 a $\dfrac{2t + 1}{t^2 + t - 1} + \dfrac{2}{t - 1}$ **b** $\dfrac{1}{z - 1} + \dfrac{2}{(z - 1)^2}$

  c $\dfrac{2x + 1}{x^2 + x + 1} + \dfrac{2x + 1}{(x^2 + x + 1)^2}$

8(f) **1 a** $2\ln|c + 2| + \ln|c + 1| + C$

  b $\ln|\lambda^2 - 1| + C$ **c** $\ln\left|\dfrac{(a - 1)^3}{a + 2}\right| + C$

  d $\ln\left|\dfrac{2y + 1}{(y - 3)^7}\right| + C$

2 a $\ln|p^2 + p - 1| + \ln(p - 1)^2 + C$

  b $\ln|z - 1| + \dfrac{2}{1 - z} + C$

3 1.79

4 0.044

7 i $x - 3 + \dfrac{7}{x + 2}$ **ii** $\dfrac{x^2}{2} - 3x + 7\ln|x + 2| + C$

8(g) **1 a** $\dfrac{(b^2 - 3)^8}{16} + C$ **b** $\dfrac{(5s - 1)^{10}}{50} + C$

  c $\dfrac{(a^3 - 2a^2 + 6)^5}{5} + C$ **d** $\dfrac{2(7q^3 - 5)^{3/2}}{3} + C$

  e $\dfrac{2}{3}(p^2 - 3p)^{1/2} + C$

  f $-\dfrac{1}{2}(\alpha^2 - 2\alpha + 10)^{-1} + C$

2 2 years

3 0.316

4 a $\dfrac{2}{3}$ (same answer as that in **Example 32**)

  b 0

5 a $\dfrac{\sec^7(\beta)}{7} + C$ **b** $\dfrac{\tan^6(A)}{6} + C$

  c $\csc(A) - \dfrac{\csc^3(A)}{3} + C$

For complete solutions see www.palgrave.com/science/engineering/singh

8(h) 2 $\dfrac{10}{\sqrt{2}}$

ME8 1 $\dfrac{mv^2}{2}$

2 $\dfrac{EAl^2}{2L}$

3 $\dfrac{1}{2}k(x_2^2 - x_1^2)$

7 $\dfrac{x^2}{2} + \ln(x^2 - 4)^2 + C$

8 7.2 kJ/kg

9 $R\left[a(T_2 - T_1) + \dfrac{b}{2}(T_2^2 - T_1^2) + \dfrac{c}{3}(T_2^3 - T_1^3)\right.$

$\left. + \dfrac{d}{4}(T_2^4 - T_1^4) + \dfrac{e}{5}(T_2^5 - T_1^5)\right]$

10 $-\cot(x) - \tan(x) + C$

13 $\dfrac{C}{0.32}[V_1^{-0.32} - V_2^{-0.32}]$

14 i 0.245025   ii 0.25   iii 0.004975

15 $-\dfrac{k}{6}\left[\dfrac{r^3}{x^6} + \dfrac{2}{x^3}\right]$

18 $0.3e^{-(2\times10^3)t} - 0.125\,e^{-(8\times10^3)t} - 0.175$

19 i $\dfrac{20}{\pi}[1 - \cos(100\pi t)]$ ii $\dfrac{[1 - \cos(100\pi t)]^2}{\pi^2}$

21 π

23 $\dfrac{EID^2\pi^4}{64L^3}$

26 ii $\dfrac{\pi}{2}$

31 7.26×10^5 J/s

32 a 1.80×10^4 b 1.84×10^4 c 2.30×10^4
   d 2.93×10^4
   (units are kJ/kmol)

9(a) 1 23 Ns

2 a 0.743   b 1.147

3 i a 0.266   b 0.254

   ii 0.25   iii a −6.4%   b −1.6%

4 53.91 m

5 69.65 m³/s

6 3.73V

9(b) 1 0.999969, exact value = 1, % error = 0.0031%

2 0.681 a i 0.6797915375   ii 0.68597084676
   iii 0.6863708136   iv 0.6865841146
   b i 0.6838407449   ii 0.6863450655
   iii 0.6865041355   iv 0.6865883836

3 175.41 Ns

4 17.62 N

5 0.3 A

6 9.58 J

7 149 m³/s

8 49.81 N

9 0.96×10^{-9}

10 0.159, 0.023, 0.001, 31.7×10^{-6}

11 a 2.553783362   b 2.553805688

9(c) 1 4 (units)²

2 Mean = 2 V, RMS = 2.31 V

3 2/π A

4 i 0 V   ii 0 V

7 Mean = 3.68 V, RMS = 4.10 V

8 2.38 mA

9 Mean = 1 V, RMS = 1.18 V, f = 1.18

10 0.75 V

11 3.86 kN

12 i 2/3   ii 2/3

9(d) 1 $x = \dfrac{2.5}{4\pi^2}[\cos(2\pi t) - 1]$

2 $x = \dfrac{t^3}{6} + 2t + 3$

For complete solutions see www.palgrave.com/science/engineering/singh

3 $a = 3.4$ m/s^2, $s = 3.6$ m

4 8.06 m/s

5 20.88 km

6 11.8 km (Simpson result)

8 2.02 s

9(e) **1** 11.885 kJ

2 −21.592 kJ

3 −11.467 kJ

5 $\dfrac{P}{6EI}(3lx^2 - x^3)$

6 $\dfrac{w}{24EI}(4L^3x - x^4 - 3L^4)$

7 $\dfrac{w}{24EI}\left[L^4 - 4L^3x - (L - x)^4\right]$

ME9 **1** 14.02 kJ

2 430 J

3 930 kJ/kg

4 393 m

5 22.1 s

6 0.92 V

7 1.49 A

8 $\dfrac{3\pi}{2}$

9 Mean = 4.01 V, RMS = 4.19 V

10 b 200 m

11 15.5 kJ

12 $y = \dfrac{1}{EI}\left[\dfrac{2.5x^4}{12} - \dfrac{11x^3}{3} + \dfrac{38.5x^2}{2} - 30.625x\right]$

13 $\dfrac{22}{7} - \pi$

10(a) **1 a** $7 + j11$ **b** $-3 - j5$ **c** $-14 + j31$
d $-14 + j31$ **e** $0.382 - j0.011$
f $2.615 + j0.077$ **g** $9 + j14$ **h** $-5 + j12$

i $-39 + j80$ **j** $-44 + j92$

2 $3 + j15, 3 - j15, -j3, -j, 0, \pi$ and e

3 $42 - j13$

4 $1, j, -j, 1, -j$

5 i $3 + j4, 2 + j11, -7 + j24, 0$ **ii** $2 + j$

6 $(0.0104 + j0.0055)$ S

7 a $a = 2, b = 3$ **b** $a = 0, b = 6$ **c** $a = 0, b = 2$
d $a = 0.12, b = 0.16$ **e** $a = -3/2, b = 4$
f $a = 1, b = 1$

8 a $20 - j15$ **b** $0.8 - j0.6$ **c** $5 - j$
d $-1.7 - j0.1$

9 $Z_t = (0.756 + j0.805)\ \Omega$

10 $I = (0.385 - j1.923)$ A

11 a $z = -1 \pm j5$ **b** $z = 1 \pm j\sqrt{2}$

c $z = \dfrac{7}{6} \pm j\dfrac{\sqrt{107}}{6}$

12 $(0.462 + j0.692)$ A

13 $I_s = (0.0843 + j0.0375)$ A, $I_p = (0.797 - j1.110)$ A

14 ii $\dfrac{R}{1 + \omega^2 C^2 R^2},\ \dfrac{-\omega C R^2}{1 + \omega^2 C^2 R^2}$

10(b) **2 a** $\sqrt{2}\ \angle 45°$ **b** $\sqrt{2}\angle 135°$ **c** $2\angle(-30°)$

d $2\angle(-120°)$ **e** $1\angle 90°$ **f** $x\angle 90°$

g $3\angle 180°$ **h** $5\angle 0°$

3 a $1.73 + j$ **b** $4.6 + j1.95$ **c** $1.41 - j1.41$
d $-0.90 - j1.07$

4 $|F| = 75.85$ N, $\arg(F) = -177.82°$

5 $F = 0$ N (equilibrium)

10(c) **1** $0.83\ \angle 51.57°$ A
2 $|z| = 0.2$ $\arg(z) = -26°$

3 a $\sqrt{3} + j$ **b** $\sqrt{3} + j$ **c** 240

5 a $6\angle(-45°)$ **b** $2\angle(-131°)$ **c** $15\angle 26°$

6 ii $\dfrac{1}{3}\left(\sqrt{8} - j\right)$

For complete solutions see www.palgrave.com/science/engineering/singh

8 Gain = $\dfrac{1000}{\sqrt{2}}$, phase = $-45°$

9 Gain = 0.11, phase = $-197.89°$

10 $I_1 = 1.27\angle(-29.93°)$ A, $I_2 = 0.94\angle(-8.13°)$ A

10(d) **1 a** $j2$ **b** $-2 - j2$ **c** -4
d $9784704 - j2063872$ **e** $-76443 + j16124$
f $12 - j316$

2 $1, j, -1$ and $-j$

3 -64

4 i $Z_0 = 1000\angle(-39.3°)$ Ω **ii** $0.01005\angle 45°$ m^{-1}
iii 884.2 m **iv** 1.407×10^6 m/s

6 i Poles: $-4, 2 + j3.46$ and $2 - j3.46$
ii Zeros: $\sqrt{2} + j\sqrt{2}, -\sqrt{2} + j\sqrt{2}, -\sqrt{2} - j\sqrt{2}$
and $\sqrt{2} - j\sqrt{2}$

7 ii 0 **iii** 1

9 $-0.42 + j0.42$

10 i $1 \pm j\sqrt{3}$ **ii** $2^{1/3}\angle 20°, 2^{1/3}\angle 100°, 2^{1/3}\angle 140°,$
$2^{1/3}\angle 220°, 2^{1/3}\angle 260°$ and $2^{1/3}\angle 340°$

10(e) **1 a** $10e^{j\pi}$ **b** $10e^{j\pi}$ **c** $2e^{j\pi/2}$ **d** $\sqrt{2}e^{j\pi/4}$
e $2e^{-j\pi/6}$

2 a je **b** $-e^\pi$ **c** $\sqrt{3}e^2 + je^2$ **d** $\dfrac{e^{1000}}{\sqrt{2}}(1 - j)$

3 All parts are equal to 1

4 All parts are equal to -1

6 $2e^{-j\frac{5\pi}{6}}$

7 $5.45e^{-j2.76}$

8 a 0 **b** $e^t(6\cos(t) - 8\sin(t))$

10 $a = 0, b = 3$

ME10 **2 a** $j32$ **b** $2^6\left(-1 - j\sqrt{3}\right)$ **c** $\cos(4\theta) + j\sin(4\theta)$

3 a $x = \cos(\theta), y = -\sin(\theta)$

b $x = \dfrac{1}{2}, y = -\dfrac{\sin(\theta)}{2(1 + \cos(\theta))}$

4 a $1 + j$ **b** $\dfrac{1}{2}\left(-\sqrt{3} + j\right)$

6 $F = 0$

7 $I_R = 12 - j12, I_Y = 11.94 + j22.46,$
$I_B = -24.83 + j7.82$ and $I_N = -0.89 + j18.28$

10 $K = 25$

11 gain = 9.81, phase = $-11.31°$

12 a stable **b** critically stable

13 i $\sigma = -1, \omega = \sqrt{11}$ **ii** $\sqrt{12}$ rad/s **iii** 0.29
iv 0.45 s

14 $\omega = 1$: gain = 0.995 and phase = $-5.71°$;
$\omega = 10$: gain = 0.707 and phase = $-45°$;
$\omega = 100$: gain = 0.0995 and phase = $-84.289°$

15 i gain = 707.11 and phase = $-45°$;
iii gain = 9.69 and phase = $-11.75°$

17 $s_1 = 1\angle 0°, s_2 = 1\angle 60°, s_3 = 1\angle 120°, s_4 = 1\angle 180°,$
$s_5 = 1\angle 240°$ and $s_6 = 1\angle 300°$

19 a $z_0\left(\dfrac{1 + k}{1 - k}\right)$ **b** $z_0\left(\dfrac{1 - k}{1 + k}\right)$

21 $\phi = \dfrac{3}{2}\left(x - \sqrt{3}y\right), \psi = \dfrac{3}{2}\left(\sqrt{3}x + y\right)$

11(a) **1 a** $\begin{pmatrix} 7 & 1 \\ 8 & 2 \end{pmatrix}$ **b** $\begin{pmatrix} 7 & 1 \\ 8 & 2 \end{pmatrix}$ **c** $\begin{pmatrix} 3 & 6 \\ 9 & -3 \end{pmatrix}$

d $\begin{pmatrix} 15 & 4 \\ 19 & 3 \end{pmatrix}$ **g** $\begin{pmatrix} 1 \\ -4 \end{pmatrix}$ **h** $\begin{pmatrix} -7 \\ -2 \end{pmatrix}$ **j** $\begin{pmatrix} 17 \\ -8 \end{pmatrix}$

e, f and **i** cannot be evaluated.

2 a $-8, -8$ **b** $-13, -13$ **c** $0, 0$
$\det A = \det B$

4 a i -9 **ii** 22 **iii** -198 **iv** -198
b i -1701.5 **ii** 624.3 **iii** -1062246.45
iv -1062246.45
c i -21.4 **ii** 79.27 **iii** -1696.38
iv -1696.38
$\det A \times \det B = \det(AB)$

5 a $\begin{pmatrix} 3 & -2 \\ -13 & 9 \end{pmatrix}$ **b** $\begin{pmatrix} 5 & -7 \\ -12 & 17 \end{pmatrix}$ **c** $\begin{pmatrix} 1 & 0 \\ 0 & 1 \end{pmatrix}$

6 a $\begin{pmatrix} -1/7 & 4/7 \\ 3/7 & -5/7 \end{pmatrix}$ **b** $\begin{pmatrix} -4/9 & 1/3 \\ 7/18 & -1/6 \end{pmatrix}$
c no inverse

11(b) **1** $i_1 = -1.25$ A and $i_2 = 2.5$ A

2 $i_1 = \dfrac{47}{95}$ A and $i_2 = \dfrac{27}{95}$ A

3 $\begin{pmatrix} 1 & -29 \\ -4 & -10 \end{pmatrix}, \begin{pmatrix} 2 & 9 \\ 8 & 39 \end{pmatrix}$

4 The system is controllable and observable

5 a Controllable and observable
b Observable but not controllable

6 a $\sqrt{7}, -\sqrt{7}$ **b** $7, -8$ **c** $6.78, -5.78$

7 $\dfrac{-RC + \sqrt{R^2C^2 - 4LC}}{2LC}, \dfrac{-RC - \sqrt{R^2C^2 - 4LC}}{2LC}$

8 $\dfrac{-c + \sqrt{c^2 - 4mk}}{2m}, \dfrac{-c - \sqrt{c^2 - 4mk}}{2m}$

9 a not stable   **b** stable   **c** not stable

10 $x = 2$ and $y = 4$ **b** $x = -1$ and $y = 1$
   c $x = \dfrac{1}{4}$ and $y = -\dfrac{1}{3}$

11(c) **1 a** -117 **b** 1288 **c** -114

2 a $\begin{pmatrix} -2 & 1 \\ 3 & 5 \end{pmatrix}$ **b** $\begin{pmatrix} 1/3 & -3/7 \\ -2/5 & \pi \end{pmatrix}$

   c $\begin{pmatrix} 7 & 2 & -3 \\ 3 & 6 & -3 \\ 4 & 1 & 1 \end{pmatrix}$ **d** $\begin{pmatrix} 1.17 & 9.39 & 2.11 \\ 1.36 & -1.45 & 5.20 \end{pmatrix}$

   e $\begin{pmatrix} -7 & 3 & 8 & -4 \\ 2 & 6 & 3 & 6 \\ 1 & -4 & -3 & 7 \\ 5 & 7 & 5 & 0 \end{pmatrix}$

   f $\begin{pmatrix} 1 & -9 & -4 & 9 \\ -3 & 4 & 2 & 19 \\ 7 & 6 & 8 & 11 \end{pmatrix}$

4 $t = 5 + j\sqrt{173}, t = 5 - j\sqrt{173}$

5 $C = \begin{pmatrix} 7 & 42 & -16 \\ -5 & -30 & 1 \\ -15 & -17 & 3 \end{pmatrix},$

   $C^T = \begin{pmatrix} 7 & -5 & -15 \\ 42 & -30 & -17 \\ -16 & 1 & 3 \end{pmatrix}$

6 $i_1 = -0.76$ A, $i_2 = -0.64$ A and $i_3 = 2.19$ A

7 a -42 **b** 44 **c** 13 **d** -85

8 a 0 **b** -509092.39

11(d) **1 a** $x = 2, y = -4, z = -3$
   b $x = 1, y = 2, z = 3$
   c $x = 1/2, y = 1/4, z = 1/8$

2 $i_1 = 1.20$ mA, $i_2 = 0.28$ mA

3 $x = -6.2, y = 30, z = -10$

4 $\ddot{x}_1 = g/3, \ddot{x}_2 = -g/3, T = 4g/3$

5 $\ddot{x}_1 = -g/5, \ddot{x}_2 = g/5, T = 12g/5$

6 $I_1 = 2.56$ mA, $I_2 = 8.58 \times 10^{-5}$ A,

   $I_3 = -0.615$ mA

11(e) **1** $x = 0, y = 0, z = 0$

2 unique, $x = 1, y = 1$

3 infinite, $x = a, y = -2a$

4 all are unique

5 a $\lambda = -4, \lambda = 2$ **b** $\lambda = 3 + j\sqrt{14}, 3 - j\sqrt{14}$

11(f) **1 a** $-4, 7$ **b** $1, 3$ **c** $-3, 3$

2 a $\lambda = -4$, eigenvector $\begin{pmatrix} -3/11 \\ 1 \end{pmatrix}$;

   $\lambda = 7$, eigenvector $\begin{pmatrix} 1 \\ 0 \end{pmatrix}$

   b $\lambda = 1$, eigenvector $\begin{pmatrix} 1 \\ 2 \end{pmatrix}$; $\lambda = 3$, eigenvector $\begin{pmatrix} 1 \\ 1 \end{pmatrix}$

   c $\lambda = -3$, eigenvector $\begin{pmatrix} -2 \\ 1 \end{pmatrix}$;

   $\lambda = 3$, eigenvector $\begin{pmatrix} 1 \\ 1 \end{pmatrix}$

3 a $j, -j$ **b** $-1 + j2, -1 - j2$
   c $1.50 + j3.57, 1.50 - j3.57$

4 $a\begin{pmatrix} -2/3 \\ 0 \\ 1 \end{pmatrix}$ corresponds to $\lambda = -5$ and $a\begin{pmatrix} 20 \\ 9 \\ 15 \end{pmatrix}$
   corresponds to $\lambda = 4$

5 $\begin{pmatrix} 0 \\ 1 \\ 2 \end{pmatrix}$ corresponds to $\lambda_1 = 2$, $\begin{pmatrix} 0 \\ 1 \\ 1 \end{pmatrix}$ corresponds to

$\lambda_2 = 3$ and $\begin{pmatrix} 1 \\ -5 \\ -4 \end{pmatrix}$ corresponds to $\lambda_3 = 5$

6 1.40 s, 2.81s

7 0.68, 14.92, 3.80

8 $0.21 + j5.73$, $0.21 - j5.73$,
$-4.01 + j2.08$, $-4.01 - j2.08$

11(g) (All temperatures are in Kelvin (K)).

1 $T_1 = 443.75$, $T_2 = 493.75$,
$T_3 = 381.25$ and $T_4 = 431.25$

2 i $T_1 = 437.5$, $T_2 = 437.5$,
$T_3 = 412.5$ and $T_4 = 412.5$

3 $T_1 = 732.14$, $T_2 = 698.21$, $T_3 = 660.71$,
$T_4 = 780.36$, $T_5 = 750$, $T_6 = 694.64$,
$T_7 = 839.29$, $T_8 = 826.79$ and $T_9 = 767.86$

4 $T_1 = 596.33$, $T_2 = 594.84$, $T_3 = 590.49$,
$T_4 = 586.71$, $T_5 = 578.91$, $T_6 = 571.03$,
$T_7 = 554.10$, $T_8 = 539.60$,
$T_9 = 497.89$ and $T_{10} = 479.16$

5 $T_1 = 350.48$, $T_2 = 350.67$, $T_3 = 351.23$,
$T_4 = 351.73$, $T_5 = 352.72$, $T_6 = 353.80$,
$T_7 = 355.85$, $T_8 = 358.02$, $T_9 = 362.66$,
$T_{10} = 366.59$, $T_{11} = 378.21$ and $T_{12} = 383.01$

ME11 1 i $\begin{pmatrix} 15 & 0 \\ 0 & 15 \end{pmatrix}$ **ii** $\begin{pmatrix} -42 & 0 \\ 0 & -42 \end{pmatrix}$ **iii** $\begin{pmatrix} ac & 0 \\ 0 & bd \end{pmatrix}$

2 $a = -2$, $b = 1$, $c = 3/2$, $d = -1/2$

3 $\begin{pmatrix} 4 & 0 \\ 0 & 4 \end{pmatrix}$, $\begin{pmatrix} 8 & 0 \\ 0 & 8 \end{pmatrix}$, $\begin{pmatrix} 1024 & 0 \\ 0 & 1024 \end{pmatrix}$

4 i $x = 5$, $y = -4$ **ii** $a = -6$, $b = 5$

iii $\begin{pmatrix} 5 & -6 \\ -4 & 5 \end{pmatrix}$

6 0

7 0

8 $I_1 = -68.45$ mA, $I_2 = -12.54$ mA,
$I_3 = 38.29$ mA

10 $i_1 = 2.19$mA, $i_2 = 0.44$mA, $i_3 = 1.28$mA

11 $2\pi\sqrt{-\dfrac{2m}{k}}$, $\pi\sqrt{-\dfrac{2m}{k}}$

12 i $\lambda_1 = 4$, $\lambda_2 = 3$ **ii** For $\lambda_1 = 4$, $\mathbf{v} = a\begin{pmatrix} 1 \\ 1 \end{pmatrix}$ and
for $\lambda_2 = 3$, $\mathbf{u} = a\begin{pmatrix} 2 \\ 1 \end{pmatrix}$

13 $C_T = \begin{pmatrix} 2 & 7 \\ 1 & 3 \end{pmatrix}$, $O_T = \begin{pmatrix} 3 & 4 \\ 2 & 29 \end{pmatrix}$. The system is
controllable and observable

14 $C_T = \begin{pmatrix} 1 & 44 & -141 \\ 5 & 29 & 364 \\ 7 & -31 & 837 \end{pmatrix}$

$O_T = \begin{pmatrix} 2 & 5 & 9 \\ 87 & 91 & -84 \\ 46 & -225 & 1450 \end{pmatrix}$

15 i $\begin{pmatrix} 1 & -7 \\ -1 & 6 \end{pmatrix}$ **ii** $\begin{pmatrix} -6 & -7 \\ -1 & -1 \end{pmatrix}$ **iii** $\begin{pmatrix} 9 & -24 \\ -6 & 21 \end{pmatrix}$
iv $(-3 \ 3)$

16 $(2 \ 3)$

17 $\dfrac{s + 1.4}{s^2 + 2s + 3}$

18 $\left(\dfrac{2777}{2560} \quad -\dfrac{47}{7680} \quad \dfrac{1953}{2560} \right)$

19 a $\begin{pmatrix} -1804 & -2583 & -2727 \\ 12 & 5685 & 2699 \\ 1269 & 6387 & 1124 \end{pmatrix}$

b $\begin{pmatrix} -99.92 & 308.63 & 61.37 \\ 357.22 & -226.62 & 498.79 \\ 340.06 & -732.14 & 56.44 \end{pmatrix}$

20

$$-0.2\left[\frac{(0.77 \times 10^9) + 515500s^3 - (0.29 \times 10^8)s^2 - (0.1 \times 10^9)s}{(1 \times 10^4)s^4 - (1.7 \times 10^4)s^3 - (2.54 \times 10^5)s^2 - (0.13 \times 10^7)s - (0.24 \times 10^8)} \right]$$

For complete solutions see www.palgrave.com/science/engineering/singh

12(a)

1 13 kN, angle 22.26° below horizontal

2 8.66 kN horizontal and 5 kN vertical

3 8.49 kN horizontal and 8.49 kN vertical

4 3.94 kN horizontal and 9.19 kN vertical

5 $\frac{1}{2}\mathbf{b} - \mathbf{a}$

7 10.85 N horizontal

8 31.80 kN

9 24.27 kN at an angle of 45.54° anticlockwise from $\mathbf{F}_1$

10 23.18 kN at an angle of 17.76° clockwise from $\mathbf{F}_1$

12(b)

1 a 43.59 kN, angle 36.59° above the horizontal
   b 10.63 kN, angle 16.40° below the horizontal

2 181.96 kN

3 5.38 kN, angle 106.95° above the horizontal

4 21.13 kN, angle 167.04° below the horizontal

12(c)

1 a $-18\mathbf{i} + 14\mathbf{j} + 11\mathbf{k}$
   b $-176\mathbf{i} + 203\mathbf{j} + 46\mathbf{k}$

2 $7\mathbf{j} + 8\mathbf{k}$

3 $12\mathbf{i} - 7\mathbf{j} + 30\mathbf{k}$

4 $(45\mathbf{i} - 21\mathbf{j} + 15\mathbf{k})$ N

5 $(-32\mathbf{i} - 16\mathbf{j} + 4\mathbf{k})$ kN

6 $(0.88\mathbf{i} + 6.58\mathbf{j} + 9.15\mathbf{k})$ kN

7 $(254.645\mathbf{i} - 672.697\mathbf{j} - 1234.618\mathbf{k})$ kN, magnitude = 1428.86 kN

12(d)

1 a 23   **b** 0   **c** −211

2 59 kJ

3 12 J

4 6.03 J

5 i $x^2 + y^2 + z^2$ **ii** $x^2 + y^2 + z^2$

6 11.31°

7 i $\mathbf{v} = (3t^2 - 2t)\mathbf{i} + 4t^3\mathbf{j}$, $\mathbf{a} = (6t - 2)\mathbf{i} + 12t^2\mathbf{j}$
   ii 4.4°

12(e)

1 i $5\mathbf{i} + 26\mathbf{j} - 17\mathbf{k}$ **ii** $-5\mathbf{i} - 26\mathbf{j} + 17\mathbf{k}$

4 a $-11\mathbf{i} - 8\mathbf{j} + 14\mathbf{k}$ **b** $3\mathbf{i} - 8\mathbf{j} + 6\mathbf{k}$

5 $20\mathbf{i} + 4\mathbf{j} + 16\mathbf{k}$

ME12

1 $\frac{5}{\sqrt{218}}\mathbf{i} + \frac{7}{\sqrt{218}}\mathbf{j} + \frac{12}{\sqrt{218}}\mathbf{k}$

2 i $\mathbf{i} + 2\mathbf{j} - 7\mathbf{k}$ **ii** $20\mathbf{i} + \mathbf{j} - 6\mathbf{k}$ **iii** 2   **iv** 2
   v 3   **vi** $29.4\mathbf{i} + 44.1\mathbf{j} - 14.7\mathbf{k}$ **vii** 20.9

3 90°

4 48.37°

5 10.84 kN at −55.46°

6 15.34 kN at 8.91°

7 $F_x = 23.77$ kN horizontal, $F_y = 31.55$ kN vertically down

8 a $63\mathbf{i} + 69\mathbf{j} + 48\mathbf{k}$ **b** $86\mathbf{i} + 225\mathbf{j} + 111\mathbf{k}$

9 35 J

10 468 J

11 a 13 J   **b** 149 J

12 $-3\mathbf{i} + 6\mathbf{j} - 21\mathbf{k}$, 22.05 N m

13 $(29.72\mathbf{i} - 4.95\mathbf{j} - 24.77\mathbf{k})$ kN

15 $\mathbf{j}(cb - ad)$

16 −3, 3

17 4

13(a) (C is the constant of integration).

1 $p = -\rho g z + C$

2 $p = -\frac{\rho \omega^2 r^2}{2} + C$

3 $y = 5x$

4 $y = e^x + C$

6 $v = 5t - 1.5t^2 + 8$, $s = 2.5t^2 - 0.5t^3 + 8t - 2.1$

For complete solutions see www.palgrave.com/science/engineering/singh

13(b) 1 $y = e^{-x} - e^{-2x}$

2 $y = e^{2x} + Ce^x$

3 $y = \dfrac{x}{1 - x^2}$

5 $v = 1 - \dfrac{1}{t}, v \to 1$

6 iii $i \to \dfrac{E}{R}$

8 $i = (3.33 \times 10^{-3})[e^t - e^{-(3 \times 10^6)t}]$ A

13(c) 1 iv $i = 0.368$ A

2 iv $v = 0.368E$

3 $v = E(1 - e^{-t/RC})$ V

4 ii $v = 15(1 - e^{-10t})$ V

iv 63.2%, 86.5% and 95.0% respectively of 15 V

13(d) 1 i $\dfrac{dh}{dt} = -\left(34.79 \times 10^{-3}\right)h^{1/2}$

ii $h = [-(17.395 \times 10^{-3})t + 1.14]^2$

iii 1.09 min

2 i $\dfrac{dh}{dt} = -(6.27 \times 10^{-3})h^{1/2}$

ii $h = [-(3.13 \times 10^{-3})t + 1.265]^2$

iii 6.73 min

3 i $\dfrac{dh}{dt} = -(17.3 \times 10^{-3})h^{-3/2}$

ii $h = [-(43.25 \times 10^{-3})t + 15.59]^{2/5}$

iii 6 min

4 $h = \left[-\left(k\sqrt{\dfrac{g}{2}}\right)t + 1\right]^2$

5 $\theta = 300 + 73e^{-(2.96 \times 10^{-3})t}$

6 $\theta = 300 + 100e^{kt}$

8 $\theta = 320 + 28e^{kt}$

13(e) 1 $y_1 = 1.1, \; y_2 = 1.231$ and $y_3 = 1.4025361$

2 At $x = 0.5, \; y = 1.8370655$

3 $y_1 = 0, y_2 = 1, y_3 = 3, y_4 = 6$ and $y_5 = 10$
Exact $y = 0, 0.5, 2, 4.5, 8$ and 12.5

4 $v(0.3) = 2.93262$

7 iv For $h = 0.25, 83.85\%$ and for $h = 0.005, 1.82\%$

13(f) 1 $y_1 = 2.215, y_2 = 2.463075$

2 $y_3 = 1.436058$

3 $y(1.6) = 1.3759$

6 Euler's and improved Euler's methods have 4.286% and 1.062% errors respectively

13(g) 1 $y_1 = 2.215513, y_2 = 2.464208$

2 $y_1 = 0.0627, y_2 = 0.1919$

3 $y(1.6) = 1.3774$

4 $v(2) = 9.1371$ m/s

ME13 2 $v = \dfrac{1}{0.1 - t}$

3 $v = \dfrac{1}{kt + 10}$ · Large k gives a large drop in velocity, v, with time t

9 i $v = 20(1 - e^{-0.1t})$ **ii** 20 m/s

10 $\dfrac{mg}{k}$

11 $\sqrt{\dfrac{mg}{k}}$

15 Runge–Kutta solution is closer to the exact solution than the Euler's solution

18 iv 9 V
v Smaller C is quicker to reach its final value of 9 V

19 iv 0.9 mA
v Larger L takes longer for the current, i, to reach its final value of 0.9 mA

14(a) 1 a $y = Ae^{-3x} + Be^{-2x}$
b $y = (A + Bx)e^{-2x}$
c $y = e^x\left[A\cos\left(\sqrt{3}x\right) + B\sin(\sqrt{3}x)\right]$

For complete solutions see www.palgrave.com/science/engineering/singh

2 a $y = Ae^{-2x} + Be^{-x}$

  b $y = Ae^{6x} + Be^{x}$

  c $y = (A + Bx)e^{3x}$

  d $y = e^{2x}[A\cos(x) + B\sin(x)]$

  e $y = (A + Bx)e^{4x}$

3 a $y = (A + Bx)e^{-3x}$

  b $y = e^{-\frac{5}{2}x}\left[A\cos\left(\dfrac{5\sqrt{3}}{2}x\right) + B\sin\left(\dfrac{5\sqrt{3}}{2}x\right)\right]$

  c $y = Ae^{m_1 x} + Be^{m_2 x}$ where $m_1 = \dfrac{-3 + \sqrt{41}}{2}$ and

  $m_2 = \dfrac{-3 - \sqrt{41}}{2}$

14(b) **1** $i = 2(e^{-2t} - e^{-5t})$

  2 a $x = A\cos(3t) + B\sin(3t)$

   b $x = A\cos(4t) + B\sin(4t)$

   c $x = A\cos\left(\sqrt{2}t\right) + B\sin\left(\sqrt{2}t\right)$

   d $x = A\cos\left(\sqrt{5}t\right) + B\sin\left(\sqrt{5}t\right)$

  7 $x = 5e^{-\omega t}(1 + \omega t)$

  9 a $v = (A + Bt)e^{-t/2RC}$

   b $v = Ae^{m_1 t} + Be^{m_2 t}$ where

   $m_1 = \dfrac{1}{2RC}\left[-1 + \sqrt{\dfrac{L - 4R^2C}{L}}\right]$ and

   $m_2 = \dfrac{1}{2RC}\left[-1 - \sqrt{\dfrac{L - 4R^2C}{L}}\right]$

   c $v = e^{\alpha t}[A\cos(\beta t) + B\sin(\beta t)]$ where

   $\alpha = -\dfrac{1}{2RC}, \beta = \dfrac{1}{2RC}\sqrt{\left(\dfrac{4R^2C - L}{L}\right)}$

  10 $L = 0.294$ H.

  11 $\omega = \dfrac{1}{\sqrt{LC}}, \zeta = \dfrac{1}{2R}\sqrt{\dfrac{L}{C}}$

14(c) **1 a** -9 **b** $-x - 1$ **c** $3\cos(3x) - 11\sin(3x)$

   d $-\dfrac{1}{10}[3\cos(x) + \sin(x)]$ **e** $\dfrac{e^{3x}}{4}$

  2 a $y = Ae^{-2x} + Be^{-x} + 3$

   b $y = Ae^{-x/3} + Be^{x} - 2x + 7$

---

c $y = (A + Bx)e^{-x} + e^{5x}$

  d $y = Ae^{-4x} + Be^{x} + 3\cos(x) + 5\sin(x)$

3 $x = A\cos(\omega t) + B\sin(\omega t) + \dfrac{mg}{k}$ where

  $\omega = \sqrt{\dfrac{k}{m}}$

4 $x = Ae^{2t} + Be^{-2t} - 0.125$

8 $i = Ae^{-(2.764 \times 10^4)t} + Be^{-(7.236 \times 10^4)t} + (50 \times 10^{-3})$

9 $y = Ae^{2x} + Be^{-x} - \dfrac{5}{3}xe^{-x}$

10 a $Y = Cxe^{-3x}$

  b $Y = x[a\cos(3x) + b\sin(3x)]$

14(d) **1** $i = -25e^{-3t} + 15e^{-5t} + 10 = 5(2 + 3e^{-5t} - 5e^{-3t})$

  2 $i = 2.5e^{-3t} + 7.5e^{-t} - 10e^{-2t}$

  3 $x = 2\cos(3t) + \sin(3t) - 2$

ME14 **1 a** $y = (A + Bx)e^{-4x}$ **b** $y = Ae^{3x} + Be^{-x}$

   c $y = Ae^{(3 + \sqrt{2})x} + Be^{(3 - \sqrt{2})x}$

  2 a $y = (A + Bx)e^{-4x} + \dfrac{1}{2}$

   b $y = (A + Bx)e^{-4x} + \dfrac{1}{4}(2x - 1)$

   c $y = (A + Bx)e^{-4x} + \dfrac{1}{128}[8x^2 + 7]$

  4 $T = Ae^{mx} + Be^{-mx}$ where $m = \sqrt{\dfrac{hP}{kA}}$

  8 i $m^2 + \dfrac{1}{RC}m + \dfrac{1}{LC} = 0$

   ii $R_{cr} = \dfrac{1}{2}\sqrt{\dfrac{L}{C}}$

  9 i $m^2 + \dfrac{R}{L}m + \dfrac{1}{LC} = 0$ **ii** $R_{cr} = 2\sqrt{\dfrac{L}{C}}$

  10 i $20 \,\Omega$ **ii** $i = (A + Bt)e^{-(10 \times 10^3)t}$

  11 $v = e^{-2000t}[9\cos(4000t) + 4.5\sin(4000t)]$

  18 $x = 2t + 0.776 - e^{-3t}\left[0.195\sin\left(\sqrt{491}\,t\right)\right.$

  $\left. + 0.776\cos\left(\sqrt{491}\,t\right)\right]$

For complete solutions see www.palgrave.com/science/engineering/singh

15(a) **1 a** $\dfrac{\partial f}{\partial x} = 2x, \dfrac{\partial f}{\partial y} = 2y$

b $\dfrac{\partial f}{\partial x} = \cos(x), \dfrac{\partial f}{\partial y} = -\sin(y)$

c $\dfrac{\partial f}{\partial x} = 3x^2 + 3y, \dfrac{\partial f}{\partial y} = 3y^2 + 3x$

d $\dfrac{\partial f}{\partial x} = \dfrac{1000}{x^2}, \dfrac{\partial f}{\partial y} = 2$

2 $\dfrac{\partial M}{\partial P} = x, \dfrac{\partial M}{\partial x} = P + wx^2$

3 a $\dfrac{PL}{AE}$ **b** $\dfrac{3PL^3}{16EI}$

4 $\sigma_x = 12Ax^2y^2 + 2Bx^4, \sigma_y = 2Ay^4 + 12Bx^2y^2$

6 $u = \dfrac{y}{x^2 + y^2}, v = \dfrac{-x}{x^2 + y^2}$

8 $c_p - c_v = T\left(\dfrac{R}{P} + \dfrac{K}{RT^2}\right)\left[\dfrac{RV + 2K/T}{(V + K/RT)^2}\right]$

9 $a = \dfrac{9RT_c V_c}{8}, b = \dfrac{V_c}{3}$

11 a $36xy - 9$ **b** $\dfrac{16 \times 10^6}{x^3 y^3} - 4$

15(b) **1** 0.00853 m^4

2 Reduction of 1.5%

3 Increased by (2.34×10^{-4}) m^3/s

4 1%

5 2.8%

6 2.3%

15(c) **1 a** (3, 4)   **b** (0.5, 1)   **c** $(-11/6, -121/12)$
and $(2, -12)$

2 a (2,1) saddle

b $\left(\dfrac{1}{\sqrt{3}}, -\dfrac{1}{2}\right)$ minimum and $\left(-\dfrac{1}{\sqrt{3}}, -\dfrac{1}{2}\right)$ saddle

c $\left(\dfrac{3}{2}, 2\right)$ minimum and $\left(\dfrac{25}{6}, \dfrac{2}{3}\right)$ saddle

3 1.216 × 1.216 × 0.609. One of the lengths is half of the other two.

4 $(2V)^{\frac{1}{3}}, (2V)^{\frac{1}{3}}$ and $\left(\dfrac{V}{4}\right)^{\frac{1}{3}}$

5 2.62 × 2.62 × 1.31

6 i 99.5

ii $\left(\dfrac{1}{\sqrt{2}}, \dfrac{1}{\sqrt{2}}\right), \left(-\dfrac{1}{\sqrt{2}}, \dfrac{1}{\sqrt{2}}\right), \left(-\dfrac{1}{\sqrt{2}}, -\dfrac{1}{\sqrt{2}}\right)$

and $\left(\dfrac{1}{\sqrt{2}}, -\dfrac{1}{\sqrt{2}}\right)$

7 $x = 3$ m and $\theta = 60°$

ME15 **1** $\dfrac{\partial M}{\partial P} = x - \dfrac{x^2}{L}, \dfrac{\partial M}{\partial x} = K + \dfrac{P}{L}(L - 2x)$

2 $\dfrac{\partial M}{\partial P} = x - \dfrac{ax}{L}, \dfrac{\partial M}{\partial x} = \dfrac{K - Pa}{L} + P$

3 $24Ax$. Ω is not an Airy stress function

4 1.5% decrease

5 7%

6 3.8% increase

7 $x = -r$

11 $\sigma_r = -2A\cos(2\theta) - 2B\sin(2\theta)$,

$\sigma_\theta = 2A\cos(2\theta) + 2B\sin(2\theta)$ and

$\tau_{r\theta} = 2A\sin(2\theta) - 2B\cos(2\theta)$

13 $x = 2^{-1/6}$ m, $l = 2^{1/3}$ m and $\theta = 45°$

14 $a = 2RVT, b = V/2$

15 $a = 1.28RVT^{3/2}, b = 0.26V$

16(a) **1 a** continuous   **b** discrete   **c** discrete
d continuous   **e** continuous

2

Temperature (°C)	15	16	17	18	19	20
Frequency	1	2	3	1	2	5
Temperature (°C)	21	22	23	24	25	26
Frequency	4	3	5	2	1	1

3 The shortest height is 1.53 m and the tallest is 2.00 m. Divide into 1.5–1.6, 1.6–1.7, ...

Height h (m)	Frequency
$1.50 \le h < 1.60$	4
$1.60 \le h < 1.70$	6
$1.70 \le h < 1.80$	7
$1.80 \le h < 1.90$	14
$1.90 \le h < 2.00$	9

12 a $\dfrac{1}{495}$ **b** $\dfrac{893}{990}$ **c** $\dfrac{19}{198}$

13 0.957

14 $\dfrac{5}{119}$

15 a 3/4   **b** 1/2   **c** 1/4

16 $\dfrac{7}{1520}$

17 0.9775

18 0.88186

16(b) $\bar{x}$ = mean and s = standard deviation

1 median = 27, mean = 24.91, SD = 13.61

2 a $\bar{x}$ = 15.20 and s = 11.46
  b $\bar{x}$ = 115.20 and s = 11.46
  c $\bar{x}$ = 1.52 and s = 1.146
  d $\bar{x}$ = 155.00 and s = 114.6

5 i $\bar{x}$ = 69.29 and s = 23.68
  ii $\bar{x}$ = 692.9 and s = 236.8
  iii $\bar{x}$ = 0.6929 and s = 0.2368
  iv $\bar{x}$ = 74.29 and s = 23.68
  v $\bar{x}$ = 7029 and s = 2368

6 mean = 159.93 kN, s = 4.66 kN

7 i good estimate is 10 kΩ
  ii small SD because data is close to 10 kΩ
  iii mean = 9.996 kΩ, s = 0.122 kΩ

8 a $\bar{x}$ = 431.35 MN/m² and s = 55.62 MN/m²
  b $\bar{x}$ = 196.29 GN/m² and s = 1.86 GN/m²

16(c) **1 a** 9/10   **b** 3/20,000

2 8×10^{-6}

3 0.93

4 i 0.3   **ii** 0.65   **iii** 0.35

5 a 7/100   **b** 93/100

6 8.4×10^{-3}

7 $\dfrac{4}{49}$

8 i 0.62   **ii** 0.38

9 0.786

10 0.955

11 $\dfrac{3}{59}$

16(d) **1** 0.0716

2 i 0.249   **ii** 3.52×10^{-9} **iii** 0.751

3 a 3.908×10^{-5} **b** £1,919,190

4 a 258,336,000   **b** 456,976,000

16(e) **1 a** 8.1×10^{-3} **b** 0.99954

2 a 0.170   **b** 1.239×10^{-5}

3 i 0.264   **ii** 0.264   **iii** 0.418   **iv** 0.582

4 0.456

5 0.6572

6 a i 0.3295   **ii** 0.0988   **iii** 0.0197
  b i 0.3293   **ii** 0.0988   **iii** 0.0198
  c The results are very similar

7 0.83521

8 $k = \dfrac{1}{17}$

9 $k = \dfrac{1}{36}$

16(f) **1** $E(X) = 3$, symmetric distribution therefore mean is 3

2 $E(X^2) = 13.42$

3 $\mu = 3$ V, $\sigma = 1.095$ V symmetric distribution therefore mean is 3 V

4 $\mu = 4.19$ V, $\sigma = 1.94$ V

5 $\mu = 1.11$ V, $\sigma = 1.05$ V

For complete solutions see www.palgrave.com/science/engineering/singh

6 i $k = \dfrac{1}{30}$ **ii** $E(X) = 3.33$ **iii** SD = 0.83

   iv variance = 0.69

7 i 2.9   **ii** 9.5   **iii** 12.4   **iv** 70.8

9 a 0.20   **b** 0.32   **c** 0.78   **d** 6.12×10^{-6}

10 a 0.0166   **b** 0.776

16(g)

1 a 0.632   **b** 0.865   **c** 0.181

2 a 0.632   **b** 0.95   **c** 0.181

3 a 0.787   **b** 0.050   **c** 0.549

4 i 10/3 years   **ii** $\dfrac{0.08t}{1 - 0.04t^2}$ $0 \le t \le 5$

5 $R(t) = e^{-0.05t^2}$

6 $f(t) = 0.1$, $F(t) = 0.1t$, $R(t) = 1 - 0.1t$,

   $h(t) = \dfrac{0.1}{1 - 0.1t}$ all defined for $0 \le t \le 10$;

   $MTTF = 5$ years

7 a 5/3 years   **b** 3 years

8 $t_{med} = 6.93$ years

16(h)

1 a 0.9332   **b** 0.95   **c** 0.975   **d** 0.05

   e 0.05   **f** 0.0228   **g** 0.9   **h** 0.1191

   i 0.1662

2 0.9655

3 0.0228

4 a 0.00135   **b** 0.00621

5 0.7592

6 a 126   **b** 45   **c** 235

7 $\mu = 68.463$ kg

8 $\sigma = 0.243$ m

9 0.57825 m to 0.62975 m

ME16 1

Capacitance (μF)	Frequency
0.900–0.930	3
0.930–0.960	6
0.960–0.990	3
0.990–1.020	16
1.020–1.050	7
1.050–1.080	7
1.080–1.110	8

2 a 0.0179   **b** 0.1841   **c** 0.18223

3 0.99

4 a 0.997   **b** 0.993

5 i 0.294   **ii** 0.006

6 a 1.024×10^{-7} **b** 0.1074   **c** 0.0881

7 i 30   **ii** 3.33   **iii** 0.83   **iv** 0.69

8 1002 hours to 1198 hours

12 Largest size is 11

13 0.4, $F(t) = 0.4t - 0.04t^2$,

   $R(t) = 1 - 0.4t + 0.04t^2$, $h(t) = \dfrac{5}{3}$ years

15 $F(t) = 1 - \dfrac{1}{(1 + t)^4}$, $R(t) = \dfrac{1}{(1 + t)^4}$,

   $h(t) = \dfrac{4}{1 + t}$, $\dfrac{1}{14641}$

16 a $F(t) = 4 - \dfrac{20}{5 + t/3}$, $R(t) = -3\left(\dfrac{t - 5}{15 + t}\right)$,

   $f(t) = \dfrac{20}{3(5 + t/3)^2}$, $h(t) = \dfrac{20}{(t - 5)(15 + t)}$

   c i 3/4   **ii** 9/17   **iii** 1/3   **iv** 3/19   **v** 0

   d a component cannot last more than 5 years

(In **16** the functions are defined for $0 \le t \le 5$.)

Appendix: Standard Normal Distribution Table

(Reproduced with permission. From Miller & Powell: *The Cambridge Elementary Mathematical Tables*, 2nd edn, 1979, Cambridge University Press.)

THE DISTRIBUTION FUNCTION $\Phi(z)$ OF THE NORMAL DISTRIBUTION N(0, 1)

z	0	1	2	3	4	5	6	7	8	9	1	2	3	4	5	6	7	8	9
											ADD								
0.0	.5000	.5040	.5080	.5120	.5160	.5199	.5239	.5279	.5319	.5359	4	8	12	16	20	24	28	32	36
0.1	.5398	.5438	.5478	.5517	.5557	.5596	.5636	.5675	.5714	.5753	4	8	12	16	20	24	28	32	36
0.2	.5793	.5832	.5871	.5910	.5948	.5987	.6026	.6064	.6103	.6141	4	8	12	15	19	23	27	31	35
0.3	.6179	.6217	.6255	.6293	.6331	.6368	.6406	.6443	.6480	.6517	4	7	11	15	19	22	26	30	34
0.4	.6554	.6591	.6628	.6664	.6700	.6736	.6772	.6808	.6844	.6879	4	7	11	14	18	22	25	29	32
0.5	.6915	.6950	.6985	.7019	.7054	.7088	.7123	.7157	.7190	.7224	3	7	10	14	17	20	24	27	31
0.6	.7257	.7291	.7324	.7357	.7389	.7422	.7454	.7486	.7517	.7549	3	7	10	13	16	19	23	26	29
0.7	.7580	.7611	.7642	.7673	.7704	.7734	.7764	.7794	.7823	.7852	3	6	9	12	15	18	21	24	27
0.8	.7881	.7910	.7939	.7967	.7995	.8023	.8051	.8078	.8106	.8133	3	5	8	11	14	16	19	22	25
0.9	.8159	.8186	.8212	.8238	.8264	.8289	.8315	.8340	.8365	.8389	3	5	8	10	13	15	18	20	23
1.0	.8413	.8438	.8461	.8485	.8508	.8531	.8554	.8577	.8599	.8621	2	5	7	9	12	14	16	19	21
1.1	.8643	.8665	.8686	.8708	.8729	.8749	.8770	.8790	.8810	.8830	2	4	6	8	10	12	14	16	18
1.2	.8849	.8869	.8888	.8907	.8925	.8944	.8962	.8980	.8997	.9015	2	4	6	7	9	11	13	15	17
1.3	.9032	.9049	.9066	.9082	.9099	.9115	.9131	.9147	.9162	.9177	2	3	5	6	8	10	11	13	14
1.4	.9192	.9207	.9222	.9236	.9251	.9265	.9279	.9292	.9306	.9319	1	3	4	6	7	8	10	11	13
1.5	.9332	.9345	.9357	.9370	.9382	.9394	.9406	.9418	.9429	.9441	1	2	4	5	6	7	8	10	11
1.6	.9452	.9463	.9474	.9484	.9495	.9505	.9515	.9525	.9535	.9545	1	2	3	4	5	6	7	8	9
1.7	.9554	.9564	.9573	.9582	.9591	.9599	.9608	.9616	.9625	.9633	1	2	3	4	4	5	6	7	8
1.8	.9641	.9649	.9656	.9664	.9671	.9678	.9686	.9693	.9699	.9706	1	1	2	3	4	4	5	6	6
1.9	.9713	.9719	.9726	.9732	.9738	.9744	.9750	.9756	.9761	.9767	1	1	2	2	3	4	4	5	5
2.0	.9772	.9778	.9783	.9788	.9793	.9798	.9803	.9808	.9812	.9817	0	1	1	2	2	3	3	4	4
2.1	.9821	.9826	.9830	.9834	.9838	.9842	.9846	.9850	.9854	.9857	0	1	1	2	2	2	3	3	4
2.2	.9861	.9864	.9868	.9871	.9875	.9878	.9881	.9884	.9887	.9890	0	1	1	1	2	2	2	3	3
2.3	.9893	.9896	.9898	.9901	.99036	.99061	.99086				0	1	1	1	1	2	2	2	2
								.99111	.99134	.99158	3	5	8	10	13	15	18	20	23
											2	5	7	9	12	14	16	18	21
2.4	.99180	.99202	.99224	.99245	.99266			.99324	.99343	.99361	2	4	6	8	11	13	15	17	19
						.99286	.99305				2	4	6	7	9	11	13	15	17
2.5	.99379	.99396	.99413	.99430	.99446	.99461	.99477	.99492	.99506	.99520	2	3	5	6	8	9	11	12	14
2.6	.99534	.99547	.99560	.99573	.99585	.99598	.99609	.99621	.99632	.99643	1	2	3	5	6	7	8	9	10
2.7	.99653	.99664	.99674	.99683	.99693	.99702	.99711	.99720	.99728	.99736	1	2	3	4	5	6	7	8	9
2.8	.99744	.99752	.99760	.99767	.99774	.99781	.99788	.99795	.99801	.99807	1	1	2	3	4	4	5	6	6
2.9	.99813	.99819	.99825	.99831	.99836	.99841	.99846	.99851	.99856	.99861	0	1	1	2	2	3	3	4	4
3.0	.99865	.99869	.99874	.99878	.99882	.99886	.99889	.99893	.99896	.99900	0	1	1	2	2	2	3	3	4
3.1	.9³032*	.9³065	.9³096								3	6	9	13	16	19	22	25	28
				.9³126	.9³155	.9³184	.9³211				3	6	8	11	14	17	20	22	25
								.9³238	.9³264	.9³289	2	5	7	10	12	15	17	20	22
3.2	.9³313	.9³336	.9³359	.9³381	.9³402						2	4	7	9	11	13	15	18	20
						.9³423	.9³443	.9³462	.9³481	.9³499	2	4	6	8	9	11	13	15	17
3.3	.9³517	.9³534	.9³550	.9³566	.9³581						2	3	5	6	8	10	11	13	14
						.9³596	.9³610	.9³624	.9³638	.9³651	1	3	4	5	7	8	9	10	12
3.4	.9³663	.9³675	.9³687	.9³698	.9³709	.9³720	.9³730	.9³740	.9³749	.9³758	1	2	3	4	5	6	7	8	9
3.5	.9³767	.9³776	.9³784	.9³792	.9³800	.9³807	.9³815	.9³822	.9³828	.9³835	1	1	2	3	4	4	5	6	7
3.6	.9³841	.9³847	.9³853	.9³858	.9³864	.9³869	.9³874	.9³879	.9³883	.9³888	0	1	1	2	2	3	3	4	5
3.7	.9³892	.9³896	.9³90	.9⁴04	.9⁴08	.9⁴12	.9⁴15	.9⁴18	.9⁴22	.9⁴250									
3.8	.9⁴28	.9⁴31	.9⁴33	.9⁴36	.9⁴38	.9⁴41	.9⁴43	.9⁴46	.9⁴48	.9⁴500									
3.9	.9⁴52	.9⁴54	.9⁴56	.9⁴58	.9⁴59	.9⁴61	.9⁴63	.9⁴64	.9⁴66	.9⁴670									

For negative values of z use $\Phi(z) = 1 - \Phi(-z)$.

* $.9^3032$ signifies .999032, and similarly for other values quoted in this form.

Index